TEST CRITIQUES COMPENDIUM

DEVIN SHAFRON

Contributors

Fred L. Adair, Ph.D.
Glenn Affleck, Ph.D.
Donald G. Barker, Ph.D.
Robert H. Bauernfeind, Ph.D.
Jack L. Bodden, Ph.D.
Merith Cosden, Ph.D.
Theodore R. Cromack, Ed.D.
Denise D. Davis, Ph.D.
George Domino, Ph.D.
Robert J. Drummond, Ph.D.
Jane Duckworth, Ph.D.
Jean C. Elbert, Ph.D.
Giselle B. Esquivel, Psy.D.
G. Cynthia Fekken, Ph.D.
Jim C. Fortune, Ed.D.
Michael D. Franzen, Ph.D.
Alan F. Friedman, Ph.D.
Kurt F. Geisinger, Ph.D.
Paul C. Hager, Ph.D.
Ronald K. Hambleton, Ph.D.
Sandra W. Headen, Ph.D.
Elaine M. Heiby, Ph.D.
Sharon Herzberger, Ph.D.
Raymond H. Holden, Ph.D.
E. Wayne Holden, Ph.D.
Raymond G. Johnson, Ph.D.
Timothy Z. Keith, Ph.D.
Grant Aram Killian, Ph.D.
Jean Powell Kirnan, Ph.D.
Allan L. LaVoie, Ph.D.

Howard D. Lerner, Ph.D.
Paul M. Lerner, Ph.D.
Eugene E. Levitt, Ph.D.
Maria M. Llabre, Ph.D.
Sheridan P. McCabe, Ph.D.
William R. Merz, Sr., Ph.D.
Kevin C. Mooney, Ph.D.
Maria Pennock-Román, Ph.D.
Rolf A. Peterson, Ph.D.
Joan M. Preston, Ph.D.
Robert C. Reinehr, Ph.D.
Richard M. Ryan, Ph.D.
Arthur B. Silverstein, Ph.D.
Jean Spruill, Ph.D.
R. Scott Stehouwer, Ph.D.
Lucille B. Strain, Ph.D.
Donald I. Templer, Ph.D.
Howard Tennen, Ph.D.
Oliver C. S. Tzeng, Ph.D.
Forrest G. Umberger, Ph.D.
Philip A. Vernon, Ph.D.
Ken R. Vincent, Ed.D.
J. Mark Wagener, Ph.D.
James A. Walsh, Ph.D.
Mary M. Wellman, Ph.D.
Judith L. Whatley, Ph.D.
Randolph H. Whitworth, Ph.D.
Brent E. Wholeben, Ph.D.
Robert E. Williams, Ed.D.
Carl G. Willis, Ed.D.

Daniel J. Keyser, Ph.D.
Richard C. Sweetland, Ph.D.
General Editors

TEST CRITIQUES COMPENDIUM

Reviews of Major Tests from the *Test Critiques* Series

TEST CORPORATION OF AMERICA

Library of Congress Cataloging-in-Publication Data
Test critiques compendium.

Includes index.
1. Psychological tests. 2. Educational tests and measurements. 3. Neuropsy-. chological tests. I. Keyser, Daniel J., 1935-   . II. Sweetland, Richard C., 1931-   . III. Tests critiques. BF176.T4195   1987   150'.28'7   87-10247
ISBN 0-933701-08-X
ISBN 0-933701-09-8 (softcover)

Test Critiques is an ongoing series, with new volumes containing approximately 100 new reviews published every 9-12 months. For a cumulative title listing of tests reviewed or other information about the series, contact Test Corporation of America at the above address or call 1-800-822-8485.

Printed in the United States of America

CONTENTS

ACKNOWLEDGEMENTS

As in each new volume of *Test Critiques*, the editors wish to acknowledge the efforts and dedication of the reviewers whose work is represented here. We are indebted especially to those whom we asked to check and update previously published reviews at a very busy time in the academic calendar; they complied graciously. We thank all of the reviewers in the *Critiques* project for the quality and professionalism that they bring so consistently to our endeavors.

We also wish to thank the test publishers whose instruments are reviewed in the series, for providing both specimen materials and support to the contributors who approach them under the auspices of *Test Critiques*.

We are grateful to the staff members at Test Corporation of America who have been involved in all the many phases of *Test Critiques*, including the compilation of this compendium. We extend our thanks to Clark Smith, Test Corporation's CEO, for maintaining his spirited commitment to this project as well as to the others in which we are involved. Our greatest indebtedness is to Jane Guthrie, senior editor of TCA, whose sense of organization, personal composure, and journalistic skill we relied upon totally both for *Test Critiques* and for the *Compendium*. Eugene Strauss and Leonard Strauss, directors of Westport Publishers, Inc., continue to give freely of their support, enthusiasm, and business advice, all of which we acknowledge with gratitude.

The editors owe a special "thank you" to S. Joseph Weaver, Ph.D., whose insight, encouragement, and expertise laid a keystone in building the *Test Critiques* series.

Finally, we wish to express our profound appreciation to our readers. Their positive response to *Test Critiques* has given impetus to this *Compendium* as well as a reassuring validity to the project as a whole. It is our most sincere desire that these books continue to have a valuable application for the many hands that use them.

A NOTE FROM THE EDITORS

The compilation of this volume was undertaken with the realization that the inclusion or exclusion of any given test could be a rather subjective decision. For this reason, we employed criteria that would remove as much bias as possible from the test selection process. Specifically, we turned to the reference literature and to professionals active in the field of assessment.

In 1984, the *American Psychologist* published the results of a survey by Lubin, Larsen, and Matarazzo indicating the 30 tests most used by their respondents. Within the introductory passages of *The Ninth Mental Measurements Yearbook*, Table 4 identifies the 50 tests in the *MMY* that generated the largest number of literature references. The contents of these two lists were combined, and to this combination we added tests that were newly published or had gained a significant following not reflected in the two sources named above (such as the Kaufman Assessment Battery for Children and the Child Behavior Checklist).

The resulting list of roughly 80 test instruments was sent in survey form to over 100 professionals—university professors who teach testing and assessment, psychologists who use tests as an integral part of their practice, and various others involved in the field of measurement and evaluation. The purpose of *Test Critiques Compendium* was described and participants were asked to indicate whether each test on the survey list should be included or excluded from the book.

Based on the survey responses, the 80 tests were ranked according to the number of votes each received for inclusion. Respondents were also asked to name tests not listed in our survey that, from their perspective, should be included in the book. This led, for example, to the inclusion of the Myers-Briggs Type Indicator, which previously did not appear on any list. Tests that received only a minimum number of inclusion votes (less than five) were eliminated from the group, and the final count of 60 tests was derived.

The statement that these are 60 "major" tests is made carefully and without implication of the worth of any tests not included. Every specific field of assessment has its own major tests. Our research was undertaken to determine, in the broadest context possible, which tests were taught the most, used the most, or, for some other reason, enjoyed a high degree of acceptance within the entire assessment field.

Because the editors could anticipate that students in graduate assessment courses would use this *Compendium*, authors of reviews published prior to 1987 (i.e., those that had appeared in Volumes I-IV, 1984-86) were asked to check their material for timeliness. Where the reviewer(s) deemed it appropriate, these previously published reviews have been updated via endnotes, citing recent literature, forthcoming revisions, and so on.

Within the descriptive copy offered at the beginning of each review, only the test publisher information has been updated (in order to facilitate purchases and user inquiries); as the *Compendium* constitutes a compilation rather than a revision, details such the academic titles or geographic locations of the participating reviewers are unchanged from the original. The editors have noted on the opening page of each review the year in which the original was published and the volume number in which it appeared.

Through the 60 test reviews assembled for this *Test Critiques Compendium* the editors wish to provide as comprehensive a resource as possible in one volume. As with our *Test Critiques* series, we sincerely hope that this volume will continue to evolve over time—as does the field of assessment itself.

TEST CRITIQUES COMPENDIUM

G. Cynthia Fekken, Ph.D.

Assistant Professor of Psychology, Queen's University, Kingston, Ontario.

ADJECTIVE CHECK LIST

Harrison Gough and Alfred Heilbrun, Jr. Palo Alto, California: Consulting Psychologists Press, Inc.

THE ORIGINAL OF THIS REVIEW WAS PUBLISHED IN TEST CRITIQUES: VOLUME I (1984).

Introduction

The Adjective Check List (ACL) is an alphabetized list of 300 adjectives commonly used to describe a person's attributes, to which subjects typically respond by marking those adjectives considered to be self-descriptive. Further applications pertain to descriptions of others, ideal self, historical figures, ideas, and so on.

The ACL was introduced by Professor Harrison G. Gough at the Institute of Personality Assessment and Research (IPAR) at the University of California, Berkeley, in 1949. Its purpose was to record staff members' observations of participants in IPAR personality studies. The technique of checking relevant adjectives from a large adjective pool was considered ideal since it is simple, time-efficient, rooted in everyday language, and has considerable descriptive scope. Originally, the ACL was comprised of 279 adjectives, either drawn from Cattell's group of 171 trait descriptors or written to reflect various theories, particularly those of Freud, Jung, Mead, and Murray. In 1951, five "overlooked" words were added. Again, in 1952, 17 words were added (and one dropped) in response to then recent work in personality theorizing and to an apparent lack of terms for describing physical characteristics. The current 300 adjective version of the ACL was published by Gough in 1952.

Professor Alfred Heilbrun, then at the State University of Iowa, developed 15 scales based on Murray's need system in 1958. The subsequent collaborative work of Gough and Heilbrun culminated in the publication of the 1965 manual for the Adjective Check List. The ACL could be scored on 24 scales, a Number of Adjectives Checked Scale, and eight additional scales developed by grouping items on either theoretical or empirical grounds. A major revision of the ACL manual, published in 1983, recommends scoring 37 scales, of which 15 are new. Two previous scales (Defensiveness and Lability) were dropped. The 1983 manual presents scoring schemes for the following: four Method of Response Scales, measuring aspects of stylistic responding; 15 Need Scales, reflecting some of Murray's needs; nine Topical Scales, measuring dimensions relevant to various research projects; five Transactional Analysis Scales, reflecting five ego states or functions recognized in Transactional Analysis; and four Origence-Intellectence Scales, measuring structural aspects of creativity and intelligence. In addition, a

3

differentiated picture of an individual's personality may be obtained simply by examining the actual adjectives checked on the ACL.

Normative data for approximately 10,000 subjects are reported in the 1983 manual. The ACL has been translated into French, German, Hebrew, Italian, Japanese, Norwegian, Portuguese, Spanish, Thai, and Vietnamese. However, only the Italian version has met Gough and Heilbrun's criteria for being fully normed and standardized. Norms for the French version are being collected.

The ACL can be obtained as a hand-scorable four-page folder on which adjectives are presented alphabetically. The examiner reads the instructions, which ask subjects to respond by placing an "X" in the box next to each self-descriptive adjective. The manual lists which adjectives are scored on which scales, although templates are recommended for hand-scoring. Scoring involves subtracting the number of contraindicative items from the number of indicative items for each scale. Raw scores are then converted to standard scores with a mean of 50 and a standard deviation of 10 for subgroups determined by sex and by total number of adjectives checked. Scale scores are strongly influenced by the number of adjectives checked and since subjects freely endorse as few or as many adjectives as they chose, raw scores must be standardized to permit accurate interpretation. Standard scores are transferred onto a profile sheet which lists the 37 scales down the left-hand margin, along with a column for recording the standard score. The remainder of the profile sheet is a grid with 37 rows and 10 columns, allowing standard scores to be plotted next to scale names. For more than a few protocols, the computerized scoring sheets available for the ACL may be faster and more convenient.

Practical Applications/Uses

The ACL appears to be appropriate for use with a wide variety of individuals. While the manual does not include particular subject specifications, it does refer to samples of high school students, undergraduate and graduate students, delinquents, psychiatric patients, and adults. The ACL can be completed in 15 to 20 minutes. Test administration is straightforward and amenable to either an individual or a group format. Acceptable examiners include psychologists, psychometrists, research assistants, proctors, etc. Computerized scoring of the ACL is infinitely preferable to hand scoring, which can take up to 30 minutes per profile. Scoring the ACL can be learned in a matter of hours, although the availability of a step-by-step outline of the scoring procedure as well as a set of templates would constitute an improvement.

Interpretation of test results requires some training in psychology. Interpretation is guided by six case illustrations presented in the manual and by empirically derived descriptions for high and low scores for each scale. External criteria used to develop descriptions were Q-sorts, adjective ratings, and ratings on trait clusters, all completed by staff members with reference to subjects participating in IPAR assessment studies.

In general, the ACL is probably best suited for research settings. This judgment is based on an examination of the dimensions reflected by most ACL scales and of the psychometric properties of the ACL. The adjectives on the ACL tend to describe normal dimensions of personality. The scope of the adjective pool has encouraged frequent use of the ACL for constructing scales by combining items either rationally (i.e., on the basis of theory) or empirically (i.e., on the basis of their relationships to external criteria) (cf. Gough, 1960). Thus, the number of scales that can be developed from the ACL is limited only by the imagination, something of a mixed blessing.

The Need Scales developed by Heilbrun were among the first rational scales. Nineteen psychology graduate students were asked to judge which ACL adjectives were related to definitions of the following 15 of Murray's needs: Achievement, Dominance, Endurance, Order, Intraception, Nurturance, Affiliation, Heterosexuality, Exhibition, Autonomy, Aggression, Change, Succorance, Abasement, and Deference. The Need Scales in the 1983 manual are modified versions of the 1965 Need Scales. Scales were shortened or lengthened to decrease item overlap and to increase item-total correlations. Other rationally developed scales include the Method of Response Scales. Scores on the Number of Items Checked Scale simply reflect the number of adjectives an individual endorses. The Number of Favorable Adjectives Checked and the Number of Unfavorable Adjectives Checked Scales were derived from the judgments of 97 college students regarding the 75 most and the 75 least favorable ACL items, respectively. Finally, the Commonality Scale was based on clusters of very frequently and very infrequently endorsed items, and may be helpful for identifying random protocols.

Adjectives comprising the other ACL scales were clustered using a more empirical approach. The Counseling Readiness Scale (Heilbrun & Sullivan, 1962) was based on selection of adjectives differentially endorsed by clients showing a more or a less positive response to counseling. This analysis yielded different sets of items for males and females. The Ideal Self Scale was developed by selecting across four samples those items more often associated with ratings of the "ideal self" than of the "real self." Adjectives for the Military Leadership Scale and the Creative Personality Scale were selected using ratings of leadership capacity and of creativity, respectively.

The 1965 ACL manual describes a number of scales, which were intended to parallel major factorial dimensions found on the California Psychological Inventory (CPI), also authored by H.G. Gough. Three of these scales are presented in the 1983 manual, although their origins are not outlined there. In fact, the Personal Adjustment Scale and the Self-Confidence Scale were developed by contrasting the self-descriptions of IPAR subjects receiving extreme ratings on "personal adjustment and personal soundness" and on "poise and self-assurance," respectively. The Self-Control Scale is described in the 1965 manual as "developed empirically." It was intended to reflect the "responsibility-socialization" factor on the CPI.

The development of the Masculine Attributes and Feminine Attributes Scales is unclear in the 1983 manual. These scales were written independently to describe masculine, forceful, and initiative individuals, and feminine, sentimental, and warm individuals. These scales have shown some validity for differentiating between male and female respondents and between individuals with heterosexual and homosexual preferences.

The Transactional Analysis Scales reflect five ego functions: Critical Parent, Nurturing Parent, Adult, Free Child, and Adapted Child. K.B. and J.E. Williams (1980) asked 15 members of the International Transactional Analysis Association to rate the relevance of ACL adjectives to ego states on a four-point scale. Judges' mean ratings were initially used to weight endorsed adjectives' contributions to total scores on each of the five ego states. Subsequently, five 44-item scales with weights of +1 and –1 were constructed, although details are reported neither in the manual nor in the original reference (Williams & Williams, 1980).

Finally, the Origence-Intellectence Scales were developed in the context of George Welsh's work on creativity and intelligence (1975). Subjects who received extreme scores on various measures of creativity and intelligence were placed in one of four quandrants: low origence-low intellectence, low origence-high intellectence, high origence-low intellectence, high origence-high intellectence. The pattern of item endorsements shown by these groups of subjects was the criterion for selecting items for the four ACL scales.

Gough and Heilbrun (1983) emphasize the ACL's potential for description of others as well as the self. Observers' ACL protocols may be compared to ACL self-descriptions either intuitively, statistically (e.g., by calculating a phi coefficient), or rationally. Rational comparison involves calculating indices such as an Insight Index, which is the total number of adjectives checked by the observer and the self divided into this total plus the number checked by the self only. The degree to which such indices reflect individual difference variables is not clear from the data reported in the manual. Other possible uses of the ACL include evaluation of pretherapy/posttherapy changes or ideal-real self differences. These comparisons may be conducted at the level of individual adjectives or of profiles based on the 37 scale scores.

The more innovative targets to which the ACL has been applied have included historical figures (e.g., Lincoln and Washington), stereotypes, and objects (e.g., cars and cities). While interesting, these applications may be overly enthusiastic. To illustrate, in comparing the uncross-validated profiles of Rome and Paris, males only are reported to rate Rome significantly higher on the Counseling Readiness Scale; females only rated Rome higher on the Number of Favorable Adjectives Checked, Intraception, Nurturance, Deference, Personal Adjustment, and Ideal Self Scales. Surely such empiricism has limited usefulness.

Technical Aspects

Ultimately, tests must be evaluated in terms of reliability and validity. Alpha estimates of internal consistency for the 37 ACL scales show acceptable median values of .76 and .75 for males and females, respectively. Reliabilities for the 15 Need Scales and three other scales may be somewhat inflated: alpha coefficients were based on the same sample used to modify these 18 scales via selection of items having high correlations with the scale total. The stability of ACL scale scores is quite strong. Six-month test-retest correlations for males showed a median value of .65; one-year test-retest correlations for females showed a median value of .71.

Based on evidence presented in the 1983 manual, the construct validity of the ACL scales appears to be modest. ACL scale scores were correlated with observer ratings on extensively defined trait descriptors; with scores on various personality inventories; with the Edwards Social Desirability Scale; and, with a vocabulary measure. Some ACL scales, such as Dominance, clearly showed an appropriate pattern of relationships to criteria, with correlations of up to .5 with observer ratings and relevant CPI scales. Validity coefficients for other ACL scales were frequently significant, although not necessarily as high. In some instances ACL scales did not correlate with criteria. For example, observer ratings of femininity correlated significantly with scores on the Femininity Scale for females but not for males; scores on the Terman Concept Mastery Test did not correlate significantly with any origence-intellectence scales for females; and the Self-Control Scale (designed to reflect the CPI Responsibility-Socialization factor) correlated only .15 and .24 with the CPI Responsibility and Socialization Scales, respectively.

Evidence for discriminant validity is weak overall. On the positive side, ACL scales tend not to be confounded by either social desirability or vocabulary level, as evidenced by the small and generally insignificant correlations reported in the manual. With the possible exception of the origence-intellectence scales, however, ACL scales correlate substantially less with validity criteria than with one another. The Self-Control Scale mentioned previously is typical. Correlations with CPI scales range from −.43 to .38, with a median absolute correlation of approximately .17. The range of intercorrelations with other ACL scales is −.78 to .73, with a median absolute correlation of .33. Such high intercorrelations may be a function of item overlap between scales, method variance, or true overlap between the constructs assessed by any pair of scales. An explicit theoretical rationale for relating scales to one another and to particular criteria is lacking and at this time evidence for discriminant validity is not particularly convincing. An evaluation of the ACL within the multitrait-multimethod paradigm is recommended not only to establish discriminant and convergence validity, but ultimately to contribute to evidence for the construct validity of the ACL.

Critique

The Adjective Check List is one of the more widely used tests in psychology (see Buros, 1978), but it is open to various criticisms. One should consider first the scope and derivation of ACL adjectives, which have been criticized previously (Rorer, 1972). The 1983 manual offers a relatively more complete outline of the origins of the adjectives, although not necessarily a better justification for inclusion. In the absence of a guiding theory or definition, adjective selection seems arbitrary and is not necessarily an improvement over a pool assembled by a researcher for a particular purpose.

The method of checking only applicable adjectives, seemingly so straightforward, involves a number of problems. While the interpretation of endorsed adjectives is fairly clear, a lack of context for the adjectives or a method for indicating the degree of adjective relevance has occasionally been noted (Spitzer, Stratton, Fitzgerald, & Mach, 1966). More serious may be the interpretive problems associated with nonendorsed adjectives. Is a nonendorsed adjective definitely not characteristic or merely irrelevant? Typically, both poles of a dimension are scored and the rejection of items sampled from one pole is considered as important for construct assessment as the endorsement of items at the other pole. The ACL ignores this aspect of measurement.

Another problem engendered by the free adjective checking technique is the strong correlation between the number of items checked and individual scale scores. Individual differences in acquiescent tendencies show correlations as high as .7 and .8 with ACL scale scores (Gough & Heilbrun, 1983). The manual proposes standardizing scale scores within categories based on the number of adjectives checked. Correlations between standard scale scores and the Number of Items Checked Scale are indeed negligible. The change from a four category (1965) to a five category (1983) system was intended to circumvent large differences in standard scores associated with small changes in the number of items checked. The middle category (for each sex) has as its midpoint the median number of adjectives checked by the normative sample. In 1965, two males with raw scores of 16 on Achievement but endorsing 75 and 76 adjectives would receive standard scores of 75 and 63, respectively; in 1983, two males receiving 16 on Achievement but endorsing 73 and 74 items would receive standard scores of 70 and 58, respectively. The criterion values for standardization categories may have changed, but the problem appears to remain. A better solution is the partialling procedure referred to in the manual and described elsewhere (Swede & Gough, 1984). When compared to the five category method in the manual, final means on 12 of the 37 scales were different for males and 21 of the 37 scales for females. While mean differences were small and correlations between scores obtained by the two methods were extremely high, users should note that the procedures do not produce identical results. The partialling procedure is much preferred. A final concern about adjusted scores: such scores must derive their meaning from correlates, lacking intrinsic or intuitive meaning. ACL scale scores have been correlated with adjectives and Q-sorts, as indicated earlier, to develop psychological profiles of extreme scorers. Replication of empirical relationships across samples and across methods other than observer ratings is not reported.

The derivation of the 37 ACL scales bears mention. An empirical approach to test construction does allow for the development of many and varied scales from the ACL item pool. By using the same group of adjectives over and over, however, scales show item overlap. At a conceptual level, this is problematic; for example, the 38-item Achievement Scale and the 40-item Dominance Scale share 13 items and standard scores on the scales correlate .77, making it difficult to argue that these scales measure distinct constructs. At an empirical level, the item overlap makes factor analysis of scales inappropriate (Nunnally, 1978). Gough and Heilbrun (1983) are certainly not the first researchers to factor analyze dependent measures but they too, need to be reminded that such solutions are potentially misleading. Finally, item factor analyses have yielded support for a relatively stable seven-factor solution (Parker & Veldman, 1969; Vidoni, 1977). The breadth of the ACL adjective pool may be considerably smaller than the 37 dimensions Gough and Heilbrun recommend.

In conclusion, the wide appeal of the Adjective Check List lies in its simplicity: it is easy to administer, straightforward to complete, and applicable to a large variety of subject types. While it may be criticized for lacking a clear rationale underlying the adjective selection, the scope of the ACL item pool has nonetheless made it popular for developing scales. The strongest criticisms of the ACL are associated with scale development. Empirically derived scales require considerable cross-validation to eliminate spurious relationships and maximizing validities. Failure to cross-validate item selection for the scales may help account for instances where ACL scales show poor convergent validity. Empirical selection of items from a single group of adjectives tends to yield scales with overlapping items, which contribute to high scale intercorrelations and make ACL scale score interpretation more difficult. The authors attempt to enhance interpretation by correlating ACL scores with external criteria, but these relationships appear not to be cross-replicated. The most serious shortcoming of the ACL scales may be the absence of a general theory as a basis for deriving and validating scales, in combination with a tendency to place unjustified weight on unreplicated empirical findings. Further data on the construct validity of the Adjective Check List are required.

References

Buros, O.K. (Ed). (1978). *The eighth mental measurements yearbook.* Highland Park, NJ: The Gryphon Press.

Gough, H.G. (1960). The Adjective Check List as a personality assessment research technique. *Psychological Reports, 6,* 107-122.

Gough, H.G., & Heilbrun, A.B., Jr. (1965). *The Adjective Check List manual.* Palo Alto, CA: Consulting Psychologists Press, Inc.

Gough, H.G., & Heilbrun, A.B., Jr. (1983). *The Adjective Check List manual* (rev. ed.). Palo Alto, CA: Consulting Psychologists Press, Inc.

Nunnally, J.C. (1978). *Psychometric theory* (2nd ed.). New York: McGraw-Hill Book Co.

Parker, G.V.C., & Veldman, D. J. (1969). Item factor structure of the Adjective Check List. *Educational and Psychological Measurement, 29,* 605-613.

Rorer, L.G. (1972). The Adjective Check List. In O.K. Buros (Ed.), *The seventh mental measurements yearbook.* Highland Park, NJ: The Gryphon Press.

Spitzer, S.P., Stratton, J.R., Fitzgerald, J.D., & Mach, B.K. (1966). The self-concept: Test equivalence and perceived validity. *Sociological Quarterly, 7,* 265-280.

Swede, S.W., & Gough, H.G. (1984). Transformations from adjectives to scales for the Adjective Check List (ACL). ACLTRANS FORTRAN IV program. *Behavior Research Methods, Instruments, & Computers, 16,* 1-2.

Vidoni, D.O. (1977). Factor analytic scales of the Adjective Check List (ACL) replicated across samples: Implications for validity. *Educational and Psychological Measurement, 37,* 535-539.

Welsh, G.S. (1975). *Creativity and intelligence: A personality approach.* Chapel Hill, NC: Institute for Research in Social Science, University of North Carolina.

Williams, K.B., & Williams, J.E. (1980). The assessment of transactional ego states via the Adjective Check List. *Journal of Personality Assessment, 4,* 120-129.

Philip A. Vernon, Ph.D.
Assistant Professor of Psychology, The University of Western Ontario, London, Ontario.

ADVANCED PROGRESSIVE MATRICES

J. C. Raven. London, England: H. K. Lewis & Co. Ltd. U.S. Distributor—The Psychological Corporation.

THE ORIGINAL OF THIS REVIEW WAS PUBLISHED IN TEST CRITIQUES: VOLUME I (1984).

Introduction

The Advanced Progressive Matrices (APM) is one of a series of nonverbal tests of intelligence developed by J.C. Raven (1962). Following Spearman's theory of intelligence, it was designed to measure the ability to educe relations and correlates among abstract pictorial forms and it is widely regarded as one of the best available measures of Spearman's *g*, or of general intelligence (e.g., Jensen, 1980; Anastasi, 1982). As its name suggests, it was developed primarily for use with persons of advanced or above average intellectual ability, but in many cases it is also a more suitable test for assessing the abilities of adults than is its predecessor, the Standard Progressive Matrices.

The initial version of the APM was prepared in 1943 for use by War Office Selection Boards in Britain. In 1947 a revised version was made available for more general use, and in 1962 the version that is presently available was issued. The 1947 test consisted of 48 items, 12 of which were removed to leave the current APM with 36 items arranged in order of increasing difficulty. The present test also consists of a set of 12 practice items, which were designed to introduce testees to the methods of working on the test. The manual (1978) suggests that the 12-item test can also be used by itself to provide a quick and rough assessment of adults' mental ability, but it is probably better used to introduce testees to the types of problems that they will encounter in the 36-item test.

Materials for the test consist of two test booklets—Set I containing the 12-item test and Set II containing the 36-item test—answer sheets, and a scoring key. Each item is printed in black on a separate page of the booklet and consists of a series of figures or forms at the top of the page and 8 response items (1 correct and 7 distractors) at the bottom of the page. The figures at the top of each page are arranged in some systematic order, or orders, and always have one figure missing. The test-taker's task is to educe the relationships or the order among the figures and to select the single figure from among the 8 response items that correctly completes the pattern. The response items are numbered (1 to 8), and in most applications the testee simply records the number of the selection on the answer sheet. The score is the number of items solved correctly; thus, possible scores range from 0 to 36.

Practical Applications/Uses

The APM may be administered individually or to groups, and can be given under timed or untimed conditions. If no time limit is used, the manual states that the test will "assess a person's total capacity for observation and clear thinking" (section 4, p. 2). If a time limit is imposed, the manual recommends 40 minutes for the 36-item test and states that it will now provide a measure of "intellectual efficiency." Section 4 of the manual contains detailed instructions for administering the APM as a group test. If it is being administered individually, the practice items in Set I can be used to demonstrate the test and explain how it is to be approached. Most people of average or above-average intelligence will understand what is required of them very quickly and, in fact, the test can be explained easily by pantomime, making it usable with deaf persons or those whose language the examiner does not speak. If it is administered without a time limit, most persons will complete the test (or as many of the items as they can) in about an hour to an hour and a half. Since the items are presented in an order designed to train examinees in the method of working, it is strongly recommended that they be instructed to attempt the items in the order in which they appear in the test booklet.

Having explained the test and established that the examinees understand what they are required to do, the examiner has no further participation in the testing process except to announce when the time limit has been reached, if one is used. Using the scoring key, the test can be scored manually in a minute or less. If required, answer forms suitable for machine scoring may be obtained from the test's publisher (H.K. Lewis & Co., Ltd.) and can be sent for processing to Document Reading Services Ltd. or to The Psychological Corporation. If the test is not being used for clinical assessment (e.g., if it is used as a measure of intelligence in a research study) it does not require a qualified psychologist for administration. With minimal training, a graduate or an undergraduate student research assistant can administer and score it competently.

The 36 items in the APM move fairly quickly from being relatively simple to extremely complex. The manual states that even persons of outstanding intellectual ability will be unable to solve all the items in less than 40 minutes and, while this may be a somewhat exaggerated claim, it is certainly true that the test proves very challenging for the majority of persons who take it. It is an ideal test for adults of above-average intelligence; the other Progressive Matrices tests (the Standard and the Coloured versions) are better suited to persons of average ability or to children. Although not specifically designed for use with special populations, the format of the test allows considerable flexibility in giving instructions and in how subjects may respond, such that it can be used with deaf persons or persons with orthopedic handicaps. It has been recommended for use in certain work settings where an assessment of a person's ability to think clearly and to make quick, accurate decisions is required. In a like manner it could also prove useful to vocational and high school counselors; the manual suggests that the APM might be used to select students for advanced science or technical courses. It is an excellent research instrument and has been used extensively in psychological research as a marker test of general intelligence. Its nonverbal nature makes the APM a good candidate for use in cross-cultural research.

Technical Aspects

The quality of the APM as a test is offset by the totally inadequate manual which accompanies it. Reliability data are presented only for the 1947 edition (the 48-item test) and, apart from the subjects' ages, little information is given about the samples which provided these data. Test-retest reliability coefficients for the 1947 edition (over an intertest interval of 6 to 8 weeks) range from .76 (among 10½-year-olds) to .91 (among adults). It is probably a reasonable inference that the 1962 test has similar reliability, but users of the test will have to look outside the manual for information pertaining to this and to the test's validity. For interpretive purposes, the manual provides "estimated norms" for the 1962 APM, which allow raw scores to be converted into percentiles (but only 50, 75, 90, and 95), and another table for converting percentiles into IQ scores. Overall, the manual must be a disappointment to many users of the APM; the sparsity of the information it provides seriously restricts the usefulness of the test as a diagnostic instrument. It has been criticized frequently in the past and the criticism bears repeating until a completely new standardization of the test is undertaken and the manual extensively revised.

A large number of publications pertaining to the Progressive Matrices is available. Interested users of the tests can purchase a *Researchers' Bibliography* (Court, 1982), containing some 1,500 references with brief abstracts and summaries, or consult *The Mental Measurements Yearbooks*, the eighth edition of which (Buros, 1978) lists 699 references to the series of Progressive Matrices tests. The APM has received somewhat less attention than have the Standard and Coloured Progressive Matrices, but a number of studies have addressed its validity as a measure of intelligence. Typically, when the APM has been factor-analyzed with other ability tests it shows a high *g* loading (i.e., a high loading on the first, unrotated factor) and negligible loadings on other factors (e.g., Vernon, 1983), supporting its construct validity as a measure of general intelligence. A recent study by Dillon, Pohlmann, and Lohmer (1981) factor-analyzed the 36 items of the test by themselves—controlling for the differential difficulty of the items—and identified two orthogonal (i.e., uncorrelated) factors underlying the test. One factor appeared to tap the ability to disembed simple figures from more complex forms, and was interpreted as being similar to Guilford's dimension of cognition of visual-figural transformations. The second factor involved mental rotations and appeared to tap certain aspects of mechanical ability.

Critique

The APM and the other Progressive Matrices are well-known tests, and research continues to add to the large body of information which already exists about them. They are among the few tests of intelligence which have their own listing in the *Psychological Abstracts*. The popularity of these tests is undoubtedly attributable to their simplicity of administration and scoring and to the fact that they will provide a valid measure of intelligence in a relatively short period of time. As has been mentioned, the APM is less useful in clinical, diagnostic situations than it is as a research instrument. In the former setting, psychologists will have to settle for a rather rough index of an examinee's level of ability and no profile of differential abilities can be obtained unless the APM is supplemented by administering additional tests, a procedure which the manual recommends. Even if adequate norms were available, it is unlikely that a clinician would choose the APM over other instruments which provide measures of a more diverse set of abilities. For research purposes, the APM is one of the best tests available if discrimination among adults of average to above-average mental ability is required. The fact that it can be group-administered also makes the APM more attractive than tests which must be given individually.

References

Anastasi, A. (1982). *Psychological testing* (5th ed.). New York: Macmillan Publishing Co.

Buros, O.K. (Ed.). (1978). *The eighth mental measurements yearbook.* Highland Park, NJ: The Gryphon Press.

Court, J.H. (1982). *Researchers' bibliography for Raven's Progressive Matrices and Mill Hill Vocabulary Scales* (6th ed.). Kent Town, South Australia: Author.

Dillon, R.F., Pohlmann, J.T., & Lohman, D.F. (1981). A factor analysis of Raven's Advanced Progressive Matrices freed of difficulty factors. *Educational and Psychological Measurement, 41,* 1295-1302.

Jensen, A.R. (1980). *Bias in mental testing.* New York: The Free Press.

Raven, J.C. (1962), *Advanced Progressive Matrices, Sets I and II.* London, England: H.K. Lewis & Co. Ltd.

Raven, J.C., Court, J.H., & Raven, J. (1978). *Manual for Raven's Progressive Matrices and Vocabulary Scales.* London, England: H.K. Lewis & Co. Ltd.

Vernon, P.A. (1983). Speed of information processing and general intelligence. *Intelligence, 7,* 53-70.

Judith L. Whatley, Ph.D.

Assistant Professor of Clinical Pediatrics and Psychiatry, University of Texas Health Science Center, San Antonio, Texas.

BAYLEY SCALES OF INFANT DEVELOPMENT

Nancy Bayley. San Antonio, Texas: The Psychological Corporation.

THE ORIGINAL OF THIS REVIEW WAS PUBLISHED IN TEST CRITIQUES: VOLUME VI (1987).

Introduction

The Bayley Scales of Infant Development (BSID) provide a three-component approach to the assessment of children's developmental status from 2 months to 2½ years of life. Two scales, the Mental Scale and the Motor Scale, result in quantitative standard scores. The Mental Scale is intended to assess sensory-perceptual abilities, object constancy, memory, learning and problem-solving ability, communication and verbal skills, and early abstracting ability. The Motor Scale is intended to measure gross and fine motor skills and control of the body. The third component, the Infant Behavior Record, provides an assessment of social and objective orientation toward the environment. Designed for use in both research and clinical practice, the scales provide an assessment of a child's current developmental status in comparison with normatively based expectations. In practice, the Mental and Motor Scales are more often employed, both by researchers and clinicians, than is the Infant Behavior Record.

The BSID culminate more than 40 years of research on the part of Nancy Bayley and her colleagues. Bayley was motivated to develop an assessment tool for infant development for which the standardization sample would be representative of normal children. Hers was a reaction against the relatively small samples of institutionalized children upon which many tests were based at the time her work began.

Bayley (1969) acknowledges the contribution of the California First-Year Mental Scale (Bayley, 1933), the California Preschool Mental Scale (Jaffa, 1934), and the California Infant Scale of Motor Development (Bayley, 1936) in the development of the BSID. An unpublished 1958 version of the BSID covered the first 15 months of life and was employed in a research program sponsored by the National Institute of Neurological Diseases and Blindness. These scales were then expanded to include the second year of life by taking items from the California Preschool Mental Scale and developing new items. This 1958-60 version was subsequently used in research sponsored by the National Institutes of Health. Using data from approximately 1,400 children aged 1 to 15 months and a sample of 160 children aged 18 to 30 months, items were selected that became the Mental and Motor Scales of the current edition.

The scales were standardized on a stratified sample of non-institutionalized children selected on the basis of the 1960 United States Census. The norms for the BSID are based upon this normal, English-speaking sample. Norms and standardized procedures have not been published for non-English-speaking populations, specific handicapped groups, or premature infants, although the scales have been used with such groups.

The Infant Behavior Record was developed from rating scales originally used with the sample from which the earlier California Mental and Motor Scales (Bayley, 1933, 1936) were developed. A 1958 unpublished version of the rating scales was administered to approximately 1,350 children; this research resulted in the present edition.

The BSID consist of 163 items on the Mental Scale and 81 items on the Motor Scale. Only those items within the range of performance, basal to ceiling, of a given child are administered. Items in the 2-to-5-month range are normed at half-month intervals; those in the 6-to-30-month range are normed at 1-month intervals. In general, test materials are presented to the child in a specified way, and the child's reactions to or behavior with the stimulus is observed and scored by the examiner. Other spontaneous behaviors that may be expected of a child in a given age range are also scored for their occurrence or nonoccurrence.

Materials for administration of the examination include a set of standardized, manipulable objects, as well as objects for visual and auditory presentation: a set of 1″ cubes, a red ring with attached string, cups, a saucer, red crayons, a rattle, a red ball, sugar pellets, a bell, a whistle doll, a picture book, puzzles, a jointed doll, a small broken doll, picture cards, a pull toy, spoons, and a non-breakable mirror. The manual also includes building specifications for a set of stairs and a walking board for assessment of motor skills for children in the older age ranges. (These standardized materials are available through The Psychological Corporation.) Other materials that must be supplied by the examiner include a table, a crib for the testing of young infants, white paper, and tissue.

The Infant Behavior Record consists of a set of 30 descriptive rating scales for behaviors characteristic of children in the first 2½ years of life. Twenty of these employ a 9-point scale, six use a 5-point scale, three use a yes/no format, and one asks for a Normal/Exceptional general evaluation. These rating scales focus on the child's social orientation, emotional tone, object orientation, attention span, goal directedness, interest focus, energy, overall evaluation of the child's performance, and representativeness of test performance. The Infant Behavior Record is based upon observations during the assessment and is completed after the exam.

Practical Applications/Uses

The Bayley Scales of Infant Development are used primarily in research settings and in clinical practice when there is a question about developmental status or when a high-risk/special population is being followed. Because of the skill level and time involved in administration, it is less likely to be used as a routine screening instrument.

The BSID were standardized on a full-term, nonhandicapped population. Since its development, it has frequently been used with children whose development is in question due to premature birth or other birth-associated risk factors. Most commonly, when BSID scores are computed for premature infants, the age of the infant is corrected for the prematurity. Rhodes, Bayley, and Yow (1984) cite Hunt and Rhodes (1977) for further information about age correction.

Use of the BSID with handicapped children is difficult and often impossible because items on the Mental Scale often require visual and auditory abilities. At all but the earliest age ranges, required responses involve reaching, gesturing, or otherwise manipulating test materials. No standardized version of the BSID has been developed for use with handicapped, blind, or deaf children, although some researchers and clinicians have employed modifications of the standardized procedure (DuBose, 1977; Kierman & DuBose, 1974). Experienced clinicians may wish to use items from the BSID as part of an assessment with these children, but no valid score is obtainable.

General instructions as well as the administration procedure for each item on the Mental and Motor Scale are detailed in the manual (Bayley, 1969). Additional clarifications are provided in the recent manual supplement (Rhodes, Bayley, & Yow, 1984). In general, the instructions are sufficiently detailed to inform the examiner of the proper procedure for administering each item. (Videotapes for use in training examiners are available through The Psychological Corporation, and training films may be rented from the Extension Media Center, University of California, Berkeley.)

The BSID are best administered in a room large enough to contain an examining table, chairs for adults, a crib or youth chair (depending on the age of the child), and free floor space for testing children who locomote. The room should be pleasant but not distracting. Optimally, other children and unnecessary adults should not be present to distract the child being examined. Rhodes, Bayley, and Yow (1984) remind examiners that the test materials should be managed in such a way that only materials being employed at the moment be visible to the child and that the child not have access to the equipment kit and its materials.

The role of the examiner is a very active and integral part of the assessment. The examiner administers test items, motivates the child to perform, observes and scores administered items and scorable spontaneous behavior, and is a stimulus toward whom the child responds. The attendant parent, at the examiner's request and direction, may administer items as instructed.

BSID norms are based on testing in a laboratory setting; however, children may be tested in a clinic or at home provided that care to limit distractions is taken. Durham and Black (1978) have reported data on infants 16 to 21 months old that showed that infants assessed at home scored higher on the Mental Scale than when they were assessed in laboratory settings. Verbal performance appeared to be especially affected.

Administration of the BSID requires training and practice. Examiners need to be able to relate well and easily to infants and their parents and should also have some background in child development and psychometrics. Most often the BSID are administered by psychologists, child development specialists, and occupational and physical therapists. The BSID are not difficult to administer, but neither can they be considered easy. This is not a test that can be administered as one reads along in the manual.

The scales are to be administered with the parent or appropriate substitute present, generally someone with whom the child is at ease. Bayley recommends that the examiner sit across from the infant rather than to the side in order to facilitate observation. Items are administered by the examiner but may be administered by the parent at the examiner's request and instruction.

The Mental Scale is to be administered before the Motor Scale due to the change of pace involved. However, many of the early items of the Motor Scale may be observed incidentally, or interspersed among items of the Mental Scale. In both the Mental and Motor Scales there is no set order for administration of items, and administration should be guided by the child's interest, energy, and attention. Basal and ceiling level criteria consist of 10 successive items passed or failed on the Mental Scale and 6 on the Motor Scale.

The Infant Behavior Record is completed following administration of the Mental and Motor Scales. The examiner selects the one most descriptive statement for each of the rating scales and may also note additional observations. Except for three items, ratings are characterizations of the child's behavior without reference to the normal distribution of the characteristic. For the three specific items, ratings are to be estimated with respect to other children of the child's age.

The time for testing will vary depending on the age and characteristics of the child. Bayley (1969) has estimated 45 minutes as the average amount of time to administer both the Mental and Motor Scales, with only 10% of cases requiring as much as 75 minutes. Completion of the Infant Behavior Record and computation of scores requires additional time.

Scoring of the BSID uses three forms, one for each of the two scales and one for the Infant Behavior Record. The Mental and Motor Scale forms provide descriptive phrases for each item and a space for noting "Pass" or "Fail" status or other notation. In addition, the record forms for these scales indicate the age at which 50% of children demonstrate an item and the range of ages that 5-95% of children pass an item. There are also situation codes noted on the form that facilitate the examiner's orderly administration of the examination. These codes draw the examiner's attention to the use of the same stimulus material or procedure for multiple items at different age levels that can be scored in one or a few presentations. The Infant Behavior Record form contains the rating scales and space for additional observations the examiner might wish to add. All three forms provide space for background information, computation of age at testing, and conversion of raw scores to standard scores.

Items on the Mental and Motor Scales are scored on a Pass/Fail basis, and only responses observed by the examiner may be scored as a "Pass." Even though there is a space provided to note responses reported by the parent, credit is not given for these responses. Scoring is intimately related to the administration of the test. Each item is scored as it is administered or observed; additionally, any scorable response is scored at any time that it is observed. Raw scores are obtained by summing items passed on the Mental and Motor Scales and adding this number to the basal level score for each scale. Using tables in the manual, raw scores are converted into standard scores known as the Mental Index and the Psychomotor Index. The child's date of birth is used as the reference from which conversions are made. Age tables making the conversion are broken down by 2-week intervals for the age range 2 to 6 months and by monthly intervals for the 6-to-30 month range. The Mental Index and the Psychomotor Index have a mean of 100 and a standard deviation of 16. Bayley (1969) cautions that these are developmental quotients and not IQs and offers instructions for individuals wishing to convert the raw score into an "age equivalent" rather than a standard score. An age equivalent score, however, does not offer the psychometric advantages of the standard score. No subtest scores are derivable from the two scales. Ratings from the Infant Behavior Record and the comparison percentages for these ratings add to the clinical evaluation of the child's social, behavioral, and emotional functioning during the test situation. These ratings may be examined using norm-referenced tables provided in the manual. The sample size for the Infant Behavior Record was relatively small for children 15 months and younger, and Bayley urges caution when interpreting these ratings (Bayley, 1969).

The BSID offer the advantages of a representative, normative sample and standardized scores with statistical properties useful for interpreting scores. Interpretation is based on reference to the norms established for the test. Clinical judgment plays a role in decision-making concerning the extent to which the child's performance is a fair representation of his or her ability. Such judgment is also called into play in developing hypotheses concerning the pattern of results that make up a given score, differences encountered between Mental and Psychomotor Indices, and the meaning of a score for a child from certain population groups. Clinical experience with children and an understanding of child development are also necessary to understand the meaning of a given score. Unfortunately, the manual offers little to educate or facilitate interpretation of scoring beyond statistical and standardization data.

Interpretation of the Infant Behavior Record makes use of percentage tables for individual rating scales. However, use of the rating scales and their interpretation calls upon the clinical skill and judgment of the examiner even more than do the Mental and Motor Scales.

Technical Aspects

The advantage of the Bayley Scales of Infant Development in comparison to other tests of infant development has been and continues to be its standardization and statistical properties. Norms of the BSID are derived from a stratified sample based upon the 1960 U.S. Census, controlling for sex, color within age group, urban/rural residence, and education of head of household. Bayley acknowledges that rural children are somewhat underrepresented but considers the effect on the norms to be negligible. The children making up the sample were tested at 14 ages: 2, 3, 4, 5, 6, 8, 10, 12, 15, 18, 21, 24, 27, and 30 months. These children were located through hospitals, well-baby clinics, municipal birth records, and social agencies, but included only "normal" children living at home. Excluded were institutional children with severe behavioral or emotional problems, those born more than 1 month prematurely, and children over 12 months of age from bilingual homes who showed significant difficulty using English. In all, 1,262 children were tested.

Bayley (1965) examined the effects of socioeconomic variables on scores using the 1958-60 precursor of the current scales. Although the difference was small, she noted significantly higher scores on the Motor Scales by black infants at all ages from 3 through 14 months. No other differences due to sex, birth order, geographic location, or parental education were found on either the Mental or Motor Scales.

Statistical properties are presented in the manual for the current scales and for research on earlier versions. Split-half reliabilities were computed for both the Mental and Motor Scales. Mutually dependent items were grouped on the same half of each scale, and reliabilities were computed separately for each age group in the standardization sample. Coefficients corrected by the Spearman-Brown formula estimate the reliability of the full-length scales. Resulting reliability coefficients for the Mental Scale ranged from .81 to .93, with a median value of .88. For the Motor Scale, values ranged from .68 to .92, with a median value of .84. The lower reliabilities were obtained on the Motor Scale when testing infants under 6 months of age. Overall, the reliability values compare quite favorably with other infant tests.

Bayley also reports standard error of measurement values for each scale by age group tested. For the Mental Scale, these range from 4.2 to 6.9. On the Motor Scale, the values range from 4.6 to 9.0.

Werner and Bayley (1966) report tester-observer and test-retest reliabilities (for a 1-week time difference) for a sample of 8-month-old infants tested with the 1958-1960 precursor of the BSID. Mean percentage of agreement between observers on the Mental Scale was 89.4 with a standard deviation of 7.1; on the Motor Scale, mean percentage agreement was 93.4 with a standard deviation of 3.2. These figures document a high level of agreement but are unfortunately limited to a restricted age range and are not based on the current version of the BSID. Test-retest reliabilities for a subsample of these babies showed mean percentage agreement on the Mental Scale of 76.4; however, the same limitations pertain. Reliability estimates are not provided in the manual for Infant Behavior Record ratings.

Reliability across test setting has been examined in a study by Durham and Black (1978) with a sample of 16- to 21-month-olds. These researchers found that children were likely to score higher when tested in the home compared to testing in a laboratory setting. Verbal performance appeared to be especially affected.

Another issue addressed in the manual is the relationship between the Mental and Motor Scales. Bayley (1969) reports that correlation coefficients by age between raw scores on these scales range from .24 to .78; the range is from .18 to .75 between standard scores. The median coefficient is .46 for both raw and standard scores. Overall, the relationship between mental and motor scores decreases with age. Bayley interprets this as being due to the increasing differentiation between mental and motor skills with development.

Researchers generally have found that infant intelligence tests in the first year of life do not accurately predict IQ performance later in childhood (Stott & Ball, 1965; McCall, Hogarty, & Hurlburt, 1972). These results are not inconsistent with Bayley's view of mental development—that it is emergent and may take different forms at different ages. Attempts to relate BSID scores to measures of intelligence in childhood have generally followed this pattern, especially for infants scoring within the normal range of functioning.

Bayley (1969) presents data correlating scores on the Mental Development Index and the IQ obtained from the Stanford-Binet Intelligence Scale for 120 children who were 24, 27, and 30 months of age. The sample was limited to children who earned basal scores on the Stanford-Binet. The coefficient of correlation obtained was .57. Although Bayley argues that this is substantial given the limited range of scores on the Stanford-Binet, the correlation still leaves much unexplained.

Cohen and Parmalee (1983) found only a moderate relation between BSID scores for preterm infants at 25 months of age and their Stanford-Binet scores at age 5 years ($r = .65$, $p < .05$). Their analysis supported social factors as being more important than any other set of variables in predicting outcome at age 5. McCall, Hogarty, and Hurlburt (1972), looking at Bayley's 1949 data, concluded that prediction of intellectual functioning in childhood was poor, although the later in infancy the developmental assessment was made and the closer in time the two assessments were made, the better the predictive ability.

The picture is somewhat different when attempting to predict the future functioning of children who receive BSID scores that are significantly lower than average. Vander Veer and Schweid (1974) related BSID scores at 18 to 30 months to tests 1 to 3 years later and found that 75% of the infants judged to be moderately to profoundly retarded remained so classified at the later testing. Additionally, none of the original group that were classified as retarded were scoring in the normal range at the subsequent testing. Ireton, Thwing, and Gravem (1970) have reported similar findings of greater predictability for low scores on the BSID. Thus, while the BSID may offer little predictability for children in the normal range or better, it does appear to be useful in predicting the course of development for those children scoring very poorly early in life.

Fewer efforts have attempted to look at the predictive value of the Infant Behavior Record. McGowan, Johnson, and Maxwell (1981) collected BSID Mental Scale scores and infant behavior ratings at 12 and 24 months and Stanford-Binet scores at 36 months on a sample of Mexican-American babies. They concluded that the behavior ratings were not helpful in adding to the predictive power of BSID Mental Scores at 12 months.

Other researchers have looked at infant behavior ratings over time, looking at the distributions of ratings and at the underlying factor structure of these ratings. Several sources (Dolan, Matheny, & Wilson, 1974; Matheny, 1983; Bayley, 1969) have noted developmental as well as individual difference changes in the distribution of ratings over the age span covered by the record. Research with twins has related these changes to genetic influences (Matheny, 1980; Matheny, Dolan, & Wilson, 1976). Results concerning factors underlying the Infant Behavior Record are mixed. Matheny (1980) has reported three factors to be recurrent across the first two years: task orientation, test affect-extraversion, and activity. McGowan, Johnson, and Maxwell (1981) failed to find the same factor structure in a study of Mexican-American 12-month-olds. It should be noted that attempts to identify factors underlying the abilities measured by the BSID Mental and Motor Scales, such as that by Hofstaetter (1954), have not been clearly successful (see Cronbach, 1967, for a critique).

Critique

Almost 20 years after their publication, the Bayley Scales of Infant Development remain one of the most used and useful assessment instruments for infant development. Bayley's original aim, to offer a well-standardized assessment making use of a representative sample of non-institutionalized children for the determination of current developmental status, has been achieved. This, indeed, is one of the greatest strengths that the BSID have to offer in contrast to many other infant assessment techniques. Additionally, the BSID offer a set of standardized scores and norms for both mental and motor development based on the same group of children. Directions for administration are clearly presented in the manual, and the more recent 1984 supplement offers additional clarification to facilitate reliable and standardized administration.

The BSID are useful in identifying infants whose development is significantly below average, and for these children it offers good predictive value for the future. The specific scale scores and the observations the test facilitates help identify neuromuscular deficits.

The BSID are also very useful in documenting, especially for research purposes, that a sample of children is developmentally within the range expected for their age. Such information provides an important base when attempting to interpret other observations made on those children. An additional advantage of the BSID is that they are one of the most frequently used infant tests and have consequently accumulated a background of data. This includes research by Bayley as well as many other researchers.

What the BSID do not do is predict future intellectual development, especially 1) within the normal range and above and 2) from the first year of life to later childhood and beyond. This shortcoming is not just a characteristic of this particular test but reflects the state of our understanding of development. Bayley and others discuss the emergent nature of intelligence and see it taking different forms at different times. However, we continue to ask, "What will this baby be like when he or she is older?" and we are not able to answer with confidence.

A second shortcoming of the BSID is the lack of reliable and valid subscores. Children appear to have strengths and deficits in particular areas, but these areas of ability are not reliably defined or measured. Individual clinicians make observations about verbal skills, problem-solving skills, fine motor skills, and so on, but no reliable or standardized subscales are available to facilitate such observations.

A related difficulty with the BSID is the lack of independence between the Mental and Motor Scales. Clearly, children with significant motor deficits will score less well on the Mental Scale as we measure "mental abilities" as manifested in motor performance.

Another problem concerns the Motor Scale at the older age range where it focuses on gross motor skills—climbing, walking, jumping, and so on. Children with fine motor deficits will fail to be identified formally by the Motor Scale as having motor problems. Clinical observations from performance on the Mental Scale can help identify such cases, but the test, and especially the Psychomotor Index, will fail to reflect this observation directly.

Finally, the manual offers little information to facilitate clinical interpretation of BSID results beyond the statistical properties of the test. Neither does it offer therapeutic or remedial interventions related to test performance or particular difficulties that a child might display. Those using the BSID in clinical settings would benefit from information concerning both these areas.

The fact that the BSID offers so much leads us to want it to do more. It does not claim to be standardized on handicapped or preterm infants, but we want to use it for these very populations and do so at the risk of misinterpreting our findings. These cannot be said to be shortcomings of the test but rather a need for better such instruments in the field of assessment, instruments for which the BSID is a model.

References

Bayley, N. (1933). *The California First-Year Mental Scale.* Berkeley: University of California Press.

Bayley, N. (1936). *The California Infant Scale of Motor Development.* Berkeley: University of California Press.

Bayley, N. (1965). Comparison of mental and motor test scores for ages 1–15 months by sex, birth, order, race, geographical location, and education of parents. *Child Development, 36,* 379–411.

Bayley, N. (1969). *Bayley scales of infant development.* San Antonio, TX: The Psychological Corporation.

Cohen, S. E., & Parmalee, A. H. (1983). Prediction of five-year Stanford-Binet scores in preterm infants. *Child Development, 54,* 1242–1253.

Cronbach, L. J. (1967). Year-to-year correlations of mental tests: A review of the Hofstaetter analysis. *Child Development, 38,* 283–290.

Dolan, A. B., Matheny, A. P., Jr., & Wilson, R. S. (1974). Bayley's Infant Behavior Record: Age trends, sex differences, and behavioral correlations. *JSAS Catalog of Selected Documents in Psychology, 4,* 9.

DuBose, R. F. (1977). Predictive value of infant intelligence scales with multiply handicapped children. *American Journal of Mental Deficiency, 81,* 388–390.

Durham, M., & Black, K. (1978). The test performance of 16 to 21 month-olds in home and laboratory settings. *Infant Behavior and Development, 1,* 216–223.

Hofstaetter, P. R. (1954). The changing composition of "intelligence": A study in t-technique. *The Journal of Genetic Psychology, 85,* 159–164.

Hunt, J. V., & Rhodes, L. (1977). Mental development of preterm infants during the first year. *Child Development, 48,* 204–210.

Ireton, H., Thwing, E., & Gravem, H. (1970). Infant mental health development and neurological status, family socioeconomic status, and intelligence at age four. *Child Development, 41,* 937–946.

Jaffa, A. S. (1934). *The California Preschool Mental Scale.* Berkeley: University of California Press.

Kierman, D. W., & DuBose, R. F. (1974). Assessing the cognitive development of preschool deaf-blind children. *Education of the Visually Handicapped, 6,* 103–105.

Matheny, A. P., Jr. (1980). Bayley's Infant Behavior Record: Behavioral components and twin analyses. *Child Development, 51,* 1157–1167.

Matheny, A. P., Jr. (1983). A longitudinal study of stability of components from Bayley's Infant Behavior Record. *Child Development, 54,* 356–360.

Matheny, A. P., Jr., Dolan, A. B., & Wilson, R. S. (1976). Twins: Within-pair similarity on Bayley's Infant Behavior Record. *Journal of Genetic Psychology, 128,* 263–270.

McCall, R. B., Hogarty, P. S., & Hurlburt, N. (1972). Transitions in infant sensorimotor development and the prediction of childhood IQ. *American Psychologist, 27,* 728–748.

McGowan, R. J., Johnson, D. L., & Maxwell, S. E. (1981). Relations between infant behavior ratings and concurrent and subsequent mental test scores. *Developmental Psychology, 17,* 542–553.

Rhodes, L., Bayley, N., & Yow, B. C. (1984). *Supplement to the manual for the Bayley Scales of Infant Development.* San Antonio, TX: The Psychological Corporation.

Stott, L. H., & Ball, R. S. (1965). Infant and preschool mental tests: Review and evaluation. *Monographs of the Society for Research in Child Development, 30* (3, Serial No. 101).

Vander Veer, B., & Schweid, E. (1974). Infant assessment: Stability of mental functioning in young retarded children. *American Journal of Mental Deficiency, 79,* 1–4.

Werner, E. E., & Bayley, N. (1966). The reliability of Bayley's revised scale of mental and motor development during the first year of life. *Child Development, 37,* 39–50.

R. Scott Stehouwer, Ph.D.

Associate Professor of Psychology, Calvin College, Grand Rapids, Michigan.

BECK DEPRESSION INVENTORY

Aaron T. Beck. Philadelphia, Pennsylvania: Center for Cognitive Therapy.

THE ORIGINAL OF THIS REVIEW WAS PUBLISHED IN TEST CRITIQUES: VOLUME II (1985).

Introduction

The Beck Depression Inventory (BDI) is a 21-item test presented in multiple-choice format which purports to measure presence and degree of depression in adolescents and adults. Each of the 21 items of the BDI attempts to assess a specific symptom or attitude "which appear(s) to be specific to depressed patients, and which are consistent with descriptions of the depression contained in the psychiatric literature" (Beck, 1970, p. 189). Although the author, Aaron T. Beck, M.D., is associated with the development of the cognitive theory of depression, the Beck Depression Inventory was designed to assess depression independent of any particular theoretical bias.

The BDI was developed by Beck and his associates at the University of Pennsylvania School of Medicine. At that time he was Professor of Psychiatry and also chief of the psychiatry section of Philadelphia General Hospital. In undertaking research on depression, Beck decided it would be useful to develop an inventory for measuring depth of depression and felt such an inventory would be particularly advantageous for research purposes.

The BDI was first developed with psychiatric patients who were drawn from routine admissions to the psychiatric out-patient department of the University of Pennsylvania Hospital and to the psychiatric out-patient department of the Philadelphia General Hospital. The test was developed over a seven-month period starting in June, 1957 with an original sample of 226 patients. A later study, which began in February, 1960, was undertaken over a five-month period with 183 patients.

The original BDI was published in 1961 (Beck et al., 1961) and was later contained in Beck's (1970) classic work, *Depression: Causes and Treatment.* The original BDI was a 21-item multiple-choice test in which the selections for each item varied from four to seven choices. Each choice was given a weight of zero to three points. A revision was undertaken in 1974, then later in 1978, standardizing each item to four possible choices, each choice still assigned a weight of zero, one, two, or three points. Additionally, a short form of the Beck Depression Inventory has been

---

The reviewer wishes to acknowledge gratefully the help of Mr. Robert Nykamp in the preparation of this review.

developed (Reynolds & Gould, 1981). This short form consists of 13 items taken from the revised 21-item test.

Although Aaron Beck is credited with the development of the BDI, credit is also given to C. H. Ward, M. Mendelson, J. Mock, and J. Erbaugh for their original work in the test's development.

Each of the inventory items corresponds to a specific category of depressive symptom and/or attitude. Each category purports to describe a specific behavioral manifestation of depression and consists of a graded series of four self-evaluative statements. The statements are rank ordered and weighted to reflect the range of severity of the symptom from neutral to maximum severity. Numerical values of zero, one, two, or three are assigned each statement to indicate degree of severity. The 21 items purport to measure the following symptoms and attitudes:

1. Sadness
2. Pessimism/Discouragement
3. Sense of Failure
4. Dissatisfaction
5. Guilt
6. Expectation of Punishment
7. Self-Dislike
8. Self-Accusation
9. Suicidal Ideation
10. Crying
11. Irritability
12. Social Withdrawal
13. Indecisiveness
14. Body Image Distortion
15. Work Retardation
16. Insomnia
17. Fatigability
18. Anorexia
19. Weight Loss
20. Somatic Preoccupation
21. Loss of Libido

Practical Applications/Uses

The BDI is designed for simple administration. A trained interviewer can read aloud each statement for the items, though usually the test is self-administered. Originally, the inventory was developed for adult and late adolescent patients, though later research indicates the BDI is appropriate for use with even young adolescents, as young as 12 years, 10 months (Strober, Green, & Carlson, 1981).

The test is very simple to administer and very simple to take. The reading difficulty level is quite low (eighth grade), the statements are easy to understand, and directions are very clear. Responses are indicated directly on the question sheet by having the subjects circle the number beside the statement they endorse. Subjects are asked to read each group of statements carefully and then select the statement from each group that best describes what they have been feeling over the past week including the day of administration. If several statements in the group seem to apply equally well, subjects are asked to circle each statement they believe fits them.

The score is obtained by taking the highest score circled for each item and adding the total number of points for all items. While the authors admit there is no arbitrary score that can be used for all purposes as a cutoff score and that the specific cutoff point depends upon the characteristics of the patients used and the purpose for which the inventory is being given, they do give the following guidelines:

    0-9   Normal Range
    10-15   Mild Depression
    16-19   Mild-Moderate Depression
    20-29   Moderate-Severe Depression
    30-63   Severe Depression

The practitioner who is in need of a simple, quick, and helpful tool in gathering information about a patient's depressive state may do well to use the BDI. The BDI is particularly useful in mental health settings as a screening device since it correlates very well with psychiatric ratings of depression. In any situation where a screening device is necessary for depression, the BDI would seem to be the test of choice. The ease of administering, taking, scoring, and interpreting the BDI makes it a very attractive test. Practitioners must be aware that the BDI should certainly not be used as the sole means of determining presence or degree of depression; as indicated above, however, it can be an excellent screening and supplementary device for individual or group administration.

Individuals who are most likely to benefit from using the BDI are mental health practitioners in a variety of settings. This would include psychiatrists, psychologists, social workers, mental health counselors, school counselors, school psychologists, and industrial psychologists. In this latter case, the BDI may be very helpful as a screening measure for industry.

The Beck Depression Inventory has been used with a wide variety of populations including psychiatric in-patients, psychiatric out-patients, general university populations, and the adolescent population. It is also appropriate for subjects with a wide range of abilities. Since the test can be given orally, all that is necessary is that the subject be able to understand the words of the test. They do not have to write a response, but merely can endorse a particular answer. This makes it particularly advantageous for individuals who have a short attention span, for individuals who have difficulty reading, and for individuals who may be less compliant with a longer test.

Because of the simplicity of the BDI, it can be administered almost anywhere by a variety of individuals, with very little training. If the BDI is administered individually, it would be quite helpful for the examiner to be aware of his or her ability to provide behavior observations that may aid in the further delineation of presence and/or degree of depression.

The time required for the test varies from individual to individual but should be no more than ten minutes. Since scoring the BDI involves simple addition, the time required is therefore minimal, usually less than three minutes per inventory. The BDI is available in computer format from National Computer Systems, Minneapolis, MN.

Interpretation is based on objective scoring and is generally straightforward. Based on the total score, individuals are categorized into five levels of depression, from normal to severe depression, as indicated previously. The amount of training for interpreting the BDI varies with the use to which the inventory is put. If it is used to diagnose a patient, the examiner should have appropriate psychological training. If the inventory is used to categorize a particular individual for research purposes only, training need not be extensive.

Technical Aspects

One of the disadvantages of the BDI is that there is no manual available for its use. Perhaps it is felt a manual is not necessary because of the simplicity of the instrument; however, it is more difficult to determine such things as reliability and validity without one.

It appears that most of the studies of BDI reliability have been undertaken with psychiatric patients. Test-retest reliability has been studied in the case of 38 patients who were given the BDI on two occasions (Beck, 1970). It was discovered that the changes in BDI scores tended to parallel changes in the clinical reading of the depth of depression, indicating a consistent relationship between BDI scores and the patient's clinical state. The reliability figures here were above .90. Item analysis also demonstrated a positive correlation between each item of the BDI and the total score. These correlations were all significant at the .001 level. Internal consistency studies demonstrated a correlation coefficient of .86 for the test items, and the Spearman-Brown correlation for the reliability of the BDI yielded a coefficient of .93.

In assessing the validity of the BDI, the readily apparent face validity of the BDI must be addressed. The BDI looks as though it is assessing depression. While this may be quite advantageous, it may make it easy for a subject to distort the results of the test.

Content validity would seem to be quite high since the BDI appears to evaluate well a wide variety of symptoms and attitudes associated with depression.

Studies of the BDI have also been undertaken with regard to concurrent validity. One study demonstrated a correlation of .77 between the inventory and psychiatric rating using university students as subjects (Bumberry, Oliver, & McClure, 1978). Beck (1970) reports similar studies in which coefficients of .65 and .67 were obtained in comparing results of the BDI with psychiatric ratings of patients. Other measures of concurrent validity have been undertaken by comparing the results of the BDI with those of other measures of depression, such as the Depression Adjective Check Lists (DACL) and the MMPI. Beck (1970) reports a correlation of .66 between the BDI and the DACL and a correlation of .75 between the BDI and the MMPI D Scale. Again, this study was undertaken with psychiatric patients as subjects.

Overall, the results of reliability and validity studies strongly support the BDI as a very useful measure for assessing depression. This has been demonstrated with a variety of subjects in a variety of situations. Additionally, cross-validation research tends to support the reliability and validity findings with regard to the BDI.

Critique

The Beck Depression Inventory is a simple, perhaps deceptively and even elegantly simple, test for the presence and depth of depression. It is extremely helpful as a screening procedure and is also of value as a validation procedure for other means of determining depression. It is difficult to determine how well the BDI has been accepted in general clinical or applied practice. Certainly the Beck

Depression Inventory has become the inventory of choice for researchers in selecting depressed subjects from a larger population (Stehouwer & Rosenbaum, 1977). It has also been useful in delineating differences in depressive symptomatology between men and women (Hammen & Padesky, 1977) and between adolescents and adults (Stehouwer, Bultsma, & Termorshuizen, in press). Nonetheless, if the clinical issue is one of determining the presence and degree of depression and if the subjects are motivated to accurately reflect their emotional status, the Beck Depression Inventory would certainly seem to be the inventory of choice for clinical as well as research purposes.

References

This list includes text citations and suggested additional reading.

Beck, A. T. (1970). *Depression: Causes and treatment*. Philadelphia: University of Pennsylvania Press.

Beck, A. T., & Beamesderfer, A. (1974). Assessment of depression: The depression inventory. In P. Pichot (Ed.), *Psychological measurements in psychopharmacology: Vol. 7. Modern problems in pharmacopsychiatry* (pp. 151-169). Basel, Switzerland: Karger.

Beck, A. T., Ward, C. H., Mendelson, M., Mock, J., & Erbaugh J. (1961). An inventory for measuring depression. *Archives of General Psychiatry, 4*, 561-571.

Bumberry, W., Oliver, J. M., & McClure, J. N. (1978). Validation of the Beck Depression Inventory in a university population using psychiatric estimate as the criterion. *Journal of Consulting and Clinical Psychology, 46*, 150-155.

Hammen, C. L., & Padesky, C. A. (1977). Sex differences in the expression of depressive responses on the BDI. *Journal of Abnormal Psychology, 86*(6), 609-614.

Reynolds, W. M., & Gould, J. W. (1981). A psychometric investigation of the standard and short form Beck Depression Inventory. *Journal of Consulting and Clinical Psychology, 49*(2), 306-307.

Steer, R. A., Beck, A. T., & Garrison, B. (1982). Applications of the Beck Depression Inventory. In N. Sartorious & T. Ban (Eds.), *Assessment of depression*. Geneva, Switzerland: World Health Organization.

Stehouwer, R. S., Bultsma, C., & Termorshuizen, I. (in press). Cognitive-perceptual distortion in depression as a function of developmental age. *Adolescence*.

Stehouwer, R. S., & Rosenbaum, G. (1977, December). *Frequency and potency of reinforcing events in anxiety and depression*. Paper presented at the 17th annual convention of the Association for the Advancement of Behavior Therapy, Atlanta, GA.

Strober M., Green, J., & Carlson, G. (1981). Utility of the Beck Depression Inventory with psychiatrically hospitalized adolescents. *Journal of Consulting and Clinical Psychology, 49*(3), 482-483.

Randolph H. Whitworth, Ph.D.

Chairman, Department of Psychology, The University of Texas at El Paso, El Paso, Texas.

BENDER VISUAL MOTOR GESTALT TEST

Lauretta Bender. New York, New York: The American Orthopsychiatric Association, Inc.

THE ORIGINAL OF THIS REVIEW WAS PUBLISHED IN TEST CRITIQUES: VOLUME I (1984).

Introduction

The Bender Visual Motor Gestalt Test (Bender-Gestalt) is the most frequently administered and thoroughly researched of all of the drawing (copying) tests. It consists of nine geometric designs (numbered A and 1-8) originally developed by Wertheimer to demonstrate the perceptual tendencies to organize visual stimuli into configural wholes (*Gestalten*). Each design, on a 4" x 6" card, is presented sequentially to the subject whose task is to reproduce them on a blank sheet of 8½" x 11" paper. Scoring systems vary, but all are in terms of the accuracy and organization of the reproductions and include such factors as the relative size of the drawings compared to the stimuli and their placement on the paper, distortions of the drawings, inversions or rotations of the figures, and other evidence of perceptual and/or fine-motor dysfunction.

The test is appropriate for all age groups, from 3 years through adult, and may be administered individually or in a group setting. The results may be interpreted in terms of maturational development in children, neuropsychologically with respect to organic brain function, or as indicators of emotional disturbance.

Two adaptations of the Bender-Gestalt, the Canter Background Interference Procedure (BIP) (Canter, 1966) and the Minnesota Percepto-Diagnostic Test (MPD) (Fuller & Laird, 1963), will also be considered in this review. Both procedures are designed to improve the efficiency of the Bender-Gestalt in neuropsychological screening. The BIP involves two administrations of the Bender-Gestalt. After the standard Bender is given, the subject reproduces the nine designs on paper covered with randomly interwoven black curved lines that cover the entire sheet. Scoring compares the two sets of drawings. The MPD uses two of the Bender-Gestalt designs (A and No. 4), presenting three different combinations of card and figure orientations to the subject, whose task is to draw these six designs on a sheet of unlined paper. Scoring involves determining the degree of rotation of the drawings measured by a protractor.

The Bender-Gestalt was developed by Lauretta Bender while a senior psychiatrist at Bellevue Hospital in New York (Bender, 1938). This pioneering work was based on the principles of Gestalt psychology utilizing nine of the figures first developed by Wertheimer, one of the founders of the Gestalt school of psychology. These geometric designs were chosen to illustrate certain Gestalt principles and Bender's classic research monograph discusses her results in these terms. In this report she included data on the administration of the test to a wide variety of clinical groups, including mentally retarded, organically brain-damaged, psychotic, and normal adults. She also gave the instrument to normal and retarded children as a method of measuring developmental level. Although she reported a variety of results involving different neuropsychological and emotional disorders, she presented no systematic data analysis nor any scoring procedures other than those associated with Gestalt psychology. She later published a "manual" for use with the test but again no reliability or validity studies nor any systematic quantitative scoring procedures were included. As she stated, "This is a clinical test and it should not be so rigidly formalized as to destroy its function which is to determine the individual's capacity to experience visual-motor gestalten in a spatial and temporal relationship" (Bender, 1946, pp. 6-7).

Nevertheless, within the next few years a number of different quantitative scoring systems were devised. Probably the best known and most widely used is that of Pascal and Suttell (1951), who developed a complete, complex scoring system that involves scoring each design separately, resulting in a total of more than 100 different scorable characteristics. Hain (1964) developed a more holistic approach, scoring the entire test performance rather than each design on the basis of 15 different categories.

Elizabeth Koppitz (1964, 1975) conducted the first comprehensive research of the Bender-Gestalt with young children. She carried out an extensive normative standardization of the test on large numbers of children, evolving a Developmental Scoring System for children aged 5-12. In addition, using the Bender-Gestalt as a type of projective technique, she described a set of ten "Emotional Indicators" for detecting psychological disturbance in children.

Hutt (1977) developed the most extensive scoring system to measure severity of psychopathology, which involves 17 different factors. Like Hain's system, this procedure treats the test performance as a whole rather than scoring each card individually.

Variations of the Bender-Gestalt that have been developed include not only the Canter BIP and Fuller's MPD but also several procedures for the measurement of memory. One such procedure by Wepman (described in Lezak, 1983) involves the presentation of each stimulus card for five seconds then having the subject attempt to reproduce it from memory. Wepman also described another procedure in which the subject attempts to recall and reproduce the designs after a regular administration.

The Bender-Gestalt is a deceptively simple test both in terms of test materials and procedures. Very likely because of this apparent simplicity, no single standardized set of stimulus materials, instructions, or procedures has ever been generally accepted and thus administrations are likely to vary from one user to the next. Despite this lack of uniformity, however, materials and procedures are generally quite similar. In addition to the nine stimulus cards, the subject is provided with one or more blank sheets of unlined paper and a soft pencil. Specific instructions do vary but most frequently the subject is simply told that he or she will be presented with several cards that are to be copied as exactly as possible. The stimulus card is then placed with its length running parallel to the top of the sheet. No instructions are given as to how the designs are to be placed on the paper, but the subject is encouraged to place all the designs on the same sheet allowing an evaluation of the subject's organizational ability. The subject is likewise discouraged from turning or moving either the stimulus or the paper, but if he or she insists on doing so the examiner should note this on the sheet.

The examiner's role in the administration is entirely passive. In response to any questions from the subject, the examiner simply reiterates that he or she is to make as good a drawing as possible. This presents the subject a minimum of structure and essentially no information about how to proceed. This increases the test's potential as projective technique as well as allowing the administrator to evaluate the subject's planning organizational ability.

The Bender-Gestalt has a distinct advantage over many other tests in that the same set of stimuli and essentially the same procedures are utilized for all age groups, from three-year-old children to geriatric adults. It is untimed, although the time for completion of the test is often recorded and may provide valuable additional information about the subject. Finally, scoring is usually relatively easy and rapid, rarely requiring more than three or four minutes, regardless of whether a formalized or intuitive scoring system is employed.

An important innovation has been the development of group administration procedures for this test. Group administration of the Bender-Gestalt has employed different techniques, including the use of enlarged stimulus cards, slide projections of the designs, or group test booklets. Research with children has shown that Bender scores derived from group administration do not differ significantly from individually administered tests (Keogh & Smith, 1961; Koppitz, 1975).

Practical Applications/Uses

Because the Bender-Gestalt is such a simple, exceptionally time-efficient procedure, it has not only become popular but has achieved the distinction of being the most frequently administered psychological test in the United States (Lubin, Wallis, & Paine, 1971). As might be anticipated in view of this enormous popularity, it has also achieved the additional, more dubious distinction of being both one of the most overrated as well as most maligned tests in current use. Realistically, however, the Bender-Gestalt is not such a simple and uncomplicated a procedure as it might appear from a casual appraisal. Evaluation and interpretation of this test, like any other psychological test, depends on the skill, training, and experience of the evaluator. In fact, Cronbach (1970) classifies the Bender as a "Complex Clinical Test" along with the Rorschach Inkblot technique! Taking even the most cautious view of the test, however, the Bender-Gestalt has proved to be a very effective and efficient psychometric tool in at least three areas of practical clinical applications.

The use of the Koppitz scoring system (or similar developmental scoring procedure) with the Bender-Gestalt to measure young children's developmental and maturational level has proved to be of great practical value for both child psychologists as well as school diagnosticians. Koppitz has demonstrated that the

Bender has a rough but direct relationship to mental ability with normal and retarded children from ages 3 to 10.

The second general area of practical application is the use of the Bender-Gestalt as a neuropsychological screening test for brain damage. While the literature reveals that the test is generally a reliable and valid instrument in the determination of organicity, caution should be exercised in using the Bender as the only neuropsychological measure since its "hit rate" for detecting organic versus nonorganic conditions is far from perfect. Research suggests that the Bender is an excellent supplemental test for brain damage, used in conjunction with other neuropsychological measures. Further, the use of the Canter BIP or Fuller MPD, rather than the standard Bender, has been shown to improve the diagnostic accuracy of detecting organic brain injury.

The third area of general application has been its use in detecting emotional and psychiatric disturbance in both children and adults. Koppitz presents a series of emotional indicators which have been shown to be related to psychological disturbance in young children. Hutt has by far the most comprehensive system to measure psychopathology, defining 17 different characteristics shown to effectively distinguish between normals, neurotics, schizophrenics, and brain-damaged patients. As with the other practical applications of the Bender-Gestalt, however, great caution should be exercised when using it as a personality instrument or projective technique and it probably never should be employed as the only personality measurement instrument. As in the other applications, the technique is especially useful as a screening or supplemental test but should not be expected to stand alone as a measure of psychopathology.

Whether administered individually or to a group, the Bender-Gestalt is one of the more nonthreatening psychological test procedures and rarely causes any degree of test anxiety in either adults or children. In young children up to about the age of ten or eleven it is viewed as a test of ability, as indeed it is for this age group, and quite similar to many school activities. With adults, although they may view the procedure as too simple or childish, they rarely find it psychologically threatening. It is thus one of the better procedures with which to begin a psychological test battery, in order to alleviate fear of testing and establish rapport. Another major advantage in the administration of this test is that the person actually giving the test need not be professionally qualified, although the interpretation clearly should be done by a technically competent professional.

Scoring procedures developed for the Bender-Gestalt are so numerous and diverse that the lack of any generally accepted scoring procedure has become one of the major problems in the use of the technique. In reviewing the literature it often appears as if almost everyone who uses the test decides to develop his or her own administration and scoring procedures, leading to an unfortunate proliferation of systems and general confusion. Regardless of which scoring system is examined, however, all have the same basic structure in that errors of reproduction of the designs and the organization of the figures on the sheet are the substance of scoring. The differences between systems thus involves what the developer chooses to score as an error and the weight that is given to each type of distortion.

Pascal and Suttell (1951), for example, developed a very elaborate and complex scoring system for adults that identifies from 10 to 13 different types of errors for each design plus seven more scores involving organization, resulting in more than 100 scorable characteristics such as asymmetry, rotations, or distortions. Koppitz's (1963, 1975) system is developmental and appropriate only for children up to the age of about 12. Again, as in all other procedures, scoring is based on errors which are then translated into age-equivalents which, in turn, can be interpreted as rough approximations of mental development. Other frequently cited scoring systems are those of Hain (1963), which employ only 15 categories of errors, and Hutt (1977), utilizing some 17 scoring factors.

The usefulness of a formal, objective scoring system as opposed to an intuitive, subjective approach to evaluating the Bender has been the subject of some debate. Arguments for the use of formal, quantitative scoring include improvement in interscorer reliability, consistency in training, improvement in diagnostic accuracy, as well as the obvious necessity for some quantitative scores for research purposes. Arguments against formal scoring procedures are generally that an experienced clinician using an intuitive approach is as accurate in diagnosis as a formal scoring procedure. Those favoring an intuitive approach often cite Goldberg's (1959) well-known study comparing groups of psychological trainees, secretaries, Pascal and Suttell scores, and Dr. Max Hutt in accuracy of diagnosis. Secretaries, objective scoring, and trainees all did equally well but Hutt was superior to everyone else. Despite these results, familiarity with at least one of the formal scoring systems is really highly desirable in allowing the evaluator to be aware of the more common distortions and perceptual aberrations that are associated with maturational and developmental problems or organic cerebral dysfunction.

Regardless of whether an objective or intuitive scoring system is employed, the interpretation of the Bender-Gestalt poses an entirely different set of problems. The Bender procedure can be viewed as a series of three steps. First the subject must be able to see the stimulus design, and thus visual acuity and sensory integrity are vital to the process. Second, the visual stimulus must be developed into a visual perception, which involves cortical associational processes in the brain. Finally, the subject must reproduce the percept, which involves motor control and dexterity. If visual acuity and motor control are intact, it is then legitimate to make the assumption that any faulty reproductions of the stimuli result either from cerebral associational or perceptual formation deficits. The degree to which that assumption is correct is generally the degree to which interpretations of organic cerebral dysfunction are accurate. Specific errors, such as rotations, inversions, omissions, and other gross distortions are certainly suggestive of cerebral impairment. Nevertheless, it should be emphasized that other factors not involving either developmental lags or organicity, such as impulsive and haphazard reproductions, severe emotional problems, or even malingering, can result in high error scores. The interpretation of the Bender thus depends not only on the number and type of errors committed, but also on the training, experience, and sophistication of the examiner. This is especially true when the Bender results are compared with other psychological, medical, and educational data which will often either confirm or rule out conclusions based on Bender errors alone.

Technical Aspects

The number and variety of reliability and validity studies on the Bender-Gestalt are of epic proportions, with *The Eighth Mental Measurements Yearbook* (Buros, 1978) listing more than 1,000 references that included studies only to 1977. In the subsequent seven years many more additional articles have been published and research continues, with scores of new studies appearing every year. With such an enormous number of publications, only an overview and summary of the reliability and validity studies can be presented. The research on reliability and validity on the Bender involving children will be presented first, followed by the data on the use of the Bender with brain damage, and finally the research on the Bender in the diagnosis of emotional disturbance.

Koppitz carried out extensive normative standardizations on large numbers of children, including a number of reliability and validity studies. Other researchers using her developmental scoring system also conducted reliability and validity research which she reports in her two volumes (Koppitz, 1964, 1975). These results involving the Bender with young children reveal interscorer reliability to be very high with correlations of .90 and above. This suggests that a formal scoring system like the Koppitz results in a high degree of consistency from one evaluator to another.

Test-retest reliability coefficients with children show great variability, ranging from low of about .50 with kindergarten children measured 8 months apart to .90 with the same age group measured two weeks apart. The majority of more than 20 different reliability studies reported by Koppitz reveal correlation coefficients in the .80 + range and suggest that normal elementary school children show relatively stable patterns of Bender-Gestalt scores from one administration to the next.

With respect to the validity of the Bender with children, Koppitz reported correlation coefficients from about .50 to as high as .80 between the Bender-Gestalt and intelligence as measured by the Stanford-Binet or Wechsler Intelligence Scale for Children up to about the age of 10. Beyond this age the correlations drop to essentially zero as most older children obtain nearly perfect scores. She also reported relatively high correlations between Bender scores and subsequent educational achievement of first-grade children. Correlations between the Bender and academic performance for older children were less impressive. Koppitz also reported a relatively high correlation between the Bender and intellectual and academic performance for retarded children as well. With children diagnosed as having minimal brain damage, she reported that the Bender is a valuable diagnostic tool but cautioned that it should not be used alone but in combination with other psychological tests and any background information available.

The use of the Bender-Gestalt in neuropsychological research has consistently proven that the technique is very effective in detecting organic brain damage in both adults and children (Lezak, 1983). Even more impressive, however, is the technique's efficiency in discriminating between psychiatric and brain-injured patients as well as normals. Studies reported by Lezak reveal that the Bender correctly identified members of each diagnostic group as well or better than any other single neuropsychological instrument, with correct diagnoses ranging from 74% to 91%. In one of these studies the Bender was equal to or better than the Halstead Neuropsychological Test Battery subtests in correctly classifying the three groups. The Bender has also proven useful in diagnosing neuropsychological changes in such conditions as chronic alcoholism, Parkinson's disease, and chronic obstructive pulmonary disease.

The use of the Canter BIP procedure in discriminating between psychiatric and organic groups has proven even more effective than the standard procedure, with Heaton and colleagues (1978) reporting a remarkably high 84% correct classification. As with the standard Bender, however, the Canter BIP is susceptible to the effects of cortical laterality and both procedures are much more likely to be effective in detecting brain damage in the right cerebral hemisphere than in the left.

Research with the Minnesota Percepto-Diagnostic Test confirms that this technique is effective in discriminating between brain-damaged, normal, and personality disturbed adults, with correct classifications ranging from 78% to 89% for these three groups (Fuller & Laird, 1963). On the other hand, the MPD has not proved effective in discriminating between organic and psychiatrically disturbed patients.

Studies involving the use of the Bender-Gestalt in the diagnosis of emotional disturbance in children and adults are relatively limited. Koppitz has developed a number of "Emotional Indicators" and presents some data of their utility in discriminating between normal and emotionally disturbed children. The numbers of children reported in these studies are relatively small and the validity of these results are somewhat questionable. Hutt's (1977) 17-factor psychopathology scale was reported to distinguish between normal, neurotic, psychotic, and organically brain-damaged patients with a high degree of statistical reliability. His interest in the Bender-Gestalt as a projective technique has led Hutt to develop another scale, which he called the "Adience-Abience Scale," to measure perceptual approach-avoidance behavior, but statistical evidence is lacking as to its effectiveness in personality evaluation. It would thus appear that the use of the Bender-Gestalt as a projective technique to determine emotional pathology or personality characteristics is much more questionable than its use to determine maturational level or organic brain dysfunction.

Critique

When the Bender-Gestalt is viewed in perspective of more than 40 years of research and clinical applications and its strengths and utility are measured against its weaknesses and limitations, the ancient and oft-castigated Bender comes out surprisingly well. As a rapid, efficient measure of perceptual-motor and cognitive development in young children the test is of enormous value for both educational and clinical work, especially if it is used as a supplemental or screening device.

The use of the Bender in neuropsychological diagnosis of organic cerebral dysfunction is also very useful, especially if the Canter BIP is employed, but it is also in this area that it has been misused most often. When employed as a screening test or in conjunction with other neuropsychological procedures, however, research has shown it to be effective in organic diagnosis and considering its efficiency in time and effort it is certainly a valuable procedure.

The test's use as a projective technique or as a measure of emotional psychopathology, again if used by itself, would appear to be its most questionable application. Under the best of circumstances caution should be exercised when interpreting the results as a measure of personality or psychopathology.

The development and especially the general acceptance of a *single* scoring system for the Bender is still a pressing need. At the very least one accepted scoring system for developmental purposes and one for diagnosis of organicity should be a goal.

There is probably no other single psychological test instrument available that requires so little time or training to administer and score and which, in turn, provides such a wealth of valuable clinical and behavioral information. These very strengths are also the bases for its all too frequent misuse or abuse by untrained, inexperienced, and even unethical examiners. This reviewer is certain that neither Bender, Koppitz, Canter, nor any other of the hundreds of researchers and proponents of the Bender-Gestalt would advocate its use as a 10-minute neuropsychological test battery or an 8-minute substitute for the WISC-R. Yet this is what is often done, and as a result not only does the reputation of the test suffer, but so does the whole field of psychological testing.

With the Bender-Gestalt, as with almost any psychological test, the criticisms of its inadequacies, misuses, and even abuses are primarily the result of deficiencies on the part of the person using the instrument and not inherent in the test itself. In the hands of a trained, experienced examiner the Bender-Gestalt is a valuable diagnostic tool and deserves to be a part of every professional psychometrician's repertoire.

References

Bender, L. (1938). A visual motor gestalt test and its clinical use. *American Orthopsychiatric Association Research Monographs*, (3).

Bender, L. (1946). *Instructions for the use of Visual Motor Gestalt Test.* New York: American Orthopsychiatric Association.

Buros, O. K. (Ed.). (1978). *The eighth mental measurements yearbook.* Highland Park, NJ: The Gryphon Press.

Canter, A. (1966). A background interference procedure to increase sensitivity of the Bender-Gestalt test to organic brain disorder. *Journal of Consulting Psychology, 30,* 91-97.

Canter, A. (1968). BIP Bender test for the detection of organic brain disorder: Modified scoring method and replication. *Journal of Consulting and Clinical Psychology, 32,* 522-526.

Cronbach, L. J. (1971). *Essentials of psychological testing* (3rd ed.). New York: Harper & Row.

Fuller, G. B. (1969). *Minnesota Percepto-Diagnostic Test* (rev. ed.). Brandon, VT: Clinical Psychology Publishing Co.

Fuller, G. B., & Laird, J. T. (1963). The Minnesota Percepto-Diagnostic Test. *Journal of Clinical Psychology, Monograph Supplement, 19*(16).

Goldberg, L. R. (1959). The effectiveness of clinicians' judgments: The diagnosis of organic brain disease from the Bender-Gestalt test. *Journal of Consulting Psychology, 23*, 25-33.

Hutt, M. L. (1977). *The Hutt adaptation of the Bender-Gestalt test* (3rd ed.). New York: Grune & Stratton.

Hain, J. D. (1964). The Bender-Gestalt Test: A scoring method for identifying brain damage. *Journal of Consulting Psychology, 28*, 34-40.

Heaton, R. K., Boade, L. E., & Johnson, K. L. (1978). Neuropsychological test results associated with psychiatric disorders in adults. *Psychological Bulletin, 85*, 141-162.

Koppitz, E. M. (1964). *The Bender-Gestalt Test for Young Children.* New York: Grune & Stratton.

Koppitz, E. M. (1975). *The Bender-Gestalt Test for Young Children: Vol. II. Research and application, 1963-1973.* New York: Grune & Stratton.

Keogh, B. K., & Smith, C. E. (1961). Group techniques and proposal scoring system for the Bender-Gestalt with children. *Journal of Clinical Psychology, 17*, 172-175.

Lubin, B., Wallis, R. R., & Paine, C. (1971). Patterns of psychological test usage in the United States: 1935-1969. *Professional Psychology, 2*, 70-74.

Pascal, G. R., & Suttell, B. J. (1951). *The Bender-Gestalt Test: Quantification and validity for adults.* New York: Grune & Stratton.

Mary M. Wellman, Ph.D.
Director of School Psychology, Department of Counseling and Educational Psychology, Rhode Island College, Providence, Rhode Island.

BENTON REVISED VISUAL RETENTION TEST

Arthur Benton. San Antonio, Texas: The Psychological Corporation.

THE ORIGINAL OF THIS REVIEW WAS PUBLISHED IN TEST CRITIQUES: VOLUME III (1985).

Introduction

The Benton Revised Visual Retention Test is a widely used instrument that assesses visual perception, visual memory, and visuoconstructive abilities. Because it measures perception of spatial relations and memory for newly learned material, it is used in clinical diagnosis of brain damage and dysfunction in children and adults, as well as in research. The Benton, as it is usually called, has three alternate forms, each of which consists of ten designs. In addition, there are four possible modes of administration.

The author, Arthur Benton, is an internationally recognized authority in the field of neuropsychology. He received his Ph.D. in clinical psychology from Columbia University in 1935. For many years he was Professor of Neurology and Psychology at the University of Iowa and is now Professor Emeritus. In addition to his well-known work with adult neurologically impaired patients, he has been intensely involved in research concerning developmental dyslexia. His long and fruitful career has yielded a plethora of publications.

The Benton was originally introduced in 1946 as a test of short-term, visual memory. In this regard, it was used as a corollary instrument with the Digit Span Test, which measures short-term auditory memory. These measures were used together as part of a clinical examination for brain damage in adults. In the years following World War II, such instruments were needed to assess neuropsychological deficits in servicemen who had suffered head trauma. The original Benton was standardized on 155 males and five females, most of whom were patients in a naval hospital. The first edition had two alternate forms, each consisting of seven stimulus items. Only one administration, that of a ten-second viewing and then immediate recall by drawing, was used. This is now known as Administration A.

In 1950 the multiple-choice version was introduced. It was designed for patients whose motoric disabilities prohibited them from drawing the designs. Thus, the task was changed from a recall memory to a recognition memory one. The patient was to indicate which of four choices was the same as the stimulus item.

The second edition of the standard 1946 Benton was introduced in 1955. Ten stimulus items were used, a copying administration was added, a more complete standardization was included, an error score and number correct score were initiated, and the directions for administration were made more precise.

The third edition was published in 1963, and the fourth, most recent edition appeared in 1974. This latest edition includes norms for the copying (C) administration and has a complete chapter that reviews recent literature concerning the test.

The fourth edition has four administrations (A, B, C, and D) and three alternate forms (C, D, and E). Each of these forms consists of ten designs printed on $5\frac{1}{2}$" x $8\frac{1}{2}$" cards, making a total of 30 cards, which are bound in a spiral booklet. Two additional forms (F and G) rely on the examinee's recognition, rather than on their reproduction ability, and each of these consist of 15 stimulus cards. The examiner uses a record form to enter the scoring. A 92-page manual details the administration, scoring, norms, interpretations, and recent literature.

The examinee is given ten blank, white $5\frac{1}{2}$" x $8\frac{1}{2}$" pieces of paper (e.g., $8\frac{1}{2}$" x 11" paper cut in half) and a pencil. Depending on which administration is used, the examinee either reproduces each design from memory or copies each design while looking at it. In administration A, each design is exposed for ten seconds, then removed, with the subject drawing by immediate recall. The same procedure is used in Administration B, except that each design is exposed for only five seconds. In Administration C, the examinee copies each design while it remains in full view, and in Administration D, each design is exposed for ten seconds, then the examinee waits 15 seconds before reproducing the design. During the viewing time, the design booklet is positioned like an easel, so that the design is at an angle of approximately 60° from the surface of the table, rather than being laid flat on the table, as in the administration of the Bender Gestalt.

Beginning with Design III, each design includes two major and one peripheral minor figures, and the examiner must direct the examinees to draw everything that they see.

Sometimes an examinee will start to draw before the viewing time limit has expired. Benton (1974, p. 2) suggests that the examiner should direct the examinee to "study the card for the full time of exposure." With impulsive children and adolescents, this reviewer has found that directing the examinee to lay the pencil on the table after the execution of each design, and then nodding to the examinee to pick up the pencil again as the new design is being removed, facilitates correct administration procedure. Also, the drawings should be numbered by the examinee to identify the spatial orientation of the drawing on the paper, as well as which design was drawn.

The Benton has norms for children and adolescents aged 8-14 years, as well as adult norms for examinees aged 15 years and older. Further, the norms include intelligence classification levels and age levels, so that the examiner can determine the expected number correct score and error score for an examinee according to both age and intellectual functioning. The number correct score has a range of 0 to 10, as each design is rated on an all-or-none basis and scored either 1 or 0. Scoring of errors enables both a qualitative and a quantitative analysis of the examinee's performance. Scoring is recorded and summarized on the one-page record form that allows the examiner to designate which designs were completely correct, as well as record, describe, and summarize types of errors for each design.

Practical Applications/Uses

The Benton is a valuable assessment tool in the diagnosis of visual perception, memory, and two-dimensional visuoconstructive abilities. As such, it is used by clinical and school psychologists in the assessment of neurological dysfunction and impairment. Because the Benton contains both drawing from memory and copying administrations, it can be used to evaluate the client's short-term visuo-spatial memory, visual perception, and graphomotor abilities. Thus, the test can be used in a variety of situations. For example, a school psychologist may need to determine if a child has a visual-motor dysfunction and needs special education services. A clinical psychologist evaluating a chronic drug-abusing adolescent may need to determine if the subject's short-term visual memory has become impaired. A clinical neuropsychologist evaluating an adult head-trauma patient may need data on the patient's current visuoconstructive abilities.

It is interesting to note that third-party payment state agencies, such as Medicaid and private insurers, often request specific tests to be used in a battery for various kinds of assessments. The Benton is listed (Commonwealth of Massachusetts, 1984, p. 6-2) as one of the "approved" tests to be used in the assessment of organicity.

As indicated above, the Benton should be used as only one instrument in a complete battery of tests to assess neurological dysfunction, rather than as the only measure of organicity. However, this caveat applies to assessment in all realms of psychological functioning. The Benton is not appropriate for use by non-psychologists. In fact, in order to purchase this test from The Psychological Corporation, the buyer must furnish a state license number or proof of membership in the American Psychological Association.

Forms F and G, which rely on the examinee's recognition rather than reproduction ability, are multiple-choice in format. Each of the 15 stimulus cards of these two forms corresponds to a four-choice response card. Designed to test the visuospatial memory of examinees with motor handicaps, these forms can be used by stroke patients with hemiparesis, individuals with cerebral palsy who can fist-point or head-point if they are unable to finger-point, or any other examinee who is unable to hold a pencil and draw the responses. Further, Form F or G can be used in conjunction with Form C, D, or E for persons without handicaps, in order to define more clearly whether the examinee's disability lies in the area of perception, memory, or motoric reproduction.

The Benton should be administered individually by a psychologist in a quiet room that is free from both visual and auditory distractions. Because this is primarily a test of memory, stimuli that might cause interference in short-term memory processing should be eliminated from the environment. Such stimuli include the examiner's verbalizations during administration of the test. In this reviewer's research, when using a tachistoscopic, visuospatial short-term memory task (Wellman & Allen, 1983; Wellman, 1985), it was noted that extraneous verbal interference reduced subjects' performance levels. When administering the Benton, this reviewer tries to give all directions at the beginning of the testing and makes any necessary comments after one design has been executed and before the next design is presented for viewing. Invariably, graduate students learning to administer the test need to be cautioned about this, as they attempt to verbalize directions to examinees while they are viewing or drawing the stimuli. The testing manual cautions the examiner to terminate conversations initiated by the examinee when using Administration D, which contains a 15-second delay period between the end of the viewing period and the onset of the drawing of the stimulus. Further, the manual directs the examiner using Administrations A, B, and C to present the cards without comment. However, the rationale for the need for silence could be given to further underscore its importance.

In general, the instructions for administration are simple, clear, and provide appropriate examiner responses to a variety of situations that may occur during testing. Although the manual states that administration takes about five minutes, it may take approximately ten minutes with compulsive or indecisive examinees. The designs are to be administered in the order presented in the spiral-bound booklet.

The scoring system, although somewhat complex, is also presented clearly in the manual. When teaching the scoring system to graduate students, usually about two hours of didactic instruction are needed, followed by sample protocols to score for practice. After administering and scoring three or four tests and receiving feedback and correction, students usually master the major categories of error scores and by their tenth protocol are usually fairly proficient in the complete scoring system.

Once the scoring system is mastered, an examiner can usually score one Benton in five to fifteen minutes, depending on the number and complexity of the errors. The only difficulties in scoring arise when an error may be scored in more than one way, but the manual gives explicit directions for resolving such problems.

Two scoring systems, a number correct score and an error score, are used. For the number correct score, each design is scored as either correct (1) or incorrect (0) for a possible total of ten points. In the error scoring system, six main categories are used to classify errors: omissions, distortions, perseverations, rotations, misplacements, and size errors. Each major category contains a variety of specific error subtypes relating to the position and degree of the error in the examinee's drawing. Each error subtype has an identifying abbreviation that is written on the scoring sheet to specify the subtypes of errors made in each drawing. The number of errors for all ten items is totaled and becomes the error score. The number of errors for each major category is also totaled and entered on the record form. The examiner then consults the norm tables in the manual to determine the expected number correct score based on the age and IQ of the examinee. The expected number correct score is then compared with the examinee's actual (obtained) number correct score. The same procedure is then followed for the error score.

One difficulty associated with the use of the norm tables for adults with brain injury is the need for estimating the examinee's "premorbid IQ." If no previous record on the patient is available, the clinician must rely on such evidence as educational and occupational history. Unfortunately, this problem occurs in doing any neuropsychological assessment of brain-injured patients when no early record is available.

The norm tables for children list expected scores based on age and "estimated

IQ." This seems unnecessarily imprecise, as an individually administered IQ test is almost always given as part of the evaluation for neurological dysfunction.

Norm tables for adults and children are available for Administrations A, B, and C, but no norms are presented for Administration D in the 1974 manual. Dr. Benton notes that the standardization data on this measure are not yet sufficient to provide suitable norms. Norms are available for Forms F and G. These norms take into account both age and intelligence level.

Machine and computer scoring are not available or warranted. The nature of the scoring is such that it can be performed easily by a trained examiner, but the judgment needed to score the designs makes scoring by a computer a complex programming task. Because no scoring instruments other than the manual and the half-page record form are needed there are no problems with the mechanics of scoring.

Test interpretation is based on an assessment of the number and types of errors made and involves several levels of analysis for diagnostic purposes. At the first level, for Administrations A and B, the examiner compares the examinee's obtained scores with the expected scores found in the norm tables. When examining the difference between these scores for the number correct, the wider the discrepancy in favor of the expected score, the more probable it is that the examinee has suffered neurological impairment. Thus, the manual (p. 43) states that for adults an obtained score of two points below the expected score raises the question of brain dysfunction, three points below suggests impairment, and four or more points below is a strong indication of such impairment. In like fashion, when interpreting the difference between the obtained and expected error scores, an obtained error score of three points greater than the expected error score raises the question of impairment, four points greater suggests such impairment, and five or more points greater is a strong indication of such dysfunction.

The same procedure is followed when interpreting children's number correct and error scores for Administrations A and B. The only difference is that the category of "strong indication" of impairment has been omitted, leaving the first two categories with the same obtained-expected score discrepancy as for adults.

Several norm tables are provided for Administration C. A table of error scores for adults provides five interpretive categories ranging from average to grossly defective. This table does not provide data by age and IQ, but is merely a general table for all adults. Children's norms for Administration C provide "critical scores" based on age, with a dichotomous split of IQ levels above and below 115. The "critical score" indicates the performance level that is at the tenth percentile or lower for children of each age level. The manual states that more precise classification is not reported because of a small sample size at some age levels.

Because no norms are presented for Administration D, only empirical observations are presented to facilitate interpretation of this procedure.

Beyond the first level of interpretation, a number of qualitative diagnostic guidelines are reported in the manual. Many of these guidelines are reported from research on the Benton, which is cited in the manual. First, the author cautions that poor performance does not necessarily signify neurological dysfunction. Rather, an examinee suffering from various psychiatric disorders, such as severe depression, schizophrenia, paranoia, or antisocial personality, will often display less than adequate performance. Further, patients with serious medical conditions that are not neurological in nature may show inadequate reproduction of the designs.

Studies of the performance of brain-injured adult patients and various samples of children on the Benton are discussed in the manual, with the aim of providing data to assist in the process of differential diagnosis. In a study cited without authors in the manual (p. 52) the scores of 100 brain-diseased patients were compared to the scores of 100 patients with no brain disease. Only 4% of the control patients, as compared to 51% of the brain-diseased group, had a number correct score of three or more points below their expected score. Thus, at that level of score discrepancy, very few false positives, but a number of false negatives, would be identified. Although it appears that the Benton does not have a high power of discrimination between brain-injured and normal adults, according to Von Kerekjarto (1962, as cited in Benton, 1974) it does discriminate better than other similar measures, such as the Bender-Gestalt using the Pascal-Suttell scoring system (Pascal & Suttell, 1951); the Graham-Kendall Memory for Designs test (Graham & Kendall, 1960); and the WAIS Block Design subtest (Wechsler, 1955).

The manual describes particular types of errors that are significant of specific kinds of cerebral dysfunction. Patients with unilateral spatial neglect often omit a figure, particularly the peripheral figure on one side of the designs, in their reproductions. The preponderance of evidence indicates that patients with parieto-occipital damage of either hemisphere are more apt to make errors than are patients whose brain damage occurs in the frontal lobes.

Some hemispheric difference seems to exist with regard to quantity of errors. Some patients with right-hemisphere lesions make more errors than patients with left-hemisphere lesions. It appears that even though the Benton contains visuo-spatial stimuli, many examinees use verbal mediation to encode and recall the designs; thus, the task is no longer strictly a visuospatial one, and less hemispheric differentiation is noted. However, when the multiple-choice form is used, differences have been reported concerning the hemisphere in which the lesion is located. Alajouanine, Castaigne, and Ribaucourt-Ducarne (1960, as cited in Benton, 1974) reported that patients with left parietal lesions performed better on the multiple-choice recognition form than on the drawing performance recall form. It is quite possible that less verbal encoding is needed when using only recognition memory. Patients with right parietal lesions had defective memory for the spatial relationships of the figures in the designs, characteristic of visuospatial dysfunction.

The Benton is not as useful in identifying children with brain injury as it is in identifying adults. However, significant discriminations which are of clinical value have been found. Thus, it is a valuable instrument to be used in combination with others in a diagnostic battery.

It can be seen from this discussion of the interpretation of the Benton that some knowledge of neuropsychology is needed by the examiner. Further, many studies (e.g., Brook, 1975; Marsh & Hirsch, 1982) in the past 20 years have reported on the use of this test with various populations, and so it behooves the clinician to be familiar with the recent literature, as the manual cannot provide a "cookbook" type of interpretation for the complexities and subtleties of neuropsychological assessment.

Technical Aspects

A number of studies on the reliability and validity of the Benton have been reported. Interrater reliability has been reported by Wahler (1956), using 29 randomly selected protocols from a pool of tests of 48 brain-injured and 50 normal adults and two scorers. Wahler states that interscorer agreement for total error score is high ($r = .95$) and for major categories of errors reliability is moderate to high ($r = .66$ to $.97$). The category of errors having the lowest interscorer reliability is substitutions ($.66$), whereas distortion and omission/addition categories produce a correlation of $.75$. All other categories have correlations over $.85$.

Alternate form reliability for Forms C, D, and E is high ($r = .85$) for Administration A (Benton, 1974). A number of studies reported in the manual indicate that Form C is slightly easier than Form D when participants are given Administration A. Using Administration C, no difference in the difficulty level of forms is observed. Weiss (1974), in a study of 106 Israeli subjects, found that, overall, although one form was no easier or more difficult than another, there was a concentration of particular types of errors on certain figures. After a complex qualitative analysis of these errors, he makes numerous recommendations for changing the scoring system. Unfortunately, these changes would make scoring extremely cumbersome and time-consuming.

No internal consistency measure has been described for these three forms, but the multiple-choice forms (F and G) are reported to have moderate internal consistency. Split-half reliability of these forms is $.76$ (Benton, 1950). To determine test-retest reliability, Brasfield (1971, as cited in Benton, 1974) gave Administration C to 194 kindergarten children twice in a four-month interval. Considering that the performance of five-year-olds is quite variable (Koppitz, 1964), the correlation of $.75$ between the two sets of scores is very promising.

Finally, in a study assessing the consistency of norms for children across samples, Brook (1975) used a cross-validation sample of subjects aged 8-17 years, presenting emotional but no organic problems, at a midwestern child guidance clinic. Brook found that the norms obtained were not significantly different from those presented in the manual for both number correct and error scores.

Criterion-related validity studies of the Benton are numerous. Largely, these are studies of concurrent validity that assess the relationship among subjects' scores on the Benton and their scores on other tests. Foremost among these have been several of the WAIS subtests, notably Block Design and Digit Span. Studies of constructional apraxia, an impairment in visual-motor synthesis ability involving mental organization of details to form a gestalt, have compared the Benton to the Block Design and other constructional tasks. Results of these studies (e.g., Benton, 1967; Benton & Fogel, 1962) show that the correlations are low among two-dimensional paper-and-pencil tasks (Benton Visual Retention Test), two- and three-dimensional constructional tasks (Block Design), and three-dimensional construction tasks (Three Dimensional Constructional Praxis Test). The authors interpret these findings as indicating that at least two types of constructional apraxia exist: a graphomotor type and an assembling type. Heilbrun (1960, as cited in Benton, 1974) found fairly low correlations between the Benton and WAIS Digit Span. The correlation was $.42$ among nonorganic subjects and $.26$ for brain-injured patients.

This is, in fact, a good test of discriminative validity because the Benton was originally designed to supplement the Digit Span in assessing brain damage. Because the Benton primarily assesses visuospatial synthesis abilities and the Digit Span assesses sequential verbal processing, low correlations between the two are expected.

Various studies have examined the ability of the Benton to assist in the diagnosis of brain injury and the differential diagnosis of brain injury as opposed to other mental disorders. Some studies suggest that the Benton has little diagnostic value. Schwerd and Salguerio-Feik (1980) administered the German edition to subjects aged 8-15 years with brain injury and suspected brain injury. Using a cutoff score of -3, they report that the test identified as brain injured 22% of the suspected brain injured, 24% of the true brain injured, and 6% of normals. An analysis of qualitative error scores was not reported in the study. Another study conducted in Germany by Velborsky (1964, as cited in Benton, 1974) concluded that the Benton had little power to discriminate among brain-injured, depressed, and schizophrenic patients. In contrast to Velborsky's results, studies summarized by Lezak (1976) indicate that the Benton has better discriminating power than other neuropsychological instruments in distinguishing brain-damaged patients from those with psychiatric disorders.

Other studies suggest more hopeful results. Marsh and Hirsch (1982) gave the Benton Form C and the Graham-Kendall Memory-For-Designs Test to 100 neurological patients diagnosed accurately by neurological examination and neurologic/radiologic laboratory tests at UCLA Medical Center. The Benton was significantly better than the Graham-Kendall in detecting neurological impairment ($p < .001$). The Benton correctly diagnosed 71% of the patients and misclassified 29%, using the number correct score, and correctly diagnosed 72% of the patients and misclassified 28%, using the error score. In contrast, the Graham-Kendall correctly identified only 35% of the neurological patients and produced 65% false negatives. Marsh and Hirsch suggest that the three geometric figures of each of the last eight Benton designs create a more demanding task and involve greater cognitive structuring than the single figures of the Graham-Kendall. The authors conclude that the Benton is valuable for use in evaluating persons with mild to moderate brain damage, although for severely impaired patients or the elderly, the Graham-Kendall is less threatening and also has norms for patients aged 70-90 years.

Petris (1980) administered the Benton in Italy to normal adults and children, as well as to brain-injured adults. Based on the results, nine indices of patients with various types of brain injury were devised by Petris for purposes of differential diagnosis.

Critique

The Benton has now been used clinically for 39 years. The fourth edition was published eleven years ago. In its metamorphosis, more elaborate directions for administration and more expanded norms have been introduced. The advantages of the Benton include short administration time and a precise and logical scoring system. Thus, it can be given and scored with little time expended. Because of its multiple-choice, drawing from memory, and copying administrations and forms, it can assist the clinician in discriminating among visual perceptual, visual memory and graphomotor visuoconstructive dysfunctions. Further, several research studies (e.g., Marsh & Hirsch, 1982) indicate that, although not highly valid, the Benton is better at determining brain injury than are several other similar instruments.

Weaknesses in the Benton include some problems in norming. The manual for the fourth edition states that over 600 subjects were used in the norming of Administration A. However, no indication is given as to the relative distribution of adults and children. As separate norms are given for each, it would be helpful to know the apportionment of the subjects. Further, the sample size of children for Administration C is 236, yet the author states that a more precise classification of performance (i.e., "critical score") was precluded because of the small sample size at some age levels. This could have been easily rectified by using larger samples. Finally, even in the fourth edition, the standardization data are not yet extensive enough to provide norms in the manual for Administration D.

In the complex field of neuropsychological assessment, no one simple instrument should be used alone. Unlike complete batteries, such as the Halstead-Reitan or Luria-Nebraska, the Benton is more valid as an instrument to detect specific dysfunction rather than to diagnose brain injury. Even the most promising research studies indicate a substantial percentage of false negatives when verified neurological patients are used. Thus, the Benton should be used to assess specific types of cognitive functioning as part of a larger battery of tests. As such, it is probably the best instrument of its kind.

References

Benton, A. L. (1950). A multiple-choice type of visual retention test. *Archives of Neurology and Psychiatry, 64,* 699-707.

Benton, A. L. (1967). Constructional apraxia and the minor hemisphere. *Confinia Neurologica, 29,* 1-16.

Benton, A. L. (1974). *The Revised Visual Retention Test* (4th ed.). New York: The Psychological Corporation.

Benton, A. L., & Fogel, M. L. (1962). Three-dimensional constructional praxis: A clinical test. *Archives of Neurology, 7,* 347.

Brook, R. M. (1975). Visual Retention Test: Local norms and impact of short-term memory. *Perceptual and Motor Skills, 40,* 967-970.

Commonwealth of Massachusetts. (1984). *Psychologist manual.* Boston: Department of Public Welfare, Medical Assistance Program.

Graham, F. K., & Kendall, B. S. (1960). Memory for Designs Test: Revised general manual. *Perceptual and Motor Skills, 11,* 147-188.

Koppitz, E. M. (1964). *The Bender Gestalt Test for young children.* New York: Grune & Stratton.

Lezak, M. (1976). *Neuropsychological assessment.* New York: Oxford University Press.

Marsh, G., & Hirsch, S. (1982). Effectiveness of two tests of visual retention. *Journal of Clinical Psychology, 38,* 115-118.

Pascal, G. R., & Suttell, B. J. (1951). *The Bender-Gestalt Test: Quantification and validity for adults.* New York: Grune & Stratton.

Petris, L. (1980). Diagnostic indices of organic cerebral damage in Benton's Visual Retention Test. *Bollettino di Psicologia Applicata, 153,* 79-87. (From *Psychological Abstracts,* 1982, *68,* Abstract No. 3939)

Schwerd, A., & Salguerio-Feik, M. (1980). Diagnostic value of the Benton Revised Visual Retention Test for children and adolescents. *Zeitschrift für Kinderund Jugendpsychiatrie, 8,* 300-313. (From *Psychological Abstracts,* 1982, *67,* Abstract No. 2517)

Wahler, H. J. (1956). A comparison of reproduction errors made by brain-damaged and control patients on a memory-for-designs test. *Journal of Abnormal and Social Psychology, 52,* 251-255.

Wechsler, D. (1955). *The Wechsler Adult Intelligence Scale.* New York: The Psychological Corporation.

Weiss, A. A. (1974). Equivalence of three alternative forms of Benton's Visual Retention Test. *Perceptual and Motor Skills, 38,* 623-635.

Wellman, M. M. (1985). Information-processing abilities in left- and right-handers. *Developmental Neuropsychology, 1,* 53-65.

Wellman, M. M., & Allen, M. (1983). Variations in hand position, cerebral lateralization and reading ability among right-handed children. *Brain & Language, 18,* 277-292.

Allan L. LaVoie, Ph.D.

Professor of Psychology, Davis & Elkins College, Elkins, West Virginia.

THE BLACKY PICTURES

Gerald S. Blum. Ann Arbor, Michigan: Psychodynamic Instruments.

THE ORIGINAL OF THIS REVIEW WAS PUBLISHED IN TEST CRITIQUES: VOLUME I (1984).

Introduction

The Blacky Pictures Test (BPT), a modified projective test consisting of a dozen carefully drawn cartoons, was designed to measure oral eroticism, oral sadism, anal sadism, oedipal intensity, masturbation guilt, castration anxiety, penis envy, identification, sibling rivalry, guilt feelings, ego ideal, love object relations, and narcissism. It is a modified projective test for several reasons, including the facts that it is so explicitly tied to psychoanalysis, that it uses such unambiguous test stimuli, and that it incorporates objective questions about each cartoon to supplement the scoring of clients' spontaneous stories.

The psychometric properties of the BPT have not been fully documented (e.g., reliability remains largely an unknown), but data collected in a variety of research settings encourage the belief that the test has construct validity. To illustrate briefly, Vernallis (1955; see Taulbee & Clark, 1982, for a complete reference) found bruxists scored higher on oral sadism as predicted, and Adams (1959) found prospective mothers who were going to breast-feed their babies had far less conflict on oral eroticism than prospective bottle-feeding mothers.

In addition to the use of the BPT in research, a recently reported survey (Lubin et al., 1984) showed that clinicians ranked it as the 30th most commonly used test in clinical settings; that figure is down slightly from 1969. Its use by clinicians has been as a diagnostic tool, to discover problem areas and to point out directions for therapy to proceed. Occasionally it seems to be used to monitor therapeutic progress. In short, the BPT is an innovative projective test that has been widely used in interesting ways. It has stood the test of nearly forty years, though it does need some work to bring it up to currently accepted standards.

The BPT was developed by Gerald S. Blum. The manual states that Blacky (the test's central cartoon character) was created in 1946 on a warm December day in California. At that time Blum was working on his Ph.D. in clinical psychology at Stanford University, a degree awarded in 1948 when he was 26. His earlier degrees were a B.S. in psychology from Rutgers and an M.A. in personality theory from Clark University. Blum's career has been spent at the University of Michigan, from 1948-1968, and at the University of California, Santa Barbara, where he has been chairman of the department. In addition he has had ten years of experience with the Veteran's Administration as a consulting clinical psychologist. Blum has received a number of research grants to advance his work in psychodynamics, was a Fulbright Scholar, and has published widely in personality and hypnosis. Perhaps his best known book is *Psychodynamics: The Science of Unconscious Mental Forces* (1966). Apparent in his many publications is the intention to discover how unconscious forces affect the mind and behavior, and the BPT seems a very natural part of that career effort.

The test's creation while Blum was still a graduate student grew out of his interest in developing measures of Freudian concepts so that psychoanalytic theory could be experimentally tested. The cartoons were explicitly designed to tap stages of psychosexual development, conflict about sexuality, and so on. This procedure made the BPT unique among projective tests in that its content was limited to the areas where interpretation was desired: other projective tests used more ambiguous stimuli, which produced much more varied subject responses and left far more scope for differences in interpretation.

While the BPT uses explicit stimuli for subjects to write spontaneous stories, Blum has failed to provide an adequate backdrop for interpretation. The manual says that a subject's becoming conscious of a cartoon's theme does not seem to affect the story much; to make that judgment requires some knowledge about normative performance. Exactly how much do scores differ as a function of awareness of purpose? Since no norms are included in the published materials, the judgment must be left to the expertise or intuition of the examiner. Surely in the more than 30 years since initial publication norms of some kind could have been developed.

The basic cartoons have not changed since the 1950 copyright. Schaeffer (1968) reports a Blacky the Cat series as well as a French translation; Blum (1956) reports an Italian version. Additional materials include the Blacky Analogies Test, a measure of cognitive performance with emotion-laden concepts, and the Defense Preference Inventory (DPI), neither readily available.

The standard BPT package comes with the 12 cartoons, 25 record blanks, male and female inquiry booklets, and the 1950 manual. The sequence of black-and-white cartoons, each roughly 8½″ by 11″, begins with the first card which simply introduces Blacky, the ageless, androgynous dog, and Blacky's family: Mama, Papa, and sibling Tippy. The remaining cartoons are arranged more or less according to the psychosexual stages from oral and anal to oedipal, latent, and genital. A professional artist seems to have prepared the cartoons, but even so there are ambiguous ones (e.g., V) that require a verbal introduction ("Here Blacky is discovering sex . . ."). There is room on the record blanks for the examiner to note the spontaneous stories. While the subject reads the questions from the inquiry booklet, the tester records the answers below the story. The main use of the blanks would be for clients who cannot read or write.

A typical BPT user would need far more information than the manual contains. At least the earlier monograph (Blum, 1949) and the later guide for research use (Blum, 1962) ought to be included with the test package. The manual also should include the DPI, and the scoring procedure that was revised in 1951 but never published. This combined information would permit the user to make informed choices about test interpretation systems, and should help make more consistent the kinds of inferences different examiners draw from the same protocol.

Practical Applications/Uses

Presently the BPT must be restricted to a very few applications until some of its flaws can be corrected. Professional psychologists with training in psychoanalysis would find it valuable in psychotherapy for revealing a client's psychosexual development, areas of particular conflict, fixations, and typical defenses. The typical therapist would thus have the benefit of considerable additional information in making interpretations of stories. All other uses of the test have been and should remain experimental. Of particular importance for researchers are these problems with the BPT: the lack of norms; no coherent (revised) scoring system for females; and no clear place in the nomological network, especially in comparison with other widely used tests.

Once the most serious limitations can be removed, there would be practically endless applications for the BPT. It could be used in an employee screening system (anal retentives in the accounting office rather than the shipping department). It could be applied in student advising and in helping students choose majors (some work along these lines has already been started). In marriage and family counseling, in milieu therapy, and in numerous other areas it may prove valuable.

Most BPT studies to date have dealt with adults, with a few assessing children down to the age of 5 years. Because of the clearer interpretive guidelines available for males some would say the test should be used only with males. When combined samples are used there are a multitude of sex differences on the standard dimensions (many of course consonant with psychoanalysis). Anecdotally, children of 5 and older enjoy taking the test, have little trouble with most of the inquiry items, and tend to get bored after the first few cartoons. Sex differences show up even in the youngest samples, as Freud might have predicted. Nonverbal patients would produce little of interpretive value.

Because the BPT requires a lot of time to administer and to score, its use as a screening device for psychiatric or personnel purposes would be restricted. It may be administered in groups, separately for males and females, to save time in administration. Nonetheless, there remains the very time-consuming hand scoring which demands a professional's expertise.

In clinical use the analyst should be the administrator as a standard procedure so that test-taking behavior can be incorporated into the scoring. In this reviewer's experience with individual testing the full session takes 45-60 minutes. After each of the spontaneous stories is written the client answers the inquiry items. After the last set of inquiries, the cartoons are sorted into "liked" and "disliked" piles and the client selects the most extreme one from each pile. If the DPI is given, more time is needed for the inquiry phase after each story.

In research applications the examiner has much more flexibility. Blum (1962) recommends using slides for group administration to reduce both experimenter effects and the amount of time subjects spend on stories and inquiry items (2 minutes each for the first, roughly 10 seconds each for the latter). In examining a variety of experimental procedures, it becomes obvious that many more administrative procedures exist. For example, Fisher (1966) did not obtain stories or use inquiry items—his subjects simply ranked their preferences for some of the cartoons. Others measure picture conflict by asking for recall of the pictures after some varying delay. Still others use semantic differential ratings of key cartoons as the scoring procedure.

In summary, the test typically takes close to an hour to administer in its fullest form, and should be administered by a professional so that test-taking behavior can be used to help with the scoring. In research settings, the test has been employed in a myriad of ways, from using individual cartoons to the full test. Group administration using timed presentation of slides seems the most economical method for large-scale research purposes.

A major decision confronts the user: should one score by the 1950 manual's instructions, supplemented with the 1949 monograph? (Clinicians probably should.) By the 1951 supplement? (Yes, if one can find a copy of it—this reviewer could not.) By the major restructuring in the 1962 article? (Definitely for research purposes, if one's design allows it.) By some other system? (As noted earlier, one would not be the first to innovate.)

Of the several techniques available for scoring this reviewer prefers the 1962 version, though it was derived exclusively from males' responses. It was derived empirically from scores obtained using the 1951 procedure and submitted to a factor analysis. The result is a 30-factor scoring system of a more refined but less immediately theoretically compelling nature. The 30 scores are organized around the original 11 dimensions. To illustrate, Anal Sadism, the original dimension, was originally scored for anal retentive or expulsive tendencies; now it can be scored for Exploitation, Neutrality, and Denial.

The chief advantages of the new system lie in the far clearer definitions and directions for scoring, and in the greater capacity for discriminating among subjects. In addition, where the old system trichotimized scores (very strong, strong, weak, or absent), the new system allows much more variation in scores (depending on how many inquiry items are answered in the keyed direction, on preference rankings, and on related themes from other stories). Scoring time for experienced examiners may be an hour or more, especially with prolific stories containing conflicted themes. If the group administration is used, with stories limited to 2 minutes, scoring will average 20-30 minutes. The time required is about the same with whichever scoring system once the examiner has committed it to memory.

A disadvantage of the 1962 scoring system is its distance from the original Freudian concepts. Formerly, if a subject had strong conflict on oral sadism, one knew fairly clearly what that meant. Currently one might question how to interpret a high score on IV-A, Undisguised Oedipal Involvement, when the client also scores high on IV-B, Disguised Oedipal Involvement. "Oedipal conflict" would obviously be the interpretation in this instance, but surely there must be something to distinguish this pattern from one where the client scores low on the first and high on the second. Much more experience with the 1962 system must be accumulated and published.

Another disadvantage concerns the meaning of the scores in an absolute sense. With the old system, a very strong score had some absolute meaning. But now what does a score of 4 mean on IX-B, Guilt-ridden Hostility toward Sibling? Does this indicate "extremely intense guilt . . ." (Blum, 1962, p. 22)? What about a score of 6? 8? A system of converting to standard scores would obviate much of this criticism. In the absence of standard scores, norms, or even large sample averages, the burden for interpreting scores rests with the examiner—a professional psychologist who has clinical experience, psychoanalytic training, BPT experience, and solid intuition. In these conditions, Zubin et al.'s (1965) criticism still applies: "A shortcoming . . . has been the subjective nature of the scoring" (p. 499). To use the DPI, one must obtain Nelson's dissertation (1955); it includes form M53 in the appendices.

Technical Aspects

A variety of experimental contexts have been used to validate the BPT, including: perceptual vigilance and defense (Blum, 1954, 1955; Nelson, 1955); international comparisons of the use of different defense mechanisms (Blum, 1956, 1964); clinical study of peptic-ulcer patients (Blum & Kaufman, 1952); academic performance (Galinsky, 1971); breast-feeding (Adams, 1959); bruxism (Vernallis, 1955); verbal conditioning (Cooperman & Child, 1971); stuttering (Carp, 1962); and castratingness (Touhey, 1977). (See Taulbee & Clark, 1982.) No critical review of this literature exists, and room prohibits the attempt here. Instead, a few observations and comments will be noted.

Most of the early studies were tentative, exploratory, and post hoc. The first research efforts attempted to match the BPT with psychoanalytic concepts rather than trying to establish its ability to predict independent complex behaviors. The early syntactical studies gave way to efforts to revise and refine the test (e.g., Blum, 1962) and to discover its strict psychometric properties (e.g., Robinson & Hendrix, 1966). Several published papers of Blum and colleagues depended on their study of 44 fraternity boys (e.g., the DPI; Blum, 1962; Cohen, 1956; Nelson, 1955); those data have probably had a disproportionate influence on the literature. If a clear conclusion emerges from the study of the literature, it echoes Beck's (1956) comment in *The Fifth Mental Measurements Yearbook:* "The test has the promise warranting its development" (p. 216). Perhaps in the intervening 28 years more of the warranted development should have taken place.

Two studies, examined in some depth, illustrate the versatility of the BPT and its potential for revealing important aspects of psychodynamics. Touhey (1977) used the BPT, including the stories and inquiry responses to cartoon VI, to classify 87 female college students into high or low penis-envy groups. These subjects were then shown a film depicting mutilation (subincision) of male genitalia. Half of each subgroup was instructed that the film showed a painless rite of passage that did no permanent damage; the other half was told that these men had been found guilty of sexual transgressions and this painful, incapacitating treatment was their punishment. The students were then invited to evaluate the film on seven scales, such as whether they approved of the practice, whether they thought it ought to be adopted in the U.S., and whether they would be willing to see the film again. Touhey predicted and found that the women classified as high penis-envy in the subgroup that received the punishment instructions were significantly more approving than the other three subgroups. This outcome is consistent with the castrating-woman stereotype. Touhey also found that regardless of condition the high-penis-envy women were more favorable towards the practice. With this study Touhey corroborated a controversial feature of psychoanalytic theory and supported the validity of the BPT.

In the second study Blatt (1964) examined how research scientists described themselves in terms of Murray's need system. The individual self-descriptions were then compared with an ideal ranking of the 20 needs as they would exist in an optimal personality. Whenever deviations from ideal showed up, they were compared with conflict as revealed by the BPT. Blatt reported that scientists with conflict on oral eroticism had deviant self-ratings on Autonomy (dependence), while those conflicted on the anal-retentive dimension had deviant needs for Order. These findings clearly help to support the conclusion that the BPT measures Freudian concepts in a meaningful way.

To summarize, the validity of the BPT seems adequate for use in groups, and probably could be developed to the point that accurate, individual predictions were possible.

No easy method of measuring the reliability of projective tests has been demonstrated. The quick and easy measures of internal consistency do not apply and when the test measures psychodynamics, which are subject to instinctual vicissitudes, even the test-retest method may not be appropriate. Charen (1956a) estimated the test-retest stability of the inquiry responses, found it was inadequate, and even found some negative correlations. Blum (1956) has objected to Charen's procedure on several grounds, some of which Charen rebutted (1956b), but the fact remains that the test author has a clear responsibility to estimate his test's reliability. This reviewer believes Blum has notably failed in this matter. A few supportive estimates of reliability exist, but in general inadequate information exists to conclude anything about BPT score stability or homogeneity.

Critique

To paraphrase the riddle of the sphinx, the BPT came in on four legs and found immediate productive applications in testing psychoanalytic theory and in clinical work. In its midlife, while it did not walk around on two legs, there still were signs of maturity as clinicians used it more and more, and burgeoning research interest led to its incorporation in an impressive array of different designs. Now, not quite on three legs, there may still be signs of age. Clinicians use it nearly as much today as in 1969, but dissertation research occurs less frequently with the BPT as a focus, and fewer research reports can be found today.

The problems with the BPT can be summarized easily. No norms exist. The preferred scoring procedure is based on males, who probably comprise the real clients for this test. Insufficient work on score stability has been published, and what has been is mixed. The inquiry items, which follow each cartoon, may very likely bias the client's story for the next cartoon. The manual is horribly inadequate. The DPI is not easily obtainable. And, what may be the most serious shortcoming, Blum's advocacy of the test has declined as his research focus has changed, and no one has come forward to replace him to use Blacky in new and dramatic ways, to defend against unfair criticism, and to continue to improve the test in response to just criticism.

Naturally the test contains some strengths. When it was introduced the Blacky Pictures Test was innovative in being so closely tied to psychoanalysis; those ties still exist. It was innovative then for its use of projective and objective material in scoring and that is probably still a strength (allowing, as it does, both unconscious and conscious material, respectively, to influence the scores). The test can clearly uncover, in at least some subjects, deep conflict that can be related to real-life behaviors. College students and children appear to enjoy taking the test, and the former enjoy seeing the results. In this reviewer's experience the test results have catalyzed important insights in clients. For these reasons clinicians are urged to use the test on at least a trial basis; and researchers are urged to use it in appropriate experiments to add to our knowledge both of this test and of psychoanalytic theory.

References

Literature cited in the text but not found here may be found referenced in Taulbee and Clark, 1982.

Blum, G. S. (1949). A study of the psychoanalytic theory of psychosexual development. *Genetic Psychology Monographs, 39,* 3-99.

Blum, G. S. (1956). Defense preferences in four countries. *Journal of Projective Techniques, 20,* 33-41.

Blum, G. S. (1962). A guide for research use of the Blacky Pictures. *Journal of Projective Techniques, 26,* 3-29.

Blum, G. S. (1966). *Psychodynamics: The science of unconscious mental forces.* Belmont, CA: Wadsworth Inc.

Blum, G. S. (1968). Assessment of psychoanalytic variables by the Blacky Pictures. In P. McReynolds (Ed.), *Advances in psychological assessment* (Vol. 1, pp. 150-168). Palo Alto: Science and Behavior Books.

Blum, G. S., & Hunt, H. F. (1952). The validity of the Blacky Pictures. *Psychological Bulletin, 49,* 238-250.

Blum, G. S., & Miller, D. R. (1952). Exploring the psychoanalytic theory of the "oral character." *Journal of Personality, 20,* 287-304.

Charen, S. (1956). A reply to Blum. *Journal of Consulting Psychology, 20,* 407.

Lubin, B., Larsen, R. M., & Matarazzo, J. D. (1984). Patterns of psychological test usage in the United States: 1935-1982. *American Psychologist, 39,* 451-454.

Massong, S. R., Dickson, A. L., Ritzler, B. A., & Layne, C. C. (1982). A correlational comparison of defense mechanism measures: The Defense Mechanism Inventory and the Blacky Defense Preference Inventory. *Journal of Personality Assessment, 46,* 477-480.

Nelson, S. E. (1955). *Psychosexual conflicts and defenses in visual perception.* Unpublished doctoral dissertation, University of Michigan, Ann Arbor.

Taulbee, E. S., & Clark, T. L. (1982). *A comprehensive, annotated bibliography of selected psychological tests.* Troy, NY: Whitston Publishing Co.

Touhey, J. C. (1977). "Penis envy" and attitudes toward castration-like punishment of sexual aggression. *Journal of Research in Personality, 11,* 1-9.

Zubin, J., Eron, L. D., & Schumer, F. (1965). *An experimental approach to projective techniques.* New York: John Wiley & Sons.

George Domino, Ph.D.
Professor of Psychology, University of Arizona, Tucson, Arizona.

CALIFORNIA PSYCHOLOGICAL INVENTORY

Harrison G. Gough. Palo Alto, California: Consulting Psychologists Press, Inc.

THE ORIGINAL OF THIS REVIEW WAS PUBLISHED IN TEST CRITIQUES: VOLUME I (1984).

Introduction

The California Psychological Inventory (CPI) is a personality assessment inventory, intended primarily for use with normal subjects. It contains 18 standard scales designed to assess those personality characteristics that are important to the understanding and the prediction of individual behavior, particularly in a social context.

The author of the CPI is Harrison Gough, a graduate of the University of Minnesota at a time when the MMPI and "dustbowl empiricism" were primary. Professor of psychology at the University of California at Berkeley since 1949, he is also the director of the Institute of Personality Assessment and Research, author of a variety of test instruments including the well-known Adjective Check List, and a prolific contributor to the professional literature.

The first CPI scale, the Capacity for Status, was published in 1948. By 1951, 15 scales were in existence, and the first copyrighted edition of the CPI came to life. In 1957 Consulting Psychologists Press published the full 18-scale CPI. The CPI manual has been revised at frequent intervals, but the 480 items making up the CPI have not, thus providing the clinical user and the researcher with both stability and current information.

As Gough states in the manual the CPI was developed with two goals in mind. The first goal, a theoretical one, was to develop descriptive dimensions with broad personal and social relevance. Gough points out that the variables assessed by our psychometric tools, especially personality inventories, come from one of three areas. The first is the development of variables that come from a specific theory. As an example, the Myers-Briggs Type Inventory is clearly embedded in a Jungian framework, and the variables assessed (intuition vs. sensation or judging vs. perceiving) are central concepts in Jungian typology. Similarly, the needs assessed by the Edwards Personal Preference Schedule have their roots in the personality theory of Henry Murray.

A second approach to the development of variables is to create new variables not tied to established theoretical frameworks. This approach is most clearly exemplified by the work of Raymond Cattell, the author of the Sixteen Personality Factor Questionnaire (16 PF). The variables on the 16 PF were developed through factor analysis, and although some of the names given to these variables are eminently sensible (e.g., intelligence), many are neologistic (e.g., parmia, alaxia) and clearly point to their "newness."

Still a third approach, the one used by Gough, is to examine the setting in which the test is to be applied and to utilize those variables that are already operational in that setting. A clear example of this approach is the Strong Vocational Interest Blank, for which the setting is the world of work and the variables operational are occupations; these occupations are not dictated by a theory, although of course they may provide some basic data for a theoretical framework. Nor are the occupations invented, although a factor-analytic approach can yield valuable information on the underlying similarities and differences.

For the CPI the setting is the normal everyday world, and the variables are what Gough calls "folk concepts," those concepts used by people to describe everyday personal and interpersonal personality characteristics, particularly as they relate to favorable and positive aspects of functioning.

The second goal in the development of the CPI was the practical one of developing "brief, accurate, and dependable" subscales that would operationalize the folk concepts identified. Gough does not claim that the 18 scales of the CPI cover all folk concepts, or even the most important ones. He does claim that the 18 scales, taken one at a time or in combination, have some relevance to most important everyday behaviors. He is at the same time fully aware of the psychological complexities of everyday life and of the psychometric limitations of our instruments, and is quite content with modest relationships.

One ramification of the notion of "folk concepts" is that CPI variables should have cross-cultural relevance. If the concept of responsibility is a true folk concept, the scale ought to be relevant in other cultures even though the scaling of this concept may be done in one particular culture. This does not imply invariance; i.e., in one culture a high degree of responsibility may lead to occupational achievement, while in another culture such a relationship may not exist. However, the folk concept of responsibility ought to be relevant in both cultures.

A second ramification lies in that the variables are meaningful and readily understood by the user. One need not have an advanced degree in clinical psychology or be trained as an analyst to understand what "tolerance," "flexibility," or "capacity for status" mean. To be sure, test interpretation is both an art and a science: some users will be virtuosos in their interpretations, and the scales themselves will contain subtleties and nuances that can only be learned through clinical experience and familiarity with the literature.

A third ramification is that the CPI variables, taken individually or in combination, analyzed either at the item level (as in the development of new scales) or at a more global level (as in the development of regression equations), ought to be relevant to problems and issues in interpersonal behavior.

Eleven of the 18 CPI scales were developed empirically. This method involves first the definition of a criterion dimension, such as dominance, tolerance, or intellectual efficiency. Secondly, a pool of items is assembled. Most of these items will be selected because of their theorized psychological relevance to the criterion dimension; some may be borrowed from other scales that presumably measure the same domain, while still others will be added because of personal preference and hunch.

In the third step the developer administers the pool of items to a sample of subjects who can be shown, by a procedure other than the test, to occupy different positions on the criterion dimension. These independent procedures can be observer or self ratings, behavioral indices such as being a scholarship recipient, scores on other tests, or composite indices that may reflect various sources of information. Typically a contrasted group approach is used, in which individuals high and low on the criterion dimension are identified and their responses to the pool of items are statistically analyzed. The goal of this item analysis is to identify those items that show a differential response pattern between the two groups, or a high correlation with the criterion dimension.

Once a group of items which statistically discriminate the two contrasted groups has been identified, these items are selected for further study and refinement. This at least should involve a cross-validation on a new sample; often the criterion-dimension is redefined operationally, using a new procedure. The keyed response for each item is thus determined empirically and usually reflects the response given by subjects high on the criterion dimension. The end result is a scale, typically consisting of 30 to 40 items, that bears a demonstrable relationship to the criterion dimension.

Four of the CPI scales were developed by the technique of internal-consistency analysis. This approach also begins with the assembling of a pool of potential items that are presumably related to the criterion dimension in question. The second step, however, requires the experimenter to determine what the scoring weights will be. For example, if one were developing a scale of depression by this method, the item "I am depressed" ought to be keyed true. The pool of items is then administered to a sample of subjects, and the protocols are scored. Subjects with the highest and the lowest total scores are identified, and an item analysis is carried out to determine which items do significantly differentiate the two subsamples.

The problem with this approach is that the experimenter may be wrong in the initial scoring decision or the item may in fact not discriminate empirically, even though logically it ought to. Gough suggests that this approach, though limited, is useful when obtaining large samples of criterion subjects is not feasible.

There are three additional CPI scales that were developed somewhat uniquely; these scales are "validity" scales, but also have personality implications. The Sense of Well-Being Scale (Wb), originally called the Dissimulation Scale, was intended to identify individuals who underestimate their well-being and exaggerate their worries and misfortunes as opposed to those individuals who portray a relatively accurate picture of their problems and concerns. The intent here is to distinguish between warranted and unwarranted complaints, and the scale was developed by comparing the responses of hospitalized psychoneurotic patients with those of normal subjects asked to feign neurosis. As with most other CPI scales, several analyses with different criterion samples were carried out to insure that the items included in the Wb scale did indeed function validly and not as a result of chance or sampling artifacts.

The Good Impression Scale (Gi) is essentially a measure of social desirability. It was constructed by administering a pool of items twice to research samples, first using standard instructions and then instructions to present the best possible portrait of oneself. Those items which showed statistically significant changes from the first to the second administration were selected.

Finally, there is the Communality Scale (Cm), originally labeled as the Infrequency Scale. This scale is composed of 28 items answered in the keyed direction by 95% or more of the subjects in the normative samples.

Although most of the scales have undergone some revisions, in the sense of fine-tuning the CPI as a published inventory has remained the same. However, Gough is currently working on a minor revision which will result in a deletion of some 20 items and the addition of two new scales: Empathy and Independence. In addition, the revised manual will discuss scales in three groups: 1) the folk concept measures (i.e., the current 18 scales and the two new ones); 2) the special purpose scales, indices, and regression equations that have been developed by Gough and others (e.g., social maturity, Type A living style); and 3) structural scales.

These last scales represent the explication of a theoretical model that has always been implicit in the CPI. Gough indicates that factor and smallest space analyses have shown that there are three major themes in the CPI: role, character, and competence, and he has developed three scales, v.1, v.2, and v.3, to assess these themes (the *v* is for vector). Role involves the interpersonal presentation of self and is currently reflected by the Dominance, Capacity for Status, Sociability, Social Presence, and Self-Acceptance scales; the new structural scale to mark this theme is v.1, composed of 31 items. Character involves intrapersonal values, as currently assessed by the Responsibility, Socialization, and Self-Control scales; the new structural scale for this theme will be v.2 (36 items). Finally, competence involves such scales as Intellectual Efficiency, Sense of Well-Being, Tolerance, Achievement via Conformance, and Achievement via Independence, and will be marked by v.3 (58 items). The three vector scales have substantial correlations with the pertinent CPI folk concept scales (e.g., v.1 correlates −.78 with the Dominance scales for a sample of 1,000 males), but have approximately zero intercorrelations with each other.

Using vectors 1 and 2 and the median as a cutoff point generates a four-fold typology: As are low on v.1 and high on v.2, and are outgoing, involved, rule-favoring persons. Bs are high on both vectors, and are described as internalizers, inwardly oriented, rule-favoring individuals. Cs are low on both vectors, and are outwardly oriented, socially adept, rule doubters. Finally, Ds are high on v.1 and low on v.2, and these are inwardly oriented, rule-doubting persons. In normative samples, close to 25% of the subjects fall in each of the four cells, although specialized samples (such as psychologists) would not show this distribution.

Vector 3 is divided into seven levels of competence, with obtained frequencies close to the theoretical distribution of 8%, 12%, 19%, 22%, 19%, 12%, and 8%. This typology then generates a total of 28 configurations (4 types X 7 levels) and provides a bridge between the nomothetic aspects of personality assessment (e.g., successful business managers are . . .) and the idiographic (Ellen is . . .).

There is currently only one form of the CPI, but a shortened version is available (Burger, 1975) consisting of 240 items. Correlations between the abbreviated and standard scales range from .78 to .93, with a median of .88. Gough has always argued that the full CPI item pool should be given so that it can be subsequently scored on new scales or in ways that had not occurred to the investigator when the data were collected. The savings in time that any short form can give must be balanced with considerations of utility, psychometric validity, and research potential.

Because of its cross-cultural focus, foreign language editions of the CPI abound. The 1984 catalog of Consulting Psychologists Press lists the availability of Italian, Spanish, and German editions, and individual researchers have translated the CPI into Chinese, Turkish, French, and other languages.

The CPI consists of a 12-page reusable booklet with 480 true-false inventory items. Twelve items are repeated, so that there are 468 different items. Of these, approximately 178 are virtually identical with MMPI items, so that the CPI has been characterized rather inappropriately as the "sane man's MMPI." From the viewpoint of the subject a number of items do appear "strange," addressing paranoid fears of picking up diseases from doorknobs or various hysterical and psychosomatic symptoms. The majority of items, however, do seem directly relevant to the concerns and behaviors of everyday life.

The test is a self-administered paper-and-pencil personality test with no time requirement. The instructions are clear, so that the role of the examiner as administrator is minimal. This does not negate the importance of establishing rapport, minimizing outside distractions, exercising appropriate control over test materials, and eliminating potential collusion in group settings.

The CPI requires a fourth-grade reading level (though the test can be administered orally) and has been administered to subjects as young as 12 as well as to elderly individuals. In general, however, the test is most appropriate for high school, college age, and young adults.

As indicated previously, the CPI currently contains 18 standard scales, and these are listed in Table 1, together with representative reliability and validity coefficients. The 18 scales are divided into four groups: 1) measures of poise, ascendancy, self-assurance, and interpersonal adequacy; 2) measures of socialization, maturity, responsibility, and intrapersonal structuring of values; 3) measures of achievement potential and intellectual efficiency; and 4) measures of intellectual and interest modes. The first two groups are largely based on the results of factor and cluster analyses. The third group was clinically defined to assemble the achievement indices together, while the fourth group includes variables with slight loadings on the first two factors.

A substantial number of factor analyses of the CPI have been carried out and as Megargee (1972) points out, there is considerable uniformity in the number of factors extracted, with most studies showing five groupings. There is a wealth of information on the CPI and the interested reader should consult three basic sources: 1) the test manual, which is exceptionally clear and well-written; 2) an *Interpreter's Syllabus* written by Gough in McReynolds (1968); and 3) *The CPI Handbook* written by Megargee (1972).

In addition to the 18 standard scales and those to be included in the planned revision, there are a number of scales and indices that have been developed for use with the CPI. Most of these appear to be research oriented and their clinical potential has not yet been tapped. Megargee (1972) discusses most of these, including estimates of MMPI scales, and factor-analytic, theoretical, and rational

Table 1

Outline of CPI Scales

Scale	How Developed*	# of Items	Reliability	Representative Evidence Validity
Class I. Measures of Poise, Ascendancy, Self-Assurance, and Interpersonal Adequacy				
1. Do Dominance	E	46	Test-retest 1-4 weeks .80	X of college leaders = 55, of non-leaders 46
2. Cs Capacity for Status	E	32	Test-retest 1 year .68	r's of .38 to .48 with scores on a Home Index
3. Sy Sociability	E	36	KR-21 .74	r = .44 with peer ratings of sociability
4. Sp Social Presence	I	56	Test-retest 1 year .65	r = .43 with staff ratings of social presence
5. Sa Self-acceptance	I	34	Split-half corrected .70	high scorers show less autonomic disturbance
6. Wb Sense of Well-being	O	44	Test-retest 1 year .71	lower scores for college students seeking counseling
Class II. Measures of Socialization, Maturity, Responsibility, and Intrapersonal Structuring of Values				
7. Re Responsibility	E	42	Test-retest 1-4 weeks .85	r = .29 with peer ratings of responsibility
8. So Socialization	E	54	Test-retest 1 year .69	male & female delinquents have lower scores
9. Sc Self-control	I	50	KR-21 .82	r's of -.21 to -.34 with staff ratings of impulsivity
10. To Tolerance	E	32	Split-half corrected .86	negative correlations with the F scale
11. Gi Good Impression	O	40	Test-retest 1 year .68	discriminates dissimulated protocols
12. Cm Communality	O	28	KR-21 .52	identifies improperly answered protocols
Class III. Measures of Achievement Potential and Intellectual Efficiency				
13. Ac Achievement via Conformance	E	38	Test-retest 1 year .69	r's of .35 to .40 with high school grades
14. Ai Achievement via Independence	E	32	Test-retest 1 year .63	r's of .19 to .44 with college grades
15. Ie Intellectual Efficiency	E	52	KR-21 .81	r = .44 with Miller Analogies Test scores
Class IV. Measures of Intellectual and Interest Modes				
16. Py Psychological-mindedness	E	22	Test-retest 1-4 weeks .53	r's of .21 to .26 with Intro. Psych. grades
17. Fx Flexibility	I	22	Split-half corrected .71	r = .18 with staff ratings of flexibility
18. Fe Femininity	E	38	Split-half corrected .73	r point biserial = .71 with sex

*E = empirical; I = internal consistency; O = other. See text for explanation.

scales developed by Hase and Goldberg of the Oregon Research Institute, as well as an anxiety scale developed by Leventhal (1966).

At a time when most clinical and personality research used a *t*-test approach (comparing group 1 with group 2 as to mean differences), Gough pioneered the use of regression equations, discriminant functions, and other more sophisticated analyses to yield a series of CPI equations to predict academic achievement and graduation in high school and college; performance in medical and dental school, psychiatric residency, and student teaching; appraisal of leadership and social maturity; severity of menstrual distress; success on parole; and estimation of locus of control. Other researchers have looked at a range of criteria-measures including preference for scientific versus humanities college majors, performance in optometry school, selection of paraprofessional telephone interventionists, police work, and Type A (prone to heart attack) living style.

There is a hand-scored answer sheet which requires the subject to place an X in either the true or false box for each item. Although all 480 responses are printed on one side of a single 8½″ x 11″ answer sheet, the sheet is well designed and highly readable. There are also several computer-scored answer sheets which are described later in this review.

There is a profile sheet where the raw scores can be plotted and converted to standard scores with means of 50 and standard deviations of 10. The graph for females is on one side and the one for males on the other. Not only are the four clusters of scales clearly indicated but the profile is marked with the 50 T-score line and standard score divisions going from 0 to 100. There is space to the right of the profile for the examiner to make notes.

Practical Applications/Uses

A basic aspect of the CPI, often neglected or misunderstood, is that the purpose of each scale is to predict what a person will do in a particular context, and/or to identify persons who will be described in a certain way. The CPI is *not* intended to measure traits. Gough often uses the analogy of the Strong Vocational Interest Blank. On the SVIB the physician scale is not intended to identify or define a trait of medicine, but rather to identify those individuals whose interests are similar to those of physicians and hence might wish to consider medicine as a possible career.

Given this approach the CPI is relevant not only to the researcher who wishes to investigate a particular phenomenon but to the practitioner who needs to make specific predictions or to understand and assist a particular client. The CPI is intended primarily for use with normal subjects, though some studies have shown its relevance to psychiatric samples. As the manual indicates, the most general use of the CPI is in educational institutions, industry, and agencies whose clientele exhibits a normal level of social functioning. *The Eighth Mental Measurements Yearbook* (1978) lists a bibliography of 1382 items, and a perusal of the titles suggests there are very few populations for which the CPI is not appropriate or useful.

In this context, note should be made that the CPI manual is one of the few test manuals that devotes considerable material to the interpretation of individual test profiles. Not only does Gough present an overall strategy of profile interpretation that is applicable to any multivariate instrument, but discusses in detail seven cases, as well as the detection of dissimulation.

The range of applications of the CPI is limited only by the user's imagination; a perusal of the literature suggests such limits have been rather broad. Not only has the CPI been used in a wide variety of studies of scholastic achievement from junior high school to professional school and in many studies of occupational groups, but also in studies dealing with creativity, leadership, marital adjustment, environmental and biographical variables, choice of contraception, time estimation, drug abuse, smoking cessation, and bulimia.

Megargee (1972) points out that there is a need for normative studies on American minority groups, and for studies that attempt to sort out the effects of race, socioeconomic level, IQ, and other demographic variables on CPI scores. He also suggests that although the CPI was designed to be used in individual interpretation, there are few studies that address the validity of the clinicians' CPI interpretations.

The CPI would certainly seem to be appropriate to those areas of psychological research that are currently growing, such as cross-cultural research and studies in gerontology, health psychology, and the relation of lifestyle to physical health.

The CPI can be administered individually or in group testing, and has been used successfully in at-home mailings. The subject may complete the test with or without supervision, at one sitting or more than one. There is no time limit, and total testing time requires the proverbial 50-minute hour. No special administration skills are needed for the CPI, and it can be administered by a proctor, graduate assistant, or clerical staff. The CPI is essentially self-administering, though professional etiquette requires that subjects be made aware of why they are being tested and what use will be made of their scores, and they should by provided with reasonable feedback.

Scoring by hand is a straightforward clerical task that requires the sequential placement of 16 plastic templates over the answer sheet and counting the Xs that show through (two templates each cover two scales). Recording the raw scores in their appropriate place is made easy by the correspondence of templates with the bottom of the answer sheet. Raw scores are then transferred to the profile sheet. Here again this clerical task is simplified by the fact that the spaces where the scores are recorded match exactly on the answer sheet and profile sheet.

The plastic templates are robust; my set is approximately 18 years old and has been used by hundreds of graduate and undergraduate students in testing and assessment courses, in research projects and dissertations, and is still in excellent condition.

For those interested in computer scoring, a number of options are available. The Consulting Psychologists Press catalog indicates the availability of computer scoring services that require special prepaid answer sheets and provide a printout with the standard 18-scale profile as well as six additional indices (Empathy, Independence, Managerial Interests, Work Orientation, Leadership Index, and Social Maturity Index).

The publisher's catalog also lists a variety of narrative computerized reports available from Behaviordyne, Inc. These reports range from comprehensive clinical reports to self-reports written in non-technical language and designed for the client. Although the reports are not a substitute for the individual clinician's judgment, they are excellent, and if used judiciously can be extremely helpful. In addition, the APA *Monitor* contains a number of advertisements from companies that provide microcomputer programs for the local administration and/or scoring of the CPI and related services.

Gough compares test interpretation to playing the piano: while just about any intelligent person can master the basic notes and perhaps even learn some triads, there are few who can transcend the mechanical aspects and become virtuosos. Similarly, any professional can do a scale-by-scale analysis of a CPI profile, but it takes a highly sensitive and skilled clinician to appreciate the nuances of a particular profile, the alterations in meaning due to both internal profile configuration and external factors, and the psychodynamic implications of configurational patterns and scale interactions. Thus the CPI is similar to the MMPI, the Rorschach, the WAIS, and other clinical tests; just about anyone can master the basics, but there are few Klopfers, Exners, Meehls, and Goughs around.

Technical Aspects

Reliability is a basic but rather complex issue. For example, the CPI is designed to assess rather enduring personality characteristics and should therefore present a rather high degree of stability. On the other hand, one would expect a certain amount of sensitivity to changes due to the effects of college, professional and personal experiences, gaining of new skills, and specific training. Many studies have not been able to disentangle the degree of instability due to low reliability and that due to actual change.

Given the large number of studies, the question of the reliability of the CPI seems to have been relatively neglected, if not by Gough then certainly by other investigators.

The manual reports two studies of test-retest reliability. In one study high school juniors (125 females and 101 males) were tested and then retested as seniors. In the second study, 100 male prisoners were retested over a period of 7 to 21 days. For the prisoner sample, the correlation coefficients range from + .49 for the Flexibility scale to + .87 for the Tolerance scale, with a median of + .80. For the high school students the coefficients are substantially lower; for females they range from + .44 for the Communality scale to + .77 for the Intellectual Efficiency scale with a median of + .68, while for males they range from + .38 for the Communality scale to + .75 for the Socialization scale, with a median of + .64.

No coefficients of internal consistency are reported in the manual, but Megargee (1972) calculated these with the Kuder-Richardson Formula 21 on the normative samples presented in the manual (3,572 high school boys and 4,056 girls). The resulting coefficients range from + .22 for the Psychological-mindedness to .94 for the Achievement via Conformance. In addition, Megargee (1972) also presents internal consistency and split-half coefficients. In general the pattern of results is adequate, with moderate stability over time and, given the empirical complexities of the CPI scales, lower internal consistency.

Perhaps because reliability is not a particularly exciting topic, it has been relatively neglected and there is need for additional information. It is of interest to note, for example, that both Megargee's (1972) excellent compendium and Gough's (1968) *Syllabus* discuss the validity of each scale in great detail but give little information on reliability.

There is a massive amount of validity information on the CPI, with some scales being better validated than others. A detailed analysis of what is available is beyond the scope of this volume, and the interested reader is referred to Megargee's book (1972) for a comprehensive review of what was available prior to 1972. In the following paragraph a few studies have been selected and discussed as either representative or illustrative of the approaches taken.

A popular research strategy has been the use of contrasted groups. Gough (1969), for example, asked the principals of 15 high schools where the CPI had been routinely administered for normative purposes to nominate boys and girls outstanding on leadership. The CPI protocols of the 179 nominated leaders were then compared with those of 2,411 students from eight other high schools. Not only were scale-by-scale analyses undertaken, but a multiple stepwise-regression analysis yielded an equation to predict leadership. This study is also illustrative of a second approach, namely the development of regression equations to forecast specific behaviors.

Another approach has been to identify rather complex behavioral criteria and analyze the CPI at both the scale and item level, both to predict the criterion and to elucidate the underlying psychodynamic meaning. An example is provided by Gough (1983) in a study of the personality correlates of the time required to complete work for the doctoral degree in psychology. Although only one CPI scale showed a significant correlation (So, $r = -.11$, $p < .05$), Gough was able to identify a cluster of 34 items significantly related to the criterion.

Another set of studies have used the CPI as a criterion itself and have looked at physiological concomitants. Hare (1978), for example, monitored the electrodermal activity of 64 prison inmates classified psychopathic and nonpsychopathic partly on the basis of the Socialization scale, and found support for the hypothesis that psychopathic inmates are electrodermally hyporesponsive to very strong stimuli, possibly because of left cerebral dysfunction.

A number of studies have looked at personality constellations related to specific criteria. Kegel-Flom (1984) studied instructors in a school of optometry and correlated self, colleague, and student ratings of teaching effectiveness with CPI profiles. As might be expected, she found that teaching effectiveness was viewed somewhat differently by each of the groups, but that there was a constellation of CPI variables related to teaching effectiveness which included self-confidence, assertiveness, drive, tolerance, and a preference for personal interaction.

Still another set of studies have looked at the cross-cultural relevance of either single scales (e.g., Nishiyama, 1974), indices (e.g., Gough, DeVos, & Mizushima, 1968), or the entire inventory (e.g., Repapi, Gough, Lanning, & Stafanis, 1983). Other studies have compared the CPI with other instruments (e.g., Nerviano &

Weitzel, 1977) or with psychometric issues of importance such as faking (e.g., Dicken, 1960).

Critique

Reviews of the CPI have run from laudatory to deprecatory, often reflecting more the reviewer's biases than the psychometric aspects of the CPI.

One type of criticism leveled at the CPI is that the results of studies comparing CPI variables with criteria are usually modest, in a range that is helpful in predicting group performance but not individual behavior. This is a complex issue, related to a variety of technical matters including issues of base rate and false positives-false negatives. Little information is available on the comparative validity of the CPI versus other instruments.

Another type of criticism has been aimed at the redundancy of the CPI, including both item overlap and the relatively high positive correlations among scales. Here again, the issue is complex and reflects the differing goals of the factor analyst who attempts to impose on a complex world a statistical model that aids in understanding that world, versus the empiricist who parallels the complexities of the real world in the inventory scales in order to more accurately predict.

Still a third type of criticism has been leveled at discrepancies between the theoretical and applied approaches of the CPI. These criticisms are directed at the lack of specific information on how some of the CPI scales were developed and what seems to be a paucity of studies on important issues such as the validity of clinical interpretations of CPI profile configurations.

On the positive side the CPI has been praised for its technical competency, careful development, cross-validation and follow-up, use of sizeable samples and separate sex norms, a highly readable manual, reporting correlations with other personality inventories, and for materials (e.g., topical bibliographies) that Gough makes available to interested researchers.

On one aspect most experts agree: the CPI is a popular test. Not only is this reflected in the ever growing number of references listed in the *Mental Measurements Yearbooks* and in its inclusion in test compendia, but also in its prominent mention in surveys. For example, Piotrowski and Keller (1983) surveyed graduate programs in clinical psychology and asked with what five objective personality measures should the clinical Ph.D. candidate be familiar. Of the 80 responding programs, 94% indicated the MMPI and 49% the CPI; the third listed was the 16 PF with 19%.

As someone who firmly believes that psychological testing is the sine qua non of psychology and that assessment is what ought to distinguish the clinical psychologist from other mental health professionals, I see the CPI in many ways as a model of psychological test development and application and a most welcome addition. At the same time much more needs to be done. Some old issues remain (how was the self-acceptance scale developed?), some key areas need further exploration (scale interactions), and some new issues will need much additional work (the new vector scales). Very clearly, the CPI has become a monumental endeavor that goes beyond the efforts of one psychologist and his students. Its impact has been more substantial than most realize, and its future seems well assured.

References

Burger, G. K. (1975). A short form of the California Psychological Inventory. *Psychological Reports, 37,* 179-182.

Dicken, C. F. (1960). Simulated patterns on the California Psychological Inventory. *Journal of Counseling Psychology, 7,* 24-31.

Gough, H. G. (1968). An interpreter's syllabus for the California Psychological Inventory. In P. McReynolds (Ed.), *Advances in psychological assessment* (Vol 1). Palo Alto, CA: Science & Behavior Books.

Gough, H. G. (1969). A leadership index on the California Psychological Inventory. *Journal of Counseling Psychology, 16,* 283-289.

Gough, H. G. (1975). *California Psychological Inventory manual.* Palo Alto, CA: Consulting Psychologists Press, Inc.

Gough, H. G. (1983). Personality correlates of time required to complete work for the Ph.D. degree in psychology. In C. D. Spielberger & J. N. Butcher (Eds.), *Advances in personality assessment* (Vol. 3, pp. 105-128). Hillsdale, NJ: Lawrence Erlbaum Associates, Inc.

Gough, H.G., DeVos, G., & Mizushima, K. (1968). Japanese validation of the CPI social maturity index. *Psychological Reports, 22,* 143-146.

Hare, R. D. (1978). Psychopathy and electrodermal responses to nonsignal stimulation. *Biological Psychology, 6,* 237-246.

Hogan, R. (1969). Development of an empathy scale. *Journal of Consulting and Clinical Psychology, 33,* 307-316.

Kegel-Flom, P. (1983). Personality traits in effective clinical teachers. *Research in Higher Education, 19,* 73-82.

Megargee, E. (1972). *The California Psychological Inventory handbook.* San Francisco: Jossey-Bass, Inc.

Nerviano, V. J., & Weitzel, W. D. (1977). The 16 PF and CPI: A comparison. *Journal of Clinical Psychology, 33,* 400-406.

Nishiyama, T. (1975). Validation of the CPI femininity scale in Japan. *Journal of Cross-cultural Psychology, 6,* 482-489.

Repapi, M., Gough, H. G., Lanning, K., & Stafanis, C. (1983). Predicting academic achievement of Greek secondary school students from family background and California Psychological Inventory scores. *Contemporary Educational Psychology, 8,* 181-188.

Judith L. Whatley, Ph.D.
Post-Doctoral Fellow, Counseling Service and Department of Psychiatry, University of Texas Health Science Center at San Antonio, San Antonio, Texas.

CATTELL INFANT INTELLIGENCE SCALE

Psyche Cattell. San Antonio, Texas: The Psychological Corporation.

THE ORIGINAL OF THIS REVIEW WAS PUBLISHED IN TEST CRITIQUES: VOLUME IV (1986).

Introduction

The Cattell Infant Intelligence Scale (Cattell, 1960) was developed with the intention of being a standardized assessment of mental ability for children aged 2-30 months. Underlying the test is a view of intelligence as being maturationally and genetically controlled. The Cattell proposes to focus on mental development and *not* on motor development, to be standardized, be objective in scoring, appeal to young children, and provide numerical rather than simply descriptive assessments of mental ability.

The Infant Intelligence Scale was developed by Psyche Cattell in the United States beginning in the 1930s and was first published in 1940. The 1960 revision is essentially a reprinting with corrections for typographical errors, but with no substantive changes. The test was developed with the Gesell tests (Gesell, 1925) as a starting point, but items were converted, omitted, added, and arranged in an age scale, with the goal being to create a test compatible with the Stanford-Binet, Form L. This was done to provide a downward extension of that test of mental ability and to consequently extend the age range covered by an assessment instrument or set of compatible instruments. Goals that Cattell attempted to achieve with the development of this new test included 1) objective procedures in administration and scoring of test items, 2) exclusion of items markedly influenced by training, 3) exclusion of items that appear related more to motor control than mental development, 4) inclusion of scaling techniques to go beyond mere descriptive ratings of mental ability, 5) extension of age range beyond that covered by tests being used in the 1930s and distribution of test items more evenly across that age range, and 6) achievement of standardized procedures. These goals reflect an appreciation of the psychometric contribution to test development, as well as Cattell's assumptions about factors thought to control "intelligence," that is, genetic or maturational rather than experiential influences.

The Cattell was developed and tested on a standardization sample of 1,346 examinations made on 274 children enrolled in the Normal Child Series study at the Center for Research in Child Health and Development at the Harvard School of Public Health. Data were collected at ages 3, 6, 9, 12, 18, 24, 30, and 36 months. Items added during the later phases of the test's development were administered to fewer, although an unspecified number, children than were items included at the beginning of the research. Subjects for the Normal Child Series were selected during the prenatal period from pregnant women attending the Boston Lying-in Hospital clinics on the basis of evidence indicating the likelihood of a normal delivery of a normal child. Other criteria for inclusion in the study were that the child's father have permanent employment and be likely to remain in residence in the area throughout the course of the study; that at least three grandparents be of Northern European stock, thus excluding minority participation; and that the mother show ability and willingness to cooperate with the study. Use of this population resulted in the majority of the participants coming from the lower-middle class, with upper-most and lower-most economic-level participants being largely excluded from the study. Socially and mentally inadequate parents' children were also excluded. As of 1937, 37 cases were lost due to unsatisfactory cooperation and 34 were dropped from the data base due to moving, death, or some other apparently nonstudy-related reason. Thus, the study was subject to an attrition rate of approximately 25%.

The sample, clearly not representative of the United States population geographically, socioeconomically, or ethnically, is one of the weakest aspects of the Cattell's development. Moreover, that original sample is by now significantly dated. Caution is consequently warranted concerning the confidence with which one can generally employ its use, even with a "normal" group of children. No known standardized versions of the Cattell exist for use with groups of children with significant visual, auditory, or major physical handicaps, or with children from other language or cultural groups. It has also been used with what are described as minor modifications with children with multiple handicaps, such as blindness and deafness (DuBose, 1977); with children with cerebral palsy (Banham, 1972); and in the diagnosis of developmental delays (Erikson, 1968).

Test materials for the Cattell consist of a set of standardized, simple, manipular objects, visual and auditory stimuli that appeal to young children (ring, teaspoons, ball, rattle, mirror, key, sugar pellets, formboard, paper, bell, pencil, jointed doll, beads, toy dog, pegboard with pegs, cubes, picture cards, bottle), the Stanford-Binet Picture Vocabulary, and the Stanford-Binet formboard. These standardized materials are available through The Psychological Corporation.

The test consists of 95 items: five for each month period from 2-12 months, five for each two-month period during the second year of life, and five for each of the two quartiles of the first half of the third year of life. At each of these age periods, one or two alternate items, which are described as acceptable but somewhat less satisfactory than the standard items, are also included for use as needed. All 95 items are not administered to a given child at a given test administration or age level; only the items necessary to establish basal and ceiling levels are presented in order to obtain a child's mental age score.

Infants aged 2-30 months are the age group for whom this test is designed. Cattell suggests that, although the test was developed and standardized for use with children at particular age points within the above range (e.g., 3, 6, 9, 12, 18, 24, and 30 months), it can be used with only slightly less accuracy with children between those age ranges. Placement of test items for the ages between the designated intervals was done less rigorously and consequently may involve greater measurement error and lower reliability.

Similar to the Stanford-Binet, items on the Cattell are grouped by age level rather than by content, and no individual subtest scores are computed. The resultant score is a single score in the form of a mental age. A ratio IQ is also obtainable.

Practical Applications/Uses

The Cattell serves as one of several test instruments that might be employed to assess infants and young children's mental development. It requires a trained examiner, time (minimally a half hour), and space for administration. Due to these requirements for administration, it is not apt to be of frequent use as a general or quick screening tool. It is more apt to be used when a more detailed assessment of mental development is required (e.g., when there is some question about a child's development or an ''at risk'' group is being evaluated). Compared to the psychometrically more sophisticated Bayley Scales of Infant Development (Bayley, 1969), it is somewhat quicker to administer and may serve successfully as an alternate to that examination under some conditions. In a study of young handicapped children, Erikson, Johnson, and Campbell (1970) report a high positive correlation ($r = 0.97$) between scores on the Cattell and scores on the Bayley. Scores on the Cattell tended to be slightly higher than on the Bayley, but these authors felt that the two tests might be considered interchangeable for use with young children referred for diagnosis of developmental problems.

One of Cattell's interests was to develop an instrument compatible with the Stanford-Binet that could be used to extend the testable age range downward. Longitudinal studies would clearly benefit from the availability of such an instrument. The Cattell achieves that aim in format and scoring procedure. The extent to which it fulfills that goal conceptually is less clear, given the low correlations obtained when comparing longitudinally obtained data on the two tests. Scores obtained after the second year of life and close in time to one another compare more favorably.

The Cattell is limited to individual administration. Examiners should be individuals who are sensitive to and comfortable with young children, as well as experienced in psychometric procedures. Their role is an active one and is an integral part of test procedure. Consequently, administration is most likely to be carried out by a psychologist or other child development specialist.

The examiner's role is to present test materials to the child and to score the child's response to each test item and the spontaneous responses related to test criteria. The examiner, therefore, must be alert to the child's behavior and note performances by the child that are relevant to items not necessarily being administered at the moment. Such ad lib performances are scorable. Test items can be administered in any order, and a good examiner will be attuned to the child's energy and interest level and will administer items in an order taking these factors into account. It is especially important that the individual who administers the Cattell be sensitive to and comfortable with young children, as it is also the task of the examiner to motivate children and elicit their cooperation and participation. Specific other tasks that the examiner must carry out include recording verbatim the verbal responses of the child on particular items, judging that items are not passed by chance but reflect intentional or purposive behavior on the part of the child, and differentiating refusal to perform from lack of ability.

The test is administered with the child either lying on or seated at a table. Some items up to and including five-month items are administered with the child lying down, whereas all items from six months onward are presented with the child seated, either in a caregiver's lap or on a chair. The presence of a mother or another attendant is also part of the general administration procedure, although that adult may not directly help the child perform items. Cattell points out that it is often helpful to inform the mother or caregiver that some things will be presented to the child that are expected to be beyond the child's current developmental level.

Cattell cautions against testing young children when they are tired, sick, or in a negative mood. Such features can easily affect the child's willingness or motivation to perform, resulting in an invalid assessment.

The setting in which the testing is administered should be free from distracting noises and sights, and materials not being used at a given time should be kept out of the child's sight. On the other hand, the setting should not be so "nondistracting" as to be sterile and uncomfortable.

Testing generally begins with items thought to be within a child's ability. A basal level is established by testing downward until all five items of a given age level have been passed. Testing continues upward, although not necessarily in order, until all five items of a level are failed.

A specific aim of Cattell's in developing this test was to provide objective scoring criteria for each test item. The test manual (Cattell, 1940) specifies in objective terms the procedure for administering each item and objective criteria for scoring on a pass/fail basis. Illustrative photographs also appear in the manual. Each item's score is recorded on a standardized record form that briefly lists a short descriptor and a space for noting whether the item was passed or not. The last part of the record form provides space for continuation of a child's score on Stanford-Binet items when the two tests are used together. This extension is provided through the fourth year of life.

The test form also contains four five-point scales on which the examiner can rate the child's test behavior in terms of willingness, self-confidence, social confidence, and attention. These ratings do not figure directly into the scoring, but provide clinical information that may be helpful in interpreting test performance.

The time required to learn scoring of the examination will vary considerably, depending on the examiner's psychometric experience and experience and knowledge regarding young children's behavior. Although the objective, behavioral criteria for a pass are generally clearly specified in the test manual, the establishment of reliable administration and scoring by comparison with an experienced examiner would be an advantage. Generally, once the scoring is mastered, it should be possible to score the examination as it is administered, with only occasional reference to the scoring manual. This is also necessary because the items to be administered will be determined by efficiently establishing basal and ceiling levels. Administering unnecessary items may occupy time that could affect the ability to complete satisfactory administration of the test.

Interpreting the obtained score requires use of both the objective score and clinical judgment. The objective score in the form of a mental age indicates the age level at which the child is performing as determined by norms established by the standardization sample. Other statistical properties of the test, such as means and standard deviations, provide other important information for test interpretation. One of the main tasks of the interpreter is to explain the extent to which the test score indicates that mental development is delayed, advanced, or consistent with expected rates. Other questions of interpretation focus on the predictability of future intellectual functioning and consideration of possible intervention. The interpreter's clinical judgment is called into play in making such decisions, as well as decisions about the extent to which the test reflects a representative sample of the child's ability (i.e., are individual item scores due to refusal, lack of ability, lack of motivation, attentional problems, or even luck). As the Cattell produces only an overall score, analysis or comment on subcomponent abilities must rely on clinical judgment and should be offered cautiously. Furthermore, interpreting scores for children at the youngest end of the scale must rely heavily on clinical judgment given the low levels of test reliability for that portion of the test. As with administration and scoring, interpretation requires training in child development, as well as psychometrics, and is most likely carried out by a psychologist or other child development expert.

Technical Aspects

Reliability studies were conducted on the standardization sample, and data are reported in the manual. Predictive validity data, comparing the Cattell scores with Stanford-Binet, Form L, scores, are also reported for that same sample. At three months, the reliability estimate obtained by the split-half method and corrected by the Spearman-Brown formula was 0.56. Correlating those scores obtained at three months with scores at 36 months on Form L of the Stanford-Binet resulted in a value of only 0.10. Thus, the scores at age three months show neither statistical reliability or predictive validity. For 6, 9, and 12 months the split-half reliability coefficients were .88, .86, and .89, respectively, and correlations with the Stanford-Binet at 36 months were .34, .18, and .56, respectively. Although reliability estimates reach acceptable levels, the predictive validity scores do not. At ages 18, 24, and 30 months the reliability coefficients were .90, .85, and .71, respectively, with correlations with the Stanford-Binet for 36 months falling at .67, .71, and .83, respectively. Again, acceptable reliability estimates are achieved with the exception of the 30-month score. The predictive validity score at 24, and possibly at 18, months appears acceptable, but the score at 30 months must be interpreted cautiously, given the low level of reliability at that age.

Some additional reliability and validity studies have been reported, usually conducted with selected samples, and generally the results have been similar to those presented by Cattell. Escalona (1950) and Gallagher (1953) each report research demonstrating the effects of nontest factors detracting from a child's optimal performance that can measurably decrease test-retest reliability. Escalona reports such data for a sample of adoptive children for whom she had judges' subjective appraisals concerning maximal performance on the Cattell; Gallagher also studied adopted infants on whom similar ratings had been made.

Use of the Cattell as a downward extension of the Stanford-Binet is not strongly supported, given the obtained correlations between the two instruments.

Although the two tests appear compatible in format and scoring, they cannot be empirically defended as being conceptually the same. Research has fairly consistently failed to find that tests of infant mental performance during the first year and one-half of life predict subsequent IQ (McCall, Hogarty, & Hurlburt, 1972; Escalona & Moriarity, 1961). This failure to predict may be due to differences in what is "intelligence" or is measurable as "intelligence" in infancy compared to later in life, due to the changeable nature of intelligence or to the unreliability of test instruments. Nonetheless, despite one's particular position in this debate, it cannot be concluded that the Cattell and the Stanford-Binet serve as extensions of one another, due to the low correlations between the two in most comparisons.

Escalona and Moriarity (1961) report results from a longitudinal study comparing the Cattell and Gesell scales with the Weschler Intelligence Scale for Children (WISC). Infants were assessed at ages one to eight months on both infant scales, whereas the WISC was employed between seven and eight and one-half years. Of interest here is the resulting correlation between the Cattell and the WISC, which was 0.05, showing virtually no predictability. Clinical appraisal data collected during infancy were considered of greater predictive ability than performance on either infant scale.

Finally, Erickson (1968) evaluated the predictive validity of the Cattell for young mentally retarded children in an effort to explore the possibility that greater predictability might be found in the case of children with subnormal ability. Her results were based on a study of children referred for diagnosis of developmental problems. IQs were measured as early as three months of age and were repeated annually through approximately age 72 months. Longitudinal comparisons of the children's IQ scores resulted in correlations ranging from .63 to .94. As with findings from other research, correlations were greater with increasing age of children at the time of initial examination, and IQs obtained during the first two years of life were less stable than those obtained later. Erikson argues that the Cattell can be used to predict subsequent performance when used with children below average in intellect, but most usefully only after the first two years of life.

Critique

The Cattell Infant Intelligence Scale is a standardized assessment instrument for the measurement of mental ability for children aged 2-30 months. It is based on the assumption that intelligence is largely maturationally and genetically controlled. Cattell developed this test with certain goals in mind, some of which were achieved, whereas others were not fully realized. Procedures for administration and scoring test items are reasonably objective and clearly specified. Cattell's goal of going beyond descriptive ratings of mental ability to the use of scaling techniques has been implemented with the use of objectively scored items with point values. However, the absence of item analyses beyond percentages passed by age, especially for items falling between major age points, somewhat reduces the reliability of the test, as well as its discrimination function. In general, test-retest reliability was found to be at acceptable levels, with the exception of the youngest ages where the examiner must rely more heavily on clinical judgment in interpreting the meaning of the individual's performance.

Cattell's attempt to provide a downward extension of the Stanford-Binet may be considered to have met some success in terms of format and scoring. There seems to be adequate predictive ability for children beyond the second year of life; for children younger than this, the predictive validity of the Cattell, based on correlations with the Stanford-Binet at later ages, fails to obtain empirical support for conceptual compatibility between the two tests. Use of the Cattell as a downward extension in these cases does not appear to be warranted and should be done cautiously.

Cattell's assumptions about intelligence as being maturationally or genetically controlled led to the exclusion of a motor development scale and to the removal of items thought to be strongly related to motor control and/or home training. However, some items that remain cannot be said to be completely free of these factors. For example, a number of items early in the scale, such as ''lifts head'' and ''head erect and steady,'' clearly involve motor ability. The absence of a separate motor scale may in some cases be a shortcoming of this test.

The greatest shortcoming of the Cattell, however, is its unrepresentative and, by now, dated standardization sample. The restricted sample, which is neither socioeconomically, ethnically, nor geographically representative of the United States population, severely restricts the confidence with which one can usefully apply the scale and interpret obtained results.

Cattell's ambitious effort in developing this test has contributed and will probably continue to contribute to our understanding of intellectual development in young children, as well as our efforts to improve our ability to measure intellectual functioning. The Cattell serves as an objective, standardized instrument that allows more objective testing of young children's functioning and facilitates clinical judgments in this area.

References

Banham, K. M. (1972). Progress in mental development of retarded cerebral palsied infants. *Exceptional Children, 39,* 240.

Bayley, N. (1969). *Bayley Scales of Infant Development.* Cleveland: The Psychological Corporation.

Cattell, P. (1940). *The measurement of intelligence of infants and young children.* Cleveland: The Psychological Corporation.

Cattell, P. (1960). *Cattell Infant Intelligence Scale.* Cleveland: The Psychological Corporation.

DuBose, R. F. (1977). Predictive value of infant intelligence scales with multiply handicapped children. *American Journal of Mental Deficiency, 81,* 388-390.

Erickson, M. T. (1968). The predictive validity of the Cattell Infant Intelligence Scale for young mentally retarded children. *American Journal of Mental Deficiency, 72,* 728-733.

Erickson, M. T., Johnson, N. M., & Campbell, F. A. (1970). Relationships among scores on infant tests for children with developmental problems. *American Journal of Mental Deficiency, 75,* 102-104.

Escalona, S. (1950). The use of infant tests for predictive purposes. *Bulletin of the Menninger Clinic, 14,* 117-128.

Escalona, S., & Moriarity, A. (1961). Prediction of schoolage intelligence from infant tests. *Child Development, 32,* 497-605.

Gallagher, J. J. (1953). Clinical judgment and the Cattell Infant Intelligence Scale. *Journal of Consulting Psychology, 17,* 303-305.

Gesell, A. (1925). *The mental growth of the preschool child.* New York: Macmillan.

McCall, R. B., Hogarty, P. S., & Hurlburt, N. (1972). Transitions in infant sensorimotor development and the prediction of childhood I.Q. *American Psychologist, 27,* 728-748.

Kevin C. Mooney, Ph.D.

Assistant Professor of Psychology, Washington State University, Pullman, Washington.

CHILD BEHAVIOR CHECKLIST

Thomas M. Achenbach and Craig Edelbrock. Burlington, Vermont: University of Vermont.

THE ORIGINAL OF THIS REVIEW WAS PUBLISHED IN TEST CRITIQUES: VOLUME I (1984).

Introduction

The Child Behavior Checklist is designed to assess in a standardized format the social competencies and behavior problems of children ages 4 through 16 as reported by their parents or others who know them well (Achenbach & Edelbrock, 1983). Social competence is assessed by up to 40 questions measuring the amount and/or quality of the child's involvement in sports, nonsport activities, organizations, chores or jobs, friendships, family, and school. The behavior problems section lists 118 childhood problems and parents are asked to rate how often each occurs. Space is also provided for parents to list and rate items that are not specifically included in the checklist.

The Child Behavior Checklist is the result of the joint efforts of Thomas Achenbach and Craig Edelbrock with financial backing from the National Institute of Mental Health. The initial work and conceptualization were primarily the efforts of Achenbach; Edelbrock appears to have had a major hand in developing the cluster analytic techniques (Edelbrock, 1979; Edelbrock & McLaughlin, 1980; Edelbrock & Achenbach, 1980) which allowed the Child Behavior Checklist to go beyond being a psychometrically sound, well-normed, descriptive instrument to an instrument capable of *categorizing* childhood syndromes or typologies. The development and norming of the Child Behavior Checklist has gone through four major stages. The first involved the initial conceptualization of the scope and content of the test. The second consisted of the psychometric development of the Behavior Problem scales and the Child Behavior Profile on a clinic population. Thirdly, normative data on behavior problems and social competencies were collected from a non-clinic population. The final stage involved the grouping or clustering of the Behavior Problem scales to arrive at a typology of childhood problems.

The behavior-problem items were derived primarily from a 1966 study by Achenbach in which he "content analyzed" over 600 clinical case histories of children. The child symptom checklist items from that study were derived from other symptom checklists and from additional problems referred to in case histo-

This reviewer wishes to acknowledge and thank Amy Harrison and Sandy Wurtele for their contributions to this review.

ries, resulting in 91 items "which seemed to involve minimal inference, which could be considered mutually exclusive with regard to specific observations, and which were not excessively molecular" (p. 7). Further research (Achenbach & Lewis, 1971), consultation with child mental health professionals, and pilot testing of draft versions with parents resulted in the final 118 behavior-problem items of the Child Behavior Checklist.

Selecting items for the positive behaviors comprising a social competency scale proved problematic since there was a much smaller developmental and clinical literature from which to pull items. Difficulties also arose during pilot testing; many items were being endorsed to some degree by all parents, there appeared to be a social desirability response bias, and some of the items seemed to be measuring merely the opposite of a behavior problem. The items found to be most useful were measures of the involvement and attainment of children in activities, social involvement, and school. These three a priori groupings of items comprise the Social Competency scales of the Child Behavior Checklist.

To arrive at the Behavior Problem scales, Child Behavior Checklists were obtained on a total of 2,300 children* as part of intake procedures at 42 mental health and related service agencies. To partially control for age and sex differences, Achenbach developed separate scales for both boys and girls for ages 4-5 years, 6-11 years, and 12-16 years. The behavior-problem items were then subjected to a principal-components factor analysis which resulted in either 8 or 9 "narrow band" behavior-problem scales for each age and sex group (see Table 1). Any behavior-problem items endorsed by less than 5% of the clinic population were excluded from the principal components analyses and were included in an "other problems" category.

The authors also obtained, for each age and sex group, second-order or "broad band" factors, based on intercorrelations among those Behavior Problem scales appearing to tap the dimensions of Internalizing and Externalizing. Thus, while the "narrow band" factor analysis of the behavior-problem items resulted in the Behavior Problem scales, the "broad band" analyses of intercorrelations among scales resulted in a higher level of analysis, placing each of the "narrow band" scales on an Internalizing-Externalizing continuum. This placement, along with the Mixed designation of those scales with a moderate loading on both Internalizing and Externalizing, comprises the Revised Child Behavior Profile, the basis for all normative comparisons of a child's behavior problem scores.

While the Child Behavior Checklist was derived primarily to determine differences *within* clinic populations, the authors also intended for the test to be able to compare a child with normative samples of agemates. This goal was accomplished by sending interviewers to randomly selected homes from census-tract blocks that approximated the socioeconomic and racial characteristics of the clinic

---

*For some categories of children, for instance girls ages 4-5, obtaining sample sizes that the authors considered adequate required up to 9 years. The First Edition Profiles for some of the categories have had to be modified in order to take into account small discrepancies between initial results and these recent analyses. The Child Behavior Checklist items remain the same; it is only some of the factors that have been slightly modified.

Table 1

Scales Found through Factor Analysis of the Child Behavior Checklist

Group	Internalizing Scales	Mixed Scales	Externalizing Scales
Boys aged 4-5	Social Withdrawal Depressed Immature Somatic Complaints	Sex Problems	Delinquent Aggressive Schizoid
Boys aged 6-11	Schizoid or Anxious Depressed Uncommunicative Obsessive-Compulsive Somatic Complaints	Social Withdrawal	Delinquent Aggressive Hyperactive
Boys aged 12-16	Somatic Complaints Schizoid Uncommunicative Immature Obsessive-Compulsive	Hostile Withdrawal	Hyperactive Aggressive Delinquent
Girls aged 4-5	Somatic Complaints Depressed Schizoid or Anxious Social Withdrawal	Obese	Hyperactive Sex Problems Aggressive
Girls aged 6-11	Depressed Social Withdrawal Somatic Complaints Schizoid-Obsessive		Cruel Aggressive Delinquent Sex Problems Hyperactive
Girls aged 12-16	Anxious-Obsessive Somatic Complaints Schizoid Depressed Withdrawal	Immature Hyperactive	Cruel Aggressive Delinquent

population. They administered the Child Behavior Checklist in interview form to parents of 50 male and 50 female children, aged 4-16. These children's scores comprise the test norms, which are grouped by sex for ages 4-5, 6-11, and 12-16 years. The criterion of "normalness" for this population was that no child in the family had received mental health or psychiatric services in the past year.

The Profile Types were derived from a sample of 2,500 clinic-referred children. These children were randomly selected from each age/sex grouping in the Child

Behavior Checklists files compiled from the previously mentioned 42 child mental health facilities. Profile Types were computed using a hierarchical cluster-analytic technique for each of the 6 age/sex groupings. The Profile Types with their corresponding frequency in Achenbach and Edelbrock's clinic-referred sample are delineated in Table 2. These Profile Types, then, represent statistically defined "nested" groupings of parent-reported behavioral problem patterns in a clinic sample. The labels for the Profile Types were chosen because they represent the highest scale or scales for each profile type but, unlike MMPI or related profiles, the Profile Types reflect the relationships among all of the scores on the Behavior Problem scales.

Materials necessary for the administration of the Child Behavior Checklist are the *Manual for the Child Behavior Checklist and Revised Behavior Profile* (Achenbach & Edelbrock, 1983); the Child Behavior Checklist; and the Revised Child Behavior Profile sheets. There is both a FORTRAN mainframe tape and an IBM-PC disk that will score the profile and give information on the child's profile pattern. Templates are also available for the hand-scoring of profiles. As the *scoring* of the profiles has been slightly modified since the first edition, any research with the profiles should use the updated (1981) version. Translations of the checklist are available from Achenbach in Chinese, Dutch, French (Canadian), German, Greek, Hebrew, Hindi, Italian, Korean, Spanish, and Swedish.

The *Manual for the Child Behavior Checklist and Revised Behavior Profile* is a detailed 230-page description of the Child Behavior Checklist. Much of the test development material briefly summarized and discussed here can be found in greater detail in this manual. The manual also contains brief descriptions of additional test forms in various stages of development—a Direct Observation Form, a Teacher's Report Form, and a Youth Self-Report Form—to supplement the parent-completed Child Behavior Checklist. They will not be discussed in this chapter except to say that they are, in combination, intended to be useful in obtaining data on children from multiple perspectives and that they are being developed in a manner similar to the Child Behavior Checklist Parental Form.

The Child Behavior Checklist contains space for demographic information on the parents and child as well as the social competency items and the behavior problem items. The first two pages comprise the demographic and social competency sections. The social competency section asks the parents to list up to three activities in which their child participates in the areas of sports, nonathletic activities, organizations, and jobs or chores. It also asks the parents to compare (on a four-item continuum from "Don't Know" to "More than Average") their child to other same-aged children on the quantity of time the child spends on each activity and the proficiency in each their child possesses. Parents are next asked to check the number and frequency of contacts their child has with close friends and to rate both how their child gets along with siblings, peers, and parents and how well the child plays and works when alone. The social competency section ends with a number of questions about the child's school performance, which include rating the child's performance in various academic subjects and indicating whether or not the child is in a special class, has ever repeated a grade, or if the child has had any academic or other school problems.

Table 2

Child Behavior Checklist Profile Types

Sample	Profile Type	Frequency (%)
Girls ages 4-5	Schizoid-Somatic Complaints	18.0
	Social Withdrawal	9.2
	Obese	12.4
	Sex Problems	9.2
	Aggressive	6.0
	Hyperactive	14.0
	Hyperactive-Aggressive	7.2
	Unclassified	4.0
Girls ages 6-11	Depressed-Social Withdrawal	12.9
	Somatic Complaints	14.0
	Schizoid-Obsessive	5.1
	Sex Problems	10.1
	Hyperactive	10.6
	Delinquent	12.9
	Aggressive-Cruel	15.4
	Unclassified	2.7
Girls ages 12-16	Anxious-Aggressive	6.7
	Somatic Complaints	9.7
	Anxious-Obsessive-Aggressive	12.2
	Hyperactive-Immature	12.9
	Delinquent	14.7
	Depressed-Social Withdrawal-Delinquent	10.3
	Aggressive-Cruel	12.4
	Unclassified	2.2
Boys ages 4-5	Depressed-Social Withdrawal	12.4
	Somatic Complaints	8.8
	Immature	10.8
	Sex Problems	10.4
	Schizoid	13.6
	Aggressive	12.8
	Aggressive-Delinquent	11.2
	Unclassified	2.2
Boys ages 6-11	Schizoid-Social Withdrawal	4.2
	Depressed-Social Withdrawal-Aggressive	7.6
	Schizoid	13.9
	Somatic Complaints	16.1
	Hyperactive	14.1
	Delinquent	22.4
	Unclassified	6.8
Boys ages 12-16	Schizoid	16.6
	Uncommunicative	12.3
	Immature-Aggressive	12.8
	Hyperactive	11.8
	Uncommunicative-Delinquent	12.0
	Delinquent	15.0
	Unclassified	3.9

A list of 118 behavior problems with reference to the child's behavior "now or in the past 6 months" comprises the Behavior Problems section of the Child Behavior Checklist. These items are rated on a 3-point scale (not true as far as the parent knows, somewhat or sometimes true, and very or often true). The items include specific observable child behaviors (e.g., talks about killing self, wets the bed), behaviors that require more general inferences (e.g., unhappy, sad, or depressed; suspicious), and behaviors that require considerable parental judgment (e.g., plays with own sex parts too much, feels too guilty, is unusually loud).

If both parents are available to complete the Child Behavior Checklist, they should do so separately. Achenbach and Edelbrock intend for any person with fifth-grade reading capabilities to be able to complete the Checklist, and most can do so in less than 20 minutes. For parents who have difficulty completing any part of the test, assistance can be provided by either a therapist or receptionist.

There are six profile sheets, provided separately for boys and girls at ages 4-5, 6-11, and 12-16 years. The scales for the social competency section appear on one side and those for behavior problems are on the reverse. The listing of the social-competency and problem-behavior items that comprise each scale allow relatively easy (if tedious) scoring without the use of the scoring templates. The scoring templates provided for each profile consist of a transparent sheet which overlays each page of the behavior problems with the designated scale(s) next to the behavior-problem item. This reviewer has found the templates not worth the effort; it is just as easy to use the Revised Child Behavior Profile Sheet to tally the ratings and to arrive at the child's scores.

To ascertain how a child's profile matches a given Profile Type, the use of one of the computer-scoring programs is necessary. Unfortunately, explanatory documentation for setting up the IBM-PC scoring disk is scanty and someone not well acquainted with computers may need assistance. Once one is set up on the computer and understands how the data is stored, the program itself is very easy to follow. Only the machine-language version is available, so customizing the format of the scoring program is impractical.[1]

The FORTRAN scoring program for mainframe computers is written for a DEC computer system. It requires extensive revision for use with other computer systems such as IBM, particularly because it is written in an old version of FORTRAN (i.e., FORTRAN IV instead of FORTRAN 77). This should be only a temporary obstacle, for as users solve these translation problems they have been asked to send their documentation to Achenbach, who is acting as a clearinghouse for such information. As with the PC program, the documentation is weak at best. Internally, however, the program is well documented and Achenbach does provide a phone number to call for consultation on the use of the scoring programs.

Practical Applications/Uses

On a clinical level the Child Behavior Checklist provides descriptive information comparing children with other children of the same age and sex. The checklist is in no way meant to supplant a thorough case history or applied behavior analysis; rather, it provides an overview of a child's social competencies and behavior problems.

The Child Behavior Checklist is of particular value to practitioners who wish to differentiate clinical from nonclinical populations and those who wish to distinguish among clinical populations. Racial or socioeconomic bias appears minimal. For inpatient populations the social competency section may not be wholly appropriate, and the behavior-problem scores on many scales may be so high as to overload the assessor with problem information. As the checklist was primarily intended as a clinical tool, those who are interested in "normal" or healthy populations may not find this instrument very useful, for most of these children will score very high on social competency and very low on behavior problems. Concomitantly, while the social competency section does provide some information on a child's social and school status, it does not measure many aspects of social competency that might be of interest with a healthy population (i.e., altruism, courage, autonomy, etc.). The test appears ideal for evaluations in child mental health programs and for clinicians who would like a broad-based measure of treatment outcome. Because of its wide problem coverage and its sound normative information, this instrument could well become a standard for use in child evaluations.

Computer scoring of the Child Behavior Checklist is probably the preferable alternative for obtaining a child's raw scores and social competence and behavior problem profiles. However, it is possible (although tedious) to hand-score the Revised Child Behavior Profile for social competence or behavior problems. Separate normed profile sheets are provided for each age and sex grouping. While there is often great similarity, the Behavior Problem Scales are different for each age and sex group based on the derived items for that group's narrow band factor analysis. As the Social Competency Scales were not empirically derived, the scales and the items that comprise each scale are identical for each group, except that the School scale is not scored for 4- and 5-year-olds. As mentioned, one cannot use hand scoring to obtain the child's Profile Type.

When scoring this instrument, each item in each behavior-problem scale is given the weight of its rating (0 = not true, 1 = somewhat or sometimes true, and 2 = very or often true). The sum of these weights is the child's raw scale score that can then be transferred to the appropriately "normed" Revised Child Behavior Profile sheet. If a parent skips more than 8 behavior-problem items, the resulting scores may not be sufficiently comparable to the standardization sample to utilize the "normed" profile.

While easy to administer and score, interpreting the Child Behavior Checklist requires a reasonable statistical background, as the statistics and bases for decision making used to develop the scales, norms, and Profile Types are quite complex. It is important to remember that high scores on a Behavior Problem Scale do not necessarily correspond to a diagnostic label; the names of these scales merely reflect their composite items. The scales were developed empirically from parental reports and while relationships between scales and DSM-III diagnoses would be heartening for construct validity, they would also be in a sense coincidental.

The Revised Child Behavior Profile sheets are formatted so one can compare a child's profile scale scores by using the percentages for age and sex that are on the left side of the sheet or by referring to the child's normalized T-score that is shown on the right side. (A normalized T-score is similar to a regular T-score in that it has a mean of 50 and a standard deviation of 10; however, for purposes of arriving at a more "normal" or bell-shaped distribution, the normalized T-score is based on the *percentiles* of raw score distribution rather than on the standard deviation of the raw scores themselves.) Because the test measures problems that may be relatively rare in a non-clinic population, the low range given on the profile sheet for each scale is 69% or a T-score of 55. In a similar vein, the high range of behavior-problem scale scores are relatively rare in a non-clinic population, making it difficult to norm very high scores. Thus for scores above the 98th percentile or a T-score of 70, the T-scores are no longer based on the percentiles from the non-clinic population. Rather, the possible raw scores are divided evenly into 30 intervals that by definition comprise the "T" scores of 71 through 100. While this does not allow for percentile-normed comparisons above the 98% score, it does permit greater differentiation among children who score exceptionally high on various scales, such as those in inpatient settings. Finally, it should be noted that both of these decisions regarding the extreme ranges of scores are present only in the Revised Child Behavior Profiles and not in the first edition.

Because of the greater range of *total* behavior-problem scores (nearly all parents endorse *some* behavior problems for their child), T-scores for the total behavior-problem scales have a low range of 30 (2%). The authors assigned scores above a T-score of 70, or the 98th percentile of non-clinic scores, as follows. First, the highest total problem score achieved on a clinic sample was given a T-score of 89. The raw scores between this and the raw score at the 98% (T-score of 70) were assigned T-scores at equal intervals from 71 to 89. Other possible raw scores above those actually found in the clinic sample were assigned T-scores in equal intervals from 90 through 100. Scores on the Internalizing dimension and the Externalizing dimension were normed in a manner similar to that for the total behavior-problem score.

Scores on the individual Social Competence scales and their total score were normed on the same population sample and in a manner similar to that used with the Behavior Problem Scales. However, with these scales low scores (i.e., indicating a lack of social competence) are clinically significant. Thus the high score for the Social Competence scales and the total score of social competence is a T-score of 55 (or 69%) and the lowest percentile score is the second percentile which corresponds to a T-score of 30. Lower social competence scores, to better differentiate among very low scorers on the Social Competence scales, were assigned T-scores at equal intervals ranging from 30 to 10.

The behavior profiles (i.e., the subject's pattern of scaled scores across the narrow band scales ranging from the most Internalizing scales to the most Externalizing scales) provide a clinician or researcher with a rather comprehensive and perspective description of a child's behavior problems. This fact alone is enough to justify the extensive use of the Child Behavior Checklist with clinic-referred children. The additional aspect of a taxonomy of Profile Types is a major step forward in the assessment of children. These profile patterns were identified by subjecting the intraclass correlations—a measure of the similarity between behavior profiles—of the clinic-referred children's profiles to a centroid cluster analysis, resulting in a number of empirically derived categories for each age and sex grouping. Two points should be emphasized. One is that these Profile Types are based on a *clinical* sample and are not likely to be appropriate for children with no serious clinical difficulties. Secondly, these Profile Types are statistically derived; they are not based on any a priori criterion or classification of disorders. The underlying assumption of such an approach is that since these types exist statistically and are "stable," they will in future research correlate with clinical and etiological variables of interest. The usefulness of such a taxonomy depends on further research and clinical experience with the Profile Types.

Technical Aspects

All tests are limited by the scope and level of their items, which are included or excluded according to the developer's paradigm of what contributes to whatever the test is designed to measure. Achenbach, in his attempt to develop a test that would differentiate among clinical populations, is clearly operating from a narrow framework of health and mental health. One should not allow the "objectivity" of the measure to obscure the fact that the test implies a psychiatric paradigm of what constitutes a problem and possibly even what should constitute the focus of concern. This is not a failing of the test, but rather is a limitation that should be clearly recognized. This is probably most evident in the social competence section where not only were the items and scales chosen with less rigor than those in the behavior problem section, but they also tap only a small portion of what most specialists in child development would deem necessary to include in social competence. Another regrettable weakness in a scale that intends to be a broad-based assessment into adolescence is scanty information on sexuality, whether conceptualized within social competence or problem behavior.

More generally, the items emphasize the description of child-centered intrapersonal and interpersonal experiences. The focus is on the child when a focus on other aspects of the environment or problems occurring therein may be more appropriate. Parents who complete the checklist and clinicians who interpret it should remember that an identified behavior problem or social incompetency is not necessarily "in the child." Care must be taken to investigate other systems such as the family, school, church, and even larger social and political systems that may be important in assessing the child but are in no way implied by the test and may even be masked.

As the Child Behavior Checklist is primarily a descriptive instrument, it is important to examine individual items closely. The scales are based on nomothetic principles and may be biased for any one parent-child pair. It may be inappropriate to assume that for any such pair each problem is as important as every other. It also is possible that two children could have the same profile scores and belong to the same Profile Type and yet have no overlap in individual items. The best defense against these and the previously mentioned concerns is what Achenbach and Edelbrock continually stress in their manual: to use the Child Behavior Checklist as a starting point and not the endpoint of a clinical evaluation.

Using parents as the test's key informants makes data on the child's behavior across time and situations quite accessible. In addition, parental perceptions of behavior typically are crucial determinants in bringing a child in for treatment, seeking continued mental health care, and defining the focus of treatment. However, parental reports also contain potential drawbacks. The most obvious difficulty arises in the assumption that parents can accurately estimate their child's behavior on a 3-point scale. Even given that ability, it is likely that parental standards or baselines contribute to specific profile types of children. As Patterson (1982) notes, "it is assumed that parents differ in what it is that they classify as deviant . . . [and this] in turn, determines which child behaviors will be monitored and which will be punished" (p. 247). For example, Patterson speculates that parents of socially aggressive children are overly inclusive in what behaviors they classify as deviant, while parents of some stealers are overly exclusive in their classification of property violations. While any apparent relationship between parental baselines and profile types does not address causation, it is particularly relevant when a therapist considers targeting not only the child's behavior but also the parents' expectations. Finally, there is the question of the degree to which parental reports reflect an accurate assessment of the child and how much they reflect instead the worry, depression, sophistication, and/or marital discord of the parent (see Gibbs [1982] for a more thorough discussion of this issue).

As previously mentioned, the clinical sample from which the scales and profile types were derived came from 42 eastern mental health and related service agencies. This sample was 81.2% white and 17.1% black, with 1.8% of the sample representing other races. Mothers completed 83% of the checklists, fathers 11.5%, and other respondents (such as foster parents or other relatives) completed 5.6% of the checklists. A large variety of occupational levels from rural, suburban, and urban areas are represented. Within the age and sex groupings, differences in scores due to age were found to range from small to nonexistent. Analyses of the items according to race and socioeconomic status revealed very few significant differences, all of which were quite small. Further research on ethnic populations, particularly Hispanics and Asian-Americans, would however be desirable.

The normative sample of children was collected by first randomly selecting blocks from census-tract data that approximated the socioeconomic and racial sample of the clinic sample; city directories were then used to select residents. Those selected were then sent a postcard indicating that an interviewer would stop by and ask them a variety of questions about their four- to sixteen-year-old child. If the home had no children living in it the residents were urged to return an enclosed postpaid card; if none was received an interviewer went to the home. If no parent was home at the time, the interviewer returned later, making at least one return visit on a weekend. If no children ages 4-16 lived in the home or if any had received mental health of psychiatric services in the past year, the interview went no further. If there was more than one child in the appropriate age range, a random number table was used to select the targeted child. A black interviewer was sent to the largely black census tracts. Of the sample contacted, 82.3% were able and willing to complete the interview. The authors do not indicate what percentage they were not able to contact, but one can assume from the procedures followed that it was quite small. From this group 50 boys and 50 girls at each age from 4 through 16 were randomly selected to comprise the normative sample. Race, socioeconomic status, and parental relationship were nearly identical to the clinical sample.

The normative sample was well selected in a careful manner and the participation rate was commendably high. The major potential bias arising from their procedures is that the normative checklists were completed in response to an interviewer. Although the respondents remained anonymous, the results may not be comparable to those compiled without such a face-to-face interaction. As normative data from other methodologies becomes available on the checklist, careful scrutiny should be given to any discrepancies. To date, normed controls used in a number of studies seem to support the validity of normative scores (see Validity section of this review).

For mothers and fathers of children seen in a mental-health setting, the authors reported interparent reliability of .985 for total behavior problems and .978 for total social competencies. The individual scale score reliabilities tended to be quite a bit lower (the median correlation for all scales was .66). As one might expect, the smaller correlations tended to occur on scales having items difficult to directly observe, such as depression and sex problems. There was an overall tendency, which amounted to very little difference in mean scores, for mothers to see more problems and less social competency in their children.

Test-retest reliability for *non-referred* samples that Achenbach and Edelbrock (1983) reported is generally very high. One-week and three-month test-retest reliabilities for the total behavior problems were .952 and .838 respectively and similar test-retest reliabilities for the total social competence was .996 and .974. If the Behavior Problem and Social Competency scales are examined individually for each age and sex grouping, one again sees very high overall reliability (the median correlation for all scales was .89 for one-week test-retest reliability) with some exceptions. For boys and girls ages 4-5 and for girls 6-11, the scale of Sex Problems had low reliabilities (from .22 to .52) as did the Obsessive-Compulsive scale for boys ages 12-16 (–.12) and the Obese scale for girls 4-5 (.42). These low correlations may be due partly to the small range and variance of the scores, reflecting the low occurrence of these problems in a non-clinical sample.

Achenbach and Edelbrock (1979; 1983) also present "stability" data comparing checklists completed before treatment began with those obtained 6 and 17 months later. Overall, the behavior-problem scores appear to be reasonably stable and yet almost all showed post-treatment decreases in problem behaviors.

The Social Competence scales have not been used as frequently in research as the Behavior Problems scales. When used, they have had generally positive but weak results in discriminating clinic-referred children from non-clinic-referred. Hodges, Kline, Stern, Cytryn, and McKnew (1982) found that all three Social Competence Scale scores were higher for normal controls than for inpatient or outpatient child psychiatric patients. Wolfe and Mosk (1983) found that abused and non-abused clinic-referred children differed from the non-referred on scores of Social and School, but not on Activities. Plaisted, Wilkening, Gustavson, and Golden (1983) cite unpublished research finding that poor performance on the Luria-Nebraska Neuropsychological Battery-Children's Revision is associated with low scores on all the Social Competence Scales.

When used to differentiate among clinical samples the findings have been more mixed. Heath, Hardesty, Goldfine, and Walker (1983), for example, found no significant differences between overall competence, social functioning, or school competence when they compared firesetters with a clinical control, but firesetters were found to score less on the Activity scale than the non-firesetters. Hodges et al. (1982) found that inpatient and outpatient child psychiatric patients did not differ on the Social Competence Scales. Similarly, Wolfe and Mosk (1983) found that abused and non-abused children in the same agency did not differ on any of the Social Competence scales. Cohen-Sandler, Berman, and King (1982) found among suicidal children higher scores on Social Involvement than depressed children at 5-month to 3-year follow-ups, which is consistent with their prediction that depressed children tend to withdraw from family and friends while suicidal children maintain their involvement with others.

While it is a global score, the total problem behavior index has been found a useful discriminator of disturbed and nondisturbed children. The authors report that using a 90% cutoff on the total behavior-problem score results in 90% of the non-referred sample being classified in the normal range and 26% of the clinical sample being classified as from a non-referred sample.

In addition, the total behavior problem score has been acquiring impressive construct validity. Weissman, Orvaschel, and Padian (1980) found that total scores from mothers' reports on the Conners Parent Questionnaire and the Child Behavior Checklist Problems correlated very highly ($r = .91$). They also found that problem behaviors correlated substantially with a depression checklist completed by mothers but found lower correlations with children's self-reports. The total behavior-problem score was related to the total pathology score of the Child Assessment Schedule—a structured interview with children (Hodges, McKnew, Cytryn, Stern, & Klein, 1982). Hazzard, Christensen, and Margolin (1983) found the Parent Perception Inventory that is completed by children correlated with the total problem behaviors. In particular they found that children who perceived their parents negatively (punitive and disciplining) had higher total behavior problems, while those who rated their parents more positively (supportive-nurturing) did not differ in behavior problems from those who rated their parents less positively. Last and Bruhn (1983) grouped children according to their Total Problem Behaviors into groups of well adjusted (T-score below 63), mildly maladjusted (T-score of 63–76), and severely maladjusted (T-score greater than 76), and compared their earliest memories. They found that the presence of a caretaker figure in a fragmented, illogical, or unbelievable memory was related to severe maladjustment, while memories did not differentiate the mildly disturbed from the well-adjusted children.

Internalizing and Externalizing categories of behavior problems have been found across many types of measures and these dimensions are an important aspect of the Child Behavior Checklist. Heath et al. (1983) investigated the difference between firesetters and non-firesetters in a child clinical population. They found that firesetters were more likely to be Externalizers than their control group. Mash and Johnston (1983) found very high correspondence between the Externalizing scale, and to a lesser but significant degree the Internalizing scale, and both the Conner's Abbreviated Rating Scale and the Werry-Weiss-Peters Activity Scale. Further, on both the Internalizing and Externalizing dimensions the group T-score means of their hyperactive sample differed significantly from those obtained from the normal samples.

One note of caution is that the high correlations between internalizing and externalizing scales make broad dichotomous conclusions risky. As the Child Behavior Checklist can be used on both sides of the next generation of longitudinal studies, more fine-grained analyses of items, scales, and Profile Types, as well as on the Internalizing and Externalizing dimensions, should prove fascinating.

While information obtained on the Profile Types would appear to have potential to contribute significantly to clinical research with children, validation studies examining the relations of the Profile Types to independent criteria of etiology, functioning, and outcome remain almost nonexistent. The only such study this reviewer found was conducted by Rapoport et al. (1981).[2] Although it is unclear how the data were analysed and grouped, these researchers found that for DSM-III diagnosed obsessive-compulsive adolescents, boys were most related to Profile Type A—Schizoid—and girls were most related to Profile Type C—Obsessive-Compulsive-Aggressive.[3] Likely reasons for this scarcity of data are that the taxonomy was only published a few years ago; that with the valuable breakdowns by age and sex, research requires a substantial population N for adequate analyses; and that to date the documentation and support for scoring the profile types have been less than optimal.

When using scales on which items appear directly related to the construct under study (e.g., the Depression scale with a depressed population), the Behavior Problem scales have proved quite effective in discriminating groups. Frame, Matson, Sonis, Fialkov, and Kazdin (1982) noted that the Depression scale of a prepubertal depressed boy was 1.5 standard deviations from the mean. Michael, Klorman, Salzman, Borgstedt, and Dainer (1981) used the Child Behavior Checklists hyperactivity scale to assist in matching different hyperactive groups. These hyperactive groups had T-scores over 70, while the normal control children had T-scores of around 50. Various Behavior Problem scales also appear able to differentiate clinically referred and non-clinically referred children. When constitutionally short children were compared to healthy normal-height counterparts they scored higher on scales of Somatic Complaints, Schizoidal tendencies, and Social Withdrawal, reaching levels typically found in psychiatrically referred children (Gordon, Crouthamel, Post, & Richman, 1982).

Within clinically referred groups and with problems not directly related to the scales under examination, the results to date have been more mixed. In the future such analyses might be better done on the Profile Types than on individual scales. Kuhnley, Hendren, and Quinlan (1982) found that within an inpatient child psychiatric sample, those who set fires scored higher on the Delinquency scale than those who did not set fires. The specific Delinquency scale items for the firesetters that were more likely to be endorsed were "destroys others' things";

"lies, cheats"; "steals at home"; and "vandalism." Other individual checklist items that differentiated the two groups were "fears he/she might think or do something wrong" and "cruel to animals" (both being scored higher for the firesetters), and "deliberately harms self or attempts suicide" and "likes to be alone" being scored higher for the non-firesetters.

Associations to unobservable, typically psychodynamic personality patterns have also been investigated, as well as relations with projective tests. With adolescents, Noam et al. (1984) found ego development was negatively correlated with scores on the scales of Somatic Complaints, Aggressive-Delinquent, Anxious-Obsessive, Depressive-Withdrawal, Immature, and Hyperactive. No behavior problem scales have been found to relate to the Egocentricity Indices of Exner's scoring system (Gordon & Tegtmeyer, 1982), nor were any checklist factors related to white-space responses in children's Rorschach protocols (Tegtmeyer & Gordon, 1983).

Achenbach and Edelbrock (1981) found that parents of a clinically referred group endorsed 116 of the 118 individual behavior problem items more often than a non-referred group, with Allergy (#2) and Asthma (#4) being the only two items that did not significantly differ between the two groups. A particularly interesting finding in this study was that the non-referred sample of girls, compared to non-referred boys, tended to score higher on Internalizing while the non-referred boys tended to score higher on Externalizing. However, the total problem behaviors of the groups were the same. This supports the notion that the much higher incidence of referral for boys in a clinical population appears to be due more to the type of problem—one that conflicts openly with official norms—rather than to the number of problems.

Critique

While many checklists have been developed, none has been done so in as careful and well-constructed a manner and is as potentially useful across a wide variety of settings as the Child Behavior Checklist. It is easy to administer, well normed, has outstanding psychometric properties, and is appropriate for a large clinic-referred child population. While still a relatively new test, clinical research with the Checklist is rapidly expanding, suggesting a promising future for the instrument in child assessment and program evaluation.

The test initially requires considerable effort in order to understand the nuances of its development. However, once one is clear on why certain statistical decisions have been made, the test is extremely logical and easy to interpret. A number of decisions in the development of the test deserve repeating. First, it is important to remember that the Child Behavior Checklist was developed with a clinically referred population in mind; the content, scales, and profile types all reflect this clinical emphasis. Users of the Child Behavior Checklist should thus be cautioned that the checklist items, scales, and profile types are all skewed to identify pathology and those children who appear to be within a broad band of healthy functioning (i.e., high scorers in social competency and low scorers in behavior problems) are generally not differentiated. A second caution is that the

Checklist focuses on an individual child's pathology and neglects other important systems (e.g., the family system) in the child's life that quite likely impact on the child's behavior problems.

Each of the behavior problem scales was empirically derived using factor analysis for the six age/sex groupings, resulting in either 8 or 9 factors for each of the groupings. While there is considerable similarity among factor items across these groupings and while the scale labels are often similar, it is important to remember how the scales were derived and that the labelling of scales is merely a shorthand for the items comprising any given scale.

The problem behavior Profile Types were empirically derived using a cluster analysis of the scale scores for each age/sex grouping and their labels merely reflect the high points of the profile pattern, when in reality the scale consists of the magnitude and relationship among all of the scale scores. The Profile Types, a taxonomy of empirical groupings of children's behavior problem profile patterns, are one of the most exciting developments in childhood assessment and broad-based behavior problem research. To quote Edelbrock and Achenbach (1980), this taxonomy "serves as a 'multivariate heuristic' for identifying *patterns of reported problems that characterize groups of disturbed children*" (p. 463). Whereas most taxonomies are narrow in focus, demand a dichotomous decision, and are quite unreliable (the childhood disorders in DSM-III are an example), these Profile Types look at a child's behavior across a wide range of behavior problems, allow a quantitative score of a given child's match to a given profile type, and appear to be quite stable. The possibilities for sophisticated etiological, outcome, and longitudinal research are vast.[4]

The weakest component of the Child Behavior Checklist is the social competency section. Its Social, Activity, and School scales are the same for the various age/sex groupings (except for 4- and 5-year-olds, where the school section may not apply) since they were not developed empirically but chosen in an a priori fashion. While the social competency section has the potential for providing useful social information in clinical cases, it should not be confused with the current social competency literature in child development that deals with issues such as coping, altruism, and positive mental health.

It is clear that as a well-normed descriptive instrument, the Child Behavior Checklist has earned a place comparable to or above any other standard assessment tool for psychologists and other health and mental health professionals who work with children. As there is to date little or no information on the correlates of children categorized according to their Profile Types, the usefulness of this aspect of the Child Behavior Checklist is left to future clinical experience and research.

In summary, the Child Behavior Checklist is a valuable addition to the field of the assessment of children. It is quite useful in distinguishing clinical from non-clinical populations and in providing a broad overview of a child's behavior problems from the parents' perspective. Its usefulness with normal groups of children appears to be quite limited. The Child Behavior Checklist is likely to be heavily used in the future, with information obtained on the Profile Types having the potential to play an important role in understanding the development of clinical problems in children.

Notes to the *Compendium:*

[1]In particular, it would be very helpful if Achenbach and Edelbrock would release the necessary information and norms for investigators to compute profile types for themselves.

[2]See also Mooney, Thompson, and Nelson (1987). Cohen, Gotlieb, Kershner, and Wehrspann obtained the narrow band Profile Types of outpatient children and then classified the children as to whether their narrow band Profile Types fell under the broader classification of an Externalizing or Internalizing Profile Type. On a battery of tests, including parent, child, and teacher ratings and a shortened form of the Wechsler Intelligence Scale for Children-Revised (WISC-R), they found significant differences between children with the Internalizing and Externalizing Profile Types.

[3]Mooney et al. (1987), using a large sample of outpatient children, found no relationship between a variety of etiological "risk factors" and the CBCL's narrow band or Internalizing and Externalizing Profile Types. However, it is unclear whether this lack of results is due to a deficiency in the validity and usefulness of the CBCL Profile Types or other problematic aspects of the study.

[4]However, to date there exists no research showing that the CBCL narrow band Profile Types are a useful or valid taxonomic system *within* a clinical population.

References

Achenbach, T.M. (1966). The classification of children's psychiatric symptoms. *Psychology Monographs, 80*(7), 1-37.

Achenbach, T.M. (1978). The Child Behavior Profile: I. Boys aged 6-11. *Journal of Consulting and Clinical Psychology, 46,* 478-488.

Achenbach, T.M., & Edelbrock, C.S. (1979). The Child Behavior Profile: II. Boys aged 12-16 and girls aged 6-11 and 12-16. *Journal of Consulting and Clinical Psychology, 47,* 223-233.

Achenbach, T.M., & Edelbrock, C.S. (1981). Behavioral problems and competencies reported by parents of normal and disturbed children aged 4 through 16. *Monographs of the Society for Research in Child Development, 46*(10, Serial No. 188).

Achenbach, T.M., & Edelbrock, C.S. (1983). *Manual for the Child Behavior Checklist and the Revised Child Behavior Profile.* Burlington, VT: University Associates in Psychiatry.

Achenbach, T.M., & Lewis, M.A. (1971). A proposed model for clinical research and its application to encopresis and enuresis. *Journal of the American Academy of Child Psychiatry, 10,* 535-554.

Cohen, N.J., Gotlieb, H., Kershner, J., & Wehrspann, W. (1985). Concurrent validity of the internalizing and externalizing profile patterns of the Achenbach Child Behavior Checklist. *Journal of Consulting and Clinical Psychology, 53,* 724-728.

Cohen-Sandler, R., Berman, A.L., & King, R.A. (1982). A follow-up study of hospitalized suicidal children. *Journal of the American Academy of Child Psychiatry, 21,* 398-403.

Edelbrock, C. (1979). Mixed model tests of hierarchical clustering algorithms: The problem of classifying everybody. *Multivariate Behavioral Research, 14,* 367-384.

Edelbrock, C., & Achenbach, T.M. (1980). A typology of Child Behavior Profile patterns: Distribution and correlates in disturbed children aged 6 to 16. *Journal of Abnormal Child Psychology, 8,* 441-470.

Edelbrock, C., & McLaughlin, B. (1980). Hierarchical cluster analysis using intraclass correlations: A mixture model study. *Multivariate Behavioral Research, 15,* 299-318.

Frame, C., Matson, J.L., Sonis, W.A., Filskov, M.J., & Kazdin, A.E. (1982). Behavioral treatment of depression in a prepubertal child. *Journal of Behavior Therapy and Experimental Psychiatry, 13,* 239-243.

Gibbs, M.S. (1982). Identification and classification of child psychopathology. In J.R. Lachenmeyer & M.S. Gibbs (Eds.), *Psychopathology in childhood*. New York: Gardner Press.

Gordon, M., Crouthamel, C., Post, E.M., & Richman, R.A. (1982). Psychosocial aspects of constitutional short stature: Social competence, behavior problems, self-esteem, and family functioning. *Journal of Pediatrics, 101,* 477-480.

Gordon, M., & Tegtmeyer, P.F. (1982). The egocentricity index and self-esteem in children. *Perceptual and Motor Skills, 55,* 335-337.

Heath, G.A., Hardesty, V.A., Goldfine, P.E., & Walker, A.M. (1983). Childhood firesetting: An empirical study. *Journal of the American Academy of Child Psychiatry, 22,* 370-374.

Hodges, K., Kline, J., Stern, L., Cytryn, L., & McKnew, D. (1982). The development of a child assessment interview for research and clinical use. *Journal of Abnormal Child Psychology, 10,* 173-189.

Hodges, K., McKnew, D., Cytryn, L., Stern, L., & Kline, J. (1982). The Child Assessment Schedule (CAS) diagnostic interview: A report on reliability and validity. *Journal of the American Academy of Child Psychiatry, 21,* 468-473.

Kuhnley, E.J., Hendren, R.L., & Quinlan, D.M. (1982) Firesetting by children. *Journal of the American Academy of Child Psychiatry, 21,* 560-563.

Last, J.M., & Bruhn, A.R. (1983). The psychodiagnostic value of children's memories. *Journal of Personality Assessment, 47,* 597-603.

Mash, E.J., & Johnston, C. (1983) Parental perceptions of child behavior problems, parenting self-esteem, and mothers' reported stress in younger and older hyperactive and normal children. *Journal of Consulting and Clinical Psychology, 51,* 86-99.

Michael, R.L., Klorman, R., Salzman, L.F., Borgstedt, A.D., & Dainer, K.B. (1981). Normalizing effects of methylphenidate on hyperactive children's vigilance performance and evoked potentials. *Psychophysiology, 18,* 665-677.

Mooney, K.C., Thompson, R., & Nelson, J. (1987). Risk factors and the Child Behavior Checklist in a child mental health center setting. *Journal of Abnormal Child Psychology, 15,* 67-73.

Noam, G., Hauser, S., Jacobson, A., Powers, S., & Miranda, D. (1984). Ego development and psychopathology. *Child Development, 55,* 184-194.

Patterson, G.R. (1982). *Coercive family process.* Eugene, OR: Castalia Publishing Company.

Plaisted, J.R., Wilkening, G.N., Gustavson, J.L., & Golden, C.J (1983). The Luria-Nebraska Neuropsychological Battery-Children's Revision: Theory and current research findings. *Journal of Clinical Child Psychology, 12,* 13-21.

Rapoport, J., Elkins, R., Langer, D.H., Sceery, W., Buchsbaum, M.S., Gillin, C., Murphy, D.L., Zahn, T.P., Lake, R., Ludlow, C., & Mendelson, W. (1981). Childhood obsessive-compulsive disorder. *American Journal of Psychiatry, 138,* 1545-1554.

Tegtmeyer, P.F., & Gordon, M. (1983). The interpretation of white-space responses in children's Rorschach protocols. *Perceptual and Motor Skills, 57,* 611-616.

Weissman, M.M., Orvaschel, H., & Padian, N. (1980). Children's symptom and social functioning self-report scales: Comparison of mothers' and children's reports. *The Journal of Nervous and Mental Disease, 168,* 736-740.

Wolfe, D.A., & Mosk, M.D. (1983). Behavioral comparisons of children from abusive and distressed families. *Journal of Consulting and Clinical Psychology, 51,* 702-708.

Donald G. Barker, Ph.D.

Professor of Educational Psychology, Texas A&M University, College Station, Texas.

THE CHILDREN'S APPERCEPTION TEST

Leopold Bellak and Sonya Sorel Bellak. Larchmont, New York: C.P.S., Inc.

THE ORIGINAL OF THIS REVIEW WAS PUBLISHED IN TEST CRITIQUES: VOLUME I (1984).

Introduction

The Children's Apperception Test (CAT-A) is a projective method of describing personality by studying individual differences in the responses made to stimuli presented in the form of pictures of animals in selected settings. The test is individually administered to children aged 3 to 10 years. The 10 items consist of 10 scenes showing a variety of animal figures, mostly in unmistakably human social settings. The use of animal rather than human figures was based on the assumption that children of these ages would identify more readily with appealing drawings of animals than with drawings of humans. The author discusses interpretation on the basis of psychoanalytic themes, but there is no compelling reason that Children's Apperception Test protocols could not be interpreted from other theoretical frameworks.

Also available, but much less used, are two other sets of items: Children's Apperception Test-Human Figures (CAT-H) and Children's Apperception Test-Supplement (CAT-S). The Human Figures set was developed in reaction to criticism of the assumption that children identify more readily with sketched animals than with sketched human beings. The human figures are placed in settings as nearly as possible identical with the corresponding scenes on the original test. The Human Figures form has been used mostly for research. The set of items termed Children's Apperception Test—Supplement contains 10 additional items dipicting animals in scenes even more structured than the original set. These seem to be intended more for therapeutic interaction than for diagnosis, although they can provoke responses that may help the therapist understand the child.

Leopold Bellak attributed his idea for the Children's Apperception Test to a remark made by a colleague, Dr. Ernst Kris, that small children could be expected to identify much more readily with pictures of animals than with pictures of humans in social settings. Already an authority on and enthusiastic user of Murray's Thematic Apperception Test (TAT), Bellak observed that the Murray test had only a few items specifically designed for children, and that even the so-called children's subset of pictorial plates did not yield rich protocols from children younger than 11 years.

Bellak planned a series of scenes involving situations vital to the 3-to-10-year developmental level. Violet Lamont, a professional illustrator of children's books, produced 18 drawings to Bellak's specifications. A number of psychologists acquainted with Bellak collected and submitted protocols of young children's responses to the sketches. The author selected the 10 plates yielding the richest content from children aged 3 to 10 years.

The Children's Apperception Test contains 10 items administered individually and orally. Each item consists of a 8½" x 11" black-and-white cardboard plate containing pictures of one or more animals in a social setting likely to be meaningful or interesting to a small child. Rapport must be established at the beginning of the session, and the examiner should describe the session as a time for interesting activities. The examiner then presents the pictures one at a time in a set order, asking the child to tell a story about the picture (i.e., what is going on, what the animals are doing). The examiner provides as much encouragement as possible and also asks the child to describe what went on before and what will happen later. After the child tells stories about all 10 pictures, the examiner conducts an inquiry, asking for elaboration, details, and clarifications. All pictures except the one being responded to should be kept out of the examinee's sight.

The examiner records pertinent features of both the stories and the inquiry on a specially prepared form, the TAT and CAT Recording and Analysis Blank (Short Form). The fold-out form is on a 24" x 11" sheet, providing for notes on all ten stories in parallel columns so that themes and patterns can be more readily construed. The form prompts for observation or judgment on 10 variables for each story, as well as across all stories: 1) main theme; 2) main hero; 3) main needs and drives of hero; 4) conception of environment (or world); 5) perception of parental, contemporary, and junior figures; 6) conflicts; 7) anxieties; 8) defenses; 9) adequacy of superego; and 10) integration of ego (including originality of story and nature of outcome). The examiner finally prepares a diagnosis or personality description of the child by synthesis of the observations recorded on the form.

The CAT-H was developed by Bellak in response to criticism (by Marcuse and others) of his hypothesis that animal pictures have more stimulus value than those of humans for young children. The procedures and forms are the same as for the CAT-A, but the pictures have been carefully composed to resemble the original plates as much as possible using human figures rather than animals. The resulting images include attempts to conceal the gender of some characters by the use of strange clothing and unusual perspectives. The manual does not seem to seriously recommend this form over the CAT-A, except perhaps for children near the upper end of the 3-to-10-year age span. The form is offered for and has been used in research.

The CAT-S is not intended as an independent assessment of personality. Instead, it contains a set of 10 supplementary pictures of animals in situations encountered by some but not all children. These plates can be presented selectively in situations in which the examiner has reason to believe that supplementation is needed. The plates include pictures of an injured kangaroo, group of mice on a playground, a cat looking at itself in a mirror, and others. The manual suggests that these supplementary pictures may be used in a sort of play-therapy context rather than a formal examination situation. The supplements are smaller than the standard CAT-A plates, and they are on irregularly shaped pieces of cardboard.

Practical Applications/Uses

The Children's Apperception Test can be used to assist in understanding the personality attributes of children too young to be tested by most other means, even other projective techniques. The test could be used in clinical settings to diagnose a child's condition and to plan treatment, but interpretation must be cautious as the test does not have and is not claimed to have psychometric precision. This test can also be administered in research settings to observe personality-relevant behaviors for the purpose of relating them to other variables. In addition, there is some indication that the use of the technique may have therapeutic value by directing attention to social situations and providing a permissive opportunity for expression. It may not be appropriate for children near the lower end of the intended age range who are immature for their age, and it may not be suitable for children at the upper end of the range who are advanced in their development.

The Children's Apperception Test must be individually administered by a psychologist specifically trained in its use. It is an oral test, requiring no written responses by the examinee. The instructions for administration allow considerable discretion on the part of examiner for establishing rapport, encouraging response, and conducting inquiries. The test is untimed, but can usually be administered in 30 minutes.

This projective technique is not "scored" in any literal, quantitative sense. The gist of stories is recorded, and the presence or absence of thematic elements is indicated on the form provided. No machine scoring is available or needed.

The author provides considerable material in the test manual and more in *The TAT and CAT in Clinical Use* about the interpretation of the Children's Apperception Test. Recommended interpretation is in terms of the 10 variables previously listed in this review. Numerous examples of sets of stories are presented, along with their interpretation. The illustrative interpretations are from a psychoanalytic point of view. Examples are also given of the story elements commonly encountered, to aid the examiner in judging whether stories are popular (in the Thematic Apperception Test sense) or novel.

Technical Aspects

As in many other projective techniques, no statistical information is provided on the technical validity and reliability of the Children's Apperception Test. This may be distressing to users, but enthusiasts maintain that projectives are not appropriate for and do not require psychometric evaluation.

Critique

Many experts are highly critical of any type of projective approach, including the Children's Apperception Test. Other critics, open to projectives, fault the test for its basic assumption that children respond more richly to animal figure stimuli than to human figure stimuli. Still others object that the settings in the plates are so obvious that the stimuli do not have sufficient ambiguity to encourage deep projection. Critics have frequently observed that the pictures are highly appealing to children ages 3-10, and some have reported that responses made by children tend to be therapeutic whether highly diagnostic or not.

The Children's Apperception Test was developed primarily as an alternative to the Thematic Apperception Test for younger children. It was based on the assumption that animal drawings would appeal to children of this age range and would encourage richer, more productive responses. The premise on which the test was based has been questioned, yet the Children's Apperception Test has remained fairly popular for many years among practitioners who deal regularly with such young clients.

References

This list includes text citations and suggested additional reading.

Anastasi, A. (1982). *Psychological testing* (5th ed.). New York: Macmillan Publishing Co.

Bell, J. E. (1970). Review of the Children's Apperception Test. In O. K. Buros (Ed.), *Personality tests and reviews* (pp. 581-582). Highland Park, NJ: The Gryphon Press.

Bellak, L. (1954). *The TAT and CAT in clinical use.* New York: Grune & Stratton.

Bellak, L., & Adelman, C. (1960). The Children's Apperception Test. In A. I. Rabin & M. R. Haworth (Eds.), *Projective techniques with children* (pp. 62-94). New York: Grune & Stratton.

Bellak, L., & Bellak, S. (1980). *A manual for the Children's Apperception Test* (7th ed.). Larchmont, NY:C.P.S., Inc.

Murstein, B. J. (1970). Review of the Children's Apperception Test. In O. K. Buros (Ed.), *Personality tests and reviews* (pp. 1228-1230). Highland Park, NJ: The Gryphon Press.

Neuringer, C. (1968). A variety of thematic methods. In A. I. Rabin (Ed.), *Projective techniques in personality assessment: A modern introduction* (pp. 224-261). New York: Springer Publishing.

Semeonoff, B. (1976). *Projective techniques.* New York: John Wiley & Sons, Inc.

Stone, L. J. (1970). Review of the Children's Apperception Test. In O. K. Buros (Ed.), *Personality tests and reviews* (pp. 582-585). Highland Park, NJ: The Gryphon Press.

Sweetland, R. C., & Keyser, D. J. (Eds.). (1983). *Tests: A comprehensive reference for assessments in psychology, education and business.* Kansas City, MO: Test Corporation of America.

Wirt, R. D. (1970). Review of the Children's Apperception Test. In O. K. Buros (Ed.), *Personality test and reviews* (pp. 1230-1231). Highland Park, NJ: The Gryphon Press.

Giselle B. Esquivel, Psy.D.
Assistant Professor of Psychology, Graduate School of Education,
Fordham University at Lincoln Center, New York, New York.

COLOURED PROGRESSIVE MATRICES

J. C. Raven. London, England: H. K. Lewis & Co. Ltd. U.S.
Distributor—The Psychological Corporation.

THE ORIGINAL OF THIS REVIEW WAS PUBLISHED IN TEST CRITIQUES: VOLUME I
(1984).

Introduction

The Coloured Progressive Matrices (CPM) is one of three in the series of Raven Progressive Matrices tests, the other two being the Standard Progressive Matrices (SPM) and the Advanced Progressive Matrices (APM). The CPM was developed by Raven in 1947 as a downward extension of the SPM for use with young children between the ages of 5 and 11 and for individuals with intellectual deficiencies or very advanced age.

This untimed, nonverbal test consists of a series of 36 problems divided into three subscales (A, Ab, B) of 12 items each. The items are arranged in order of difficulty both within and between scales. Each item consists of a colored abstract design (matrix) with a missing piece to be completed by selecting the best alternative choice, based on characteristics ranging from discrete form to more complex logical relationships.

The CPM is based on Spearman's theory of g as a factor associated with analogical reasoning. Its purpose is to assess a child's observational ability and developmental readiness for analogical reasoning and the extent to which this capacity is impaired in adults. As such, the CPM is not a test of either innate capacity or global intelligence. However, if used in conjunction with the Crichton Vocabulary Scale or another standardized measure of acquired knowledge, it may provide an estimate of general intellectual ability.

Accumulated research studies over the years indicate that the CPM is highly reliable and valid in terms of what it portends to measure. It is especially useful as a supplement for assessing children of limited English proficiency and those with oral communication problems. Recent findings suggest that the test may be used equally well with children of both sexes and from different ethnic and socioeconomic backgrounds. Major efforts toward restandardization in the United States and other countries will serve to further enhance its validity and wide usage in cross-cultural settings.

The Coloured Progressive Matrices (CPM) was developed by John C. Raven, a

The reviewer wishes to acknowledge and express appreciation to Scott Sigmon for some technical aspects, to William Summers, and to John Raven for historical facts and other specific information.

British psychologist who lived between the years of 1902 and 1970. In his early days, Raven was associated with many significant persons in psychology and came especially under the influence of Spearman and the British school of general factor analysis. After spending several years as teacher and headmaster, Raven began research on the origins of mental deficit with Lionel Penrose, a geneticist at the Colchester Institute (Court, 1971). It was through their work that Spearman's theory of a general factor (*g*), associated with the capacity for analogical reasoning and relatively independent from acquired knowledge, first began to be applied in the development of a practical measure.

Raven's publication of the Standard Progressive Matrices (SPM) in 1938 was timely with the start of World War II and became widely accepted in his clinical work at the Mill Hill Emergency Hospital, and in its use with British Army recruits, because of its "apparent" nonverbal nature and lack of educational bias (Burke, 1958). In order to provide a measure of acquired information, Raven developed the Mill Hill Vocabulary Scale, to be used as a supplement to the Standard Matrices.

In 1943, Raven was invited to establish a new Department of Psychological Research at the Crichton Royal in Dumfries, where he taught and continued his work (Court, 1971). Raven continuously refined his instrument into various editions which reflected the new data available from such studies as that conducted on Colchester children between 1943 and 1944. In 1947, the Advanced Progressive series came into restricted circulation for use with intellectually gifted individuals.

During the same year the Coloured Progressive Matrices (CPM) became available for use with young children, those individuals with intellectual impairments, and the very old. The original CPM (Sets A, Ab, B) was a modification of the SPM (Sets A, B, C, D, E), which included an interpolated set of problems of intermediate difficulty (Set Ab), consistent with the sequential cognitive development of children. The Crichton Vocabulary Scale was published in 1949 as a verbal supplement to the CPM for use with young children.

In 1956, the CPM was revised slightly through rearrangement of a few items to provide for a more uniform increase in order of difficulty. Alternative choices were also analyzed and rearranged to account for positional effects where necessary. The test was standardized on a representative sample of 608 school children from the Burgh of Dumfries in Scotland between the ages of 5 and 11½. The adult group sample consisted of 271 people aged 60 to 89 years from the Rutherglen Centre, a place of social contact and a clinic for clients who were free from mental illness and senile dementia.

Until his retirement in 1964 Raven continued to work on the development of the Matrices, including the CPM, and on their advancement in clinical and cross-cultural application. After his retirement he proceeded more slowly, yet continued to develop ideas which left the basis for useful clinical measures of spatial orientation and concept formation. Following Raven's death in 1970, collaborative efforts were continued by his colleagues.

Since that time, the various Matrices manuals have been revised and compiled into a single volume. An extensive bibliography on significant international research studies has been developed by J. H. Court (1972, 1974, 1976, 1980, 1982).

Major efforts led by John Raven (son and colleague of J. C. Raven) are being undertaken towards restandardization in the United States (John Raven, personal communication, June, 1984).[1]

The CPM set is published by H. K. Lewis & Company in London and distributed in the United States by The Psychological Corporation. Manuals are also available in various languages through foreign publishers containing norms for non-English-speaking examinees and information for researchers and psychologists. The basic set contains a manual, test booklet, answer sheets, and scoring key. The CPM manual consists of a general overview section on all the Progressive Matrices and Vocabulary Scales and a separate section specifically on the CPM. This last section provides a description of design and use, instructions, standardization, norms, and an evaluation of test responses of children and elderly persons in terms of developmental expectancies. A useful description is also provided regarding the performance of emotionally disturbed children and other clinical groups. The 1984 manual contains updated norms and a summary of reliability and validity data.

The Book Form of the test consists of a reusable booklet with 36 consecutive items presented separately by page and divided into three sets (A,Ab,B) of 12 items each. Most CPM items are designed in bright colors to attract a child's attention. Each matrix is missing a piece which needs to be completed by the selection of the correct one from among six alternative pieces placed below the matrix. The items are arranged in order of difficulty within and between sets.

In the first twelve items of set A, the matrix forms a continuous pattern with a gap which needs to be completed on the basis of such characteristics as design, size, and orientation. Items on set Ab consist of three discrete figures which need a fourth in order to form a spatially related whole. In set B, the figures have to be analyzed and completed on the basis of analogical relationships.

The Board Form of the test is recommended for routine work with very young children, in the clinical assessment of intellectual deficiency or decline, and in cross-cultural studies. It has been chiefly used in psychological research. Since it is not available commercially, a detailed description of construction is provided in the manual (Court & Raven, 1984) for interested workers and researchers. An advantage of the board form over the book form is that problem-solving approaches by trial and error can be observed and compared with solutions by direct perception and inference. It is also useful for training and for persons who need concrete and physically guided instructions. Results have been found to be approximately the same as those in the book form. A major disadvantage of the board form, however, is its limited availability and construction expense.

Practical Applications/Uses

As mentioned previously, the matrices were developed on the basis of Spearman's g factor theory. Spearman (1923) linked this general factor to analogical reasoning, a process characterized primarily by the ability to "educe relations and correlates." This ability is distinct from "reproductive ability," which involves the recall of acquired information. Raven (1948) viewed general intelligence as a combination of both types of ability.

The CPM was specifically designed to determine the extent to which a child has developed the capacity for analogical reasoning, or "eductive ability," as one aspect of intelligence. Initially in this process the child is able to perceive similarities or differences between patterns in a somewhat "passive" or direct manner. This is followed by the apprehension of simultaneous attributes, such as size or orientation of discrete figures in the perceptual field. Gradually, discrete figures may be conceived as forming a gestalt or spatially related whole. Eventually the child develops the ability to discern the relationships that exist between discrete pairs and to infer new correlates from these (Raven, Court, & Raven, 1984). This cognitive development process forms the basis for nonverbal abstract reasoning.

As an assessment instrument, the CPM may not be used as a substitute for a standardized intelligence test, even if the norms were appropriate, primarily because it measures only one aspect of intelligence. Nevertheless, the test serves as a valuable tool when used as a supplement to estimate the intellectual ability of illiterate, bilingual, deaf, language-impaired, and limited English-proficient children (Jensen, 1980). Moreover, as Sattler (1974) has suggested, a testing of the limits procedure may be applied to the CPM to make qualitative observations which have diagnostic significance. In school settings, the use of the CPM in this supplemental manner may serve to prevent the misclassification of children and their inappropriate placement in special classes (Sigmon, 1983; Valencia, 1984).

In terms of intervention, the CPM has been used effectively as a model in training and developing problem-solving strategies in children. Budoff and Corman (1976) trained children whose abilities in analogical reasoning tasks were deficient by providing them with more meaningful and concrete cues as a basis for solving tasks of a similar nature. Feurstein (1979) used a similar approach as the basis for his original work in the dynamic assessment of learning potential and in the development of cognitive strategies in children and adolescents.

In clinical settings, various modifications of the CPM may be applied, such as tactile versions for blind children (Rich & Anderson, 1965). The board form, slide presentations, and short forms of the CPM have also been used in assessing persons with cerebral palsy, sensory motor deficits, and other types of impairment.

The CPM may be administered by psychologists, educators, or researchers. Administration procedures allow the examiner to provide very specific and clear instructions or to demonstrate, if necessary, without any verbal instruction. This makes it a test especially suitable for children with limited language proficiency, deafness, aphasia, and other communication problems. Although task requirements are relatively easy to understand, the examiner may allow ample time when explaining these to very young or less able children. Moreover, redemonstrations and verbal explanation of errors may be provided for a few initial items. This flexibility helps to insure that performance is actually reflective of the examinees' ability when they attempt to solve the tasks.

Another characteristic of the CPM is that since it is untimed, assessing power of thinking rather than speed, it does not penalize those children with slower and/or more reflective styles. The test may be timed, however, when an assessment of speed of accuracy is desired. Younger and less able children who are unable to record their responses on the answer sheet are normally tested individually. Older children and adults usually do better on their own and may be tested in groups, thus decreasing the time of administration. The test usually takes from 10 to 20 minutes to administer.

An examinee who solves most of the problems on set B can proceed without interruption to sets C, D, and E of the Standard Scale. This procedure helps to assess the extent of intellectual maturity in analogical reasoning and to determine possible giftedness of young children in this area.

Scoring is simple and facilitated through the use of the scoring key. The number of correct responses is added and converted into percentile points using the norm tables based on the British samples. Extrapolated norms are also provided for children down to the age of 3½ and for adults up to the age of 100, based on the finding that both the decline of scores in old age and increase in childhood follow a linear trend. The utility of these norms, however, is quite limited for use in the United States, given the standardization samples on which they are based. The use of extrapolated hypothetical norms based on samples from American school children (Sigmon, 1983) and results from studies on specific reference groups may serve as viable alternatives pending restandardization of the Matrices in the United States.

In summarizing quantitative test results, an IQ score is not calculated, in keeping with Raven's view that intelligence in children is characterized by leaps and bounds and that the CPM is not a test of intellectual ability. Comparisons are made instead on the basis of developmental age expectancies. Tables are provided for the qualitative analysis of each consecutive problem (i.e., major operations in apprehending the problem and selecting choices as well as the type of erroneous choices made). This kind of qualitative interpretation serves to enhance the diagnostic value of the test.

Technical Aspects

The reliability data, which are available in the 1977 CPM manual, are based primarily on a study conducted between 1952 and 1954 comparing a clinical group of emotionally disturbed children and normal school children between the ages of 6½ and 12½. The results are suggestive of the test's high internal consistency and test-retest reliability. The manual also reports reliability data from an earlier study, showing a test-retest reliability of .65 for children under 7, of .80 for children older than 9, and of .90 for the whole age range for which the test was constructed.

A growing number of studies has confirmed that the reliability of the CPM is very satisfactory. Split-half reliability coefficients range between .80 and .90. Test-retest reliability scores fall between .80 and .87, although these are somewhat lower (.71) for periods over one year. A recent summary of studies with ethnic minority children also suggests high reliability with this group (Valencia, 1984). Although some age differences have been reported, showing higher reliability for children between 9 and 11½ and lower for those under 7, these differences may reflect the fluctuating nature of cognitive development during earlier years rather than psychometric flaws in the scale in this respect.

Concurrent validity studies show a moderate correlation between the CPM and conventional tests of intelligence (Birkmeyer, 1964). This finding is consistent with the fact that it is not a test of general intellectual ability. As a nonverbal test the CPM has higher correlation with other nonverbal tests, such as the Leiter International Performance Scale (Musgrove & Counts, 1975), and may be used interchangeably with similar tests in assessing reasoning ability, particularly with deaf children. Although the CPM is not a test of nonverbal reasoning alone, it correlates more highly with performance than with verbal scales and subtests. Correlations with Wechsler Intelligence Scale for Children, for example, range between .47 for Information and .74 for Block Design (Martin & Wiechers, 1954).

The CPM has relatively low correlations with tests of academic achievement, but it appears to have a relationship with some areas of learning such as reading comprehension and problem-solving tasks (convergent-cognitive figural). Furthermore, Stallings (1975) found a correlation between increased scores on the CPM and children's involvement in exploratory and independent activities in flexible classroom environments. Although a direct relationship has not been found between higher academic scores and the CPM, this study suggests that there may be some association between analogical reasoning and creative kinds of learning.

In terms of predictive validity, Hausman (1973) found the CPM to be a meaningful evaluator of learning potential with Spanish-speaking educable mentally retarded children. Wiedl (1978) found that while intelligence tests are better predictors of academic performance under regular teaching, learning test versions of the CPM have superior predictive validity under conditions of adaptive teaching.

Construct validity for the CPM rests partly in its high loading with g as the logical reasoning factor it purports to assess (Orpet, Yoshida, & Meyers, 1976). However, it is difficult to measure such a pure construct independent from the modality through which it is expressed or the specific process used by children in approaching the tasks. This is reflected by the small loading also found with the k factor, which involves visuospatial skills. Additional factorial differences between sets have been found that may reflect distinct stages in the developing process toward analogical reasoning in children. Thus, set A taps a more concrete perceptual dimension based on completion of patterns, set Ab involves reconstruction and closure of patterns, and set B requires the capacity for reasoning by analogy (Corman & Budoff, 1974; Carlson & Jensen, 1980).

Finally, cross-cultural validation studies show mostly positive results. The CPM has been widely used in places such as Australia, Africa, Canada, France, Germany, New Zealand, Spain, Belgium, Italy, and Venezuela. Where differences between groups have been found, these have been more reflective of qualitative than quantitative differences in intelligence. For example, spatial aspects of the test were influential in the Indian context in producing lower scores than those found in Britain (Sinha, 1979). Kirby and Das (1978) have associated cultural differences with the different strategies employed for CPM solution.

Critique

The Coloured Progressive Matrices has made a significant contribution to the area of assessment. It has been established as a reliable and valid test with wide applicability in cross-cultural settings. Although the lack of representative norms in some countries, including the United States, poses a problem, recent efforts towards restandardization[2] will help to further increase its usage. As a nonverbal test, the CPM has been valuable in evaluating children with sensory deficits and oral communication problems. As a relatively nonbiased instrument the CPM has provided supplemental information for making more appropriate decisions involving ethnic and language minority students. It has served as a screening procedure for identifying disadvantaged gifted children. Its major strength lies in its lasting influence and relevance for these various groups of children. Moreover, extensive research studies on the CPM have helped to clarify our understanding of children's thinking abilities and other aspects of information processing and cognitive reasoning. Finally, the CPM has served to broaden the view of assessment as a more dynamic process, and some critical dimensions of intelligence as amenable to educational training.

Notes to the *Compendium:*

[1] See Raven (1986).
[2] See Raven (1986).

References

Birkmeyer, F. (1964). The relationship between the Coloured Progressive Matrices and individual intelligence tests. *Psychology in the Schools, 1,* 309-312.

Budoff, M., & Corman, L. (1976). Effectiveness of a learning potential procedure in improving problem-solving skills of retarded and non-retarded children. *American Journal of Mental Deficiency, 81,* 260-264.

Burke, H. R. (1958). Raven's Progressive Matrices: A review and critical evaluation. *The Journal of Genetic Psychology, 93,* 199-228.

Carlson, J. S., & Jensen, M. (1980). Reliability of the Raven Coloured Progressive Matrices Test: Age and ethnic group comparisons. *Journal of Consulting and Clinical Psychology, 49,* 320-322.

Corman, L., & Budoff, M. (1974). Factor structures of retarded and nonretarded children on Raven's Progressive Matrices. *Educational and Psychological Measurement, 34,* 407-412.

Court, J. H. (1971). John C. Raven, M.Sc., F.B.Ps.S. (1902-1970). *British Psychological Society Bulletin, 24*(82), 47-48.

Court, J. H. (1972, 1974, 1976, 1980, 1982). *Researchers' bibliography for Raven's Progressive Matrices and Mill Hill Vocabulary Scales* (eds. and supplements). Adelaide, South Australia: Flinders University.

Feurstein, R. (1979). *The dynamic assessment of retarded performers: The learning potential assessment device, theory, instruments, and techniques.* Baltimore: University Park Press.

Hausman, R. M. (1973). Efficacy of three learning potential assessment procedures with Mexican-American educable mentally retarded children. *Dissertation Abstracts International, 33,* 3438A.

Jensen, A. R. (1980). *Bias in mental testing.* New York: The Free Press.

Kirby, J. R., & Das, J. P. (1978). Information processing and human abilities. *Journal of Educational Psychology, 70,* 58-66.

Martin, A. W., & Wiechers, J. E. (1954). Raven's Coloured Progressive Matrices and the Wechsler Intelligence Scale for Children. *Journal of Consulting Psychology, 18*(2), 143-144.

Musgrove, W. J., & Counts, L. (1975). Leiter and Raven performance and teacher ranking: A correlation study with deaf children. *Journal of Rehabilitation of the Deaf, 8*(3), 19-22.

Orpet, R. E., Yoshida, R. K., & Meyers, C. E. (1976). The psychometric nature of Piaget's conservation of liquid for ages six and seven. *Journal of Genetic Psychology, 129,* 151-160.

Raven, J. (1986). *Manual for Raven's Progressive Matrices and vocabulary scales. Research supplement no. 3: A compendium of North American normative and validity studies.* London: H. K. Lewis & Co.

Raven, J. C. (1948). The comparative assessment of intellectual ability. *British Journal of Psychology, 39,* 12-18.

Raven, J. C., Court, J. H., & Raven, J. (1984). *Coloured Progressive Matrices.* London: H. K. Lewis and Co.

Rich, C. C., & Anderson, R. P. (1965). A tactual form of the Progressive Matrices for use with blind children. *Personnel and Guidance Journal, 43*(9), 912-919.

Sattler, J. M. (1982). *Assessment of children's intelligence and special abilities* (2nd ed.). Boston: Allyn and Bacon Inc.

Sigmon, S. B. (1983) Performance of American schoolchildren on Raven's Coloured Progressive Matrices Scale. *Perceptual and Motor Skills, 56,* 484-486.

Sinha, D. (1979). Cognitive and psychomotor skills in India. *Journal of Cross Cultural Psychology, 10,* 325-355.

Spearman, C. (1923). *The nature of 'intelligence' and the principles of cognition.* London: Macmillan.

Stallings, J. (1975). Implementation and child effects of teaching practices in follow through classrooms. *Monographs of the Society for Research in Child Development, 40,* 1-33.

Valencia, R. R. (1984). Reliability of the Raven Coloured Progressive Matrices for Anglo and for Mexican-American children. *Psychology in the Schools, 21,* 49-52.

Wiedl, K. H. (1978). Ecological aspects of differential predictive validity. *Psychologie in Erziehung und Unterricht, 25,* 369-371.

Fred L. Adair, Ph.D.

Professor of School Education, College of William and Mary, Williamsburg, Virginia.

COOPERSMITH SELF-ESTEEM INVENTORIES

Stanley Coopersmith. Palo Alto, California: Consulting Psychologists Press, Inc.

THE ORIGINAL OF THIS REVIEW WAS PUBLISHED IN TEST CRITIQUES: VOLUME I (1984).

Introduction

The Coopersmith Self-Esteem Inventories (CSEI) are multiform, paper-and-pencil instruments designed to measure in any individual those evaluative attitudes toward the self that one holds in social, academic, family, and personal areas of experiences. In the context of measuring self-esteem it is herein defined as ". . . the evaluation a person makes, and customarily maintains, of him- or herself; that is overall self-esteem is an expression of approval or disapproval, indicating the extent to which a person believes him- or herself competent, successful, significant and worthy" (Coopersmith, 1981, p. 1).

The antecedents and consequences of self-esteem were of lifelong interest to Stanley Coopersmith. Beginning with his dissertation in social psychology, "Self-Esteem as a Determinant of Selective Recall and Repetition" (Coopersmith, 1957), he continued throughout his career to refine and measure the concepts surrounding self-esteem. He began his teaching and research career at Wesleyan University in 1957, moving to the University of California at Davis in 1963 as chairman of the psychology department. At the time of his death he was director of the Self-Esteem Institute, Lafayette, California, had been a member of Sigma Xi, and was a Fellow of both the Social Science Research Council and the National Science Foundation.

With the publication of the results of an intensive study of the antecedents and consequences of self-esteem in 1967 (reprinted by Consulting Psychologists Press in 1981), Coopersmith published the first edition of the CSEI (acquired by Consulting Psychologists Press in 1981). Originally published as Form A or the Long Form, currently it is published as the School Form. Later he published Form B or the Short Form, which consists of the first 25 items of the School Form. Both of these forms are designed to measure self-esteem in school children ages 5 to 15. Form C or the Adult Form was also published, designed for use with persons ages 16 and older.

The School Form consists of 50 items measuring self-esteem and an 8-item Lie Scale. Most of the items measuring self-esteem were selected from the Rogers and Dymond (1954) Scale with the author writing several original items. The items making up the Lie Scale were included as an index of defensiveness and are all answered in the defensive mode ("Unlike Me"). In its final form, the School Form was administered to two classes (Grades 5 and 6; N = 86) including both males and females whose scores ranged from 40 to 100, with a mean score of 82.3 and a standard deviation of 11.6. The mean score for 44 males was 81.3 with a standard deviation of 12.2 and the mean score for 43 females was 83.3 with a standard deviation of 16.7. Differences between the two sexes were not significant. Subsequently the inventory was administered by the research staff to 1,748 public school children in central Connecticut with the mean for females of 72.2, standard deviation of 12.8, and the mean for males of 70.1, standard deviation of 13.8. Both populations produced a score distribution that was skewed in the direction of high self-esteem. The School Form is designed for children ages 8-15 and is scorable on five scales: General Self, Social Self-Peers, Home-Parents, School-Academic, Total Self, and Lie Scales.

The School Short Form was developed through an item analysis of the School Form and consists of the first 25 items of the School Form. Designed primarily as a time-limited version of the longer inventory, it correlates .86 with the School Form. The Adult Form is an adapted version of the School Short Form and includes language related to older persons. Wording was changed in 8 items to reflect adult lifestyles and experiences. The total score of the Adult Form correlates with the School Form in excess of .80 for three samples of high school and college students (N = 647). The Adult Form is scorable on one scale—General Self—and the reading level is for individuals ages 8 and over.

The School Form of the CSEI is furnished on three pages (each approximately 5½″ x 8½″) of a four-page, folded, expendable booklet, printed in green ink on white paper stock. The first page contains the name of the inventory, the name of the author, space for the subject's name, age, school, sex, and grade, and the date of taking the inventory. It also contains directions for completing the self-description which are complete and make no reference to the true mission of the inventory. Items 1-29 appear on the second page, as well as boxes for marking "LIKE ME" or "UNLIKE ME" and one in the lower left corner for a score on the Short Form. While not designated as such on the School Form, items 1-25 make up the Short Form and the publisher states this in the manual. Presumably an administrator could give appropriate instructions and have students complete only the first 25 items and score the test accordingly. The third page contains items 30-58, boxes for marking "LIKE ME" or "UNLIKE ME," and six boxes at the bottom in which to mark scores for the five scales and a total score. The fourth page is left blank. Although no profile for the scales is provided for an individual display of scores, one could easily be made.

The Adult Form is printed on the front and back of a page (approximately 5½″ x 8½″) printed in black ink on white paper stock. On the first page are the name of the inventory, the name of the author, space for the subject's name, age, institution, sex, and occupation, and the date. Directions for completing the self-description are directly below the space for name, age, etc., and are complete, making no reference to the true nature of the inventory. On the lower right side there is a box in which to write the individual's score. The second page contains the Adult Form items and the boxes for marking "LIKE ME" or "UNLIKE ME."

This inventory may be administered to either individuals or groups with ease. The directions on each form are explicit enough that little else need be said. The words "self-esteem," "self-concept," and "self-evaluation" are not included in either the title or the body of the three forms and examiners are advised not to use them when giving oral directions.

It is recommended that administrators read the directions aloud and ask examinees to follow along as they are read. (Earlier forms of the inventory included practice items, as is noted in the manual, however none were included on the forms currently in use.) Once this is done examinees are directed to open their booklets and begin. While clarification of word meanings may be necessary for younger persons, caution is urged to avoid influencing their responses. Adult Forms are essentially self-administering, though if there is a question about a person's ability to understand the directions, administrators should proceed as with the School Form. When inventories have been completed they are collected for scoring.

Practical Applications/Uses

The Coopersmith Self-Esteem Inventories had their genesis in one man's curiosity about the relationship between the antecedents and consequences of self-esteem. His general notions on the subject led him to test his theories in an extensive study (Coopersmith, 1981) that took six years to complete. Coopersmith's work led naturally to the question that if high self-esteem was good, what could be done to effect change in an individual whose self-esteem was low? The CSEI provide a well-accepted, thoroughly researched and validated measure of the concept.

The School Form of the inventory may be used by counselors, researchers, or teachers to provide an initial baseline measure of self-esteem prior to initiating a program to enhance self-esteem in children. High self-esteem has been positively correlated with creativity, academic achievement, resistance to group pressure, willingness to express unpopular opinions, and effective communication between parents and youth. The manual reports successful use of the CSEI with many groups, including educable mentally retarded and learning disabled individuals and American Indians, as well as in program evaluation.

Counselors might use responses on the inventory as a means of eliciting information on problem areas from a counselee. This is more applicable to persons of low self-esteem since these subjects are typically nonverbal, often shy and withdrawn, and have difficulty affirming their identity. School psychologists could include a self-esteem score in a test battery to offer a possible explanation for many school- or home-related deficiencies. As with counselors, answers to individual items might prove illuminating for specific deficits. Vocational and career counselors, when assessing traits known to be effective in career decision-making, may find a measure like the CSEI valuable in the pursuit of certain careers. High self-esteem is a very desirable quality in many professions (e.g., sales, management, teaching, communications) and a person seeking to enter a field where high self-esteem is almost essential for success could thus become aware of any personal deficiencies before entering into such a venture.

Clinical use of such an inventory is probably unnecessary because persons seeking psychological assistance are invariably feeling low self-esteem, whether or not they are aware of it. Within the first few minutes of a therapeutic interview clients demonstrate their lack of self-esteem and measurement would be redundant.

The manual mentions populations for whom the inventories might be inappropriate. Among them are persons whose judgment about themselves is at variance with the way they are viewed by teachers and peers. As a means of establishing validation criteria for a measure of self-esteem the observations of significant others are important, especially in cases where high self-esteem is not paired with high performance. In addition, some cultural, ethnic, or religious groups, or subgroups, may have values and perceptions that differ significantly from the statements in the inventories.

Since the CSEI was designed to measure self-esteem in grade school children and later adapted for use with older students and adults, it would seem to be appropriate for almost any age group. The manual, however, presents normative data only on the School Form, presumably with school-aged populations. There are only three citations to older-than-pupil populations in the manual and no normative data. In a personal communication with Raynor Gilberts, the current director of the Center for Self-Esteem Development, this reviewer was informed that data are currently being collected to establish adult norms. Users are urged to develop local norms and since the inventories are relatively easy to administer and quick to score, local norming would be easily accomplished.

There is no provision made for administering the CSEI to non-English-speaking subjects since it is only published in English. It has been administered extensively to Spanish-surname groups with normative data reported. Reference is made to some difficulty measuring self-esteem in other ethnic groups, but no research is cited to back up the contention. The CSEI could be read aloud to a visually impaired subject, though normative data are not available for comparison purposes.

The inventories can be administered to groups or individuals with equal facility. Classrooms or psychometric centers are both appropriate settings for taking the inventories. Persons trained in testing and measurement would be qualified to administer the inventories since the manual is explicit on how administration is accomplished. The time required to complete the inventory will likely vary with the age of the examinee, but normal time rarely exceeds 10 minutes.

A set of hand-scoring keys made of heavy white stock is provided for each form. The keys are printed in the same color as the ink on each form of the test. The boxes on each scoring key conform to those on the answer sheet so that scoring takes a minimum amount of time. By actual timing it was discovered that the School Form could be hand scored and the total score derived in less than two minutes and the Short Form and Adult Form in less than one minute.

Machine or computer scoring is available through the Center for Self-Esteem Development, 669 Channing Avenue, Palo Alto, California, 94301. Individual and group profiles are also available with this service. A score is derived by multiplying the raw score by 2 on the School Form and by 4 on the Short Form and Adult

Form. The basis for scores is that a totally positive self-esteem score is 100 and a totally negative one is 0.

Interpretation of the inventories, beyond statements regarding high, medium, or low self-esteem, may prove difficult since no exact criteria are provided. The manual wisely states that only general guidelines are available and even they should be used with caution. These guidelines recommend "two procedures when the inventory is used in a school setting: the supplemental use of a behavior observational rating and the development of local norms. It is assumed that in clinical or treatment settings supplemental measures or observations will always be used" (p. 8). Chapter 6 of the manual offers several normative populations from which estimates can be made. A high score on the School Form Lie Scale may mean that the examinee answered defensively or responded positively to all items. In either instance a supplemental observational rating or teacher report will validate the score. Since the Lie score may invalidate the performance, further evaluation is usually necessary.

Technical Aspects

Predictive validity (the prediction of the future existence of the construct self-esteem) was estimated by correlating subscale scores of the CSEI through regression analysis (Donaldson, 1974). It was discovered that scores on Reading Achievement could best be predicted by the Lie Scale, but a General Self subscale multiple r of .53 ($p < .01$) was also quite high (N=643).

Another way of estimating the validity of the constructs for any testing instrument is to demonstrate that the subscales are measuring what they purport to measure. Using a group from a study previously cited, Kokenes (1974) performed a factor analysis of the CSEI responses of 7,600 children in Grades 4 through 8 and discovered that the four bipolar dimensions obtained were highly congruent with the test's subscales.

The manual reports several studies demonstrating the reliability of this test at several grade levels. To estimate the internal consistency of the instrument Spatz and Johnson (1973) administered the School Form to over 600 students in Grades 5, 9, and 12. With an N = 600 they randomly selected 100 subjects from each grade level and calculated Kuder-Richardson reliability estimates (KR-20). At all three levels they obtained coefficients in excess of .80, considered adequate for the instrument. Reported Short Form reliabilities were somewhat lower with a group of 103 college students. Bedian, Geagud, and Zmud (1977) reported KR-20s of .74 for males and .71 for females.

Test-retest and alternate form reliability estimates are also presented on several different populations, but because of the questioned appropriateness of such statistics, caveats are expressed. Coopersmith attempts to measure the enduring traits subsumed under the construct self-esteem with his inventories. Since all affective states are subject to significant and often unpredictable change over short periods of time, interpretation of a score on the CSEI must be approached with caution. Behavioral observations by significant others are suggested as a supplement to the measurement.

Critique

The Coopersmith Self-Esteem Inventories appear to be well researched, well documented, and widely used. The School Form is used much more often than are the Short or Adult Forms, and this fact is well documented by the populations reported in the literature. The author was candid about the inventories and how they were developed and are best used. The manual is excellent and appears to have been prepared with all of the thoroughness one could ask. The writing style is easily understandable and follows the APA *Standards for Educational and Psychological Tests*.

The manual and the book from which it is in part derived, *The Antecedents of Self-Esteem* (Coopersmith, 1981), portray self-esteem as a trait not evenly distributed in the population but highly desirable to have and to cultivate. Chapter 4 of the manual, "Building Self-Esteem in the Classroom," was adapted from *Developing Motivation in Young Children* (Coopersmith, 1975) and offers the prospective teacher or counselor specific suggestions for enhancing self-esteem in children. Though not identified as such it follows the affective teaching espoused by several other authors but especially Alschuler, whose *Teaching Achievement Motivation* (1970) inspired many in the 1970s.

The CSEI may be used with the confidence that their development has been well thought out and researched from the beginning by a competent developmental psychologist who literally worked himself to an early grave in the endeavor. Coopersmith was frank about what he found and careful to give caveats where necessary. Though the concept of self-esteem is not altogether specific, neither are many other well-developed constructs in the measurement of personality. That fact does not keep researchers from measuring these constructs, however carefully or not they do so. By using the CSEI judiciously one can achieve a measure of self-esteem that is as reasonable as possible with self-report instruments.

References

Alschuler, A. A. (1970). *Teaching achievement motivation*. Middletown, CT: Educational Ventures.

Bedian, A. G. (1977). Some evidence relating to convergent validity of Form B of Coopersmith's self-esteem inventory. *Psychological Reports, 40*, 725-726.

Coopersmith, S. (1958). Self-esteem as a determinant of selective recall and repetition. *Dissertation Abstracts, 18*, 1119.

Coopersmith, S. (1975). *Developing motivation in young children*. San Francisco: Albion Publishing Company.

Coopersmith, S. (1981). *The antecedents of self-esteem*. Palo Alto, CA: Consulting Psychologists Press, Inc.

Coopersmith, S. (1981). *Self-esteem inventories*. Palo Alto, California: Consulting Psychologists Press.

Donaldson, T. S. (1974, February). *Affective teaching in the Alvin Rock Voucher schools*. (Rand Corporation, 1700 Main Street, Santa Monica, CA 90406)

Gilberts, R. (1981). *Review of research on the self-esteem inventory*. Unpublished manuscript.

Gilberts, R. (1983). The evaluation of self-esteem. *Family and Community Health, 6*, 29-49.

Kokenes, B. (1973). *A factor analytic study of the Coopersmith Self-Esteem Inventory.* Unpublished doctoral dissertation, Northern Illinois University, Dekalb.

Kokenes, B. (1974). Grade level differences in factors of self-esteem. *Developmental Psychology, 10,* 954-958.

Kokenes, B. A. (1978). A factor analytic study of the Coopersmith Self-Esteem Inventory. *Adolescence, 13,* 149-155.

Rogers, C. R., & Dymond, R. F. (Eds.), *Psychotherapy and personality change: Coordinated studies in the client-centered approach.* Chicago: University of Chicago Press.

Simon, W. E., & Simon, M. G. (1975). Self-esteem, intelligence and standardized academic achievement. *Psychology in the Schools, 31,* 97-100.

Spatz, K., & Johnson, F. (1973). Internal consistency of the Coopersmith Self-Esteem Inventory. *Educational and Psychological Measurement, 33,* 875-876.

R. Scott Stehouwer, Ph.D.
Associate Professor of Psychology, Calvin College, Grand Rapids, Michigan.

DETROIT TESTS OF LEARNING APTITUDE-2

Donald D. Hammill. Austin, Texas: PRO-ED.

THE ORIGINAL OF THIS REVIEW WAS PUBLISHED IN TEST CRITIQUES: VOLUME II (1985).

Introduction

As I searched for the materials on the Detroit Tests of Learning Aptitude (DTLA), I soon discovered that things have changed for the DTLA. The original publisher, Bobbs-Merrill Company, no longer owns the test and in many respects the DTLA no longer exists. What does exist, however, is the Detroit Tests of Learning Aptitude-2 (DTLA-2). More than just a revision, the DTLA-2 could better be called "Son of DTLA" since in many respects it is a new test. Revisions that have taken place and the added and refined materials seem on the whole to have produced a significantly improved version of the DTLA.

In one of the more recent reviews of the DTLA, Silverstein (1978) criticized the standardization and statistical evaluation of the DTLA. Further, little material had been available regarding this standardization and evaluation. In fact, well over 40 years ago reviewers commented that the DTLA appeared to have been published before it was properly standardized and before adequate data were available regarding reliability and validity. Reviewers consistently called for remediation of this situation. The DTLA-2 is an attempt not only to correct this deficit but also to create a more useful and widely utilized test, one which holds true to the demographic characteristics of today's children and adolescents.

As the DTLA-2 is an aptitude test, it is concerned with measuring ability or capability on the part of a certain population, specifically the ability to learn new tasks. The DTLA-2 was developed by Donald D. Hammill in order to "determine strengths and weaknesses among intellectual abilities, . . . identify children and youths who are significantly below their peers in aptitude, and . . . serve as a measurement device in research studies investigating aptitude, intelligence, and cognitive behavior" (Hammill, 1985, pp. 10-11).

Donald Hammill, Ed.D., is retired from the department of special education at Temple University and is the owner of PRO-ED, the current publisher of the DTLA-2. He has authored or coauthored a number of assessment devices in the field of special education and his background includes extensive training in education, special education, and speech pathology.

The main purpose of the DTLA-2 is to focus on intra-individual strengths and weaknesses with regard to learning aptitude. In other words, this test provides an

This reviewer wishes to acknowledge gratefully the help of Mr. Robert Nykamp in the preparation of this review.

opportunity to compare the individual with him- or herself on various aspects of intellectual and cognitive abilities.

The potential for the DTLA (and in essence the DTLA-2) has long been evident (Anastasi, 1938; Wells, 1949). Reviewers focused particularly on the asset of the test's ability to evaluate intra-individual strengths and weaknesses. Another asset of the DTLA was its physical layout and the relative ease of administering and scoring the test, a test that was not threatening to the children who were tested. The DTLA-2 has attempted to maintain these strengths and where possible to improve upon them. An example of this is the fact that the use of item analysis has resulted in shortening many subtests without any ill effect.

Each review of the DTLA, as indicated previously, was particularly critical of the poor standardization and lack of appropriate reliability and validity information. The last DTLA handbook (Baker & Leland, 1967) had failed to discuss such aspects as the interpretation of scores and profiles and the theoretical issues underlying the test. Another consistent criticism was the fact that demographic variables had changed a great deal from the 1930s when the test was developed and standard-ized. In fact, the test itself had an outdated appearance.

The DTLA-2 represents the result of a major effort to develop an appropriate level of standardization. Studies were also undertaken to establish an appropriate level of reliability and validity. The material has been extensively revised and updated, including the development of a presentation of test items that appear appropriate to the 1980s and beyond.

Consulting on the revision were 100 professionals who had used the test frequently. These professionals were asked to choose the most useful and appro-priate subtests of the DTLA and they indicated a clear preference for seven of them. Although some were retitled, these subtests were retained in the DTLA-2: Word Opposites, Sentence Imitation, Oral Directions, Word Sequences, Design Reproduction, Object Sequences, and Letter Sequences. In addition, four new subtests were developed: Story Construction, Symbolic Relations, Conceptual Matching, and Word Fragments.

No other forms of this test have been developed. However, as with the DTLA, the author of the DTLA-2 recommends selecting certain subtests depending on the capabilities or physical impairments of the person to be tested. The DTLA-2 comes in a box comparable in size to the various Wechsler tests which contains the test manual, examiner record forms, student response forms, summary and profile sheets, and a spiral-bound picture book.

Because this is an individually administered test the examiner's participation is crucial. As with all such tests, the examiner sets the tone for the examination procedure. The directions are read to the child being tested and, as with all individually administered tests, the examiner is in a position not only to score the tests but also to observe the child's response to the test itself as well as to his or her successes and failures. The test is intended for persons from the age of 6 years, 0 months through 17 years, 11 months. This would include most persons from first through twelfth grade.

One real advantage of this test concerns the use of basal and ceiling scores. This format, very similar to that of tests such as the Peabody Individual Achievement Test (PIAT), not only saves a good deal of time and effort on the part of the examiner, but also saves time for the child being tested. It lessens both boredom due to overly simple items and frustration due to too many failures in any individual subtest.

As indicated, the DTLA-2 consists of 11 subtests, which yield nine composite scores. Examinee responses vary from answering questions directly to drawing pictures. The subtests and variables that each subtest is intended to measure are as follows (adapted from Hammill, 1985):

Word Opposites: A measure of vocabulary, specifically a test of antonym knowledge.

Sentence Imitation: A measure of rote sequential memory which is influenced by competence in standard English grammar.

Oral Directions: A series of complex tests that measures listening comprehension, spatial relations, manual dexterity, short-term memory, and attention.

Word Sequences: A test of short-term verbal memory and attention.

Story Construction: A verbal subtest that measures story-telling ability and that depends on creating and telling a logical story.

Design Reproduction: A measure of attention, manual dexterity, short-term visual memory, and spatial relations.

Object Sequences: A measure of attention and visual short-term memory.

Symbolic Relations: A measure of problem solving and visual reasoning.

Conceptual Matching: A measure of the ability to observe theoretical or practical relationships between objects.

Word Fragments: A measure of the ability to form closure and recognize partially presented familiar words in printed form.

Letter Sequences: A measure of short-term visual memory and attention.

The nine composite scores were developed as an aid to delineating intra-individual strengths and weaknesses. Discrepancies between the composites are particularly important in determining these. Most normal individuals will show little discrepancy. Discrepancies can be particularly useful in determining abilities on which the individual can capitalize as well as specific areas of disability that may create frustration and that the individual and significant others must address.

The nine composites (General Intelligence, Verbal Aptitude, Non-Verbal Aptitude, Conceptual Aptitude, Structural Aptitude, Attention-Enhanced Aptitude, Attention-Reduced Aptitude, Motor-Enhanced Aptitude, and Motor-Reduced Aptitude) are described as follows (adapted from Hammill, 1985):

The General Intelligence (GIQ) score represents a person's overall, global aptitude. Comprised of the standard scores of all the DTLA-2 subtests, it is considered the best predictor of achievement. Hammill also considers it the best estimate of the individual's ability to handle the intellectual demands of the environment.

In the *Linguistic Domain,* two composites are considered important: Verbal Aptitude (VBQ) and Non-Verbal Aptitude (NVQ). The Verbal Aptitude score is a measure of the examinee's ability to understand, integrate, and use spoken and written language. Non-Verbal Aptitude represents the ability to use spatial relationships and nonverbal symbolic reasoning.

The *Cognitive Domain* is determined by abilities involving conceptual meaning (abstract thinking, reasoning, and problem solving) and those which involve structural knowledge (recognition of physical properties). The Conceptual Aptitude (COQ) and Structural Aptitude (STQ) composites measure these abilities.

In the *Attentional Domain* two composites attempt to guage differences between short-term memory and attentional abilities on the one hand and long-term memory on the other. Attention-Enhanced Aptitude (AEQ) is a measure of immediate recall, short-term memory, and focused concentration. Attention-Reduced Aptitude (ARQ) measures long-term memory.

The ability to utilize particularly fine motor skills is assessed within the *Motoric Domain*. The Motor-Enhanced Aptitude (MEQ) composite is a measure of the ability to utilize complex motoric abilities, particularly hand-eye fine motor coordination. The second composite, Motor-Reduced Aptitude (MRQ), shows the examinee's ability to perform tasks that are relatively free from fine motor abilities.

The answer forms provided include the Examiner Recording Form on which the examiner notes whether an item was passed or failed. Most of the subtests are presented orally; however, a number of other subtests involve motor activity such as reproducing designs and following directions to perform certain motor tasks. These latter two items are completed on the Student Response Form.

Results of the test are transcribed to a Summary and Profile Sheet. There are five sections to this sheet: Section I contains a record of subtest scores including the raw scores, percentiles, and standard scores for each of the 11 subtests; Section II allows for a listing of other tests administered prior to or along with the DTLA-2 and for indicating the date, standard score, and the DTLA-2 equivalent of the score for each of the other tests; Section III contains a profile of subtests scores that is completed on a graph representing the subtests and the standard scores for each subtest; Section IV contains workspace for computing composite quotients; and Section V presents a profile of composite quotients and space to present a profile of the results of other tests which were administered.

Practical Applications/Uses

The DTLA-2 has a number of practical uses for persons between the age of 6 and 18 years. The Detroit Tests of Learning Aptitude have long been used in connection with school systems and in fact were developed originally with school children and their academic needs in mind. The DTLA-2's ability to test for intraindividual strengths and weaknesses and to delineate specific areas of learning ability and disability make it particularly well suited for academic planning. The results of the DTLA-2 can be very beneficial in planning an appropriate course of study and method of teaching for specific students. Realizing that a child has problems with auditory attention but is capable of using motoric ability, for example, may be particularly beneficial in developing a more "hands-on" approach to the educational experiences of this child.

The DTLA-2 can also be beneficial in vocational assessment and planning. For example, determining that an individual has strong ability in the motoric area or a strong ability in the linguistic area can be very helpful in outlining specific careers and/or training the individual might pursue.

The DTLA-2 offers a number of uses in mental health settings. In fact, this reviewer first became acquainted with the DTLA in such a setting. The DTLA has been a useful component in a battery of tests focusing on intellectual/academic/cognitive abilities or disabilities of child and adolescent patients. It was particularly well suited, for example, in determining attentional deficits children may evidence. The DTLA proved valuable in detecting auditory attention deficits, which have profound implications for a child's ability to follow directions or orders from parents. Anecdotally, the DTLA was helpful in behavior management in this regard, helping parents to understand their child's limitations in understanding verbal commands and aiding them in the development of healthier interaction and more appropriate methods of child rearing. The DTLA-2 holds perhaps even greater promise for use in mental health settings.

The DTLA-2 can be quite helpful as a neurological screening device. The DTLA has been a useful device in neurological assessment of children and presumably DTLA-2 will be better. Using the discrepancy formula of the composites, an examiner can very quickly and precisely pinpoint areas of deficit in which there may be some neurological impairment. Thus, the DTLA-2 can be helpful either as a screening device for further neuropsychological testing or as part of a neuropsychological battery for children and adolescents.

The DTLA-2 is particularly well suited for use by psychologists and educators. Persons in the fields of special education, vocational education, counseling, clinical psychology, and neuropsychology can benefit from using this test. Additionally, because the DTLA-2 focuses on grammatical and speech structure it may be potentially useful for professionals in the field of speech pathology.

The DTLA-2 can be used with persons from age 6 years, 0 months to 17 years, 11 months. It has been standardized on boys and girls from both city and rural areas in a representative sample of geographic areas in the United States. The education level of the parents in the sample varied from less than high school graduate through postgraduate training. The standardization sample included white and black children as well as American Indians and those of Hispanic and Asian ethnic backgrounds.

In terms of test administration, the DTLA-2 is to be given on an individually administered basis. Certain subtests could be appropriately administered to subjects with physical impairments (e.g., for the deaf, an examiner could select Design Reproduction, Conceptual Matching, and Letter Sequences; for the blind, Word Opposites, Sentence Imitation, and Word Sequences). The testing environment should be free from distractions, well ventilated, well lighted, quiet, private, and comfortable. Some training is necessary to administer the test. A psychometrist or technician should be easily able to master the requisite skills for administering the test. However, interpretation of the test is another matter and should be left to a trained professional. The time necessary for administering the DTLA-2 depends on the number of subtests used, though the full test should take between 50 minutes and 2 hours.

The manual for the DTLA-2 is one of the best this reviewer has ever seen for any test. It includes a thorough review of issues regarding aptitude testing, description of the test, rationale for the test, its appropriate and inappropriate uses, interpretation data as well as methods for controlling bias, sharing results, and testing the limits. These are all presented in a very readable format. Dr. Hammill is to be commended for the excellent, clear, and well-developed manual.

Scoring methods and instructions are presented clearly in the manual. Each subtest is scored differently and the procedures for each are explained. The test is for the most part very objective and scoring is therefore relatively easy to determine. The manual presents sample responses for each subtest that are quite helpful in learning how to administer and score the tests. The author suggests administering the DTLA-2 at least three times as adequate practice.

In order to shorten the testing and scoring time, basal and ceiling scores are used. Answers are checked against the manual and a designation of correct or incorrect is given for each. At the time of this writing there was no machine or computer scoring available. Because of the simplicity of the scoring procedure, few if any scoring problems should arise.

Interpretation of the test is based on the objective scoring. Though scoring is a relatively simple matter, interpreting the DTLA-2 is more difficult. Interpretation is based on the subtest scores, the total score, and the composite scores, and the use of scoring profiles eases the process. Effective interpretation depends, however, on the interpreter's qualifications, not only regarding the specific test but also in terms of what the interpreter knows about development, childrens' and adolescents' cognitive abilities, etc. This takes a relatively high degree of training and therefore interpretation is best left to individuals who are well acquainted with individually administered psychological tests in general and issues in learning abilities and disabilities, intellectual assessment, aptitude assessment, etc., specifically.

Technical Aspects

Because the DTLA-2 is in essence a new test, the only available studies of reliability and validity are those of Dr. Hammill and his associates. The results of these studies are presented in the manual.

In terms of reliability, internal consistency for test scores was analyzed for 300 subjects in six age groups ranging from six- to seven-year-olds through 16- to 17-year-olds. Coefficients ranged from .81 to .95 for the subtests and from .95 to .96 for the composites. Thirty-three subjects ranging in age from six to 17 were used in a study of test-retest reliability. Subjects were retested after a two-week interval. With a correction factor for the restricted range of the sample, all composites yielded a coefficient greater than .80 (range .80 to .93). Seven of the 11 subtests yielded a coefficient greater than .80 (range .82 to .91). Oral Directions, Object Sequences, Conceptual Matching, and Letter Sequences yielded coefficients ranging from .63 to .78.

The manual also reports studies regarding content, criterion-related (concurrent), and construct validity. For content validity, Hammill states the DTLA-2 fits well with Wechsler's (1974) verbal-performance dichotomy and Kaufman and Kaufman's (1983) sequential-simultaneous processing dichotomy. The DTLA-2 also fits well (as it should) with Hammill's taxonomy (Hammill, Brown, & Bryant, in press).

An investigation of the criterion-related (concurrent) validity of the DTLA-2 was conducted using 76 students either currently enrolled in or referred for special education programs. All subjects had received the Wechsler Intelligence Scale for Children-Revised (WISC-R) and 25 of them had also been administered the Peabody Picture Vocabulary Test (PPVT). Results were corrected for reduced reliability due to restricted variance. The subtest comparisons yielded coefficients ranging from .38 to .76 with a median of .55. The composites yielded coefficients ranging from .54 to .84 with a median of .71.

The construct validity of the DTLA-2 has been researched regarding five issues. First, the subtests have been demonstrated to relate to chronological age and to conform to developmental patterns known to exist for aptitude measures. Second, the subtests have been shown to be highly intercorrelated, suggesting they all measure a similar trait (i.e., learning aptitude). Third, the test scores correlate well with tests of academic achievement, confirming the idea that the DTLA-2 measures abilities related to academic performance. Fourth, the DTLA-2 differentiates between individuals known to be of normal learning aptitude and those of poor learning aptitude. Finally, the items of each subtest correlate highly with the total score of the subtest, suggesting the items of each subtest measure similar traits. While the results of these studies support the construct validity of the DTLA-2, studies utilizing a multitrait-multimethod matrix (Campbell & Fiske, 1967) could be most beneficial.

Taken together, these few studies suggest adequate reliability and validity for the DTLA-2. However, all of the information available on reliability and validity is found only in the test manual. These data are for the most part clearly given, but certainly need cross-validation. Also, further delineation of the subjects employed in reliability and validity testing would be helpful. The information provided on these subjects leads one to wonder whether they are representative of the larger population on which the DTLA-2 is to be employed.

Critique

Because of the very recent publication of the DTLA-2 (two months prior to the preparation of this review), other reviews are not yet available. As mentioned previously, there were numerous reviews of the DTLA—one of the major reasons for the development of the DTLA-2 was in order to deal with the identified weaknesses of its predecessor.

Personally, this reviewer found the DTLA very useful and believed it had great potential, while remaining aware of its limitations. The DTLA-2 may be in a good position to fulfill the potential of the original. It seems to address some unique issues, particularly those relative to intra-individual strengths and weaknesses. As educational planning becomes more individualized not only for students in general but specifically for those with special needs, the DTLA-2 may become even more useful. As neuropsychological assessment of children advances, the DTLA-2 may help lead the way. As clinicians seek a better understanding of the cognitive functioning of children and adolescents and as vocational planning takes on an ever-increasing importance in society, the DTLA-2 is at least potentially in a position to help guide the effort.

However, much more work is necessary in order to assess the usefulness of the new test. Research must be undertaken to cross-validate the reported studies of reliability and validity. Further research will determine whether the DTLA-2 can take its expected place in clinical, academic, and neuropsychological assessment.

References

Anastasi, A. (1938). Detroit Tests of Learning Aptitude. In O. K. Buros (Ed.), *The 1938 mental measurements yearbook* (pp. 108-109). Highland Park, NJ: The Gryphon Press.

Baker, H. J., & Leland, B. (1967). *Detroit Tests of Learning Aptitude: Examiner's handbook.* Indianapolis: Bobbs-Merrill.

Campbell, D. T., & Fiske, D. W. (1967). Convergent and discriminant validity by the multitrait-multimethod matrix. In D. Jackson & S. Messick (Eds.), *Problems in human assessment* (pp. 124-132). New York: McGraw-Hill, Inc.

Hammill, D. (1985). *Manual of the Detroit Tests of Learning Aptitude-2.* Austin, TX: PRO-ED.

Hammill, D., Brown, L., & Bryant, B. (in press). *A consumer's guide to tests in print.* Austin, TX: PRO-ED.

Kaufman, A. S., & Kaufman, N. L. (1983). *Kaufman Assessment Battery for Children.* Circle Pines, MN: American Guidance Service.

Silverstein, A. B. (1978). Detroit Tests of Learning Aptitude. In O. K. Buros (Ed.), *The eighth mental measurements yearbook* (pp. 214-215). Highland Park, NJ: The Gryphon Press.

Wechsler, D. (1974). *Wechsler Intelligence Scale for Children-Revised.* Cleveland: The Psychological Corporation.

Wells, F. L. (1949). Detroit Tests of Learning Aptitude. In O. K. Buros (Ed.), *The third mental measurements yearbook* (p. 356). Highland Park, NJ: The Gryphon Press.

Maria Pennock-Román, Ph.D.
Research Scientist, Division of Measurement Research and Services, Educational Testing Service, Princeton, New Jersey.

DIFFERENTIAL APTITUDE TESTS

George K. Bennett, Harold G. Seashore, and Alexander G. Wesman. San Antonio, Texas: The Psychological Corporation.

THE ORIGINAL OF THIS REVIEW WAS PUBLISHED IN TEST CRITIQUES: VOLUME III (1985).

Introduction

The Differential Aptitude Tests (DAT) are currently among the most widely used measures of multiple abilities. This battery is primarily intended for use in educational and vocational counseling with students in Grades 8-12, although it is also appropriate for young adults not in school. In addition, it has been used for a variety of basic research studies on effective teaching methods and the nature of cognitive abilities, especially sex differences. The original Forms A and B, developed in 1947, have been revised and restandardized in 1962 (Forms L and M), in 1972 (Forms S and T) and in 1982 (Forms V and W). A special, large-print edition is available from the publisher for the visually impaired, and several different versions have been developed in other languages for use outside the United States (DAT Program Office, The Psychological Corporation, personal communications, March, 1985).

The three authors of the tests, George K. Bennett (1904-1975), Harold G. Seashore (1906-1965), and Alexander G. Wesman (1914-1973), were all distinguished psychologists with much experience in test development; they held high offices in The Psychological Corporation and professional organizations. In addition to the DAT, Bennett developed the Tests of Mechanical Comprehension, the Short Employment Tests, the Stenographic Aptitude Test, and the Stenographic Proficiency Test, and Wesman authored the Wesman Personnel Classification Tests and Personnel Tests for Industry. Although The Psychological Corporation staff has made changes in the recent revision, the DAT still maintains its original format, the same content areas, and many of the first items. Additionally, the *Administrator's Handbook, Directions for Administration and Scoring,* and *Technical Supplement* still carry the authors' names (i.e., Bennett, Seashore, & Wesman, 1982a, 1982b, 1982c, respectively).

The rationale for the development of the DAT is described briefly in the current handbook, but is discussed with more detail in the fifth edition of the DAT manual (Bennett, Seashore, & Wesman, 1974). Drawing on the work of Kelley, Thurstone, and others, the authors approached the measurement of aptitude from a multifaceted point of view. They criticized early group tests of mental ability that computed only one total score after sampling mental behavior with various types of content and items. According to Bennett, Seashore, & Wesman (1974, pp. 1-3), "a student may have excellent verbal facility, yet lack numerical or mechanical aptitude. A test which contains items that measure several of these aptitudes, but yields only a total score, obscures almost as much as it reveals of the true potentiality of the student."

In the 1940s when the DAT and other batteries of this type emerged there was a growing interest in the measurement of intraindividual variation in performance on intelligence tests. Anastasi (1982) describes the situation as follows:

> Crude attempts to compare the individual's relative standing on different subtests or item groups antedated the development of multiple aptitude batteries by many years. Intelligence tests, however, were not designed for this purpose. The subtests or item groups were often too unreliable to justify intraindividual comparisons. In the construction of intelligence tests, moreover, items or subtests are generally chosen to provide a unitary and internally consistent measure. In such a selection, an effort is therefore made to minimize, rather than maximize, intraindividual variation . . . . (pp. 357-358)

According to Anastasi, the development of multiple aptitude batteries was due in part to the gradual realization that "general" intelligence tests, rather than being general, were "primarily measures of verbal comprehension" that rarely touched on some areas (e.g., mechanical abilities).

Counselors with an interest in obtaining measures of different skills began to string together various unrelated tests. However, the DAT authors recognized that this practice presented problems in understanding the meaning of scores because the tests were standardized on different populations, at different times, and for different purposes, and therefore made broad interpretive leaps in assuming that the rating of "average" meant the same in the various tests. (Bennett et al., 1974, p. 3). The authors facilitated the interpretation of scores on the DAT by having norms that were based on the same examinees who had taken all of the separate tests. Thus, as pointed out by Cronbach (1984, p. 301), "the DAT was the pioneer among integrated batteries."

The intention of the authors was to include a manageable number of tests that would be practical and effective predictors of how well people perform in academic studies and in the jobs that they later undertake. They wanted the content of the measures to be applicable to many fields of endeavor and the score reports to be readily interpretable by informed counselors and teachers. In addition, although the set of tests was designed to represent an integrated whole, they wanted each test to have sufficient psychometric reliability and practical utility so that it could be administered and interpreted independently of the others. The objective was to have power tests that were relatively unspeeded for the majority of examinees, except for aptitudes in which speed was an important part of the skill being measured, as in the Clerical Speed and Accuracy Test. The authors also believed that it was desirable to have two alternate forms for each test for use in large testing sessions and retesting. When alternate forms are distributed to adjacent students in large sessions, the examinees are discouraged from copying from their neighbors. In retesting, the use of a different form of the test minimizes the effects of memory and practice. For the most part, the authors succeeded in designing an instrument that met these goals; the psychometric quality and many practical features of the battery ensured the popularity of the DAT to the present time.

The basic test specimen set includes one test booklet for all eight tests, answer sheets, order forms for selecting the type of scoring service desired, and three slender manuals: the *Administrator's Handbook, Directions for Administration and Scoring,* and *Technical Supplement.* All materials are attractive and of durable quality. In the tests, important directions are highlighted in colors, and the print is easy to read.

One of the most attractive features of this battery is that in addition to the basic scoring and reporting services, a school can select the Career Planning Program service. This service includes a career planning questionnaire, which assesses the examinees' interest in activities and occupations, with the tests. It also provides the *Counselor's Manual* (Super, 1982) and an optional orientation booklet that contains test-taking hints and practice exercises for each test, information on the program, and the choices to be marked in the questionnaire. Students who answer the questionnaire and complete at least seven tests can receive a DAT Career Planning Report. These interpretive reports are produced by a computer analysis of the students' occupational choices in terms of their individual abilities, interests, and plans.

Each form has eight tests designed to tap aptitudes that are relevant to particular academic or occupational choices. The eight tests, their number of items and testing time, plus comments from this reviewer and from the Individual Report that students are given with their profiles of test scores follow:

Verbal Reasoning (VR): (50 items, 30 minutes) Consists of verbal analogies items. According to the publisher, it measures the "ability to reason with words, to understand ideas expressed in words."

The vocabulary terms and type of background information needed to understand the analogies have recently been revised and simplified; the items were found to include material that most students have encountered by sixth grade. Hence, the test does measure reasoning skills, provided that the examinee has been exposed to at least a sixth-grade level curriculum. On the other hand, for seriously educationally disadvantaged examinees, low scores may reflect poor vocabulary and general knowledge rather than a deficiency in reasoning per se.

Numerical Ability (NA): (40 items, 30 minutes) Consists of problems that require some arithmetic computation, with many assuming knowledge of very basic concepts of first-year algebra (e.g., exponents, negative numbers), but no geometry. In some instances the student is told what operation to perform (e.g., multiply two given numbers). In others, the operations needed to solve the posed problem or to complete a series are not explicitly expressed. Most items have relatively little verbal content. A few ask to find a number that is "evenly divisible" by another, which may confuse some students. Specifically, the items require knowing how to compute percentages, take the square root of fractional numbers, multiply and divide numbers with decimals and negative signs, sum quantities of time expressed in years and months or hours and minutes, do arithmetic operations with fractions, solve for unknown quantities in simple equations, and compute fractions raised to the power of given exponents.

The profile report claims that the test measures the "ability to reason with numbers, to understand and work with ideas expressed in numbers." This reviewer finds that the phrase "ability to reason with numbers" is somewhat inaccurate in that so much arithmetic computation is involved and the type of reasoning required does not demand much ingenuity. This test should be considered a test of mastery of junior-high-school mathematics, which is relevant to the occupations listed in the profile description.

Abstract Reasoning (AR): (45 items, 20 minutes) Consists of items having no words and requiring an understanding of the rule governing a series of abstract designs and identifying which one comes next. It measures reasoning skill and careful attention to detail. The description of this measure in the profile report seems to this reviewer to be quite appropriate: "ability to understand ideas that are not presented in words or numbers; to see relationships among things, such as objects, patterns, diagrams, or designs."

Clerical Speed and Accuracy (CSA): (two parts of 100 items each, 3 minutes per part) Consists of items that measure how quickly and accurately the examinee can compare letter and number combinations and mark identical ones.

Mechanical Reasoning (MR): (70 items, 30 minutes) Consists of pictures of mechanical devices or persons working with tools. The examinee must read and answer a question about the picture. In the profile report, it is described as measuring an "understanding of mechanical principles and devices, and of the laws of everyday physics." To some extent, one would expect that extensive experience with machines and tools or a course in mechanical physics would improve performance on this test. A study of male adolescents in Sweden with similar tests (Balke-Aurell, 1982) shows that spatial and mechanical skills increased with training in technical areas. Some items require knowledge about specific facts in science; nevertheless, there are a sufficient number of items that draw from experiences common to most persons' everyday lives outside of school to justify classifying it as an aptitude test.

Space Relations (SR): (60 items, 25 minutes) Consists of patterns that can be folded into figures. The examinee is shown a picture of the object, which has been laid flat and unfolded, and must identify which figure among the choices can be made from the pattern. It also requires paying attention to details such as the location of shading and dark dots in the figures. The publishers describe it as a measure of the "ability to 'think in three dimensions,' to picture mentally the shape, size, and position of objects when shown only a picture or pattern."

Spelling (Sp): (90 items, 20 minutes) Consists of words, some of which are spelled correctly and others incorrectly. The words are not read aloud to the examinees. They must recognize whether the spelling is correct on sight without hearing how the word is supposed to sound.

Language Usage (LU): (50 items, 20 minutes) Consists of items, whereby the examinee reads a sentence and identifies what part of it, if any, contains a mistake in grammar, punctuation, or capitalization.

Revisions and Norming Samples: In each revision of the DAT, the first letter designates the form that is the updated version of Form A, and the second corresponds to Form B. In other words, there is a rough correspondence between Forms A, L, S, and V, and between Forms B, M, T, and W, although occasionally some items have been reassigned to different forms.

The frequent revisions of the tests reflect the care and conscientiousness of the publisher in maintaining the content and form of the test current with changes in society and language use. For example, in the 1972 edition the major changes reflected a concern with reducing the "culture loading" of test content in order to make it more acceptable to minority groups. In addition, the LU Test was modernized to conform to the newer, more permissive standards of writing. The book containing illustrative case histories for evaluating DAT profiles was updated in 1977 to include more modern examples of career counseling situations. The changes in the latest revision of the test were intended to achieve greater fairness to both sexes, better readability, and other improvements. The art work for MR, which contains people and vehicles, has been frequently revised in order to maintain a contemporary appearance. In response to the expressed preferences of school personnel, the format of the tests has evolved from eight booklets (one per test) for Forms A and B to a two-booklet format in 1962 and later a single booklet with the eight tests in 1972 (Form S only) and 1982 (both forms).

The original standardization sample and the subsequent samples for the revisions have all been very large. The norming groups for Forms A and B (1947) and Forms L and M (1962) each included approximately 50,000 cases. The norms for Forms S and T (1972) were based on more than 64,000 students in Grades 8-12 from 76 public and parochial school systems throughout the United States. The norming sample for the 1982 revision (Forms V and W) was slightly smaller than the previous one—61,000 cases and 64 school systems. Norms continue to be provided separately for males and females because some of the tests reveal sizable sex differences. Although the provision of separate-sex norms has been criticized in recent years, the publishers have argued that using both sets of norms leads to less stereotyping of careers by sex.

The changes in test content, directions, and test lengths of the 1982 revision are presented in outline form in Table 1. The major changes involve the VR, NA, and MR Tests, for which a substantial number of items are new or modified. In general, these changes appear to be improvements. For example, in the analogy test items for the VR Test, words such as "modiste" and references to poets and painters have been replaced with more commonly known words and cultural background information.

During the revision, the "readability level" of the directions and test items were analyzed separately for each test, using the Dale-Chall (1948) and the EDL Core Vocabularies (1979). For four out of the five tests containing items with words, it was found that 93 to 100% of the words occurred in the curriculum by the sixth grade. The Spelling words were a little more difficult; 68% were in the curriculum by Grade 6 and 94% by Grade 8. The directions to seven out of the eight tests were readable at the sixth-grade level, and those for the CSA were readable at the eighth-grade level. Therefore, the analyses confirm that the tests are, on the whole, appropriate for grades 8-12.

However, some problems in test-item content remain. There continue to be a few items in the MR Test that Mastie (1976) has criticized because "they seem to require knowledge of learned facts and become science achievement measures" (p.

Table 1

**Norming Samples and Changes in Test Directions and Test Items
in the 1980 DAT Revision**

Norming Sample	61,000 students, 64 schools throughout U.S.
Directions	Modified slightly to increase "readability."
Order of Items	Items arranged in ascending order of difficulty within each test.
Test Content	
Verbal Reasoning	Replaced and modified some items. Difficult words replaced and incorrect options revised to be more plausible. New items written.
Numerical Ability	Replaced and modified some items. Revisions intended to widen range of content and to make some of incorrect options more attractive and plausible. New items written.
Abstract Reasoning	Reduced from 50 items to 45. Time shortened.
Clerical Speed & Accuracy	Some items transposed from one form to the other.
Mechanical Reasoning	Changes in the art work, to enhance fairness to members of both sexes and remove dated pictures. Balanced number of male and female figures depicted.
Space Relations	None.
Spelling	Reduced from 100 items to 90. Left the time limits the same.
Language Usage	Minor wording changes to enhance fairness to both sexes (e.g., balance in male and female pronouns). Reduced from 60 items to 50. Time shortened.

91). Those items, which involve the height of spurting water and the absence of air on the moon, are still included. In addition, there are some new items on the NA Test that may be introducing irrelevant sources of difficulty because they are expressed verbally and can be puzzling to some. Specifically, two items on both Form V and Form W require knowing what it means for one number to be "evenly divisible" by another. Because verbally loaded items on the test battery for the High-School and Beyond Study have been found to be differentially more difficult for black examinees than for white examinees at the same ability level (Shepard, Camilli, & Williams, 1983), future studies should address the impact of this new type of item on minority students taking the DAT.

Given that all but the CSA are intended to be power tests, the speededness for the other seven tests was checked in the eighth-grade sample. The results showed that between 88 to 99% of the females and 82 to 96% of the males answered one or more of the last five questions. Therefore, the tests allow sufficient time for eighth-graders; one can assume that they are essentially unspeeded at the higher grades also.

The sampling design for the 1982 standardization is described in greater detail than for earlier editions (Bennett et al., 1982a, pp. 33-34). A two-stage procedure was followed in order to obtain representative groups of students for each of the Grades 8-12. In addition to geographic region, the selection of schools in the first involved stratification on two variables: enrollment size of the school district and the socioeconomic status of the community served by the district. After selection of participating school systems was completed, a small number of additional systems were included to improve the correspondence between population and sample characteristics. In the second stage, all students enrolled in Grades 8-12 in small districts were included, and a representative sample was chosen for the intermediate-sized districts. Representativeness was determined on the basis of ethnic composition and median scores for each school on a standardized achievement test. After the testing was completed, differential weighting of cases was carried out to ensure that the obtained norms reflected the appropriate proportions of students from each type of school district.

In the resultant sample, the estimated proportion of black children for all grades combined was 14.1% as compared to the census figure of 15.1% for Grades 9-12. In the handbook, the lower figure for the standardization is explained on the basis of possible inaccuracies in the estimates given by schools and the reliance on socioeconomic data and enrollment size rather than race as stratification variables. No information is given on the racial composition of the sample by grade level, and the percent of students in other minority groups is not reported. What is more troubling is that, despite the careful sampling procedure, the median family income and median years of education for the school districts in the norming group given by the 1970 census are somewhat lower than the 1970 census figures corresponding to the overall U.S. population ($8,919 and 11.6 years for the school districts vs. $9,590 and 12.1 years for the U.S. population as a whole). These puzzling discrepancies are not discussed in the handbook; it is possible that a decade after the census these school districts became more representative of the U.S. population as a whole.

Users should also be aware that the 1982 spring norms are still based on interpolation of values rather than by actually testing students in the spring. The same caution that Mastie (1976) advised for the 1972 spring norms is still warranted if students are tested late in the school year; it is preferable to test students at the beginning of the year and use the fall norms.

Practical Applications/Uses

In vocational and educational counseling, the handbook cautions that the tests should be considered measures of *developed* abilities, i.e., they assess "a capacity to learn, given appropriate training and environmental input" (Bennett, Seashore, &

Wesman, 1982a, p. 5). Nevertheless, some researchers believe that it would be more appropriate to call some of these tests achievement tests. For instance, Mehrens and Lehmann (1984, p. 388) point out that the two tests Sp and LU could be called measures of achievement because they are more closely tied to a curriculum in spelling and language arts. The diversity of the DAT has enabled researchers to measure reasoning and other aptitudes with less emphasis on verbal content, which is particularly appropriate in studies on the deaf (Arnold & Walter, 1979; Parasinis & Long, 1979). In addition, some studies have investigated the administration and appropriateness of the DAT for deaf students (Garrison & Coggiola, 1980) as well as for schizophrenic patients (Rutgers University, 1980).

For the visually impaired, a large-print edition has been developed by the publishers. Other special editions include ten licensed adaptations of the DAT for use outside the United States (DAT Program Office, The Psychological Corporation, personal communications, February, 1985). Three different versions in Spanish have been adapted for use in Guatemala, Spain, and Argentina. There are also Dutch, Hindi, Italian, Brazilian Portuguese, French Canadian, British English, and Filipino English versions that are licensed. Some of the research literature on foreign editions of the DAT since 1953 is referenced in Buros' *Seventh* and *Eighth Mental Measurements Yearbook* (1972; 1978), and *Tests in Print III* (Mitchell, 1983). Not included in these sources are studies in Argentina (Fogliatto, 1972), East Pakistan (Baroya, 1967; Hashmi, 1967), India (Deb, 1980), and Brazil (De Santis-Feltran, 1982, 1983). The DAT has also been used by researchers in Korea (Korean Institute for Research, 1970), Yugoslavia (Makarovic, 1970), and Czechoslovakia (Hlavenka & Rapos, 1973-74).

The DAT's variety of measures give comprehensive diagnostic information on students who drop out of school (Cumming et al., 1966; Fuller & Friedrich, 1972; Garrison, 1983; Sewell, Palmo, & Manni, 1981) and are useful in the evaluation of teaching methods in science and mathematics education (Crosby, Freemont, & Mitzel, 1960; Duke, 1966; Geiger, 1968; Lee, 1966; Tanner, 1968).

In addition, the DAT has been used in an impressive amount of basic research on the nature of achievement and aptitudes, including the relationship between DAT scores and visual short-term memory and age (Adamowicz & Hudson, 1978); visual imagery (McKelvie & Rohrberg, 1978); cross-modal spatial ability (Kumar, 1975); neural impulse (Glover, 1974); brain laterality (Meiners & Dabbs, 1977); handedness (Gregory, Alley, & Morris, 1980); performance on gross-motor tasks (Beitel, 1980); field differentiation (Cooperman, 1980); critical thinking (Brown, 1967); writing competence (Quellmatz, Capell, & Chih-Ping, 1982); and reduction of test anxiety (Goldsmith, 1980). It has also been used to study the information-processing demands of test items (Whitely, 1977) and strategies or components in the process of reasoning (Eastman & Behr, 1977; Sternberg, 1980; Sternberg & Weil, 1980). Several researchers have examined sex differences in aptitudes using the DAT (Fennema & Sherman, 1977, 1978; Newcombe, Bandura, & Taylor, 1983; Sherman, 1979, 1980a, 1980b; Sherman & Fennema, 1977; Tapley & Bryden, 1977). Others have investigated the educability of spatial ability (Carpenter, Brinkman, & Lirones, 1965) and its relationship to performance in sports among highly skilled women squash players (Graydon, 1980).

Administration: The battery is designed for ease in group administration, in settings such as classrooms. Less than two hours of training is required to familiarize examiners with administration procedures, and answer sheets can be scored by hand or machine. It is recommended that testing be carried out in two or more sessions because the total procedure, including distribution and collection of materials, giving directions, and answering questions, takes approximately four hours.

The directions for administration are clear and complete with good attention to important details. For example, students are instructed to "mark the choice that is your best guess" if they are not sure of an answer. Three alternative testing schedules are outlined, but these do not exhaust the possibilities. Because the eight tests can take about four hours to administer, and the questionnaire takes an additional half hour, the suggested schedules divide the testing into two, four, or six sessions.

The only thing that this reviewer would add to the instructions is to make sure that the students understand the uniqueness of the answer sheet for the CSA Test. For the other tests, students mark the letter corresponding to the answer option that they choose from the problems in the test booklet. However, for the CSA Test the answer choices do not appear on the booklet but on the answer sheet itself. Thus, the answer options are not labelled A, B, C, D; they have combinations of two letters or numbers. This test is probably the hardest one to administer well because its time limits are so short. The publishers have admitted (Handbook, p. 36) that in the standardization for both the 1972 and 1982 editions, time limits were not well adhered to in some schools and some adjustments were necessary in the norms.

As stated earlier, this reviewer also recommends testing in the fall rather than the spring because of the way in which the spring norms have been derived by interpolation.

Answer Sheets and Scoring: Answer sheets have been prepared for two types of test-scoring machinery—MRC and NCS; both can also be scored by hand. The MRC format is more condensed than the NCS, which makes it more difficult for students to erase mistakes without disturbing other answers, but it is faster to score by hand with fewer scoring keys. The NCS format has larger print but has a greater number of different pages to be scored. Scoring templates can be obtained from the publisher for scoring the MRC sheets by hand, but the school must make up its own templates based on the list of keyed answers for the NCS. If the questionnaire is administered, only a special MRC sheet and machine-scoring can be used. Hence, this reviewer recommends that the NCS sheet be used when a machine-scoring service is desired and the questionnaire is not administered. For hand-scoring, the MRC is preferable; if the DAT Career Planning Program is chosen, the MRC sheet is required.

Aids to the Interpretation of Test Scores: The battery yields nine scores, one for each test with an additional VR + NA Score that correlates highly with measures of intelligence and represents overall scholastic aptitude because it correlates highly (usually above .70) with group tests of ability yielding a single score. Schools or counselors have several options concerning the way scores are reported. They may choose a report that lists students and their scores, sets of press-on labels for each student, or individual reports for each student that have the scores plotted on a profile chart by a computerized scoring service. The percentiles can be based on either female or male norms and both may be requested for each student. If the individual reports are not requested, the scores can still be plotted by hand on profile sheets available from the publisher.

In the profile, a one-inch bar is drawn around each score so that the actual percentile falls in the middle of the bar. No test is free of errors of measurement and these bars reflect the range of values that the individual's true score may actually have. According to the Handbook (p. 43), for all scores except the MR scores for females, each half-inch of the bar represents at least 1.5 units of standard error of measurement. Hence, users can have more than 90% confidence that each one-inch band contains the true score.

This reviewer strongly urges users to order or plot the profile charts because single raw scores or percentiles without graphical representation can be quite misleading, even for trained personnel. Small fluctuations in raw score due to chance errors of measurement can produce very large differences in percentiles at the middle of the scale. Without the profile, many persons would consider a difference between the 40th and 60th percentile a large gap when, in fact, both scores are very close to average and the difference could be due to chance. On the other hand, the difference between the 1st and 20th percentiles or the 80th and 99th percentile, are truly quite large and probably will hold up if the test is repeated. The profile chart has the percentiles in the middle ranges (20th to 80th) close together and the percentiles at the extremes further apart, which scales the differences accordingly and prevents misinterpretation of scores. Because plotting scores can be quite tedious, the computerized service generating the profiles seems most desirable.

The Interpretative Report provides a useful rule of thumb for deciding when scores on two tests differ significantly: if the bars for the two tests do not overlap subjects are probably better in the kind of ability in which they scored higher; if the bars overlap by more than half their length, the difference between the scores can probably be ignored; if they overlap by not more than half their length, the subjects should consider other things that they know about themselves in order to decide if the difference may be important.

If the Career Interest Questionnaire has been administered, students who complete at least seven of the eight tests will receive an individual report. These computer-generated Career Planning Reports evaluate the first, second, and third vocational choices of the students in light of their aptitude test scores and stated interests. It may confirm the appropriateness of the choices or suggest alternative occupational areas. In addition to the *Counselor's Manual*, materials include *Counseling from Profiles: A Casebook for the Differential Aptitude Tests* (Myers & Thompson, 1977), which provides complete information to aid the counselor in the interpretation of reports.

The publishers provide much useful information concerning the interpretation of the scores; the use of percentile bands in the profiles, with their one-half-inch rule, is simple and practical. The information on individual case histories, based on the 1972 and earlier editions, needs updating for the 1982 edition; nevertheless, it is still useful because the changes have not been dramatic, and there is a high correlation between corresponding forms in the 1972 and 1982 versions.

The main problem this reviewer sees with the battery is the temptation to overinterpret the importance of tests other than VR and NA. These different measures do have much clinical appeal because they allow a counselor to identify sources of strengths for students of overall low ability and weak areas for the otherwise strong student. However, the evidence overwhelmingly indicates that the overall VR + NA Score is the best predictor of grades even in areas such as science and industrial arts. Some studies suggest that spatial ability and mechanical reasoning may affect the pattern of occupational choice in the long run, but the evidence is scanty and weak at best for the specific tests of the DAT. In the explanations of the student profiles, the publishers make sharp differentiations among the occupations listed as relevant to each skill without a strong body of research to justify each statement. Linn (1978, p. 659) cautions that "differential predictions and profile interpretations are generally not defensible in terms of the existing evidence of differential validity. If they are to be made at all, they must be based primarily on theory and clinical judgment."

Although some researchers have urged that the norms for male and females be combined (e.g., Gold, 1977), the publishers only provide separate-sex norms. Bennett et al. (1982a) and Herman (1977) recommend that students compare their scores against male norms if they are interested in a career dominated by men, and against female norms if they are interested in a female-dominated career. In this way, students can evaluate their standing relative to probable competitors. For three tests (VR, NA, and AR), sex differences are small and it does not matter which norms are used. In MR and SR, males tend to have higher scores; in LU, Sp, and CSA females tend to have higher scores. This reviewer has no objection to the use of separate-sex norms, provided that a counselor keeps in mind two caveats: 1) the VR + NA Score should be the primary score as a basis for advising, with other scores having lesser importance; and 2) specialized aptitudes are not fixed and can be improved if the student is exposed to relevant tasks and courses.

Specifically, counselors should avoid the tendency to regard the MR Test as a measure of aptitude that is less modifiable by training than other skills, such as Sp and LU. The items on the MR Test require knowledge of specific subject matter in science and experience with tools. Because sex differences on this test are quite large, counselors should be particularly wary not to discourage girls from entering male-dominated fields if they score poorly on this test but not on other relevant tests. The case of Susan Tenney in *Counseling from Profiles* (Myers & Thompson, 1977) illustrates this point.

Susan, a ninth-grader, was considering eventually entering a career in architecture or engineering. She had very high scores in MA, AR, and SR; her MR score was high for a girl but in the 60th percentile for boys. The manual states that this low score "should constitute a warning since the average engineering student tends to be well above the median for all boys on this test" (p. 47). Considering that her SR score was in the 90th percentile for *boys*, and that the correlation between SR and MR is generally quite high, this lower score should not be given much emphasis. It could reflect psychometric error or a lack of background experience with the information on the tests. Some of the items depend on specific knowledge of science and her scores could go up if she were to take more science courses or gained more experience with tools. This type of remediation was not pointed out to her; in contrast, the weaknesses of a male student in LU and Sp were viewed as "signals for remedial attention" (p. 23).

The distinction between aptitude and achievement tests is quite subtle at this age level, not only for the specialized tests but for the more generalized ones as well. In the interpretation of subtests such as VR, counselors must keep in mind that for seriously educationally disadvantaged examinees, low scores may reflect poor vocabulary and general knowledge rather than a deficiency in reasoning per se. Although the recent edition of this subtest has reduced the difficulty of the words used in the analogies, vocabulary level and educational background still play an important role in this test. It does measure reasoning skills, provided that the examinee has been exposed to at least a sixth-grade level curriculum. Similarly, students who have not had junior-high-school level mathematics—covering exponents, extensive work with fractions, and negative numbers—will score poorly on the NA Test, even if they have the potential to master this material.

One of the problems identified by Linn (1978, p. 485) with the 1972 edition is that the manual did not "specifically address issues in the use of the test with members of minority groups. . . . The lack of attention to possible issues in the use and interpretation of the DAT with minority groups seems particularly surprising in view of the stated objective of reducing the 'culture loading' in the revision of the Verbal Reasoning test." The same criticism still holds for the 1982 edition. The book of case studies (Myers & Thompson, 1977) includes some students with Spanish surnames; however, no students are identifiably black and no mention is made of the possible effects of bilingualism, cultural differences, and inequality in the adequacy of schooling on the test scores and their interpretation.

Despite the shortcomings pointed out here, which represent departures from an ideal, the interpretive materials provided for the DAT are truly of a very high quality; they compare well with those of competing batteries.

Technical Aspects

Most of the technical information on the DAT is found in the technical supplement, although the reliabilities and norms are reported in the administrator's handbook.

The reliability coefficients for all of the tests, except the CSA, represent internal consistency measures for the tests using the split-half procedure (odd- vs. even-numbered items) and the Spearman-Brown formula. For the CSA, which has two separately timed parts, alternate-form reliability coefficients were computed; this method is more appropriate for highly speeded tests than the split-half procedure. Most tests for both forms had generally high reliability coefficients in the low .90s for all grade levels and both sexes. The most consistent exceptions occurred for the MR Test; it had reliabilities that dipped to .88 for males at some grade levels and ranged from a low of .83 in Grade 8 (Form V) to a high of .88 in Grade 11 (Form W) for females. In general, the tests tended to have adequate but slightly lower reliabilities for Grade 8. There do not appear to be marked differences in reliabilities between Forms V and W. It should be noted that for CSA the estimates of reliability were computed in Grade 8 and Grade 10 in only school, whereas for other tests the reliabilities were estimated for all grades using subsamples of the standardization sample.

In one school district, ninth-grade students took both Form V and Form W in a counterbalanced order, one week apart. The interform correlations for corresponding tests were generally quite high, ranging from .69 to .93 for males and .64 to .92 for females. The VR + NA Score had the highest inferform correlation for both males and females. These data indicate that corresponding tests of Forms V and W are essentially equivalent and that the scores are quite stable over short intervals of time.

In order to compare the 1982 and 1972 editions, students in two school districts in the standardization sample were administered the new and older versions of the tests about two to three weeks apart. They took Form V and T or Forms W and S because these pairs had the fewest items in common, and it was desirable to avoid giving the same items twice. (Form S is the earlier version of Form V, and T is the earlier version of Form W.) The correlations between corresponding subtests in Forms V and T were generally in the .80s and high .70s, with the exception of the CSA, which had correlations ranging from .39 to .52. Similar results were found when comparing Forms W and P, but the correlations for the CSA were higher (.50 and .82 for males and females).

It is interesting to compare the interform correlations for the tests that underwent the most substantial revisions: VR, NA, MR, and LU. Apparently, the correlations between the newer and older versions are not markedly different than the correlations between corresponding subtests in Forms V and W. Hence, the changes in these tests did not alter radically the estimates of abilities of the students, and there is substantial continuity with earlier versions of the tests.

As pointed out by Anastasi (1982, p. 375), "the amount of validity data available on the DAT is overwhelming, including several thousand validity coefficients." There are, indeed, numerous reports of the DAT's correlations with standardized achievement test batteries, other aptitude tests, course grades in specific subjects in Grades 8-12, college level courses, and success in post-high-school job training programs. However, past reviewers of the DAT have criticized the paucity of data on the differential validity of the DAT for predicting success in different occupations (e.g., Bechtoldt, 1953; Bouchard, 1978; Carroll, 1959; Linn, 1978; Quereshi, 1972; Schutz, 1965). Unfortunately, this weakness remains in the otherwise very comprehensive collection of psychometric information assembled for these tests. The publishers report only one study of actual job performance, which involved only selected tests (Daniel et al., 1982).

It has been found that the VR + NA Score has very high correlations (.70s to .80s) with tests of general ability that yield one score. Also, it has correlations above .60 with college admissions tests. The correlations with achievement tests are quite high and in the expected patterns, e.g., NA is the best predictor of mathematics achievement scores.

On the other hand, when it comes to grades and job training, the best overall predictor for most criteria is the VR + NA Score; other tests do not contribute a differentially better prediction in the expected directions. For example, among the median correlations between tests and science grades for males, the highest were .53 and .48 for NA and VR + NA Scores, respectively. Sp and LU correlated more highly with science grades than the AR, MR, and SR tests. Similar problems exist with other batteries of this type (Toronto Board of Education, 1973).

The intercorrelations among the tests tended to be moderate to quite high, ranging from .46 to .80, with the exception of the Sp and CSA measures. The highly speeded CSA Test had low correlations with the other tests, usually below .3; the Sp had somewhat lower correlations with MR and SR (.27 to .43). Past reviewers have questioned whether these correlations are sufficiently low to justify the claim that each test measures a relatively distinct ability.

The lack of differential prediction of grades and high intercorrelations among the tests have led reviewers to debate whether the tests really measure different abilities. Bouchard (1978, p. 485) concludes that "the DAT achieves very little differential validity and measures predominantly general intelligence. . . . Dependence on subtest scores as the primary predictor would be misleading." Bechtoldt (1953, p. 678) attributes the disappointing empirical evidence to "the complete inadequacy of school grades as criterion measures." Anastasi (1982, p. 379) comments that lack of differential validity is not unique to the DAT since most "multiple aptitude batteries have fallen short of their initial promise" in this regard. On the other hand, Cronbach (1984, p. 309) defends the vocational relevance of traits such as spatial ability:

> Measures of specialized abilities fail to account for differences within a profession. But selection by institutions and self-selection operate steadily over many years to encourage the person to pursue whichever lines of work he or she handles well. . . . [The results of Project TALENT showed that] high spatial ability [among males] greatly increased the probability of gravitating into a career as architect, engineer, or physicist and, among females, of gravitating into high school mathematics teaching, physical therapy, or the visual arts.

Bennett (1955) found a similar pattern of self-selection for students who participated in a longitudinal study of the DAT.

Because a substantial number of items were changed on the VR, NA, MR, and LU Tests to update them and remove sexist or "culturally loaded" material, this reviewer compared the technical information for current and 1972 forms for these tests. The validities discussed are median validities for predicting grades in academic, nontechnical courses.

The norms for MR do show a small narrowing of sex differences in the means for this test of about one point. This difference is not an artifact of a change in units between the older and newer versions; the number of items were the same in both editions and the standard deviations for both sexes were approximately equal in 1972 and 1982. At all grade levels, the means increased for both males and females in the new edition. In comparing the validity coefficients for both editions, this reviewer found that for the MR Test the validities did not seem to change consistently across age groups from 1972 to 1982 for either males or females. However, sex differences in validity coefficients narrowed. Whereas in 1972 the test was slightly more valid for females, in 1982 the validities were about the same for males and females, except in English and Literature where the test was more valid for females.

On the LU Test, the difference in the means of boys and girls also narrowed by

0.7 to 1.8 points for the various grade levels. On the other hand, these means were lower in 1982 than in 1972 for both sexes, and the standard deviations were about one point smaller. The median validities for this test also dropped for both sexes. These coefficients were approximately the same for males and females in both editions.

For the VR and NA Tests, the means and standard deviations did not change appreciably and there were no sex differences as before. However, the VR Test had slightly lower median validities in 1982 for predicting grades in traditional academic subjects for both males and females. In 1972 this test had generally higher validities for females; in 1982 there were smaller differences in validities for the two gender groups.

The validities for the NA Test stayed about the same in 1982 as in 1972 for males; for females there was a small decrease in 1982. Whereas previously there were no consistent differences in validities between the sexes, in 1982 this test was slightly more valid for males.

Critique

There is no question that the DAT is a psychometrically sound instrument that is practical, easy to administer, and useful for academic counseling and research on cognitive aptitudes. Its popularity is well deserved. The changes in the 1982 edition have been relatively subtle; correlations between the new forms and the older ones are quite high. However, the Mechanical Reasoning Test has definitely improved. The publishers managed to reduce the sex differences in the means a little without affecting test validity. On the other hand, the validities of the Verbal Reasoning Test across subject areas appear to be consistently lower. A statistical analysis should be done to determine if the decrease is greater than expected by chance sampling variations. The test is now, in a sense, less of an achievement test because some items that contained achievement-type material (e.g., difficult vocabulary, knowledge of poets) were removed. Because achievement tests generally give a better prediction of grades than do aptitude tests, it may be that the elimination of these items reduced the correlations with grades. Unfortunately, one cannot evaluate whether the "cultural loading" of the tests has been, in effect, reduced because the publishers do not provide technical information on racial or ethnic differences in the tests. This reviewer would like the DAT publishers in future editions to set an example for other batteries of this type by discussing the use and interpretation of test scores pertaining to members of racial minorities and other groups that are outside the cultural mainstream of the United States.

As in other batteries measuring a variety of aptitudes, the differential validity of the DAT's separate tests for predicting success in different occupations or academic courses is poor. The evidence suggests that overall academic ability (VR + NA) is the best predictor of success in most courses and jobs, although aptitude in other areas influences self-selection for some careers. Unfortunately, there are too few studies on the relationship between occupational choices, job success, and DAT scores; more research needs to be done. Thus, many psychometricians question the appropriateness of making fine discriminations in terms of occupational counseling. In fact, DAT materials with overstated claims of differential prediction may mislead students who receive low scores in some subtests but high overall scores for VR + NR. Given the available evidence, this reviewer would prefer that the publishers make the description of the vocational relevance of tests such as Spatial Relations, Mechanical Reasoning, and Clerical Speed and Accuracy more tentative.

The danger in overinterpreting the importance of the specialized tests is aggravated by the sex differences in some of these tests—differences that are *not found* in the Verbal Reasoning and Numerical Ability Tests. Hence, special caution should be exercised in the vocational counseling of young persons interested in careers that are dominated by the opposite sex. Because few males are apt to be interested in female-dominated occupations due to generally lower salaries than those of male-dominated careers, the sex differences in the norms impact more frequently on females. In particular, counselors should think twice before discouraging girls with high overall aptitude but lower scores on Mechanical Reasoning and Spatial Relations Tests from considering technical, mathematical, or science careers. These tests are related to self-selection for careers, but generally add very little to differential prediction of grades or success in these occupations; moreover, these aptitudes can be improved with training.

When students obtain lower scores on a test than are desirable for a particular occupational choice, perhaps the best strategy would be to suggest remedial courses rather than dissuading them from considering that occupation, particularly if their overall score is adequate. On the other hand, high scores in the specialized tests could be used to encourage the pursuit of careers that a student may not have considered initially. In vocational counseling, tests should open doors for students, not close them.

Regardless of the lack of evidence for differential validity, this reviewer prefers the DAT to other group tests that yield a single score. The variety of abilities measured and the deemphasis on verbal test content make the DAT particularly appropriate for measuring aptitudes in poor readers or persons with limited English proficiency. Nonverbal tests cannot substitute entirely for verbal tests because they do not measure exactly the same skills, although they are substantially correlated. Nevertheless, nonverbal tests can predict school performance to some extent and may help to identify talent that has gone unnoticed because of poor reading or communication skills.

Moreover, the DAT and similar tests are indispensable tools for research on cognitive abilities; the DAT's availability in foreign editions makes it a powerful instrument for cross-cultural research. The study of variation in the pattern of strengths and weaknesses of students in the diverse tests across cultures can help us to understand individual differences in the process of learning and lead us to the development of more adequate cognitive psychological theories.

References

Adamowicz, J. K., & Hudson, B. R. (1978). Visual short-term memory, response delay, and age. *Perceptual and Motor Skills, 46,* 267-270.

Anastasi, A. (1982). *Psychological testing* (5th ed.). New York: Macmillan Publishing Company.

Arnold, P., & Walter, C. (1979). Communication and reasoning skills of deaf and hearing signers. *Perceptual and Motor Skills, 49,* 192-194.

Balke-Aurell, G. (1982). *Changes in ability as related to educational and occupational experience.* Gothenburg: Acta Universitatis Gothoburgensis.

Baroya, G. M. (1967). Reliability, validity, and comparability of Forms L and M of the "Verbal Reasoning" and the "Numerical Ability" Subtests of the Differential Aptitude Tests for use in East Pakistan. *Dissertation Abstracts, 27*(9-A), 2865-2867.

Bechtoldt, H. (1953). Review of the Differential Aptitude Tests. In O. K. Buros (Ed.), *The fourth mental measurements yearbook* (pp. 676-680). Highland Park, NJ: The Gryphon Press.

Beitel, P. A. (1980). Multivariate relationships among visual-perceptual attributes and gross-motor tasks with different environmental demands. *Journal of Motor Behavior, 12,* 29-40.

Bennett, G. K. (1955). The D.A.T.—A seven-year follow-up. *The Psychological Corporation Test Service Bulletin, 49,* 11-15. (ERIC Document Reproduction Service No. ED 079348)

Bennett, G. K., Seashore, H. G., & Wesman, A. G. (1974). *Manual for the Differential Aptitude Tests, Forms S and T* (5th ed.). Cleveland: The Psychological Corporation.

Bennett, G. K., Seashore, H. G., & Wesman, A. G. (1982a). *Administrator's handbook, Differential Aptitude Tests, Forms V and W.* Cleveland: The Psychological Corporation.

Bennett, G. K., Seashore, H. G., & Wesman, A. G. (1982b). *Directions for administration and scoring, Differential Aptitude Tests, Forms V and W.* Cleveland: The Psychological Corporation.

Bennett, G. K., Seashore, H. G., & Wesman, A. G. (1982c). *Technical supplement, Differential Aptitude Tests, Forms V and W.* Cleveland: The Psychological Corporation.

Bouchard, T. J. (1978). Review of the Differential Aptitude Tests. In O. K. Buros (Ed.), *The eighth mental measurements yearbook* (pp. 655-658). Highland Park, NJ: The Gryphon Press.

Brown, T. R. (1967). *Attitudes toward science and critical thinking abilities of chemistry and nonchemistry students in the Tacoma public schools.* Corvallis, OR: Oregon State University. (ERIC Document Reproduction Service No. ED 022671)

Buros, O. K. (Ed.). (1972). *The seventh mental measurements yearbook.* Highland Park, NJ: The Gryphon Press.

Buros, O. K. (Ed.). (1978). *The eighth mental measurements yearbook.* Highland Park, NJ: The Gryphon Press.

Carpenter, F., Brinkmann, E. H., & Lirones, D. S. (1965). *Educability of students in the visualization of objects in space.* Ann Arbor, MI: Michigan University. (ERIC Document Reproduction No. ED 003271)

Carroll, J. B. (1959). Review of the Differential Aptitude Tests. In O. K. Buros (Ed.), *The fifth mental measurements yearbook* (pp. 670-673). Highland Park, NJ: The Gryphon Press.

Cooperman, E. W. (1980). Field differentiation and intelligence. *Journal of Psychology, 105*(1), 29-34.

Cronbach, L. J. (1984). *Essentials of psychological testing* (4th ed.). New York: Harper & Row Publishers.

Crosby, G., Freemont, H., & Mitzel, H. E. (1960). *Mathematics individual learning experience.* New York: City University of New York, Queens College. (ERIC Document Reproduction Service No. ED 003558)

Cumming, J. F., Davies, M. M., Nagle, J. P., & Thompson, G. M. (1966). *Project 13—Outreach counseling, Minneapolis public schools. Progress report.* Minneapolis: Minneapolis Public Schools. (ERIC Document Reproduction Service No. ED 011676)

Dale, E., & Chall, J. S. (1948). A formula for predicting readability: Instructions. *Educational Research Bulletin, 27,* 11-28.

Daniel, M., Wiesen, J. P., Trafton, R. S., & Bowker, R. (1982, September). *The relationship of aptitudes to the performance of skilled technical jobs in engine manufacturing* (Technical Report 1982-5). Fort Worth, TX: Johnson O'Connor Research Foundation, Human Engineering Laboratory.

Deb, M. (1980). DAT as predictor of students' success in psychology. *Psychological Research Journal 4*(1), 23-28.

DeSantis-Feltran, R. C. (1982). Validade preditiva da Bateria De Aptidoes Especificas DAT—Revisao bibliografica. *Arquivos Brasileiros De Psicologia, 34*(4), 66-87.

DeSantis-Feltran, R. C. (1983). Estudo da validade preditiva da Bateria De Aptidoes Especificas DAT. *Arquivos Brasileiros De Psicologia, 35*(1), 110-112.

Duke, R. D. (1966). *The development of new supplementary teaching materials and an analysis of their potential use in the high school biology curriculum.* Austin, TX: University of Texas. (ERIC Document Reproduction Service No. ED 024561)

Eastman, P. M., & Behr, M. J. (1977). Interaction between structure of intellect factors and two methods of presenting concepts of logic. *Journal for Research in Mathematics Education, 8,* 379-381.

EDL core vocabularies in reading, mathematics, science, and social studies. (1979). New York: EDL/McGraw-Hill.

Fennema, E., & Sherman, J. (1977). Sex-related differences in mathematics achievement, spatial visualization and affective factors. *American Educational Research Journal, 14,* 51-71.

Fennema, E. H., & Sherman, J. A. (1978). Sex-related differences in mathematics achievement and related factors: A further study. *Journal for Research in Mathematics Education, 9,* 189-203.

Fogliatto, H. M. (1972). Factorial stability of two psychological tests: D.A.T. and Kuder. *Revista Interamericana De Psicologia, 6*(3-4), 213-223.

Fuller, G. B., & Friedrich, D. (1972). Rural high school dropout: A descriptive analysis. *Perceptual and Motor Skills, 35*(1), 195-201.

Garrison, D. R. (1983). Psychosocial correlates of dropout and achievement in an adult high school completion program. *Alberta Journal of Educational Research, 29*(2), 31-39.

Garrison, W. M., & Coggiola, D. C. (1980). *Time limits in standardized testing: Effects on ability estimation.* Rochester, NY: National Technical Institute for the Deaf. (ERIC Document Reproduction Service No. ED 209905)

Geiger, H. B. (1968). *Effectiveness in learning Newton's second law of motion in secondary school physics using three methods of learning.* Philadelphia: Temple University. (ERIC Document Reproduction Service No. ED 028095)

Glover, H. (1974). A study of the relationship between scores on a visualization test and neural impulse on a visual task. *Dissertation Abstracts International, 35*(3-A), 1497.

Gold, A. M. (1977). The use of separate-sex norms on aptitude tests: Friend or foe. *Measurement and Evaluation in Guidance, 10*(3), 162-170.

Goldsmith, R. P. (1980). The effect of training in test taking skills and text anxiety management on Mexican American students' aptitude test performance. *Dissertation Abstracts International, 40*(11-A), 5790.

Graydon, J. (1980). Spatial ability in highly skilled women squash players. *Perceptual and Motor Skills, 50,* 968-970.

Gregory, R. J., Alley, P., & Morris, L. (1980). Left-handedness and spatial reasoning abilities: The deficit hypothesis revisited. *Intelligence, 4*(2), 157-159.

Hashmi, S. A. (1967). Effect of previous academic achievement on the performance of first-year college students of East Pakistan on the "Verbal Reasoning" and the "Numerical Ability" subtests of the Differential Aptitude Tests. *Dissertation Abstracts, 27*(8-A), 2391-2392.

Herman, D. O. (1977). A reply to Gold's article on separate-sex norms. *Measurement and Evaluation in Guidance, 10*(3), 172-174.

Hlavenka, V., & Rapos, I. (1973-1974). Studies of the Differential Aptitude Tests. *Psychologica: Zbornik Filozofickez Fakulty, U. Komenskeho, 23,24*(12-13), 133-173.

Korean Institute for Research in Behavioral Sciences. (1970). *A study on the aptitude structures of freshmen in Seoul National University.* Seoul: Author. (ERIC Document Reproduction Service No. ED 128353)

Kumar, S. (1975). Reliability and validity of a cross-modal test of spatial ability. *Perceptual and Motor Skills, 41*(3), 805-806.

Lee. A. E. (1966). *The development of new supplementary teaching materials and an analysis of their potential use in the high school biology curriculum. Final report.* Austin: Texas University. (ERIC Document Reproduction Service No. ED 015138)

Linn, R. L. (1978). Review of the Differential Aptitude Tests. In O. K. Buros (Ed.), *The eighth mental measurements yearbook* (pp. 658-659). Highland Park, NJ: The Gryphon Press.

Makarovic, J. (1970). *Prediction of scholastic achievement in gymnasium for testing of elementary school pupils, 3*(3), 24-33.

Mastie, M. (1976). Review of the Differential Aptitude Tests. *Measurement and Evaluation in Guidance, 9*(2), 87-95.

McKelvie, S. J., & Rohrberg, M. M. (1978). Individual differences in reported visual imagery and cognitive performance. *Perceptual and Motor Skills, 46*, 451-458.

Mehrens, W. A., & Lehmann, I. J. (1984). *Measurement and evaluation in education and psychology* (3rd ed.). New York: Holt, Rinehart and Winston.

Meiners, M. L., & Dabbs, J. M., Jr. (1977). Ear temperature and brain blood flow: Laterality effects. *Bulletin of the Psychonomic Society, 10*, 194-196.

Mitchell, J. V., Jr. (Ed.). *Tests in print III.* Lincoln, NE: The Buros Institute of Mental Measurements.

Myers, R. A., & Thompson, A. S. (1977). *Counseling from profiles: A case book for the Differential Aptitude Tests* (2nd ed.). New York: The Psychological Corporation.

Newcombe, N., Bandura, M. M., & Taylor, D. G. (1983). Sex differences in spatial ability and spatial activities. *Sex Roles: A Journal of Research, 9*(3), 377-386.

Parasinis, I., & Long, G. L. (1979). Relationships among spatial skills, communication skills, and field independence in deaf students. *Perceptual and Motor Skills, 49*, 879-887.

Quellmalz, E. S., Capell, F., & Chih-Ping, C. (1982). Effects of discourse and response mode on the measurement of writing competence. *Journal of Educational Measurement, 19*(4), 241-258.

Quereshi, M. Y. (1972). Review of the Differential Aptitude Tests. In O. K. Buros (Ed.), *The seventh mental measurements yearbook* (pp. 1049-1051). Highland Park, NJ: The Gryphon Press.

Rutgers, The State University. (1980). *Final report of the Vocational Assessment Project, 1979-80.* New Brunswick, NJ: Rutgers, The State University. (ERIC Document Reproduction Service No. ED 190811)

Schutz, R. E. (1965). Review of the Differential Aptitude Tests. In O. K. Buros (Ed.), *The sixth mental measurements yearbook* (pp. 1005-1007). Highland Park, NJ: The Gryphon Press.

Sewell, T. E., Palmo, A. J., & Manni, J. L. (1981). High school dropout, psychological, academic, and vocational factors. *Urban Education, 16*, 65-76.

Shepard, L., Camilli, G., & Williams, D. M. (1983, April). *Accounting for statistical artifacts in item bias research.* Paper presented at the annual meeting of the American Educational Research Association, Montreal.

Sherman, J. (1979). Predicting mathematics performance in high school girls and boys. *Journal of Educational Psychology, 71*, 242-249.

Sherman, J. (1980a). Mathematics, spatial visualization, and related factors: Changes in girls and boys, grades 8-11. *Journal of Educational Psychology, 72*, 476-482.

Sherman, J. (1980b). Predicting mathematics grades of high school girls and boys: A further study. *Contemporary Educational Psychology, 5*(3), 249-255.

Sherman, J., & Fennema, E. (1977). The study of mathematics by high school girls and boys: Related variables. *American Educational Research Journal, 14,* 159-168.

Sternberg, R. J. (1980). Representation and process in linear syllogistic reasoning. *Journal of Experimental Psychology: General, 109,* 119-159.

Sternberg, R. J., & Weil, E. M. (1980). An aptitude X strategy interaction in linear syllogistic reasoning. *Journal of Educational Psychology, 72,* 226-239.

Super, D. E. (1982). *DAT career planning program: Counselor's manual* (2nd ed.). New York: The Psychological Corporation.

Tanner, R. T. (1968). *Expository-deductive vs. discovery-inductive programming of physical science principles.* Stanford, CA: Stanford University. (ERIC Document Reproduction Service No. ED 027207)

Tapley, S. M., & Bryden, M. P. (1977). An investigation of sex differences in spatial ability: Mental rotation of three-dimensional objects. *Canadian Journal of Psychology, 31,* 122-130.

Toronto Board of Education. (1973). *Aptitude testing: A critical examination of the Differential Aptitude Tests, alternative batteries, and problems in prediction* (Report No. 17). Ontario: Author. (ERIC Document Reproduction Service No. ED 068487)

Whitely, S. E. (1977). Information-processing on intelligence test items: Some response components. *Applied Psychological Measurement, 1,* 465-476.

Robert J. Drummond, Ed.D.

Interim Chairperson, Division of Educational Services and Research, University of North Florida, Jacksonville, Florida.

EDWARDS PERSONAL PREFERENCE SCHEDULE

Allen L. Edwards. San Antonio, Texas: The Psychological Corporation.

THE ORIGINAL OF THIS REVIEW WAS PUBLISHED IN TEST CRITIQUES: VOLUME I (1984).

Introduction

The Edwards Personal Preference Schedule (EPPS) is a forced choice, objective, non-projective personality inventory, derived from the theory of H. A. Murray, which measures the rating of individuals in fifteen normal needs or motives. The fifteen variables selected are: Achievement, Deference, Order, Exhibition, Autonomy, Affiliation, Intraception, Succorance, Dominance, Abasement, Nurturance, Change, Endurance, Heterosexuality, and Aggression. On the EPPS there are nine statements used for each scale. Social Desirability ratings have been done for each item, and the pairing of items attempts to match items of approximately equal social desirability. Fifteen pairs of items are repeated twice for the consistency scale.

The author of the test is Allen L. Edwards, a professor of psychology at the University of Washington. He has authored several widely used texts in the areas of statistics and psychological measurement. His publications include a number of articles on the issue of social desirability in personality and a book, *The Social Desirability Variable in Personality Assessment and Research* (1957). Dr. Edwards has also developed another test, the Edwards Personality Inventory (1967).

The EPPS is a 225-item paired comparison test contained in an eight-page test booklet. The front cover includes the directions for the test, and there are seven pages of items presented in double columns. Each item consists of two statements, and test-takers must decide which statement is most characteristic of them.

Example:
A. I like to do things by myself.
B. I like to help others do things.

An item representing each need is paired twice with one from every other need. It takes 105 items to compare each need once, and 210 items for it to be compared twice.

Since the test is of the objective type and has a separate answer sheet, the

139

examiner can administer the test to an individual or to a large group. Specific guidelines for the examiner are contained in the manual. The manual recommends that an individual taking the test be allowed to read the directions on the test booklet and begin taking the test without examiner assistance. For large groups, it is suggested that the examiner read the directions aloud and direct the examinees to read along silently. When the directions have been read, if the examinees have any questions the direction and illustrations give guidelines on how questions should be answered by the examiner. The test is untimed, but it usually requires about forty minutes for college students to complete. The author suggests that the answer sheet be checked for possible omissions before being collected; it is important that a response be recorded for each item.

The EPPS was initially designed for use with college students and adults. Edwards presents norms for both a college sample and a general adult sample. The test has also been used with high school students. Separate norms were developed for this group, but are not reported in the manual. The test is most commonly used with subjects eighteen to sixty-five years of age.

Evaluation of difficulty level in the classical sense is not appropriate for this type of test. More appropriate issues are the readability of the test and the nature of the tasks given. Review of several samples of items indicates they are at the seventh-grade reading level. Adults with a high-school education should not have trouble reading the items or understanding the directions, but examinees with less education or with reading problems will probably experience difficulty.

There are no subtests, but the EPPS contains sixteen scales—fifteen personality scales and a consistency score.

Two types of answer sheets have been developed for the test. There are hand-scoring sheets available, with a scoring template to help identify items belonging to the consistency scale. There are also different types of machine-scorable answer sheets available, including IBM 805, IBM 1230, and Op Scan, for individuals interested in machine scoring.

On the profile sheet for the EPPS the raw scores for each of the sixteen scales are translated into percentile ranks using the appropriate norm tables. The ranks for each scale are then plotted on a graph, which displays quartiles along with the 1, 5, 10, 20, 30, 40, 60, 70, 80, 90, 95, 99 percentiles.

Practical Applications/Uses

The EPPS Test can be utilized for research, teaching, and counseling purposes, and the results can be used to stimulate interaction between the counselor and the client when the discussion relates to needs or motives. The test is based upon Murray's Need System, a very useful and understandable conceptual framework for examining the motives of normal people.

The test is a good teaching tool for professors or teachers presenting courses in personality theory, personality assessment, or measurement. It is applicable in general and applied psychology classes since it is based on the work of Murray, a major personality theorist. Because the test utilizes the method of paired comparison, and illustrates ipsative measurement (the examinee ranks the variables internally) and the concept of social desirability, it is also appropriate in measurement and psychometric theory courses.

The test has been widely used in research studies. When researchers are interested in an experimental instrument that measures the needs or motives of individuals, the test may be of value. Because questions have been raised concerning the validity of the scale, and the interpretability of the type of scores generated, caution should be exercised when using the EPPS.

Edwards states that the EPPS was designed as a research and counseling tool to provide a quick and convenient measure of normal personality variables. He feels that the EPPS is useful in stimulating discussions about dimensions of interpersonal relations, levels of aspiration, and vocational and educational plans and goals. The test can be a tool which helps the counselor assist the client in the assessment and diagnosis of problems.[1]

Edwards states that the EPPS may be of value to individuals wanting to differentiate among groups in different occupations. The test could also be of use in assessing the differences between successful and unsuccessful workers in a given job.

The EPPS has become one of the most widely used personality inventories. By the time *The Eighth Mental Measurements Yearbook* (Buros, 1978) was published, 1,640 references of studies that had used the test were listed. The test has been used with numerous occupational groups for descriptive purposes, for construct validity studies, for criterion-referenced validity studies, and in all types of clinical and counseling studies. Psychologists, counselors, and educational researchers have been primary users of the test. This test was designed for use with normal adults and college students. It is not appropriate for clinical populations, nor for adults who have reading disabilities. As the EPPS has been on the market since 1953, new insights concerning its usage are not expected.

Since the EPPS is a personality test and students may have questions about procedures, the examiner should be a psychologist or a counselor who has had coursework in tests and measurements and training in tests and administration and interpretation. It would be classified as a Level B under the APA standards for types of tests.

The directions for administration of the EPPS are clear, and it is considered easy to administer. The author gives the test taker a number of sample items to illustrate how to answer the paired comparison situations. Edwards suggests that the administrator tell the test takers at the end of twelve-minute segments how many items should have been completed (Edwards, 1959, p.6). This procedure would not appear to be helpful, however, since adults vary in reading speed and the test is supposed to be untimed.

There are scoring procedures outlined for both machine- and hand scoring of the IBM answer sheets, and procedures concerning how to hand score regular answer sheets. A professional should require only a few minutes to learn how to score the test. There are different types of machine- and computer-scoring services available. In hand scoring the regular answer sheet, one must be careful to place the template over the answer sheet correctly. There are checks to use to determine if the scales were added correctly. If a counselor or psychologist has a large number of tests to score, it would be easier to use the machine- or computer-scoring services. Test takers can also be guided on how to score their own tests.

Test interpretation is based on objective scoring. Each of the needs is paired twice with every other need. The highest possible score for each scale is 28, occurring if the examinee chooses the statement for a given need as being more personally characteristic than all the other needs it was paired with. The lowest possible score for each scale is zero, given when the subject fails to choose the items representing a particular need in the 28 comparisons presented.

Even though the scaling is ipsative, norm-referenced interpretations are given. Normative data are presented for college students and adult heads of households. The raw scores can be translated into percentile and standard scores.

One can certainly translate the results of the EPPS into a rank order since paired comparison method leads inherently to a rank order of the variables. The interpretation would then be based on an ordinal scale, and problems of interpretation rise from the nature of the ordinal scale. The scale does not have equal units of measurement. The distance between ranks may be slight or great, and the ranks of one individual may be incomparable with those of another person. Edwards (1959) recommends that the first interpretation should utilize the rank ordering of the raw scores of the client. The counselor should then move to the norm-referenced interpretation using the percentile norms. He further states that the examinee should be given a copy of the descriptions of each of the scales.

The examiner needs a thorough knowledge of psychometric theory as well as personality theory prior to any attempt at interpreting the EPPS. On the surface, the interpretation appears to be simplistic; however, the ipsative nature of the scores clouds the normative type of interpretation.

Technical Aspects

Edwards suggests that one definition of validity might be viable for the interpretation of the validity of the EPPS. He states that "validity is the extent to which the test or inventory measures what it purports to measure" (Edwards, 1959, p.21). He discusses the concurrent validity of the test, which was obtained by comparing ratings of the strength of Murray's needs with scores on the EPPS. Mixed results have been found in these studies, with some examinees showing a high degree of agreement between their rankings of the fifteen variables and their scores on the EPPS, while others found little or no agreement.

There are numerous studies comparing the EPPS with other personality inventories. The manual reports studies comparing the EPPS with the Guilford-Martin Personality Inventory and the Taylor Manifest Anxiety Scale. Other researchers have correlated the California Psychological Inventory, the Adjective Check List, the Thematic Apperception Test, the Strong Vocational Interest Blank, and the MMPI with the EPPS. In these studies there are often statistically significant correlations among the scales of these tests and the EPPS, but the relationships are usually low-to-moderate and often are difficult for the researcher to explain. These kinds of comparisons do little to demonstrate the construct validity of the EPPS.

There have also been predictive validity studies comparing scores on an EPPS scale such as Achievement with grades at the end of the semester. There have been mixed results in these studies, and correlations range from zero to a moderate relationship.

Split-half reliability coefficients, or coefficients of internal consistency for 1,509 students in the college normative group, are presented by the author. They range from .60 to .87 with a median of .78. He also presents test-retest stability coefficients with a one-week interval. These are based on a sample of 89 students and range from .55 to .87 with a median of .73 (Edwards, 1959). Other researchers have reported similar results over a three-week period, showing correlations of .55 to .87, with a median of .73.

The split-half coefficients were acquired by dividing each 28-item scale into two scales of fourteen items each. Since nine items are rotated for each variable in the test, each part-scale would have some of the same items. The overlap of items might overinflate the reliability of a scale. The coefficients would also give a picture of the equivalence of the two sets of items for each variable. The uncorrected coefficients would indicate that the items may not be highly correlated with each other. These articafts of the forced choice technique make interpretations of reliability data difficult.

Critique

The experts' evaluations of the EPPS can be best summarized by Heilbrun's opening statement about the test in *The Seventh Mental MeasurementsYearbook:* "Sixteen years and three *Mental Measurements Yearbooks* later, a reviewer is left with much the same critique of the EPPS as before. Edwards' publication of this instrument in 1954 came more as an exercise in test construction than as a serious entry into the market of validated tests" (Heilbrun, 1978, p. 149). In reviewing the assets of the EPPS, Baron (1959) felt that the test showed promise as a system to measure manifest needs. Fiske (1959) mentioned that the test was based on sophisticated theoretical formulation and showed promise in measuring personality. Stricker (1965) pointed out that the test has stimulated a tremendous amount of research. Heilbrun (1972) concluded that the test has been a useful catalyst for research, and has also stimulated psychometric debate.

In reviewing some of the areas of concern about the EPPS by the critics, Baron (1959), Radcliff (1965), and McKee (1972) point out that the EPPS has failed to perform its function of controlling the factor of social desirability. All of the critics raise questions about the validity of the scale. Radcliff (1965), Stricker (1965), and Heilbrun (1972) call attention to the problems caused by the ipsative type of measurement used by the test. Stricker (1965) felt that the authors should have reported evidence on the item validities, while the reliability of the test is questioned by McKee (1972). Borden (1959), Fiske (1959), and Radcliff (1959) felt that the norms were inadequate and unrepresentative. McKee (1972) points out that, as with many personality inventories, the EPPS can be faked.

Essentially, the critics concluded that the test had potential for use and research, but needed more studies on the validity and on the social desirability factor of the test. They point out that even though there is some evidence of the construct validity of several scales, the EPPS is primarily a research tool and should be used only for experimental purposes. McKee (1972) concludes that the ipsative nature of the scale, and the attempt at control of social desirability have taken too great an expense from the technical quality of the test; he feels that there are other (better) tests to measure dimensions of normal personality, such as the California Psychological Inventory. Critics in general indicate that because there are so many technical problems with the test, and so little supporting evidence of its validity, other proven instruments should be considered before selecting the EPPS.[2]

I have used the EPPS for years. I initially used it in my doctoral dissertation at Teachers College, Columbia University, to compare forced choice and ranking methods of personality assessment. Professor Edwards was very cooperative and sent me his information on the social desirability of the items on the EPPS, and he and The Psychological Corporation allowed me to use items from the EPPS to make four rank order forms. I have used the EPPS in my classes in General Psychology, Personality Theory, Psychological Measurement, and Learning Theory and Guidance Practices, and in a number of research studies and in counseling situations.

Overall, I have found the test to be a useful tool to stimulate discussion, not only about need theory but also concerning methods of psychological assessment and the qualities needed in a psychological test. In counseling situations, the EPPS has helped to trigger client's insight into their own personality structure. I would not use the test for screening or placement purposes, or as a clinical test. I am quite reluctant to use it for research purposes because I have found that many individuals become frustrated and bored when taking the test, due to the item repetition and to dislike of the forced choice format.

We have become more concerned about standards for educational and psychological tests, and have become more sophisticated in our awareness of these standards. The author and The Psychologcial Corporation have not placed a high priority on the updating and upgrading of this test and its manual; the test still contains the same technical problems identified by the previous reviewers.

In the selection of any test, one must consider the purpose behind its selection and its validity for this purpose. The EPPS has been utilized in numerous research and experimental studies for over a period of thirty years. It is doubtful that new research would dramatically change the findings already reported in the literature. Today, the EPPS has primarily instructional and historical value; it is time for psychologists and researchers to consider other and newer approaches to measure the personality structure of normal adults.

Notes to the *Compendium*:

[1]Helms (1983) presents in review literature on the use of the EPPS in vocational and educational counseling and found just a limited number of studies completed in these areas.

[2]Helms (1983) provides a comprehensive review of the uses and research on the EPPS and is a good source of information on the test.

References

Baron, F. (1959). Edwards Personal Preference Schedule. In O.K. Buros (Ed.), *The fifth mental measurements yearbook* (pp. 114-118). Highland Park, NJ: The Gryphon Press.

Borden, E.S. (1959). Review of 1959 revision of the manual of the EPPS. *Journal of Counsulting Psychology, 23,* 471.

Buros, O.K. (Ed.) (1978). *The eighth mental measurements yearbook.* Highland Park, NJ: The Gryphon Press.

Edwards, A.L. (1957). *The social desirability variable in personality assessment and research.* New York: Dryden Press.

Edwards, A.L. (1959). *Edwards Personal Preference Schedule.* New York: The Psychological Corporation.

Edwards, A.L. (1967). *Edwards Personality Inventory.* Chicago: Science Research Associates, Inc.

Fiske, D.W. (1959). Edwards Personal Preference Schedule. In O.K. Buros (Ed.), *The fifth mental measurements yearbook* (pp. 118-119). Highland Park, NJ: The Gryphon Press.

Heilbrun, A.B. (1972). Edwards Personal Preference Schedule. In O.K. Buros (Ed.), *The seventh mental measurements yearbook* (pp. 148-149). Highland Park, NJ: The Gryphon Press.

Helms, J. E. (1983). *A practitioner's guide to the Edwards Personal Preference Schedule.* Springfield, IL: Charles C. Thomas.

McKee, M.G. (1972). Edwards Personal Preference Schedule. In O.K. Buros (Ed.), *The seventh mental measurements yearbook* (pp. 149-151). Highland Park, NJ: The Gryphon Press.

Radcliff, J.A. (1965). Edwards Personal Preference Schedule. In O.K. Buros (Ed.), *The sixth mental measurements yearbook* (pp. 195-200). Highland Park, NJ: The Gryphon Press.

Shaffer, L.F. (1955). Review of the Edwards Personal Preference Schedule. *Journal of Consulting Psychology, 19,* 156.

Stricker, L.J. (1965). Edwards Personal Preference Schedule. In O.K. Buros (Ed.), *The sixth mental measurements yearbook* (pp. 200-207). Highland Park, NJ: The Gryphon Press.

Allan L. LaVoie, Ph.D.

Professor of Psychology, Davis & Elkins College, Elkins, West Virginia.

EMBEDDED FIGURES TEST

Herman A. Witkin. Palo Alto, California: Consulting Psychologists Press, Inc.

THE ORIGINAL OF THIS REVIEW WAS PUBLISHED IN TEST CRITIQUES: VOLUME I (1984).

Introduction

The Embedded Figures Test (EFT), along with a host of other tests in the Witkin stable, was designed to measure the individual differences in perceptual style that came to be called *field dependence-independence* (Witkin, 1950; this article contains all of the items plus the practice materials). Clients who can quickly find a simple figure hidden, or embedded, in a larger complex geometrical figure tend to be more field-independent than those who disembed the simple figure more slowly. In turn, field-independent people are more likely to impose structure on their experiences, to have a more differentiated cognitive system, to assign more weight to vestibular than to visual cues, and to be less interpersonally sensitive and competent.

By September of 1981 more than 3,800 articles had reported research results using the EFT or one of the more than 25 related tests. Experimental uses of the test have clustered in the areas of education, social psychology, personality, learning, clinical, and perception/cognition. Clearly the EFT and its cousins have been often and widely used, ranking among the most popular tests (e.g., the MMPI and its direct offshoots have appeared in about twice as many articles).

Unlike its predecessors, such as the Rod and Frame Test (RFT), the EFT is scored for, and subjects are urged to obtain, correct answers. This forces subjects to show their capacity for disembedding rather than their habitual practice. Strictly speaking, then, the EFT measures not perceptual style but perceptual ability.

The late Herman Allen Witkin (see Messick, 1980), developer of the test, was born August 2, 1916, and died July 8, 1979. In 1939, he earned his doctorate in psychology from New York University, specializing in animal behavior (his dissertation was on maze-learning in rats). He then took a one-year position as research associate to Wolfgang Kohler, the Gestalt psychologist who had been studying the effects of the perceptual field on perception. In the mid-1940s he worked with Max Wertheimer, another of the three great Gestalt psychologists. When Wertheimer died before their work ended, Kohler took over the direction of it. Witkin (1954) mentioned Wertheimer as the individual ". . . who gave warm encouragement to the first efforts to investigate the personalities of subjects with different modes of space orientation" (p. 496, n. 5); this book was dedicated to Wertheimer. Under his direction then, Witkin's research program for the next 25 years was laid out: to identify individual differences in perception of space orientation and to find the corresponding personality differences. His career took him from Brooklyn College, to the State University of New York, and finally to Educational Testing Service as chair of the Personality and Social Research Group.

Witkin and his colleagues developed a large number of tests to measure the ability to maintain a vertical posture despite conflicting visual cues (see, e.g., Witkin, 1949; Witkin et al., 1954/1972). In one test the subject was placed in a tiltable chair in a tiltable room (the TR-TC test) and asked to restore the chair to upright. When the room was tilted away from the upright, incredible individual differences emerged in the ability of subjects to ignore the visual cues in restoring the chair to verticality. Some subjects could consistently ignore the room's visual cues, placing the chair within a few degrees of verticality through the use of bodily and vestibular cues. Others made errors as large as 56 degrees! The latter, when asked to adjust the chair with eyes closed, were able to achieve verticality. But with eyes reopened, the chair was again tilted away from the upright to make it consistent with the visual cues.

Another of these tests is the RFT, especially Series 3, in which the subject adjusts a luminous rod to verticality while a luminous frame is tilted, in an otherwise dark room. The average error for RFT3 is on the order of 10 degrees. Among other tests are the Body Adjustment Test (BAT), Articulation of Body Concept (or one of the other draw-a-person scoring systems), Wechsler analytic triad, and the moving-room test. The perceptual ability common to these tests has been called field dependence-independence, for scores reflect one's ability to overcome the effects of a visual field (independence) or to be influenced by it (dependence).

Field dependence corresponds to an intellectual style called *global-articulated*. The *global* person has trouble distinguishing self from field, makes less structured judgments, and discriminates social cues better than *articulated* persons. These are all field-dependent behaviors. The articulated person has a more stable internal frame of reference, makes finer distinctions between self and field, and is less social. These are field-independent behaviors.

A later theoretical extension of field dependence has been to a concept labelled *degree of differentiation*. The field-dependent individual has a less differentiated personality as a consequence of perceptual and cognitive tendencies. The more differentiated individual has a clearer sense of self and takes a more active role in imposing structure on the field. There has been a strong tendency to call the cognitive and personality dimensions field-dependence despite the separate definitions and labels offered by Witkin.

The last theoretical extension (Witkin & Goodenough, 1981) aligns field dependence-independence with *degree of autonomous functioning*. The field-independent person has a greater degree of autonomous functioning; that is, this person functions more in tune with internal frames of reference than external ones. Such a person also has greater cognitive restructuring skills, less interpersonal competency, and more specialized (lateralized) brain functioning. The field-dependent person has less capacity for autonomous functioning, less skill in cognitive restructuring, greater interpersonal competency, and less brain specialization.

The EFT can be conceptualized in terms of autonomous functioning: since it measures the ability to disembed, or restructure, one's experiences, it measures at the second level in the newer theory. If subjects who score best on the EFT are better at autonomous functioning (the highest level of the new theory), it follows that they are better at differentiating self from field, are more field independent, and are less interpersonally competent. How well this theoretical picture fits what actually goes on in people must be determined by research; little has been done on this question as yet. Until then it remains safer to think of the EFT as a traditional measure of field dependence-independence.

The EFT was developed to test the consistency of the perceptual style identified in the earlier tests, and was then called the Imbedded Figures Test (Witkin, 1949). Witkin hypothesized that the field-dependent subject would have trouble disembedding a simple figure from the complex one containing it, and the independent subject would do it more quickly. Generally, that relationship has been obtained.

Since the EFT requires only the test paper and a stopwatch, in contrast to the elaborate and non-portable equipment of some of the other tests, it began to be used more frequently. Gradually under the pressure of increased use, it went from the original 24 figures, with 5 minutes per figure (a maximum of 2 hours testing per person), to 12 items with 3 minutes per figure (a 36-minute limit). Some subjects finished the test in under 5 minutes.

The EFT is too difficult for children much below age 12. One can use the children's EFT (CEFT, ages 5 to 11 or 12; the norms go up to age 12) or the Preschool EFT (ages 3 to 5). A group version (the GEFT) has also been created (Witkin et al., 1971), which can be used with large numbers of subjects and takes about 15-20 minutes. In some studies, the analytic triad from the WAIS or WISC has been used to classify subjects into field dependence-independence (using scores on Block Design, Object Assembly, and Picture Completion subtests).

In a recent field-dependence bibliography (Cox & Gall, 1981) this reviewer counted the frequency of use of the various tests in 51 randomly selected studies. The GEFT was used in 29 (57%), the EFT in 8 (16%), the CEFT in 5 (10%), the portable RFT in 2 (4%), and some other in the last 7 reports. Clearly the ease of using the GEFT has led to its wider use.

Incidentally, the EFT has not been universally adopted as published. Jackson (1956) created one grouping of figures consisting of 12 of the original 24; scores on the 12 correlated .99 with the full scale scores. His version is sometimes used (e.g., Ihilevich & Gleser, 1971), though it differs by three items from Witkin's EFT, Form A (which is simply the first 12 items of the original EFT). Competing versions of other tests also exist.

Finally, a series of specialized versions of the original test has been created to deal with special populations, such as the Tactile EFT and Auditory EFT for use with the blind (see Witkin & Oltman, 1967).

The EFT kit comprises two sets of 12 complex-figure cards (3"x5"), Forms A and B; a set of eight simple-figure cards; two practice cards; a stylus to be used by the subject to trace the simple figure in the complex; and a pad of recording sheets on which the examiner may note times, errors, or comments. With one exception the complex figures are colored in ways that help to mask the embedded simple figures while the simple-figure cards are printed black on white. The manual must be ordered separately; it also serves the GEFT and CEFT.

To preserve the paper cards they should be encased in plastic, for some examinees have trouble keeping the stylus off the figures. This is especially true with younger subjects for whom thinking seems to be facilitated by physical contact. The minimum examinee age has been indicated as 10, however norms indicate that 10-year-olds have a very difficult time with this test. Further, some when they fail get discouraged and stop trying. The EFT has, though, been used successfully with retarded children.

Some of the items are very difficult, others very easy. Witkin (1950) found item A-2 (now 6A) was failed by 20% of the initial sample, and the average search time was 144 seconds. In contrast, item A-4 (now 17A) was failed by no one, and average search time was 15 seconds. The range of individual subject performance was even greater, with the fastest man averaging under 5 seconds and the slowest woman more than 177 seconds.

Practical Applications/Uses

The primary purpose of the EFT must be considered as an alternate, easier way to measure field dependence-independence. Used for this purpose, it will allow one to tap into that very large body of literature. In applied settings, the following are some of the ways the EFT may be used:

1. As a supplement to verbal IQ tests, it will provide a reliable estimate of analytic IQ. This may be helpful in classifying the retarded (see, e.g., Witkin et al., 1966).

2. In the clinic, it may be used diagnostically to help differentiate among forms of psychosomatic illness, to predict responsivity to directive vs. nondirective therapy, or to identify the kind of defense mechanisms a client would employ.

3. In personnel screening the EFT can be an aid to selecting appropriate candidates. For example, fast disembedders will probably work better in isolation than those who disembed slowly; the latter will do better in social settings where sensitivity and skill are required. Similarly, the fast disembedders will be better at analytic tasks, the slower at tasks requiring accumulation of information (see, e.g., Witkin & Goodenough, 1977).

4. In guidance counseling, scores can help advisors place students in the correct major or appropriate career tracks (see Witkin, Moore, Goodenough, & Cox, 1977; Witkin, Moore, Owen, Raskin, Oltman, Goodenough, & Friedman, 1977). For example, field-dependent students are more likely to graduate from college with majors in education, while independents are likely to choose the sciences.

5. Field dependence is more typical of females than males, so the EFT may be used in studying the basis of sex differences in perceptual and cognitive styles.

6. Incomplete research suggests the EFT could aid in understanding the roles of day residues and stress in dream formation and recall (e.g., Goodenough, Witkin, Lewis, Karlack, & Cohen, 1974).

7. Physiological psychologists may find uses for the EFT in understanding arousability, autonomic control, cerebral dominance, and similar variables (see Cox & Gall, 1981).

Other topics to which the EFT has been related include aesthetic preferences, learning and instructional preferences, expression of aggression, and sex role behavior, among others (Cox & Gall, 1981).

As the EFT is an individual test it must be administered one-on-one. The examiner need not be extensively trained; the only skills needed are in using a stopwatch and being able to recognize the simple figure when successfully traced. If the client works for 3 minutes without finding the figure, the search is interrupted and the next trial started. Total time for Form A or B may reach 45 minutes, but for the fastest subjects 10 minutes will suffice.

To insure that the examinee takes in the complex figure, it is presented alone for 15 seconds and the subject describes it. Then the examiner covers it with the simple figure for 10 seconds to allow the subject to absorb the shape of the simple figure. When the complex card has again been uncovered the search begins, but with the simple figure card out of sight. They are never visible together, as the disembedding is to take place in the subject's mind.

An investigation by Jackson, Messick, and Myers (1964) has made it clear that there are practice effects, so that item statistics depend on where in the test the item appears. This was true when calculating the proportion of subjects who successfully disembedded that item and for average search time. Hence, one who wishes to use the norms in the manual must take care not to alter the item order.

Scoring is even easier than administering the test. By adding up the search times and dividing by 12, the average time for successful disembedding is produced. If the simple figure was not found, the default time of 3 minutes is assigned.

The distribution of scores approximates normality. Hence, a client's score can easily be converted to standard scores by using the means and standard deviations reported in the manual. As there are strong age and sex differences in EFT performance, scores have to be matched with the corresponding means in the norm table. Slightly different procedures are recommended for young children, though again this reviewer would recommend using the CEFT for 10- to 12-year-olds rather than the EFT.

Much of the information about the meaning of field dependence-independence remains tentative. Interpretation as to the degree of dependence is straightforward given the standard score, but interpretation of the meaning of degree of dependence requires a thorough knowledge of the experimental literature. Even then, interpretation and prediction are easy only for extreme scores or in a few selected settings (e.g., in educational guidance).

Technical Aspects

The EFT has good internal-consistency reliability. The scores on the odd items typically correlate about .80 with the scores on the even items with a range of .61 to .93 in several studies. Test-retest stability has been reported at .89 for an interval of more than three years (Bauman, 1951; see the Manual) and .92 for a one-week interval (Witkin et al., 1966).

While these results are good, there are some factors which introduce doubt. Consistency for females almost always falls lower than for males (the average difference reported in the manual is about .06). A second point arises from the studies of developmental influences on scores. Clearly scores change, sometimes very significantly (e.g., around age 12, and then in the opposite direction around age 40). Learning effects have also been well documented. Taking all the data together, however, a fair conclusion is that the EFT measures reliably. The two forms, A and B, also correlate well, about .80.

The real validity question concerns convergent validity. How well do scores on the EFT converge on, or measure the same thing as, scores from the other measures of field dependence? Extensive information is available; to summarize it briefly, the EFT has greater validity for males than for females. It correlates significantly with the RFT3, averaging about .45 to .60, and with the BAT, averaging about .35 to .50. The EFT correlates very highly with the GEFT, ranging between .60 and .85; with the tactile EFT at .78; and with the auditory EFT at .63. Correlations with Body Drawing are lower, sometimes at zero or less, and with the TR-TC coefficients range from .20 to .45.

In summary, the EFT obviously cannot be considered a parallel form of the RFT, the BAT, and the others. It differs from them in several ways, including its reliance on the visual and its emphasis on successful performance. Users should be impressed that it correlates as well as it does, especially with the RFT where correlations occasionally reach into the .80s. The overall pattern gives this reviewer confidence that the EFT does tap the field-dependence dimension, and hence may be used to access the vast literature on psychological differentiation (see also Goodenough, 1976; Witkin & Goodenough, 1977).

Critique

The EFT can take its place near the front of the line of tests that measure field dependence-independence. It requires no special apparatus, is easy and quick to administer and score, gives reliable, stable results, and correlates well with the other leading tests of differentiation. A very large body of relevant literature has accumulated, numbering around 4,000 articles (though one of the peculiarities of this literature is the disproportionately large number of master's and doctor's theses).

The manual accompanying the EFT is now 13 years old and in need of revision. It should be expanded to review more of the recent literature, and the inclusion of more normative data would be very useful.

Overall, the EFT can be considered a solid test with many immediate and potential applications.

References

Cox, P.W., & Gall, B.E. (1981). *Field dependence-independence and psychological differentiation* (Supplement No. 5). Princeton, NJ: Educational Testing Service.

Goodenough, D.R. (1976). The role of individual differences in field dependence as a factor in learning and memory. *Psychological Bulletin, 83,* 675-694.

Goodenough, D.R., Witkin, H.A., Lewis, H.B., Karlack, D., & Cohen, H. (1974). Repression, interference and field dependence as factors on dream forgetting. *Journal of Abnormal Psychology, 83,* 32-44.

Ihilevich, D., & Gleser, G.C. (1971). Relationship of defense mechanisms to field dependence-independence. *Journal of Abnormal Psychology, 77,* 296-302.

Jackson, D.N. (1956). A short form of Witkin's Embedded Figures Test. *Journal of Abnormal and Social Psychology, 53,* 254-255.

Jackson, D.N., Messick, S., & Meyers, C.T. (1964). Evaluation of group and individual forms of embedded-figures measures of field-independence. *Educational and Psychological Measurement, 24,* 177-192.

Messick, S. (1980). H.A. Witkin (obituary). *American Psychologist, 35,* 99-100.

Witkin, H.A. (1949). The nature and importance of individual differences in perception. *Journal of Personality, 18,* 145-170.

Witkin, H.A. (1950). Individual differences in ease of perception of embedded figures. *Journal of Personality, 19,* 1-15.

Witkin, H.A., Dyk, R.B., Faterson, H.F., Goodenough, D.R., & Karp, S.A. (1962). *Psychological differentiation.* New York: John Wiley & Sons.

Witkin, H.A., Faterson, H.F., Goodenough, D.R., & Birnbaum, J. (1966). Cognitive patterning in mildly retarded boys. *Child Development, 37,* 301-316.

Witkin, H.A., & Goodenough, D.R. (1981). *Cognitive styles: Essence and origins.* New York: International Universities Press.

Witkin, H.A., Goodenough, D.R., & Oltman, P.K. (1979). Psychological differentiation: Current status. *Journal of Personality and Social Psychology, 37,* 1127-1145.

Witkin, H.A., Lewis, H.B., Hertzman, M., Machover, K., Meissner, P.B., & Wapner, S. (1972). *Personality through perception: An experimental and clinical study.* Westport, CT: Greenwood Press Publishers. (Original work published 1954)

Witkin, H.A., Moore, C.A., Goodenough, D.R., & Cox, P.W. (1977). Field-dependent and field-independent cognitive styles and their educational implications. *Review of Educational Research, 47,* 1-64.

Witkin, H.A., Moore, C.A., Owen, D.R., Raskin, E., Oltman, P.K., Goodenough, D.R., & Friedman, F. (1977). Role of the field-dependent and field-independent cognitive styles in academic evolution: A longitudinal study. *Journal of Educational Psychology, 69,* 197-211.

Witkin, H.A., & Oltman, P.K. (1967). Cognitive style. *International Journal of Neurology, 6,* 119-137.

Witkin, H.A., Oltman, P.K., Raskin, E., & Karp, S.A. (1971). *A manual for the Embedded Figures Test.* Palo Alto, CA: Consulting Psychologists Press, Inc.

Robert Drummond, Ph.D.

Interim Chairperson, Division of Educational Services and Research, University of North Florida, Jacksonville, Florida.

EYSENCK PERSONALITY INVENTORY

H. J. Eysenck and Sybil B. G. Eysenck. San Diego, California: Educational and Industrial Testing Service.

THE ORIGINAL OF THIS REVIEW WAS PUBLISHED IN TEST CRITIQUES: VOLUME II (1985).

Introduction

The Eysenck Personality Inventory (EPI) is an objective 57-item paper-and-pencil inventory that measures Extraversion-Introversion and Neuroticism-Stability. There is also a Lie (or falsification) Scale that provides for detection of response distortion. The test items are in the form of questions to which the examinee gives a "Yes" or "No" answer.

The authors of the test are H. J. Eysenck and Sybil B. G. Eysenck. H. J. Eysenck is a prominent British psychologist who has contributed significant books and tests on dimensions of personality and personality measurement, as well as having authored several personality inventories. He served as Professor of Psychology at the Institute of Psychiatry at the University of London and was director of the psychology departments at Maudsley and Bethlehem Royal Hospitals.

Sybil B. G. Eysenck also has coauthored books and tests on personality with her husband as well as having authored the Junior Eysenck Personality Inventory.

The EPI was based on the theories of research on extraversion and neuroticism that H. J. Eysenck began in the 1940s. EPI is a refinement of the Maudsley Personality Inventory (MPI) (Eysenck, 1962) used initially to measure the constructs. The EPI is similar to the MPI but improved. The items on the EPI were carefully reworded in order to make them more understandable by subjects of low intelligence and/or education. A Lie Scale was developed for the test in order to help identify subjects who put themselves in a very positive or desirable light. The MPI did not have such a scale. The construction of the test led to increased reliability and validity of the scale over the MPI.

There are three forms of the EPI, Forms A & B and an Industrial Form A1, which are also available in Spanish from the publisher. In addition, microcomputer software programs are available for Forms A & B for the Commodore PET 4000 or 8000 series.

The EPI consists of 57 questions in which the respondent is asked to mark "Yes" or "No." For example, an item might read: "Would you call yourself a calm person?" The items are presented in two columns on the back side of the test booklet. The directions are on the front side of the inventory. The test is self-administered, but the examiner can read the instructions to the examinee or group if necessary.

The test is designed for high school and college students and adult populations.

The readability level is, in general, appropriate for adults but the language is somewhat dated.

There are three subscales on the test: an Extraversion-Introversion Scale consisting of 24 items, a Neuroticism Scale consisting of 24 items, and a Lie Scale consisting of nine items.

The test booklet is consumable and serves as the answer sheet. Examinees record their answers in the spaces provided under the yes/no statements by each item. There are percentile norms available for American college students. Computerized scoring is available from the publisher and scores can be compared to different age, sex, and occupational groups.

Practical Applications/Uses

Eysenck reports the use of the test in employee selection and placement, educational guidance and counseling, in clinical diagnosis, and for experimental studies. The test is short, reliable, and a valid measure of two major dimensions of personality. It is convenient to use in research studies and in undergraduate and graduate classes in personality theory or personality assessment to illustrate Eysenck's theory. Overall, this reviewer has found that other tests, such as the California Psychological Inventory or the MMPI, provide more useful information to assist diagnosis in clinical situations. Such tests as the Myers-Briggs Type Indicator give a fuller picture of the extraversion/introversion dimension.

The test, however, can be used by the counselor or the psychologist as a quick screen of the two personality dimensions assessed by the inventory because it can be given and scored within 10 to 15 minutes.

The test measures Extraversion/Introversion and Neuroticism. Extraversion on the EPI refers to the outgoing, uninhibited, impulsive, and sociable inclinations of a person (Eysenck & Eysenck, 1968a, p. 5). Neuroticism on the EPI is defined as emotional overresponsiveness and liability to neurotic breakdown under stress (Eysenck & Eysenck, 1968a, p. 5). The two dimensions of personality were found to be independent of each other.

The test has been used in educational settings for guidance and research purposes. Both extraversion and neuroticism have been found to be related to academic attainment. The EPI has been used extensively for clinical diagnostic purposes and found to differentiate among different types of psychopathologies.

The EPI has been widely used in many contexts, including to study personality structure and correlates, because it is a short, reliable, and valid inventory. It has been used in schools, industrial settings, mental hospitals, and clinics by psychologists, counselors, researchers, and college professors. However, overall, this reviewer does not see any major new uses of the EPI because it is dated and has been superseded by the Eysenck Personality Questionnaire (Eysenck & Eysenck, 1975).

The EPI is appropriate for use with high school and college students and adults with normal intelligence or above, as well as those with lower educational levels. It is appropriate for use with different clinical populations. Data were collected from the EPI of different neurotic and psychotic groups and normal individuals from different occupational levels. American college students were also assessed. The test would be inappropriate for certain types of handicapped adults, such as the visually handicapped and some physically handicapped, because it would have to be orally administered. Additionally, the small-sized print might cause some adults problems in reading the items.

The test can be either administered individually or in large groups. The instructions are printed in full on the front side of the questionnaire. Normally, the administrator is directed to read these directions aloud to the subjects. Because it is a personality test, it would be best administered by a psychologist or counselor, especially when questions arise in the test session. It is a simple test to administer and the directions take only a minute or two to read. If both the test and directions are read by the examiner, the administration time is roughly about ten minutes.

The test takes approximately 30 seconds to handscore. There are three hand overlay stencils for scoring purposes. They are placed on top of the answer sheet and one point is given for each blackened answer space showing through the holes. The test can also be scored on the microcomputer.

The interpretation is based on the objective scores derived from the scoring of the test. There are norms presented for American college students only. Although sex differences have been found on the EPI scales, only combined sex groups are presented. Although means and standard deviations are presented for different types of occupational and clinical groups, the number and description of these groups are limited. In order to interpret the test correctly, the EPI manual, as well as some of the other books on personality that Eysenck has written on his personality theory and assessment (e.g., Eysenck, 1952, 1970), should be read carefully. Overall, one of the major problems is the limited number of reference groups and norm tables available.

One needs to note the definitions of the constructs measured. Although many tests measure constructs such as introversion/extraversion, not all test authors define the constructs in the same way. It is important for the user to read the items belonging to each scale.

Technical Aspects

Extensive validity and reliability studies have been conducted on the EPI. Test-retest and split-half reliabilities are reported on the scale.

The test-retest reliabilities range from .84 to .94 for the complete test and between .80 and .97 for the separate scales (Eysenck & Eysenck, 1968a, p. 15). The split-half reliabilities, A vs. B for 1,655 normals, 210 neurotics, and 90 psychotics, range from .74 to .91. No reliability information is presented for the Lie Scale. Eysenck suggests that if the test is to be used for individual decisions, both forms should be used, whereas, if it is to be used for experimental studies, one form would be sufficient. Overall, the Extraversion and Neuroticism Scales have fairly substantial reliability for personality tests.

Eysenck provides information on the factorial and construct validity of the test as well as the concurrent validity. Much evidence has been collected of the existence and orthogonality of the Extraversion and Neuroticism dimensions. The scales differentiate between different types of neurotics and psychotics and normals. Correlation studies were reported between the EPI and Taylor Manifest

Anxiety Scale, Cattell's IPAT Anxiety Scale, the Multiple Affect Adjective Check List, and the California Personality Inventory. The correlational studies provide some evidence of the concurrent validity of the scales on the EPI. Eysenck has conducted extensive factor analytic studies which demonstrate the independence of the Extraversion and Neuroticism diversions (Eysenck & Eysenck, 1968b).

The test has been studied by other researchers and psychologists and, in general, there is support for the validity and reliability of the scale. The manual of the EPI does not contain emperical studies comparing the EPI and the Maudsley Personality Inventory, the test from which the EPI was generated.

Critique

Lingoes (1965), Cline (1972), Lanyon (1972), and Tellegen (1978) have presented critical reviews of the EPI in editions of the *Mental Measurements Yearbooks.*

Tellegen (1978) and Lanyon (1972) question the conceptualization of the two dimensions of the test and the value of the narrow conceptualization. Lingoes (1965) and Lanyon (1972) have problems evaluating the validity and usability of the Lie Scale. Tellegen (1978) has identified problems with the dimension framework, provided by the EPI of the two constructs.

One major area of criticism is the manual. Tellegen (1978) criticizes the test for not discussing explicitly the content and conceptualization of each scale. Lanyon (1972) feels there is a deficiency in the reporting of norms for the test.

Overall, the critics feel that the EPI is psychometrically well constructed but probably of limited value to practitioners. This criticism is based partially on the conceptualization of constructs measured and partially on the deficient information reported in the manual on the interpretation of the test. They feel that the manual ought to be updated and improved by including better normative and interpretative information as well as a more complete review of the research conducted using the EPI. The test is perceived as a gross screening device or a test that might supplement the information from other more appropriate and global measures such as the MMPI and CPI.

This reviewer has used the EPI over a 20-year period for clinical, educational, and research uses. Adults tend to have a negative reaction to the items as well as to the test format. Most of the items on the Neuroticism Scale focus on negative behaviors and this stimulates discussion and a negative reaction to the test. Examinees do not like the appearance of the test, the size of print, and the language used in phrasing the questions. The items on the Lie Scale are of questionable value. The EPI is easy to administer and score and has sufficient reliability for research purposes. The test has been a good instrument to illustrate Eysenck's theory of personality in psychology classes and to stimulate discussion on the structure of personality.

Generally in counseling I have found global measures of personality such as the 16PF, CPI, Myers-Briggs, and MMPI to be more useful in diagnoses and assessment. Nevertheless, the EPI still has value as a research tool and a supplemental instrument, especially if the user reads the test manual and booklet carefully and understands the conceptualization of the constructs measure.

References

Cline, V. B. (1972). The Eysenck Personality Inventory. In O. K. Buros (Ed.), *The seventh mental measurements yearbook* (pp. 161-163). Highland Park, NJ: The Gryphon Press.

Eysenck, H. J. (1952). *The scientific study of personality.* London: Routledge and Kegan Paul.

Eysenck, H. J. (1962). *The manual of the Mandsley Personality Inventory.* San Diego: Educational and Industrial Testing Service.

Eysenck, H. J., & Eysenck, S. B. (1968a). *The manual of the Eysenck Personality Inventory.* San Diego: Educational and Industrial Testing Service.

Eysenck, H. J., & Eysenck, S. B. (1968b). *Personality structure and measurement.* San Diego, CA: Knapp.

Eysenck, H. J. (1970). *The structure of human personality.* London: Methuen.

Eysenck, H. J., & Eysenck, S. B. G. (1975). *Eysenck Personality Questionnaire.* San Diego: Educational and Industrial Testing Service.

Eysenck, S. B. G. (1965). *Junior Eysenck Personality Inventory.* San Diego: Educational and Industrial Testing Service.

Lanyon, R. I. (1972). The Eysenck Personality Inventory. In O. K. Buros (Ed.), *The seventh mental measurements yearbook* (pp. 163-164). Highland Park, NJ: The Gryphon Press.

Lingoes, J. C. (1965). The Eysenck Personality Inventory. In O. K. Buros (Ed.), *The sixth mental measurements yearbook* (pp. 215-217). Highland Park, NJ: The Gryphon Press.

Tellegen, A. (1978). The Eysenck Personality Inventory. In O. K. Buros (Ed.), *The eighth mental measurements yearbook* (pp. 802-804). Highland Park, NJ: The Gryphon Press.

Alan F. Friedman, Ph.D.

Assistant Professor of Psychology, Northwestern University Medical School, Chicago, Illinois.

EYSENCK PERSONALITY QUESTIONNAIRE

H. J. Eysenck and S. B. G. Eysenck. San Diego, California: Educational and Industrial Testing Service.

THE ORIGINAL OF THIS REVIEW WAS PUBLISHED IN TEST CRITIQUES: VOLUME I (1984).

Introduction

The Eysenck Personality Questionnaire (EPQ) (Eysenck & Eysenck, 1975) is the latest in a series of questionnaires designed to measure the major factors of normal and abnormal personality. Based on a factor-analytic and experimental research program of about 40 years duration, the EPQ consists of scales for measuring Psychoticism (P), Extraversion (E), and Neuroticism (N), and contains a Lie (L) scale to ascertain the validity of an individual's scores. The EPQ can be used with persons aged 16 and older. The Junior EPQ is appropriate for ages 7 to 15. The adult EPQ contains 90 statements (81 for the Junior EPQ) to be answered "yes" or "no." The subject can respond directly to the items on the questionnaire or the items can be given in an interview format.

The EPQ was developed by H. J. Eysenck and S. B. G. Eysenck who are, respectively, Professor and Senior Lecturer in Psychology at the University of London Institute of Psychiatry. Professor Eysenck is one of the most productive and influential figures in psychological research. His studies have included factor-analytic investigations of normal and clinical populations in an attempt to identify the major personality factors, as well as laboratory and applied experiments to establish correlates of the factors and to validate treatment programs.

The EPQ was preceded by two questionnaires developed by Eysenck: the Maudsley Personality Inventory (Eysenck, 1959) and the Eysenck Personality Inventory (EPI) (Eysenck & Eysenck, 1968). These questionnaires measured Eysenck's two most researched personality dimensions, E and N. These personality scales were revised on the EPQ and the P scale was added as a result of more recent research on this dimension (Eysenck & Eysenck, 1976).

Practical Applications/Uses

The EPQ was designed to measure the primary factors of personality, identified as P, E, and N. These factors are generally orthogonal (i.e., separate and unrelated

---

This reviewer wishes to acknowledge and thank James A. Wakefield, Jr., Ph.D., Professor of Psychology, California State University-Stanislaus, and Nancy A. Goad, M.A., school psychologist, Mono County Office of Education, Bridgeport, California, for their significant contributions to the writing of this review.

to each other) and account for most of the nonintellectual personality variance in a wide variety of subject populations that have been investigated (Kline & Barrett, 1983). While these factors are generally uncorrelated with IQ, they have been shown to be related to a variety of learning, attentional, and school achievement variables (Eysenck, 1976).

High scores on the P scale are related in part to empathy defects, aggressive and hostile traits, a preference for liking odd or unusual things, impulsivity, and general socialization difficulties. *Schizoid, psychopathic,* and *behavior disorders* are psychiatric terms often used to describe high scorers on this dimension. Low scorers on P are seen as empathic, sensitive, and not hostile or aggressive toward others. They are generally well socialized and have few, if any, problem behaviors.

High scorers on the E scale are described as extraverts with tendencies to seek out others and enjoy their company. Others see extraverts as likable, but sometimes too loud and outgoing. It appears that the E scale on the EPQ is more a measure of sociability and less a measure of impulsivity than was the E scale on the EPI (Rocklin & Revelle, 1981). Low scorers on this scale are referred to as introverts and are more reserved than extraverts. They tend to work more slowly and accurately and to persist longer at tasks than do extraverts.

The N scale reflects the principal dimension of emotional normality-abnormality. Those scoring high on N are generally nervous, maladjusted, and overemotional. Their emotional overresponsiveness may dispose them to develop neurotic disorders under stress.

The EPQ is not limited to a particular population of subjects. It can be administered to groups, individuals, or even read to subjects when necessary (e.g., for individuals with reading deficits or for telephone surveys). Less than 30 minutes are required for administration. Hand scoring keys are available, which are easy to use, and scoring takes only about 10 minutes. Machine scoring and reports for both adult and junior forms are available through PsychSystems of Baltimore. Interpretations of test results are based on objective scoring; however, implementing effective clinical interventions based on these scores will depend on the clinician's knowledge of personality theory and research.

In a recent attempt to implement clinical interventions based on the results of research on Eysenck's personality factors, Wakefield (1979) has generated prescriptive-teaching and behavioral-control strategies for school children covering all the combinations of P, E, and N. Although the EPQ has not been as popular as the MMPI, the CPI, or the 16PF, the wealth of research, ease of administration, and the readily available clinical intervention strategies keyed to the EPQ should contribute to its future increase in popularity.

Technical Aspects

Test-retest reliabilities and internal consistencies for several samples, including students, social workers, normal adults, and prisoners, are reported in the manual. Over one month, the overall test-retest reliabilities were .78 (P), .89 (E), .86 (N), and .84 (L). Internal consistencies were also in the .70s for P and the .80s for the other three scales. The reliabilities for the Junior EPQ were somewhat lower, especially below about age 9. In general, these reliabilities indicate that the variance of the scales ranges from approximately 75% consistent (for P) to approximately 90% consistent (for E). These figures are much better than typical reliabilities for projectives, interviews, and behavioral ratings, and compare favorably with reliabilities of other much longer objective personality questionnaires such as the MMPI and the 16PF.

Normative information for the EPQ is also presented in the manual. Subjects for the adult norming of the EPQ were drawn from the United Kingdom and included most social classes. The manual indicates that data based on American samples are presented in separate supplements although no specific reference is given. The United Kingdom norming of the adult P, E, and N scales included 2,312 men and 3,262 women. The L scale was normed on 1,624 men and 2,462 women; this is due to the fact that a number of subjects were tested on a version of the questionnaire not containing the L scale. The L scale sample, however, does not differ significantly from the previous sample in composition or mean scores.

The manual reports an age trend for P, E, N, and L with men having a higher P score than women and both sexes showing a decline with age. On the E scale, both men and women become more introverted with age. For women, however, this trend is much less pronounced and by age 50 women's E scores were higher than those of men. On the N scale, women had higher scores. Both men and women declined on N with age at a similar rate. Women also had higher L scores. Both sexes showed a rapid increase on L with age. Information on the effects of social class and occupation on P, E, N, and L are included, but the manual indicates that more work is necessary to establish the relationships among these variables.

Information about abnormal groups such as psychotics, neurotics, endogenous depressives, and prisoners is also included in the manual. Psychotics and prisoners were found to have the highest P scores while the scores of drug addicts, alcoholics, and individuals with personality disorders and sex problems were also elevated. Endogenous depressives, while their P scores were higher than the normal population, were well below the other groups on P. Psychotic groups and criminals who were classified as neurotics and personality-disordered were found to have elevated N scores. Psychiatric groups were found to have elevated L scores, but prisoners did not, suggesting that those subjects in the psychiatric group (both psychotics and neurotics) would have higher P and N scores had they not dissimulated.

Subjects for the norming of the Junior EPQ included over 3,000 children from different kinds of schools and regions of the U.K. No obvious age trend was reported for P, although both boys and girls in the 10- and 11-year-old age category were lower than those children of other ages. This difference, however, was not large. E scores increased with age, with girls showing the more rapid increase. N scores increased only for girls while the scores for boys were irregular. Both boys and girls showed a marked decline with age on their L scores. Sex differences on the Junior EPQ were similar to those found on the adult EPQ, with boys having higher P and E scores and girls higher N and L scores.

The construct validity of the EPQ scales has been extensively investigated against behavioral, emotional, learning, attentional, and therapeutic criteria. P is related to aggression, impulsiveness, sensation-seeking, and tough-minded attitudes, among others. E is related to sociability, global response style, and responsiveness to reward/punishment, among others. N is related to worry, anxiety, depression, and emotional reactivity, among others. L measures dissimulation. The following present more specific validation studies: Eysenck (1967) for E and N; Eysenck and Eysenck (1976) and Friedman et al. (1976) for P and L; Eysenck (1976) for original technical articles; Wakefield and Goad (1982) for a less technical overview; Wakefield (1979) for practical suggestions for interventions; Friedman et al. (1983) and Eysenck et al. (1983) for discussions of Eysenck's factors in connection with the MMPI and the DSM-III.

Critique

The EPQ is a short questionnaire designed to measure three fundamental dimensions of personality. Research on the EPQ and its predecessors is voluminous and has covered several decades, while its use as a clinical instrument is relatively recent. Although practical and technical characteristics of the EPQ are strong, other personality tests with more scales have generally been more popular with clinicians. Empirically supported interpretive material and intervention strategies keyed to the EPQ are readily available.

References

Anderson, K. A., & Revelle, W. (1983). The interactive effects of caffeine, impulsivity and task demands on a visual search task. *Personality and Individual Differences, 4,* 127-134.

Eysenck, H. J. (1959). *The Maudsley Personality Inventory.* London: University of London Press. (Also published by Educational and Industrial Testing Service, Inc. [EdITS], San Diego, 1962)

Eysenck, H. J. (1967). *The biological basis of personality.* Springfield, IL: Charles C. Thomas.

Eysenck, H. J. (1976). *The measurement of personality.* Baltimore: University Park Press.

Eysenck, H. J., & Eysenck, S. B. G. (1968). *The manual of the Eysenck Personality Inventory.* San Diego: Educational and Industrial Testing Service, Inc.

Eysenck, H. J., & Eysenck, S. B. G. (1975). *Manual: Eysenck Personality Questionnaire (Junior & Adult).* San Diego: Educational and Industrial Testing Service, Inc.

Eysenck, H. J., & Eysenck, S. B. G. (1976). *Psychoticism as a dimension of personality.* London: Hodder & Stoughton.

Eysenck, H. J., Wakefield, J. A., Jr., & Friedman, A. F. (1983). Diagnosis and clinical assessment: The DSM-III. *Annual Review of Psychology, 34,* 167-193.

Friedman, A. F., Gleser, G. C., Smeltzer, D. J., Wakefield, J. A., Jr., & Schwartz, M. S. (1983). MMPI overlap scales for differentiating psychotics, neurotics, and nonpsychiatric groups. *Journal of Consulting and Clinical Psychology, 51,* 629-631.

Friedman, A. F., Wakefield, J. A., Jr., Boblitt, W. E., & Surman, G. (1976). Validity of psychoticism scale of the EPQ. *Psychological Reports, 39,* 1309-1310.

Kline, P., & Barrett, P. (1983). The factors in personality questionnaires among normal subjects. *Advances in Behavior Research and Therapy, 5,* 141-202.

Rocklin, T., & Revelle, W. (1981). The measurement of extraversion: A comparison of the EPI and the EPQ. *British Journal of Social Psychology, 20,* 279-284.

Wakefield, J. A., Jr. (1979). *Using personality to individualize instruction.* San Diego: Educational and Industrial Testing Service, Inc.

Wakefield, J. A., Jr., & Goad, N. A. (1982). *Psychological differences: Causes, consequences, and uses in education and guidance.* San Diego: Educational and Industrial Testing Service, Inc.

Jean Powell Kirnan, Ph.D.

Assistant Professor of Psychology, Trenton State College, Trenton, New Jersey.

Kurt F. Geisinger, Ph.D.

Associate Professor and Chairperson of Psychology, Fordham University, Bronx, New York.

GENERAL APTITUDE TEST BATTERY

U.S. Employment Service. Washington, D.C.: United States Department of Labor.

THE ORIGINAL OF THIS REVIEW WAS PUBLISHED IN TEST CRITIQUES: VOLUME V (1987).

Introduction

The General Aptitude Test Battery (GATB) was developed by the United States Employment Service (USES) in response to the growing need for a comprehensive instrument for use in occupational counseling. Utilizing tests that measure a variety of basic aptitudes, the GATB evaluates an applicant's ability to perform successfully in a wide variety of occupations. As such, its introduction in 1947 was welcomed by job counseling and selection professionals faced with the task of placing the growing labor market of the post-World War II era.

Since its inception in 1947, the GATB has undergone continuous revision and modification. Its development spans a 50-year period as today it continues to be the object of considerable research. The groundwork for the GATB was laid in the mid-1930s when the development of tests of aptitudes and skills was widespread among psychologists and others. Tests that could quickly and accurately match applicant abilities with job requirements were greatly needed both during World War II and in the years following. Both the Minnesota Stabilization Research Institute and the Occupational Research Program in the USES pioneered this research. By the 1940s, over 100 aptitude tests had been developed for use in placing workers. However, this growing number of tests made the task of efficient and accurate placement almost impossible for general counseling purposes. In order to be counseled for more than one occupation, individuals seeking vocational guidance would have to be administered several tests. Researchers recognized that similarities existed in the factors being measured by the varying tests as well as the basic aptitudes required by different occupations. Factor-analytic techniques then were used to identify basic aptitudes and to reduce the number of tests needed to measure them.

The factor analysis performed by USES was conducted during the period 1942 to 1944 and employed several experimental groups, totalling 2,156 individuals. Each of these individuals had taken some combination of 59 tests—54 of which were Employment Service tests representative of the 100 tests then in use by the Service. While all subjects were male and applicants for or incumbents in defense training courses, the sample was otherwise representative of the Employment Service's counselees in terms of age, education, and experience.

163

The analysis identified ten factors: Intelligence, Verbal Aptitude, Numerical Aptitude, Spatial Aptitude, Form Perception, Clerical Perception, Aiming, Motor Speed, Finger Dexterity, and Manual Dexterity. The tests to measure these factors were selected on the basis of internal or factorial validity (magnitude of the factor loading) as well as external or practical validity (based on a review of previously conducted criterion-related validity studies in the occupations). It must be noted, however, that the criteria of occupational success are not provided in the manual. It is also not reported whether the studies were concurrent or predictive in nature. The resulting 11 paper-and-pencil tests and four apparatus or performance tests comprised the first form of the GATB, B-1001. Thus, from over 100 available tests, 15 were chosen that in a single administration could measure the basic aptitudes required of most of the jobs for which USES employed ability tests. The aptitudes and the tests that measure them are provided in Table 1.

A second version of the instrument, B-1002 Form A, was introduced in 1952. Several enhancements were made in this revision: B-1002 provided examinees with a separate answer sheet (for this reason, items in B-1001 that were not in multiple-choice format were revised accordingly), new experimental items were added, provision was made for an alternate form (Form B), and new time limits were set to accommodate the new items as well as the use of a separate answer sheet. The new items allowed for the replacement of current B-1001 items that were no longer effective as well as the construction of the alternate form. In order to accomplish all the above, an experimental test consisting of the revised B-1001 items, new items, and a separate answer sheet was administered under untimed conditions to 10 samples of approximately 200 subjects each. The tests were administered in an untimed fashion to help establish new time limits and to facilitate the subsequent item analysis by ensuring that each examinee had the opportunity to attempt every item.

The final items for the B-1002 forms were selected on the basis of item difficulty and discriminability. Alternate forms were made as equivalent as possible. Several easy items were included in the beginning of each test to allow the examinee to "warm-up," and all items were arranged in order of increasing difficulty. Efforts were made to reduce testing time by eliminating tests that were redundant. Three tests—Two-Dimensional Space, H Markings, and Speed—were dropped from the original B-1001. These were eliminated for one of two reasons: either they were not weighted heavily in determining an aptitude score, or ongoing validity studies had shown that they did provide incremental validity. The revision also resulted in the elimination of one aptitude, Eye/Hand Coordination, which was found to be highly correlated with and thus redundant of another aptitude, Motor Speed. With this change, Motor Speed was renamed Motor Coordination.

Thus, the current version of the GATB assesses nine aptitudes through the use of 12 tests, as listed in Table 2.

Numerous variations on the answer sheet were to follow the revision to B-1002. Typically, these were minor changes to allow for machine scoring on different optical scanning equipment. In every case, comparability studies were conducted to ensure the new answer sheets did not affect applicants' scores.

No information on the development of Forms C and D, the most recent versions of the battery, were included in the development manual (Section III) provided by USES. It certainly must be hoped that similar comparative studies were conducted

Table 1

Original Version of the GATB

Aptitude	*Test*
Intelligence	Three-Dimensional Space
	Vocabulary
	Arithmetic Reasoning
Verbal Aptitude	Vocabulary
Numerical Aptitude	Computation
	Arithmetic Reasoning
Spatial Aptitude	Two-Dimensional Space
	Three-Dimensional Space
Form Perception	Tool Matching
	Form Matching
Clerical Perception	Name Comparison
Aiming or	H Markings
Eye/Hand Coordination	Mark Making
Motor Speed	Speed
	Mark Making
Finger Dexterity	Assemble and Disassemble
Manual Dexterity	Place and Turn

to equate these new versions with Forms A and B. A subsequent research report (Kolberg, 1977) states that Forms C and D had been developed as alternates in response to a liberal retesting policy in local Employment Service offices.

Initially, norms were established based on test results of 519 employed workers who took the GATB. However, it was necessary to update these data because the sample of 519 did not include an adequate sampling of occupations. In 1952, a special stratified sample of 4,000 cases was extracted from 8,000 records of job incumbents tested while they were working in a wide variety of occupations. These were selected in a stratified procedure to be representative of the 1940 U.S. census data of the working population. The primary variable in the stratification was Occupation. Sex, age, and geographic location were also considered as factors. Because of the changing role of women in the work force—both during World War II and in the years following—it was arbitrarily decided that the sample be half male and half female. The geographic distribution of the 4,000 cases was found to differ significantly from the U.S. census data; the North Central region was overrepresented because there was a high degree of diversification of occupations as well as many

Table 2

Current Version of the GATB

Aptitude	*Test*
Intelligence	Three-Dimensional Space Vocabulary Arithmetic Reasoning
Verbal Aptitude	Vocabulary
Numerical Aptitude	Computation Arithmetic Reasoning
Spatial Aptitude	Three-Dimensional Space
Form Perception	Tool Matching Form Matching
Clerical Perception	Name Comparison
Motor Coordination	Mark Making
Finger Dexterity	Assemble and Disassemble
Manual Dexterity	Place and Turn

large centralized industries in that region. The authors, however, did not consider these to be of sufficient magnitude to threaten the usefulness of the data.

Thus, norms for the GATB version B-1001 were calculated on the sample of 4,000. When B-1002 Form A was introduced, new conversion tables were needed. The formulae for these tables were derived from a study of 585 high school and junior college students tested in three different states. Form B of B-1002 also has separate conversion tables; again, these were derived from a study of 412 high school juniors and seniors in two states. In 1966, an analysis of data collected in a variety of studies (1950-1966) reaffirmed the stability of these norms.

Only 2 of the 10 aptitudes in B-1001 were based on a single test score. The other eight were derived via weighted linear combinations of scores. The justification for combining these measures was obtained from the earlier factor analysis conducted on the sample of 2,156 examinees, as well as the initial normative sample of 519. Factor loadings and intercorrelations among the tests were used to determine how these tests would be weighted additively to arrive at a single aptitude score. Data from the normative sample were also used to aid in standardizing these scores so that each of the resulting GATB aptitudes had a mean of 100 and a standard deviation of 20.

Similarly, conversion tables were needed to convert raw scores to aptitude scores for the subsequent versions of the GATB. Normative data from high school comparability studies cited above were used to derive equivalent aptitude scores for B-1002 Forms A and B.

Not all aptitudes are required in every occupation. Therefore, the identification of those aptitudes necessary for a given occupation and the determination of the predicted levels of competency was required. Groups of aptitudes essential for a given occupation were combined and called Specific Aptitude Test Batteries or SATBs. A total of more than 450 SATBs have been developed, each composed of two, three, or four GATB aptitudes.

Aptitudes were evaluated as being important to job performance (and thus selected for a SATB) on the basis of job analysis information and statistical data described below. When possible, predictive validity studies were used. When this type of design was not feasible, the concurrent approach was employed. The data for these analyses were collected at various state employment offices and consolidated for analysis at the national office. The statistical data considered in addition to the job analysis information were 1) the three aptitude tests yielding the highest mean scores within an occupation, 2) aptitude tests with low standard deviations (evidence of relative homogeneity), and 3) aptitudes demonstrating significant correlations ($p < .05$) with the criterion (job or training performance).

Once selected, the aptitudes were used to arrive at "trial norms," which were evaluated in the sample for their ability to discriminate "successful" and "unsuccessful" workers. The resulting norms were subsequently cross-validated on new occupational samples.

To calculate all SATBs for any one applicant would be extremely time-consuming; on the other hand, computation of a select few would be too specific for general counseling purposes. For this reason, families of jobs called Occupational Aptitude Patterns or OAPs were constructed. By grouping occupations into families and assessing each candidate on all the job families, the OAPs make maximum use of the data gathered by the GATB. The OAPs are especially useful in counseling because most applicants are being counseled for a large number of occupations.

The initial decision as to which occupations would comprise an OAP was determined by grouping the SATBs according to the aptitudes that they had in common. These trial OAPs were then evaluated and finalized based on their correlation with job success. The cut scores for the OAPs were determined from the cut scores of the SATBs that comprise them. The OAPs have been revised on a number of occasions and currently there are 66 OAPs.

A major criticism of the SATB development is that a number of different aptitude combinations were equally effective in predicting success. This lack of stability further affects the OAPs, which rely heavily on SATB validity information for their development. For this reason, the development of the OAPs has shifted from a heavy reliance on the SATB validity studies to an emphasis of job analytic data because these data are considered to be more reliable.

This new emphasis on job analysis in OAP development corresponds with an overall effort to integrate the GATB and its associated measures with the *Dictionary of Occupational Titles* (DOT) (U.S. Department of Labor, 1977). The DOT provides a detailed description of the 12,000 occupations in the U.S. economy. The usefulness of the GATB can be maximized by coordinating the job descriptions and required aptitudes with those described in the DOT.

More recently, a comprehensive system for job counseling called the Counselee Assessment/Occupational Exploration System has been introduced. The most recent edition of the *Dictionary of Occupational Titles* (4th edition) serves as the core of this system. The primary components of the system are 1) the Guide for Occupational Exploration, 2) measures of occupational interest (USES Interest Inventory and the Interest Check List), and 3) measures of the necessary aptitudes (the GATB). The guide classifies the occupations in the DOT into 12 Interest Areas and 66 Work Groups. As stated previously, the OAPs have been developed to correspond to these work groups and cover 97% of the non-supervisory occupations listed in the DOT. Thus, applicants can be measured on interests as well as aptitudes.

Much concern has been raised over the use of tests requiring reading and arithmetic calculations with educationally and culturally disadvantaged persons. In 1965, the USES introduced the Nonreading Aptitude Test Battery (NATB) as an alternative to the GATB for use with educationally deficient individuals. Designed for those with low levels of literacy skill, the NATB utilizes 14 tests to arrive at measures of the same nine aptitudes as the GATB.

The GATB was relied on heavily in the development of this new instrument. The rationale for this dependence was that the GATB measures the most important aptitudes for the occupations under consideration, and it had already been validated. Thus, the GATB norms could be used cautiously for the NATB. In fact, correlation with the GATB was one of the criteria used in the selection of the 14 NATB tests. The strategy of high correlation with an existing, validated, in-use instrument is an acceptable and common means of initially demonstrating validity. However, the fact that the NATB was specifically designed for individuals for whom the GATB was not a valid measure raises concern over the use of this technique in the NATB's development. USES recognizes this shortcoming and refers to it as "indirect" validity.

Tests 8 through 12 of the GATB—Mark Making and the Finger and Manual Dexterity measures—are included in the NATB. The nine other tests that comprise the NATB are: Picture Word Matching, Oral Vocabulary, Coin Matching, Matrices, Tool Matching, Three-Dimensional Space, Form Matching, Coin Series, and Name Comparison.

However, a relatively recent report (Angrisani, 1982) indicates that the NATB has undergone at least one revision. In a description of the NATB, this report, like other articles, states that 14 tests are used to measure the 9 aptitudes. However, when giving an in-depth description of the NATB, only 11 tests were listed and these were referred to as the 1982 edition. No indication was given as to how or why three tests were dropped from the original version.

In conjunction with the NATB, USES designed a screening procedure to identify disadvantaged individuals. Using this procedure, a counselor can determine in about five minutes if an applicant will be able to understand the instructions and test items on the GATB. It is suggested that those who do not pass the exercises should probably be tested with the NATB. However, because the screening test has not been validated, it is advised that it be used only as an aid and not as the sole basis for such a testing decision.

The documentation on the prescreening instrument is confusing. It is referred to as the Wide Range Scale, the BOLT Wide Range Scale, and the GATB-NATB Screening Device. It is assumed that all these names refer to the same instrument, which consists of eight vocabulary and eight arithmetic items. The use of this pre-screening device appears to be twofold: 1) to determine whether the GATB or the NATB should be administered; and 2) to determine what level of the Basic Occupational Literacy Test (BOLT) should be given. The BOLT was designed by USES as a measure of the applicant's literacy skills and relates these to literacy requirements of jobs. It is recommended that the BOLT be given to anyone who takes the NATB. The use of the prescreening device with BOLT may explain this variation on its name.

Additional aids to disadvantaged applicants are the Pretesting Orientation Techniques. These procedures were designed to orient and provide test-taking practice for disadvantaged individuals through the use of booklets and sample items available in "Doing Your Best on Aptitude Tests," "Group Pretesting Orientation on the Purpose of Testing," "Doing Your Best on Reading and Arithmetic Tests," and "Pretesting Orientation Exercises."

A Spanish version of the GATB, the Bateria de Examenes de Aptitud General (BEAG), has been available since 1978. Two of the Prescreening Orientation Techniques—"Doing Your Best on Aptitude Tests" and "Group Pretesting Orientation on the Purpose of Testing"—are also available in Spanish.

A number of modifications to the GATB have recently been made in an effort to accommodate handicapped applicants better. These include special administration methods for the testing of deaf applicants and special norms for the manual dexterity tests administered to seated applicants (those confined to wheelchairs or otherwise unable to stand for extended periods of time).

The GATB consists of 12 tests (eight paper-and-pencil and four performance tests) that yield measures of nine aptitudes. The paper-and-pencil tests are presented to the applicant in one of two booklets, and responses are recorded on a separate answer sheet. Book 1 contains the first four paper-and-pencil tests; Book 2 contains tests 5-7. Test 8, Mark Making, requires a separate response sheet, and the performance tasks in tests 9-12 require special materials. Descriptions of the 12 tests and some sample items follow. (Actual items from the GATB are not presented for reasons of confidentiality. Rather, a representative item from each of the four tests was chosen and then modified for inclusion.)

1. *Name Comparison*—determining if name pairs are similar or different.

Example: H. W. Longfellow & Co.——H. W. Longfellow, Co.
(*Answer:* different)

2. *Computation*—arithmetic exercises utilizing addition, subtraction, multiplication, and division.

Example:	MULTIPLY (X)	A	3822
		B	3972
	427	C	3843
	9	D	3783
	────	E	none of these

(*Answer:* C)

3. *Three-Dimensional Space*—selecting the proper three-dimensional representation of a flat, two-dimensional drawing. Lines appear on the two-dimensional stimulus figure indicating where it is to be folded or bent.

4. *Vocabulary*—from four words, select a word pair that constitutes either a synonym or an antonym.

Example: a. kind   b. empathetic   c. cruel   d. good
(*Answer:* a and c)

5. *Tool Matching*—identifying the exact likeness of a stimulus tool from four drawings that differ only slightly from each other.

6. *Arithmetic Reasoning*—solving of mathematical word problems.

Example: John can run 1/5 as far as Bill. Bill runs 12 miles. How far can John run?

A   2.4 miles
B   2 miles
C   3 miles
D   1 mile
E   none of these
(*Answer:* A)

7. *Form Matching*—matching similar shapes and forms in two different groups.

8. *Mark Making*—require making specified marks in a series of squares preprinted on a special response sheet.

The four performance tests, 9 through 12, are conducted using boards and are designed as measures of manual and finger dexterity. Tests 9 and 10 are referred to as *Place* and *Turn*, respectively. A rectangular pegboard divided into two sections, each containing 48 holes, is used for both tests. The examinee is required to move pegs from one side of the pegboard to the other for "Place." For "Turn," examinees must pick up each peg and turn it, or invert it, before returning it to the same hole in the pegboard. Test 9, Place, is performed three times, and then Test 10, Turn, is performed three times. The final score for each test is the number of pegs attempted summed across the three timed trials.

Tests 11 and 12, the measures of finger dexterity, are referred to as *Assemble* and *Disassemble*, respectively. The apparatus differs slightly, depending on the form of the test. For all forms, the major work part is a rectangular board containing 50 holes. What varies across forms is the location of the washers and rivets, which the individual is required to either assemble or disassemble. For Forms A, B, and D, the rivets are to be picked up (Assemble) or returned to (Disassemble) the upper portion of the board, where 50 additional holes are located. The rivets are to be taken from or returned to a vertical rod located at the mid-line of the board. However, for Form C, both washers and rivets are taken from and returned to two shallow cups built into the top of the board. No documentation was provided as to why this difference exists between the forms. Regardless of form, the basic tasks were the same. On the Assemble test, examinees place one washer and one rivet into each hole in the pegboard. On the Disassemble test, they remove the washers and rivets and return them to their original locations. Unlike Place and Turn, there is only one trial for Assemble and Disassemble.

While the administration of the battery does not require professional expertise, the role of the examiner is an active one. In addition to the normal administration duties of reading instructions, timing the tests, answering questions, and reporting any unusual conditions, the examiner must also demonstrate examples, set up equipment for performance tests, and score the performance tests.

Practical Applications/Uses

Tests developed by USES are intended for use by the public employment system in occupational counseling. Thus, they are available to the United States Employment Service as well as all State Employment Services. However, it is recognized that these instruments may prove useful to other counselors in both employment and educational settings. Use by other organizations requires prior approval from the State Employment Service. While the GATB could be used for any applicant to a State Employment Service, the usual age range of such applicants is 18 to 54, and most of the subjects in the standardization and validation samples fell into this age range.

The Administration and Scoring manual reports that only the earlier Forms, A and B, are available for release to other agencies. Form D is the primary test for use in government agencies, with Form C as the first retest. Forms A and B are also used as subsequent retests by government users.

In addition to its ability to predict job success, the GATB has been examined for its usefulness in predicting college success, especially for those occupations that require college training. In this capacity the GATB might help those who counsel high school students unsure of whether or not they should attend college. Because of time and cost considerations, most of the studies of the GATB with college students were restricted to using academic performance in college (e.g., grades) as the criterion of success rather than subsequent job performance. A consistent result was that Intelligence, Verbal Aptitude, and Numerical Aptitude were most frequently found to have significant correlations with this form of success. Because Intelligence correlates most highly, a minimum cutoff score for this aptitude was developed for use in counseling. In fact, three different cutoffs for this aptitude were derived for use with different types of colleges: junior colleges, four-year colleges, and four-year professional colleges (defined as colleges offering highly specialized professional courses, such as medicine, dentistry, and engineering). However, if the counselor requires information regarding a specific occupation or field of study, the GATB norms developed for that occupational area should be used.

Consideration was given to the use of the GATB with younger high school students. Studies of Grades 9-12 revealed substantial maturational increases in aptitude scores. Additionally, the rate of maturation was quite varied for the individual subjects. For this reason, separate norms were derived for students enrolled in 9th and 10th grades. Reliability studies showed a good deal of instability and thus special bands were established. The purpose of these bands is to identify marginal scorers (marginal in the sense of qualifying for an OAP). If a student was a marginal scorer on an OAP, no interpretation was given for that particular occupational group. If the purpose of the GATB is to direct students to occupational groups for which they are qualified, it would seem that the entire battery would be invalidated if one OAP could not be interpreted. Four aptitudes—Intelligence, Verbal Aptitude, Numerical Aptitude, and Clerical Perception—were most predictive of success in six different high school subject areas: science, English, math, social studies, foreign language, and communication. In terms of validity, however, when used to predict college success, none of the aptitudes was as effective as a measure of academic standing in high school. Thus, because the OAPs are not reliable for this age group and academic standing in high school is superior in validity, there seems no reason to use or promote the use of the GATB for 9th and 10th graders.

Instructions for the administration of the GATB are clear and comprehensive. The administrator is advised as to materials needed, proper test environment, and preparation tips. The manual includes highlighted verbatim instructions, examples of common misunderstandings on the part of examinees, and suggestions for the testing of handicapped individuals. The instructions are so thorough that they provide a model for the administration of any standardized instrument. However, the booklet is lengthy, and the danger lies in the fact that many administrators may not have the time or inclination to read it thoroughly.

The aptitude battery requires approximately two-and-one-half hours for its administration. However, actual working time is only 48 minutes. The additional time is required for instruction, distribution of materials, and scoring of the performance tests. All tests are timed and provide for practice exercises in Tests 1-7, examples for Test 8, and demonstrated examples for Tests 9-12.

The test is most efficient if administered to groups. The manual provides suggested ratios of examiners to examinees. These ratios are: 1 to 10 for Tests 1-8 and 1 to 5 for the performance tests, Tests 9-12.

Normally, the 12 subtests are administered in numerical order. However, some deviations are acceptable. These are provided in the Administration manual. No indication is given as to whether the test may be given in two different administrations.

Scoring services are provided by Intran Corporation (INTRAN) and National Computer Systems (NCS), or the examiner may elect to hand score the battery. Stencils are available to facilitate hand scoring. Two interesting points are raised in the scoring section of the manual. The manual suggests, to those who hand score the test, that if parts of the GATB are administered for specific batteries, only the appropriate parts of the battery should be scored instead of scoring the entire test. This statement is the only reference in the manual to the administration of specific parts of the GATB rather than the entire battery. Its placement in the hand-scoring section of the Administration and Scoring Manual is obscure at best. There are no clear instructions as to when this type of administration would be appropriate or how it should be undertaken.

Additionally, the hand-scoring instructions alert the examiner to unusually low scores. When such scores appear, the examiner is instructed to determine if the examinee made an error in recording his or her responses on the answer sheet. This often occurs if the examinee skips a question. When this appears to have happened, the examiner is instructed to give credit for correct answers that are recorded incorrectly. However, the manual does not indicate if low scores are given the same special consideration when the tests are machine scored.

For each examinee the scoring services provide a special report containing actual aptitude scores, aptitude scores plus one standard error of measurement, and a rating of H, M, or L (assumed to mean High, Medium, or Low) for each of the 66 OAPs. For hand scoring, the Test Record Card is provided as an easy method of recording aptitude scores and evaluating the candidate on the OAPs. The examiner begins by recording raw scores on this card. A table (referred to as a "conversion table") is used to translate the raw scores into converted scores. Each of the 12 tests has its own conversion table, and separate tables are provided for each form of the GATB—A, B, C, and D. Additionally, simple linear equations are provided to convert the raw scores for users whose raw data have been computerized. These converted scores are then added together (where more than one test contributes to the measure of an aptitude) to arrive at the nine aptitude scores.

These aptitude scores are then compared to the cut scores established for each of the 66 OAPs. Three norm tables are available for this evaluation—9th grade, 10th grade, and adult. The examinee receives a rating of H, M, or L for each OAP. An H indicates a high probability that the individual will do well in that particular occupational group (OAP). An H is assigned if scores meet or exceed the cut scores for all the aptitudes associated with that particular OAP.

A rating of M indicates that the individual's score is similar to the scores obtained by workers judged as satisfactory in job performance. These individuals will do well on the job but probably not as well as someone judged to be an H. The M is assigned if the individual's scores plus one standard error of measurement meet or exceed the appropriate cut scores.

An L is assigned if an individual's scores are more than one standard error of measurement below the cut scores for the aptitudes identified as essential. This rating indicates that the probability of satisfactory performance on the job is very low. Other occupations should be considered, and counselors should suggest other possible avenues to explore.

Thus, through machine scoring or hand scoring, a rating of H, M, or L is assigned for each of the 66 OAPs. The same procedure can be followed for determining the applicant's qualifications on the SATBs. However, these determinations must always be made by the examiner, as the machine-scoring services do not provide this information. It is recommended that an applicant be evaluated on a SATB only if the individual is applying for a specific job opening. The SATBs, therefore, are used selectively, while every applicant is evaluated on all the OAPs. One of the reasons for this selectivity is that there are over 450 SATBs. Thus, evaluation on all SATBs would be rather cumbersome and overwhelming for the counselee.

In Section II of the manual, Occupational Aptitude Pattern Structure, over 2,500 occupations are listed within their respective OAP categories. These are a subset of the over 10,000 occupations listed in the Guide for Occupational Exploration. Counselors are instructed to refer to this Guide for occupations not listed in the GATB manual. It should be recalled that the OAPs apply only to nonsupervisory occupations.

While the instructions for hand scoring are clear and easy to follow, the process is quite time-consuming, particularly if one were evaluating several candidates. Additionally, the sample record presented in Section I, Administration and Scoring, of the manual is incorrect. The sample indicates that the applicant took Form D of the GATB when in fact the raw score conversions are based on those for Form B, a quite confusing problem for an examiner engaged in hand scoring, especially if he or she does not have access to both the Form B and Form D conversion tables.

Some of the advantages of the GATB are that the battery enables the counselor to evaluate the candidate on a large number of occupations through the use of one exam. Also, only those aptitudes essential for successful performance on the job are utilized. The use of multiple cutoffs prevents any one aptitude from compensating for the lack of another, a practice that may be beneficial or detrimental.

The test authors caution against the use of the GATB alone and stress that counselors should use other information at their disposal such as school records, the interview, background data, and other standardized tests. Specifically mentioned are interests, leisure activities, physical capacity, personal traits, social and economic factors, acquired skills, and education and training.

Two instruments have been developed by USES to assess an applicant's occupational interests. It is suggested that the Interest Check List and the USES Interest Inventory be used in conjunction with the GATB. However, the manual provides very little direction for the measure of these other variables or how they should be incorporated into either counseling or selection decisions. Other USES reports make reference to the previously mentioned Guide for Occupational Exploration. Descriptions of the Guide refer to it as useful to counselors in coordinating the interests and aptitudes of applicants.

Technical Aspects

The reliability of the GATB was assessed using the test-retest method. Time periods between testing administrations ranged from 1 day to 3 years. Retest studies have been conducted using the same test as well as an alternate form. Reliability coefficients for the aptitudes typically ranged from .80 to .90. A practice effect, common in most retesting situations, is typically observed on the GATB. The manual is confusing in its treatment of retests; counselors are warned to be cognizant of the practice effect in a retest situation, but statistical corrections are not provided. On the other hand, the manual suggests that because of the high reliability observed over several years, there is no need for retesting unless the applicant has experienced significant training. Additional evidence against retesting is found in a subsequent discussion in the manual in which training on the job did not increase aptitude scores. However, a research report states the reason behind the development of Forms C and D was to provide alternate forms for retesting situations. Thus, it is rather confusing when retesting is warranted, and it is not clear how a counselor should view the scores of a retest.

The reliabilities for Finger and Manual Dexterity tests were lower than for the other aptitudes, an expected finding in that performance tasks typically show greater contamination from retests than do paper-and-pencil tests.

Numerous studies have been conducted in a variety of state employment offices over a period of time. Several studies examined males and females separately and found consistently significant reliability coefficients for both groups. Subjects included high school and college students, local office applicants, and current employees.

The validation research of the GATB has been, and continues to be, ongoing. Most notably, this research has resulted in the development of the Specific Aptitude Test Batteries (SATB) and the Occupational Aptitude Profiles (OAP) discussed earlier. The validation of the SATBs and OAPs was not conducted in a single time or place; studies were carried out at the various state employment offices over a period of time. The manual reports the results of these studies in table form and cites a few in detail as examples. For this reason, discussion of the validation can only be carried out on a general level.

The effectiveness of the GATB as a selection tool is continually being measured by follow-up studies of job success for applicants tested with the GATB. Results of these studies indicated that those individuals who met the norms of an OAP or SATB were more successful as evidenced by lower turnover rates, lower training costs, lower make-up costs, and higher productivity (the manual does not define what "make-up" costs entail). Additionally, from the counselee's perspective, some studies showed that individuals who followed counseling advice based on GATB results were more satisfied with their occupations than those who did not follow the counselor's recommendations. Only one study was cited that reported no differences in job success for those who met OAP norms as compared to individuals who did not meet the norms. The manual mentions problems with the sample and criteria but is not specific.

A number of studies have been conducted on both the correlation of the GATB with other tests as well as the intercorrelations of the GATB subtests. In general, the GATB aptitudes correlate highly with other measures of similar aptitudes. Additionally, relatively high intercorrelations were found among the measures of cognitive abilities, and lower intercorrelations were found between measures of cognitive and motor abilities.

The manual provides a wealth of statistical data on the specific occupations for both counseling and research needs. Some limited research results are also presented on a variety of topics, such as training effects and GATB differences for sex, age, minority, and disabled groups. More research is clearly needed in these areas.

A series of research studies were recently conducted by John Hunter for the USES. These provide up-to-date information on a variety of topics including validity, utility, fairness, and the replacement of the SATB cutoff scores with a regression model. This line of research has primarily addressed two critical issues: 1) expanding the use of the GATB to other occupations; and 2) increasing its utility by replacing the ratings of H, M, and L with percentile rankings. The first step in this process was the application of the validity generalization model to over 500 validity studies already conducted on the GATB. This research effort demonstrated that the GATB is valid and that this validity can be generalized to the over 12,000 occupations listed in the *Dictionary of Occupational Titles*. Additionally, the findings suggested the use of an alternative scoring method, which utilizes general aptitudes specifically keyed to a job family structure based on the job complexity levels listed in the DOT.

Such a method involved the reduction, using factor analysis, of the nine aptitudes to three general abilities similar to Fleishman and Quaintance's (1984) formulations. Each of the three general abilities is composed of three specific abilities, as shown in Table 3.

Table 3

Abilities Tested by the GATB

General	*Specific*
Cognitive	General Intelligence
	Verbal
	Numerical
Perception	Spatial
	Form Perception
	Clerical Perception
Psychomotor	Motor Coordination
	Finger Dexterity
	Manual Dexterity

The use of different multiple regression equations, each utilizing a different combination of abilities, would enable researchers to maximize high validity for a large number of occupations. For example, high complexity jobs were better predicted by cognitive abilities, while low complexity jobs were more accurately predicted by psychomotor abilities. Hunter provides five different regression equations for use with the three general abilities of the GATB.

The equations were designed for use with the DOT. As previously described, the DOT is part of a job analysis system that provides job descriptions for 12,000 occupations. Each of the occupations is rated for complexity on a scale of one to five. The five regression equations correspond to the five complexity levels. Thus, given the level of complexity of the job, a measure of predictive job performance could be obtained using the appropriate equation. (Five separate regression equations are also provided to predict training success in the different complexity levels).

The proposed regression model is contradictory to the multiple cutoff technique currently used to arrive at the ratings of H, M, and L. The USES manual specifically states that the relationship between aptitudes and job success is nonlinear and cites this finding as a major reason behind the use of multiple cutoff scores. However, this claim of nonlinearity has not been substantiated by other researchers. Hunter cites studies of Employment Service data and employment industry data as a whole that contradict the USES claim by providing support for the existence of a linear relationship between test scores and job performance.

If one accepts Hunter's findings and assumes the relationship between test scores and job performance to be in fact linear (or at least well-represented by a linear relationship), then the designation of individuals as H, M, or L would not prove optimal. That is, when trying to select the best applicants, an employer is unable to differentiate between a high H and a low H and thus selects randomly within a given category. Similarly, in a tight labor market, an employer is unable to identify the high M or high L for employment.

Substantial losses in productivity using the multiple cutoff method have been documented by Hunter through utility analysis. The importance of this loss is especially great when one considers the application of the GATB on a national level. Hunter anticipates increases in worker productivity through the introduction of the multiple regression model at 50 to 100 billion dollars a year.

Hunter calls for the use of optimal selection to maximize the utility of the GATB. Optimal selection is defined as using the most predictive ability test for a given job and selecting the top scorers on that test(s). He cites benefits for the employer such as increased productivity, increased promotional pool, and a competitive edge. Additionally, there may be psychological and economic benefits to applicants who are properly placed in jobs.

However, optimal selection usually results in adverse impact and the subsequent failure to meet Affirmative Action goals because ethnic minority groups historically average lower scores on cognitive ability tests. This finding is also true of the GATB and raises concern over its fairness. However, evidence provided by Hunter indicates that these lower average scores on the GATB correspond to lower average job performance. In fact, Hunter provides evidence of a slight tendency for overprediction of minority job performance by the GATB. For this reason, the GATB may be considered a fair instrument when used properly. To substantiate this finding, a validity generalization study by Hunter of over 51 validity studies found no evidence of either differential validity (a significant difference between the correlations obtained for two ethnic groups) or single group validity (a statistically significant correlation for one ethnic group but not for the other). The problem then becomes one of maximizing utility while still hiring representative numbers of minorities. Hunter suggest the use of optimal selection within each protected group as a possible solution.

Interestingly, minorities score higher on the measures of psychomotor abilities. As a major battery measuring psychomotor abilities, the GATB may be a fairer and more valid predictor of occupations that require such skills. As stated earlier, lower complexity jobs are frequently more accurately predicted by tests of psychomotor ability. Thus, for these occupations, use of the GATB would result not only in better qualified hires but also in a more racially balanced workforce.

Critique

The GATB is a comprehensive test battery designed to meet the occupational counseling needs of a specific time period. It has been the object of much research, and a wealth of data has been collected on this instrument. Despite its extensive use and research, the GATB does have some psychometric shortcomings.

Perhaps the two gravest criticisms are directed at the development of the SATBs and the use of multiple cutoff scores. First, the factor analysis that identified the aptitudes, the tests used to measure them, and the weights used for this purpose were based in part on a normative sample of only 519. The reader will recall that no women were included in this sample, and all examinees were either trainees or applicants for training courses. Additionally, this was the same sample referred to subsequently as nonrepresentative of the working population because it did not include a wide range of occupations. In fact, this was the reason that the second normative sample of 4,000 was analyzed. Thus, the sample that was inadequate for general working norms was somehow adequate for factor analysis.

Humphreys (1985) was among a number of critics who pointed out that the 59 tests used in the factor analysis did not include measures of a number of traits found by other researchers to be effective, such as measures of mechanical information and comprehension. Fleishman and Quaintance's (1984) current model of abilities includes 52 different abilities for example. One must wonder also, with computerization and other technological advances in the workplace, if skills necessary for new occupations, as well as redefined existing jobs, are not being overlooked. In conjunction with this, Weiss (1972) points out that the original occupations were overwhelmingly blue collar while today jobs are predominantly of a white collar nature.

Several critics have commented on the high intercorrelations among the aptitudes, particularly Intelligence, which is composed of three measures that are used in estimating other aptitudes as well (Keesling, 1985). Perhaps most disconcerting is the high intercorrelations among the factors that comprise the SATBs. Keesling notes that because of these high correlations a variety of "aptitude-cutting score" combinations exist that would be equally valid to those finally decided upon. This instability in the SATBs is recognized in the manual. The heavy reliance of the OAP development on the SATBs serves to carry these errors into other aspects of the GATB.

As stated earlier, multiple cutoffs were defended in part on the nonlinear relationship of the criterion and the predictor. However, this logic contradicts the statistical procedures employed in the construction of the GATB, many of which assume or imply that a linear relationship exists between the test and the criterion. Such a relationship is implied in the use of high mean score and low standard deviation as two of the criteria considered in determining which aptitudes would constitute a SATB (Weiss, 1972).

Additionally, the validity studies usually reported a phi correlation measuring the relationship between the dichotmized variables of predictor—exceeded or met the norm or failed to meet the norm, and criterion—successful on the job or not successful. It is unclear whether "met the norm" refers to applicants rated as H or as both H and M.

The continual reliance on high school samples for major statistical work is questionable. For example, the conversion of B-1001 to B-1002, Form A, was based on a study of high school students in only four states. Similarly, the conversion study of B-1002, Form A, to Form B was based on a sample of high school students in three states.

There were three major reasons cited in the manual for the use of multiple cutoff scores: 1) ease in computation for clerical staff in local state offices; 2) (apparent) benefits of a noncompensatory model; and 3) the lack of a linear or "straightline" relationship between aptitudes and job success. Prior to 1945, a total weighted score was in use, a method that was abandoned because the norms did not demonstrate a "straightline" relationship with job success.

In addressing the first rationale above, this benefit would hardly seem to be an issue today in light of technological advancements. The machine-scoring services provided by NCS and INTRAN could easily be programmed to provide predictors based on multiple regression equations. Many offices now have personal computers, so that even if hand-scoring was elected, the data could easily be run through a program to calculate predicted scores.

Several critics have referred to the other two reasons as "pseudo-problems." Claims of nonlinearity and the need for a noncompensatory model are not supported empirically and are actually contrary to the assumptions in the test's construction cited earlier (Weiss, 1972). The regression model suggested by Hunter responds to the criticisms aimed at the multiple cutoff score. Additionally, it bypasses the errors inherent in the instability of the SATBs and their subsequent effect on the OAPs. In fact, by utilizing three general aptitudes, the regression model agrees with the basic assumptions underlying the initial development of the GATB: the existence of certain basic aptitudes common to many occupations.

However, the introduction of such a model would probably change the way in which the GATB is used. This model provides five regression equations, each designed to predict job performance for the five different job complexity levels identified in the *Dictionary of Occupational Titles*. As such, an applicant can be evaluated for any occupation by referring to the job complexity level listed in the DOT. The developmental history of the GATB suggests that its primary use is for general occupational counseling and not specific job placement. The new model would be useful on a general level only as a means of indicating to which job complexity levels an applicant should be directed. Whether complexity level alone is adequate for differentiating occupations remains to be seen, but this seems unlikely to the current authors.

Overall, the manuals are comprehensive but overwhelming to the layperson and the professional alike. The manual on test development in particular is confusing because of numerous references to all the different versions and forms of the GATB. Instead of presenting the history of the instrument in a chronological manner, it is presented chronologically within subsections such as item analysis, factor analysis, validity, and so on. It is often unclear as to which sample and what version is being used in a particular analysis.

USES should be commended for its noble efforts in providing assessment techniques sensitive to the special needs of the educationally and economically disadvantaged applicants. The NATB, BEAG, BOLT, and Prescreening Techniques are all evidence of USES' commitment to fairness and equal opportunity. However, the heavy reliance in the NATB's development on the GATB and the lack of separate validation studies for the NATB leave too many unanswered questions to warrant recommendation of its use.

The GATB has proven useful in its application and, relative to other tests, its technical development is good. However, USES should certainly correct the shortcomings cited by researchers over the years. The more recent research efforts in validity generalization, fairness, and regression-based scoring are a first step in addressing these concerns.

References

This list includes text citations and suggested additional reading.

Angrisani, A. (1982). *U.S. Employment Service tests and assessment techniques* (USES Test Research Report No. 32). Washington, DC: U.S. Employment Service, U.S. Department of Labor.

Droege, R. C., Ferral, M., & Hawk, J. (1977). *The U.S. Employment Service occupational test development program* (U.S. Employment Service Test Research Report No. 31). Washington, DC: U.S. Employment Service, U.S. Department of Labor.

Fleishman, E. A., & Quaintance, M. K. (1984). *Taxonomies of human performance.* Orlando, FL: Orlando Academic Press.

Humphreys, L. G. (1959). General Aptitude Test Battery. In O. K. Buros (Ed.), *The fifth mental measurements yearbook* (pp. 698-700). Highland Park, NJ: Gryphon Press.

Hunter, J. E. (1983a). *The dimensionality of the General Aptitude Test Battery (GATB) and the dominance of general factors over specific factors in the prediction of job performance* (USES Test Research Report No. 44). Washington, DC: U.S. Employment Service, U.S. Department of Labor.

Hunter, J. E. (1983b). *Fairness of the General Aptitude Test Battery: Ability differences and their impact on minority hiring rates* (USES Test Research Report No. 46). Washington, DC: U.S. Employment Service, U.S. Department of Labor.

Hunter, J. E. (1983c). *Overview of validity generalization* (USES Test Research Report No. 43). Washington, DC: U.S. Employment Service, U.S. Department of Labor.

Hunter, J. E. (1983d). *Test validation for 12,000 jobs: An application of job classification and validity generalization analysis to the General Aptitude Test Battery* (USES Test Research Report No. 45). Washington, DC: U.S. Employment Service, U.S. Department of Labor.

Keesling, J. W. (1985). General Aptitude Test Battery. In J. V. Mitchell, Jr. (Ed.), *The ninth mental measurements yearbook* (pp. 1645-1647). Highland Park, NJ: Gryphon Press.

U.S. Department of Labor. (1977). *Dictionary of occupational titles* (4th ed.). Washington, DC: U.S. Government Printing Office.

U.S. Department of Labor. (1970). *Manual for USES General Aptitude Test Battery, section III: Development.* Washington, DC: Government Printing Office.

U.S. Department of Labor (1979). *Manual for the General Aptitude Test Battery, section II: Occupational aptitude pattern structure.* Washington, DC: Government Printing Office.

U.S. Department of Labor. (1980). *Manual for the General Aptitude Test Battery, section II-A: Development of the occupational aptitude pattern structure.* Washington, DC: Government Printing Office.

U.S. Department of Labor. (1982). *Manual for the General Aptitude Test Battery, section I: Administration and scoring (Forms A and B).* Minneapolis, MN: Intran Corporation.

U.S. Department of Labor. (1983). *Manual for the General Aptitude Test Battery, section I: Administration and scoring (Forms C and D).* Salt Lake City, UT: Utah Department of Employment Security.

Weiss, D. J. (1972). General Aptitude Test Battery. In O. K. Buros (Ed.), *The seventh mental measurements yearbook* (pp. 1058-1061). Highland Park, NJ: Gryphon Press.

Arthur B. Silverstein, Ph.D.

Professor of Psychiatry, University of California, Los Angeles, California.

GOLDSTEIN-SCHEERER TESTS OF ABSTRACT AND CONCRETE THINKING

Kurt Goldstein and Martin Scheerer. San Antonio, Texas: The Psychological Corporation.

THE ORIGINAL OF THIS REVIEW WAS PUBLISHED IN TEST CRITIQUES: VOLUME I (1984).

Introduction

As the name indicates, these tests are based on a distinction between two modes of behavior, abstract and concrete, which are thought to depend on two corresponding "attitudes," conceived of as capacity levels of the total personality. Abstract behavior has the following characteristics: 1) detaching one's ego from the outer world or from inner experiences; 2) assuming a mental set; 3) accounting for acts to oneself and verbalizing the account; 4) shifting reflectively from one aspect of a situation to another; 5) holding in mind various aspects simultaneously; 6) grasping the essentials of a given whole, breaking the whole up into parts, isolating and synthesizing the parts; 7) abstracting common properties reflectively and forming hierarchic concepts; and 8) planning ahead ideationally, assuming an attitude toward the "merely possible," and thinking or performing symbolically. In contrast, concrete behavior lacks these characteristics.

Normal adults are said to be capable of both modes of behavior, whereas abnormal (brain-damaged, psychotic, or mentally retarded) individuals are confined to just one, the concrete. The Goldstein-Scheerer Tests were designed to determine whether a subject can or cannot assume the abstract attitude, and to assess the nature and degree of an existing impairment (i.e., Is the subject limited to the lowest level of concrete behavior? able to succeed on only some of the tests? capable of learning from a series of aids?).

The battery had its origin in the work of Kurt Goldstein (1878-1965), a neurologist. He and his colleagues were concerned with evaluating the status of patients suffering from brain injuries during and after the first World War. Years later Goldstein formed an association with Martin Scheerer (1900-1961), a psychologist. Both men had fled Nazi Germany to settle in the United States, and their collaboration resulted in the monograph that also serves as the manual for their revisions of the tests (Goldstein & Scheerer, 1941). The test materials and record forms were subsequently made available by The Psychological Corporation.

Five tests comprise the battery and the authors recommend that all of them be administered to every subject suspected of impairment, since a subject need not perform consistently from one test to another and intertest comparisons provide a broader and sounder basis for evaluating the degree and nature of a defect. However, the literature suggests that this recommendation is not generally followed, and that some of the tests (e.g., the Object Sorting Test) are much more widely used than others (e.g., the Stick Test).

Descriptions of these five tests follow:

1. *Goldstein-Scheerer Cube Test.* The purpose of this test, a modification of the Kohs Block Design Test, is to determine whether the subject is able to copy colored designs with blocks. The materials are four identical wooden blocks with colored sides and two booklets, each containing six printed designs. If the subject does not succeed in correctly reproducing the original design, the examiner presents a series of modifications of decreasing difficulty: 1) an enlargement of the design; 2) the design in original size, but divided by lines which break up the figure into four squares; 3) an enlargement of the divided design; 4) a model of the design built of four blocks; and 5) three models, only one of which is an exact reproduction of the original design. At whatever step the subject succeeds the examiner again presents the original design. To be credited with success the subject must not only provide a correct solution, but must also acknowledge that the solution is correct.

2. *Gelb-Goldstein Color Sorting Test.* The purpose of this test is to determine whether the subject is able to sort a variety of colors according to definite color concepts. The materials are 61 woolen skeins of different hues and shades. In the first of a series of "experiments" the skeins are placed in a random heap and the subject is instructed to pick out all those that can be grouped with a sample skein. In the next experiment the subject is asked to match a skein either with one of the same hue or with one of the same brightness. The third experiment requires the subject to pick out all the skeins of a given hue. In the fourth experiment the subject must state which of two rows of skeins is more alike: six of the same hue but differing in brightness, or six of different hues but of the same brightness. In each of the experiments the subject is asked to state the reason for a particular grouping.

3. *Gelb-Goldstein-Weigl-Scheerer Object Sorting Test.* The purpose of this test is to determine whether the subject is able to sort a variety of objects according to general concepts and to shift from one basis of sorting to another. The materials are some 30 common objects that can be sorted by use, situation, color, form, occurrence in pairs, or material. In the first experiment the objects are placed in a seemingly random order and the subject is instructed to hand over all those that can be grouped with a sample object. The next two experiments require the subject first to sort all the objects into separate groups, and then to sort them in a different way. In the fourth experiment the examiner presents groups that the subject has not yet formed and that represent conceptual categories or classes (e.g., all metal, all round, all red, etc.). In every experiment the subject is asked to explain why the objects were grouped as they were.

4. *Weigl-Goldstein-Scheerer Color Form Sorting Test.* The purpose of this test is to determine whether the subject is able to sort a set of figures according to both color and shape. The materials are 12 plastic figures of three different shapes and four different colors; the reverse sides of all the figures are white. In the first two experiments the subject is instructed to sort the figures that belong together, and then to sort them again in a different way. If the subject is unable to shift, the

examiner attempts to induce a shift in a third experiment, for instance by turning the figures with their white sides up, and if the subject then succeeds, the first two experiments are repeated. To be regarded as successful, the subject must not only sort and shift without difficulty, but must account verbally for the principle of the sorting.

5. *Goldstein-Scheerer Stick Test.* The purpose of this test is to determine whether the subject is able to copy figures composed of sticks and to reproduce them from memory. The materials are 30 plastic sticks of four different lengths. On the first part of the test the examiner constructs samples of 36 increasingly intricate figures and the subject is instructed to reproduce them. On the second part of the test the procedure is repeated, only this time each figure is exposed for 5-10 seconds, depending on the subject's condition, after which it is removed and the subject must copy it from memory.

Practical Applications/Uses

A recent national survey of patterns of psychological test usage (Lubin, Larsen, & Matarazzo, 1984) found that the Goldstein-Scheerer Tests were still, though just barely, among the 30 most frequently used tests. These findings did not differ greatly from those of a survey conducted in 1969, but they indicated that the popularity of the battery had declined considerably since a 1959 survey. None of these surveys shed any light on the specific purposes for which the tests are employed, but presumably they are still being used and of some value for the purpose for which they were originally intended (i.e., detecting and defining one form of psychological deficit: an impairment of abstract thinking).

The literature indicates that although the battery has been used primarily with adult brain-damaged and schizophrenic patients of many types and subtypes, it has also seen some use with many other groups of subjects: other psychiatric populations; normal children, adults, and elderly persons; mentally retarded, deaf, and reading-retarded individuals; and children and adults from other cultures. The tests are (at least apparently) so elementary and simple that there is virtually no limit to the types of subjects with which they can be used.

The procedure for administering the tests was described in some detail earlier, although it should be noted that after having given them according to instructions, the examiner is permitted to vary the procedure according to the need for further clarification of the subject's status. For this reason, and also because the examiner is supposed to record all the subject's comments, questions, verbal responses, and in some cases every move that the subject makes with the test materials, the administration of the tests is quite demanding. The examiner should be thoroughly knowledgeable, not only about the theoretical distinction between abstract and concrete behavior, but also about the various ways in which the two modes of behavior are exhibited in performance on the tests.

In The Psychological Corporation's current clinical catalog, the working time for the battery is given as about 20-30 minutes, but it would seem that in some cases it might take that long to administer just one or two of the tests. One other caution is in order: between the time the monograph was published and the time the tests became commercially available, a number of changes were made in the test materials, either as the result of later work by the authors or with their approval. These changes are said not to affect the clinical usefulness of the tests, but they may prove confusing to the unwary.

Although the monograph provides criteria of success (abstractness) and failure (concreteness) on each of the tests, and many specific examples of the latter, the authors offer no formal quantitative scoring system. (A number of other investigators, among them Boyd, 1947; Rapaport, Gill, & Schafer, 1945; and Weiss, 1964, have attempted to provide such systems for some of the tests.) They acknowledge that there are gradations of both concrete and abstract behavior, but maintain that the abstract and concrete attitudes differ not only in degree but in kind, and conclude that only a qualitative analysis can be instrumental in determining the nature and degree of a defect.

In the absence of a generally accepted scoring system it is obvious that there can be no norms, and so the interpretation of performance on the battery is based on clinical judgment, hence relatively subjective. A good deal of training and experience is therefore required of the examiner to insure that the interpretations offered are adequate and proper.

Technical Aspects

Consistent with the authors' belief that qualitative analysis has to precede quantitative analysis, the monograph (Goldstein & Scheerer, 1941) presents no data whatever on the reliability and validity of the tests. Reliability studies by other investigators appear to be virtually nonexistent, and the bulk of the research on validity has regarded the tests as measures of organicity, seeking to differentiate brain-damaged patients from clinically normal controls and/or schizophrenic patients. As a rule these studies have not used the entire battery, have sometimes introduced changes in procedure, and perforce have employed ad hoc scoring systems. Moreover, as noted in an early review (Yates, 1954), the studies have generally suffered from a number of methodological and conceptual flaws. Thus it is not surprising that the results have at best been mixed. All too often, research reporting positive findings (e.g., Leventhal, McGaughran, & Moran, 1959; McGaughran & Moran, 1957) has been followed by research reporting a failure to replicate (e.g., Sturm, 1964).

In summary, patients with moderate to severe brain damage tend to do poorly on the tests (but so do some types of schizophrenics), whereas patients with mild or subtle defects (and other types of schizophrenics) may perform quite adequately. It appears to be a telling comment on the validity of the tests as measures of organicity that they are not even mentioned in a number of reviews of the literature on the assessment of psychological deficit in brain damage (and schizophrenia) that have been published in the last two decades.

Critique

When this battery was first introduced it was greeted with considerable enthusiasm for it offered a means of evaluating psychological functions that were not adequately assessed by other tests. Even in the early days, however, the tests were severely criticized on various grounds: technical (the lack of standardization), theoretical (uneasiness with the conception of abstract and concrete attitudes as capacity levels), and empirical (the conflicting findings on validity). Forty years later, it seems fair to conclude that although they are of limited value for diagnosis or differential diagnosis, they may still serve a useful purpose in simply describing how a patient thinks. None of the previously mentioned criticisms disqualify them for that modest but potentially important purpose. Whether or not they remain in use, the Goldstein-Scheerer Tests deserve an honored place in the history of clinical psychology and clinical neuropsychology.

References

Boyd, F. (1949). A provisional quantitative scoring system with preliminary norms for the Goldstein-Scheerer Cube Test. *Journal of Clinical Psychology, 5,* 148-153.

Goldstein, K., & Scheerer, M. (1941). Abstract and concrete behavior: An experimental study with special tests. *Psychological Monographs, 53*(2, Whole No. 239).

Leventhal, D. B., McGraughran, L. S., & Moran, L. J. (1959). Multivariable analysis of the conceptual behavior of schizophrenic and brain-damaged patients. *Journal of Abnormal and Social Psychology, 58,* 84-90.

Lubin, B., Larsen, R. M., & Matarazzo, J. D. (1984). Patterns of psychological test usage in the United States: 1935-1982. *American Psychologist, 39,* 451-454.

McGraughran, L. S., & Moran, L. J. (1957). Differences between schizophrenic and brain-damaged groups in conceptual aspects of object sorting. *Journal of Abnormal and Social Psychology, 54,* 44-49.

Rapaport, D., Gill, M., & Schafer, R. (1945). *Diagnostic psychological testing: The theory, statistical evaluation, and diagnostic application of a battery of tests* (Vol. 1). Chicago: Year Book Publishers.

Sturm, I. E. (1964). "Conceptual area" among pathological groups: A failure to replicate. *Journal of Abnormal and Social Psychology, 69,* 216-223.

Weiss, A. A. (1964). The Weigl-Goldstein-Scheerer Color-Form Sorting Test: Classification of performance. *Journal of Clinical Psychology, 20,* 103-107.

Yates, A. J. (1954). The validity of some psychological tests of brain damage. *Psychological Bulletin, 51,* 359-379.

Lucille B. Strain, Ph.D.

Professor of Education, Bowie State College, Bowie, Maryland.

GRAY ORAL READING TESTS-REVISED

J. Lee Wiederholt and Brian R. Bryant. Austin, Texas: PRO-ED.

The original of this review was published in Test Critiques: Volume V (1987).

Introduction

The 1986 revision of the Gray Oral Reading Tests (GORT-R) comprises individually administered standardized tests designed to measure several aspects of oral reading ability. The tests yield information regarding oral reading speed and accuracy, oral reading comprehension, total oral reading ability, and oral reading miscues. Results of the tests are reported as Passage Scores, Comprehension Scores, and Oral Reading Quotients. Provisions are made for recording five types of oral reading miscues: meaning similarity, function similarity, graphic/phonemic similarity, multiple sources, and self-correction.

For nearly three quarters of this century, the Gray Oral Reading Tests have been preeminent among the relatively few widely used tests designed to measure oral reading ability. Developed by educator, scholar, researcher, and writer Dr. William S. Gray, the tests were originally published in 1915 as the Standard Oral Reading Paragraphs (SORP). Revised in 1963 by Dr. Helen M. Robinson, editions of the Gray Oral Reading Tests (GORT) were published in 1963 and again in 1967.

Dr. William S. Gray, to whom a tribute is paid by J. Lee Wiederholt and Brian R. Bryant, authors of the current GORT-R, was eminently qualified by education and experience for establishing a prototype for the measurement of oral reading ability. Dr. Gray's educational career spanned many years and included many levels of education. His teaching career, which began in the rural schools of Adams County, Illinois, in 1904-1905, subsequently included several years of teaching and research at the University of Chicago. His educational administration career included service as director of the Training School at Illinois State Normal University and dean of the College of Education at the University of Chicago. A proponent of the eclectic method of reading instruction and an innovator, Dr. Gray initiated the first course in reading education at the University of Chicago and the annual reviews of the literature of reading research. He was originator of the first annual reading conference at the University of Chicago. Not only was he one of the founders of the International Reading Association, but he was also its first president in 1956.

Prolific in writing and research, Dr. Gray authored approximately 500 books, articles, and research reports. These included the monographs *Studies of Elementary School Reading through Standardized Tests* and *Remedial Cases in Reading: Their Diagnosis and Treatment.* Known widely as the "father of the Dick and Jane books," he was co-author of the Scott-Foresman Basic Reader Series and the Elson-Gray readers. During the years 1925, 1937, and 1948, respectively, Dr. Gray served as chairman of the Yearbook Committees of the National Society for the Study of Education (NSSE). The Standard Oral Reading Paragraphs was among the then-available standardized tests described in the 1925 NSSE Yearbook.

Available in one form and consisting of only four pages, the SORP contained a series of independent paragraphs designed to measure oral reading ability from first through eighth grade. The test combined oral reading rate and oral reading accuracy in order to arrive at what was called a "B" score, similar to a grade-equivalent score. The limited purpose of the "B" score was to determine whether a student should be promoted to the next school grade. Materials accompanying the paragraphs were directions and a form for recording test results.

The 1963-1967 Gray Oral Reading Test (GORT) appeared in four equivalent forms designed for student use. Each form contained 13 independent paragraphs of ascending levels of difficulty. Each paragraph was followed by four questions for assessing literal comprehension. The first paragraph in each form was preceded by a picture. Three measures were recorded for each passage: errors, time, and comprehension performance. A grade-equivalent score was determined by the amount of time required for reading a passage along with the number of errors made. Passage Scores were converted to tentative grade norms for boys and girls, respectively. Oral reading errors included words aided, gross and partial mispronunciations, omissions, insertions, substitutions, repetitions, and reversals. Other observations for which provisions were made included word-by-word reading, poor phrasing, lack of expression, monotonous tone, overuse of phonics, loss of place, and the like. Materials accompanying the GORT included a *Manual of Directions for Administering, Scoring and Interpretation, Revised* and an *Examiner's Record Booklet* for each of the forms of the tests.

Two estimates of reliability were reported in the manual for the GORT: Coefficients of Equivalence and the Standard Error of Measurement. Validity was based solely on the nature of procedures used in construction of the tests. The GORT was normed on a sample of 502 subjects from public schools in two districts in Florida and in several schools in metropolitan and suburban Chicago. Sex of the students was the demographic factor emphasized, and attempts were made to include only "average" students at each grade level.

The GORT-R is available in two alternate, equivalent forms: Form A and Form B. Both Form A and Form B are included in a single, spiral-bound book identified as the Student Book. Each of the forms contains 13 independent paragraphs of progressively ascending levels of difficulty. Each paragraph is followed by five multiple-choice test items designed to measure a variety of types and levels of comprehension.

The Passage Score of the GORT-R measures a student's ability to read a paragraph with speed and accuracy. The score results from determination of the amount of time required for reading the paragraph and the number of deviations made from the print. A Conversion Matrix, appropriate for the particular passage, is then used to determine the applicable Passage Score. Each conversion matrix provides a vertical axis for Deviations from Print (an index of accuracy) and a horizontal axis for Rate (an index of speed). The point at which these two indexes intersect for a student's performance yields a Passage Score somewhere within the range from 0-9. A Passage Score is generated for each paragraph read and the number of these is summed after testing is completed.

The GORT-R Comprehension Score is derived from a student's responses to the five multiple-choice test items following each paragraph. The questions are designed to assess comprehension at literal, inferential, and critical levels and to focus variously on cognitive and affective outcomes. The sum of all questions answered correctly provides this raw score. A combination of the Passage Score and the Comprehension Score produces an Oral Reading Quotient (ORQ), which provides an overall index of the student's total ability to read orally.

The quantitative data produced by the GORT-R are interpreted as standard scores and percentiles. Thus, provisions are made for comparing GORT-R scores with test scores resultant from administration of other tests. Ease of communication and understanding is facilitated by the use of percentiles.

In addition to the quantitative data generated by the GORT-R, provisions are made for analyzing and recording five types of oral reading miscues: (a) meaning similarity, (b) function similarity, (c) graphic-phonemic similarity, (d) multiple sources, and (e) self-correction. Analysis of miscues related to meaning similarity (replacing a printed word with another word similar in meaning) permits assessment of a student's use of comprehension strategies. Analysis of function similarity (substitution of a printed word with a word of similar syntactic function) permits assessment of a student's use of appropriate grammatical forms in reading. Analysis of graphic-phonemic similarities (occurring when printed words are replaced with words similar in appearance or sound) permits assessment of a student's use of word-attack strategies. Analysis of multiple sources permits an examiner to determine the nature of the miscues made by a reader that fit several categories at the same time. Analysis of self-correction miscues sensitizes an examiner to the self-correction strategies used by a student during oral reading.

In addition to the five categories of miscues for which special provisions are made in the GORT-R, attention can be given to the miscues of omissions, additions, dialects, and reversals (at the option of the examiner).

The 54-page GORT-R manual contains information pertaining to test administration, scoring procedures, interpretation, uses of test results, and development of the tests. In the overview of the manual, a comparison is made between the earlier versions of the GORT and the GORT-R. Emphasis is placed on the new normative, reliability, and validity data supporting the GORT-R, the new comprehension questions, development of the paragraphs, and modifications of the scoring criteria and analysis of miscues.

According to directions given in the manual, administering, scoring, and interpreting the results of the GORT-R are relatively simple procedures. The tests are designed for students between the ages of 6 years, 6 months and 17 years, 11 months. The time required for administration of each form of the GORT-R ranges from 15 to 30 minutes. Although it is recommended that the tests should be administered in one session, suggestions are also given for modifying testing time if essential. An explanation is given of the uses of basals and ceilings and their influence on reduction of testing time. Instructions are given for interpreting results of the tests, for completing the Profile/Examiner Record Form, and for pursuing additional assessment and instructional strategies.

Appendix 1 of the GORT-R manual contains standard scores and percentiles for Passage Scores and Comprehension Scores for both forms and a table for conversion of standard scores to quotients and percentiles. Appendix 2 contains four Passage-Score Conversion Matrices for paragraphs 1 through 13, useful with either of the test forms. Appendix 3 presents samples of student performance on the tests, and Appendix 4 is a reproducible Examiner's Worksheet for analyzing and recording oral reading miscues.

The Profile/Examiner Record Form is used for recording and summarizing a student's oral reading performance on either Form A or Form B. The record form is designed for specifying pertinent information about the examiner and the student, recording the GORT-R scores, noting the results of other tests, profiling the student's scores, summarizing oral reading miscues, and making comments and recommendations. The form is used also for recording the rate for each paragraph read and for scoring the comprehension items.

All of the paragraphs contained in Form A and Form B are reproduced in the Profile/Examiner Record Form. Each paragraph in the Profile/Record Form is preceded by a "prompt," a statement to be read by the examiner to the student prior to the reading of the paragraph. The prompt serves to provide the student with an immediate purpose for reading the paragraph. Each paragraph is followed by blanks for scoring pertinent aspects of the student's performance.

Practical Applications/Uses

The purposes served by the GORT-R are such that the practical applications and uses of the tests serve a wide range of practitioner needs in a variety of professional settings. The tests are useful to teachers at all levels of schooling, to counselors of various types, to graduate students and other researchers, and to administrators in educational and other settings.

Specific purposes for which the tests are designed are stated in the GORT-R manual: (a) to help identify those students who are significantly below their peers in oral reading proficiency and who may profit from supplemental help, (b) to aid in determining the particular kinds of reading strengths and weaknesses that individual students possess, (c) to document students' progress in reading as a consequence of special intervention programs, and (d) to serve as a measurement device in investigations in which researchers are studying the reading abilities of school-age children.

All the purposes for which the GORT-R is designed are, at various times, of interest to teachers both at the elementary-school level and at the level of secondary education. In elementary-school classrooms, identification of students who need special help in reading either in the classroom or in clinical settings is frequently a concern. The classroom teacher who expects to use effective and efficient techniques of corrective reading needs easily acquired data for making as precise a diagnosis of a student's reading difficulties as possible. In addition, dependable data are needed by classroom teachers who wish to identify those students whose reading difficulties are severe enough to merit recommendation for further diagnosis and remediation in a clinical setting. The ease with which the GORT-R can be administered and the relatively short time required for its administration make it especially useful for gathering initial data for corrective reading or for making referrals of students for remedial reading by reading specialists.

In secondary schools, in which the major concerns for reading regard content areas such as mathematics, natural or social sciences, and the like, a short time spent by a teacher administering the GORT-R to an individual can lead to the use of those instructional strategies of most value in achieving objectives of a content area. Although oral reading and silent reading are not entirely synonymous, they require enough of the same types of abilities that knowledge of an individual's capabilities in one mode of reading can offer some insights into the other. Oral reading assessment can help in the identification of some of the most debilitating problems that may be hampering the reading of a student. As in the case of the classroom teacher at the level of elementary education, a teacher at the secondary level needs data to substantiate referrals of students to sources for particular assistance.

Because of the nature of the GORT-R, a student's specific reading difficulties can be identified. Not only do the results of the GORT-R yield information regarding the relative standing of a student in comparison with pertinent peers, but the precise nature of the student's difficulties can be identified.

Reading specialists in clinical settings need baseline or initial data prior to engaging in more detailed diagnostic procedures or to implementing a remedial reading program for a student. Results of the GORT-R can suggest direction for subsequent diagnostic procedures needed for acquiring in-depth understanding of a student's reading problems. The results of the GORT-R are also useful for determining the types of remediation approaches that may be effective initially for a student. Because it can produce baseline data as a result of pretesting and subsequent data through posttesting, the GORT-R is a convenient instrument through which the effectiveness of any intervention program can be determined.

The GORT-R is particularly suitable for types of action research frequently implemented by teachers and administrators in classroom or school settings. Because of its two equivalent forms, the GORT-R can be used for assessment purposes before and after the use of instructional materials or procedures. Similarly, the tests are useful to graduate students and other researchers interested in experimental research in reading.

For reading diagnosticians seeking to determine patterns of a student's reading behaviors, the GORT-R is useful as an instrument by which results of other types of assessment measures can be compared. Results of informal assessment techniques such as informal reading inventories and cloze procedures can be compared with results of the GORT-R for a more certain evaluation of a student's needs in oral reading instruction. Because of its standard scores, results of the GORT-R can be compared to results of standardized measures of reading behaviors other than those required for oral reading.

Uses of the GORT-R, however, are not confined to educational settings. The tests can be used in any setting in which there may be a need to gain some insight into the reading ability of individuals of the relevant ages served by the tests.

Technical Aspects

Information regarding the technical development of the GORT-R is available in the test manual. Detailed accounts are given about the construction of the paragraphs and test items, standardization procedures, and studies related to reliability and validity of the tests.

Both text structure and content of the paragraphs were given careful attention and subjected to a number of criteria to determine suitability for inclusion in the GORT-R tests. The organization and complexity of the text as well as the interrelationship of ideas in a passage were taken into account. Text content was selected according to difficulty level of the vocabulary and interest level of the topic.

Prepared by a professional writer, the paragraphs in the GORT-R were subjected to some of the same criteria used to determine difficulty and suitability for the various levels of the 1967 GORT. In light of newer research results, however, less reliance was placed on use of readability formulas for determining difficulty levels in the GORT-R. While paragraphs and test questions in the GORT-R were subjected to analysis by appropriate readability formulas (Flesch, Fry, Dale-Chall, Farr-Jenkinson-Patterson, and Danielson-Bryan), concerns were broadened to include attention to several other factors. These factors included sentence structure, logical connections between sentences and clauses, and the coherence of topics. Three word lists comprising words to which students are exposed during the school years were used as primary sources for controlling the vocabulary in GORT-R paragraphs.

Some of the paragraphs in the GORT-R are modifications of content used in the 1967 GORT while others are based on fables, current events, unusual situations, or satirical social interactions. In accordance with the intent of the GORT-R to assess general reading ability and skills, general content rather than content emphasizing specific academic subjects was utilized in the paragraphs.

Unlike the 1963-67 GORT, in which comprehension questions were limited to literal comprehension, the GORT-R utilizes comprehension questions representing all levels of comprehension. For example, questions that assess comprehension at inferential and critical levels are also included. In addition to focusing on cognitive concerns, affective questions requiring personal and emotional reactions to the text are also taken into consideration. Most of the questions are passage-dependent because comprehension of particular paragraphs is being tested. Questions were avoided that could be answered through recall of similar text features, such as matching shapes or sounds of words in the questions with those in the paragraphs. In addition, care was taken to ascertain that the vocabulary used in the questions was not more difficult than that used in the paragraph on which the questions were based.

Unlike the open-ended questions posed in the 1963-67 GORT in which students were asked to make free responses, the GORT-R utilizes multiple-choice test items to assess oral reading comprehension. In the GORT-R, the student selects the response choice that best completes the idea expressed by the stem. Examiners are no longer required to judge the correctness of answers on the basis of suggested answers given in the record booklet.

Statistical criteria related to item discrimination (or discriminative power of the test items) are presented in the GORT-R manual. Item discrimination was interpreted as "the degree to which an item discriminates correctly among examinees in the behavior that the test is designed to measure" (Anastasi, 1982, cited in the GORT-R manual, p. 29). The point-biserial correlation technique, in which each item is correlated with the total test score, was used to determine the discriminative power of the test items. It is reported that, in view of lack of guidance in the literature concerning the magnitude of acceptable coefficients for item discrimination, the conventions governing interpretations of validity coefficients were utilized. On this basis and on the average, test items were found to be satisfactory in terms of discriminative power.

The GORT-R was standardized on a sample involving 1,401 students representing 15 states: Alabama, California, Iowa, Kansas, Louisiana, Maine, Massachusetts, Mississippi, Montana, New Mexico, New York, Texas, Virginia, Washington, and Wisconsin. The states represented four regions of the United States: Northeast, North Central, South, and West. Other demographic characteristics of the standardization sample included sex, type of residence (i.e., urban or rural), race, ethnicity, and age. Sample percentages for each of the characteristics were representative of percentages reported in the *Statistical Abstract of the United States* (1965) and were considered nationally representative.

Studies were undertaken during development of the GORT-R to determine its reliability with respect to three types of reliability: internal consistency, alternate forms, and standard error of measurement.

Internal consistency pertains to the degree to which individual items of a test intercorrelate and measure the same construct. In the case of the GORT-R, the construct under question was reading comprehension. Coefficient alpha, a statistical technique that shows how well scores obtained under just one condition represent universe (true) scores (Cronbach, 1970), was used to estimate the internal consistency reliability of each of the two GORT-R test forms. Using 25 protocols selected at random from the normative sample representing each one-year interval between 7 and 17 years of age, coefficients alpha were developed for Passage Scores and Comprehension Scores for each test form. Coefficients reported in Table 4 of the manual range from .83 to .98 for Passage Scores and Comprehension Scores across all levels of both forms. Guilford's (1978) formula for computing internal consistency coefficients for composites was used in calculations for the Oral Reading Quotients. All coefficients obtained across both forms of the GORT-R range from .92 to .95, indicative of a satisfactory degree of internal consistency.

Assessment of alternate-form reliability of the two forms of the GORT-R produced satisfactory results. To investigate the alternate-form reliability, correlation coefficients were computed across both forms to determine the degree to which each test related to its counterpart. Calculations on the means of the scores of 100 protocols selected across ages from the normative sample yielded the following coefficients: Passage Scores, .80; Comprehension Scores, .81; and Oral Reading Quotients, .83. All three coefficients were found to be significant at the .001 level. Results of the analysis of student-performance means were all within one standard error of measurement of each other.

The standard error of measurement for the GORT-R tests was calculated with reliability coefficients from the internal consistency research and is reported both for raw scores and standard scores in the GORT-R manual. Thus, the examiner can determine the probability of the range of scores within which the true score of an individual may lie.

The GORT-R was subjected to validity studies to assess the degree to which the tests are indeed measures of oral reading ability. Content, criterion-related, and construct validity were investigated.

For content validity, the paragraphs and test items were systematically examined to determine whether the content and skills required covered a representative sampling of behaviors involved in oral reading ability. This examination, along with the care used in construction of the paragraphs and test items, indicated appropriate content validity according to the purposes of the tests.

The GORT-R was examined for criterion-related validity by using concurrent measures as the relevant criterion. Results of the GORT-R were compared to the results of achievement tests related to reading performance and with ratings by teachers of students' overall reading ability. Specifically, GORT-R scores of 100 students (Grades 9-12) of the norming sample were correlated with the students' scores on the Iowa Tests of Educational Development (ITED). This process resulted in coefficients of .28 (Comprehension Scores), .47 (Passage Scores), and .38 (Oral Reading Quotients). For some of the students from the norming sample, scores on Form C of the Formal Reading Inventory (FRI) by Wiederholt (1986) were correlated with the scores for Form A and Form B of the GORT-R. The resultant coefficients were Form A—CS = .66, PS = .44, ORQ = .54; Form B—CS = .59, PS = .59, and ORQ = .48. For 98 of the students, scores from the GORT-R were also correlated with the ITED Total Reading Scores, resulting in the following coefficients: CS = .37, PS = .38, and ORQ = .39. Correlations made between the scores of 74 students of the sample (Grades 3 and 4) on the Comprehensive Test of Basic Skills (CTBS) and their scores for Form A of the GORT-R yielded the following coefficients: CS = .49, PS = .40, and ORQ = .47.

Additionally, ratings by three elementary teachers (Grades 3, 4, and 5) of 37 of their students' overall reading ability using a Likert-type rating scale (1-poor to 5-excellent) were correlated with the GORT-R scores of the students with the resultant coefficients: Form A: CS = .72, PS = .47, ORQ = .63; Form B: CS = .78, PS = .67, ORQ = .74.

Construct validity was examined according to certain traits generally recognized as related to reading ability. These included age, language abilities, total school achievement, and intelligence factors. With respect to age, investigation showed an increase of the GORT-R scores with increase in age, as expected. Furthermore, it was shown that the difficulty level of successive paragraphs in the GORT-R increased. Investigation of the relationship between the GORT-R scores and language ability of a group selected from the norming sample produced coefficients indicative of the construct validity of the GORT-R scores. Support for the construct validity of the GORT-R scores was also found in correlations involving the GORT-R scores of 206 students in the normative sample and their total school achievement. The results of two analyses run to investigate the relationship between the GORT-R scores and results of intelligence measures supported the construct validity of the GORT-R scores. Other evidence of construct validity of the GORT-R scores was apparent in the degree to which the scores discriminated between groups of readers and in the degree of discriminative power of the test items.

Critique

The Gray Oral Reading tests, for many years popular and highly satisfactory for their intended purposes, promise even greater effectiveness as a result of the 1986 revision by J. Lee Wiederholt and Brian R. Bryant. Available from PRO-ED publishers under the new title GORT-R (Gray Oral Reading Tests-Revised), the tests represent a continuation of the best features of previous editions and improvement in the aspects in which changes were desirable.

The GORT-R continues use of the traditional format of 13 paragraphs of difficulty ascending levels. Continued as well is the practice of presenting one paragraph per page in the examinee's book. The revised questions for assessing comprehension of the paragraphs are characterized by significant changes in both form and role. Rate and accuracy continue to play important parts in evaluation of oral reading ability, but these are only aspects of a wide complex of behaviors examined by the GORT-R.

Several positive changes are apparent in the physical properties of the GORT-R. For example, by shortening the manual and the Profile/Examiner Record Form by one inch, materials for this text can now be stored in files and notecovers of conventional sizes. In addition, deleting the picture preceding the initial paragraphs of the two forms will enhance the quality of testing and prevent early "datedness" of the tests. Finally, the "prompts" provided for the examiner to use prior to having a student read a paragraph is in keeping with what is known about the importance of establishing a purpose for reading.

The expansion of purposes increases uses for which the GORT-R is suitable. The content of the paragraphs is in keeping with events and ideas timely in terms of student's experiences and interests. Use of multiple-choice test items with specific correct answers improves objectivity in assessing comprehension of the paragraphs. Questions pertaining to a particular paragraph are placed on the reverse side of the page on which the paragraph appears, which keeps the questions immediately available to the examiner and makes it easy for the student to read along as the questions are read. Rather than testing literal comprehension solely, the GORT-R questions utilize a variety of comprehension forms; testing, then, of oral reading comprehension is more closely related to the nature of comprehension as it is conceived for silent reading or for listening. Because the raw scores achieved from the comprehension tests are converted to standard scores and percentiles using normative data, several types of comparisons are therefore possible.

The GORT-R reflects excellent technical preparation. Many of the criticisms levelled at the previous editions have been taken into consideration and used to improve technical aspects of development. Appropriate norms, no longer "tentative," are available in the manual. Normative scores are presented as standard scores and percentiles for Comprehension Scores, Passage Scores, and Oral Reading Quotients for Form A and Form B. The norming sample of 1,401 students, representing a variety of demographic characteristics, was to a large extent a nationally representative group of students in the United States. The tests will be appropriate for an expanded population of users. It should be noted that the tests continue to be focused on students of average abilities. Modifications are recommended for use of the tests by individuals who are outside the ranges of the ages or other specific characteristics of the norming sample.

Basic testing procedures of the GORT-R are simple, easy to understand, and easy to follow even by examiners who may lack experience in testing. Steps in administration and scoring are clearly explained in the manual. Testing time of 15 to 30 minutes is relatively short for an individually administered test. Ages of students served by the GORT-R are inclusive of students throughout the elementary and secondary levels of school. In some cases, the tests will be appropriate for college-level students, including those who require some type of reading remediation or those who may serve as subjects in research projects related to improvement of reading ability or instruction.

The GORT-R is designed for measurement of oral reading ability. It makes no claims for measurement of silent reading ability or for interpretation of its results in terms of other aspects of reading ability. Developers of the GORT-R express recognition of the contradictory results of research investigating relationships between oral and silent reading. In view of these results, they recommend administration of silent reading tests when warranted. They also offer the suggestion that further testing may not be essential when a student's silent reading comprehension is satisfactory for his or her age and ability.

The GORT-R thus is an essential and timely addition to the relatively few tests available for measuring oral reading ability. Not only is it important as a measurement device, but it focuses attention on—and offers insights regarding—the importance of oral reading instruction per se.

References

This list includes text citations and suggested additional reading.

Allington, R. I., Chodos, L., Domaracki, G., & Truex, P. (1977). Passage dependency: Four diagnostic reading tests. *The Reading Teacher, 30,* 369-375.

Anastasi, A. (1982). *Psychological testing* (5th ed.). NY: Macmillan.

Buros, O. K. (Ed.). (1965). *The sixth mental measurements yearbook.* Highland Park, NJ: The Gryphon Press.

Cook, R. C. (Ed.). (1941-1942). *Who's who in American education* (4th Ed., Vol. 10). Nashville, TN: Who's Who in American Education, Inc.

Cronbach, L. J. (1970). *Essentials of educational testing.* NY: Harper & Row.

D'Angelo, K., & Mahlios, M. (1983). Insertion and omission miscues of good and poor readers. *The Reading Teacher, 36,* 778-782.

Durkin, D. (1978-1979). What classroom observations reveal about reading comprehension instruction. *Reading Research Quarterly, 14,* 482-533.

Goodman, K. S. (1969). Analysis of oral reading miscues: Applied psycholinguistics. *Reading Research Quarterly, 5,* 9-30.

Gray, W. S. (1967). *Gray Oral Reading Tests.* Austin, TX: PRO-ED.

Guilford, J. P., & Fruchter, B. (1978). *Fundamental statistics in psychology and education.* NY: McGraw-Hill.

Wiederholt, J. L., & Bryant, B. R. (1986). *GORT-R (Gray Oral Reading Tests-Revised) manual.* Austin, TX: PRO-ED.

Randolph H. Whitworth, Ph.D.

Chairman, Department of Psychology, The University of Texas at El Paso, El Paso, Texas.

THE HALSTEAD-REITAN NEUROPSYCHOLOGICAL BATTERY AND ALLIED PROCEDURES

Ward Halstead and Ralph M. Reitan. Tucson, Arizona: Reitan Neuropsychology Laboratory.

THE ORIGINAL OF THIS REVIEW WAS PUBLISHED IN TEST CRITIQUES: VOLUME I (1984).

Introduction

The Halstead-Reitan Neuropsychological Test Battery and Allied Procedures is the best known and most widely used of the formalized sets of clinical and psychometric procedures designed to assess the relationship between brain function and behavior. One of the first comprehensive neuropsychological test batteries developed, it was designed to provide a global assessment of impairment (or integrity) of the verbal and nonverbal cognitive, auditory, visual, and perceptual-motor abilities for the diagnosis, treatment, and rehabilitation of both adults and children with organic brain lesions (brain damage). The Halstead-Reitan Neuropsychological Battery is, in reality, not a single battery but consists of three separate batteries: Adult, Intermediate Children (9-14 years), and Young Children (5-8 years). Each battery includes a minimum of 14 separate neuropsychological tests, measuring up to 26 different aspects of brain-behavior relationships. In addition, the appropriate Wechsler Intelligence Scale, a broad-range academic achievement measure (usually the Wide Range Achievement Test), and (for adults) the Minnesota Multiphasic Personality Inventory are routinely administered as a part of the test battery.

The Halstead-Reitan battery is based on the pioneering work of Ward Halstead, who in 1935 established one of the first neuropsychology laboratories at the University of Chicago. He originally developed and validated a series of neuropsychological tests to assess the impact of impairment in brain function on the entire range of human abilities. Halstead's original neuropsychological battery consisted of 7 tests yielding a total of 10 scores, which in turn could be combined into a single measure of the degree of organic damage, an Impairment Index. One of Halstead's first graduate students, Ralph M. Reitan, modified the Halstead battery and then extended it with the addition of other test instruments, including developing instruments appropriate for the diagnosis of brain damage in children.

Reitan also conducted numerous validation studies with the revised battery in his own neuropsychological laboratory, established at Indiana University Medical Center. Reitan's approach to the validation of neuropsychological tests is of as much interest and importance as the development of the tests themselves. All patients admitted to the neurology and neurosurgical center at the Medical Center were routinely administered The Halstead-Reitan Battery. Reitan then evaluated the test results and wrote a report in which he gave a diagnosis regarding the presence or absence of brain lesions, the location and extent of any such lesions, and extent of impairment of such an individual, and included his prognosis for recovery or rehabilitation. These results were then compared with data obtained from medical neurological or neurosurgical procedures, such as electroencephalographic studies, surgical procedures or, in some cases, autopsy results.

The test battery has thus been subjected to extensive research and undergone revisions over the past 25 years, both by Reitan and others. The battery has now gained considerable acceptance not only within the neuropsychological community but increasingly in the entire field of psychology, medicine, and even courts of law.

As noted previously, the Halstead-Reitan Battery is composed of three separate batteries, each of which will be described in turn. The Adult Neuropsychological Battery consists of five of the original Halstead tests, plus a number of additional neuropsychological measures which have been added by Reitan. Descriptions of the five Halstead tests follow.

1. *Category Test.* This is a test of abstracting ability, requiring the subject to evolve a single principle (a "category") from visual presentations of geometric shapes and colors. The stimuli consist of 208 photographic slides projected one at a time onto a screen. The subject has a response apparatus consisting of four lights, numbered one through four, with levers below each. On the presentation of a stimulus, the subject responds by pressing any one of the four levers. If the response is correct, the subject is rewarded by a chime; if the response is incorrect, the subject is informed of this by the sound of a rather unpleasant buzzer. The 208 slides are grouped into six sets of items, each organized on the basis of different principles or categories, followed by the seventh subtest composed of the previously shown items.

Scoring is on the basis of the subject's number of errors. The cutoff score recommended by Reitan (derived from Halstead's original work) is 50 or 51 errors, which has been demonstrated to discriminate well between brain-damaged and neurologically intact groups of younger patients. In addition to measuring abstract concept formation, this test measures the ability to concentrate and maintain attention over a protracted period of time. It also involves a visuospatial component which correlates with both the Picture Arrangement and Block Design subtests of the Wechsler Scales (Lansdell & Donnelly, 1977).

One unique aspect of this concept formation task that makes it extremely effective in discriminating organically impaired individuals from normals is that although the principle to be evolved remains constant, the form of the stimuli changes on every presentation, which is particularly difficult for persons who have sustained even mild organic injury.

2. *Tactual Performance Test* (Time, Memory, and Localization). The Tactual Performance Test (TPT) employs a modification of the Seguin-Goddard Form Board, using 10 blocks of different shapes. The subject is blindfolded prior to being presented with the board and the task is to place the varied shapes into their

respective similarly shaped holes as rapidly as possible. The subject uses only the preferred hand on the first trial, the nonpreferred hand on the second trial, and both hands on the third trial. Because learning does take place, it is anticipated that approximately 30-40% improvement will occur from one trial to the next.

Scoring is on the basis of the time taken for each individual trial as well as the total time. After completion of the three trials the board is removed and the blindfold taken off of the subject, who is then instructed to draw a diagram of the board with the blocks in their proper place as best the patient can remember them. A score for Memory (the number of blocks remembered) and Localization (the number of blocks placed in their correct order) is then derived. The TPT assesses motor speed, tactile and kinesthetic ability, learning, and incidental memory (ability to remember things when not explicitly directed to do so) (Boll, 1981).

3. *Speech Sounds Perception Test.* This test consists of 60 sets of 4 four-letter nonsense words, beginning and ending with different consonants but with the vowels "ee" between them (e.g., "geep," "keet," etc.), administered via tape recording. The subject responds by marking a multiple-choice answer sheet with what is believed to have been heard. Scoring is on the basis of the number of errors, with the cutting score being 7 or more errors. This test is very sensitive to brain damage generally, and to left-hemisphere lesions in particular. It not only measures forms of receptive aphasia but also the capacity to maintain concentration and attention to a rather boring task.

4. *Rhythm Test.* This is a subtest taken from the Seashore Test of Musical Talent, with the subject's task being to discriminate between like and unlike pairs of rhythmic beats presented via tape recording. The task is simply to identify whether the series of rhythmic beats are the same or different, scoring being based on the number of correct identifications. Errors are then translated into a scaled score, with "1" the best and "10" the worst possible scores. This subtest measures nonlanguage auditory perception, attention, and concentration, susceptible to organic brain injury generally and right hemisphere involvement specifically.

5. *Finger Tapping Test* (FTT). This is the most widely employed test of manual dexterity and was originally called the "Finger Oscillation Test" by Halstead. The subject's task is simply to use the index finger to tap a telegraph-type key with a recording device that measures the number of taps for a 10-second period. Each hand makes five 10-second trials, with brief rest periods between, and the average mean number of taps for each hand are the criterion scores. The average for normal young adults is about 55 taps with the preferred hand and approximately 10% less in the nonpreferred hand. Lateralized brain lesions usually result in slowing the tapping speed in the contralateral hand, and this test is especially effective in determining which cerebral hemisphere, if any, has been damaged.

These five tests result in a total of seven scores (the TPT yielding 3 scores of Time, Memory, and Localization) which are used to compute the Impairment Index. This is a simple procedure wherein the total number of scores falling into the organically impaired range is divided by 7, then rounded off. Thus a person who scored in the impaired range on two of the seven subtest scores would yield an Impairment Index of 0.3; five such scores would yield an Impairment Index of

0.7; and all seven, an Impairment Index of 1.0. An Impairment Index of 0.5 or greater is considered highly diagnostically significant for brain injury.

The neuropsychological examination also includes additional tests, added by Reitan to the original Halstead battery, and referred to as "Allied Procedures." They include the following:

6. *Trail Making Test.* This test, like most of the tests in the battery, was derived from other sources, in this case from the Army Individual Test Battery. The Trail Making Test is an easily administered test of visual, conceptual, visuomotor tracking. It consists of two parts, A and B, which involve connecting 25 circles numbered from 1 to 25 randomly placed on a sheet of paper (Part A) and connecting in alternating sequence numbers and letters, that is, 1-A-2-B-3-C, etc. (Part B). The score is the combined total time required to complete both tasks. Like most other tests involving both motor speed and attention, the Trail Making Test is extremely sensitive to the effects of organic brain injury. This task also provides a measure of how a person responds to complex visual array and the ability to mentally sequence stimuli.

7. *Strength of Grip Test* (Hand Dynamometer). This is a simple task involving two trials, alternating between nonpreferred and preferred hands, using a plunger-type dynamometer to measure grip strength. Again, this test gives a good measure of absolute strength in either hand as well as providing a good left/right comparison measure. Score is simply the average grip strength (in kilograms) in each hand.

8. *Sensory Perceptual Examination* (Tactile, Visual, and Auditory). This procedure determines the individual's sensitivity to the perception of unilaterally presented stimuli to each side of the body. With the subject's eyes closed, the back of each hand is touched lightly in randomly alternating sequences. This same procedure is done to the left and right sides of the face. In addition to unilateral tactile stimulation, the subject is also presented simultaneous stimulation to both hands and face. The score is the number of times the subject reports that only one side is touched when bilaterally stimulated. These errors (referred to as "suppressions") and the side of the body on which the stimulation was *not* noted are recorded, which gives an indication of the lateralization of any organic injury affecting the tactile sensory processes.

A similar procedure is performed visually, with the patient instructed to look at the examiner's nose as the examiner moves his extended right or left hand, and the subject indicates which hand he sees moving. The upper, median, and lower visual fields are measured again using both unilateral and simultaneous (bilateral) visual presentations. The score is again the errors made when the subject reports only one side to bilateral stimulation.

Finally, the examiner measures auditory preception by rubbing his fingers together near the subject's right and left ears while the subject's eyes are closed. The patient reports whether the sound is heard on the right or left side. Again, simultaneous and unilateral presentations are randomly made and the score is the number of times the client reports hearing only on one side during bilateral presentation.

These measures provide information concerning sensory-perceptual involvement generally and lateralization of brain injury specifically.

9. *The Aphasia Screening Test* (Halstead & Wepman, 1959). This is a procedure that has been modified several times and in its present form consists of 32 items designed to screen all the elements of receptive and expressive aphasic disabilities, as well as other commonly associated communication problems. It is a relatively brief test, usually requiring less than 30 minutes to administer, and it provides a wealth of information involving language and communication function. There are no specific scoring standards employed but erroneous responses are recorded in a diagnostic profile. It is important to recognize that in The Aphasia Screening Test the interpretation is based on training and clinical judgment, rather than any specific psychometrics or cutting scores.

10. *Tactile Form Recognition.* Here the subject is asked to identify, by tactile apperception ("feeling" with the hands), four common geometric plastic shapes (circle, square, triangle, cross) without the use of vision. The subject does not name the objects or, in fact, make any verbal response, but simply points with the other hand to the corresponding stimulus displayed on the testing apparatus. Each stimulus is presented twice to each hand, and the score is the total number of errors made by each hand. This task is sensitive to any generalized disturbance in tactile form recognition as well as useful in identifying any lateralization of injury.

11. *Tactile Finger Localization and Fingertip Number Writing Perception.* Both of these tests involve tactile perception, are both brief, and provide a significant amount of information. In tactile finger localization, the subject's eyes are closed and the examiner lightly touches each of the subject's fingers in a predetermined order, with the subject responding which finger was touched. In fingertip writing, the subject's eyes are again closed and the examiner, using a stylus, writes the numbers "3," "4," "5," or "6" four times on each of the subject's fingertips on both hands. The subject identifies which number he perceives that the examiner has written. These subtests are sensitive to any damage in the sensory areas of the brain, and again, provide additional information about lateralization of injury.

The Halstead-Reitan Neuropsychological Test Battery for Children is the downward extension of the battery to include children between the ages of 9 and 14 and was developed by Reitan while at Indiana University Medical Center. The battery is very similar to the adult form with the following modifications:

Category Test. The intermediate version utilizes only 168 slides (rather than 208 of the adult battery) distributed into only 6 tests. Further, some of the items of the adult test were reordered to obtain more consistent levels of difficulty.

Tactual Performance Test. A six-block formboard, rather than the ten-block used in the Adult form, is employed with the intermediate children.

Trail Making Test: The number of items in both forms A and B are reduced from 25 to 15.

Speech Sounds Perception Test. The stimulus is presented by tape as in the adult form. The number of items remain the same but the number of multiple-choice nonsense words on each line has been reduced from 4 to 3.

The Reitan-Indiana Neuropsychological Test Battery for Children is a further downward extension for children between the ages of 5 and 8. The changes from the adult form are as follows:

Category Test. This contains a major change, with the stimulus items being reduced from 208 to 80. Further, the child is not required to respond to number stimuli (1, 2, 3, or 4) but to colors (red, green, blue, and yellow) attached to the levers. This alteration was necessary since many preschool children tested did not know their numbers but testing revealed that they could match colors.

Aphasic and Perceptual Examination Test. The aphasia test is reduced to 22 from 32 items. In the perceptual and motor tests, much simpler sensory stimuli (such as substituting Xs and Os for numbers) is employed.

In addition, several additional neuropsychological tests were developed for use with younger children as well. While the children's batteries are often administered, they have never reached the same level of general use and acceptance as the adult batteries.

Practical Applications/Uses

When administering the Halstead-Reitan Battery, the competence of the test administrator is vital to achieving valid results. According to Boll (1981), the training of a neuropsychology technician-examiner requires at least 4 to 6 weeks of full-time training and he strongly recommends that the entire procedure, including all instructions, be memorized. Boll's rationale is that the examiner's role is to focus full and total attention on the subject being evaluated in order to insure the patient's understanding, attention, and concentration.

Administration of the Halstead-Reitan Battery requires considerably more time than the usual psychological evaluation. Administration time varies, of course, with the subject's age, education, general health, degree of organic damage, and other variables. In general, the adult battery requires about five hours of testing that includes the administration of the WAIS (or WAIS-R) and a test similar to the Wide Range Achievement Test. In some cases the battery may be administered in as little as four hours but generally not much less than that. Only very rarely does testing require more than a single day for completion.

Individual test scores are recorded on protocol sheets, with standardized or cutoff scores recorded as well. Although there is no single standard form or diagnostic profile employed with the Halstead-Reitan Battery, most psychologists using the battery follow Reitan's procedure and record all of the various test scores on a single summary sheet, which allows the interpreter to survey the results of the entire battery.

As is obvious from the foregoing, this is an elaborate, lengthy, rather complicated set of procedures requiring extensive training not only for interpretation but even for test administration. As such, only those individuals with considerable training in the field of neuropsychology would routinely employ these procedures. Nevertheless, the field of neuropsychology is one of the most rapidly expanding areas in psychology and in this reviewer's opinion some training, knowledge, and expertise in the administration and interpretation of neuropsychological procedures will very likely soon become a part of any clinical psychologist's repertoire. Although relatively few psychologists in private practice today were trained as neuropsychologists due to the relatively recent development of the field, more and more of these psychologists are achieving their training and expertise through self-study, neuropsychological workshops, convention symposia on neuropsychology, and the like. Workshops in the administration and interpretation of the Halstead-Reitan Neuropsychological Test Battery are frequently presented by Ralph Reitan himself and his colleagues.

The purpose of the Halstead-Reitan Neuropsychological Test Battery is not simply to diagnose the presence or absence of brain injury but to determine the extent or severity of such organic involvement, lateralization to either left or right hemisphere, the specific localization of such injury, and some estimation of general prognosis for recovery or rehabilitation. This is accomplished by using four different inferential procedures to derive information from the Halstead-Reitan Battery test data.

First, the absolute level of performance of the subject is determined by comparing his or her test scores with normative data both on brain-injured and normal controls. A second level of evaluation is the interpretation of patterns of performance. In this case, the variations in performance both between and within the various subtests provide information about the type and location of brain injury. A very simple and obvious example of patterns of performance might be seen by looking at the differences between Verbal and Performance IQs on the Wechsler Scales and relating this to left (Verbal) or right (Performance) cerebral involvement.

A third technique is the identification of specific behavioral deficits which are often referred to as "neurological pathognomic signs." Again, the most obvious examples would be receptive or expressive aphasias identified by the Aphasic Screening Test employed in the test battery. The fourth procedure compares performance between the left and right sides of the body with a view to determining either organic involvement or the functional integrity of the contralateral cerebral hemisphere.

Utilizing all of these procedures, the neuropsychologist is able to make very sophisticated evaluations and diagnoses not only involving gross organic injury, such as that resulting from head trauma, but much more subtle conditions such as the early onset of such degenerative conditions as Alzheimer's disease or specific language learning disabilities. With respect to the latter group, in view of the recent interest in the identification and treatment of learning disabilities in the public schools, it is quite likely that some sophistication in neuropsychological procedures and diagnosis will become an integral part of the school diagnostician's training.

The administration and scoring of the Halstead-Reitan Battery is relatively straightforward and no difficulty should be encountered in following the directions in each of the subtest manuals. It should be noted that the test manuals are published privately by Reitan (1969) and the test apparatus is also available only through Reitan. Because the test apparatus is quite extensive, really much too bulky to be portable and quite expensive, testing is almost always conducted in either a neuropsychological clinic, neuropsychological laboratory, or the psychologist's office.

Because the test apparatus and procedures are so extensive and lengthy, some revisions and alterations of the test battery have been published. The Category

Test, for example, is available on individual stimulus cards rather than slides (Adams & Trenton, 1981). It is also available in a single booklet form which preserves the color and integrity of the stimulus shapes while being portable and much less expensive (DeFillipis, McCampbell, & Rogers, 1979). The correlations between administration of the slide presentation and the card and/or booklet forms are very high, essentially the same as the original reliability data for the slides. Other modifications of the battery have included shortening the Category Test by about half (Calsyn, O'Leary, & Chaney, 1980) and correlations between the shortened and full versions are again very high. Other modifications have involved dropping some of the subtests or administering only parts of the more time-consuming tests. While these modifications do in fact reduce the administration time, the potential danger lies in that much valuable neuropsychological information may be lost as well.

Technical Aspects

Compared with other fields of psychology involved in the development and application of psychometric instruments, neuropsychological evaluation has rarely emphasized technical test construction problems, as Davison (1974) points out. This is in sharp contrast to other fields of psychological measurement which often involve themselves in technical measurement problems to the exclusion of substantive psychological content. Again according to Davison, the Halstead-Reitan Battery has had relatively few test construction problems. This is due in great part to the reliance on already well-standardized tests, as has been noted in this review in the description of the tests themselves.

Further, the emphasis has been on concurrent, rather than predictive, validity, which is generally somewhat easier to establish. The test criterion of brain damage, while presenting very real technical difficulties itself, is nevertheless a much more objective criterion in assessment than other psychological or clinical criteria, such as functional psychoses or legal insanity. As a result, the validation of the Halstead-Reitan Battery has been meticulous and extensive. The initial validation of a full set of neuropsychological procedures was first reported by Reitan in 1955, which also included cross-validation of Halstead's work. To briefly summarize this extensive study, it demonstrated that the test battery was highly effective in identifying brain-damaged and non-brain-damaged controls beyond the .001 level of significance. Vega and Parsons (1967) did the first major cross-validation of the Halstead-Reitan Battery and although their results were somewhat poorer than those reported by Reitan and his associates, they also demonstrated a highly significant separation of brain-damaged and non-brain-damaged patients. More recent validation studies of the Halstead-Reitan (Filskov & Goldstein, 1974) have confirmed the statistical clinical validity and utility of these procedures for well-trained neuropsychologists in the diagnosis and prognosis of brain damage.

The issue of reliability of neuropsychological evaluations has been secondary in the field of neuropsychological evaluation. As Boll (1981) so succinctly states, "If the procedures measure validly, it is reasonable to assume an adequate degree of reliability." As with most clinical procedures, test-retest reliability may be spuriously low if the patient has shown significant improvement or, conversely, deterioration from one test evaluation to the next. To elaborate, accepted neuropsychological procedures now suggest that after obtaining a neuropsychological evaluation on a patient, the neuropsychologist should perform follow-up evaluations periodically every six to nine months. Clearly, significant changes from one administration to the next, while producing what might be considered "low reliability," comprise vital information. Finally, with respect to overall reliability, Matarazzo, Matarazzo, Weins, Gallo, and Klonoff (1976) reported that neuropsychological examinations provide consistent, reliable classifications of patients as organic or non-organic, despite changes in their physical status, practice effects, and any other variables that influence test performance.

In short, despite some methodological problems involved, the reliability and validity of the Halstead-Reitan Neuropsychological Test Battery, as well as other similar neuropsychological batteries, are higher than for almost any other type of psychometric procedure. This is due to the relatively larger amount of information obtained by the more extensive, lengthy evaluation than is usual in the routine clinical evaluation. It is also due to the reliance on objective, behavioral measurements rather than purely subjective clinical inference. Finally, the validity of the Halstead-Reitan Battery is directly related to organic cerebral brain damage which, while not always easy to objectify, is clearly more objective than the often vague, ill-defined hypothetical constructs the psychologist is often asked to diagnose.

Critique

The Halstead-Reitan Neuropsychological Test Battery is really the prototype for a comprehensive professional neuropsychological evaluation. As a standardized procedure, however, it has a number of deficiencies. The apparatus is bulky, clumsy, and very expensive, due in part to the fact that it is not mass produced. As noted, a number of modifications and alterations have been developed that in some ways compensate for this deficiency. Further, the administration of the test battery is very long and at times can be quite stressful for the patient, especially one with brain injury. This is particularly true of the Tactile Performance Test, requiring a protracted period of time with the patient blindfolded (Lezak, 1983). The Halstead-Reitan procedures require a highly trained professional both to administer as well as to interpret the results. Although there have been some attempts at computer scoring and interpretation, the results have been disappointing compared to individual evaluations by highly skilled neuropsychologists.

Another major criticism of the battery is the use of rather simple scoring and extensive application of cutoff scores. The use of normed, standardized scores which would allow some estimation of the degree or extent of impairment, rather than simply signifying "organic-not organic" would be much preferable. As a result of these shortcomings, very few neuropsychologists administer the entire Halstead-Reitan Battery exactly as it was developed. Almost every neuropsychologist (including the reviewer) substitutes other standardized procedures or develops their own. Despite the criticisms and its shortcomings, however, most retain at least part of the Halstead-Reitan Neuropsychological Test Battery in their neuropsychological evaluations. Perhaps most important, they also rely on the procedures and techniques of neuropsychological evaluation, developed by Halstead and Reitan, to whom the whole field of neuropsychology owes an enormous debt. Thus, despite some inadequacy and inefficiencies, the Halstead-Reitan Test Battery remains the model (and in some ways, the ideal) for comprehensive neuropsychological assessment.

References

Adams, R.L., & Trenton, S.L. (1981). Development of a paper-and-pen form of the Halstead Category Test. *Journal of Consulting and Clinical Psychology, 49,* 298-299.

Boll, T.J. (1981). The Halstead-Reitan Neuropsychology Battery. In S.B. Filskov & T.J. Boll (Eds.), *Handbook of clinical neuropsychology.* New York: Wiley-Interscience.

Calsyn, D.A., O'Leary, M.R., & Chaney, E.F. (1980). Shortening the Category Test. *Journal of Consulting and Clinical Psychology, 48,* 788-789.

Davison, L.A. (1974). Current status of clinical neuropsychology. In R.M. Reitan & L.A. Davison (Eds.), *Clinical neuropsychology: Current status and applications.* New York: Hemisphere Publishing Corp.

DeFillippis, N.A., McCampbell, E., & Rogers, P. (1979). Development of a booklet form of the Category Test: Normative and validity data. *Journal of Clinical Neuropsychology, 1,* 339-342.

Filskov, S.B., & Goldstein, S.G. (1974). Diagnostic validity of the Halstead-Reitan Neuropsychological Battery. *Journal of Consulting and Clinical Psychology, 42,* 382-388.

Halstead, W.C., & Wepman, J.M. (1959). The Halstead-Wepman Aphasia Screening Test. *Journal of Speech and Hearing Disorders, 14,* 9-15.

Lansdell, H., & Donnelly, E.F. (1977). Factor analysis of the Wechsler Adult Intelligence Scale subtests and the Halstead-Reitan Category and Tapping tests. *Journal of Consulting and Clinical Psychology, 45,* 412-416.

Lezak, M.D. (1983). *Neuropsychological assessment* (2nd ed.). New York: Oxford University Press.

Matarazzo, J.D., Weins, A.N., Matarazzo, R.G., & Goldstein, S.G. (1974). Psychometric and clinical test-retest reliability of the Halstead impairment index in a sample of healthy, young, normal men. *Journal of Nervous and Mental Disease, 158,* 37-49.

Reitan, R.M. (1955). Investigation of the validity of Halstead's measure of biological intelligence. *AMA Archives of Neurology and Psychiatry, 73,* 28-35.

Reitan, R.M. (1969). *Manual for administration of neuropsychological test batteries for adults and children.* Indianapolis: Author.

Reitan. R.M. (1974). Methodological problems in clinical neuropsychology. In R.M. Reitan & L.A. Davison (Eds.), *Clinical neuropsychology: Current status and applications.* New York: Hemisphere Publishing Corp.

Vega, A., Jr., & Parsons, O.A. (1967). Cross-validation of the Halstead-Reitan tests for brain damage. *Journal of Consulting Psychology, 31,* 619-623.

Grant Aram Killian, Ph.D.

Assistant Professor of Psychology, Nova University, Fort Lauderdale, Florida.

HOUSE-TREE-PERSON TECHNIQUE

John N. Buck. Los Angeles, California: Western Psychological Services.

THE ORIGINAL OF THIS REVIEW WAS PUBLISHED IN TEST CRITIQUES: VOLUME I (1984).

Introduction

The earliest written article on analyzing drawings was by Cook (1885), who recognized that there were successive stages in the development of children's drawings. During the turn of the century and the 40 years following Cook's publication, numerous articles on drawings appeared and were reviewed by Goodenough (1926), who wrote the first text using a scoring procedure for assessing children's intelligence by adding up the number of appropriate human details. The total number of appropriate details was then converted to an IQ score.

The House-Tree-Person (H-T-P) projective technique developed by John Buck (1948) was originally an outgrowth of the Goodenough (1926) scale utilized to assess intellectual functioning. Like Machover (1949), who was also interested in projective drawings as an appraisal of children's intelligence, Buck felt artistic creativity represented a stream of personality characteristics that flowed onto graphic art. They believed that through drawings, subjects objectified unconscious difficulties by sketching the inner image of primary process. By allowing subjects to respond by their own construction to stimuli that are familiar and ambiguous, they assumed that subjects would project a self-portrait that could be used to assess personality dynamics. Since it was assumed that the content and quality of the H-T-P was not attributable to the stimulus itself, they believed it had to be rooted in the individual's basic personality. Thus, in the interpretative process, the three objects assume symbolic aspects of the subject's world: the *House* mirrors the subject's home life and intrafamilial relationships; the *Tree* reflects the elemental relationships that the subject experiences within his or her environment; and the *Person* echoes the subject's interpersonal relationships. Since the H-T-P was an outcropping of an intelligence test, Buck (1948) developed a quantitative scoring system to appraise gross classification levels of intelligence along with a qualitative interpretive analysis to appraise global personality characteristics. Unfortunately, the original standardization research falls short of today's standards for acceptable methods of test development and construction.

The sample size used in developing norms for the quantitative scoring system

---

This reviewer wishes to acknowledge and thank Evelyn Villafana for her assistance in the preparation of this review.

was most meager (N = 140) and no attempt was made to select randomly a stratified sample of subjects from the general population (Buck, 1981). Criteria for inclusion in the study were nebulous. Twenty adults were selected for each of seven intellectual levels (imbecile, moron, borderline, dull average, average, above average, and superior) based on an assessment of the person's "complete clinical picture . . . and not a score on one or more standard intelligence tests" (Buck, 1981, p. 8). Furthermore, no attempt was made to select an equal number of males and females (e.g., the sample at the imbecile level consisted of 5 males and 15 females, and the sample at the superior level consisted of 19 males and 1 female) and no attempt was made to control for age across and within the seven intellectual levels (ages ranged from 13.6 to 29 years). Finally, two separate collection methods were used during the standardization study: the drawings of subjects who had less than above average intelligence (high school or less) were obtained by individual examination, whereas the drawings of subjects who had above average intelligence and more (college students) were obtained by group examination.

The standardization data for the qualitative analysis are equally suspect (Buck, 1981). The normative sample consisted of 150 adult subjects: 52 patients who were seen at the University of Virginia Hospital and 98 who were seen at the Lynchburg State Colony Hospital or at the Colony's Mental Hygiene Clinics in other cities. Sex, age, race, IQ scores, and other demographic information about the sample were not presented. Eight gross classification groups of unequal numbers were formed (adult maladjustment = 10, epilepsy with personality maladjustment = 29, psychopathic personality = 22, psychoneurosis = 53, prepsychotic state = 3, mental deficiency with psychosis = 6, organic psychosis = 11, and functional psychosis = 16); however, criteria for inclusion in the various groups were not reported.

To date, no new normative studies on the H-T-P have been conducted. But despite the questionable underpinnings of the instrument, research on the H-T-P has expanded into several different countries and a variety of modifications or variations have emerged over the years, as well as a Post-Drawing Interrogation Form for children under age 15.

Diamond (1954) developed a projective test that combined aspects of the Thematic Apperception Test (TAT) and the H-T-P. The subject is instructed to make up (i.e., write) a story involving a *House,* a *Tree* and a *Person,* and is told that the three objects have personality as well as the ability to speak to one another. The subject is further instructed to describe what kind of house, tree, and person the characters are and how they feel about each other.

The Draw-A-Family test, which according to Hammer (1978) has no known authorship, is a projective technique that is often used in combination with the H-T-P. In this test, the subject is simply asked to draw a picture of his or her family. The drawing is used to assess the individual's perception of him- or herself in the family system and/or the relationship to parental and sibling figures. Interpretations focus on omitted figures, exaggerated figures, insignificant small figures, spatial placement of figures, activity of figures, facial expressions, and content and movement of all the figures. Figures in the Draw-A-Family test are analyzed in isolation and in relation to the overall gestalt. Both views are presumed to reflect subjects' perceptions of their relationships within the family.

The Man-Woman Drawing, a modification of the Draw-A-Person technique developed by Machover (1949), has now also become by tradition part of the H-T-P administration. After drawing a person the subject is then requested to draw a person of the opposite sex. Hammer (1978) was the first to combine both techniques into what he has called the H-T-P-P which includes a drawn person of each sex.

According to Hammer (1978) the Draw-A-Person-In-The-Rain also has no known authorship and is a technique that assesses adaption and withdrawal under unpleasant environmental conditions. Subjects are told to draw a person in the rain, and interpretations focus on the use of protective shields, the position of an umbrella, the amount of figural exposure with shields, facial expressions, additional environmental phenomena (e.g., clouds or rainbows), and other figures, as well as content and movement within the drawing.

Hammer (1978) developed a test called Draw-A-Member-Of-A-Minority Group to assess the projection of negative traits in oneself onto a member of a minority group. In this drawing, projected attributes and prejudices are assumed to reflect unconscious negative traits in oneself and not attributes of the minority group.

The Rosenberg Draw-A-Person Test (1948) is conducted with carbon paper for both male and female drawings. Once the drawing is completed, the examiner retains the carbon copy and asks the subject to alter, erase, or cross out the original drawing. After changes have been completed, a post-drawing inquiry focuses on the modifications.

Caligor (1951) developed an Eight-Figure Redrawing technique based on the Rosenberg test. In this test, subjects make a series of eight human figure drawings and are then instructed to change each of them. Instead of using carbon, subjects are given onionskin paper that they can see through which is then placed on each previous drawing. Interpretations are based on serial change and deeper personality levels are assumed to underlie each progressively repeated sketch. The principle assumption underlying this method is that repetition of task will lead to deeper unconscious identifications.

The Animal-Drawing-Story technique developed by Levy and Levy (1978) requires 1) the drawing of any animal, 2) a pet name for the animal, 3) naming the kind of animal drawn, and 4) an optional imaginative story about the animal. Normative data was based on 7,346 drawings obtained from selected adult males and females, institutionalized male and female psychotics, male and female adolescents, and male prisoners. After completion of the test, the animal drawings are analyzed normatively, formally, and symbolically.

Harrower (1978) developed the Most-Unpleasant-Concept-Test, another projective technique, that requires subjects to graphically sketch the most unpleasant thing that they can imagine. Subjects are allowed to pencil draw the image in actuality, schematically, or symbolically, and after completion subjects describe and give free associations to the drawing. Carbon paper provides additional information about erasures, pressure, and shading. Results are interpreted in terms of the subject's reactions, content, and whether the concept was an internal or external event. The technique is based on the qualitative analysis of the results of 500 subjects in psychotherapy who were given a full battery of psychological tests.

Although the Goodenough (1926) Draw-A-Man Test was developed 22 years before the H-T-P, it served as the groundwork. Harris (1963) revised the Goodenough test in order to develop an alternate form to the Man scale, to extend the scale to adolescents, to develop norms on a more representative sample, to establish a quality scale for quick approximation for point scores, and to extend the test to a Self drawing. The text of this revision (Harris, 1963) includes the manual, general scoring instructions, scoring and practice examples, a short scoring guide, conversion tables, plates for quality scales, percentile ranks, and a test booklet.

The H-T-P is a projective technique that utilizes pencil and crayon freehand drawings of a House, a Tree, and a Person. During the administration of the drawings the subject is given almost complete freedom in sketching the three objects.

The test materials suggested by Buck (1981) to administer the H-T-P include:

1. A four-page 7" x 8½" Drawing Form. The first page provides for identification information, the second has *House* printed at the top, the third has *Tree* printed at the top, and the fourth has *Person* printed at the top. Two drawing forms are needed for each subject: one folder is used for the freehand pencil drawings and the other is used during the crayon drawings. A Two-Copy Drawing Form is available for group testing which records erasures. The chemically treated paper provides a second copy that records all lines drawn even if they are erased on the original paper, and may thereby provide further clinical application for the practitioner;

2. A four-page H-T-P Post-Drawing Interrogation (P-D-I) Folder which is used after the achromatic and chromatic drawing phase (an abbreviated version of the P-D-I can be used after the chromatic drawings). In addition, there is a set of revised questions that are recommended for subjects under 15 years of age;

3. A four-page H-T-P Scoring Folder that is used for quantitative scoring;

4. Several pencils with erasers that are recommended along with a set of 8 or more crayons (red, green, blue, yellow, brown, black, purple, and orange);

5. The H-T-P Manual; and,

6. A stopwatch.

In most settings, all of these materials are not used and the majority of clinicians agree with Ellis (1970), who "sees no particular value in employing the standard H-T-P booklets for the actual drawing. Any ordinary size white paper serves the same purpose" (p. 592). In terms of the P-D-I, Ellis goes on to state that "The time and effort spent in making this postdrawing interrogation on an individual basis is questionably expended, as against utilizing this time for a general psychological interview" (p. 107).

The H-T-P test consists of two drawing phases with each phase followed by a structured interview. The first step consists of having the subject sketch a freehand pencil drawing of a House, a Tree, and a Person, which is then followed by the examiner asking 60 questions from the P-D-I. After the subject answers the questions about the three achromatic pencil drawings, the subject is again requested to produce a freehand drawing of the three objects, but this time with the eight or more crayons. This is followed by the same structured questions of the P-D-I or a shorter version if the subject is fatigued or if undue time has elapsed.

The P-D-I consists of 60 questions varying from direct and concrete to indirect and abstract. Preceding the number of each question are the letters H, T, or P, which respectively deal with the House, the Tree, and the Person. In addition, the questions are followed by one or more letters to indicate A for Association, P for Pressure, and R for Reality-Testing, which are used for scoring purposes. To prevent the possibility of an answer set and reduce the chance of a subject remembering a response to a previous drawing, the different types of questions have been intermittently spaced.

During the drawing phase the examiner's participation is minimal and is primarily focused on recording 1) the order in which the subject sketches parts of each drawing; 2) all spontaneous comments as they relate to details; 3) any emotions expressed by the subject; 4) the initial latency period; 5) any intra-whole pauses; and 6) the total time used to sketch each object. During the questioning of the P-D-I, which is not intended to be a rigidly structured procedure, the examiner becomes more active, asking the subject to respond to the 60 questions on the H-T-P and to any additional questions that may seem clinically pertinent.

Although Buck's original normative group was arbitrarily set at 15 years or older, more recent research on the H-T-P has found that children as young as 4 can perform the task (Jolles, 1952; Beck, 1955; Bieliauskas, 1960); hence, no specific age limitations are given in the revised manual (Buck, 1981). In addition, since the difficulty level of the drawing is set by the subject, no specific mental limitations are given in the revised manual (Buck, 1981). Consequently, the test can be used for ages as young as 4 and over, and can be used with hearing-impaired (Davis & Hoopes, 1975) and handicapped children (Johnson & Wawrzaszek, 1963). In the original normative group, subjects ranged from "imbecile" with an IQ score of 25 to "superior" with an IQ score of 140.

Once the P-D-I has been administered and the interview has been completed, the examiner records items of detail, proportion, and perspective in the Scoring Folder. After completing the elaborate scoring tables by examining the plates containing sketches that illustrate quantitative scoring points, the examiner derives an IQ figure for the percentage of raw G, a net weighted score, a weighted "good" score, and a weighted "flaw" score, which then comprise the items for the profile configuration.

The H-T-P requires few materials, minimal space, and a small flat desktop surface; consequently, any quiet setting with illumination can be appropriate as long as the subject is comfortably seated with sufficient room to draw. The H-T-P can be administered in an individual or group format. For group testing the same materials are used with the possible exception of the "Two-Copy Drawing Form," which permits permanent recording of erasures. After completing the achromatic drawings, the P-D-I is distributed and the examinees are instructed to answer all 60 questions, and the procedure then continues as in the individual examination method. Empirical studies are scant comparing the two methods. According to Buck (1981) group administrations may have merit, but are less informative than individual testing. In contrast, Cassel et al. (1958), Cowden et al. (1955), Ellis (1970), and Hammer (1978) suggest that group testing is less time-consuming and seems to provide richer diagnostic and prognostic data.

The testing procedure is simple and straightforward. A trained psychologist is not necessary to administer the separate phases of the test and either a secretary, an aide, a teacher, or a mental health counselor can be quickly trained to administer the test. The examiner first presents the drawing form sheet to the subject with only the second page showing with the word *House* at the top and states:

> Take one of these pencils, please. I want you to draw me as good a picture of a house as you can. You may draw any kind of house you wish, it's entirely up to you. You may erase as much as you like, it will not be counted against you. And you may take as long as you wish. Just draw me as good a house as you can. (Buck, 1981, p. 18)

When a subject seems concerned about his or her drawing ability, the examiner should assure the person that it is not a test of artistic ability. Once completed, the examiner then repeats the drawing instructions for the Tree and the Person. After the drawing phase is terminated, the examiner administers the P-D-I in order to determine what the House, Tree, and Person meant to the subject. Since this part of the test is not intended as a rigid structured interview, the examiner may always conduct further questions that seem clinically appropriate.

In general, normal adults will use between 30 and 50 minutes for the drawing of the H-T-P series (either the achromatic or chromatic) and a similar amount of time in discussing the P-D-I; however, these time ranges depend on the subject's degree of adjustment and level of intelligence. In extreme cases, all six drawings could be done in less than 5 minutes or in the case of obsessive-compulsive styles, more than an hour could be taken for each series.

In most settings, clinicians do not administer the chromatic drawing since there is no scientific evidence that it provides any additional clinical information. Moreover, the standard procedure does not provide for the drawing of persons of the opposite sex or family. Greater insight could be achieved by eliminating the chromatic phase and including sketches of the family and the opposite sex. Ellis (1970) strongly advises "clinical psychologists who use the *H-T-P* to take Buck's administration and scoring procedures with decided skepticism and to adopt testing methodology to their own realistic work schedules" (p. 593).

The H-T-P technique was designed to foster projection, and through the interpretations of drawings to assess an individual's efficiency, sensitivity, maturity, flexibility, personality integration, and level of interaction with the environment by allowing subjects to paint a picture of their world where each drawing is assumed to represent aspects of a self-portrait. As a projective prognostic tool, the drawings are intended to lay bare symbolically conscious and unconscious mental processes that aid in the identification of suppressed or repressed dispositions and conflicts and outline the various resistances and defense mechanisms to these dispositions. As a therapeutic tool, the drawings are intended to function as a springboard for the elaboration of fruitful associations, furnishing a broader picture of the personality to further psychological insight. The House, the Tree, and the Person were selected because they are familiar, can be easily drawn, and promote open dialogue with all ages and personality types.

Practical Applications/Uses

The H-T-P has become one of the standard projective tools in the psychological test battery and frequently serves as the introductory test in the battery by providing a minimally threatening and maximally absorbing beginning to psychological assessment. Serving as an easy bridge to the psychological examination with minimal contact with the examiner, the subject is not intimidated by specific questions concerning intelligence or by threatening inkblots, which may disturb unconscious conflicts.

As a diagnostic tool, the H-T-P through tradition has been given a secure position in a prognostic battery, although there is question about its contribution to the entire clinical picture. Wyatt (1949) maintains that the H-T-P harnesses deeper and more primary conflicts that are less differentiated than those obtained through other projective tests, while Hammer (1955) argues that the H-T-P provides a "grosser personality picture" (p. 17). For him, the Rorschach provides the "richer personality picture" (Hammer, 1978, p. 600) except when subjects are guarded, then the H-T-P proves to be the more revealing device (Hammer, 1954; Landesberg, 1953). On the other hand, as a screening tool for detecting the onset of incipient psychopathology, Hammer (1955) and Zucker (1948) see the H-T-P as a more sensitive prognostic tool than the Rorschach. However, this is refuted by recent research, which suggests that the H-T-P is one of the least useful tools for discriminating between normal subjects and psychiatric patients, and that if employed in a test battery the H-T-P is more likely to increase error and reduce the probability of making an accurate diagnosis (Wildman & Wildman, 1975).

As a screening device, Buck (1981) and Boring and Topper (1949) recommend the H-T-P be employed for measuring therapeutic change; as treatment progresses the size of the drawings change appropriately and intra-whole proportions improve. As an initial screening device (Buck, 1981) the H-T-P can also be used 1) in a group testing format to identify adjustment levels; 2) to evaluate personality integration and adjustment prior to training programs, therapy (Hammer, 1978), employment, and school enrollment (Beck, 1960); and 3) as an evaluation tool to assist research.

As a therapeutic tool, the H-T-P can facilitate free associations that will enhance increasingly deeper levels of insight (Buck, 1981). In art therapy it can be employed with adolescents or children and can be a supplementary tool in analytic group therapy (Naumberg, 1978).

In short, as a diagnostic, therapeutic and screening device, the H-T-P seems to have many practical uses and is easy to administer; however, because of the extremely complex quantitative scoring system and the subjective nature of the qualitative analyses, psychological interpretation of the technique requires a trained and experienced clinician. Although a useful therapeutic tool which can be used in numerous settings, the scoring method and the sophistication required to interpret the H-T-P seem to preclude its diagnostic utility in non-clinical settings unless a trained professional is available for interpretation. Thus, despite its value and ease of administration in numerous settings, scoring and interpretation constraints preclude its use as a prognostic or diagnostic device by school counselors, educators, or social workers who are not trained in quantitative and qualitative scoring. Its utility as a diagnostic or screening tool is thus limited to the private practitioner or clinician in a mental health service. Moreover, if the research findings of Wildman and Wildman (1975) and others persist, it seems the H-T-P will have even less value since in a test battery it decreases the probability of making an accurate diagnosis. In the future, unless there is research support, the H-T-P may be relegated to the position of only a therapeutic tool for art therapy, or be limited as a diagnostic tool for nonverbal patients. For more frequently than not, the H-T-P tells the clinician what he or she already knows, or other instruments are used to develop or support hypotheses to interpret the H-T-P drawings.

The elaborate quantitative scoring system presented by Buck (1981) involves several phases before the final IQ scores are calculated. Initially, the examiner uses the scoring tables, item by item, assessing spontaneous and omitted items for all the achromatic and chromatic drawings in terms of detail, proportion, and perspective. Two major classes ("flawed" score or "good" score) are subdivided into intelligence levels totaling eight possible scores (each with a symbolic designation) that can be given to each characteristic of a drawing. "Flawed" scores from most to least flawed are rated as "very inferior (D3)," "imbecile (D2)," and "moron (D1)," while "good" scores are "borderline (A1)," "dull average (A2)," "average (A3)," "above average (S1)," and "superior (S2)." After scoring all sketches for all possible drawn or omitted characteristics, the eight grand raw scores (D3, D2, D1, A1, A2, A3, S1, S2) are converted to grand total weighted scores. The examiner then calculates the percentage of raw G and enters all possible scores onto the tabulation sheet, where IQ figures are derived from the percentage of raw G score, the net weighted score, the weighted "good" score, and the weighted "flawed" score. If there is significant scatter or more than one intelligence classification level between "good" and "flawed" scores, then this is suggestive of repression or deterioration.

This extensive quantitative scoring of the H-T-P, however, is rarely if ever used because it is cumbersome and scorer reliability is less than adequate (Bieliauskas, 1956). Despite a clear need for revision in the quantitative scoring, Bieliauskas's (1956) suggestion for refinements in aspects of the H-T-P scoring have been disregarded. Scoring instructions in the manual continue to be complex, ambiguous, and lacking in clarity, and "in any set of drawings the examiner may find items for which no scoring or only some scoring is provided" (Buck, 1981, p. 34). In confounding cases, Buck (1981) suggests, "If it is learned that a drawn whole is a stereotype or a reproduction of a learned figure ('Teacher makes us draw them that way'), the examiner may treat the figure qualitatively only" (p. 35). But this is untenable; if a drawing has been invalidated for quantitative scoring because it is a learned whole that has no psychological precursors, how can the same sketch have valid psychological meaning for projective interpretation? Further, Buck's (1981) own caveats bring to question the validity of the quantitative scoring: "In constructing this relatively objective quantitative scoring system, it was very difficult to divorce the measurement of 'architectural artistry' which is presumed to be a highly specialized and specific ability from the appraisal of good proportional relationships" (p. 35). On this point, Bieliauskas and Bristow (1959) found that the drawings of art-trained subjects significantly received more favorable scores. In another study evaluating the feasibility of the quantitative system for children in Grades 2-5, Bieliauskas and Moens (1961) found negative results, indicating that the scoring was not applicable to children.

Because of these proceeding points, the value and meaning of the H-T-P IQ scores seem questionable. Consequently, a longer than average training period is required to learn the quantitative scoring method, and the "technique cannot be properly mastered from manuals and journal articles" (Harriman, 1970, p. 860); however, once mastered, according to Buck (1981) an "experienced examiner usually can score and interpret a full achromatic-chromatic H-T-P in one hour and a half or less" (p. 251). Krugman (1970) and others feel "the method of scoring the H-T-P for intelligence is so complicated . . . that it seems doubtful whether anyone but the author of the test can achieve so high a correlation with a standardized intelligence scale. Furthermore, the time required for scoring the objective part seems prohibitive . . ." (p. 345). Hayworth (1970) also feels the "scoring criteria (descriptive and diagrammatic) are so detailed, qualified, and ambiguous that the reliability of scoring is questionable, and no data are offered on this aspect. The time spent in such scoring would be better spent in administering a standard intelligence test" (p. 1240).

Unfortunately, due to these complexities and the continual need for plate comparisons in the manual, computer scoring is impossible. In addition, three inherent difficulties in the scoring involve the use of the manual. First, the comparison plates needed to illustrate scoring points are extremely small. Second, assessing the relative quality value based on the eight possible intelligence levels for each characteristic of the House, Tree, and Person for detail, proportion, and perspective is cumbersome and time-consuming. And third, the manual does not clearly explain the tabulation sheet used to plot mean raw scores, percentage scores, and derived IQ scores. Faced with all these shortcomings led Ellis (1970) to state, "Considering the unknown reliability and, especially, validity of the H-T-P intelligence estimations and personality interpretations at the present time, and considering the time available to the psychologist in a normal clinical situation, it is unlikely that the H-T-P is normally worth this many hours of a busy psychologist's time" (p. 592). According to Harris (1963):

> Buck's manual is not clear as to procedure of evaluation, or wholly satisfactory as a guide to interpretation . . . Buck's own statistical criteria for denoting certain characteristics as unusual while consistently applied, appear to have no basis in statistical logic. (p. 49)

The interpretation of the H-T-P is based on both the quantitative scoring and qualitative analysis. In terms of the quantitative scoring, interpretations involve five steps (Buck, 1981): 1) differences between the IQ scores; 2) appraisal of mean score patterns; 3) evaluation of detail, proportion, and perspective; 4) comparison of the "good" and "flawed" scores; and 5) comparison of the achromatic and chromatic drawings. In terms of the projective analysis of the drawings, interpretations involve the evaluation of: 1) detail, proportion, and perspective; 2) elapsed time, line quality, attitude toward the task, color choice, and drive; 3) clinical analysis of the P-D-I; and 4) the subject's concepts of each sketch encompassing both graphic and verbal productions.

Several limitations affect the reliability and level of training needed to interpret the H-T-P (Buck, 1981):

1) No single sign itself is an infallible indication of any strength or weakness in the S.
2) No H-T-P sign has but one meaning.
3) The significance of a sign may differ markedly from one constellation to another.
4) The amount of diagnostic and prognostic data derivable from each of the points of analysis may vary greatly from S to S.
5) Colors do not have any absolute and universal meaning.
6) Nothing in the quantitative scoring system can be taken automatically at face value. (p. 80)

Moreover, each sketch can symbolize multiple concepts. For example, the House could represent (Buck, 1981): 1) home as it is now, or as the subject 2) would like it to be; 3) an unsatisfying or 4) a satisfying past home. As a self-portrait, aspects of the house hypothetically can represent: 1) the subject's psychosexual maturity; 2) the subject's accessibility; 3) the subject's contact at the level of reality; 4) the subject's intra-personal balance; 5) the degree of rigidity of the subject's personality; 6) the relative roles of the psychological past and future in the subject's psychological field; and 7) the subject's attitude toward his or her family and/or the subject's interpretation of the family's feeling toward him or her.

The Person may potentially represent an individual in the subject's environment whom the subject most likes, dislikes, or feels ambivalent towards. As a self-portrait, aspects of the Person hypothetically may represent: 1) the subject as he or she is now, 2) feels now, 3) would like to be; 4) the subject's concept of his or her sexual role; 5) the subject's attitude toward interpersonal relationships in general or 6) toward a specific relation; and 7) certain specific fears and/or obsessions.

The Tree could represent either the subject or some other person, and as a self-portrait aspects of the Tree could represent: 1) the subject's subconscious picture of self in relation to his or her psychological field; 2) the subject's subconscious picture of his or her development; 3) the subject's psychosexual level; 4) the subject's contact with reality; or 5) the subject's feeling of intra-personal balance.

Because of these complexities and the multiple meanings that can be given to various aspects of a drawing, a high degree of training and experience is required to properly interpret the sketches. To simplify the task, Wenck (1984) and Jolles (1983) have considerably reduced the task by providing a catalog of examples for various interpretative points. Although most of the interpretations are hypothetical, anyone interested in using the H-T-P from a projective standpoint will find the task of projective interpretation less overwhelming. However, beginners should use caution with these texts since there is little experimental support for the interpretations of these signs, and these interpretations should only be considered as tentative hypotheses that need further support from other sources. It is worth quoting Ellis's (1970) observation that "Buck frequently contradicts his own warnings and makes rash general and specific interpretative statements about the H-T-P which, as yet, are not backed up by any factual evidence whatever. He presents, in fairly dogmatic form, hypothesis after hypothesis which may logically seem to be true but which have not yet been psychologically and scientifically established" (p. 593).

Technical Aspects

In both the original manual (Buck, 1948) and the revised manual (Buck, 1981) evidence regarding reliability and validity are conspicuously missing. According to Ellis (1970) the original manual "displays incredible naivete, fanaticism, and arrant disregard for any attempt at scientific validation of the material presented" (p. 592). In terms of the revised manual, perhaps the most accurate denunciation comes from the developer himself (Buck, 1981):

> There is almost no statistical proof of the validity of the qualitative scoring points and their interpretations which would satisfy . . ." the tenets of research design and scientific method. . . ." And it appears unlikely, in fact, that such evidence will be available for several reasons, two of which are: 1. The fact that almost no *H-T-P* scoring point has a single implication. . . . 2. The fact that a given characteristic or trait may be expressed in the *H-T-P* in many ways. . . . It is the author's belief that the validity of the principle of the *H-T-P* method as a whole has been satisfactorily established (although the evidence is almost wholly clinical). The evidence of the validity of the individual differential items and their interpretations is less well-established but is certainly sufficient to justify the conclusion that the *H-T-P* is a mature clinical instrument. . . . As for reliability, no significant data are offered at this time. . . . The *H-T-P* does not possess a high order of statistically defined reliability. . . . (p. 164)

One wonders how a test can be valid and not reliable: if a test is not accurate or consistent (reliability) how can a test actually measure what it purports to measure (validity)! A test is simply not valid in general; tests are valid for a specific purpose. In test construction the types of reliability (test-retest, alternate form, and internal consistency) and validity (content, criterion-related, and construct) should be presented. For the H-T-P, reliability studies are meager and are limited to test-retest reliability or interrater reliability. Although alternate form reliability is not possible with the H-T-P, internal consistency reliability could be performed, but as yet no studies are available to assess the extent to which items on a drawing correlate among themselves.

Reliability studies using the H-T-P IQs would be relatively easy to perform; unfortunately, most clinicians do not use these scores and question their validity. Since IQ scores are frequently not utilized, investigations on test-retest reliability lack objective criteria in determining similarity of reproduction between different administrations. In addition, even if IQ scores were used to assess reliability, scores could show a positive correlation yet form could be radically different, producing alternate clinical interpretations. The most reasonable solution would seem to be internal consistency reliability; however, this requires interrater reliability which currently seems difficult to achieve on the H-T-P.

In evaluating the drawings of 32 paranoid schizophrenics, Fisher and Fisher (1950) found little interrater reliability. Their results indicated that trained psychologists had no greater interrater reliability than untrained raters. Lehner and Gunderson (1952) used the Draw-A-Person technique with normal subjects on 21 graphic traits and found a "relatively higher" agreement between raters on these 21 traits. However, with 43 college students, Bieliauskas (1956) compared judges on "flaw score" and "percent of raw G" and found lower than required correlations for interrater reliability.

In terms of test-retest reliability, Gasorek (1951) found conflicting evidence in children's drawings for consistency and reliability of formal and structural properties. Lehner's and Gunderson's (1952) study on test-retest reliability found results similar to Gasorek's. When limiting interrater reliability to just 21 items, they found interrater reliability to be greater than test-retest reliability.

If Buck (1981) is correct that "the H-T-P does not possess a high order of statistically defined reliability" (p. 164), then validation studies seem to have little if any meaning. If this projective test is not consistent in its measurement, and if raters cannot consistently agree on scoring, then it cannot possibly measure what it purports to measure. Buck incorrectly assumes that validity can exist without reliability. Thus, the following review of validation studies seems to have little value, except from a historical perspective.

The validity studies conducted by Buck (1981) show evidence that the projective technique indeed reflects intellectual functioning and not just non-intellective factors. The correlation coefficients between the H-T-P percent of raw G IQ and IQs of other tests are as follows: Otis, Higher Examination (.41); Stanford-Binet, Forms L-M (.45); Wechsler-Bellevue, Verbal (.70), Performance (.72), Full Scale (.75). However, more recently Hellkamp and Johnson (1970) found nonsignificant correlations between the H-T-P IQs and the Wechsler Adult Intelligence Scale and Raven's Coloured Progressive Matrices. Since this last study used psychometrically sound instruments, it raises serious questions concerning the actual meaning of these IQs.

The most frequent validation studies attempt to correlate graphic traits of a particular group to those of another defined group. Typically, groups are defined in terms of psychotic disorders, personality disorders, organic disorders, or physical characteristics.

Singer (1950) found inconclusive results when comparing 40 college students and 34 schizophrenic patients given the H-T-P. On the other hand, Holzberg and Wexler (1950) using a 174-item checklist found statistically significant differences between 38 schizophrenic women and 78 nurses. More recently, Wildman and Wildman (1975) showed that the H-T-P, out of three projective tests, discriminated the least between 10 nurses and 10 female patients. Twenty sets of protocols were given to 6 blind clinical psychologists; the MMPI had the highest hit rate, discriminating with 88% accuracy, while the H-T-P discriminated with only 53% accuracy.

Demming (1949) found no statistically significant differences between the H-T-Ps of 20 psychopathic patients and 20 normal controls that were matched for intelligence. Similarly, Royal (1949) and Blum (1954) did not find a significant difference between normals and neurotics on the Draw-A-Person test. Giedt and Lehner (1951) also found no significant difference between normals and neurotics in terms of the age assigned to the Person drawing. Gravitz (1969) grouped 200 normal adult males and females into those high and low on depression. Using the Depression (D) score of the Minnesota Multiphasic Personality Inventory it was hypothesized, based on Buck's theory, that subjects with high D scores would tend to draw smaller figures than those with low D scores. Statistical analyses failed to show any significant differences in the size of drawings based on D scores. Marzolf and Kirchner (1972) gave 1,054 college students the H-T-P and the 16 Personality Factor Questionnaire. All drawings were scored based on Buck's 108 drawing characteristics, but there were no significant relationships between drawing characteristics and personality traits.

Beck (1955) investigated the H-T-Ps of 25 organic and 13 non-organic mentally retarded children and found no significant differences in the drawings of the House. Michael-Smith (1953) studied 25 matched pairs of children with normal and abnormal EEG patterns. H-T-P signs indicative of organicity were negative except for "line quality." A follow-up study was recommended and was conducted by Bieliauskas and Kirkham (1958), who used 18 criteria to examine the H-T-P drawings of 20 organic and 20 non-organic subjects matched for sex, age, and IQs, but found that the H-T-P signs of organicity were not valid since the H-T-P signs failed to hold for either group. Williams (1964) matched 20 schizophrenic, 20 organic, and 20 normal controls on the same 18 criteria for organicity and 32 signs for schizophrenia, and again no statistically significant differences were found.

Waxenberg (1955) compared three groups of females, comprised of 20 asthmatics, 20 non-psychosomatics, and 20 with histories of ulcerative colitis, on the H-T-P drawings, the Thematic Apperception Test, the Bender-Gestalt, and the Rorschach, and found no significant differences on all tests between the three groups. Silverstein and Robinson (1956) compared 22 children with orthopedic disabilities with 44 healthy children on the H-T-P; findings showed no significant differences between the two groups. Wawrzaszek et al. (1958) compared 41 matched pairs of handicapped and non-handicapped children, and no significant differences were found between the two groups. Davis and Hoopes (1975) compared the Human Figure Drawings of 30 deaf and 80 hearing children between ages 7.5 - 10.5 on 19 items concerning the characteristics of the ear and mouth. No significant differences between deaf and hearing children were found related to the ear and mouth drawings, except that there was more frequent shading around the mouth for the hearing children. There were however significant differences in the branch structure of the Tree: deaf children tended to imply a branch system, while hearing children drew out the branch system.

Other validity studies have been conducted in an attempt to verify certain specific hypotheses that Buck (1981) formulated. The overwhelming majority of the research has shown that many of these clinical interpretations are not valid. According to Hayworth (1970), "The most recent research studies designed to test various hypotheses connected with the *H-T-P's* rationale have generally reported non-significant findings" (p. 1241).

Critique

From the present review, it seems that the psychometric properties of the H-T-P fall short of today's standards and the test seems to lack the required focused relevance for the practicing clinician or experimental psychopathologist. Psychological studies should be directed at valid assessment techniques that can reliably differentiate processes and functions that may be clearly implicated in various disorders (Killian et al., 1984). The use of unreliable techniques that are recommended only by their availability and familiarity should be abandoned. Instead, reliable and valid tests such as the Stroop Color and Word Test could be easily administered and scored, yet provide significant information about processes and functions in various disorders (Killian, in press). On one final note:

> The "figure drawing" approach, as loosely described by Machover, Buck, Jolles, and others, appears more simple and direct, but permits the interpreter to "project" as much as his subject! (Harris, 1963, p. 51)

References

Beck, H. S. (1955). A study of the applicability of the H-T-P with respect to the drawn house. *Journal of Clinical Psychology, 11,* 63-66.

Beck, H. S. (1960). The house-drawing test as a predictor of first-grade achievement. *Journal of Experimental Psychology, 2,* 197-200.

Bieliauskas, V. J. (1956). Scorer's reliability in the quantitative scoring of the H-T-P technique. *Journal of Clinical Psychology, 12,* 366-369.

Bieliauskas, V. J. (1960). Sexual identification in children's drawings of the human figure. *Journal of Psychology, 16,* 42-44.

Bieliauskas, V. J., & Bristow, R. B. (1959). The effect of formal art training upon the quantitative scoring on the H-T-P. *Journal of Clinical Personality, 15,* 57-59.

Bieliauskas, V. J., & Kirkham, S. L. (1958). An evaluation of the "organic" signs in the H-T-P drawings. *Journal of Clinical Psychology, 14,* 50-54.

Bieliauskas, V. J., & Moens, J. F. (1961). An investigation of the validity of the H-T-P as an intelligence test for children. *Journal of Clinical Psychology, 17,* 176-180.

Blum, R. (1954). The validity of the Machover DAP technique. *Journal of Clinical Psychology, 10,* 121-125.

Boring, R. O., & Topper, R. C. (1949). *A psychodiagnostic screening technique.* Tuscaloosa, AL: Veterans Administration Hospital.

Buck, J. (1948). The H-T-P technique, a qualitative and quantitative scoring method. *Journal of Clinical Psychology Monograph Supplement No. 5,* 1-120.

Buck, J. (1981). *The House-Tree-Person technique: A revised manual.* Los Angeles: Western Psychological Services.

Caligor, L. (1951). The determination of the individual's unconscious concept of his masculinity-femininity identification. *Journal of Projective Technique, 15,* 494-509.

Cassel, R. H., Johnson, A. P., & Burns, W. H. (1958). Examiner, ego defense, and the H-T-P test. *Journal of Clinical Psychology, 14,* 157-160.

Cook, E. (1885). Art teaching and child nature. *London Journal of Education.*

Cowden, R. C., Deabler, H., & Feamster, J. H. (1955). The prognostic values of the Bender-Gestalt, H-T-P, TAT, and Sentence Completion. *Journal of Clinical Psychology, 11*(3), 271-275.

Davis, C. J., & Hoopes, J. L. (1975). Comparison of H-T-P drawings of young deaf and hearing children. *Journal of Personality Assessment, 39,* 28-33.

Demming, J. A. (1949). *The H-T-P test as an aid in the diagnosis of psychopathic personality.* Unpublished master's thesis, Kent State University.

Diamond, S. (1954). The house and tree verbal fantasy. *Journal of Projective Technique, 18,* 316-325.

Ellis, A. (1970). H-T-P: A projective device and a measure of adult intelligence. In O. K. Buros (Ed.), *Personality tests and reviews* (pp. 591-594). Highland Park, NJ: The Gryphon Press.

Fisher, S., & Fisher, R. (1950). Test of certain assumptions regarding figure drawing analysis. *Journal of Abnormal Social Psychology, 45,* 727-732.

Gasorek, K. (1951). *A study of the consistency and reliability of certain of the formal and structural characteristics of children's drawings.* Unpublished doctoral dissertation, Teacher's College, Columbia University.

Giedt, F. H., & Lehner, G. F. (1951). Assignment of ages on the DAP by male NP patients. *Journal of Personality, 19,* 440-448.

Goodenough, F. (1926). *Measurement of intelligence by drawings,* New York: World Book Company.

Gravitz, M. A. (1969). Figure drawing size as an index of depression and MMPI scores in normal adults. *Journal of Clinical Psychology, 25,* 77-79.

Hammer, E. F. (1954). A comparison of H-T-P's of rapists and pedophiles. *Journal of Projective Techniques, 18,* 346-354.

Hammer, E. F. (1955). *The H-T-P clinical research manual.* Los Angeles: Western Psychological Services.

Hammer, E. F. (1978). *The clinical application of projective drawings.* Springfield, IL: Charles C. Thomas.

Harriman, P. (1970). H-T-P Projective Technique. In O.K. Buros (Ed.), *Personality tests and reviews* (pp. 858-860). Highland Park, NJ: The Gryphon Press.

Harris, D. B. (1963). *Children's drawings as measures of intellectual maturity.* New York: Harcourt Brace Jovanovich, Inc.

Harrower, M. R. (1978). The Most-Unpleasant-Concept Test, a graphic projective technique for diagnostic and therapeutic use. In E. F. Hammer (Ed.), *The clinical application of projective drawings* (pp. 365-390). Springfield, IL: Charles C. Thomas.

Hayworth, M. (1970). H-T-P: A projective device. In O. K. Buros (Ed.), *Personality tests and reviews* (pp. 1240-1241). Highland Park, NJ: The Gryphon Press.

Hellkamp, D. T., & Johnson, J. E. (1970). Actuarial and clinical estimates of present and premorbid intelligence on the H-T-P, Raven CPM, and WAIS. *Ohio Research Review, 3,* 307-314.

Holzberg, J. O., & Wexler, M. (1950). The validity of human form drawings of personality deviations. *Journal of Projective Techniques, 14,* 343-361.

Johnson, O. G., & Wawrzaszek, F. (1963). Psychologist's judgment of physical handicap from H-T-P drawings. *Journal of Consulting Psychology, 25,* 284-287.

Jolles, I. (1952). A study of the validity of some hypotheses for the quantitative interpretation of the H-T-P for children of elementary school age: I. Sexual identification. *Journal of Consulting Psychology, 8,* 113-118.

Jolles, I. (1983). *A catalog for the qualitative interpretation of the H-T-P.* Los Angeles: Western Psychological Services.

Killian, G. A. (in press). The Stroop and Color Word Test. In D. J. Keyser & R. C. Sweetland (Eds.), *Test Critiques* (Vol. II). Kansas City, MO: Test Corporation of America.

Killian, G. A., Holzman, P. S., Davis, J. M., & Gibbons, R. (1984). Effects of psychotropic medication on selected cognitive and perceptual measures. *Journal of Abnormal Psychology, 93*(1), 58-70.

Krugman, M. (1970). H-T-P: House, Tree, and Person. In O. K. Buros (Ed.), *Personality tests and reviews* (pp. 345-346). Highland Park, NJ: The Gryphon Press.

Landesberg, S. (1953). Relationship of Rorschach to the H-T-P. *Journal of Clinical Psychology, 9,* 179-183.

Lehner, G. F., & Gunderson, E. K. (1952). Reliability of graphic indices in a projective test (DAP). *Journal of Clinical Psychology, 8,* 125-128.

Levy, S., & Levy, R. (1978). Symbolism in animal drawings. In E. F. Hammer (Ed.), *The clinical application of projective drawings* (pp. 311-343). Springfield, IL: Charles C. Thomas.

Machover, K. (1949). *Personality projection in drawings of a human figure.* Springfield, IL: Charles C. Thomas.

Marzolf, S. S., & Kirchner, J. H. (1972). H-T-P drawings and personality traits. *Journal of Personality Assessment, 36*(2), 148-165.

Michael-Smith, H. (1953). The identification of pathological cerebral function through the H-T-P technique. *Journal of Consulting Psychology, 9,* 293-295.

Naumberg, M. (1978). Art therapy: Its scope and function. In E. F. Hammer (Ed.), *The clinical application of projective drawings* (pp. 511-517). Springfield, IL: Charles C. Thomas.

Rosenberg, L. (1948). *Modification of Draw-A-Person test.* Unpublished master's thesis, New York University.

Royal, R. E. (1949). Drawing characteristics of neurotic patients using a drawing of a man and a woman technique. *Journal of Clinical Psychology, 5,* 392-393.

Silverstein, A. B., & Robinson, H. A. (1956). The representation of orthopedic disability in children's figure drawings. *Journal of Consulting Psychology, 20,* 333-341.

Singer, R. H. (1950). *A study of drawings produced by a group of college students and a group of hospitalized schizophrenics.* Unpublished master's thesis, The Pennsylvania State College.

Wawrzaszek, F., Johnson, O. G., & Sciera, J. L. (1958). A comparison of the H-T-P responses of handicapped and non-handicapped children. *Journal of Clinical Psychology, 14,* 160-162.

Waxenberg, S. E. (1955). Psychosomatic patients and other physically ill persons: A comparative study. *Journal of Consulting Psychology, 19,* 163-169.

Wenck, S. (1984). *H-T-P drawings: An illustrated diagnostic handbook.* Los Angeles: Western Psychological Services.

Wildman, R. W., & Wildman, R. W., II. (1975). An investigation into the comparative validity of several diagnostic tests and test batteries. *Journal of Clinical Psychology, 31,* 455-458.

Williams, L. A., Jr. (1964). *An evaluation of selected schizophrenic signs in the H-T-P drawings.* Unpublished master's thesis, Xavier University, Cincinnati.

Wyatt, F. (1949). The case of Gregor: Interpretation of test data. *Journal of Projective Techniques, 13,* 155-205.

Zucker, L. A. (1948). A case of obesity: Projective technique before and after treatments. *Journal of Projective Techniques, 12,* 202-215.

Joan M. Preston, Ph.D.
Associate Professor of Psychology, Brock University, St. Catharines, Ontario.

ILLINOIS TEST OF PSYCHOLINGUISTIC ABILITIES-REVISED

Samuel A. Kirk, James J. McCarthy, and Winifred D. Kirk. Champaign, Illinois: University of Illinois Press.

The original of this review was published in Test Critiques: Volume I (1984).

Introduction

The Revised Illinois Test of Psycholinguistic Abilities (ITPA) is an individually administered test for children aged 4-8 years, measuring 12 functions employed in the acquisition and use of language. The test consists of 10 main subtests (Auditory Reception, Visual Reception, Auditory Association, Visual Association, Verbal Expression, Manual Expression, Grammatic Closure, Visual Closure, Auditory Sequential Memory, and Visual Sequential Memory) and two supplementary subtests (Auditory Closure and Sound Blending). Raw scores for each subtest are used to derive Scaled Scores for each subtest, a Composite Score, Psycholinguistic Age Scores for subtests and Composite, and Psycholinguistic Quotients for subtests and Composite. The ITPA was conceived as a diagnostic test and can be administered in approximately one hour.

The ITPA was developed by S.A. Kirk, Ph.D., currently Professor of Special Education, University of Arizona, and former director of the Institute for Research on Exceptional Children, University of Illinois; J.J. McCarthy, Ph.D., currently Professor in the Department of Studies in Behavioral Disabilities, University of Wisconsin; and W.D. Kirk, Ph.D., formerly Adjunct Professor, Department of Speech and Hearing, University of Arizona.

Test development began in 1950 when various methods were explored for evaluating receptive and expressive language in children. Later McCarthy and S. Kirk modified Osgood's model of the communication process to generate a model of children's communication skills. Like Osgood, McCarthy and Kirk hypothesized three dimensions: 1) Processes (receptive process, organizing process, and expressive process), 2) Levels of Organization (representational level and automatic level), and 3) Channels of Communication (auditory-vocal and visual-motor). In 1961, an experimental edition of the Illinois Test of Psycolinguistic Abilities was published. It consisted of nine tests based on the three-dimensional model. The development of the experimental ITPA is described by Kirk and McCarthy (1961). The experimental edition was used extensively in both research and clinical studies (see Bateman, 1965; McCarthy & Kirk, 1968). In 1965, work on a revised test was begun to improve the tests in the battery and to add tests not included in the experimental version.

The Revised Illinois Test of Psycholinguistic Abilities was published in 1968. The 12 subtests were designed to reflect particular compartments of the three-dimensional model. For example, the Visual Reception subtest reflects the visual-vocal channel, receptive process, representational-level compartment. Three subtests, the Grammatic Closure, and the two supplementary tests, Auditory Closure and Sound Blending, reflect auditory-vocal channel, organizing process, automatic-level compartment.

The normative group consisted of 962 children aged 2 to 10, each of whom met five specific criteria: 1) average intellectual functioning, 2) average school achievement, 3) sensory-motor integrity, 4) at least average characteristics of personal-social adjustment, and 5) English spoken as the family language.

The ITPA consists of a variety of materials and books. These include one copy each of the *Examiner's Manual* (Kirk, McCarthy, & Kirk, 1968), *The Development and Psychometric Characteristics of the Revised Illinois Test of Psycholinguistic Abilities* (Paraskevopoulos & S. Kirk, 1969) and *Aids and Precautions in Administering the Revised Illinois Test of Psycholinguistic Abilities* (W. Kirk, 1974). The test materials are: Picture Book 1, containing photographs of stimuli for the Visual Reception subtest; Picture Book 2, containing line drawings for the Visual Association, Grammatic Closure, and Sound Blending subtests, and photographs of stimuli for the Manual Expression subtest; a booklet, rubber tray, and plastic chips for the Visual Sequential Memory subtest; an envelope, green wooden block, rubber ball, button, and nail for the Verbal Expression subtest; a hammer for the Manual Expression subtest; five sets, 25 each, of picture strips and five scoring keys for the Visual Closure subtest; a phonograph record containing items for the Auditory Sequential Memory, Auditory Closure, and Sound Blending tests; and a set of 25 record forms. All the materials are of good quality and designed to facilitate testing.[1]

The ITPA is an individual test that can not be administered by an untrained person. However, a person with a relevant master's degree and experience in administering individual tests to children could learn to administer and to score the ITPA in a few days. The test is appropriate for children aged 4 to 8. Norms are provided for the range from 2 to 10 years to permit interpretation of results for children scoring below or above the 4-8 year norms.

The 12 subtests include:

1. *Auditory Reception,* which measures the ability to gain meaning from auditorily received stimuli. This subtest consists of demonstration items and 50 test items, each requiring the child to reply "Yes" or "No" (e.g., Do rocks play?). Items increase in difficulty and testing continues until the child fails 3 items in any block of 7 consecutive items.

2. *Visual Reception* refers to the ability to gain meaning from visually received stimuli. This subtest consists of demonstration items and 40 test items. The child sees a photograph of an object and must select the most conceptually similar item from among photographs of 4 options. Items increase in difficulty and testing continues until the child fails 3 consecutive items.

3. *Visual Sequential Memory* involves the ability to reproduce from memory sequences of visually received stimuli. This subtest consists of demonstration items and 25 test items. The child sees a card showing 2 or more squares containing line patterns and is required, after removal of the sequence card, to replicate the sequence of squares in the same order, using plastic chips and a tray. Items increase in difficulty and testing continues until the child fails 2 consecutive items.

4. *Auditory Association* is the ability to relate auditorily received material in a meaningful way. The subtest consists of demonstration items and 42 test items. The administrator reads an incomplete analogy and the child is required to supply the final term (e.g., A daddy is big, a baby is _____). Items increase in difficulty and testing continues until the child fails 3 consecutive items.

5. *Auditory Sequential Memory* involves the ability to reproduce from memory, immediately after presentation, sequences of stimuli that are auditorily received. The subtest consists of demonstrations and 28 test sequences of 2 or more digits presented at a rate of 2 per second. Later items contain longer sequences. Testing continues until the child fails 2 consecutive items.

6. *Visual Association* refers to the ability to relate visually received stimuli in a meaningful way. The subtest consists of demonstration items and 42 test items. Each item consists of 5 line drawings, one object within a center circle and one of the remaining objects in each of 4 corners of a surrounding rectangle. The child is required to indicate which of the 4 alternate drawings most meaningfully relates to the object in the circle. Items increase in difficulty. Testing continues until the child fails 3 consecutive items.

7. *Visual Closure* assesses the child's ability to identify a common object from an incomplete visual presentation. The test consists of one demonstration and four test picture strips. The administrator shows a sample picture of the object or objects to be located. Then the picture strip is exposed and the child points to all of the specified objects he or she can find in 30 seconds. Each child completes all test items.

8. *Verbal Expression* is the ability of the child to express his or her own concepts vocally. The test materials consist of 5 objects including a demonstration item. The administrator asks the child to "tell all about this." Each child completes all test items.

9. *Grammatic Closure* assesses the child's acquisition of automatic habits for handling syntax and grammatic inflections. This subtest consists of a demonstration item and 33 test items. The child sees 2 line drawings side by side. The administrator points to the drawing on the left and makes a statement about the object. The administrator then points to the diagram on the right using an incomplete statement. The child is required to provide the missing word (e.g., Here is a bed, here are two _____.).

10. *Manual Expression* refers to the child's ability to express ideas manually. This subtest consists of a toy hammer (for demonstration) and photographs of 15 common items. The child is asked to show what is done with the photographed objects by pretending to use real objects. Each child completes all test items.

11. *Auditory Closure* assesses the child's ability to produce a complete word when parts of the word are deleted from the auditory presentation. This supplementary subtest consists of demonstration items and 30 test items (e.g., child

hears "da____y," and must complete the word "daddy"). Items increase in difficulty. Testing continues until child fails 6 consecutive items.

12. *Sound Blending* measures the child's ability to produce an integrated whole word after hearing the single sounds in the word. Sounds are spoken singly, one every half second. Items increase in difficulty and include both English words and nonsense words (e.g., D__OG, child says "dog"). This supplementary subtest includes demonstration items and 32 test items. The first 7 items are used with pictures. The last 8 items are nonsense words and demonstration nonsense words are presented before these items are given. Testing on English words stops when the child fails 3 consecutive items. If the third consecutive error occurs after item 18, the administrator continues testing using nonsense words until the child fails 3 consecutive nonsense items.

On several subtests (Auditory Reception, Visual Reception, Visual Sequential Memory, Auditory Association, Auditory Sequential Memory, Visual Association, and Sound Blending), children aged 6 or older begin with later items than younger children. Demonstration items are given to older children and if they are unable to pass the first few items, the administrator gives additional demonstration items and starts testing at item one. The child's responses to each item are recorded on the Record Form by the administrator. The Record Form also contains a Summary Sheet and a Profile of Abilities for plotting scaled scores.

Practical Applications/Uses

The ITPA may be used for assessment and diagnosis in educational, clinical, and research settings where knowledge of a variety of language-related processes in children is required. A particular advantage of the ITPA is the measurement, on separate subtests, of the child's ability to understand, to organize, and to express information, and to assess these processes separately for the auditory-vocal channel and the visual-motor channel. The test should be administered to English-speaking children aged 4 to 8 years. It should not be used with children who are unable to hear or understand spoken instructions. Additional factors to be considered before administering the ITPA are discussed later in this review.

The ITPA is designed to diagnose specific language difficulties in children. Each subtest measures one language function to facilitate the identification of specific strengths and weaknesses in the child's acquisition and use of language. In addition, research studies suggest the automatic-level subtests of the ITPA are related to reading and spelling achievement.

The ITPA is appropriate for children aged 4 to 8 years. Because it requires children to comprehend verbal instructions and to use language in various ways, the test is unsuitable for young preschool children. The content of the items is appropriate for primary grade children, and this makes the test unsuitable for adolescents or adults. While the normative group consisted of "average" children, the ITPA has been used successfully to test children ages 6-8 years having severe oral language disorders (see Luick, Kirk, Agranowitz, & Busby, 1982). The ITPA has been shown to differentiate between groups of children as a function of several characteristics. Black children score higher than white and Mexican-

American children on Auditory Sequential Memory but lower than these two groups for Visual Sequential Memory. The pattern is reversed for Mexican-American children who score higher than the other two groups on Visual Sequential Memory and lower for Auditory Sequential Memory (Cicerelli, Granger, Schemmel, Cooper, & Holthouse, 1971). Disadvantaged bilingual children score lower on auditory-vocal subtests compared to visual-motor subtests (Jorstand, 1971; Michaelson, 1970). Aboriginal children score lower than the normative population on most subtests (Kuske, 1969; Lombardi, 1970; Norcombe & Moffitt, 1970). Such findings do not reflect test weaknesses but rather confirm that the ITPA is sensitive to differences in children's opportunities to acquire and use language.

Some characteristics of children are related to performance on particular subtests. Paraskevopoulos and Kirk (1969) present data concerning sex differences, social class, number of siblings, position among siblings, and intelligence. They also indicate the kinds of perfomance deficits typical for children who have special problems including dyslexia, mental retardation, mongolism, blindness or partial sight, deafness or hearing deficiencies, cerebral palsy, and articulatory speech defects. The ITPA is an individually administered test requiring about one hour to complete. The examiner should hold a relevant postgraduate degree, be experienced in testing children, and be trained specifically in the administration of the ITPA. The manual clearly sets out the materials and instructions for each subtest but an alert, well-trained examiner is required to insure that standardized procedures are followed exactly.

Scoring of the ITPA is done by hand and can be completed, with profile, in less than one hour by an experienced examiner, although a training period of 2 or 3 days is needed to learn how to score the test. The scoring instructions are clearly presented with both correct and incorrect responses indicated for many subtests to insure standardized scoring. In addition to the Examiner's Manual, W. Kirk's *Aids and Precautions in Administering the Illinois Test of Psycholinguistic Abilities* (1974) is useful in avoiding difficulties and clarifying procedures for scoring.

Once raw scores have been obtained for each subtest, Scaled Score Norms (for the child's chronological age) can be obtained using the appropriate tables. Scaled Scores are used to prepare a Profile of Abilities (provided in Record Form). Scaled Scores permit the child's performance to be compared with norms of same-age peers. For each referral group, the mean Scaled Score is 36 with a standard deviation of 6. When a child's Scaled Score is 10 points above or below the Mean or Median Scaled Score (36), the child is considered to have a substantial discrepancy. Differences of 7-9 points suggest borderline discrepancies while differences of 6 points or less are considered to be within the range of average or typical performance. Scaled Scores, because they take into consideration both mean performance and variability within the group, are normally the basis for interpretation of a child's performance.

Psycholinguistic Age scores can be obtained using the appropriate table. Psycholinguistic Age norms indicate the chronological age for which a particular raw score is typical and thus reflects the child's level of ability.

The Psycholinguistic Quotient is calculated using the formula Composite Psycholinguistic Age Score divided by Chronological Age, the result multiplied by 100. The child's Psycholinguistic Quotient is interpreted in relation to a group mean score of 100 and reflects rate of psycholinguistic development. Comparable Psycholinguistic Quotients may also be calculated separately for each subtest.

Interpretation of the ITPA should be done by an individual with a postgraduate degree in a relevant discipline and both training and experience in interpreting test scores. Guidelines for evaluating both interindividual differences and differences in subtest performances for a particular individual are provided by Paraskevopoulos and Kirk (1969).

Technical Aspects

Validity indicates the extent to which a test measures what it is built to measure. Construct validity involves the evaluation of the theory underlying the test. Cohen (1973) used the Guttman-Lingoes Smallest Space Analysis, a non-metric procedure, to analyze intercorrelations among the 10 main subtests of the ITPA administered to 569 first-grade children. A reasonably good fit was obtained in two dimensions. The results supported the channel and process dimensions of the communications model of the ITPA. Elkins (1973) also used the Guttman-Lingoes Smallest Space Analysis to examine the representations of the ITPA subtests in 2, 3, and 4 dimensions. The data were obtained from the ITPA standardization sample (reported in Paraskevopoulos & Kirk, 1969) of 962 children. The best fit was obtained for the three-dimensional representation supporting the three dimensions of the ITPA communications model. The results also showed clear separation of auditory and visual channels as well as proximity for Manual and Auditory Expression and for Auditory and Visual Association.

A factor analysis of ITPA data obtained for 237 children, ages 6.0-8.3 years, having severe oral language disorders but at least average WISC Performance IQ scores, showed two clear factors: auditory-vocal and visual-motor (Luick, Kirk, Agranowitz, & Busby, 1982). Cluster analysis of this data revealed several clusters, all of which showed that scores for visual-motor subtests were superior to auditory-vocal channel subtests and, within the auditory-vocal channel, the lowest scores were on the Auditory Association and Grammatic Closure subtests (Luick, Kirk, Agranowitz, & Busby, 1982). These findings support the two levels of channel dimension and also offer evidence of validity for the Auditory Association and Grammatic Closure subtests. Roe (1977) found significant correlations between Gesell scores in infancy and visual-motor subtests at age 5 but not auditory-vocal subtests at age 5. This study also supports the two levels of the channel dimension. Ninety-seven children with learning disorders were administered the ITPA by Pirozzolo, Obrzut, and Hess (1983). Their factor analyses supported the independence of both the channel and levels of organization dimensions. Elkins (1974) administered the ITPA to a total of 234 Australian children in Grades 1-3 and obtained scaled scores that approximated the American norms for the 10 main subtests. Results for the two supplementary tests, Auditory Closure and Sound Blending, differed from the American norms, a finding that may reflect pronunciation differences between American and Australian English. The experimental ITPA was administered to 100 four-year-old British children by Mittler and Ward (1970), who obtained scores that correspond closely to the

American norms. These studies confirm the stability of the American norms for the main subtests. McCarthy (1977) analyzed ITPA scores for 100 kindergarten children using the Campbell and Fiske multitrait-multimethod approach. Her results demonstrated both convergent and discriminant validity for the ITPA subtests.

Several studies have obtained correlations between ITPA scores and intelligence measures. Paraskevopoulos and Kirk (1969) report significant correlations of ITPA Composite scores and Psycholinguistic Quotients with Stanford-Binet Mental Age, IQ, and Vocabulary scores across several age levels (standardization data). Huizinga (1973) tested 100 six-year-old children and found ITPA Total Scaled Scores correlated significantly with Stanford-Binet IQ (.88), as well as WISC Verbal, Performance, and Full Scale IQs (.75, .58 and .80). Guest (1970) obtained significant correlations for ITPA Total Score with Wechsler Preschool and Primary Scale of Intelligence Verbal, Performance, and Full Scale IQs (.69, .47, and .67) for 47 kindergarten children. Wechsler Full Scale IQ was significantly correlated with ITPA Total scores (.61) for 73 third-grade children (Bartin, 1971) and with ITPA Psycholinguistic Quotients (.87) for 136 children aged 4-10 years (Humphrey & Rice, 1973). Similar findings were obtained by Polley (1971) using 160 children with learning disabilities in Grades 1-3. ITPA Total Scaled Score correlated significantly with WISC Full Scale IQ (.49). In addition, the ITPA score for auditory-vocal subtests correlated significantly with WISC Verbal IQ (.28) and the ITPA score for the visual-motor channel tests was related to WISC Performance IQ (.20). The correlations in this study were much lower than the other studies. The children with learning disabilities may have had a restricted range of abilities which depressed the correlations.

Studies of reading show that poor readers obtain low scores on automatic processing tests (Celebre, 1971; Dees, 1971; Macione, 1969; Richardson, DiBenedetto, Christ, & Press, 1980; and Ruhly, 1970). Poor spellers obtain low scores on the Sound Blending subtest (Bannatyne & Wichiarajote, 1969). Elkins and Sultman (1981) obtained different performance levels on the ITPA for two groups of children that differed on measures of reading and spelling. In addition, significant group separations were obtained for six out of seven discriminant analyses. Both Auditory Sequential Memory and Auditory Reception loaded heavily in the discriminant functions performed. Yom, Wakefield, and Doughtie (1975) found that scores on the Auditory Association and Grammatic Closure subtests were related to performance of five-year-old children on six conservation tasks. In a study of predictive validity, Dennis (1980) found that Grammatic Closure, Auditory Association and Auditory Sequential Memory were the best predictors of achievement for normal children, while Visual Sequential Memory and Auditory Association predicted best for learning disabled children. When IQ was controlled, Auditory Sequential Memory added significantly to the prediction equations for all measures of reading and language, and Grammatic Closure contributed significantly to Word Study and Language. Such findings provide evidence of validity of various subtests.[2]

Reliability reflects the consistency of test scores and is demonstrated by internal consistency and stability of measurement. Internal consistency assesses homogeneity of measurements and three types of homogeneity are presented for the ITPA by Paraskevopoulos and Kirk (1969). Homogeneity of items within a subtest is reported for each subtest across 8 age levels. Internal consistency coefficients range from .60-.96, with a median of .88, except for one coefficient of .45 for Sound Blending for the age level 8.7-9.1. High internal consistency was confirmed (range .63-.93) for a group of educable mentally retarded children. Internal consistency coefficients for difference scores among the ITPA subtests range from .67-.91 with a median of .81. The correlations of subtests with Composite scores by age level range from .18-.85 with a median of .56. Taken together these measures demonstrate excellent reliability for the ITPA.

Stability of measurement is reflected in test-retest coefficients. Five-month test-retest coefficients for 4-, 6-, and 8-year-old children ranged from .28-.90 with a median of .71 for subtests, .87-.93 for Composite scores, and .86-.91 for Psycholinguistic Quotients (Paraskevopoulos & Kirk, 1969). These findings were confirmed by Hatch and French (1971), who obtained slightly higher test-retest coefficients after a three-month retest period for educable mentally retarded children.

Standard error of measurement is another index of stability and reflects the degree of change expected when a series of measurements are made for the same individuals. It is dependent on both the standard deviation of test scores and test-retest reliability. A small standard error of measurement relative to the standard deviation of the test indicates stability of individual scores. Paraskevopoulos and Kirk (1969) report standard errors of measurement by age level for the 12 ITPA subtests and Composite score using raw scores, scaled scores, and Psycholinguistic Age levels. The standard deviations tend to be 2 or 3 times greater than the standard errors of measurement, indicating excellent stability of individual scores. ITPA reliability was confirmed by Mittler and Ward (1970), who observed standard errors of measurement for British four-year-old children that were very similar to the American norms as well as high split-half reliability coefficients for subtests.

Critique

The Revised Illinois Test of Psycholinguistic Abilities is a diagnostic test measuring 12 aspects of children's language processing. It is based on a useful three-dimensional communications model. Research evidence indicates that both validity and reliability of the ITPA are good.[3] The test was carefully constructed by individuals who possess expertise in test development as well as considerable experience in diagnosing language problems in children. The considerable care with which the test was constructed is documented by Paraskevopoulos and Kirk (1969), who present the details and criteria for selecting and testing the standardization groups, a description of the construction of the subtests, information about the derivation and interpretation of scores, measures of reliability, the effects of several characteristics of children on performance, interscorer reliability, several tables of norms and test statistics, and the complete raw data for the standardization group. An individual who intends to administer the ITPA must invest several days in learning to use and score this test, but the usefulness of the subtests in diagnosing specific strengths and weakness in the language use of individual children outweigh this disadvantage. In short, the ITPA is a well-designed, useful test for assessing children's language.

Notes to the *Compendium:*

[1]A cassette tape containing items for the Auditory Sequential Memory, Auditory Closure, and Sound Blending tests now replaces the phonograph record. In addition, Paraskevopoulos and Kirk's (1969) *The Development and Psychometric Characteristics of the Revised Illinois Test of Psycholinguistic Abilities* has been replaced by a 1985 version that includes an extensive list of ITPA validity studies published between 1965 and 1983.

[2]Validity studies done in the 1980s confirm and extend the findings of earlier research.

Language Delay. Studies of language delay show that language-delayed children score lower than normal-language children on various ITPA subtests. Rizzo and Stephens (1981) observed lower scores on the Auditory Reception subtest for language-delayed children aged 4-6 years. Similar results were obtained by Paul and Cohen (1985) in their longitudinal study of serious language delays. At initial evaluation, the subjects had a mean age of 6 years and a language delay of at least 9 months. Scores for Auditory Reception increased with age and, with Performance IQ, were a good predictor of school placement. Silva, McGee, and Williams (1983) measured language delay at ages 3, 5, and 7 years. The Auditory Reception and Verbal Expression tests were administered at age 7. General language delay at age 7, the combined total for Auditory Reception and Verbal Expression, was related to language delay at both age 3 and age 5. Expression delay at age 3 was correlated with low scores for Verbal Expression at age 7, and comprehension delays at age 5 were related to low scores for Auditory Reception at age 7.

A longitudinal investigation by Largo, Molinari, Pinto, Weber, and Duc (1986) studied Swiss preterm and term infants from birth to age 5. A Swiss-German version of the ITPA was administered at age 5. Children who were slow to reach language milestones between ages 1-3 had lower ITPA scores at age 5. This relationship held for both term and preterm children. At age 5, later-born children performed better on the ITPA than firstborns. Three subtests received extensive analysis. Auditory Association, Auditory Sequential Memory, and Grammatic Closure each correlated significantly with a score for Perinatal Optimality, reflecting lack of complications during pregnancy, birth, and the neonatal period for preterm children. These three tests also correlated with birth weight for preterm children, while Auditory Sequential Memory was related to birthweight for term children. For preterm subjects, Auditory Association and Grammatic Closure scores were positively related to gestational age and socioeconomic status. For term children, Auditory Sequential Memory correlated with gestational age, while all three ITPA subtests were associated with socioeconomic status. Some sex differences were observed. Term girls scored higher than term boys on the ITPA. However, preterm boys outperformed preterm girls, who weighed an average of 206 g less than the boys at birth. Some preterm children had cerebral palsy. Preterm males without CP had higher total combined scores for Auditory Association, Auditory Sequential Memory, and Grammatic Closure than preterms with CP. In addition, preterm males without CP scored higher than preterm males with CP on the Auditory Association and Grammatic Closure tests. These four studies indicate that various ITPA subtests not only measure language delays but are correlated with a variety of factors associated with language delay, including low IQ, low SES, prematurity, perinatal complications, and birth defects.

Intelligence/Reasoning. Although language performance tends to be related to intelligence, McLesky, Kandaswamy, and Colarusso (1980), in a canonical correlation analysis of ITPA and WISC scores for learning disabled children, found that 24% of the WISC subtest variance is redundant given the ITPA, while 22% of the ITPA subtest variance is redundant given the WISC. Only WISC Digit Span and ITPA Auditory Sequential Memory were significantly related. Similarly, Weed and Ryan (1983) argue that Visual Sequential Memory is different from visual reasoning because this subtest was unrelated to the Raven's Progressive Matrices test, a Piagetian stick seriation task, an alphabetic seriation task, and the Gates-MacGinitie Work Recognition Test. Nelson-Schneider (1983) observed that Verbal Expression was associated with induction acuity, while Grammatic Closure, which reflects automaticity, is unrelated to induction. Those studies indicate that, while some similarities exist, the ITPA is not a measure of intelligence.

Reading. Johnson, Shelton, and Arndt (1982) used cluster analysis to identify children with articulation difficulties. In a discriminant function analysis, the Grammatic Closure and Auditory Closure tests contributed to the differentiation of individuals in terms of reading and language performance. In a cross-sectional study, Fletcher, Satz, and Scholes (1981) tested children aged 5.5, 8.5, and 11 years. Scores for Grammatic Closure increased significantly as a function of age. These findings were obtained for both reading-disabled children and control subjects. Stanley, Smith, and Powys (1982) report that dyslexic children obtain lower scores than controls for the Visual Association test but not for the Manual Expression test. Auditory Sequential Memory was associated with reading scores for Chinese children in Grades 4 and 6, who were tested using a Chinese language version of this test (Leong, Cheng, & Das, 1985).

Intervention Studies. Several studies have used the ITPA to assess the effects of therapeutic intervention. Evans (1984) identified disadvantaged Welsh school children approximately 4½ years of age who were high versus low risk in terms of educational handicap. Two years later, high-risk children who participated in an intervention program performed better than high-risk controls on the ITPA. In a study by Naylor and Pumfrey (1983), poor readers received either training to alleviate specific deficits (overlearning auditory and visual sequences), general language training (language reception, expression, and conceptualization), or were placed in a control group. Both training groups obtained higher composite Psycholinguistic Age scores on the ITPA than controls. Only the specific training group showed changes in the pattern of deficits (i.e., improved performance for auditory closure and visual sequential memory). Hemming (1983) found that retarded adults moved from large institutions to smaller units showed improved ITPA scores compared to those remaining in large institutions. These findings suggest that the ITPA is sensitive to a variety of interventions, contrary to Clark's (1982) contention, based on a selective review of literature, that it is not sensitive to therapeutic changes.

General Studies. In a factor-analytic study, Sutter and Battin (1984) found that the ITPA provides appropriate information relevant to language instruction. Green (1980) has shown that, for preschool children, performance on the Visual Association test is a function of response latency, which is related to task difficulty. Chapman, Silva, and Williams (1984) found that Auditory Reception and Verbal Expression scores correlated with 9-year-old children's self-perceptions of general ability and reading performance.

[3]Recent research on the ITPA has tended to focus on language delays, reading difficulties, or intervention programs. Specific subtests, rather than the entire ITPA, are employed to investigate particular rather than general language questions. A few studies of intelligence and the ITPA have been conducted, focusing on differences rather than similarities between IQ and ITPA subtests.

References

Bannatyne, A.D., & Wichiarajote, P. (1969). Relationships between written spelling, motor functioning and sequencing skills. *Journal of Learning Disabilities, 2,* 4-16.

Bartin, N.G. (1971). The intellectual and psycholinguistic characteristics of three groups of differentiated third grade readers (Doctoral dissertation, State University of New York at Buffalo). *Dissertation Abstracts International, 32,* 228A.

Bateman, B. (1971). *The Illinois Test of Psycholinguistic Abilities in current research: Summaries of studies.* Urbana, IL: University of Illinois Press.

Celebre, G. (1971). Psycholinguistic abilities and oral word recognition associated with relative level of personality adjustment in primary school-age children with minimal brain dysfunction (Doctoral dissertation, Temple University, Philadelphia). *Dissertation Abstracts International, 32,* 781A.

Chapman, J., Silva, P., & Williams, S. (1984). Academic self-concept: Some developmental and emotional correlates in nine-year-old children. *British Journal of Educational Psychology, 54,* 284-292.

Cicirelli, F.G., Granger, R., Schemmel, D., Cooper, W.C., & Holthouse, N. (1971). Performance of disadvantaged primary-grade children on the revised ITPA. *Psychology in the Schools, 8,* 240-246.

Clark, F. (1982). The Illinois Test of Psycholinguistic Abilities: Considerations of its use in occupational and physical therapy practice. *Physical and Occupational Therapy in Pediatrics, 24,* 29-39.

Cohen, A. (1973). Smallest space analysis of the revised Illinois Test of Psycholinguistic Abilities. *Psychology in the Schools, 10,* 107-110.

Deese, J.C. (1971). A study of the discrimination by the subtests of the revised Illinois Test of Psycholinguistic Abilities between successful and unsuccessful readers of normal intelligence (Doctoral dissertation, Memphis State University, Memphis). *Dissertation Abstracts International, 32,* 4483A.

Dennis, S. (1980). The predictive and diagnostic validity of the ITPA. *Dissertation Abstracts International, 40,* 5786A.

Elkins, J. (1973). *A Guttman-Lingoes non-metric representation of the subtests of the revised ITPA.* Unpublished manuscript, University of Queensland, Department of Education.

Elkins, J. (1974). The use of the revised ITPA with some Queensland children. *Australian Psychology, 9,* 71-77.

Elkins, J., & Sultman, W. (1981). ITPA and learning disability: A discriminant analysis. *Journal of Learning Disabilities, 14,* 88-92.

Evans, R. (1984). Children with special needs in the ordinary school: An approach to intervention using "remedial" teachers in a preventive role. *Early Child Development and Care, 18,* 61-104.

Fletcher, J., Satz, P., & Scholes, R. (1981). Developmental changes in the linguistic performance correlates of reading achievement. *Brain and Language, 13,* 78-90.

Green, H. (1980). Preschool children's conceptual tempo and performance on visual discrimination tasks. *Journal of Psychology, 106,* 21-25.

Guest, K.E. (1970). Relationships among the ITPA, receptive and expressive language tasks, intelligence and achievement (Doctoral dissertation, University of Wisconsin). *Dissertation Abstracts International, 31,* 5845A.

Hatch, E., & French, J. (1971). The revised ITPA: Its reliability and validity for use with educable mentally retarded. *Journal of School Psychology, 9,* 16-23.

Hemming, H. (1983). The Swansea relocation study of mentally handicapped adults. *International Journal of Rehabilitation Research, 6,* 494-495.

Huizinga, R.J. (1973). The relationship of the Illinois Test of Psycholinguistic Abilities to the Stanford-Binet Form L-M and the Wechsler Intelligence Scale for Children. *Journal of Learning Disabilities, 6,* 451-456.

Humphrey, J., & Rice, A. (1973). An evaluation of several methods of predicting full-scale IQ from the ITPA. *The Journal of Special Education, 7,* 133-140.

Johnson, A., Shelton, R., & Arndt, W. (1982). A technique for identifying the subgroup membership of certain misarticulating children. *Journal of Speech and Hearing Research, 25,* 162-166.

Jorstad, D. (1971). Psycholinguistic learning disabilities in twenty Mexican-American students. *Journal of Learning Disabilities, 4,* 143-149.

Kirk, S.A. (1968). The Illinois Test of Psycholinguistic Abilities: Its origin and implications. In J. Hellmuth (Ed.), *Learning disorders* (Vol. 3, pp. 395-427). Seattle: Special Child Publications.

Kirk, S.A., & McCarthy, J.J. (1961). The Illinois Test of Psycholinguistic Abilities—An approach to differential diagnosis. *American Journal of Mental Deficiency, 66,* 399-412.

Kuske, I.I., Jr. (1969). Psycholinguistic abilities of Sioux Indian children (Doctoral dissertation, University of South Dakota, Vermillion). *Dissertation Abstracts International, 30,* 4280A.

Largo, R., Molinari, L., Pinto L., Weber, M., & Duc, G. (1986). Language development of term and preterm children during the first five years of life. *Developmental Medicine and Child Neurology, 28,* 333-350.

Leong, C., Chang, S., & Das, J. (1985). Simultaneous-successive syntheses and planning in Chinese readers. *International Journal of Psychology, 20,* 19-31.

Lombardi, T.P. (1970). Psycholinguistic abilities of Papago Indian children. *Exceptional Children, 36,* 485-494.

Luick, A., Kirk, S., Agranowitz, A., & Busby, R. (1982). Profiles of children with severe oral language disorders. *Journal of Speech and Hearning Disorders, 47,* 88-92.

Macione, J. (1969). Psycholinguistic correlates of reading disabilities as defined by the Illinois Test of Psycholinguistic Abilities (Doctoral dissertation, University of South Dakota, Vermillion). *Dissertation Abstracts International, 30,* 3817A.

McCarthy, J.F. (1977). Convergent and discriminant validity of the ITPA. *Dissertation Abstracts International, 38,* 6077A.

McCarthy, J.J., & Kirk, S.A. (1963). *The Construction standardization and statistical characteristics of the Illinois Test of Psycholinguistic Abilities.* Urbana, IL: Institute for Research on Exceptional Children.

McLesky, J., Kandaswamy, S., & Colarusso, R. (1980). A canonical correlation analysis of the WISC and ITPA for a group of learning-disabled children. *Journal of Special Education, 14*(2), 253-259.

Michaelson, A.W. (1970). *Psycholinguistic differences of monolingual and bilingual culturally-deprived children.* Unpublished master's thesis, Colorado State University, Fort Collins.

Mittler, P., & Ward, J. (1970). The use of the Illinois Test of Psycholinguistic Abilities on British four-year-old children: A normative and factorial study. *British Journal of Educational Psychology, 40,* 43-54.

Naylor, J., & Pumfrey, P. (1983). The alleviation of psycholinguistic deficits and some effects on the reading attainments of poor readers: A sequel. *Journal of Research in Reading, 6,* 129-153.

Nelson-Schneider, A. (1983). Inductive acuity and the acquisition of language. *Journal of Psycholinguistic Research, 12,* 263-273.

Norcombe, B., & Moffitt, P. (1970). Cultural deprivation and language defect: Project Enrichment of Childhood. *Australian Psychologist, 5,* 249-259.

Paraskevopoulos, J., & Kirk, S. (1969). *The development and psychometric characteristics of the revised Illinois Test of Psycholinguistic Abilities.* Urbana, IL: University of Illinois Press.

Paul, R., & Cohen, D. (1985). Outcomes of severe disorders of language acquisition. *Annual Progress in Child Psychiatry and Child Development,* 413-429.

Pirozzolo, R., Obrzut, J., & Hess, D. (1983). Construct validity of the Illinois Test of Psycholinguistic Abilities for a clinical population. *Psychology in the Schools, 20,* 146-152.

Polley, D. (1971). *The relationship of the channels of communication of the ITPA and the WISC.* Unpublished doctoral dissertation, University of Northern Colorado, Greeley.

Richardson, E., DiBenedetto, B., Christ, A., & Press, M. (1980). Relationship of auditory and visual skills to reading retardation. *Journal of Learning Disabilities, 13,* 77-82.

Rizzo, J., & Stephens, M. (1981). Performance of children with normal and impaired oral language production on a set of auditory comprehension tests. *Journal of Speech and Hearing Disorders, 46,* 150-159.

Roe, K. (1977). Correlations between Gesell scores in infancy and performance on verbal and non-verbal tests in early childhood. *Perceptual and Motor Skills, 45,* 1131-1134.

Ruhly, V. (1970). A study of the relationship of self-concept, socioeconomic background, and psycholinguistic abilities to reading achievement of second grade males residing in a suburban area (Doctoral dissertation, Wayne State University, Detroit). *Dissertation Abstracts International, 31,* 4560A.

Silva, P., McGee, R., & Williams, S. (1983). Developmental language delay from three to seven years and its significance for low intelligence and reading difficulties at age seven. *Developmental Medicine and Child Neurology, 25,* 783-793.

Stanley, G., Smith, G., & Powys, A. (1982). Selecting intelligence tests for studies of dyslexic children. *Psychological Reports, 50,* 787-792.

Sutter, E., & Battin, R. (1984). Using traditional psychological tests to obtain neuropsychological information on children. *International Journal of Clinical Neuropsychology, 6,* 115-119.

Weed, K., & Ryan, E. (1983). Alphabetical seriation as a reading readiness indicator. *Journal of General Psychology, 109,* 201-210.

Yom, B., Wakefield, J., & Doughtie, E. (1975). The psycholinguistic and conservation abilities of five-year-old children. *Psychology in the Schools, 12,* 150-152.

Ronald K. Hambleton, Ph.D.

Professor of Education and Psychology, School of Education, University of Massachusetts at Amherst, Amherst, Massachusetts.

IOWA TESTS OF BASIC SKILLS, FORMS G AND H

A.N. Hieronymous and H.D. Hoover. Chicago, Illinois: The Riverside Publishing Company.

THE ORIGINAL OF THIS REVIEW WAS PUBLISHED IN TEST CRITIQUES: VOLUME VI (1987).

Introduction

The Iowa Tests of Basic Skills (ITBS) are achievement tests intended to measure fundamental skills in the areas of word analysis, vocabulary, reading, language, work-study, and mathematics for students in Grades K to 9. In addition, there are supplemental subtests for measuring listening, social studies, science, and writing skills. All of the ITBS use a multiple-choice item format except the writing skills subtest, which employs a writing sample instead.

Broadly speaking, the tests that make up the core of the ITBS can be used to assess students' current skills and growth in the areas of reading, language, and mathematics. The ITBS are intended to provide scores applicable for individualizing instruction, providing guidance, assessing group (class, school, or district) performance, and evaluating school programs. Scores for individuals and groups may be interpreted from a criterion-referenced as well as a norm-referenced perspective.

The documents reviewed by this writer provided almost no discussion of the history of the Iowa Testing Program or the test authors. This omission is unfortunate, because both the program and the authors have outstanding reputations in the testing field. The earliest edition of the ITBS was published in 1935, when the Iowa Testing Program was under the direction of E. F. Lindquist, one of the major contributors to the field of psychometrics in the 1930s, '40s, '50s, and '60s. Over the more than 50 years of its history, the ITBS have become widely regarded as among the very best educational achievement tests in the testing field. Major authors of the current edition of the ITBS, Al Hieronymous and H. D. Hoover, not only have directed work on the ITBS through several highly rated previous editions, but they also have contributed important papers to the field of testing in the areas of test development, item bias, test score scaling, norming, and test score equating. Both authors are Professors of Education at the University of Iowa and have outstanding reputations in the psychometrics field.

None of the ITBS documents that the publisher provided this reviewer addressed the main changes in the 1985 edition from the previous 1978 edition (Forms 7 and 8). In addition to the obvious updates associated with any new achievement test battery, it would appear, however, that new tests in Social Studies and Science for Grades 1 to 9 were added to the ITBS along with Listening and Writing subtests for Grades 3 to 9. Since 1963, Riverside Publishing Company has also published the Cognitive Abilities Test. This set of tests for Grades K to 12 measures verbal, quantitative, and non-reasoning abilities and was normed on the same student population as the ITBS. With the availability of both achievement and ability test results on students, the use of the test data for diagnosing student strengths and weaknesses and prescribing instruction should be enhanced.

Forms G and H of the ITBS consist of 10 levels, organized into three batteries:

Battery	Level	Average Age	Development Level Grade
Early Primary	5	5	K.1-1.5
	6	6	K.8-1.9
Primary	7	7	1.7-2.6
	8	8	2.7-3.5
Multilevel	9	9	3
	10	10	4
	11	11	5
	12	12	6
	13	13	7
	14	14	8-9

Users of the ITBS undoubtedly will appreciate its comprehensiveness and flexibility. However, with these features go a very large number of products that can be confusing. This reviewer counted well over 100 different products in the material he was sent to review.

What follows is a list of most of the main documents associated with the ITBS:

Early Primary Battery, Levels 5-6

1) one form, denoted G;
2) two levels, denoted Levels 5 and 6;
3) two editions, machine scorable and hand scorable
   a) for the machine-scorable edition, two types of answer sheets are possible: MRC, which are scored by the Riverside Scoring Services, and NCS, which are scored by schools with National Computer System equipment;
4) Level 5, which measures Listening, Word Analysis, Vocabulary, Language, and Mathematics;
5) Level 6, which adds a Reading subtest;
6) a teacher's guide, with complete information about test purposes, test preparation, test administration directions, scoring directions, test score interpretations, uses of test results, and norms tables; tests are untimed (except for Reading) and orally administered;
7) scoring keys for the hand-scorable booklets;
8) NCS directions for administration;
9) practice tests;
10) parent/teacher report forms; and
11) special norms booklets.

Primary Battery, Levels 7-8

1) all of the information previously described for the Early Primary Battery for two forms, denoted G and H;
2) Basic and Complete Battery, and Complete Battery plus Social Studies and Science supplements (Form G only)
   a) the Basic Battery includes tests to measure Word Analysis, Vocabulary, Reading Comprehension, Spelling, and Mathematics Concepts, Problems, and Computation;
   b) the Complete Battery includes the Basic Battery plus tests to measure Listening, Capitalization, Punctuation and Usage and Expression in the language arts area, and Work-Study Skills;
   c) all tests except Vocabulary, Reading, Capitalization and Punctuation are untimed and orally administered;
3) four available editions: Complete, machine scorable; Complete plus Social Studies and Science, machine scorable; Basic, machine scorable; and Basic, hand scorable.

Multilevel Battery, Levels 9-14

1) two forms, denoted G and H;
2) all of the information previously described for the Early Primary Battery;
3) Basic and Complete Batteries with Social Studies, Science, Listening, and Writing supplements
   a) the Basic Battery has six tests to measure Vocabulary, Reading Comprehension, Spelling, and Mathematics Concepts, Problem-Solving, and Computation;
   b) the Complete Battery includes the Basic Battery plus five tests to measure Capitalization, Punctuation, Usage and Expression, and Work-Study (visual and reference materials);
4) spiral-bound versions of the Basic and Complete Multilevel Batteries; spiral-bound versions for each level within the Basic and Complete Multilevel Batteries; Social Studies and Science, Listening, and Writing supplements are packaged separately (only Form G is available for the supplements); and
5) five different types of answer sheets.

Basically, the foregoing descriptions of these batteries are not totally accurate or complete because of many additional variations, options, exceptions, and features that are omitted. Readers will need to refer to the complete set of ITBS documents for the detailed information. Table 1.8 in the *Manual for School Administrators* will be helpful to users interested in a summary. Table 1 below provides an overview of the content of the ITBS at each level and the number of test items in each subtest.

Practical Applications/Uses

According to the authors, the three ITBS batteries were designed to address nine main uses:

1) to determine the developmental level of students in order to better adapt materials and instructional procedures to individual needs and abilities;
2) to diagnose specific qualitative strengths and weaknesses in students' educational development;

Table 1[1]

Content for the Iowa Tests of Basic Skills, Forms G and H

Tests					*Number of Items*						
	Level	5	6	7	8	9	10	11	12	13	14
Practice Page		10	10								
Li: Listening		31	31	32	32						
WA: Word Analysis		35	36	47	50						
V: Vocabulary		29	29	30	30	30	36	39	41	41	41
R: Reading/Reading Comprehension			58	56	61	44	49	54	56	57	58
Words			13								
Word Attack			7								
Pictures			13	23	23						
Sentences			13	14	14						
Picture Stories/ Stories			12	19	24						
L: Language		29	29								
L1: Spelling				27	29	30	36	40	41	41	41
L2: Capitalization				60	66	28	29	30	30	31	32
L3: Punctuation				46	60	28	29	30	30	31	32
L4: Usage and Expression				27	27	33	36	38	40	43	43
W: Work Study											
W1: Visual Materials				28	29	33	36	43	45	49	51
W2: Reference Materials				30	33	33	39	40	41	42	43
M: Mathematics		33	33								
M1: Mathematics Concepts				33	36	28	32	35	39	41	42
M2: Mathematics Problem Solving				22	28	24	26	27	29	30	32
M3: Mathematics Computation				27	32	34	37	39	41	42	43
Social Studies				39	39	38	40	42	43	45	45
Science				35	35	38	40	42	43	45	45
Listening Supplement						31	33	34	36	38	40
Writing Supplement						5*	5*	5*	5*	5*	5*

[1]This table is adapted from Table A in the 1987 edition of the Riverside Publishing Company *Test Resource Catalog* (p. 5).

*The optional writing test has a choice of five different prompts (directions to the student) for Levels 9-11 and the same number for Levels 12-14.

3) to indicate the extent to which individual students have the specific readiness skills and abilities needed to begin instruction or to proceed to the next step in a planned instructional sequence;
4) to provide information that is useful in making administrative decisions in grouping or programming to better provide for individual differences;
5) to diagnose strengths and weaknesses in group performance (class, building, or system) that have implications for changes in curriculum, instructional procedures, or emphasis;
6) to determine the relative effectiveness of alternate methods of instruction and the conditions that determine the effectiveness of the various procedures;
7) to assess the effects of experimentation and innovation;
8) to provide a behavioral model to show what is expected of each student and to provide feedback that will indicate progress toward suitable individual goals; and
9) to report performance in the basic skills to parents, students, and the general public in objective, meaningful terms.

All nine uses are generally of interest to school districts, though uses 1, 2, 4, 5, and 9 are probably the most popular. Validity data to support the recommended uses of the ITBS will be discussed later in this review.

The ITBS seem well suited for regular students in Grades K through 9 in the United States. In addition, the publishers report that braille and large-print editions can be prepared. They further note that revised directions may be used to administer the ITBS (or parts thereof) to students with severe reading disabilities and other handicaps.

Apart from these special administrations, the ITBS are intended to be administered by teachers (with the help of proctors, if possible) to students in their classrooms. Groups of 25 to 30 students, with one or two proctors or aides, and a room with good lighting and desk space are the physical arrangements needed for a good test administration. The *Teacher's Guides* are clear. However, test administrators should be very familiar with the directions and tests before they begin. Advanced planning is essential. A teacher's checklist for testing is available and would be invaluable.

The time required to administer an ITBS battery depends upon the level and the choice of options. The Basic Battery administration time is approximately 2 hours. The Complete Battery administration time is about 4 hours. Roughly speaking, the supplemental tests in social studies, science, listening, and writing require another 40 minutes each. Practice tests are also available, and these tests are strongly recommended for students who have little experience in taking tests.

Scoring options and methods are well documented in the various teacher's guides. Hand scoring is possible but very tedious, and many of the group score summaries so important to accomplishing most of the main uses are not feasible with hand scoring. The hand-scoring option seems most useful when only a few students are involved in the testing and only individual reports (needed quickly) are of interest. In addition to hand scoring by the school district and machine scoring by the test publisher, local scoring with an optical scanner is also possible and feasible. With the correct equipment, personnel, and planning, test scoring could be done very quickly at the local level with fast returns of individual and group reports. Cost savings could also be realized in this way. However, ensuring quality control can be a problem with local scoring.

A complete list of score reports and scoring options is described in Riverside Publishing Company's 1987 *Test Resource Catalog*. A school district would be hard pressed to come up with a score-reporting request that is not included among the 26 main options (which the publisher refers to as services).

A listing and brief description of the main services follows:

1) Student criterion-referenced skills analysis
   —permits criterion-referenced interpretations of student performance on the skills measured in the battery
   —up to 85 skill scores may be reported on a typical student report
   —normative basic skill scores also appear on the report
   —areas of relative strength and weakness are highlighted

2) Pupil profile narrative
   —profiles student performance on the basic skills
   —both student and parent reports are available

3) Building criterion-referenced skills analysis
   —for each grade in each building, group performance on the skills and tests in the battery are reported along with district and national averages

4) Group narrative report
   —highlights group performance for all tests and all grades in each building and in the district via the use of graphic profiles and verbal descriptions

5) Class, building, and system summaries
   —reports various summary derived score statistics for each grade in a building (reporting each class average and the building average) and for each grade in the system (reporting each building average and the system average)

6) Group item analysis report
   —for items and skills, percent correct is provided for each class for each grade in each building, and for each grade for the district as a whole

7) Test results by class section
   —reports student performance in configurations of interest (and different from classrooms)
   —are of interest when students have different teachers for different subjects

8) Student press-on labels
   —includes the normative scores for a student requested in Service 9
   —on the right of the label is a portion of the label that can be forwarded to parents on one of several available reports for parents

9) List reports of student scores
   —are available by homeroom, in alphabetical order, or in ranked order
   —many options are available for organizing and reporting student normative scores from the test administration
   —district has some choice of normative scores to be reported

10) Frequency distributions
   —provides easy to read information about score distributions and descriptive statistics on the basic skills tests

—report serves to highlight the diversity of skills in grades and schools of a school district

11) Pre/post reporting
   —provides pre- and posttest scores for students
   —data are useful, for example, in Chapter 1 reporting or program evaluation studies

12) Individual performance profile
   —provides skills information for each student (and national norms) along with a graphical representation of the student skill performance data
   —provides GEs and percentile ranks on the basic skills
   —provides similar information to Service 1 but with the aid of graphics

13) Individual item analysis
   —main part of report provides student percent correct and national pecent correct for each test and skill, and the student's response to each item
   —normative scores on the basic skills test also are printed

21) Class/building/system diagnostic report
   —is similar to the Student Criterion-Referenced Skills Analysis but instead of one page per student, the skills for all students in a class (classes in a school, or schools in a district, depending upon the district's preference) are presented side by side.

In addition, each of the available services is cross-classified with the audience (or audiences) that would be interested in the service. Seven audiences are used in cross-classifying the services: students, parents, classroom teachers, curriculum coordinators, building or district administrators, counselors, and researchers. Overall, the set of reports appears to be clearly presented, comprehensive, and responsive to the various information needs of school districts. In fact, the score reporting capabilities are nothing short of outstanding.

National percentile norms tables at both the student and school level are available for fall, midyear, and spring administrations. Large-city school norms, Catholic school norms, international school norms, and low and high socioeconomic norms are also available. Local school system norms make up another one of the options provided by the test publisher. In addition, raw score to grade-equivalent score, grade-equivalent to percentile rank and developmental standard score, and percentile rank to stanine and normal-curve equivalent score conversion tables are available.

Those who wish can obtain an analysis of student performance on the skills measured in the ITBS. Using norm-referenced achievement tests (NRT) to provide test data at the skill level for absolute score interpretations is not without problems, however. For example, the specifications of the skills measured by the ITBS are not as complete as they are with high-quality criterion-referenced tests (CRTs; see, for example, Berk, 1984; Popham, 1978). Without more content specifications, there is also the danger that schools may do a skills match using the skills names only. This could result in erroneous matches and ultimate misuse of the CRT results obtained from the ITBS. Further, as item selection criteria for CRTs are very different from those used with NRTs, the representativeness of the ITBS items as measures of the skills of interest is unknown. With these cautions in mind, the available criterion-referenced ITBS information can be used to identify student, classroom, building, and district strengths and weaknesses. At the student level it is noted that the number of items per skill is typically low, and information on the reliability of these skill scores is not reported in the teacher's guides, though confidence bands (reflecting reliabilities) do appear in the Individual Performance Profile Reports. Users should proceed with caution until they have established the reliability and validity of these scores for themselves.

Technical Aspects

The *Manual for School Administrators* and the accompanying *Teacher's Guides* are clearly written, and the technical discussions are excellent. The authors offer rationales for important decisions that were made in the test development process (e.g., decisions about test content), they offer complete details about their actual developmental procedures (e.g., content reviews, item writing and reviews, item bias studies), and they were willing to draw attention to possible weak areas in the ITBS (e.g., content match to particular school curricula; various ITBS derived scores).

The *Manual for School Administrators* and the *Teacher's Guides* are among the best in the testing field. In fact, the manual's "Part 6—Technical and Other Considerations" should be required reading for students studying educational tests and measurements. The authors provide an eminently sensible statement about the roles of the test authors, test publishers, and potential users of the ITBS in test validation, as well as some guidelines for conducting validity investigations and interpreting the results. Ultimately, as they note, the final determination of test validity is a judgment based on a review of technical evidence compiled by the test authors, other considerations such as the norms, and factors unique to the school district, such as their informational needs and curricula.

Among the types of technical evidence reported in the administrator's manual are development steps for each subtest, test and subtest reliabilities, item statistics, item bias results, and correlations between test scores and a variety of criterion measures (e.g., school grades).

Sufficient information about test reliability has been reported to keep even the severest critic happy. Estimates of equivalent form and internal consistency reliability (along with other descriptive statistics) are reported for each form of each subtest at each level in both the fall and spring of the grades in which the forms are scheduled for administration. A number of coefficients representing the stability of test scores over the period of a year are also reported. However, the value of this information, as noted by the authors, is very limited for achievement tests; the information seems to say more about the nature of instruction than the tests themselves.

One special feature of the reliability reporting is the inclusion of information pertaining to the precision of score estimation at various ability levels. Such data are becoming available for testing programs built within an item response theory framework, but they are seldom, if ever, reported when test reporting is done within a classical measurement framework. Test reliabilities for the subtests tend to be in the .80s and .90s, though they are somewhat lower for the Level 5 and 6 subtests and for the Listening subtest at all four levels in which it is included.

Predictive validity evidence for the new forms is confined to six studies carried out on earlier editions. Fortunately, 1) the usefulness of predictive validity evidence is limited for achievement tests, and 2) the studies reported were based on large samples and were supportive of the ITBS. Among the validity results reported are those that show strong relationships between 1) fall ITBS scores for kindergarten and first-grade students and their grades at the end of the school year, and 2) ITBS scores in Grades 4, 6, 8 and high school and first-year college GPAs. Also, data addressing ceiling and floor effects, test completion rates, readability levels, relationships between achievement and ability, sex differences, and socioeconomic, sex, and cultural bias provide additional evidence for the validity of scores obtained from the ITBS.

The writing assessment seems to be one part of the ITBS especially in need of additional technical support. Questions about the reliability of scoring across judges as well as about the consistency of student performance over various writing prompts are especially interesting and need to be fully answered before the writing scores can be used with confidence. Steps were taken by the authors to standardize the administration and scoring. Interrater reliabilities are high, though reliability of writing scores over similar prompts are only modest, ranging from about .39 to .70.

In view of the long history of the ITBS and its careful development, one would expect more validity evidence to support the various recommended uses of the ITBS. Certainly content validity evidence for achievement tests is important, and this evidence is well documented by the authors and highly supportive of the ITBS. However, additional evidence to show that the ITBS scores, for example, are useful for "grouping . . . students to better provide for individual differences" and other recommended uses would be helpful. Each application of the ITBS scores is associated with a different type of inference, and therefore evidence for the validity of each recommended use should be reported in the technical documentation.

Test norming for the ITBS was carried out in the fall of 1984 and the spring of 1985. Three stratifying variables were used to classify public school districts and played a key role in the school sample selection: geographic region, school district size, and socioeconomic status of the community. The main goal was to draw a sample that could represent the national population with respect to both ability and school achievement; the use of previous years' achievement data was helpful in achieving the goal. The number of students per grade was never less than 10,000 for the fall administration and, except for kindergarten, was never less than 12,800. In total, over 126,000 participated in the fall norming. About a third of the sample also participated in the spring 1985 norming. Weights were used to adjust the actual data for minor departures from the desired national demographic statistics.

Baglin (1981) was one of the first researchers to draw attention to the possible bias in norms tables due to the non-participation of selected school districts, as well as to the prevalence of users of the test battery in the norming group. Unfortunately, information with which to address these two possible sources of bias was not reported in the technical documentation. It is true, however, that the authors did replace non-participating school districts with similar school districts that were willing to participate.

Critique

The ITBS present an outstanding example of a nationally normed achievement test battery. In every respect, these tests appear to be built upon modern measurement principles and practices. In some areas, such as the assessment of item bias and reliability and the scale development, the authors actually developed the methodologies they used, and these methodologies influence the measurement field more broadly than through the ITBS only. The concurrent availability of the Tests of Achievement and Proficiency (for Grades 9 to 12, or Levels 15 to 18) and the Cognitive Abilities Test (for Grades K to 12), along with the ITBS measures of science, social studies, listening, and writing, provides school districts with a comprehensive Grades K to 12 achievement and ability testing program.

About the only criticism this reviewer has of the ITBS concerns its packaging. Accompanying the very large number of options in test booklets, supplemental tests, levels, forms, answer sheets, score reporting, and so on are complexity in packaging and possible user confusion about what one actually needs. The extensive options obviously have been made available to meet broad user needs as well as to provide flexibility to schools in designing sensible school testing programs; therefore, perhaps the time has arrived for the authors to prepare a detailed flowchart representing the desired sequence of discussions and decisions for working through the myriad documents, options, and considerations encountered in building a school standardized achievement testing program. Possibly such a system could even be prepared for microcomputers, allowing users to work interactively.

More validity evidence addressing the authors' recommendations for possible uses of the ITBS would also be desirable. Certainly the recommended uses seem reasonable, but evidence to support them would further enhance an outstanding testing package.

References

Baglin, R.F. (1981). Does "nationally" normed really mean nationally? *Journal of Educational Measurement, 18,* 97-107.

Berk, R. A. (Ed.), (1984). *A guide to criterion-referenced test construction.* Baltimore, MD: Johns Hopkins University Press.

Popham, W. J. (1978). *Criterion-referenced measurement.* Englewood Cliffs, NJ: Prentice-Hall.

William R. Merz, Sr., Ph.D.

Professor and Coordinator, School Psychology Training Program, California State University-Sacramento, Sacramento, California.

KAUFMAN ASSESSMENT BATTERY FOR CHILDREN

Alan S. Kaufman and Nadeen L. Kaufman. Circle Pines, Minnesota: American Guidance Service.

THE ORIGINAL OF THIS REVIEW WAS PUBLISHED IN TEST CRITIQUES: VOLUME I (1984).

Introduction

The Kaufman Assessment Battery for Children (K-ABC) is an individually administered battery of tests designed to measure the intelligence and achievement of children 2½ to 12½ years old. It consists of 16 subtests of which no child takes more than 13. Ten of the subtests are classified as Mental Processing Subtests and six as Achievement Subtests. Subtest scores are aggregated into four global scales: Sequential Processing, Simultaneous Processing, Achievement, and Nonverbal. The Sequential Processing and Simultaneous Processing Scales are added to obtain the Mental Processing Composite. Three subtests make up the Sequential Processing Subscale, and seven subtests form the Simultaneous Processing Subscale. Six Mental Processing Subtests may be administered nonverbally to assess fairly the intellectual function of children who have communication impairments. Six subtests comprise the Achievement Subscale. The Mental Processing Scale measures problem solving in novel situations, while the Achievement Scale assesses factual knowledge usually acquired in school or from attending to the environment in which the child lives.

The K-ABC is a clinical instrument designed for evaluating preschool and elementary school children. The authors claim that the type of intelligence measured by this battery is based on the findings of neuropsychology and cognitive psychology. The work of Luria (1966) and Das (Das, Kirby, & Jarman, 1979) provide the theoretical framework of the battery.[1] The definition of intelligence addresses level of skill and is grounded in the "individual's style of solving problems and processing information" (Kaufman & Kaufman, 1983c, p.2). Further, the authors claim that the Mental Processing Scales measure fluid abilities while the Achievement Scale measures crystallized abilities.

The authors, Dr. Alan S. Kaufman and Dr. Nadeen L. Kaufman, are eminently qualified to devise a new measure of intelligence and achievement. Alan Kaufman is now on the faculty at the University of Alabama. Both are professionally active in the American Psychological Association, National Association of School Psychologists, the American Educational Research Association, and the Council for Exceptional Children.

Dr. Alan Kaufman was a student of Robert L. Thorndike at Columbia University. He was employed by The Psychological Corporation from 1968 through 1974; during that time he worked with David Wechsler and Dorothea McCarthy in developing and standardizing the WISC-R and the McCarthy Scales. He trained school psychologists at the University of Georgia and at the National College of Education in Evanston, Illinois. He serves on the editorial boards of the *Journal of Consulting and Clinical Psychology,* the *Journal of Learning Disabilities,* and the *Journal of School Psychology.* He wrote *Intelligent Testing with the WISC-R* (1979), which many consider to be a classic on the clinical use of the WISC-R.

Dr. Nadeen L. Kaufman has graduate degrees in psychology, reading and learning disabilities, and special education. As a certified teacher and school psychologist she has had extensive background working with learning disabled children. She is associate editor of *School Psychology Review* and co-authored with Alan the book *Clinical Evaluation of Young Children with the McCarthy Scales* (1977).

The test itself was developed from 1978 to 1983. The authors list six primary goals for the K-ABC:

1. to measure intelligence from a strong theoretical and research basis
2. to separate acquired factual knowledge from the ability to solve unfamiliar problems
3. to yield scores that translate to educational intervention
4. to include novel tasks
5. to be easy to administer and objective to score.
6. to be sensitive to the diverse needs of preschool, minority group, and exceptional children (Kaufman & Kaufman, 1983b, p. 5)

Tasks and items for the Mental Processing Scales were developed during 1978 and 1979 from the sequential-simultaneous information processing model of Das (Das, Kirby, & Jarman, 1979). During this stage more than 50 mental processing and achievement tasks were devised. These tasks were tested empirically in ten separate studies during 1980. From these studies and the investigators' clinical judgment 20 of the most promising subtests were developed and tried. The subtests were divided into four overlapping forms, which were tested on a sample of 794 children ages $2\frac{1}{2}$ to $12\frac{1}{2}$. To insure comparable ability among different groups, the Peabody Picture Vocabulary Test-Revised (Dunn & Dunn, 1981) was administered to every child as well. Data obtained in these trials were submitted to item analyses using both traditional methods and the Rasch-Wright one-parameter latent trait model. Internal consistency was computed utilizing Cronbach's coefficient alpha (1970). Analysis for item bias was accomplished employing the Rasch-Wright method (Wright & Stone, 1979) and the Angoff and Ford (1973) transformed item-difficulty method. Construct validity was examined employing factor analysis; both principal component analysis and principal factor analysis were utilized. In addition, there was an editorial review to identify items that might be biased on the basis of sex or subculture. Examiners who administered the tryout editions rated each subtest for ease of administration, ease of scoring, interest to children, clarity of instruction, and design of materials. Seventeen subtests containing over 400 items were included in the standardization edition.

The K-ABC was standardized from April to September 1981 on more than 2,000 children at 34 test sites in 24 states. Sampling was based on the most recent population reports available from the U.S. Census Bureau, stratifying within each age group by gender, geographic region, socioeconomic status, race or ethnic group membership, community size, and educational placement of the child in either regular or special classes. Information on the standardization contained in the *K-ABC Interpretive Manual* (Kaufman & Kaufman, 1983c) is well documented. Steps taken to insure the representativeness of the sample to population parameters are clear.

Norms were developed during 1982 from the performance of 1,981 children in the K-ABC standardization sample, after it was determined that the age groups included did not differ systematically from each other on important background variables or in cognitive ability as measured by other intelligence and achievement tests. Normalized standard scores were computed for each age group. For the Global Scales and the Achievement Subtests the mean is 100, and the standard deviation is 15; for the Mental Processing Subtests the mean is 10, and the standard deviation is 3. National percentile ranks and stanines, subtest age equivalents, arithmetic and reading grade equivalents, and sociocultural percentile ranks were prepared, and supplementary norms for out-of-level testing of $4\frac{1}{2}$- and 5-year-old children were constructed. In the opinion of this reviewer, the section on norming is one of the clearest that has been presented in test manuals of this kind.

The Nonverbal Subtests enable a child with communication problems to be assessed and to attain an estimated Mental Processing Subscore. Thus, children who are hearing impaired, speech disordered, or language disordered may be fairly evaluated. Non-English-speaking children may also be assessed more fairly than with other major intelligence tests. A Spanish version of the Mental Processing Subscales has been devised for use in Mexico and in other countries where Spanish is spoken. Children in the United States speaking Spanish who cannot be evaluated fairly in English may be tested with this version; however, the authors strongly recommend that Spanish-speaking children in the United States be tested with the English version of the K-ABC if they are sufficiently bilingual as the English norms are more appropriate.[2] The test was not designed for use with nor standardized on children older than $12\frac{1}{2}$. It probably would not be useful with the blind because of the reliance on sight for performance, nor would it be useful with children who have receptive language disorders.[3]

The K-ABC includes three Easel-Kits which allow the examiner to see the test directions and scoring key while the child sees the item being administered. Easel 1 contains six Mental Processing Subtests: Magic Window, Face Recognition, Hand Movements, Gestalt Closure, Number Recall, and Triangles. Easel 2 contains four Mental Processing Subtests: Word Order, Matrix, Analogies, Spatial Memory, and Photo Series. Easel 3 contains the six Achievement Subtests: Expressive Vocabulary, Faces and Places, Arithmetic, Riddles, Reading/Decoding, and Reading/Understanding. Five sets of manipulatives are involved: the Magic

Window Disk, Triangles, Photo Series, Cards, and Matrix Analogies Chips. There are two manuals: *K-ABC Administration and Scoring Manual* and *K-ABC Interpretive Manual*. Finally, the K-ABC Individual Test Records for entering information and graphing performance are included. There is an optional carrying kit, a sturdy purple plastic case which has been dubbed "The Grape" by many examiners. The two test manuals are well prepared, presenting detailed information in a lucid style. The administration and scoring manual contains explicit directions for both administering and scoring the battery; tables for obtaining derived scores and for evaluating significant diferences among scores are included. The interpretive manual provides clearly written information on the standardization and development of the tasks, items, and norms, and presents the results of reliability and validity studies. In addition, there is an extensive section on interpreting the results and planning for educational intervention.[4]

The administration manual states that the examiner must be a competent, trained individual with a good knowledge of psychology and individual intellectual assessment. It states too that those who are legally and professionally competent to administer tests like the Wechsler scales and Stanford-Binet are qualified to give the K-ABC. Some of the prerequisite knowledge listed for competent examiners are "a good understanding of theory and research in areas such as child development, tests and measurements, cognitive psychology, educational psychology, and neuropsychology, as well as supervised experience in clinical observation of behavior and formal graduate-level training in individual intellectual assessment" (Kaufman & Kaufman, 1983c, p. 4).[5] Examiner involvement is extensive. The section of the administration manual on general testing procedures gives suggestions for establishing and maintaining rapport; states the importance of following directions, giving feedback, and objective scoring; and describes clinical observation of behavior. Excellent guidelines for testing limits are provided; here, testing limits mean nonstandardized administration as a follow-up to standard administration in order to gain more clinically significant information. Excellent cautions on the use of this information are presented. The Mental Processing Subtests contain teaching items so that the child may be taught how to take the test with the first three items, while the Achievement Subtests do not have such items. Teaching or practice items have been used with good results in paper-and-pencil testing programs.

The purposes of the battery are psychological and clinical assessment, educational evaluation of learning disabled and other exceptional children, educational planning and placement, minority group assessment, preschool assessment, and limited neuropsychological assessment. The authors clearly indicate that the K-ABC is not a measure of innate ability, nor are the derived scores unchangeable. They state that it is not a complete neurological test battery and they do not maintain that it is "the" complete test battery. They do maintain however that it is an excellent partial assessment that will probably require other systematic information-gathering techniques such as observation and testing.

The individual test record provides a systematic method for recording a child's responses. On the face sheet, raw scaled scores are recorded for each of the subtests. National percentile ranks are entered, strengths or weaknesses are indicated, and other derived scores are entered. A summary of Global Scale

Scores is made, including a standard error band, national percentile ranks, sociocultural percentile ranks, and other derived scores. Score profiles employing scaled scores may be displayed and graphed in such a way that percentile ranks, deviations from the mean, and standard error bands may all be entered. The way information is displayed assists in interpreting information to parents and teachers. The test itself is attractive, the manuals very well written, and individual test records well layed-out. Great care has been taken to make a set of materials which facilitate administration and interpretation.

A microcomputer program titled *ASSIST for the Kaufman Assessment Battery for Children* (Ingram, 1984) is available to enter and analyze data obtained from testing. Raw scores for each subtest are entered onto a data disk and may be stored or analyzed. The program is available for the Apple II +, the Apple IIe, or DOS 3.3 capable microcomputers. Results may be printed on dot matrix printers or may be copied by hand from the screen. An additional diskette is necessary for letter quality printers and this involves additional cost. The programs are relatively easy to use and the manual is clearly written. Essentially, this set of programs takes some of the clerical drudgery out of scoring. This process is more accurate and allows for easy editing. The manual is like a recipe book, taking the reader step-by-step through the programs. It is a "user friendly" system and an extensive background in using microcomputers is not necessary. One must have available a computer capable of accepting the programs and must be familiar with turning on the machine, putting the diskette into the appropriate drive, and booting up the program. The ASSIST program is a valuable adjunct for using the K-ABC.

Practical Applications/Uses

The K-ABC appears to be most useful as a tool for psychoeducational assessment and for educational planning and placement. The Luria-Das (Luria, 1966; Das, Kirby, & Jarman, 1979) sequential-simultaneous processing model differs significantly from the theoretical constructs of other instruments used for these purposes, notably the WPPSI, the WISC-R, or the Stanford-Binet. Claims made by the authors and studies cited in the interpretive manual support these contentions. In fact, evidence cited indicates that differential diagnosis is possible with this tool, and thus the K-ABC would be of interest to psychologists, audiologists, speech and language specialists, occupational therapists, school resource specialists, teachers, and others who work with children who have difficulty with school learning.[6]

Because the intellectual and achievement measures are included in the same battery, they have been normed on a single group of students. The norm groups include special-education students, something that is not the case with many of the other test batteries. By design, the K-ABC is especially appropriate for those children with communication problems, who have limited English proficiency, or who speak a nonstandard English dialect. Efforts taken to minimize cultural bias have produced a battery which by the authors' claim functions well with children whose performance may be systematically under- or overestimated with other tests usually employed with this age group.[7]

The battery is designed to measure intelligence and to separate the acquisition of factual knowledge from the ability to solve unfamiliar problems. The authors suggest that the battery would be useful if it were included as part of a systematic neuropsychological assessment. They suggest also that it would be useful in psychological and clinical assessment. This reviewer would agree that including the battery as part of the assessment process would be appropriate and useful. Suggestions that it be used to evoke "clinical responses" is problematic; when used in that way a subtest or an item becomes an unstandardized, informal observation, which was criticized earlier by the authors when they reviewed the consequences of dismissing standardized, norm-referenced instruments as biased. Including this battery as part of a neuropsychological assessment is reasonable and beneficial, though to replace instruments designed to measure neuropsychological function with the K-ABC would be inappropriate until it has been demonstrated that the battery measures similar constructs.

The K-ABC is an individually administered battery. The examiner must have had extensive experience with individual testing and intellectual assessment. Thus, the administration and scoring manual states that "examiners who are legally and professionally deemed competent to administer existing individual intelligence tests . . . are qualified to give the K-ABC" (Kaufman & Kaufman, 1983b, p. 4). The test is easy to administer with the well-written directions in the administration manual. Easel-Kits are easy to use, as are the manipulatives. Practice with the manipulatives is crucial to smooth administration. The manual is clear, concise, and understandable. Test times vary from approximately 35 minutes for the seven subtests given to 2½-year-olds to 75 to 80 minutes for the 13 subtests given to 7- through 12½-year-olds.

Scoring instructions are clear and scoring tables are easy to use. The Individual Test Record is well done and is easily used to record responses, score subtests, and display results. Once scoring has been mastered, approximately 15 to 30 minutes are required. The ASSIST microcomputer program reduces scoring time. Individual information on the child being assessed is filled out on the front page of the individual test record. Sociocultural information including race and socioeconomic background is taken so that scores may be referenced to appropriate sociocultural percentiles. Responses are scored as the child gives them. Starting points for the various age groups are indicated on the individual test record by a hand with a pointing index finger and the age level noted next to it. Stopping points are indicated by an upheld hand with the age level next to it. Correct responses are coded as 1, errors as 0. The total raw score is computed from the ceiling item (the highest item administered) minus the number of errors. Raw scores are converted to scaled scores and standard scores from tables available in the administration and scoring manual. Scaled scores are entered onto the first page. National percentile ranks and other derived scores are entered from appropriate tables. Strengths and weaknesses are indicated by referring to a table that compares each scaled score with the mean scaled score of the mental processing composite and with the achievement composite. Standard score bands of error are obtained from an appropriate table as well. Scores may be profiled and displayed for easier interpretation.

Interpretation is based on objective scoring and on clinical judgment, and an extensive section on interpretation is presented in the interpretive manual. The five empirical steps to interpretation are objective and well explained. Translating the objective findings into educational intervention requires experience and skill, so the individual making recommendations must have a good knowledge of curriculum, individual learning, and instruction. This comes from both academic training and experience; the cases presented in the interpretive manual illustrate this very well. The chapter on interpretation employs the work of Das, Gunnison, Kaufman, and Kirby applying the sequential-simultaneous processing dichotomy to remediation in education.This section clearly illustrates the ways in which the authors suggest that information derived from the battery may be translated into educational intervention.[8] In order to become proficient in using the K-ABC one must be thoroughly grounded in the theoretical underpinnings of the battery. An understanding of the strengths and weaknesses and advantages/disadvantages of the theoretical position must be grasped by those attempting to formulate intervention strategies. This requires training, study, and experience.[9]

Technical Aspects

The interpretive manual devotes 76 pages to the topics of reliability, validity, and profiles. The section on reliability deals with split-half reliabilities, test-retest reliabilities, alternative-levels reliabilities, standard errors of measurement, and intercorrelations among scales and subtests. The section on validity deals with construct validity, predictive validity, and concurrent validity. Profiles of exceptional children and profiles for key background variables are also presented. In the opinion of this reviewer after examining a number of other individually administered intelligence and achievement measures, this section is one of the most clearly written and complete technical presentations available to those administering and using a battery.

The reliabilities presented compare favorably with those reported in the administrator's manuals for the WPPSI, the WISC-R, the PIAT, and other individually administered tests. Split-half reliabilities for subtests were computed employing W ability scaled scores generated by the Rasch-Wright model. Split-half coefficients were corrected using the Spearman-Brown formula for the Global Scales. The Guilford method for determining the reliability of a composite was employed. Mean reliability coefficients were computed for both subtests and Global Scales employing Fisher's z transformation. Technically, the methods employed are the most appropriate to minimize assumptions and maximize accurate information. For the Mental Processing subtests, split-half reliabilities range from a low .62 for the Gestalt Closure for the 7-0 to 7-11 age group to a high of .92 for Triangles for the 5-0 to 5-11 age group. For preschool children, mean coefficients for the Mental Processing subtests range from .72 to .89; for school-aged children, from .71 to .85. For the Achievement subtests, coefficients range from a low of .70 for Faces and Places for the 3-0 to 3-11 age group to .97 for Reading/Decoding for the 6-0 to 6-11 subgroup. For preschool children, mean coefficients range from .77 to .87; for school-aged children, from .84 to .92. Internal-consistency reliability coefficients for the Global Scales range from a low of .84 for Simultaneous Processing for the age groups 2-6 to 2-11 and 3-0 to 3-11, to .97 for Achievement for all age groups from 7-0 to 7-11 through 12-0 to 12-5. Mean internal-consistency reliabilities for preschool children range from .86 to .93; for school-age children, from .89 to .97.

Test-retest reliability coefficients were computed for three age groups for the Global Scales. For ages 2-6 to 4-11 (N = 84) coefficients range from .77 to .95; for ages 5-0 to 8-11 (N = 92) from .82 to .95; and for ages 9-0 to 12-5 (N = 70) from .87 to .97. Means and standard deviations of the first and second testing for each of the three groups, as well as difference between means, are reported. Test-retest reliability coefficients for the Mental Processing subtest score for the same age groups range from .62 to .86, and for the Achievement subtests, from .72 to .98.

Alternate-levels reliabilities were computed for the 4- and 5-year levels because these two levels differ the most in subtests administered. Counterbalanced order was employed to negate practice effects with half the children tested in reverse order. For the 41 children tested the correlation coefficients ranged from .83 for the Nonverbal Scale to .95 for the Achievement Scale.

Standard errors of measurement are reported by age for the Global Scales and the subtests for the standardization sample. At this point the authors introduced the notion of using the standard error to create a band around the obtained score to state a confidence of having captured the true score. Standard errors all appear to be reasonable and within the range one would expect given reliabilities, length of test, and standardization sample performance.[10]

Intercorrelation among Global Scales and among subtests are presented next. The authors state that "relationship among the components of the test battery plays an important role in determining the reliability of an instrument, and affects interpretation of profile fluctuations" (Kaufman & Kaufman, 1983c, p. 90). The correlations between the Sequential Global Scale and the Simultaneous Global Scale are of a magnitude one would expect if the two scales measure different attributes, yet high enough to show some relationship, exactly what Thorndike recommends for batteries of tests in *Personnel Selection* (1950). For all but one age level correlations between each of the Mental Processing Scales and achievement are larger than between the Global Scales with each other or with achievement; again, exactly what would be desired. The Nonverbal Scale correlates with the Achievement Scale at approximately the same magnitude that the Simultaneous Scale correlates with the Achievement Scale. Mean intercorrelations among Mental Processing subtests is of a lower magnitude than intercorrelations between Mental Processing subtests and Achievement subtests for both preschool and school-aged children. Again, from a psychometric standpoint, this is a desirable attribute for a battery.

Validity is demonstrated with 43 studies cited and summarized in the interpretation manual. Samples studied include children who are normal, dyslexic, learning disabled, hearing impaired, educably mentally retarded, trainably mentally retarded, behaviorally disordered, and physically impaired. Approximately equal amounts of space are given to construct validity and criterion-related validity. Both are extremely important aspects in light of *Larry P. v. Riles* (1979). Construct validity is divided into five portions: developmental changes, internal consistency, factor analysis, convergent and discriminant validation, and correlations with other tests. Criterion-related validity is divided into two portions: predictive validity and concurrent validity. Concurrent validity is divided into five sections: individual achievement tests, group achievement tests, test of general cognitive ability, Luria-Nebraska Children's Battery, and brief tests. There is no treatment of content validity; this may be an important oversight for a new test that seeks novel tasks with which to measure traditional concepts.

A very well-documented section on construct validity shows developmental changes as increases in mean score by age level by subtests. The increases between contiguous age groups are large enough to argue cogently that the test is measuring a developmental process. Under the section on internal consistency the authors demonstrate that subtests making up a specific global processing scale correlate reasonably with that total scale, again good evidence of construct validity. Two approaches to factor analysis, principal-components analysis and principal-factor analysis, were employed. Both analyses of the Mental Processing subtests for each of the age levels supported the contention that there are two factors corresponding to Simultaneous Processing and Sequential Processing. Factor analysis of all subtests yielded three factors with appropriate subtests loading on the appropriate global scales. Data reported in the manual indicates a sufficiently large sample size to draw reliable conclusions from the factor analysis. The confirmatory analysis employed yielded equally promising results. However, this reviewer would have liked to see cluster-analytic methods used as well as the confirmatory analysis which was selected.

Convergent and discriminant validation includes two notions: a test should correlate with variables to which it should relate theoretically, and it should not correlate highly with variables from which it should be theoretically different. This was done in two studies where scores on the K-ABC were correlated with factor scores on the Das-Kirby-Jarman Successive-Simultaneous Battery (1979). A group of 37 trainably mentally retarded children and a group of 53 learning-disabled children were tested. In both studies the K-ABC Sequential Processing Scale correlated more highly with the Das-Kirby-Jarman successive-simultaneous factor while the K-ABC Simultaneous Processing subtests correlated more highly with the Simultaneous Battery.

Correlations between the K-ABC subtest scaled and standard scores and the WISC-R and Stanford-Binet IQs are reported. These are all of reasonable magnitude, save for the Gestalt Closure subtest. K-ABC Global Scale scores and the Stanford-Binet IQ were correlated for various samples of students ranging from high-risk preschool through normal to gifted. Only five of the 30 correlations fell below .45.

Criterion-related validity may be broken into two segments: predictive validity and concurrent validity. The section on predictive validity shows the correlation between K-ABC Global Scale standard scores and four different achievement tests administered to different groups of students from 6 to 12 months apart. Correlations in most instances range from .35 through .68, with 155 correlations below .20 and 15 above .65. Concurrent validities are of greater magnitude, with correlations for various racial or ethnic groups of very similar magnitude for different achievement tests. In all instances the Mental Processing composite scores and the

Achievement scaled scores correlate more highly with subtests of a variety of achievement tests. Correlations of the K-ABC Global Scale scores and the Luria-Nebraska for two samples of learning-disabled students show correlations ranging from .406 to .863. Correlations with brief tests such as the Columbia Mental Maturity Scale and the Slosson Intelligence Test, with error scores on the Bender-Gestalt, and with raw scores on the VMI are reported.[11]

Profiles for combined groups of learning disabled students and for dyslexic students are presented, as are profiles for educably mentally retarded, trainably mentally retarded, behaviorally disordered, physically impaired, high-risk preschool, hearing-impaired, and gifted students. The profiles show differences in global scales and differences in subtest scores. It appears to this reviewer that a good deal more work must be done to demonstrate the stability of these findings across groups. The replication of profile results is crucial because of the recency of publication and the large number of studies presently in press or being completed. Much more data will be required before a judgment can be made about the usefulness of the profiles in dicriminating among pupils with different learning characteristics.

The profiles are also examined against the background variables of gender, socioeconomic status, and race or ethnic group. Females achieved higher scores at the preschool level than males, with this difference disappearing at the school-age levels. The authors conclude that gender differences are minor and unimportant for school-aged children but support better performance by preschool girls than preschool boys. There appear to be differences related to parental education, with preschool and school-aged children of individuals with less than a high school education achieving lower scores than those with parents who graduated from high school. As the level of parental education increased, so did the children's mean scores on the Global Scales and K-ABC subtests. When differences in the mean performance of various ethnic goups were examined it was clear that the school-aged children from the various ethnic groups had lower mean scores than children from the majority group. Navajo and Sioux school-aged children preformed comparably to black and Hispanic school-aged children. When the differences between blacks and whites and Hispanics and whites were compared on the K-ABC and the WISC-R, smaller mean differences between blacks and whites and Hispanics and whites were noted on the K-ABC than on the WISC-R IQs. Thus, there is evidence that the ethnicity of an individual may be a smaller moderator variable in performance on the K-ABC. Socioeconomic status reflected by level of parental education is a larger contributor to differences than is ethnicity. Gender appears to be an important variable favoring females at the the preschool level.

Thirteen invited articles and a response by Alan Kaufman will appear in a forthcoming issue of *Journal of Special Education*. Two of the articles which address questions of validity (Jensen, in press; Keith & Dunbar, in press) along with Kaufman's response were available to this reviewer. Several excellent points in these papers pertain to this review. Jensen concludes that the K-ABC does not measure something different from tests such as the WISC-R or the Stanford-Binet. From this reviewer's point of view this is much to the credit of the K-ABC, lending credence to the contention that this battery, in fact, is a new measure of intelligence and achievement. Keith and Dunbar conclude that an alternate model to the Luria-Das model may explain the factor-analytic findings; these authors conclude that three factors labeled Verbal Memory, Nonverbal Reasoning, and Verbal Reasoning explain the analysis rather than the Kaufman's Simultaneous, Sequential, and Achievement factors. Another article by Reynolds, Willson and Chatman (1983) supports the construct validity of the K-ABC as does another factor-analytic study by Kaufman and Kamphaus (in press).[12]

These challenges to the interpretation of data presented in the K-ABC interpretive manual are presented well. They are answered directly by Kaufman (in press). He points out that the challenges are a necessary and important process in developing a research literature which will allow clinicians to test wisely and to interpret findings more accurately and usefully with the new battery. Specific efforts have been made to lessen the effects of bias. The contention of less bias is confirmed in one instance and is challenged to some extent in another. Kaufman cites research that shows smaller mean differences among ethnic groups than other tests of the same kind; however, examination of mean differences is a weak method for investigating test bias. Data presented by Reynolds, Willson, and Chatman (1983) show some bias which would over-predict for blacks. Here, regression is used and investigates what this reviewer has labelled "test fairness" (Merz, 1977) rather than what Reynolds and the others call test bias.

Critique

The Kaufman Assessment Battery for Children is an exceptionally well designed, attractive set of tasks intended to assess the intelligence and achievement of children 2½ to 12½ years of age. The manuals for administration and interpretation are well written, presenting information in a straightforward, clear fashion. The report of technical considerations such as reliability, validity, and influence of moderator variables is complete and understandable. Only an intuitive grasp of correlation would be necessary to derive useful information from the tables presented in those sections. The one failing is in not presenting information on content validity. The authors state that the "battery includes a blend of novel subtests and adaptations of tasks with proven clinical, neuropsychological, or other research-based validity" (Kaufman & Kaufman, 1983c, p. 1); a much clearer explication of the rationale behind some of the novel subtests would be quite helpful.

The administration and scoring manual is exceptionally well prepared. The design and editorial work show attention to detail. The amount of research presented in the interpretive manual and the section on hypothesis generation are unusual. They provide an excellent basis to begin prescriptive teaching aimed at remedying deficiencies by utilizing strengths. Because the K-ABC has been so recently developed there has been insufficient time for the battery to have developed a track record. A number of empirical studies on this instrument are in progress at the present time; hopefully, this indicates that its usefulness will grow in identifying ways to work with children who have special needs.[13]

All in all, the battery appears to meet the goals set out for its design. It measures intelligence from a strong theoretical and research basis if the numbers of papers and quality of investigation are used as indications. The 43 studies cited in the construct validity section of the interpretive manual, the studies cited in the reliability section of that same manual, the papers which will appear in the special edition of the *Journal of Special Education* devoted to the K-ABC, and papers presented at professional association meetings all indicate a strong research base. The authors' attention to the sequential-simultaneous model and alternative interpretations presented by Keith and Dunbar (in press) are testimony to the theoretical basis for the test. The battery was constructed to separate acquisition of factual knowledge from ability to solve unfamiliar problems; however, how well this is done has been questioned by Jensen (in press) and by Keith and Dunbar (in press). The positions taken by each have been rebutted by Kaufman (in press) and by Kaufman and Kamphaus (in press). A good deal of additional research will be required to amass enough replicable information from which to make a final judgment.

A large portion of the interpretive manual is devoted to showing how scores may be translated into educational intervention. How effective these interventions are in bringing about change is yet to be determined. Insufficient time has passed to evaluate the efficacy of the match between test performance and instructional strategy; this is, however, the crux of program planning for individual students. If the test meets this goal, it will be an extremely important addition to education. Novel tasks have been used in the battery and greater emphasis on a description of content validity would have been helpful for at least two of the least traditional tasks, Hand Movements and Faces and Places. An analysis of the administration and scoring manual and the ASSIST microcomputer manual leads this reviewer to judge the battery easy to administer and objective to score.

The battery's sensitivity to the diverse needs of preschool, minority group, and exceptional children must be judged from several perspectives. Certainly, the age span covered by the subtests and norms meets this goal. The efforts at making the test less biased have produced equivocal results from the interpretation of at least two reviewers (Jensen, in press; Chatman, Reynolds, & Willson, in press). On the face of the analyses, a good deal of effort has been addressed to creating a test fair to minority groups. The Spanish version and the Nonverbal Scale are significant contributions to this effort. As far as meeting the diverse needs of exceptional children, certainly those with language handicaps will be more fairly assessed.

On balance, it appears that the goals listed in the interpretive manual have been met and that this battery is a valuable addition to the assessment effort.

Notes to the *Compendium*:

[1]Das (1984) objects to stating that the K-ABC is based solidly on the work of Luria when it addresses only one of the three blocks posited by Luria (1966). He feels that its items can be classified as verbal and nonverbal performance as well as simultaneous and sequential, challenging the construct being measured. He states as well that the test does not measure sequential processing adequately. Sternberg (1984) concludes, too, that the sampling of behaviors is too narrow to represent adequately the sequential-simultaneous processing

model posited by the Kaufmans (1983c). He sees difficulties with the evidence presented for the construct validity of the test. In his reply, Alan Kaufman (1984; previously cited as in press) rebuts these criticisms with theoretical bases and practical considerations for viewing the simultaneous-sequential dichotomy as a logical organizational base for the instrument.

[2]The Spanish version may not be much help to individuals working with Cuban, Puerto Rican, Central American, or South American Spanish-speaking children. First, the dialects may be so different that the children may miscue with Mexican Spanish directions. Secondly, the norms may not be appropriate because they are developed for Mexican groups.

[3]In fact, Kaufman, Kamphaus, and Kaufman (1985) state that the dependence on visual stimuli make the K-ABC unsuitable for use with visually handicapped children.

[4]Unfortunately, little on planning interventions from data obtained with the K-ABC appears in the literature. The interpretive manual and Gunnison's article in the fall 1984 *Journal of Special Education* are the only materials on the topic that this reviewer found.

[5]This statement still leaves unclear in whose province the use of this instrument falls. With the concern about roles in multidisciplinary teams of professionals, it would be helpful to identify more clearly whether the K-ABC requires the same level of training as the Wechslers or the Binet in order to designate it as the responsibility of one member of the team rather than another.

[6]However, much more work on differential diagnosis is required before the test can be used with a high degree of confidence.

[7]While efforts to lessen bias are claimed effective by the Kaufmans, the literature is less clear on the matter. Four studies found differing levels of performance among ethnic groups. Naglieri (1984, 1986) reported that the K-ABC was appropriate for Navajo children but perhaps not for black children. The means for blacks were lower than they were for whites; the magnitude of the difference was approximately what it was on the WISC-R. Reynolds and others (1985) found that the K-ABC overpredicted the academic levels of black children. Finally, Valencia and Rankin (1986) found support for similar factor structure for Anglo and Mexican-American children on the K-ABC.

[8]While this translation is the crux of the process, it lacks empirical validation. Only Gunnison's (1984) work illustrating the translation process has appeared in the literature. At the present time there are little hard data with which to examine the efficacy of interventions based on the analysis of performance on the K-ABC.

[9]Much more work must be done with interpretations and formulating intervention strategies. Additional work with scores and contrasting groups will assist with interpretation; that will take time. Work with formulating intervention strategies is another matter. This is probably the weakest area of utilizing test scores and profiles of test scores and not exclusively a problem of the K-ABC; the same is true of the Wechslers, the Binet, and many other tests being used regularly to assess the learning problems of school-aged children.

[10]As Mehrens (1984) points out, these reliabilities and standard errors are quite reasonable. There is a need for long-term stability studies and additional reliability studies; however, this is no different from any other new test, especially one doing what this one is purported to do.

[11]The studies on criterion-related validity fare better than those on construct validity. Sternberg (1984), Das (1984), and Jensen (1984) comment primarily on content validity. Studies since that time reveal conflicting results and interpretations. In short, the area of construct validity is the one most open to criticism. Some sort of meta-analysis approach might help to clarify the conflicting findings. In addition, much more discussion of the hypothetical constructs on which the K-ABC is based will be necessary to clarify the meaning of the construct validity studies. Mehrens (1984) attributes the Kaufmans' confident statements about validity to overly optimistic interpretation of data analyses; Sternberg

(1984) is less kind in his views. Truly, this is an area for serious investigating, for it is upon construct validity that a test wed to theory must stand. Much more remains to be done.

[12]Since 1984, additional studies have been undertaken. Studies supporting Kaufman's position can be found as can those that challenge his position. On the whole, research tends to support the reliability of the K-ABC. Criterion-related validity seems to fare well, too. The area of construct validity is controversial; it is here that attention must be focused.

[13]In preparing this update, 39 citations on the K-ABC were found. Research seems to be directed primarily at reliability, validity, and bias rather than toward the more theoretical issues of how the underlying constructs are being measured. Save for articles in the 1983 special issue of the *Journal of Special Education* devoted entirely to the K-ABC, little else was found that addressed the nature of what the test measures. Although still a young instrument, the whole issue of prescriptive teaching from the K-ABC has been examined insufficiently. In California, with the recent order (*Larry P. v. Riles,* 1986) modifying judgment in the 1979 *Larry P. v. Riles* decision, the use of this instrument and others like it have been curtailed not only by the decision of the court but also by the broader reaction of the California State Department of Education and by the caution of local school districts. In short, minority children being assessed for special education placement in California will not be tested with an intelligence measure.

References

Angoff, W.H., & Ford, S.F. (1973). Item-race interaction on a test of scholastic aptitude. *Journal of Educational Measurement, 10,* 95-106.

Chatman, S.P., Reynolds, C.R., & Willson, V.L. (1983, August). *Multiple indexes of test scatter on the Kaufman Assessment Battery for Children.* Paper presented at the annual meeting of the American Psychological Association, Anaheim, CA.

Cronbach, L.J. (1970). *Essentials of psychological testing* (3rd edition). New York: Harper & Row.

Das, J.P. (1984). Simultaneous and successive processes and the K-ABC. *Journal of Special Education, 18,* 229-238.

Das, J.P., Kirby, J.R., & Jarman, R.F. (1979). *Simultaneous and successive cognitive processes.* New York: Academic Press.

Dunn, L., & Dunn L. (1981). *Peabody Picture Vocabulary Test-Revised (PPVT-R).* Circle Pines, MN: American Guidance Service.

Golden, C.J., Hammeke, T.A., & Purisch, A.D. (1980). *Luria-Nebraska Neuropsychological Battery.* Los Angeles: Western Psychological Services.

Gunnison, J.A. (1984). Developing educational intervention from assessments involving the K-ABC. *Journal of Special Education, 18,* 325-343.

Ingram, R. (1984). *ASSIST for the Kaufman Assessment Battery for Children.* Circle Pines, MN: American Guidance Service.

Jensen, A.R. (in press). The black-white difference on the K-ABC: Implications for future testing. *Journal of Special Education.*

Kaufman, A.S. (1979). *Intelligent testing with the WISC-R.* New York: John Wiley & Sons.

Kaufman, A.S. (in press). K-ABC and controversy. *Journal of Special Education.*

Kaufman, A.S., Kamphaus, R.W., & Kaufman, N.L. (1985). The Kaufman Assessment Battery for Children (K-ABC). In C.S. Newkirk (Ed.), *Major psychological assessment instruments.* Boston: Allyn & Bacon.

Kaufman, A.S., & Kaufman, N.L. (1977). *Clinical evaluation of young children with the McCarthy Scales.* New York: Grune & Stratton.

Kaufman, A.S., & Kaufman, N.L. (1983a). *K-ABC: Kaufman Assessment Battery for Children.* Circle Pines, MN: American Guidance Service.

Kaufman, A.S., & Kaufman, N.L. (1983b). *K-ABC: Kaufman Assessment Battery for Children, administration and scoring manual.* Circle Pines, MN: American Guidance Service.

Kaufman, A.S., & Kaufman, N.L. (1983c). *K-ABC: Kaufman Assessment Battery for Children, interpretive manual.* Circle Pines, MN: American Guidance Service.

Kaufman, A.S., & Kamphaus, R.W. (in press). Factor analysis of the Kaufman Assessment Battery for Children (K-ABC) for ages 2½ through 12½ years. *Journal of Educational Psychology.*

Keith, T.Z. (in press). Questioning the K-ABC: What *does* it measure? *School Psychology Review.*

Keith, T.Z., & Dunbar, S.B. (in press). Hierarchial factor analysis of the K-ABC: Testing alternate modes. *Journal of Special Education.*

Larry P. v. Riles, 495 F. Supp. 926 (N.D.Cal., 1979).

Larry P. v. Riles, Slip Op. 71-2270 (N.D.Cal., 1986).

Luria, A.R. (1966). *Higher cortical functions in man.* New York: Basic Books.

Mehrens, W.A. (1984). A critical analysis of the psychometric properties of the K-ABC. *Journal of Special Education, 18,* 297-310.

Merz, W.R. (1977). Test fairness and test bias: A review of procedures. In M.J. Wargo & D.R. Green (Eds.), *Achievement testing of disadvantaged and minority students for educational program evaluation.* Monterey, CA: CTB/McGraw-Hill.

Naglieri, J.A. (1984). Concurrent and predictive validity of the Kaufman Assessment Battery for Children with a Navajo sample. *Journal of School Psychology, 22,* 373-379.

Naglieri, J.A. (1986). WISC-R and K-ABC comparison for matched samples of black and white children. *Journal of School Psychology, 24,* 81-88.

Reynolds, C.R., Willson, V.L., & Chatman, S.P. (1983, August). *Regression analysis of bias on the Kaufman-Assessment Battery for Children.* Paper presented at the annual meeting of the American Psychological Association, Anaheim, CA.

Reynolds, C.R., Willson, V.L., & Chatman, S.P. (1985). Regression analyses of bias on the Kaufman Assessment Battery for Children. *Journal of School Psychology, 23,* 195-204.

Reynolds, C.R., Willson, V.L., & Chatman, S.P. (in press). Relationship between age and raw score increases on the Kaufman Assessment Battery for Children. *Psychology in the Schools.*

Sternberg, R.J. (1984). The Kaufman Assessment Battery for Children: An information-processing analysis and critique. *Journal of Special Education, 18,* 269-279.

Thorndike, R.L. (1949). *Personnel selection: Test and measurement techniques.* New York: John Wiley & Sons, Inc.

Valencia, R.R., & Rankin, R.J. (1986). Factor analysis fo the K-ABC for groups and Anglo and Mexican American children. *Journal of Educational Measurement, 23,* 209-219.

Wright, B.D., & Stone, M.H. (1979). *Best test design.* Chicago: MESA Press.

Michael D. Franzen, Ph.D.

Director of Neuropsychology, West Virginia University Medical Center, Assistant Professor of Behavioral Medicine and Psychiatry, School of Medicine, and Assistant Professor of Psychology, West Virginia University, Morgantown, West Virginia.

LURIA-NEBRASKA NEUROPSYCHOLOGICAL BATTERY

Charles J. Golden, Thomas A. Hammeke, and Arnold D. Purisch. Los Angeles, California: Western Psychological Services.

THE ORIGINAL OF THIS REVIEW WAS PUBLISHED IN TEST CRITIQUES: VOLUME III (1985).

Introduction

The Luria-Nebraska Neuropsychological Battery (LNNB) is an attempt to provide a comprehensive neuropsychological assessment technique that combines features of Aleksandr Luria's neuropsychological theory of brain organization and function with the North American psychometric tradition. Although administration of the LNNB can be learned in approximately one to two weeks, interpretation is extremely complex, relying on complicated notions of test theory and on an understanding of Luria's theory. Although the items in the LNNB are based on actual procedures used by Luria and published by Anne Lise-Christensen (1978), it should not be confused with Luria's Neuropsychological Investigation or with Lurian methodology.

Lurian theory specifies that observable behaviors are the result of combinations of molecular skills and that the successful performance of a behavior is dependent on cooperation among several brain areas. These brain areas are linked in a functional chain that eventuates in the behavior. Therefore, according to Lurian theory, there is no such thing as a pure point-to-point, cortico-behavioral relation. By extension, there is no task that can be strictly localized. Failure to successfully perform a certain behavior can be the result of dysfunction in any of the links that comprise the functional chain. It is by comparing performance across similar items that differ only in a single task requirement that we can isolate the dysfunctional unit.

The LNNB is divided into eleven clinical scales: Motor Functions, Tactile Functions, Rhythm, Visual Functions, Receptive Speech, Expressive Speech, Reading, Writing, Arithmetic, Memory, and Intellectual Processes, plus a twelfth scale (Intermediate-Term Memory Scale) for Form II only. Consistent with Lurian theory, none of the items are presumed to measure only a single function. Therefore, in more recent versions of the test protocol, the names of the scales have been dropped in favor of number identifications. The point remains that the items in Scale C1 (formerly the Motor Function Scale) measure mainly motor functions.

260

The scales are described as follows:

Motor Scale (C1): The Motor or C1 Scale has 51 items, which measure basic and complex motor skills. There are timed and untimed procedures to measure motor ability of the hands, arms, face, and mouth. Some of the items require activity of only one side of the body at a time, while others require coordinated activity from both sides of the body. There are items that require imitation as well as items requiring motor activity in response to verbal instructions. Three items evaluate drawing ability and speed from verbal instructions and three others measure drawing ability and speed from copying visual stimuli.

Rhythm Scale (C2): The 12-item Rhythm or C2 Scale evaluates the subject's ability to perceive and produce rhythm and pitch relationships. As with all of the scales the items are arranged from simple to complex. The first items require the subject to determine whether the tones and pitch patterns heard are the same or different. Later items require the subject to repeat rhythmic patterns, to count the number of tones in a sequence, and to reproduce short melodies.

Tactile Scale (C3): The Tactile or C4 Scale contains 22 items, measuring ability in different kinesthetic and tactile skill areas. It includes measures of location by tactile sensation, discrimination between pressure and pain, direction of movement by tactile sensation, two-point discrimination, and stereognosis.

Visual Processes Scale (C4): The 14-item Visual Processes or C4 Scale has 14 items and includes measures of object recognition with different degress of clarity, identification of overlapping objects in a line drawing, items similar to Raven's Progressive Matrices, telling time from printed clocks, accurate perception of the number of objects in a three-dimensional object represented in a two-dimensional drawing, and a block rotation task.

Receptive Speech Scale (C5): The Receptive Speech or C5 Scale has 33 items, which measure the ability of a subject to perform auditory discrimination of phonemes and words both by repeating the stimuli and by writing down the stimuli, phonemic discrimination when pitch changes, the ability of a subject to follow verbal auditory commands, the comprehension of spatial relationships by verbal auditory information, the understanding of complex logical relationships, and the understanding of complex and inverted grammatical structure.

Expressive Speech Scale (C6): The 42-item Expressive Speech or C7 Scale measures the ability of the subject to repeat phonemes, words, phrases, and sentences. In some of the items the stimuli are given in an auditory, verbal modality and in some items the information is given in a written modality. In addition, the subject is asked to name objects from pictures, fill in the missing word in a sentence, generate a sentence that contains three given words, rearrange words into a sentence, produce speech in overlearned tasks (such as counting forwards and backwards), and produce spontaneous speech from pictorial, written, and spoken stimuli.

Writing Scale (C7): The Writing or C7 Scale has 13 items. In this scale the subject is asked to state the number of letters in given words, to spell words, to copy letters and words from printed stimuli and from memory, and to provide spontaneous writing on a given topic. Scoring takes into account spelling, grammar, content, and motor writing errors.

Reading Scale (C8): The 13-item Reading or C8 Scale asks the subject to identify letters, morphemes, and words from dictation and from written material. In addition, the subject is asked to read sentences and a short story.

Arithmetic Scale (C9): The Arithmetic or C9 Scale has 22 items, requiring the subject to read and write simple and complex arabic and roman numerals. This scale also requires the subject to perform simple and complex arithmetic operations mentally and on paper. The last two items require serial subtractions of 7 and 13 respectively.

Memory Scale (C10): The Memory or C10 Scale contains 13 items, which measure both immediate memory and slightly delayed memory with interference. Memory is assessed on the areas of pictorial representation and verbal material. In addition, one item requires the memorization of a rhythmic pattern. Memory is assessed both when there is a meaningful context and when unrelated lists are used as stimuli. The last item in this scale requires the subject to remember words that are paired with pictures.

Intellectual Processes Scale (C11): The 34-item Intellectual Processes or C11 Scale measures thematic understanding of pictures, the ability of subjects to arrange pictures into a meaningful narrative, identification of humorous aspects of pictures, simple word definitions, abstraction abilities, comprehension of analogies, understanding of proverbs, and verbal arithmetic.

Intermediate Memory Scale (C12): The Intermediate Memory or C12 Scale exists only on Form II of the LNNB. It has 10 items, which ask the subject to recall activities and procedures that occurred earlier in the test situation. Therefore, it is also a measure of incidental memory.

The LNNB is available in two parallel forms: Form I and an alternate form, Form II. There were two main thrusts to the development of Form II. The first thrust of the development was to create an alternate from that could be used in retest situations, minimizing the risk of practice effects. The second thrust was to meet some of the criticisms raised against the LNNB. A few items had increased numbers of trials in order to increase their reliability. Of the remaining items, 21% (56 of the items) were exactly the same as in Form I. The identical items are largely found on the Rhythm, Motor Functions, and Receptive Speech Scales. These items were judged not to have a significant practice component in performance. The remaining items were written to have the same intent but different content than the items in Form I. The stimulus cards are printed in larger and clearer print in order to minimize the effect of peripheral visual impairment on the test results. The stimulus items are also collected in a single, wire-bound booklet to help ease administration. Administration of Item 235 in the Memory Scale was changed in order to rule out possible confounds from serial memorization instead of association memorization as is the intent of the item.

In addition to the twelve clinical scales there are two sensorimotor scales—a Right Hemisphere Scale, a Left Hemisphere Scale—and a Pathognomonic Scale. The Right and Left Scales can be used to help lateralize the area of dysfunction. The Pathognomonic Scale is composed of items that best differentiate between neurologic and psychiatric subjects. Recent work has developed two additional scales (Profile Elevation and Impairment) that are comprised of items drawn from items on the original clinical scales (Sawicki & Golden, 1984). The Profile Impairment Scale consists of 22 items which correlated most highly with the number of scales above critical level. The Profile Elevation Scale consists of 28 items which correlated most highly with the mean scale elevation. The relative elevations of these two scales can provide useful information. For example, if both scales are elevated above critical level and the Profile Elevation Scale is elevated relative to the Profile Impairment Scale, then one can expect a fair amount of compensation for deficits to occur. As compensation occurs, the two scales tend toward equal elevations. These scales can provide useful information about the recency of the injury, the severity of impairment, and the degree of predictable compensation.

The LNNB material consists of several visual stimuli and a set of audiotaped rhythm and pitch stimuli. In addition, the examiner will need to provide a few common items, such as a rubber band, key, nickel, protractor, rubber eraser, and large paper clip, for use in the tactile and visual recognition items. The procedures are interactive with the examiner who gives verbal instructions, observes and, in some cases, times the responses of the subject. The LNNB is portable and can be performed at a subject's bedside. Forms I and II of the LNNB are intended to be used with subjects over the age of 14 years. There is another version for use with children aged 8-13 years. The LNNB can be used with most levels of neuropsychological functioning, although the reliability of the test is attenuated in the normal range of ability. Because of ceiling effects, the accuracy of the LNNB is limited in the superior range of functioning.

The test answer protocol also contains the instructions for administration for each item. There is also a patient response booklet in which the subject can write and draw for those items that require these activities.

All of the scales are presented as standardized T-scores with a mean of 50 and a standard deviation of 10. Interpretation is partly conductd by comparing the scale score to the critical level. The critical level is individually determined through a regression formula that uses the subject's age and level of education. In several, scale scores that exceed the critical level can be viewed as reflective of impaired performance. However, there are several other interpretative strategies that increase the accuracy of the instrument. These strategies include examination of the scatter of the scores, examination of the factor analytic derived scales, examination of the empirically derived localization scores, item analysis, pattern analysis, and consideration of the qualitative aspects of performance.

Practical Applications/Uses

The LNNB has been useful in the diagnosis of brain impairment, in the localization of neuropathological processes, and in the differential diagnosis of psychiatric and neurological disorders. Additional uses have been suggested in rehabilitation planning for brain-impaired individuals, in vocational placement, and in forensic applications including disability determination.

Interpretation of the results of the LNNB must occur in the context of a complete history because there are multiple reasons why a subject may fail a given item. The battery provides the most information in the hands of a trained clinical neuropsychologist. However, it can be used for screening purposes or for gross identification of behavioral-cortical deficits when used by the general psychologist who has had some training in test theory, neuropsychology, and in the administration of the LNNB. Caution must again be raised here: training in the administration of the LNNB does not make one an expert in the interpretation of the LNNB.

The original purpose of the LNNB is to identify cortico-behavioral impairment. It does this by assessing the ability of the subject to perform certain behaviors in general areas thought to be mediated by brain function (e.g., memory, motor skills, visual acuity and integration, tactile sensation). However, recent work has suggested the possibility of using the battery to assess subcortical deficits and as an aid to the assessment of multiple sclerosis.

The LNNB is appropriate for adolescent and adult subjects aged 15 years and older. Because of its reliance on verbal instructions and verbal responses, it may have limited utility in assessing aphasics. Because some of the items require the subject to describe visual stimuli, parts of the test are not applicable to blind subjects. However, because of the test's background in Lurian theory, many tasks are given in more than one modality, allowing for assessment of individuals with different peripheral impairments. For example, at different points in the assessment of expressive speech, subjects are asked to read certain words and to repeat those same words from auditory stimuli so that subjects with either an auditory or a visual impairment would be able to demonstrate expressive speech capacity in one of those tasks.

The LNNB is administered in a one-to-one setting with the examiner and the subject sitting across from each other at a desk. The test can be administered by a B.A. level technician who has received training specific to the LNNB. Most experienced test technicians can learn to administer the battery in 30-40 hours of training. Most of the items involve a straightforward administration technique. However, some of the items require behavioral skills that will need practice before they can be successfully completed. For example, in the expressive speech section, the subject is asked to give a spontaneous speech about the conflict between the generations. The examiner is expected to record the time until the response begins and then count the number of words produced in the first five seconds following the initiation of the response. This necessitates recording time latency and frequency simultaneously, a task that requires practice.

The LNNB requires about two and one half hours to administer to an intact subject. Depending on the type and degree of impairment, administration to a brain-impaired subject can take longer. The LNNB is best administered in a single session and in the order given in the protocol. Breaks can be taken at the end of one scale and before beginning another scale. If the subject fatigues easily, testing can be terminated for the day at the end of a scale and resumed the next day. It is important not to break in the middle of a scale as this will cause too great of a deviation from standardized procedures.

Each item is given a score of 0, 1, or 2, where 0 corresponds to a normal performance, 1 corresponds to a borderline performance, and 2 corresponds to an impaired performance. The manual contains explicit rules for the assignment of item scores. Item scores are then summed to provide scale raw scores that are translated to standardized T-scores with a mean of 50 and a standard deviation of 10 points.

In addition, items can be summed to provide factor scale scores and localization scale scores, which are also standardized T-scores. The factor scales were derived from principal components factor analyses of each of the clinical scales separately. The localization scales were empirically derived from examination of those items that were associated with known discretely localized lesions. Western Psychological Services, the publisher of the LNNB, markets a computer-scoring program that calculates the standardized T-scores for the 14 clinical scales, the eight localization scales, and the 30 factor analytic scales, saving as much as 90 minutes in the scoring procedure.

The first step in interpreting the LNNB is in calculating the critical level. The critical level is that score beyond which obtained scores represent impaired performance (Golden, Hammeke, & Purisch, 1980). The critical level is one standard deviation above the average scale score expected with a person with a given age and level of educational attainment. The critical level is determined substituting values for age and education in a regression formula that was empirically derived. In this way interpretation is more individualized than would be the case with absolute cutoff points. The critical level can be adjusted to reflect the amount of scatter in the scale scores. If there is more than thirty points difference between the highest and the lowest scale score, then the critical level can be lowered to 25 points above the lowest scale score (Moses, Golden, Ariel, & Gustavson, 1983).

Interpretation of the LNNB occurs at five levels. First, the clinical scales are examined in order to determine which are above critical level. Next the scatter among scale scores is examined. The localization scales are also examined against the critical level, as are the factor analytic scales, although there are no rules regarding the interpretation of scatter on these two sets of scales. Next, performance on items across scales is examined in order to determine if any patterns consistent with neuropsychological theory exist. Last, the qualitative aspects of performance are examined. Although interpretation can be terminated at any level (and should be terminated early in the case of the inexperienced examiner), the battery gives the most amount of information and the most accurate information when the five interpretative steps are followed (Golden, Ariel, Moses, Wilkening, McKay, & MacInnes, 1982).

Technical Aspects

The technical aspects of the LNNB have been evaluated in approximately 130 studies to date. Constraints of space dictate that only a cursory review of selected studies can be possible in this review.

Golden, Berg, and Graber (1982) report the results of a study designed to evaluate the test-retest reliability of the LNNB. Twenty-seven patients with diagnoses of static injuries were tested twice with an average test-retest interval of 167 days. Resulting correlation coefficients ranged from 0.77 for the Right Hemisphere Scale to 0.96 for the Arithmetic Scale. These results were subsequently replicated in Plaisted and Golden (1982), who used 30 psychiatric subjects.

Interscorer reliability was evaluated on a sample of five patients. For each patient a different pair of examiners scored the responses while observing the patient. The agreement between raters ranged from 92% agreement for one patient to 98% for another patient. Overall, there was a 95% agreement rate (Golden, Hammeke, & Purisch, 1980). It must be kept in mind that these percentages may be inflated by the fact that about two thirds of the items are tri-chotomously scored and the remaining one third of the items are dichotomously scored.

Bach, Harowski, Kirby, Peterson, and Schulein (1981) attempted to remediate the deficiencies of the above study by evaluating the interrater agreement in a sample of borderline performances in two subjects and five raters. Bach et al. used Cohen's kappa as an uninflated measure of agreement and reported values ranging from 0.79 to 1.00. The subjects used in this study were normal confederates who were instructed how to respond to the items so as to produce borderline performance on a random set of items. The identities of the items that were performed in a bor-derline manner were unknown to the examiners.

In order to evaluate and extend these findings Moses and Schrefft (1985) used two examiners to evaluate 16 neurological patients, 15 psychiatric subjects, three subjects with a history of substance abuse, and two normal medical patients for a total of 36 subjects. The authors found that correlation coefficients for summary score ranged from 0.97 for the Intellectual Processes Scale to 0.99 for the Receptive Speech Scale, with mean differences in total scores of 1.47 and 0.00 T-score points, respectively. Localization scales' correlation coefficients ranged from 0.96 to 0.99. The authors also evaluated the agreement for decisions regarding whether the scales exceeded critical level. Agreement levels ranged from 33/36 for the Recep-tive Speech and the Writing Scales to 36/36 for the Expressive Speech, Tactile Func-tions, Arithmetic, and Intellectual Processes Scales. Analyses of the same data using chi-square and Kendall's tau resulted in similar results. In examining agree-ment in individual item scores, the authors found complete agreement in 95.71%, disagreement of one-point value in 3.5%, and disagreement of two points in 0.78% of the individual item scores.

Under Lurian theory, the battery was constructed such that the scales are not homogeneous. The items in a single scale may require different discrete skills in order to produce a behavior that is related to the skill area purported to be assessed by the scale. For example, the Memory Scale requires visual skills for the suc-cessful completion of some items and auditory receptive skills for others, but all of the items involve some form of memory. Because of this fact, one would expect moderate degrees of internal consistency in the scales.

Golden, Fross, and Graber (1981) evaluated the split-half reliability of the clinical scales using an odd-even split in a sample of 338 patients. The sample was con-stituted of 74 normal controls, 83 psychiatric patients, and 181 neurological patients. The split-half reliability correlation coefficients ranged from a high of 0.95 for the Reading Scale to a low of 0.89 for the Memory Scale. Once again these values may be inflated by the categorical nature of the scoring procedure. How-ever, the values do represent acceptable degrees of consistency.

Internal consistency reliability of the summary, localization, and factor-analytic scales was evaluated for a heterogeneous group of subjects by Moses, Johnson, and Lewis (1983). These evaluations were also conducted separately for a sample of 451 brain-impaired, 414 schizophrenic, 128 mixed-psychiatric, and 108 normal sub-jects. Values ranged from 0.54 to 0.78 for the normal subjects and 0.83 to 0.93 for the remaining subjects. The relatively lower values for normal subjects is probably due to restriction of range in those subjects and underlines the problems attendant on using the LNNB to describe neuropsychological functioning in normal subjects.

The original attempts at providing validational evidence for the LNNB focused on its accuracy in discriminating neurologic patients from normal subjects. Hammeke, Golden, and Purisch (1978) report that all of the summary scales were successful in discriminating brain-damaged from normal subjects at the 0.001 level of significance. Other studies have investigated the discriminative ability of the LNNB in elderly subjects (Spitzform, 1982), psychiatric populations (Golden, Graber, Moses, & Zatz, 1980; Golden, MacInnes, Ariel, Ruedrich, Chu, Coffman, Graber, & Bloch, 1982; Puente, Heidelberg-Sanders, & Lund, 1982; Shelley & Goldstein, 1983), alcoholics (Chmielewski & Golden, 1980), and epileptics (Berg & Golden, 1981). These studies report varying levels of success, but generally support the utility of the LNNB.

Perhaps one of the more significant studies was reported by Malloy and Webster (1981). A difficult question in neuropsychological assessment is the differential diagnosis of patients who present ambiguous or inconclusive symptoms and therefore represent diagnostic dilemmas. Using the rule of classifying subjects who have three or more clinical scales above critical level as impaired, the authors found 80% agreement between the LNNB and noninvasive neurodiagnostic techniques in subjects with mild cortical impairment.

In order to evaluate the validity of the LNNB as a technique to measure specific neuropsychological tasks, a series of factor analyses was undertaken. The several studies that were conducted present general support for this idea, but too little is known about the temporal stability of these scales to recommend them for clinical use in retest situations (Golden, Hammeke, Purisch, Berg, Moses, Newlin, Wilkening, & Puente, 1982).

Concurrent validity of the LNNB has been assessed by correlating its results with that of the Halstead-Reitan Neuropsychological Battery with encouraging results. Golden, Kane, Sweet, Moses, Cardellino, Templeton, Vicente, and Graber (1981) report that when the summary scores of the LNNB are correlated with the summary scores of the Halstead-Reitan, the minimum value was 0.71. Chelune (1982) objected that the large correlations may have been due to shared overlap of the two batteries with the Wechsler Intelligence Scale. Golden, Gustavson, and Ariel (1982) respond to this criticism with a conceptual argument, namely, that postmorbid IQ will reflect brain impairment. The authors then partialled out level of education as an estimate of premorbid IQ and still found moderate correlations between the two batteries. Shelley and Goldstein (1982) report that in a sample of 137 subjects, correlations between the LNNB and Halstead-Reitan scores ranged from 0.60 to 0.96. They also conducted a factor analysis and found that both batteries contributed substantially to the extracted three-factor solution, indicating shared sensitivity to the same neuropsychological variables.

Another factor to consider is whether the use of the LNNB and the Halstead-Reitan will result in a similar classification of subjects. Golden et al. (1980) found that a discriminant function analysis of the two batteries resulted in approximately 85% accurate classification for each. Goldstein and Shelley (1984) report similar rates of accuracy in diagnosis for the two batteries.

Other attempts to evaluate the concurrent validity of the LNNB have correlated the results of the Peabody Individual Achievement Test with appropriate scales of the LNNB in 100 neurological, psychiatric, and normal subjects with acceptable results (Gillen, Ginn, Strider, Kreuch, & Golden, 1983). Prifitera and Ryan (1982) have compared the performance of the LNNB Memory Scale with other memory assessment techniques. Prifitera and Ryan (1981) compared the LNNB Intellectual Processes Scale with the WAIS. These studies represent a start at the necessary task of evaluating the specific concurrent validity of the components. However, more research needs to be done before definite conclusions can be reached.

Studies evaluating the accuracy of the localization scales fo the LNNB have been conducted resulting in generally accurate identification of the site of lesion. The use of the localization scales resulted in a 89% correct classification rate in the derivation sample and 74% correct classification rate in the cross-validation sample (Golden, Moses, Fishburne, Engum, Lewis, Wisniewski, Conley, Berg, & Graber, 1981). However, these results need to be replicated because of the inclusion of closed-head-injury subjects in the sample. Closed-head-injured patients are possibly injured in sites other than the site of the main injury due to contre-coup effects and may be contaminating these characteristics of the experimental sample.

The first empirical investigation of Form II administered both Form I and II to the same set of subjects (Ariel, Knippa, Strider, Golden, & Sawicki, 1984). These subjects consisted of 51 normal subjects, 14 psychiatric in-patients, three subjects with traumatic head injury, one retarded subject, and six neurologic in-patients. The correlations between the two sets of clinical scales ranged from 0.77 for the Rhythm Scale to 0.91 for the Arithmetic Scale. Correlations between the factor scales ranged from 0.40 to 0.88. There were significant differences on eight of the 16 clinical scales, ten out of the 30 factor scales, and three out of the eight localization scales. This underlines the need for the computation of T-scores using a separate standardization sample. Currently, computation of T-scores is conducted using a regression formula predicting Form II scores from Form I scores, and work is being done to provide T-scores on the basis of a separate normative sample.

Critique

As with many new tests, the LNNB has not been without its critics. And as with many new tests, there have been replies to these criticisms in the literature. The exchange of criticisms and answers have helped to place the LNNB in sharp relief against the criteria of test construction and neuropsychological assessment method. Adams (1980a) criticizes the LNNB on the basis of the methodology of its development, stating that multiple *t*-tests were an inappropriate means of deciding which items to retain in the battery. He similarly criticizes the use of a small subject sample in a discriminant function analysis. Golden (1980) replies unconvincingly that the derived discriminant function had an acceptable ratio of predictors to subjects. However, the fact remains that the number of tested potential predictors exceeded the number from which safe conclusions could be reached, given the sample size. Golden's (1980) reply to the issue of multiple *t*-tests is more convincing. *T*-tests were used in an apparent attempt to minimize the probability of Type I errors. The items have subsequently been evaluated and this methodology has apparently had little effect on the accuracy of the battery as a whole.

Adams (1980b) replies by questioning the veracity of the reports of the accuracy of the battery, calling into question the scientific honesty of the test's developers. These doubts are in need of empirical evidence to validate their relevance.

Spiers (1981) repeats some of the criticisms of Adams (1980a, 1980b) and adds a few conceptual criticisms. He states that the idea of a battery was antithetical to the methods of Luria and that the availability of MMPI-like graphs on which to plot results will increase the probability that insufficiently trained psychologists will use the battery clinically. He also states that the battery is not really comprehensive. The first criticism is certainly true; the LNNB is not really the Lurian method. However, the authors have been at pains to state that the name Luria-Nebraska is honorific because of its reliance on the contributions of Luria to the field of neuropsychology and the reliance of the battery on his theory. The second criticism is directed more to the users of the battery than to the battery itself and is well taken. Unless the users of the LNNB have had specific training in neuropsychological assessment, they are better to limit themselves in interpretation.

Spiers (1982) further criticizes the LNNB on the basis of the intent of the items and criticizes the standardized, quantified approach in general because of its reliance on average scores, which mask individual differences. Hutchinson (1984) criticizes Spiers (1982) on a factual basis stating that there are discrepancies between Spiers' reading of the manual and verbatim quotes from the manual. He further accuses Spiers of misunderstanding the intended use of the battery. Spiers (1984) replies that Hutchinson (1984) misinterpreted his (Spiers') statements and accuses Hutchinson of misreading his article. Stambrook (1983) has also criticized the LNNB in the literature, however his article does little more than repeat the criticisms of Adams (1980a, 1980b) and Spiers (1981, 1982). By now, two things should become clear: 1) part of the controversy surrounding the LNNB is actually related to the disagreement between proponents of the standardized, quantified approach to neuropsychological assessment and proponents of the intuitive, qualitative approach and 2) part of the controversy has degenerated into personal accusations. Although theoretical arguments are necessary to the evaluation of any assessment techniques, they are insufficient without direct empirical evidence, and ad hominem arguments have no place in the exchange of ideas.

At this point it would be useful to briefly discuss the drawbacks of the LNNB. It does not have a universally valid utility in all populations. Use of the LNNB should be done with reference to the experimental literature regarding the population in question. Further, the LNNB is not totally comprehensive as it does not assess reading comprehension or memory other than short-term (verbal) memory.

Due to its reliance on Lurian theory and to Luria's emphasis on the verbal mediation of behavior, the LNNB is dependent on language skills. Crosson and Warren (1982) criticize the use of the LNNB in aphasic populations. Delis and Kaplan (1982) present a case of an aphasic where the use of only the clinical scales would result in a misunderstanding of the subject's deficits. Golden, Ariel, Wilkening, McKay, and MacInnes (1982) reexamined the data regarding the case reported by Delis and Kaplan (1982) and found that the LNNB was inappropriately used because the primary language of the subject was Spanish. They (Golden et al., 1982) also found evidence in the history of chronic alcohol abuse, a seizure disorder, bilateral atrophy, and a left temporal-parietal infarct, indicating that the conclusions of the

LNNB were in closer agreement with the actual deficits of the subject than was implied by Delis and Kaplan (1982). Despite this foray into personal aspects of the controversy, the heavy dependence on language may limit the utility of the LNNB for language-impaired subjects.

Because the LNNB cannot distinguish between average and superior performance, it may be inappropriate in the assessment of normal neuropsychological functioning, except to rule out the possibility of impairment. As stated earlier, the reliabilities of the scale scores are attenuated in normal subjects.

Finally, although the LNNB relies on an understanding of Lurian constructs for its most effective use, the validity of these constructs have yet to be conclusively validated. In short, despite the many studies conducted to date, due to the complexity of the battery, there is still need for more research to fully evaluate each aspect of the LNNB. The LNNB can be recommended for clinical use, but only in the hands of a trained psychologist who is familiar with the research regarding the question to the answered and the population from which the subject is drawn. The LNNB seems to have wide applicability in both research and clinical settings. Given its young age and the amount of research conducted already, the LNNB is likely to further our understanding of brain and behavior relations.

References

Adams, K. M. (1980a). In search of Luria's battery: A false start. *Journal of Consulting and Clinical Psychology, 48,* 511-516.

Adams, K. M. (1980b). An end of innocence for behavioral neurology? Adams replies. *Journal of Consulting and Clinical Psychology, 48,* 522-524.

Ariel, R., Knippa, J., Strider, M. A., Golden, C. J., & Sawicki, R. F. (1984). *Alternate form reliability of the Luria-Nebraska Neuropsychological Battery.* Unpublished manuscript.

Bach, P. J., Harowski, K., Kirby, K., Peterson, P., & Schulein, M. (1981). The interrater reliability of the Luria-Nebraska Neuropsychological Battery. *Clinical Neuropsychology, 3,* 19-21.

Berg, R. A., & Golden, C. J. (1981). Identification of neuropsychological deficits in epilepsy using the Luria-Nebraska Neuropsychological Battery. *Journal of Consulting and Clinical Psychology, 49,* 745-747.

Chelune, G. J. (1982). Reexamination of relationship between Luria-Nebraska and Halstead-Reitan batteries: Overlap with WAIS. *Journal of Consulting and Clinical Psychology, 50,* 578-580.

Chmielewski, C., & Golden, C. J. (1980). Alcoholism and brain damage: An investigation of the Luria-Nebraska Neuropsychological Battery. *International Journal of Neuroscience, 10,* 99-105.

Christensen, A.-L. (1975). *Luria's neuropsychological investigation.* New York: Spectrum.

Crosson, B., & Warren, R. L. (1982). Use of the Luria-Nebraska Neuropsychological Battery in aphasia: A conceptual critique. *Journal of Consulting and Clinical Psychology, 50,* 22-31.

Delis, D. C., & Kaplan, E. (1982). Use of the Luria-Nebraska Neuropsychological Battery in aphasia: A case critique. *Journal of Consulting and Clinical Psychology, 50,* 32-39.

Gillen, R. W., Ginn, C. E., Strider, M. A., Kreuch, T., & Golden, C. J. (1983). The Luria-Nebraska Neuropsychological Battery and Peabody Individual Achievement Test: A correlational analysis. *International Journal of Neurosciences, 21,* 51-62.

Golden, C. J. (1980). In reply to Adams' "In search of Luria's battery: A false start." *Journal of Consulting and Clinical Psychology, 48,* 517-521.

Golden, C. J., Ariel, R. J., Wilkening, G. N., McKay, S. E., & MacInnes, W. D. (1982). Analytic techniques in the interpretation of the Luria-Nebraska Neuropsychological Battery. *Journal of Consulting and Clinical Psychology, 50,* 40-48.

Golden, C. J., Berg, R. A., & Graber, B. (1982). Test-retest reliability of the Luria-Nebraska Neuropsychological Battery in stable, chronically impaired patients. *Journal of Consulting and Clinical Psychology, 50,* 452-454.

Golden, C. J., Fross, K. H., & Graber, B. (1981). Split-half reliability of the Luria-Nebraska Neuropsychological Battery. *Journal of Consulting and Clinical Psychology, 49,* 304-305.

Golden, C. J., Graber, B., Moses, J. A., & Zatz, L. M. (1980). Differentiation of chronic schizophrenics with and without ventricular enlargement by the Luria-Nebraska Neuropsychological Battery. *International Journal of Neuroscience, 11,* 131-138.

Golden, C. J., Gustavson, J. L., & Ariel, R. (1982). Correlations between the Luria-Nebraska and Halstead-Reitan batteries: Effects of partialling out education and postmorbid I.Q. *Journal of Consulting and Clinical Psychology, 50,* 770-771.

Golden, C. J., Hammeke, T. A., & Purisch, A. D. (1980). *Luria-Nebraska Neuropsychological Battery manual.* Los Angeles: Western Psychological Services.

Golden, C. J., Hammeke, T. A., Purisch, A. D., Berg, R. A., Moses, J. A., Newlin, D. B., Wilkening, G. N., & Puente, A. E. (1982). *Item Interpretation of the Luria-Nebraska Neuropsychological Battery.* Lincoln, NE: University of Nebraska Press.

Golden, C. J., Kane, R., Sweet, J., Moses, J. A., Cardellino, J. P., Templeton, R., Vicente, P., & Graber, B. (1981). Relationship of the Halstead-Reitan and Luria-Nebraska Neuropsychological Batteries. *Journal of Consulting and Clinical Psychology, 49,* 410-417.

Golden C. J., MacInnes, W. D., Ariel, R. N., Ruedrich, S. L., Chu, C., Coffman, J. A., Graber, B., & Bloch, S. (1982). Cross-validation of the Luria-Nebraska Neuropsychological Battery to discriminate chronic schizophrenics with and without ventricular enlargement. *Journal of Consulting and Clinical Psychology, 50,* 87-95.

Golden, C. J., Moses, J. A., Fishburne, F. J., Engun, E., Lewis, G. P., Wisniewski, A. M., Conley, F. K., Berg, R. A., & Graber, B. (1981). Cross-validation of the Luria-Nebraska Neuropsychological Battery for the presence, lateralization, and localization of brain damage. *Journal of Consulting and Clinical Psychology, 49,* 491-507.

Goldstein, G., & Shelley, C. (1984). Discriminative validity of various intelligence and neuropsychological tests. *Journal of Consulting and Clinical Psychology, 52,* 383-389.

Hammeke, T. A., Golden, C. J., & Purisch, A. D. (1978). A standardized, short, and comprehensive neuropsychological test battery based on the Luria neuropsychological evaluation. *International Journal of Neuroscience, 8,* 135-141.

Hutchinson, G. L. (1984). The Luria-Nebraska Neuropsychological Battery controversy: A reply to Spiers. *Journal of Consulting and Clinical Psychology, 52,* 539-545.

Malloy, P. F., & Webster, J. S. (1981). Detecting mild brain impairment using the Luria-Nebraska Neuropsychological Battery. *Journal of Consulting and Clinical Psychology, 49,* 768-770.

Moses, J. A., Golden, C. J., Ariel, R. N., & Gustavson, J. L. (1983). *Interpretation of the Luria-Nebraska Neuropsychological Battery* (Vol. 1). New York: Grune and Stratton.

Moses, J. A., Johnson, G. L., & Lewis, G. P. (1983). Reliability analyses of the Luria-Nebraska Neuropsychological Battery summary, localization and factor scales. *International Journal of Neuroscience, 20,* 149-154.

Moses, J. A., & Schrefft, B. K. (1985). Interrater reliability analyses of the Luria-Nebraska Neuropsychological Battery. *International Journal of Clinical Neuropsychology, 7,* 31-38.

Plaisted, J. R., & Golden, C. J. (1982). Test-retest reliability of the clinical, factor, and localization scales of the Luria-Nebraska Neuropsychological Battery. *International Journal of Neuroscience, 17,* 163-167.

Prifitara, A., & Ryan, J. J. (1981). Validity of the Luria-Nebraska Neuropsychological Battery

Intellectual Processes Scale as a measure of adult intelligence. *Journal of Consulting and Clinical Psychology, 49,* 755-756.

Prifitera, A., & Ryan, J. J. (1982). Concurrent validity of the Luria-Nebraska neuropsychological Battery Memory Scale. *Journal of Clinical Psychology, 38,* 378-379.

Puente, A. E., Heidelberg-Sanders, C., & Lund, N. L. (1982). Discrimination of schizophrenics with and without nervous system damage using the Luria-Nebraska Neuropsychological Battery. *International Journal of Neuroscience, 16,* 59-62.

Sawicki, R. F., & Golden C. J. (1984). The Profile Elevation Scale and Impairment Scale: Two new summary scales for the Luria-Nebraska Neuropsychological Battery. *International Journal of Neuroscience, 23,* 81-90.

Shelley, C., & Goldstein, G. (1982). Psychometric relations between the Luria-Nebraska and Halstead-Reitan Neuropsychological batteries in a neuropsychiatric setting. *Clinical Neuropsychology, 4,* 128-133.

Shelley, C., & Goldstein, G. (1983). Discrimination of chronic schizophrenia and brain damage with the Luria-Nebraska Neuropsychological Battery: A partially successful replication. *Clinical Neuropsychology, 5,* 82-85.

Spiers, P. A. (1981) Have they come to praise Luria or to bury him?: The Luria-Nebraska Neuropsychological Battery controversy. *Journal of Consulting and Clinical Psychology, 49,* 331-341.

Spiers, P. A. (1982). The Luria-Nebraska Neuropsychological Battery revisited: A theory in practice or just practicing? *Journal of Consulting and Clinical Psychology, 50,* 301-306.

Spiers, P. A. (1984). What more can I say? In reply to Hutchinson, one last comment from Spiers. *Journal of Consulting and Clinical Psychology, 52,* 546-552.

Spitzform, M. (1982). Normative data in the elderly on the Luria-Nebraska Neuropsychological Battery. *Clinical Neuropsychology, 4,* 103-105.

Stambrook, M. (1983). The Luria-Nebraska Neuropsychological Battery: A promise that may be partly fulfilled. *Journal of Clinical Neuropsychology, 5,* 247-269.

Michael D. Franzen, Ph.D.

Director of Neuropsychology, West Virginia University Medical Center, and Assistant Professor of Behavioral Medicine and Psychiatry, West Virginia University School of Medicine, Morgantown, West Virginia.

LURIA-NEBRASKA NEUROPSYCHOLOGICAL BATTERY, FORM II

Charles J. Golden, Thomas A. Hammeke, and Arnold D. Purisch. Los Angeles, California: Western Psychological Services.

THE ORIGINAL OF THIS REVIEW WAS PUBLISHED IN TEST CRITIQUES: VOLUME IV (1986).

Introduction

The Luria-Nebraska Neuropsychological Battery (LNNB), Form II is an attempt to provide both an alternate form to Form I and some improvements over the original form. The second test is organized in the same way as the first, with the exception of an additional scale, C12 (Intermediate Memory). In response to criticism regarding the content validity of the scales, the names have been removed and replaced with numbers, C1 through C11 for the clinical scales, S1 through S5 for the supplementary scales, L1 through L8 for the localization scales, and numbers for the factor scales. The LNNB is suggested for use with subjects over the age of fifteen years. Because of its reliance on verbal instructions, the LNNB may have limited utility in assessing subjects with receptive language impairments. Revised administration instructions, which allow for alternate response modalities, extend the utility of the instrument for subjects with expressive language impairments.

Consistent with the aims of the LNNB, Form I, Form II combines features of Alexandr Luria's theories of brain function with a standardized approach to assessment. Therefore, many of the comments made with regard to Form I in an earlier review (Franzen, 1985) would also apply here and will not be repeated in great depth. Lurian theory specifies that observable behavior is the result of multiple areas of the brain working in cooperation. No single task can be strictly localized. It is by comparing performance on multiple items that differ in small aspects of stimulus characteristics or task demands that the evaluator can make statements regarding a dysfunctional process. For the sake of psychometric convenience, items on the LNNB are grouped into scales that share a central set of functions. For example, all items from the first scale require mainly motor operations for their successful completion. These item scores are then summed and transformed to provide T-score measurements of the function that underlies the respective scales.

The stimulus materials for Form II are similar to those of Form I. The visual stimuli have been bound into a single booklet arranged in the chronological order required for administration, an improvement over the separately bound portions of Form I. Additionally, the stimulus print has been enlarged so as not to penalize subjects with poor vision for peripheral impairment reasons. The audiotape stimulus for the C2 (Rhythm) Scale is identical to the audiotape for Form I.

In some instances, entirely new stimuli have been devised for Form II, using the same rationale as was used in devising the original items. For example, the CB (Reading) Scale shares no items across the two forms, nor does the C11 (Intellectual Processes) Scale. As mentioned above, the C12 (Intermediate Memory) Scale was a new development in Form I. In other instances, items were retained across the two forms, as in the case of the C2 (Rhythm) Scale, where 92% of the items originally appeared in Form I.

Form II is divided into 12 scales, which are similar to the original scales in Form I. In each of the scales the items are arranged from simple to more complex procedures. The Form II scales are described as follows:

C1 (Motor Functions): Measures basic and complex motor skills. Motor ability is measured in timed and untimed procedures sampling from the hands, arms, face, and mouth. Some items require coordinated movements across both sides of the body, whereas other items require isolated movements. Motor activity is assessed both from imitation and verbal instructions.

C2 (Rhythm): Measures the accurate perception and production of pitch and rhythm relationships. The first few items require the subject to determine whether the pitches and rhythmic patterns heard are the same or different, whereas later items require the subject to reproduce rhythms and pitches heard and to produce rhythms and melodies from verbal description.

C3 (Tactile Functions): Measures ability in different kinesthetic and tactile skill areas. It includes measures of location by tactile sensation, discrimination of pressure and pain, determination of movement, two-point discrimination, and stereognosis.

C4 (Visual Functions): Measures object recognition with different degrees of clarity and identification of overlapping objects, items similar to Raven's Progressive Matrices, a block rotation task, and other visual skills.

C5 (Receptive Speech): Measures auditory discrimination of phonemes and words either by repetition or dictation, the ability of the subject to follow auditory instructions, the comprehension of verbal spatial relations, and the comprehension of complex and inverted grammatical structure.

C6 (Expressive Speech): Assesses the ability of the subject to repeat phonemes, words, phrases, and sentences. In some of the items, the stimuli are presented in an auditory modality, and in some the stimuli are presented in a written modality. Confrontation-naming tasks are also included in this scale. Additionally, the production of spontaneous speech is evaluated as is the production of overlearned speech.

C7 (Writing): Measures the ability to state the number of letters in given words, to spell words, to copy words and letters from written stimuli and from memory, and to produce spontaneous writing on a given topic. As mentioned above, this scale has been decomposed into Motor Writing and Spelling subscales.

C8 (Reading): Measures the ability to read letters, morphemes, and words from written stimuli and to identify words when they are spelled auditorily. The subject is also asked to read sentences and a short paragraph.

C9 (Arithmetic): Assesses number recognition and the ability of the subject to write numbers from dictation. Simple and complex arithmetic operations are also assessed. Finally, the subject is asked to perform serial subtractions.

C10 (Memory): Assesses immediate memory and memory after a slight delay. Pictorial and verbal memory, as well as one instance of rhythmic memory, are assessed. Memory is assessed in a meaningful context, and unrelated objects are used as stimuli. However, this scale should not be construed as a comprehensive memory assessment, but rather as a screen for memory problems.

C11 (Intellectual Processes): Measures thematic understanding of pictures, the identification of humorous aspects of a picture, simple word definitions, proverb understanding, abstraction abilities, the comprehension of analogies, and verbal arithmetic.

C12 (Intermediate-Term Memory): Measures the ability to recall aspects of items that were given previously in the test. As such it can be construed as a test of incidental memory, as well as of a test of delayed memory with interference.

Practical Applications/Uses

Because of the great similarity between the two forms of the LNNB, information regarding administration and scoring can probably be assumed to be equivalent. Research regarding the amount of training necessary to achieve reliable scoring indicates that approximately 40 hours of training will produce adequate administration by a B.A.-level technician. The tests are administered from test protocols that contain the administration instructions and basic scoring criteria. More detailed scoring instructions are contained in the manual. In addition, a patient response booklet contains space for the written and drawn responses of the subject and the printed stimuli for the C4 Scale. In both forms, scores are represented in the form of standardized T-scores, with a mean of 50 and a standard deviation of ten points. A critical level is computed with a regression formula using the age and level of education of the subject as predictors.

Western Psychological Services has published a microcomputer scoring program for the LNNB. Thus far the program is available only for IBM personal computers. The program is copyrighted and an internal disk counter keeps track of each time the program is run. When the purchased number of runs is completed, the portion of the program that provides the score is deleted. Also available is a mail-in computer scoring service. Both of these scoring systems calculate the scale scores for basic clinical and supplementary scales and provide information regarding the ipsative comparison of scale scores of the same individual. A sample report is contained in the manual. Although computer scoring may save clerical time, the mail-in service may actually increase the latency between test and report.

The publisher also publishes a new manual to accompany Form II (Golden, Purisch, & Hammeke, 1985). Actually, it is a combined manual for both forms, and at 323 pages, plus various appendices, it is a substantial addition to the original manual. The new manual includes information regarding new derived scales, such as the Impairment and Elevation Scales. These scales contain items that are drawn from across the various scales, the patterns of which provide information regarding the acuteness of an injury and the probability of behavioral compensation following injury. There is also a power and a speed scale composed of items that were theoretically hypothesized to be related to those constructs. T-score conversions are available for the Impairment and Elevation Scales, based on a normative sample of approximately 800 cases. T-score conversions for the Speed and Power Scales are based on a normative sample of 45 medical patients. Initial empirical studies (Sawicki & Golden, 1984) suggest that a relatively higher Profile Elevation Scale (over the Impairment Scale) is related to recency of injury and the probability of compensation and that a relatively higher Speed Scale (over the Power Scale) is related to a posterior focus for the injury. More research is needed before these rules can be strongly recommended for clinical use.

Other changes in the manual include decomposing the C7 (Writing) Scale into the Spelling and Motor Writing subscales. Although users of the LNNB were probably already making that distinction in formulating an interpretation of results, this change helps insure that process. A major change involves the development of a set of qualitative categories that can be scored independently of the quantitative scores. These categories can be summed and the frequency counts compared to the frequency of occurrence in a normative sample (N = 48). Because of the limited size of the normative sample, interpretation of the qualitative information is best conducted with reference to neuropsychological theory. However, the qualitative categories serve a useful function by helping assessors to systematize their observations. In order to make the qualitative scores more useful, one would need a larger normative sample and a comparison sample of impaired subjects.

Expanded administration instructions in the new manual provide guides for administering the LNNB to individuals with peripheral sensory losses. The manual also contains a set of 25 cases written by James A. Moses. These are helpful and can serve as a beginning step in learning some aspects of interpretation. There are also discussions of illustrative profiles for different localizations. Although these features can be an aid in learning interpretation, they should not be used as a substitute for supervised learning experiences.

Because of the heterogeneity of the scale composition, interpretation of the LNNB must not be limited to scale interpretation, although that is a necessary first step. Full interpretation must also take into account scores on the empirically derived localization scales, the factor-analytic scales, comparisons of items across scales, and a consideration of the qualitative aspects of the subject's performance. Interpretation is best left to individuals who have received specialized training in neuropsychological assessment and who are familiar with Lurian theory.

The appendices to the manual contain information regarding the factor structure of Form I, item difficulty indices, and information regarding mean performances for each item. Similar information regarding Form II is needed.

Technical Aspects

Form II was developed and standardized on a sample of 73 subjects who were given both forms of the test. There were no significant differences in raw scores for any of the scales. Scale scores for Form II were then derived by regressing the raw scores for Form II against the raw scores for Form I and applying the linear solution to T-score transformations for Form II. Form II was then administered to a sample of 125 normal subjects, 140 subjects with central nervous system dysfunction, and 34 schizophrenic subjects. A MANOVA conducted on the data resulted in significant differentiation among the groups. Subsequently, Form II was administered to 100 normal subjects and 100 neurologically impaired subjects. Use of the clinical rules stated in the manual resulted in classification accuracy rates equivalent to that of Form I. Although these studies help us to evaluate the validity of Form II, much of the research originally conducted on Form I now needs to be conducted on Form II. Also needed is a separate normative sample for the derivation of T-scores for Form II. Although the above mentioned research indicates that Form I and II appear to be equivalent, a separate normative sample will help insure the accuracy of estimates of the precision of scores obtained from the use of Form II.

Critique

Overall, Form II of the LNNB shows promise as a clinical neuropsychological assessment instrument. Initial research supports its equivalency with Form I. There are approximately 150 published studies using the LNNB Form I. Not all of this research needs to be replicated with Form II, but the basic reliability and validity studies are in need of replication.

References

Franzen, M. D. (1985). Review of the Luria-Nebraska Neuropsychological Battery, Form I. In D. J. Keyser & R. C. Sweetland (Eds.), *Test Critiques* (Vol. III, pp. 402-414). Kansas City, MO: Test Corporation of America.

Golden, C. J., Purisch, A. D., & Hammeke, T. A. (1985). *Luria-Nebraska Neuropsychological Battery: Forms I and II.* Los Angeles: Western Psychological Services.

Sawicki, R. F., & Golden, C. J. (1984). The Profile Elevation Scale and the Impairment Scale: Two new summary scales for the Luria-Nebraska Neuropsychological Battery. *International Journal of Neuroscience, 23,* 81-90.

Timothy Z. Keith, Ph.D.

Associate Professor of School Psychology, University of Iowa, Iowa City, Iowa.

McCARTHY SCALES OF CHILDREN'S ABILITIES

Dorothea McCarthy. San Antonio, Texas: The Psychological Corporation.

THE ORIGINAL OF THIS REVIEW WAS PUBLISHED IN TEST CRITIQUES: VOLUME IV (1986).

Introduction

The McCarthy Scales of Children's Abilities (McCarthy) is designed to assess a variety of intellectual and motor abilities for children aged 2½ to 8½ years. The McCarthy takes about an hour for administration by a trained psychologist and seems to provide an enjoyable and valid method for assessing young children's abilities.

The McCarthy was authored by the late Dorothea McCarthy and published by The Psychological Corporation in 1972; it was designed to measure children's cognitive (McCarthy avoided the term *intelligence*) and motor abilities. The McCarthy consists of 18 short mental and motor tests grouped into five scales: Verbal, Perceptual-Performance, Quantitative, Memory, and Motor. The first three are non-overlapping and are further combined into the General Cognitive Index (GCI), a measure of overall cognitive functioning that is similar to an overall IQ.

The tests are grouped in a variety of combinations, with several appearing on two of the five scales, which are described as follows:

Verbal: Consists of five measures of verbal expression and verbal concept formation, including Pictorial Memory that asks the child to recall a series of pictures named by the examiner; Word Knowledge consisting of two parts: receptive language and picture vocabulary (part one) and defining words (part two); Verbal Memory requiring the child to repeat a series of words or sentences (part one) and retell a story after the examiner has told it (part two); Verbal Fluency in which the child names objects in a category within a time limit; and Opposite Analogies where the child completes sentences with an appropriate opposite word.

Perceptual-Performance: Consists of seven measures of perceptual and spatial abilities and nonverbal reasoning including Block Building in which the child copies formations of blocks; Puzzle Solving requiring the child to put together a series of simple, colorful puzzles; Tapping Sequence in which the child copies a series of notes on a toy xylophone; Right-Left Orientation given only to children above five years who are asked to differentiate right and left on oneself and on a picture of a boy; Draw-A-Design asking the child to copy a series of geometric designs; Draw-A-Child where the child draws a picture of a child who is the same sex as self; and Conceptual Grouping, a logical classification task on which the child sorts brightly colored blocks on the basis of size (large and small), shape (circle and square), and color (three colors).

Quantitative: Consists of three measures of facility with numbers, basic pre-arithmetic concepts, and arithmetic reasoning, including Number Questions requiring the child to solve oral arithmetic problems; Numerical Memory in which the child recalls simple digits, including digits forward (part one) and digits reversed (part two); and Counting and Sorting requiring the child to count blocks and sort them into equal groups, and display knowledge of such concepts as "each" and ordinal numbers.

Memory: Consists of four measures of short-term auditory and visual memory from the first three scales (Pictorial Memory, Tapping Sequence, Verbal Memory, and Numerical Memory).

Motor: Consists of five measures of fine and gross motor coordination, including two tasks from the Perceptual-Performance Scale (Draw-A-Design and Draw-A-Child), plus Leg Coordination requiring the child to perform gross motor tasks such as walking a straight line, standing on one foot, and skipping; Arm Coordination requiring the child to bounce a ball (part one), catch a bean bag (part two), and throw the bean bag at a target (part three); and Imitative Action requiring the child to copy a series of the examiner's movements, such as twiddling thumbs and looking through a tube.

The standardization of the McCarthy, which was carried out from October 1970 through September 1971, was excellent. It consisted of a nationally representative sample of children aged 2½-8½ years, with approximately 50 girls and 50 boys included at each of ten age levels (every six months between 2½ and 5½ years and every year from 5½ to 8½ years). It was stratified based on age, sex, color, geographic region, and father's occupational status in accordance with 1970 census estimates and informally stratified as to rural versus urban residence. Although the sample was composed of primarily "normal" children and excluded institutionalized and other severely handicapped children, it did include those with suspected or presumably mild handicaps. The final standardization sample included several extra cases for a total of 1,032 children.

The McCarthy test materials are all attractive and functional. And, unlike many tests for children this age, the McCarthy materials are interesting enough, when used by a skilled examiner, to retain the interest of all but the most distractable children. The manual (McCarthy, 1972) is well organized and generally complete, although frequent users would be advised to supplement the manual with Kaufman and Kaufman's (1977) excellent text on the test's development and use. Although the record forms, which have ample space for notes and observations, are easy to use, they do require considerable transferring of scores and clerical accuracy. However, the main drawbacks of the kit are the carrying case, which is made of the usual flimsy material, and the high cost of the kit.

Practical Applications/Uses

The McCarthy provides a general measure of cognitive development for young children. As such, its most frequent use will probably be with children suspected of having problems in learning, behavior, or development. The McCarthy should also be considered for inclusion in any standard psychological assessment battery for children in this age group, and, thus, could replace the Wechsler Preschool and Primary Scale of Intelligence (WPPSI), the Wechsler Intelligence Scale for Children-Revised (WISC-R), or the Stanford-Binet Intelligence Scale: Form L-M. (For reviews of these instruments see Elbert & Holden, 1985; Vernon, 1984; and Holden, 1984, respectively.) The McCarthy is most appropriate for clinical, counseling, and school psychologists in educational or clinical settings. As with most individually administered intelligence tests, the McCarthy requires training and practice by qualified examiners—generally psychologists—in order to be administered and interpreted properly. With practice, the McCarthy's administration is smooth and natural and takes an experienced examiner approximately 45-60 minutes to complete. By way of comparison, the McCarthy is probably slightly more difficult to learn than the WISC-R or the Kaufman Assessment Battery for Children (K-ABC; for a review see Merz, 1984), but somewhat easier than the Stanford-Binet.

The directions for scoring individual tests are generally clear and straightforward. A few of the tests (e.g., Word Knowledge and the drawing tests) are subjective and somewhat more difficult to score, but again the directions in the manual are fairly complete and easy to follow. However, the process of converting from raw scores to derived scores is considerably more difficult. Raw scores, some weighted by a factor of two or one-half, are transferred to the back of the record form and are added together in various combinations into the five scales, then transferred to the front of the record form, where the Verbal, Perceptual-Performance, and Quantitative Scales are further summed into the raw General Cognitive Index (GCI) score. Tables are then used to convert these raw scores into standard scores, with the resulting GCI having a mean of 100 and a standard deviation of 16, while the standard scores of the other scales have means of 50 and standard deviations of 10. The standard scores may then be plotted on a profile on the front of the cover sheet as a means of visual comparison of strengths and weaknesses. Obviously, such score conversion is somewhat cumbersome, and the juggling required can easily cause mistakes. This reviewer highly recommends double and triple checking the scoring.

Although somewhat involved, scoring the McCarthy is not difficult with training and practice. What is more unfortunate is that there is no method provided to prorate spoiled or invalid tests, a not-uncommon occurrence when working with children at this age level. In this reviewer's experience, for example, the Right-Left Orientation Test often gives unsatisfying results because children who have no understanding of the concepts of right and left occasionally score better by random guessing than those who are just beginning to develop an understanding of the concepts and answer consistently, but incorrectly. (Indeed, Right-Left Orientation is probably the weakest of all the McCarthy tests and if the McCarthy is ever revised, the publishers should consider discarding this test.)

Fortunately, instructions for prorating, along with tips on administration and scoring, are provided by Kaufman and Kaufman (1977). Anyone who regularly uses the McCarthy should have access to this book, which is also invaluable for interpretation of the test results. Although the manual's discussion of the tests and scales is helpful in interpretation, the book aids immensely in this task by presenting a logical system for comparing scores on scales and individual tests and by presenting a number of alternative methods of interpreting the resulting combinations of scores. As with any such intelligence test, it is in the interpretation that training and supervised practice are most important.

Technical Aspects

Evidence to support the reliability and validity of the McCarthy is presented in the manual. In addition, there has been considerable research performed with the McCarthy in the 13 years following its initial publication; much of this research is reviewed carefully in Kaufman (1982). Considering the volume of research that has been performed on the McCarthy Scales, this brief review will draw heavily from these two existing sources (Kaufman, 1982; McCarthy, 1972; for a review of early research on the McCarthy see Kaufman & Kaufman, 1977), and will not discuss studies in detail.

The McCarthy GCI seems to provide a reliable measure of overall intellectual ability: internal consistency estimates are generally reported in the .90s and stability estimates in the .80s over a variety of test-retest intervals. There is also evidence to support the reliability of the other McCarthy scales, with stability and internal consistency estimates (when applicable) generally in the .70s to .80s.

The McCarthy GCI seems to correlate with Weschler and Binet IQs as well as these measures correlate with each other (.70s-.80s), thus supporting the validity of the GCI. Similarly, with correlations of .55-.68, the GCI correlates substantially with the newer K-ABC Mental Processing Composite (Kaufman & Kaufman, 1983). The McCarthy GCI and other scales also correlate substantially with measures of achievement, an important criterion for a measure of intelligence. Although the coefficients vary considerably, research suggests that the GCI correlates significantly with a variety of measures of achievement when administered concurrently (.27-.84, with most in the .5 to .75 range) or when the achievement tests are administered later (.48-.76), and further that the GCI correlates about as highly with achievement measures as do other tests of intelligence (Kaufman, 1982).

There is mixed support for the construct validity of the various McCarthy scales in factor analyses of both normal (e.g., Kaufman, 1975) and exceptional (Keith & Bolen, 1980; Naglieri, Kaufman, & Harrison, 1981) children. Such analyses have offered strong support for the General Cognitive Index as a measure of overall functioning. In addition, factor analyses have offered support for the Verbal, Perceptual-Performance, and Motor Scales of the McCarthy, with memory factors appearing less often and quantitative factors appearing rarely. For example, in their analysis of referred children, Keith and Bolen (1980) found no quantitative factor, but did find that the quantitative tests were among the best measures of the general cognitive factor found.

There are some problematic aspects of the McCarthy, however. There are few items to assess social comprehension, judgment, or abstract reasoning, and especially verbal abstract reasoning (Kaufman & Kaufman, 1977). The ceiling on most tests is inadequate to assess school-aged gifted children, although most scales seem to work well with low achieving school-aged children, the type of child most frequently referred for psychoeducational assessment. Finally, there is evidence that the McCarthy yields slightly lower scores for learning disabled (LD) children than does the WISC-R, although this difference appears to be nowhere near as large as was initially reported. For example, Kaufman (1982) estimates a GCI/Weschler Full Scale difference of about six points with LD children, as compared to three points for normal children. However, the differences may be larger with retarded children, with the McCarthy producing substantially lower scores. (For more information about such studies, including a critique of the earlier studies, see Kaufman, 1982.)

Critique

The McCarthy Scales of Children's Abilities is a well-developed, extremely well-standardized test of ability for children ages 2½ to 8½ years. Its weaknesses include a paucity of items to assess social comprehension, judgment, and verbal abstract reasoning; an inadequate ceiling for gifted children above about six years and normal children above about seven years; and the possible underestimation by the GCI of WISC-R or Binet IQs for exceptional children. However, the McCarthy's strengths seem to outweigh its weaknesses. It has excellent norms and standardization. It assesses a variety of cognitive and motor skills and provides a diagnostically useful profile of scores; with the addition of Kaufman and Kaufman's (1977) book, the McCarthy is a flexible assessment device and can be interpreted from a variety of perspectives. Finally, the test is attractive and, perhaps most importantly, children find it interesting, a characteristic lacking in most other tests for this age group. Thus, although not appropriate for older normal or gifted children, the McCarthy should probably be considered the test of choice for preschool children and for children up to about age eight who are referred for learning problems.[1]

Notes to the *Compendium:*

[1]With the publication of the Stanford-Binet: Fourth Edition (Thorndike, Hagen, & Sattler, 1986), the choice among tests may not be so clear cut. The new Binet overcomes many of the deficiencies of earlier editions, but many of the tests may prove to be difficult and less interesting for young children. Similarly, the K-ABC is a better-normed alternative, but shares some of the McCarthy's weaknesses. For example, the K-ABC does not assess verbal abstract reasoning, at least not on the Mental Processing Scales. In sum, examiners now have a choice of assessment instruments for young children; they will need to become familiar with available instruments and choose an instrument to match a particular case.

References

Elbert, J. C., & Holden, E. W. (1985). Review of the Wechsler Preschool and Primary Scale of Intelligence. In D. J. Keyser & R. C. Sweetland (Eds.), *Test critiques* (Vol. III, pp. 697-709). Kansas City, MO: Test Corporation of America.

Holden, R. H. (1984). Review of the Stanford-Binet Intelligence Scale: Form L-M. In D. J. Keyser & R. C. Sweetland (Eds.), *Test critiques* (Vol. I, pp. 603-607). Kansas City, MO: Test Corporation of America.

Kaufman, A. S. (1975). Factor structure of the McCarthy Scales at five age levels between 2½ and 8½. *Educational and Psychological Measurement, 35,* 641-656.

Kaufman, A. S. (1982). An integrated review of almost a decade of research on the McCarthy Scales. In T. R. Kratochwill (Ed.), *Advances in school psychology* (Vol. 2, pp. 119-169). Hillsdale, NJ: Erlbaum.

Kaufman, A. S., & Kaufman, N. L. (1977). *Clinical evaluation of young children with the McCarthy Scales.* New York: Grune & Stratton.

Kaufman, A. S., & Kaufman, N. L. (1983). *K-ABC: Kaufman Assessment Battery for Children: Interpretive manual.* Circle Pines, MN: American Guidance Service.

Keith, T. Z., & Bolen, L. M. (1980). Factor structure of the McCarthy Scales for children experiencing problems in school. *Psychology in the Schools, 17,* 320-326.

Merz, W. R., Sr. (1984). Review of the Kaufman Assessment Battery for Children. In D. J. Keyser & R. C. Sweetland (Eds.), *Test critiques* (Vol. I, pp. 393-405). Kansas City, MO: Test Corporation of America.

McCarthy, D. (1972). *Manual for the McCarthy Scales of Children's Abilities.* New York: The Psychological Corporation.

Naglieri, J. A., Kaufman, A. S., & Harrison, P. L. (1981). Factor structure of the McCarthy Scales for school-age children with low GCIs. *Journal of School Psychology, 19,* 226-232.

Thorndike, R. L., Hagen, E. P., & Sattler, J. M. (1986). *The Stanford-Binet: Fourth edition: Guide for administering and scoring.* Chicago: Riverside Publishing Co.

Vernon, P. A. (1984). Review of the Wechsler Intelligence Scale for Children-Revised. In D. J. Keyser & R. C. Sweetland (Eds.), *Test critiques* (Vol. I, pp. 740-749). Kansas City, MO: Test Corporation of America.

J. Mark Wagener, Ph.D.

Clinical Psychologist, Mental Health Clinic-Student Health Center, Oregon State University, Corvallis, Oregon.

MEMORY-FOR-DESIGNS TEST

Frances K. Graham and Barbara S. Kendall. Missoula, Montana: Psychological Test Specialists.

THE ORIGINAL OF THIS REVIEW WAS PUBLISHED IN TEST CRITIQUES: VOLUME II (1985).

Introduction

The Memory-For-Designs Test (MFD) is an instrument designed to screen for impairment of brain functioning by measuring the ability to copy simple geometric designs from immediate memory. It is meant to be used as a tool to behaviorally assess the presence of central nervous system impairment in children and adults. The ability to perceive geometric shapes, retain these perceptions in memory, and reproduce them manually has long been considered to be related to the intactness of brain functioning.

In the MFD the subject is requested to draw a reproduction of each of 15 designs immediately after it has been presented for five seconds. The quality of the reproductions is then evaluated according to standards presented in the test manual. A scoring system is used to categorize performance levels into critical, borderline, and normal classifications of functional intactness. Raw scores can be modified to control for variation associated with age and vocabulary level. The test is used to differentiate individuals with cerebral dysfunctions from normals and from those with psychiatric dysfunctions.

This test was developed by Frances K. Graham and Barbara S. Kendall in order to test the assumption that perceptual motor memory was impaired by brain damage. It was the result of the first well-developed attempt to scientifically demonstrate the relationship between these functions and brain damage. The authors wished to develop a highly reliable instrument that could differentiate brain-damaged subjects from others on the basis of empirical criteria. It was originally designed as a research instrument in the 1940s (Graham & Kendall, 1946; Kendall & Graham, 1948). Since then it has come into widespread clinical use as a simple method of screening for brain damage.

The 15 most promising designs from a group of 40 were selected in a preliminary study to be used in the validation procedure. A group of 70 mixed brain-disordered patients and a group of 70 controls were matched for age, education, and occupation. Various types of drawing errors were weighted to reflect the frequency with which they occurred in the brain-disordered group as compared to the controls. The scoring was then cross-validated by use of an additional group of 33 brain-disordered and 168 control subjects. Additional normative data are presented in the most recent manual (Graham & Kendall, 1960). The test remains in its original form with little change in the scoring procedure other than incorporation of the new normative data.

The test materials consist of fifteen 5" x 5" white cardboard squares, each of which is printed with a black design. All of the designs are made of straight lines and are oriented squarely with the card. The client is presented with a pencil with an eraser and a sheet of 8½" x 11" paper. Clients are told that the cards will be presented individually for a 5-second interval and that they are to draw the design from memory after the card has been removed from sight. Clients are prevented from copying the design directly and are not encouraged to guess or complete partially remembered designs.

Each design is scored using an empirically derived scale from 0 to 3, with higher scores reflecting types of errors most indicative of brain damage. Satisfactory reproductions, including omitted or incomplete drawings, are given a score of 0, while those with rotation errors are given the maximum score. Specific verbal scoring criteria and sample scoring of examples of drawings are available in the manual. Scores are then totaled for the fifteen drawings in order to obtain a performance rating. The authors recommend that clients obtaining raw scores of 12 and above be placed in the brain-damaged category, 5-11 be placed in a borderline category, and 4 and below be considered normal.

The authors provide tables for conversion of raw scores into difference scores in order to partial out the effects of age and vocabulary level. This provides a means of relating performances among groups of varying age and verbal ability levels. The use of this difference score is necessary for children as well as for adults of low intelligence or those of advanced age. Although the use of difference scores was originally recommended for all, research has suggested that it is unnecessary with adults of normal intelligence and may even slightly reduce the effectiveness of the test (Hunt, 1952).

Practical Applications/Uses

The task involved is easy for most children and adults. The test is considered to be appropriate for individuals 8.5 through 60 years of age. Recently, normative data have become available for subjects over 60 years (Kendall, 1962; Riege, Kelly, & Klane, 1981). Performance levels of children less than 8½ years have been found to be too variable to make this a useful clinical tool with them.

This test is intended to be used by professionals interested in screening individuals for possible brain damage. It is likely to find its greatest usage in clinical settings where in- or out-patients are exhibiting symptoms of impaired functioning that may be the result of cerebral dysfunction. It also may be useful in educational settings as part of a diagnostic workup for students demonstrating learning disabilities or other cognitive impairments. Its simplicity of administration and nonintrusive nature make it quite appealing as a first-step procedure for gathering information in order to make a decision about referral for more extensive neurological evaluations.

This test was designed to measure the presence or absence of disorder associated with impairment of cerebral brain tissue function. The test authors indicate that they presume that completion of the MFD requires the ability to perform a sequence of behaviors. These include attending to and perceiving a patterned visual stimulus, retaining the perception in memory briefly, and reproducing the stimulus by executing a relatively complex motor act. They also assume that success is related to the ability to inhibit drawing when memory is imperfect. The test was designed to be a global measure of organic impairment, rather than to be used to localize lesions.

The MFD has been used primarily in psychiatric settings. It is frequently used in an attempt to distinguish between organic brain damage and functional conditions. Its greatest potential usefulness lies in identifying individuals with relatively mild organic impairment rather than those with gross deficits, which could obviously be demonstrated by other means.

The administration of this test should take place on an individual basis in a setting without distractions. The instructions and procedures are clear, and anyone trained in individual psychological test administration could be expected to administer it without difficulty. It is important to attend to the orientation of the presentation of the designs so that rotation errors can be appropriately scored. The total administration time varies from 5 to 10 minutes. Since it is a simple, short-term task, modification of standard testing procedure should be unnecessary.

Since the instructions for scoring are clearly presented and numerous examples are provided, learning to score appropriately is not difficult. Though scoring judgments must be made, neophyte scorers typically do not demonstrate significant errors. Designs are scored individually by comparing the client's drawing with the standards described in the manual. It is important that the scorer be familiar with the general principles of scoring and supplement these with the examples provided.

Technical Aspects

Research data indicate that interrater reliability of the scoring of the MFD is very strong. The authors reported a correlation of .99 between the total raw scores they independently assigned for the 140 original validation subjects. Because they had devised the scoring system together, one would anticipate high agreement. However, other investigators consistently report interrater reliabilities ranging from the upper .80s to mid .90s.

A significant level of sophistication is required in order to interpret the meaning of the scores. Highly conservative cutoff points were deliberately chosen by the authors in order to minimize the possibility of false-positives. Use of the suggested cutoff scores provides results that are highly indicative of brain damage when the scores fall into the critical area, but they are not indicative of the absence of brain damage when they fall below the critical scores. Therefore, the actual hit rate (true-positive and true-negative categorizations) is substantially lower than would be the case for a "best fit" cutoff score. An appreciation of the relative cost involved from a false-positive as opposed to a false-negative is necessary to choose the optimal cutoff scores. Contrary to the authors' choice of minimizing false-positives, some writers such as Krug (1971) have suggested that it is much less desirable to diagnose an organic patient as functional than vice versa. Heaton, Boode, and Johnson (1978) make the point that particularly in acute treatment settings, most clinicians would probably find the cost of an unproductive neurologic workup much more acceptable than the consequences of misdiagnosing brain disease and treating it as if it were a functional disorder.

Extensive reliability and validity data are available. The index of reliability using the split-half method was 0.92 for the original subjects. The authors reported test-retest reliabilities in the 0.80s. Researchers typically report improvement in scores in a test-retest situation, indicating that there is a practice effect. The data indicate that there is a trend for the improvement to be greater with organic subjects than non-brain-damaged subjects.

The manual lists a total of 535 control subjects and 243 brain-disordered subjects derived from a variety of groups used for validation. The brain-disordered groups included both acute and chronic conditions for which there was clear evidence of tissue damage. Control groups included psychotics, psychoneurotics, personality disorders, adjustment reactions, somatic disorders, and a small number of normals. Also included was a group of idiopathic epileptics (whose scores, incidentally, placed them in the normal category).

The authors employed a multiple regression technique to arrive at difference scores which partialed out the effects of age and vocabulary, thus making it unnecessary to equate groups. The difference scores of the combined controls and combined brain-disordered groups clearly discriminated. Using a difference score cutoff of 6.5, 46.5% of the brain-disordered subjects were correctly diagnosed, while 5% of the control subjects were categorized as brain-damaged. While 75.7% of the controls were placed in the normal category, 31.3% of the brain-disordered subjects were placed in the normal category. The remainder of the subjects fell into the borderline area.

Other studies provided evidence of the MFD's ability to discriminate between organic and non-organic psychiatric patients. Shearn, Berry, and Fitzgibbons (1974) found that MFD scores obtained when patients initially were hospitalized predicted more accurately eventual psychiatric diagnosis of suspected brain disorder than did initial psychiatric impressions. McManis (1974) found that 20 brain-damaged psychiatric patients scored significantly higher than 20 non-brain-damaged psychiatric patients. He found that he could improve his hit rate by dichotomizing on a raw score cutoff of 6.5. With this cutoff, he accurately categorized 89% of his female and 82% of his male subjects.

In addition to the question as to whether a test discriminates between diagnostic groups, an important practical consideration is how well this is done in comparison to other tests designed to make similar distinctions. In studies in which the MFD was used along with other organic screening instruments, the MFD generally compares favorably, particularly when a "best fit" cutoff score was used. Pullen and Games (1965) found no significant difference between the Bender Gestalt and MFD accuracy rates in classifying 60 first admissions to a state mental hospital. However, they recommended the use of the MFD over the Bender, particularly when used by relatively inexperienced clinicians, due to its higher interrater reliability. Korman and Blumberg (1963) administered a battery of organicity screening instruments to a group of 40 cerebral-damaged subjects matched with 40 undamaged subjects. They found the MFD to have the highest accuracy of all of their measures. The reported hit rate of 90% for the MFD is very impressive, and higher than those typically found for organicity screening measures. Thus, though there are some exceptions, the literature indicates that the MFD not only significantly discriminates between diagnostic categories, it does this as well as other measures available.

Investigators have explored some nontraditional uses of the MFD with mixed results. No clinically significant relationship has been demonstrated between MFD drawing styles and personality traits, nor has it been shown to correlate with delinquency. It also has not differentiated consistently between good and poor readers.

Though the norms indicate only a relatively low correlation between intelligence and MFD scores, the MFD may be reflective of intelligence in a highly selected group. Ong and Jones (1982) found a correlation of −0.96 between the MFD and WISC Full Scale IQs in a class of educable mentally retarded children. They suggested that the MFD may be used as a supplement to other ability measures in the placement of these children. However, the sample in this study was quite small and the high correlation was found among students already placed in this class. Additional data demonstrating the ability of the MFD to discriminate between educable retarded students and normal students would be necessary before it could be used as a placement tool in classroom settings.

Critique

There is much to be said in favor of the MFD as a screening device to make global decisions regarding the presence or absence of cerebral damage. It is quickly and easily administered. The manual is well written, and procedures for administering and scoring the instrument are clearly elaborated. Interrater reliability is quite high for an instrument of this nature, and test-retest reliability is acceptable.

Studies generally demonstrate that it discriminates well between brain-damaged and non-brain-damaged groups. This remains the case in situations where the brain damage is not readily apparent and where psychiatric conditions of comparison groups make distinctions difficult. Most studies indicate that scores on the MFD can discriminate as well or better than other global measures of organicity.

The MFD is also one of those rather rare psychological instruments that requires subjects to perform the same task and uses the same criteria for scoring for all individuals from age 8 through adulthood. Using difference scores, which control for age and vocabulary level variance, allows comparisons along an exceptional age range. This feature could be useful in future longitudinal studies.

Although the test is easy to administer and the scoring can be mastered fairly quickly, sophistication is required in choosing the appropriate cutoff scores. Acceptance of the authors' suggested cutoff will result in minimizing false-positives and maximizing false-negatives. Most studies indicate that reducing the adult raw score cutoff by approximately 5 points and eliminating the borderline category results in the most accurate overall classification. In a clinical setting the examiner must take into account the relative cost to the patient of over-diagnosing as well as under-diagnosing the probability of brain damage.

While the MFD clearly is most useful as a clinical tool, continued research may demonstrate its utility in educational settings. There is some suggestion that the MFD may measure the intellectual ability of educable mentally retarded students. Attempts at relating reading ability to MFD performance have produced mixed results, and this does not appear to be a particularly promising area for the use of this test.

References

Graham, F. K., & Kendall, B. S. (1946). Performance of brain-damaged cases on a Memory-for-Designs Test. *Journal of Abnormal and Social Psychology, 41,* 303-314.

Graham, F. K., & Kendall, B. S. (1960). Memory-for-Designs Test: Revised general manual. *Perceptual and Motor Skills, 11,* 147-188. (Monograph Supplement 2-VII)

Heaton, R. K., Boode, L. E., & Johnson, K. L. (1978). Neuropsychological test results associated with disorders in adults. *Psychological Bulletin, 85,* 141-162.

Hunt, H. F. (1952). Testing for psychological deficit. In D. Brower & L. E. Abt (Eds.), *Progress in clinical psychology. Vol. 1.* New York: Grune and Stratton.

Kendall, B. S. (1962). Memory-for-Designs performance in the seventh and eighth decades of life. *Perceptual and Motor Skills, 14,* 399-405.

Kendall, B. S., & Graham, F. K. (1948). Further standardization of the Memory-for-Designs Test on children and adults. *Journal of Consulting Psychology, 12,* 349-354.

Korman, M., & Blumberg, S. (1963). Comparative efficiency of some tests of cerebral damage. *Journal of Consulting Psychology, 27,* 303-309.

Krug, R. S. (1971). Antecedent probabilities, cost efficiency, and differential prediction of patients with cerebral organic conditions or psychiatric disturbance by means of a short test of aphasia. *Journal of Clinical Psychology, 27,* 468-471.

McManis, D. L. (1974). Memory-for-Designs performance of brain-damaged and non-damaged psychiatric patients. *Perceptual and Motor Skills, 38,* 847-852.

Ong, J., & Jones, L., Jr. (1982). Memory-for-Designs, intelligence, and achievement of educable mentally retarded children. *Perceptual and Motor Skills, 55,* 379-382.

Pullen, M., & Games, P. (1965). Comparison of two tests of brain damage. *Perceptual and Motor Skills, 20,* 977-980.

Riege, W. H., Kelly, K., & Klane, L. T. (1981). Age and error differences on Memory-for-Designs. *Perceptual and Motor Skills, 52,* 507-513.

Shearn, C. R., Berry, D. F., & Fitzgibbons, D. J. (1974). Usefulness of the Memory-for-Designs Test in assessing mild organic complications in psychiatric patients. *Perceptual and Motor Skills, 38,* 1099-1104.

Jim C. Fortune, Ed.D.
Professor of Educational Research and Evaluation, Virginia Polytechnic Institute and State University, Blacksburg, Virginia.

Theodore R. Cromack, Ed.D.
Director, Commodity Donation Demonstration Study, Virginia Polytechnic Institute and State University, Blacksburg, Virginia.

METROPOLITAN ACHIEVEMENT TESTS: 5th EDITION

Irving H. Balow, Roger Farr, Thomas P. Hogan, and George A. Prescott. San Antonio, Texas: The Psychological Corporation.

THE ORIGINAL OF THIS REVIEW WAS PUBLISHED IN TEST CRITIQUES: VOLUME III (1985).

Introduction

The Metropolitan Achievement Test: 5th Edition (MAT) is a group-administered battery of general purpose achievement tests designed to assess school curriculum from kindergarten through twelfth grade. The content of the tests was based on expressly formulated instructional objectives that had undergone a validation process with a sample (N = 3,047) of teachers so as to carry out the belief that achievement tests should assess that which is taught. The MAT, which is divided into instructional levels, measures facts, skills, concepts, and applications in language, mathematics, and reading at each level and includes science and social studies at appropriate levels.

The original Metropolitan Achievement Test, developed by The Psychological Corporation, a subsidiary of Harcourt Brace Jovanovich, Inc., dates back to the early 1930s when the test was designed to meet the curriculum assessment needs of New York City. Later editions of the test were expanded to better reflect a more national curriculum. In the 1970 Edition of the MAT, both fall and spring empirical norms and norming data for every form were introduced as new services. The 1978 Edition has continued these services and has introduced a new, two-component system of achievement assessment to better accommodate the testing needs of its users.

The new two-component system is designed to meet two needs of test users. One component, the Instructional Tests, is designed to provide in-depth, specific achievement information on the objectives tested for the purpose of analysis of curriculum and instruction at a given level. The second component, the Survey Tests, is designed to provide a more global evaluation of student performance in the major areas of a curriculum at a given level. This component requires less testing time and is designed to serve as a vehicle for administrative use in program

290

planning and evaluation. Both components are interrelated through item-sampling procedures and can be used in concert. Each of the components is designed to yield both norm-referenced and criterion-referenced information, as well as a full range of scoring and analysis services.

The Instructional Tests focus on reading, mathematics, and language at six instructional levels: 1) Primer, Grades K.5-1.4; 2) Primary, Grades 1.5-2.4; 3) Primary 2, Grades 2.5-3.4; 4) Elementary, Grades 3.5-4.9; 5) Intermediate, Grades 5.0-6.9; 6) Advanced 1, Grades 7.0-9.9. Each Instructional Test is made up of subtests facilitating the address of curricular objectives. The subtest composition of the Instructional Tests is shown in Table 1.

No provision is made for out-of-level testing by design because the authors feel that curricular considerations outweigh the advantages of this practice (Prescott, Balow, Hogan, & Farr, 1978). Practice tests are provided and the teacher's manual for administering and interpreting is clear and comprehensive and includes information concerning administration procedures, scoring, test construction, and interpretation. Administration time requirements for the Instructional Tests include 1) Primer: Reading—118 minutes, Mathematics—90 minutes, Language—55 minutes; 2) Primary 1: Reading—140 minutes, Mathematics—110 minutes, Language—115 minutes; 3) Primary 2: Reading—122 minutes, Mathematics—105 minutes, Language—130 minutes; 4) Elementary: Reading—130 minutes, Mathematics—125 minutes, Language—150 minutes; 5) Intermediate: Reading—134 minutes, Mathematics—165 minutes, Language—120 minutes; and 6) Advanced 1: Reading—74 minutes, Mathematics—170 minutes, Language—110 minutes. Two forms, JI and KI, are available for the Instructional Tests at each level.

The second component, the Survey Test, is available at eight levels: Preprimer, Grades K.0-K.5; Primer, Grades K.5-1.4; Primary 1, Grades 1.5-2.5; Primary 2, Grades 2.5-3.4; Elementary, Grades 3.5-4.9; Intermediate, Grades 5.0-6.9; Advanced 1, Grades 7.0-9.9; and Advanced 2, Grades 10.0-12.9. The Survey Tests for Preprimer, Primer, and Advanced 2 include only the three basic curriculum areas of reading, mathematics, and language. Science and social studies are also included in the other five levels of the Survey Tests. Two forms, JS and KS, are available at every level except the Preprimer. The test administration times for the total test battery are as follows: Preprimer, 100 minutes; Primer, 98 minutes; Primary 1, 115 minutes; Primary 2, 113 minutes; Elementary, 120 minutes; Intermediate, 120 minutes; Advanced 1, 115 minutes; and Advanced 2, 110 minutes. Well-developed teacher manuals that contain approximately the same information as is provided in the teacher manuals for the Instructional Tests are also provided for the Survey Tests.

Practical Applications/Uses

The authors of the Metropolitan Achievement Test suggest that the two-component test is designed to serve two purposes: 1) to provide a detailed, highly specific evaluation for use in the planning of instructional programs for individual students and for the class and for use in detailed study of the curriculum, and 2) to provide a global, overall evaluation of students' performance in major areas of the curriculum for administrative use in program evaluation and achievement

Table 1

Subtest Composition of the Instructional Tests by Level

Subtest	Primer	Primary 1	Primary 2	Elementary	Intermediate	Advanced 1
READING						
Visual Discrimination	X					
Letter Recognition	X					
Auditory Discrimination	X	X				
Sight Vocabulary	X	X	X	X		
Phoneme/Grapheme Consonants	X	X	X	X	X	
Phoneme/Grapheme Vowels			X	X	X	
Vocabulary in Context		X	X	X	X	X
Word Part Clues		X	X	X	X	
Rate of Comprehension				X	X	X
Skimming & Scanning					X	X
Reading Comprehension		X	X	X	X	X
MATHEMATICS						
Numeration	X	X	X	X	X	X
Geometry & Measurement	X	X	X	X	X	X
Problem Solving	X*	X	X	X	X	X
Operations: Whole Numbers	X*	X	X	X	X	X
Operations: Laws & Properties				X	X	X
Operations: Fractions & Decimals					X	X
Graphs & Statistics					X	X
LANGUAGE						
Listening Comprehension	X	X	X	X	X	X
Punctuation & Capitalization		X*	X	X	X	X*

*Administered as a single test

Table 1 *(continued)*

Subtest	Primer	Primary 1	Primary 2	Elementary	Intermediate	Advanced 1
			MATHEMATICS *cont.*			
Usage		X*	X	X	X	X
Grammar & Syntax		X	X	X	X	X
Spelling	X	X	X	X	X	X
Study Skills	X	X	X	X	X	X

monitoring (Prescott, Balow, Hogan, & Farr, 1978). The care that has been taken to establish the linkage of the Instructional Tests to the curriculum that currently is being taught in the schools, the provision of the instructional objectives used in the table of specifications for the test in the teacher's manual, the use of clustering to provide a criterion-referenced score along with the norm-referenced score, and the emphasis placed on item analysis in the manual make the test especially appropriate for curriculum study.

Although these reviewers believe that the test also serves ably as a general achievement assessment instrument, they believe that more information on the program that is being evaluated is needed before they can recommend the test for program evaluation. It is clear that the Survey Test provides an efficient, overall measure of the whole content in a curriculum area, but in these reviewers' opinion, a stronger tie to the specific curriculum of the program that is being evaluated is needed. Additionally, the level of performance of the students in the program being evaluated should be taken into consideration. Although the Survey Test has been developed on a representative, general curriculum, there is no reason to assume that it includes the principal objectives of a given program that is to be evaluated. If the students in such a program are in either performance extreme in regard to achievement, either floor or ceiling effects could render the Survey Test inappropriate as an evaluative instrument.

Technical Aspects

The *Metropolitan Achievement Test Special Report Number 11* (The Psychological Corporation, 1978) presents reliability estimates in the form of Kuder-Richardson Formula 20 coefficients and standard errors of measurement for each form, level, subject, and norming period. The reliability estimates were calculated on both the Instructional Tests and on the Survey Tests for all subject areas tested. The Reading Test was also subjected to a test-retest reliability study (fall and spring) and the subtests of the Instructional Test were studied for internal consistency and for intercorrelation. In Table 2, results of the internal consistency study are reported for one form of the basic battery of the Survey Test for the fall administration. The other administrations produced nearly identical coefficients. In Table 3, the range of reliability estimates for the subtests of the Instructional Tests is reported. Table 4 shows the results of the test-retest (fall and spring) study of the Reading subtests of the Instructional Tests. Generally, the reliability appears adequate, but not outstanding for a standardized test.

Table 2

Kuder-Richardson Formula 20 Reliability Coefficients for the Survey Tests

Level	Range of N	Reading	Mathematics	Language
Preprimer	N/A	.94	.77	.72
Primer	669-680	.85	.86	.81
Primary 1	707-723	.96	.86	.87
Primary 2	698-711	.95	.88	.91
Elementary	665-679	.96	.90	.88
Intermediate	729-742	.95	.90	.92
Advanced 1	851-878	.94	.89	.90
Advanced 2	N/A	.92	.90	.86

Table 3

Range of Kuder-Richardson Formula 20 Reliability Coefficients across Subtests of the Instructional Tests

Level	Range of N	Reading	Mathematics	Language
Primer	669-680	.84-.92	.72-.87	.66-.92
Primary 1	707-723	.85-.93	.79-.88	.66-.92
Primary 2	698-711	.85-.94	.81-.90	.66-.94
Elementary	665-679	.82-.95	.85-.92	.68-.94
Intermediate	729-742	.76-.95	.80-.91	.82-.92
Advanced 1	851-878	.77-.90	.75-.91	.79-.91

Table 4

Test-Retest Reliabilities for Selected Metropolitan Reading Tests

Level	Subtest	Sample Size	Reliability Coefficient
Elementary	Comprehension	271	.92
	Rate of Comprehension	268	.84
Intermediate	Comprehension	221	.92
	Rate of Comprehension	221	.73
	Skimming & Scanning	221	.79
Advanced 1	Comprehension	169	.87
	Rate of Comprehension	168	.78
	Skimming & Scanning	168	.74

The primary evidence that the publishers present for validity is that of content validity. Included in the argument for content validity are the following: descriptive information on how the table of specifications was developed from instructional objectives for each content area; teachers' reports on the number of the objectives and which ones that they actually taught; textbooks that were reviewed to establish the levels and scope of content covered in each test; and evidences of content validity that are specific to the subject area, such as correlations of the test performance to selected reading placement inventories. The study that involved teacher ratings of the instructional objectives represents a commendable practice in achievement test development; however, the number of teachers participating in the study at each level is not impressive. Ratings of the reading tests were completed by 33 teachers at the Primer level, 229 at the Primary 1 level, 195 at the Primary 2 level, 366 at the Elementary level, 324 at the Intermediate level, and 200 at the Advanced level.

Evidence for content validity of the tests for the instructional objectives listed in the administration manuals appears convincing. Yet, evidence of the content validity for a given program or use of the test is not conclusive without additional information. Two other evidences to support validity arguments for specific uses of the test can be found in the publisher's special report series for the Metropolitan. These are the correlational study between performance on the test and performance on the Otis-Lennon School Ability Test (OLSAT) reported in the *Special Report Number 19* (The Psychological Corporation, 1978) and the correspondence between grade-equivalent scores on the 1978 Metropolitan and the 1973 Stanford Achievement Test reported in the *Special Report Number 21* (The Psychological Corporation, 1978). The correlations between the Metropolitan and the OLSAT range from .55 to .89 with the majority of subscores being correlated at .73 or better for sample sizes exceeding 650. These results can be interpreted as evidence of construct validity for some uses of the test. The close correspondence between the grade-equivalent scores on the Metropolitan and the Stanford can be used as evidence of concurrent validity. The relative low intercorrelations between performances on the subtests for the same sample used in the study of the relationship between performance on the Metropolitan and the Otis-Lennon ability tests attest to the utility of the subtest scores as contributing some unique information.

Critique

Earlier versions of the MAT were criticized for three weaknesses: items were more appropriate for urban students than for rural (Gronlund, 1978; Wolf, 1978); the test manual reported internal consistency but no test-retest reliability (Gronlund, 1978; Winegard & Bentler, 1974); and use of the tests as criterion referenced when they were standardized as norm referenced was not warranted (Gronlund, 1978; Wolf, 1978). These weaknesses have largely been overcome. Examination of items fails to support the first criticism in that no items could be found which would be particularly familiar to urban youth as opposed to rural. Secondly, the newest revision reports both internal reliability and test-retest reliability (see Table 4). Lastly, it appears that there are sufficient cautions in the teacher's manual to warrant utilizing scores as criterion referenced. Teachers (and administrators) are cautioned to assure linkages of curriculum objectives and the items are defined in clusters to provide a criterion-referenced score.

The Psychological Corporation (1978) has conducted other studies in the *Special Report* series which offer the potential of increasing the utility of the Metropolitan Test. Among these supportive studies is a report that establishes a correspondence between the instructional reading level of the Metropolitan and the Basal Reader Series. Additionally, the publishers have had the test reviewed by a panel to reduce the potential of ethnic bias and have reviewed the test for balance of sex-group relevant terms and for wordings that may foster a sex bias. Other *Special Reports* provide detailed descriptions of the development of performance scores such as grade equivalent scores, national item analysis results, studies relating test performance to socioeconomic status, and studies relating the two components of the testing program. The publishers appear to have made a sincere effort to increase the utility of the test.

References

Gronlund, N. E. (1978). Review of the Metropolitan Achievement Test. In O. K. Buros (Ed.), *The eighth mental measurements yearbook,* (pp. 65-67). Highland Park, NJ: The Gryphon Press.

Prescott, G. A., Balow, I. H., Hogan, T. P., & Farr, R. C. (1978). *Metropolitan Achievement Tests: Teacher's manual for administering and interpreting.* Cleveland: The Psychological Corporation.

The Psychological Corporation. (1978). *Metropolitan Achievement Tests special report numbers 1-21.* Cleveland: Author.

Wingard, J. A., & Bentler, P. M. (1974). Review of the Metropolitan Achievement Test. *Measurement and Evaluation in Guidance, 7*(3), 204-208.

Wolf, R. M. (1978). Review of the Metropolitan Achievement Test. In O. K. Buros (Ed.), *The eighth mental measurements yearbook* (pp. 67-69). Highland Park, NJ: The Gryphon Press.

Denise D. Davis, Ph.D.

Assistant Professor of Clinical Psychology, Indiana University School of Medicine, Indianapolis, Indiana.

MILLON BEHAVIORAL HEALTH INVENTORY

Theodore Millon, Catherine J. Green, and Robert B. Meagher, Jr. Minneapolis, Minnesota: National Computer Systems/PAS Division.

THE ORIGINAL OF THIS REVIEW WAS PUBLISHED IN TEST CRITIQUES: VOLUME III (1985).

Introduction

The Millon Behavioral Health Inventory (MBHI) is a self-report inventory designed to provide a psychological profile of general medical patients undergoing medical evaluation or care. The purpose of the MBHI is to provide health-care professionals with patient information that will facilitate the development and delivery of comprehensive treatment plans. The MBHI profile offers information about four dimensions of psychological functioning: Basic Coping Styles, Psychogenic Attitudes, Psychosomatic Correlates, and Prognostic Indices. The test does not confirm or support specific medical diagnoses or prognoses, but does appraise the potential impact of psychosocial factors on the course and treatment of designated medical disorders. Interpretation of the MBHI profile is based on a combination of empirically based, actuarial descriptions and a theoretically derived substrate of personality traits. Use of MBHI results and interpretive reports requires the involvement of a health-care professional with education in psychometric logic and methods, as well as relevant psychological practice experience. Profiles and narrative interpretations of the MBHI are specifically intended for confidential professional consultation and are not intended to provide information directly to patients or their families.

The MBHI was developed in 1982 by Theodore Millon, Ph.D., Catherine J. Green, Ph.D., and Robert B. Meagher, Jr., Ph.D. Millon, the principal author, is a clinical psychologist and distinguished personality theorist. Prior to the MBHI publication, Millon developed the concept of basic personality style with regard to emotional and behavioral traits (Millon, 1969, 1981) and served as senior consultant for personality disorders in the publication of the third edition of the *Diagnostic and Statistical Manual of Mental Disorders* (DSM-III) by the American Psychiatric Association (1980). In 1977, he also developed (and recently revised) the Millon Clinical Multiaxial Inventory (Millon, 1977, 1982) to measure personality patterns and disorders among mental health patients. The MBHI appears to be a logical extension of Millon's conceptualization of personality functioning into the area of behavioral medicine and general medical care. The development of a specific instrument to serve the needs of psychodiagnosticians working with physically ill patients was based on the rationale that medical populations are different from psychiatric populations. Therefore, traditional mental health constructs and standard psychological tests may not be valid or useful for medical populations. Millon specifically designed the MBHI to provide essential psychological information in a way that is useful and relevant in medical diagnostic and treatment settings.

The population involved in construction of the MBHI included a total of 2,113 medical patients and over 2,500 nonclinical reference subjects. Norms for the Basic Coping Styles and Psychogenic Attitudes scales are based on a selected sample of 300 patients (130 males and 170 females) and 452 nonpatients (212 males and 240 females). Six additional scales of psychosomatic correlates and clinical prognostic indices were developed and cross-validated with a series of medical patients, including 437 males and 482 females. The normative sample appears to be reasonably representative of designated age groups ranging from 18 to over 65 and includes a diverse ethnic and socioeconomic distribution, proportionally similar to the general population. As such, it is primarily a white, middle-class, adult normative sample. The need for normative adjustments on the basis of certain demographic factors, such as race, ethnic group, geographic region, extreme age, or education levels, has not yet been demonstrated, although data accumulated over time may justify some normative changes. Thus far, sex is the only moderator variable that has necessitated normative distinctions. Specific research concerning extensions or modifications of the MBHI is not yet available.

The MBHI test material consists of a two-page, printed questionnaire that should be filled out with a No. 2 pencil. Both questions and responses are contained in this test booklet. Instructions for completion of the inventory, printed on the front of the test booklet, are simple, concise, and complete. Patients endorse the 150 statements regarding thoughts, feelings, and actions as either true or false in their self-description. Space for coding the response is provided next to each statement. Also on the front of the test booklet are spaces for assigning a confidential identification number and indicating the respondent's sex, age, and major medical problems. In addition, there is a special code section that designates 15 columns for specific clinical information regarding diagnosis, health service setting, current therapy, socioeconomic level, ethnic racial background, and the examiner's own special code category.

Practical Applications/Uses

Administration of the MBHI is quite simple and poses no particular problems. The manual for the MBHI provides a clear description of the test development and instructions for its use. The MBHI is intended for individuals between the ages of 17 and 70, with a minimum of an 8th-grade reading level. It can be completed in approximately 20 minutes, and this greatly enhances its value in a medical setting where patients must often undergo tedious and painful procedures. The brevity of its administration also serves to minimize patient resistance and fatigue. It is important that recipients of the MBHI are not severely fatigued, apprehensive, medically debilitated, or highly sedated, as these states could interfere with adequate responses.

The examiner's participation in the testing process primarily involves explaining the purpose of the testing to the patient, supervising test administration, and interpreting test results. Beyond that, instructions for administration of the test suggest that the most reliable results can be obtained by allowing patients to simply follow the instructions printed on the test booklet. Thus, office personnel, such as nurses, secretaries, or hospital receptionists, can administer the MBHI, and the instrument can be completed in a waiting or examining room. Under certain circumstances, it may be acceptable to allow patients to complete the MBHI at home or to mail the test booklet to them with sufficient written notification of the nature and purpose of the inventory.

Results of the MBHI form a profile consisting of 20 scales that are grouped into four categories: Basic Coping Styles, Psychogenic Attitudes, Psychosomatic Correlates, and Prognostic Indices. Scale elevation is indicative of an increasing degree of associated features.

Basic Coping Styles has 8 scales (styles)—Introversive, Inhibited, Cooperative, Sociable, Confident, Forceful, Respective, and Sensitive. These features are typically interpreted in terms of a configural pattern of several scales.

Psychogenic Attitudes has 6 scales (attitudes)—Chronic Tension, Recent Stress, Premorbid Pessimism, Future Despair, Social Alienation, and Somatic Anxiety. These measure personal feelings and perceptions of the patient with respect to psychological stressors that potentially increase psychosomatic susceptibility or aggravate an ongoing disease.

Psychosomatic Correlates (applicable only to patients who have previously been medically diagnosed with a specific disease syndrome) has 3 scales—Allergic Inclination, Gastrointestinal Susceptibility, and Cardiovascular Tendency. These indicate the degree of psychosomatic complication for someone with an allergic, gastrointestinal, or cardiovascular disorder, respectively.

Prognostic Indices (also applicable only to diagnosed medical patients) has 3 scales—Pain Treatment Responsivity, Life Threat Reactivity, and Emotional Vulnerability. These seek to pinpoint future treatment problems. Of the three, Pain Treatment Responsivity identifies patients who are similar to patients who have shown a poor response to medical treatment for pain. Life Threat Reactivity identifies those who are likely to deteriorate more rapidly than typical for patients with a chronic or progressive life-threatening illness. Emotional Vulnerability pinpoints those patients who may be more likely to have a severe negative emotional reaction to major surgery or other life-dependent treatment programs.

The MBHI is a potentially useful tool for any psychologist or psychiatrist who works in a medical setting or who consults with medical practitioners. It provides a wealth of psychological information that is presented in a concise format and can be readily applicable to the development of a comprehensive health-care plan. Because it is brief and easily administered, the MBHI apparently can elicit psychological information in a nonintrusive and cost-effective manner. With an evaluation based on the MBHI, psychiatric clinicians can provide vital psychological information that can assist medical staff in decision making as well as treatment delivery and follow-up aspects of medical intervention. Although the MBHI potentially could be of practical use with any medical patient, it seems to be particularly useful with patients who have disorders strongly associated with psychogenic factors, including allergic problems, gastrointestinal problems, and chronic pain. Other potential recipients of the MBHI include behavioral medicine patients and any patient who is facing a potentially life-threatening disease (e.g., cancer, congestive heart disease) or life-threatening intervention (e.g., surgery, chemotherapy). In addition, the MBHI may be useful in facilitating the adjustment of patients for whom there has been considerable body alteration precipitated by illness or injury (e.g., spinal injury, mastectomy). Private practitioners who deal with patients who are undergoing any of these procedures or who have had illnesses with a psychosomatic feature may wish to utilize the MBHI for their own therapeutic purposes. Because the MBHI was not intended for nonmedical patients it is of limited value with individuals not engaged in some aspect of the medical process. It would not be appropriate to use it as a routine screening device in a nonmedical setting.

The MBHI is designed to be scored via computer service provided by National Computer Systems (NCS). Scoring sheets can be mailed to the central office in Minnesota for machine scoring, which includes a computer-printed profile and interpretive report. Electronic teleprocessing is also available to expedite the machine-scoring process. With access to a terminal and a printer, patients' responses can be transmitted electronically to a computer via a telephone modem connection. The computer will score and interpret the results and transmit a clinical profile and narrative report on the user's in-office printer either instantly or overnight. NCS (personal communication, May 1, 1985) also plans to make available, as of mid-1985, a software package for an IBM PC or XT that practitioners may purchase for scoring and producing a profile and narrative report with their own computer facilities.[1] Although scale items are listed in the test manual, it would be exceedingly difficult to hand-score this instrument. In addition, there are certain scale adjustments that are made on the basis of corrections for psychological defensiveness or exaggeration that are not easily incorporated into a hand-scoring attempt. Therefore, use of this instrument relies on the use of NCS's interpretative scoring systems or one's own IBM computer and appropriate software.

Interpretation of the MBHI is based on transformation of raw scale scores into base rate scores that indicate the relative prevalence of features common to a designated classification. Thus, the base rate scores indicate the likelihood that a patient does or does not exhibit certain features, rather than locating the relative position of a patient on a normal distribution. Coping Styles scales indicate the base rates of personality features found in nonpsychiatric populations. The Psychogenic Attitudes scales are transformed into T-scores, which suggest a normalized distribution because data as to the base rates of these characteristics were unavailable. The Psychosomatic Correlates and Prognostic Indices scales are interpreted in light of base rate prevalence among specific groups of patients. Two base rate cutting lines were arbitrarily drawn to indicate the relative salience of scale characteristics. A base rate cutting line for "presence" of features has been set at 75 and "prominence" of features is indicated by a base rate score of 85 or above.

Further narrative interpretation of the MBHI is accomplished by analysis of the profile configuration. Scoring of the MBHI by NCS typically includes an automated narrative report that is based on actuarial data and theoretical conceptualization. A sample report is included in the instruction manual and appears to be readily understandable to clinicians with expertise in psychological evaluation and patient management. However, the automated report needs to be interpreted in the context of relevant personal details, including demographic factors, behavioral observations, and other distinguishing features or test data. Clinicians with experience in psychodiagnostic interpretation may readily generalize those skills in constructing their own interpretive reports once they have familiarized themselves with Millon's theoretical schema. With sufficient understanding of the test and the theoretical substrate, the MBHI could easily be incorporated into most psychodiagnostic repertoires. Because adequate and proper use of this test does require a sophisticated level of clinical judgment, distribution of the MBHI is limited to psychologists, psychiatrists, and other practitioners who are approved by an NCS review committee.

Technical Aspects

Validation of the MBHI has apparently been incorporated into all developmental phases of the test completed thus far. A theoretical-substantive validation was accomplished by initial selection of items for the Coping Styles scales on the basis of their congruence with Millon's (1969) theoretical schema of personality style. The Psychogenic Attitudes scale items were substantively linked to relevant findings in empirical literature. The remaining psychosomatic and prognostic scales were derived completely from empirical correlations. The initial item pool was reduced by eliminating items that were too complex, biased, or unclear. In addition, clinicians with medical-psychodiagnostic experience sorted items into their theoretically appropriate scale, and only items with a 70% correct inclusion rate were retained.

Validation of the MBHI continued with administration of a provisional form of the inventory to a nonmedical population, calculating internal consistency correlations and evaluating the overlap of scale structures. Overlapping of scale items was specifically planned for theoretically related scales because the characteristics being measured are conceptually interrelated. However, items were deleted at this stage if their highest correlation was with a scale other than the one originally assigned, or if there was a significant correlation with a theoretically incompatible scale. Items with insufficient scale correlations (.30 or less) were also eliminated. External, criterion validity was incorporated by administration of a final form of the inventory in a large number of medical settings. In addition, medical personnel rated specific patients on certain criterion variables such as the probable effects of psychosocial factors.[2] Items that discriminated between patients on the basis of these criterion ratings were selected for the psychosomatic and prognostic indices. The final prognostic scale, Emotional Vulnerability, was constructed with items from the MCMI (Millon, 1977) which have the strongest empirical discrimination for psychological disturbance. Point-biserial correlations for all items and all scales were then recalculated. Items were added to any scale, other than a theoretically incompatible one, to which a correlation above .30 was evidenced. The final number of items in each scale ranges from 20-48. Each of the empirically derived scales has been evaluated in at least one cross-validational study.

Postconstruction empirical data on the MBHI (Millon et al., 1982) indicate satisfactory test-retest reliability, with a mean correlation coefficient of .82 for Coping

Styles scales, a mean coefficient of .85 for Psychogenic Attitudes scales and a mean coefficient of .80 for the empirical scales, with the exception of Emotional Vulnerability ($r = .59$). The K-R 20 measure of internal consistency indicated a median coefficient of .83 for all scales. From this, it appears that the MBHI scales are sufficiently reliable, despite the inherent fluctuation of the emotional content that they intend to measure.

Each of the Coping Styles and Psychogenic Attitudes scales has been assessed in relationship to several measures of personality and symptomatology. Correlations presented in the test manual generally reflect theoretically compatible relationships, thus offering empirical evidence of convergent validity. A factor analysis conducted with a mixed sample, which consisted primarily of medical patients, produced four major factor groupings. These factors have apparently contributed empirical support to configural interpretations. Further categorization via statistical cluster analysis has revealed differences in the frequency of configurations as a function of sex and patient status.

Critique

Literature that presents empirical support of the usefulness of the MBHI with specific subgroups of patients is limited at this point to unpublished manuscripts and results summarized in the test manual. These reports generally indicate that the MBHI shows promise in identifying psychological characteristics that can mediate the development of symptoms (Head, 1979) or the response to treatment in a variety of patient groups such as coronary bypass surgery patients (Levine, 1980) or chronic pain patients (Green et al., 1980; Murphy et al., 1983).[3] Lists of recently completed and ongoing research are available from the publisher. The absence of independently published validity data has been noted as a major limitation to the usefulness of this instrument (Lanyon, 1984).

The MBHI generally seems to be an innovative instrument that promises clinical utility. Further research on the validity of the automated interpretive report is needed, as the predictive accuracy of these reports has not yet been clearly substantiated. Expanded research may be more feasible with the availability of onsite software for scoring purposes because rapid turnaround time is a primary consideration in most medical settings. The MBHI is brief, easily administered, sufficiently reliable, and well-suited to medical settings. Most of all, it expedites the process of gathering and interpreting psychological information that can facilitate planning medical treatment. The MBHI offers potential for meeting an important need in the expanding field of behavioral medicine.

Notes to the *Compendium:*

[1]MICROTEST software is now available for IBM PCs, XTs, and compatibles.

[2]Interrater reliability on this dimension was not reported.

[3]Literature presenting empirical support of the MBHI's usefulness with specific patient subgroups is still very limited. Two published studies report MBHI results that distinguish certain medical patients. Katz, Martin, Landa, and Chadda (1985) found that cardiac patients with arrhythmia and no history of myocardial infarction scored higher on MBHI

Inhibited Coping Style and Social Alienation attitudes and lower on the Respectful Coping Style when compared to medical/surgical control patients. MBHI results also distinguished chronic headache patients from normal controls and from patients with other forms of chronic pain (Gatchel, Deckel, Weinberg, & Smith, 1985).

References

American Psychiatric Association. (1980). *Diagnostic and statistical manual of mental disorders* (3rd ed.) Washington, DC: Author.

Gatchel, R., Deckel, W., Weinberg, N., & Smith, J. (1985). The Millon Behavioral Health Inventory in the study of chronic headaches. *Headache, 25,* 49-54.

Green, C., Meagher, R., & Millon, T. (1980, November). *Patients' social responses in a pain program: Uses of the MBHI.* Paper presented at the annual meeting of the Society of Behavioral Medicine, New York.

Head, R. (1979). *The impact of personality on the relationship between life events and depression.* Unpublished doctoral dissertation, University of Miami, Coral Gables.

Katz, C., Martin, R., Landa, B., & Chadda, K. (1985). Relationship of psychological factors to frequent symptomatic ventricular arrhythmia. *The American Journal of Medicine, 78,* 589-594.

Levine, R. (1980). *The impact of personality style upon emotional distress, morale, and return to work in two groups of coronary bypass surgery patients.* Unpublished master's thesis, University of Miami, Coral Gables.

Lanyon, R. (1984). Personality assessment. *Annual Review of Psychology, 35,* 667-701.

Millon, T. (1969). *Modern psychopathology.* Philadelphia: W. B. Saunders.

Millon, T. (1977). *Millon Clinical Multiaxial Inventory manual.* Minneapolis: National Computer Systems, Inc.

Millon, T. (1981). *Disorders of personality—DSM III Axis II.* New York: John Wiley & Sons.

Millon, T. (1982). *Millon Clinical Multiaxial Inventory manual (2nd ed.).* Minneapolis: National Computer Systems, Inc.

Millon, T., Green, C. J., & Meagher, R. B., Jr. (1982). *Millon Behavioral Health Inventory.* Minneapolis: National Computer Systems.

Murphy, J., Sperr, E., & Sperr, S. (1983, March). *Comparisons and clinical utility of MMPI and Millon Behavioral Health Inventory profiles in a chronic pain population.* Paper presented at the convention of the Southeastern Psychological Association, Atlanta.

Sheridan P. McCabe, Ph.D.

Associate Professor of Psychology, University of Notre Dame, Notre Dame, Indiana.

MILLON CLINICAL MULTIAXIAL INVENTORY

Theodore Millon. Minneapolis, Minnesota: National Computer Systems/PAS Division.

THE ORIGINAL OF THIS REVIEW WAS PUBLISHED IN TEST CRITIQUES: VOLUME I (1984).

Introduction

The Millon Clinical Multiaxial Inventory (MCMI) was developed by Theodore Millon to provide a personality inventory useful for clinical diagnosis. The Minnesota Multiphasic Personality Inventory (MMPI) was the long established clinical instrument which the MCMI was designed to replace, at least for diagnostic testing. Millon refers to a number of shortcomings in the MMPI that include its length, the logic of its development, and its utility for formulating diagnoses, which the MCMI would be designed to remedy. The MCMI is a 175-item personality inventory with a true/false response format that yields scores on twenty scales. These scales are organized into three broad categories to reflect distinctions between persistent personality features, current symptom states, and level of pathological severity. Scoring was designed to be done on a computer with a program that could produce either an interpretive report or merely the scores with a profile.

The author of this test, Theodore Millon, is already well known for his work on psychopathology and his theoretical formulations in this area are described in two of his books (Millon, 1969; Millon, 1981), which represent an important basis for the rationale of this test. Perhaps of even greater significance is Millon's contribution to the development of the third edition of the *Diagnostic and Statistical Manual of Mental Disorders* (DSM-III), which has made fairly substantial changes in the diagnostic nomenclature, especially with regard to neurotic conditions and personality disorders. The important role of Millon's theoretical work in this area is readily evident in the new approach of the DSM-III. The MCMI is planned and organized to identify clinical patterns in a manner easily related to the DSM-III categories.

The test was envisioned as an updated MMPI, much shorter than its predecessor and yielding results in terms of personality styles and dimensions of pathology based on the DSM-III. The criticism of the MMPI was that it was too exclusively empirical and lacked a clear theoretical basis; Millon responded to this and the MCMI was based on a definite theory of personality and psychopathology. In addition, the MCMI supports the DSM-III distinction between enduring personality patterns (DSM-III Axis II) and more acute clinical disorders (DSM-III Axis I). Another difference between the MMPI and the MCMI is the logic inherent in the development of the test and its items. Rather than using items that differentiated between specific diagnostic groups and normals, Millon chose to use items that differentiated the diagnostic groups from a general psychiatric patient population. This is designed to enhance the discrimination efficiency of the scales and improve differential diagnosis. In addition, actuarial base rate data were employed in calculating and quantifying scale measures. Millon made a strategic decision to work toward diagnostic efficiency rather than psychometric elegance, and deliberately accepts a considerable degree of item overlap among the scales.

Millon uses an elaborate three-stage approach in test development beginning with a theoretical-substantive stage in which items are generated, progressing through an internal-structural stage, in which the operational characteristics of the items are studied and many items discarded on technical grounds, and arriving finally at an external-criterion stage in which empirical tests of the scales are made. It is through this three-stage strategy that the theoretical basis of the test, as well as practical clinical considerations, are permitted to operate and guide the development along with the technical and empirical features emphasized in more traditional empirical test development. Millon argues that this strategy provides the best basis for validity and for the generalizability, dependability, and accuracy of the results.

The development of the MCMI took place through a long process of item formulation, editing, testing, and revising. In the course of this process, a pool of over 3,500 items was reduced to 175. In the external-criterion phase of validation, the test was empirically evaluated based on its administration to 682 psychiatric patients.

The test itself consists of a relatively simple combined question-and-answer test booklet that includes directions and space for subject-identifying information. There is a special grid where the clinician may record clinical information. On the four-sided sheet, three sides are devoted to the questions and answers. The entire form is designed for optical scanning, i.e., the questions are listed with circles for marking true or false answers next to each item. The instructions are simple and direct. The test is designed to be given to patients over seventeen-years-old with reading skills at or above the eighth-grade level. There is only one form of the MCMI, although Millon has devised other tests for individuals under eighteen and for medical patients. The manual suggests that the test can be administered in groups, can be given by secretarial or nursing staff, and can even be given to reliable patients on an unsupervised basis.

Practical Applications/Uses

The growing need for clinicians to provide DSM-III diagnoses will probably lead to increased use of the MCMI. The information that this test yields appears to be highly relevant and very useful to those who work with people experiencing emotional or interpersonal difficulties. The fact that it requires relatively little sophistication to administer and is rather short and convenient to take, plus the ready availability of computer scoring, make it attractive to the practitioner. The

MCMI will be particularly helpful in clinical screening, treatment planning, and in arriving at diagnostic formulations. Following an initial screening with this instrument, the clinician can more carefully target a more comprehensive and thoroughgoing assessment program.

The purpose of the MCMI is succinctly stated in the manual:

> The primary intent of the MCMI is to provide information to clinicians—psychologists, psychiatrists, counselors, social workers, physicians, and nurses—who must make assessments and treatment decisions about persons with emotional and interpersonal difficulties. The Profile Report of scale scores can serve as a screening device to identify those who may require more intensive evaluations or professional attention. The clinical narrative Interpretive Report provides a detailed analysis of personality and symptom dynamics as well as suggestions for therapeutic management. (Millon, 1983, p. 2)

The test is designed to facilitate diagnosis and give particular attention to personality disorders. The scales of the test are based on the categories outlined by Millon in his theory of psychopathology. This theory posits eight basic styles of personality functioning that can be derived from a 4 x 2 matrix. The first dimension of this matrix refers to the primary source of positive reinforcements: detached (those who experience few rewards); dependent (those who derive reinforcement from the reactions of others); independent (those who derive reinforcement from their own values and desires without reference to others); and ambivalent (those who experience conflict between their own desires and the expectations of others). The second dimension of the matrix deals with the basic pattern of coping behavior and has two major divisions: the active pattern, characterizing those who seem aroused and attentive, arranging and manipulating life events to achieve gratification; and the passive pattern, characterizing those who seem apathetic, restrained, yielding, or content to allow events to take their own course without personal regulation or control.

The eight positions in this matrix form the basis of the eight personality dimensions of the test. The relationship of the eight scales to the matrix cells is indicated in the following illustration:

	Detached	Dependent	Independent	Ambivalent
Pas.	Schizoid (Asocial)	Dependent (Submissive)	Narcissistic	Compulsive (Conforming)
Act.	Avoidant	Histrionic (Gregarious)	Antisocial (Aggressive)	Pass. Aggress. (Negativistic)

In the case of each scale, the name of the scale and its significance is identical to the corresponding DSM-III category. Millon regards these scales as defining the basic types or styles of personality. Pathology is understood as an extreme deviation of one's basic personality orientation.

The other scales of the test measure the direction and severity of pathological manifestations of these basic patterns. The three patterns of personality pathology are regarded as extensions and distortions of the basic personality style, and as such are chronic and dysfunctional. The clinical syndromes on the other hand are also distortions of the basic personality style but are distinct or transient states that are brought on by stressful situations. The latter therefore are represented as DSM-III Axis I. The test is thus carefully built up from a coherent theory of psychopathology and closely consistent with the DSM-III nomenclature.

The next three scales deal with more serious personality disturbances and are basically elaborations of the eight basic styles just described. They represent enduring and persistent pathology, are representative of Axis II in the DSM-III, and are to be distinguished from the more transitory clinical syndromes of Axis I. These scales are as follows:

> *Scale S Schizotypal (Schizoid).* This represents a deterioration in patients characterized by one of the two basic detached patterns, the schizoid type and avoidant type. It corresponds to Schizotypal in the DSM-III.
>
> *Scale C Borderline (Cycloid).* This is a more severe variant of the basic dependent and ambivalent patterns. It corresponds to the Borderline classification in the DSM-III.
>
> *Scale P Paranoid.* This is most associated with the two independent basic personality styles (the narcissistic and the antisocial), and to a lesser degree the compulsive and the passive-aggressive. It corresponds to the Paranoid classification in the DSM-III.

The remaining scales of the test are concerned with clinical syndromes. Millon also sees these as extensions or distortions of basic personality styles, but different in that they are transient states generally in reaction to stress. As such they all represent DSM-III Axis I disorders. The first six of these are generally of moderate severity, while the last three are more severe and correspond to psychotic conditions. These scales and their DSM-III equivalent categories are as follows:

> *Scale A Anxiety.* High scores on this scale are associated with anxious or phobic feelings in a pathological degree. The DSM-III equivalent is Anxiety Disorder, panic, phobic and obsessive types.
>
> *Scale H Somatoform.* People scoring high on this scale express psychological difficulties through somatic complaints and are preoccupied with concerns about health and physical symptoms. The DSM-III equivalent is Somatoform Disorder, somatization, pain and hypochondriacal types.
>
> *Scale N Hypomanic.* High scorers manifest periods of superficial, elevated but unstable moods, restless overactivity and distractability, and impulsiveness. This scale corresponds to Manic Disorder, moderate in the DSM-III.
>
> *Scale D Dysthymic.* A person scoring high on this scale is depressed, preoccupied with feelings of discouragement or guilt, and exhibits a lack of initiative. The DSM-III equivalent is also Dysthymia.

Scale E Alcohol Abuse. High scores on this scale are associated with alcoholism and frequently high scorers have a history of alcohol abuse, with disruption of their career and family life. This is also termed Alcohol Abuse in the DSM-III.

Scale T Drug Abuse: High scores on this scale are associated with a recurrent or recent history of drug abuse. It is equivalent to the category of Drug Abuse in the DSM-III.

Scale SS Psychotic Thinking. High scorers are generally characterized by incongruous, disorganized, and regressive behavior, and by confusion and disorientation. This scale corresponds to Schizophrenic Disorder in the DSM-III.

Scale CC Psychotic Depression. A person scoring high on this scale would ordinarily be incapable of functioning because of severely depressed mood and would be characterized by a dread of the future and sense of hopeless resignation. In the DSM-III, this syndrome is referred to as Major Depression.

Scale PP Psychotic Delusions. High scorers would be considered paranoid, and would be characterized by delusions of persecutory or grandiose nature. In the DSM-III, this condition is called Paranoid Disorder.

Although in the manual Millon de-emphasizes the importance of measures of test-taking attitudes and response styles, he does include two scales on the MCMI to deal with what he calls the "denial versus complaint" attitude and the problem of random or confused responding. These scales are:

Scale W Weight Factor. This scale detects excessive degrees of either psychological defensiveness and self-enhancement or emotional complaining and self-deprecation, and then adjusts the personality pathology and symptom disorder scales accordingly.

Scale V Validity Index. This scale detects respondents who fail to cooperate, are unable to comprehend the items, or for other reasons do not answer relevantly. The resulting profiles are then flagged as invalid.

The character of the MCMI's scales presents a clear indication that the primary use of the test is in the diagnosis of emotional disorders; thus, this instrument will be of interest primarily to those concerned with the use of tests in clinical diagnosis. Because of the way in which the test was developed and the basis of the scale and derived scores, it is important that the test be used only with those who are drawn from a clinical population (i.e., have some emotional or interpersonal dysfunction that is important to classify or describe). It would be inappropriate to administer the test to individuals drawn from a population which is normal with respect to emotional problems. The results of such an administration would be subject to misinterpretation.

The normative group for the MCMI included 297 non-clinical subjects and 1,591 subjects drawn from a patient population. The subjects were obtained from 27 states in the U.S. and from Great Britain. The author provides good details on the characteristics of the norm group, including the distribution on variables such as sex, race, and socioeconomic level. Millon however was more concerned about drawing subjects from diagnostic categories that reflected the actual incidence of the various clinical disorders in the general population. It would appear to be a well-selected and carefully considered norm group.

When the test first appeared on the market, the only scoring method available was the computer scoring service offered by the test publisher. Only the interpretive report was available, and consequently scoring was an expensive proposition. It is likely that this policy discouraged more widespread experimentation with the test and was a barrier to more systematic research using this instrument. More recently the publisher announced the availability of a Profile Report, which gave the scores and presented a graphic profile. This form of report could be purchased for far less than the interpretive report.

Initially it was necessary to mail in the answer sheet for computer scoring. While the publisher is to be commended for its established reputation for efficient operation and the prompt return of the results, it would still take about a week to obtain the average results under this system. More recently, a new service is available that may be utilized by anyone who has a modem and either a terminal or a microcomputer with a printer. By becoming a subscriber one can, with a local telephone call, log on to the host computer, enter the test responses, and have printed out on one's own printer either an interpretive report or a profile report. For very little more than the cost of a mailed-in answer sheet the results are instantly available. In the manual, the author gives reasons why hand scoring is not encouraged and he emphasizes the advantages of computer scoring. However, in the most recent catalog, the publisher indicates that hand-scoring templates are now available for those who wish to do their own scoring. This move should encourage wider utilization of the test and promote more research of the kind that has characterized the MMPI in its long history.

The manual provided for this test is exceptionally complete and clearly written. It presents a thorough background in the rationale of the test, the steps involved in its development, and the technical data necessary for its proper use. A good deal of very helpful information is provided covering interpretation of the results. However, because of the complexity of the results, one would do well to follow the advice in the manual and acquire a thorough familiarity with the author's book on personality disorders (Millon, 1981) to be fully at ease using this test in a clinical setting.

It is apparent that considerable effort went into the development of the interpretive report. The reports that this reviewer has seen have been clear, well written (considering that they were authored by a computer), and useful in terms of understanding the client. This is a highly subjective impression, but one that is further supported by the author's data in the manual. Millon subjected the interpretive report to an evaluative study in which computer-generated MCMI reports were rated by clinicians along with two versions of interpretive reports on the MMPI. There were 23 clinicians in the study who had had considerable experience with the patients given both the MCMI and the MMPI. The interpretive reports were rated on a number of dimensions, which were grouped into the three categories of Information Adequacy, Accuracy of Descriptions, and Report Format and Utility. One of the MMPI scoring programs was generally given unsatisfactory ratings on most scales; the other MMPI scoring service and the MCMI report were both given quite satisfactory ratings. In general, however, the MCMI interpretive report received the highest ratings. While this reviewer was not one of the raters in the study, these results do conform to his impressionistic evaluation based on experience with each of the scoring services.

While the availability of a comprehensive and adequate computer-generated report, especially one which relates to the specific demands of the DSM-III diagnostic system, makes the task of test interpretation far easier, it must be emphasized nonetheless that the MCMI is a very complex test and appropriate interpretation requires a high level of training and sophistication in personality testing.

Technical Aspects

The MCMI represents a departure from the traditional manner of development characterizing most personality inventories. As mentioned earlier in this review, Millon has made a studied effort to proceed along a three-stage validation process rather than to assemble the test and check out its criterion-related validity later. The first stage of Millon's process, the theoretical-substantive stage, represents a theory-driven approach to the writing of scale items. Items were written to characterize each of the personality styles and syndromes inherent in his theory. The items prepared for each of the tentative scales ranged between 177 and 432 in number for a total of over 3,500. This item pool was then reduced, judged on the basis of factors such as clarity, content validity, etc. These judgments were aided by clinician and patient ratings. This stage resulted in approximately 1,100 items, which were arranged in two forms of 566 items each.

These two forms were then used in the next validation stage, which Millon calls internal-structural validation. The objective here was to achieve a high degree of internal consistency in the scales. Millon makes a strong case for allowing a relatively high degree of item overlap to exist among the scales (i.e., a given item might appear on two or more related scales). This stage was pursued by administering the two forms to groups of patients and intercorrelating the items and computing the item endorsement frequencies. To maximize scale homogeneity, only those items were retained for additional study which showed their highest correlation with the scale to which they were originally assigned. Items that showed extreme endorsement frequencies were dropped. Next, items whose pattern of intercorrelation was inconsistent with the structural model of the theory were screened out. This stage resulted in 289 remaining items.

The final stage of the validation process was the external-criterion phase. This stage was pursued in a way that provided for both convergent and discriminant validity. A study was conducted in which 682 patients of over 200 clinicians were administered the 289-item research form. Twenty criterion groups of these patients were formed, based on the ratings of the clinicians who knew them. Each item was tested against the criterion of endorsement by the criterion group that corresponded to the scale on which that item was placed, as well as the relative nonendorsement by other criterion groups. Scales were thus constructed to support the structural model and to optimize differential diagnostic efficiency. Items which proved less effective were eliminated, and 150 items remained. In this original study, three clinical disorder scales (hypochondriasis, obsession-compulsion, and sociopathy) were found to be of minimal use. Therefore, Millon decided to replace these scales with the scales of hypomanic, alcohol abuse, and drug abuse. These three new scales were developed following the same three-stage procedure. This resulted in adding 25 additional items to the test, giving the final 175 items of the MCMI.

The final stage in the development of this test was the development of empirically validated configural patterns. The data from the 682 psychiatric patients (on which the development of the items was based) were subjected to a cluster analysis. An a priori decision was made about the number of clusters to derive based on the theoretical rationale of the test. Millon concluded that "the overall correspondence between the results of statistical clustering and criterion-theoretical groupings was extremely high for both the configural types that were produced and the individual MCMI patient assignments within them" (Millon, 1983). Thus the meaning of the statistically derived clusters is provided by the clinical theory on which the test is based.

In the manual, Millon reports a good deal of information on the reliability, the internal structure of the test, and external validation both of the test scores and the configural information yielded by the test. Data on reliability are presented in terms of test-retest reliability over one-week and five-week intervals. In both instances, clinical populations were used to provide the data. These coefficients range between a high of .91 and a low of .61. The stability of the personality pattern scales is somewhat higher than that of the pathological scales. In general the median stability coefficient is around .80, a level that might be considered both quite acceptable for an inventory of this type and impressive given the length of the test. One should note, however, that the time interval is not great and that sample sizes of 59 and 86 are somewhat small. The author also reports KR-20 coefficients computed over the standardization sample of almost 1,000. These data indicate a high degree of internal consistency of the scales since the coefficients range between .95 and .70, with the exception of the psychotic delusion scale which attained a coefficient of only .58.

As noted before, Millon has departed from the more traditional psychometric approach to test development in his willingness to accept a relatively high degree of item overlap among the scales of the test. In his technical discussion of the internal structure of the test, Millon provides considerable data on item overlap and item intercorrelation. He reports that the item overlap was planned and stems from the character of the theory on which the test is based. However, it is substantial, ranging from 0 to 65%. In addition to the extent of item overlap, a preponderance of the items are keyed in the "true" direction, raising the question of the operation of response bias. Table 1 indicates the number of keyed items on each of the scales along with the number of these items keyed in the "false" direction.

Item intercorrelation is generally high, reaching a level of .96 in one case and with nine coefficients of .90 or greater. Millon also reports a factor analysis of these intercorrelations, in which four factors were obtained. These factors might be characterized as: 1) depressive and labile emotionality; 2) paranoid behavior and thinking; 3) schizoid behavioral detachment and thought; and 4) social restraint

Table 1

Number of Keyed Items for MCMI Scales		
Scale	Number of items	False items
1 Schizoid (Asocial)	37	11
2 Avoidant	41	4
3 Dependent (Submissive)	33	7
4 Histrionic (Gregarious)	30	13
5 Narcissistic	43	16
6 Antisocial (Aggressive)	32	11
7 Compulsive (Conforming)	42	30
8 Passive Aggressive (Negativistic)	36	3
S Schizotypal (Schizoid)	44	4
C Borderline (Cycloid)	44	0
P Paranoid	36	0
A Anxiety	37	1
H Somatoform	41	2
D Dysthymic	36	4
B Alcohol Abuse	35	3
T Drug Abuse	46	5
SS Psychotic Thinking	33	0
CC Psychotic Depression	24	0
PP Psychotic Delusion	16	0
Validity	4	0

and conformity as opposed to social aggression. Millon observes the parallel between the first three factors and the classical distinction among affective disorders, paranoid disorders, and schizophrenic disorders.

Finally, Millon provides data on the relationship of both the clinical scales of the MCMI and its configural scores and patterns with other diagnostic inventories, particularly the MMPI. In general, inspection of these data suggests that the MCMI scales have the predicted positive or negative relationships with the appropriate MMPI scales. However, the most relevant data on the validity of the MCMI for clinical applications are the data on the cross-validation studies of the configural interpretation of the test. Millon made the decision to use a cross-validation sample that was highly similar to the original validation sample in terms of the base rates of the clinical categories. The procedures in which clinical judgments and ratings were obtained were identical. Nonetheless, the results of the reported cross-validation study are impressive. The overall pattern of valid-to-false positives ratios proved to be remarkably robust.

The technical and research data presented in the manual are thorough, impressive, and clearly explained. The clinical utility of this relatively short and convenient inventory is in no small part due to the innovative and effective approach to test development used by the author. It is however regrettable that almost all of the available research and data on this very promising inventory is that supplied by the author. The preeminence among personality inventories that the MMPI has enjoyed over a number of decades is due in very large measure to the impressive array of available research that has accumulated on this test. A bibliography of thousands of items and the development of hundreds of scales, such as in the case of the MMPI, has been exceeded by no other test. The MMPI has stood the test not only of time, but of the effects of myriads of researchers both friendly and inimical. By contrast, only the most meager collection of research by individuals other than the test author has accumulated on the MCMI. Perhaps this is due in part to the only recent availability of economic scoring techniques, which may have rendered the MCMI in the past a far less attractive option for research endeavors than a number of alternative inventories. Hopefully, the fact that hand-scoring procedures are now available may lead to the undertaking of considerably more research investigations by independent investigators.

Critique

From both a practical and theoretical point of view, this instrument appears to be promising and innovative. The development of a test that can effectively compete with the MMPI is a very ambitious endeavor to say the least. However, given the scarcity of research support from independent investigators and the lack of a well-established clinical track record, it is far too soon to render a judgment on the success of this endeavor. Based on the evidence and arguments which Millon presents in the well-written manual, though, it would appear that this test shows excellent promise to become the major clinical screening inventory of the future.

From a theoretical or psychometric point of view, the innovative development of this instrument is intriguing. The thoroughness and tight conceptual rationale for the three-stage process of test development satisfies many of the criticisms leveled at more traditional approaches to the development of personality invento-ries. Casual initial use of the instrument in clinical practice suggests that it has a great deal to offer, perhaps due to its careful, albeit atypical, psychometric devel-opment, the author's concern about appropriateness of clinical norm groups, and his utilization of base-rate scores. Of particular note is the utilization of a well-developed conceptual framework for the basic approach of the test. This makes interpretation more directed and rational and less a blindly empirical task. The idea of construct validity in terms of the MCMI has a more comprehensive and practical contribution to make. Of course, to the extent that the theory is lacking the test may be less useful.

From the practical point of view, this test has a great deal to offer. Its intimate relationship to the DSM-III is one important feature in this connection. The basic diagnostic groups on which the MMPI was originally developed are somewhat far removed from the schema inherent in the DSM-III. The fact that not only do the MCMI scales correspond to the DSM-III categories, but the normative groups and the basis for the derivation of the base-rate scores stem directly from the DSM-III render the MCMI particularly useful in approaching the diagnostic classification of a patient. Secondly, the length of the test, 175 true/false items, as well as the simplicity of responding make this a very useful test. The length of the MMPI frequently made its administration difficult to patients who were infirm or who had difficulty in attending for long periods of time. The MCMI will be much easier to administer to an elderly patient, for example. The ready availability of a variety of scoring methods will also contribute to the ease of use of the MCMI. Hand scoring as well as mail-in or remote-entry computer scoring, with a choice of either profile or interpretive reports, give the clinician a useful range of scoring options.

There is only one major category in which this test can be faulted. This is the lack of available research and clinical literature. Although this test has been available for approximately seven years, there are very few references to it in the published literature. While the number of articles which continue to appear on the MMPI each year reach the hundreds, all of the articles published on the MCMI since its introduction can be counted on one's fingers. In addition to the published articles, there are also a dozen or so unpublished dissertations and articles which the test publisher can supply. However, one does note that most of these research studies have appeared in the past two years. Now, with the availability of hand scoring, it is not unlikely that the research activity with this test will increase substantially in the near future. Should this research support and extend the promise of this instrument as suggested by the manual, then this test may well become the primary clinical personality inventory of the next decade.[1]

Notes to the *Compendium:*

[1]A revision of this test, termed MCMI-II, was announced in 1987. The principal changes incorporated into this revision include 45 new or reworded items. These replace existing items, so that the length of the test remains the same. A new scoring method has been devised that employs item weighting, intended to provide greater differentiation between diagnostic syndromes. In addition, two new personality pattern scales have been added to the original eight: Aggressive/Sadistic and Self-Defeating. Finally, some scoring adjustments that were somewhat awkward in the original MCMI are now incorporated into the test as Modifier Indices and function like validity scales (Disclosure, Desirability, and Debasement). The preliminary description provided by the publisher suggest that these changes will make possible a revised interpretive report providing a more comprehensive analysis of the subject's results.

References

American Psychiatric Association. (1980). *Diagnostic and statistical manual of mental disorders* (3rd ed.). Washington, DC: Author.

Flynn, P.M., & McMahon, R.C. (1983). Stability of the drug misuse scale of the Millon Clinical Multiaxial Inventory. *Psychological Reports, 52,* 536-538.

Flynn, P.M., & McMahon, R.C. (1983). Indicators of depression and suicidal ideation among drug abusers. *Psychological Reports, 52,* 784-786.

Gilbride, T.V., & Hebert, J. (1980). Pathological characteristics of good and poor interpersonal problem-solvers among psychiatric outpatients. *Journal of Clinical Psychology, 36,* 121-127.

Green, C.J. (1982). The diagnostic accuracy and utility of MMPI and MCMI computer interpretive reports. *Journal of Personality Assessment, 46,* 359-365.

Gynther, M.D., & Gynther, R.A. (1983). Personality inventories. In I.B. Weiner (Ed.), *Clinical methods in psychology* (2nd ed., pp. 152-232). New York: John Wiley & Sons, Inc.

Millon, T. (1969). *Modern psychopathology.* Philadelphia: W.B. Saunders.

Millon, T. (1981). *Disorders of personality: DSM-III, Axis II.* New York: John Wiley & Sons, Inc.

Millon, T. (1983). *Millon Clinical Multiaxial Inventory manual* (3rd ed.). Minneapolis: Interpretive Scoring Systems.

Snibbe, J.R., Peterson, P.J., & Sosner, B. (1980). Study of psychological characteristics of a workers compensation sample using the MMPI and Millon Clinical Multiaxial Inventory. *Psychological Reports, 47,* 959-966.

Eugene E. Levitt, Ph.D.
Professor of Clinical Psychology, Indiana University School of Medicine, Indianapolis, Indiana.

Jane C. Duckworth, Ph.D.
Professor of Counseling Psychology, Ball State University, Muncie, Indiana.

MINNESOTA MULTIPHASIC PERSONALITY INVENTORY

Starke Hathaway and J. Charnley McKinley. Minneapolis, Minnesota: University of Minnesota Press—Distributed exclusively by National Computer Systems/PAS Division.

THE ORIGINAL OF THIS REVIEW WAS PUBLISHED IN TEST CRITIQUES: VOLUME I (1984).

Introduction

The Minnesota Multiphasic Personality Inventory (MMPI) is an objective verbal inventory consisting of 550 statements, 16 of which are repeated, making a total of 566 in the complete test format. The replicated statements were originally included to facilitate the first attempt at scanner scoring. Though they are no longer needed for this purpose, they persist in the inventory.

The statements are keyed "True" and "False." In the original card deck form, the respondent was permitted a third grouping of statements to which he or she did not wish to respond for whatever reaon. The standard Psychological Corporation booklet form of the inventory urges the respondent not to "leave any blank spaces if you can avoid it" (i.e., try to respond to every statement). The vast majority of all types omit fewer than five statements (Clopton & Neuringer, 1977). The frequency of omissions appears as an index called *Qu* or *?*.

The inventory is scored in subunits, eight of which are conventionally termed *clinical scales* and provide the so-called clinical profile. The clinical scales are Scale 1 (Hypochondriasis), 33 statements; Scale 2 (Depression), 60 statements; Scale 3 (Hysteria), 60 statements; Scale 4 (Psychopathic Deviate), 50 statements; Scale 6 (Paranoia), 40 statements; Scale 7 (Psychasthenia), 48 statements; Scale 8 (Schizophrenia), 78 statements; and Scale 9 (Hypomania), 46 statements.

Two other scales were added from within the original item pool. Neither was thought to directly reflect a diagnostic entity. Scale 5 (Masculinity-Feminity) with 60 statements was developed along with the eight clinical scales. Shortly thereafter, Scale 0 (Social Introversion) with 51 statements was added (Drake, 1946).

Three additional measures were designed to estimate the validity of the clinical profile. The L (Lie) Scale has 15 statements, each dealing with a common, relatively insignificant weakness to which most people are willing to confess. The F (Infrequency) Scale is made up of 64 statements that were answered in the keyed direction by less than 10% of the inventory's original standardization group. The K Scale (30 statements) was designed to trap the respondent who attempts to conceal actual psychopathology and to present him- or herself in a speciously favorable light.

The score value on the respondent's K scale, or a fraction of it, was recommended in order to improve discriminability of Scales 1, 4, 6, 8, and 9 (McKinley et al., 1948). This practice has become established in the use of the MMPI even though the original finding has never been adequately tested independently.

Qu is also sometimes regarded as a validity check, though there are multiple reasons why Qu may be high without suggesting a response set. There is some independent evidence, however, that the clinical profile is distorted when Qu exceeds 30 (Clopton & Neuringer, 1977).

Over the years since the development of the original 13 scales, many additional or "special" scales have been derived from among the MMPI's 550 items. This tactic has been so popular that the *MMPI Handbook* (Dahlstrom et al., 1975) lists over 550 special scales. Most of them have been little used, but a handful (viz., the Wiggins Content Scales [Wiggins, 1966] and the Harris-Lingoes Subscales [Harris & Lingoes, 1968]) have been employed with a reasonable frequency.

The MMPI takes a fair amount of time to administer, anywhere from 40 minutes to one and a half hours, depending on reading speed and response tendencies. The inventory's developers were aware of this shortcoming and suggested that a 366-statement short form of the inventory could be substituted, losing only the K Scale and Scale 0 in addition to many special scales (Hathaway & McKinley, 1951). Since that time, seven other short forms ranging in length from 71 to 168 statements have been published (Newmark & Faschingbauer, 1978). These require estimates of clinical scale scores. To our knowledge none is in common use.

The MMPI was published in 1943 by Starke Hathaway, Ph.D., and J. Charnley McKinley, M.D., of the University of Minnesota. The authors wanted a test to use for routine diagnostic assessment, mostly to help in assigning an appropriate psychodiagnostic label.

Hathaway and McKinley collected a large group of items from psychological and psychiatric case histories, textbooks, and earlier published scales of personal and social attitudes. Around 1,000 items were so identified and eventually reduced to 504 relatively independent items.

Scales were derived by empirically determining those items that differentiated between a group of normals and a criterion group such as depressives or schizophrenics. Most inventories before the MMPI had used theoretical scoring, deciding which item belonged in a scale by a judgment based on a theory.

The group of normals used in the comparison with the criterion groups consisted primarily of friends and relatives of patients in the University of Minnesota hospital as well as a few other groups of normal persons such as precollege high school graduates, WPA workers, and general medical patients. The criterion groups were selected from patients at the University of Minnesota hospitals. Those items that differentiated between the normal group and a criterion group became the scale that identified that criterion group.

The scales thus derived were cross-validated on a new group of normals and new members of the criterion group. Those items that did not continue to differentiate between the two groups were dropped and only those that successfully differentiated between the new criterion group and the normal group were retained.

Almost immediately after the MMPI was published, it became apparent that many patients did not obtain a clearly high score on only a single clinical scale, thereby rendering diagnosis simple. Rather, more often, more than one scale was elevated, even if one followed the usual procedure of considering a scale with a T-score of 70 or more (two standard deviations above the mean for the standardization group) to be indicative. Hathaway (1947) proposed a coding system for the MMPI clinical profile to account for multiple elevations. The code typology proposed a few years later by Hathaway and Meehl (1951) set a standard for interpretive strategy that is still widely followed.

The most common procedure is to base MMPI interpretation on two-point code types (i.e., 1-3, 4-9, 8-6, etc.), where the numbers refer to the scales. Several perennially popular volumes contain descriptions of the behaviors and personality traits associated with the various code types (Marks & Seeman, 1963; Marks et al., 1974; Gilberstadt & Duker, 1965). Less often used but equally valid are more modest presentations by Duckworth (1979), Graham (1972), and Greene (1980). These "cookbooks" are all based partly on empirical data and partly on the authors' clinical experience.

The intent of the MMPI appeared to be to facilitate diagnosis; however, the word "personality" in the inventory's title has been taken literally for decades. Indeed, there has been more research on personality using the MMPI than psychopathology per se.

Early on, Black (1953) found that various personality characteristics were associated with each scale using a sociometric technique involving college students. Very recently, deMendonca et al. (1984) analyzed the content of ten books on the MMPI published between 1960 and 1977. They found that certain key descriptor terms were commonly associated with a high score on the various scales:

Scale 1	Immature, self-centered, complaining, demanding
Scale 2	Pessimistic, withdrawn, slow, timid, shy
Scale 3	Immature, egotistical, suggestible, friendly
Scale 4	Rebellious, resentful, impulsive, energetic, irresponsible
Scale 5 (male)	Fussy, idealistic, submissive, sensitive, effeminate
Scale 5 (female)	Aggressive, dominant, masculine
Scale 6	Suspicious, hostile, rigid, distrustful
Scale 7	Worrying, anxious, dissatisfied, sensitive, rigid
Scale 8	Confused, imaginative, individualistic, impulsive, unconventional
Scale 9	Energetic, enthusiastic, active, sociable, impulsive
Scale 0	Aloof, sensitive, inhibited, timid

Research indicates that all three of the validity scales are more complex than originally intended and probably do not have a great deal to do with actual validity. On the other hand, they have been used in personality assessment. DeMendonca et al. (1984) label high scorers on the L Scale as conventional, rigid, self-controllers; on the F Scale, as restless, changeable, dissatisfied, and opinionated; on the K Scale, as defensive and inhibited.

The inventory was intended to be used with adults. Recent research suggests that a seventh-grade reading level is required to respond validly on all scales (Ward & Ward, 1980). The inventory has been used fairly widely with adolescents and separate norms for this age group are available (Marks et al., 1974).

In many agencies, the MMPI is still scored by hand but there are a number of different scoring services available including at least eight commercial programs that print out a narrative personality assessment from the answer sheet.

Practical Applications/Uses

There is general agreement among clinicians, supported by every survey of test usage, that the MMPI is and has been the most widely used verbal inventory of all time. Though it was designed merely to provide psychiatric diagnoses, its applications have ranged far beyond this limited role. A plurality of administrations are doubtlessly done in psychiatric hospitals, mental health clinics, college counseling centers, and similar agencies. In those settings, the MMPI is usually part of a test battery that yields a report on patient symptomatology, diagnosis, and prognosis, but the MMPI has also been widely employed in personality research. Among the major clinical groups that have been studied are alcoholics and drug addicts, delinquents, criminals, the brain-damaged, suicides, and medical patients of all kinds. Personality studies have ranged from those afflicted with acne, asthma, and aphasia, through cancer, dermatitis, homesickness, multiple sclerosis, stuttering, vitamin deficiency, and underachievers, to such groups as salesmen, teachers, student nurses, clergymen, firemen, policemen, and ethnic and religious groups. It would not be far amiss to speculate that the MMPI has been used in some fashion in almost every area of modern human endeavor.

Psychologists are the primary users of the MMPI, but most psychiatrists, psychiatric social workers, and psychiatric nurses are acquainted with it and its utility.

Technical Aspects

Measuring the validity and reliability of the MMPI is complicated by the fact that the inventory is comprised of many different scales each with its own validity and reliability. Nevertheless, the inventory as a whole has been found to be reasonably effective for its original task of diagnosing problem behavior and emotions.

Only three reliability studies are reported in the MMPI manual (Hathaway & McKinley, 1951). However, many studies are reported in *The Eighth Mental Measurements Yearbook* (Buros, 1978). These show that there is considerable evidence for the MMPI's reliability depending on the group measured. Test-retest reliabilities reported in the manual range from the .50s to the low .90s. Certain scales reflecting mood, such as Scale 2, are quite variable over time, whereas other scales supposedly more "characterological," such as Scale 1, have much higher test-retest reliability. Some split-half reliabilities are especially low, which is not surprising in view of the heterogeneity of some of the scales.

The validity of the MMPI varies with the population examined and the questions to be answered. The inventory has been the subject of thousands of studies (Taulbee et al., 1977; Buros, 1978) and seemingly works best with diagnosing those who are severely disturbed and are demographically most like the original Minnesota normative sample (i.e., white and middle-class). The inventory seemingly is less valid with groups divergent from this population, such as those from different races or cultures.

The normative data on which MMPI clinical interpretation is based are more than 40 years old, a criticism that has been heard not infrequently. Very recently, a new set of norms has been published (Colligan et al., 1983). These data were collected by telephone interviews from a large Midwestern sample that resembled the original standardization group demographically. In general, the findings are that the contemporary sample scores on the average from 3 to 7 T-score points above the original mean scores on the clinical scales. It remains to be seen whether the Colligan et al. (1983) data will be trusted, a circumstance that should lead to some modifications in currently used MMPI forms.

Critique

Despite its enormous popularity, the MMPI has hardly been immune to criticism. An excellent summary is provided by Faschingbauer (1979):

> Its scales are heterogenous, redundant, and overlapping. At times, subsets of items that are known to be inversely correlated are summed in what can only be a cancellation approach to a scale score. Over 100 items are not even scored . . . . Additionally, the MMPI scales were based on Kraepelinian nosology, which is considered by many MMPI users to be somewhat anachronistic. Furthermore, Kraepelin's system was used to develop clinical scales that differentiate "normals" from groups with known types of psychopathology. Yet users of the MMPI are usually asked to differentiate among diagnoses, something the MMPI scales were never developed to do . . . . From a theoretical perspective, traits seem to be mixed with states in the MMPI code type in a haphazard manner . . . . The MMPI has mixed bipolar and unipolar traits in an apparently thoughtless manner . . . . Some MMPI items are anachonisms of the 1930s, using words with which many patients are unfamiliar . . . other items are worded so complexly . . . that they are often answered in error . . . . How can we . . . justify the common use of norms developed on white, rural, married persons with an elementary education in the 1930s? (pp. 374-375)

Faschingbauer (1979) notes that despite this impressive roster of defects,

. . . the MMPI has been, and continues to be, useful. At present, the MMPI is the most widely used objective personality test in the world. Its major use is in the screening, diagnosis, and clinical description of psychopathology. Rarely a week passes when I and my colleagues are not awed by the power of the MMPI in this area. I believe that most other clinical psychologists and psychiatrists also consider the MMPI superior to other tests for this purpose. But the usefulness of the MMPI has expanded beyond the original intentions of its developers. Nearly 500 scales provide for a variety of screening, assessment, selection and prediction applications . . . . the research utility of the MMPI seems to equal or even surpass its clinical worth. (pp. 375-376)

A final accolade to this ageless instrument is found in a critique by King (1978): ". . . The MMPI remains matchless as the objective instrument for the assessment of psychopathology . . . . [It] still holds the place as the *sine qua non* in the psychologist's armamentarium of psychometric aids."[1]

Notes to the *Compendium:*

[1]Presumably in response to methodological criticisms, the MMPI currently is undergoing formal revision, despite its manifest success over the years. The University of Minnesota Press has not yet announced a publication date for the revision, nor has it issued any statements concerning the nature and scope of the item alterations, eliminations, and/or additions.

References

Black, J.D. (1953). *Adjectives associated with the various MMPI codes.* Unpublished doctoral dissertation, University of Minnesota, Minneapolis.

Buros, O.K. (Ed.). (1978). *The eighth mental measurements yearbook.* Highland Park, NJ: The Gryphon Press.

Clopton, J.R., & Neuringer, C. (1977). MMPI Cannot Say scores: Normative data and degree of profile distortion. *Journal of Personality Assessment, 41,* 511-513.

Colligan, R.C., Osborne, D., Swenson, W.M., & Offord, K.P. (1983). *The MMPI: A contemporary normative study.* New York: Praeger Publishers.

Dahlstrom, W.G., Welsh, G.S., & Dahlstrom, L.E. (1975). *An MMPI handbook: Volume II. Clinical interpretation* (rev.). Minneapolis: University of Minnesota Press.

deMendonca, M., Elliott, L., Goldstein, M., McNeill, J., Rodriguez, R., & Zelkind, I. (1984). An MMPI-based behavior descriptor/personality trait list. *Journal of Personality Assessment, 48,* 483-485.

Drake, L.E. (2946). A social I.E. scale for the MMPI. *Journal of Applied Psychology, 30,* 51-54.

Duckworth, J.C. (1979). *MMPI interpretation manual for counselors and clinicians.* Muncie, IN: Accelerated Development, Inc.

Faschingbauer, T.R. (1979). The future of the MMPI. In C.S. Newmark (Ed.), *MMPI: Clinical and research trends.* New York: Praeger Publishers.

Gilberstadt, H., & Duker, J. (1965). *A handbook for clinical and actuarial MMPI interpretation.* Philadelphia: W.B. Saunders Co.

Graham, J.R. (1972). *The MMPI: A practical guide.* New York: Oxford University Press.

Greene, R.L. (1980). *The MMPI: An interpretive manual.* New York: Grune & Stratton.

Harris, R.E., & Lingoes, J.C. (1968). *Subscales for the Minnesota Multiphasic Personality Inventory: An aid to profile interpretation* (rev. version). Berkeley: University of California, Department of Psychiatry.

Hathaway, S.R. (1947). A coding system for MMPI profile classification. *Journal of Consulting Psychology, 11,* 334-337.

Hathaway, S.R., & McKinley, J.C. (1951). *Manual for the Minnesota Multiphasic Personality Inventory* (rev.). New York: The Psychological Corporation.

Hathaway, S.R., & Meehl, P.E. (1951). The Minnesota Multiphasic Personality Inventory. In *Military clinical psychology* (Department of the Army Technical Manual), TM 8:242.

King, G.D. (1978). Review of the MMPI. In O.K. Buros (Ed.), *The eighth mental measurements yearbook* (pp. 935-938). Highland Park, NJ: The Gryphon Press.

Marks, P.A., & Seeman, W. (1963). *The actuarial description of personality: An atlas for use with the MMPI.* Baltimore: Williams & Wilkins.

McKinley, J.C., Hathaway, S.R., & Meehl, P.E. (1948). The MMPI:VI. The K Scale. *Journal of Consulting Psychology, 12,* 20-31.

Newark, C.S., & Faschingbauer, T.R. (1978). Bibliography of short forms of the MMPI. *Journal of Personality Assessment, 42,* 496-502.

Taulbee, E.S., Wright, H.W., & Stenmark, D.E. (1977). *The Minnesota Multiphasic Personality Inventory (MMPI): A comprehensive annotated bibliography (1940-1965).* Troy, NY: Whitson Publishing Co.

Ward, L.C., & Ward, J.W. (1980). MMPI readability reconsidered. *Journal of Personality Assessment, 44,* 387-389.

Wiggins, J.S. (1966). Substantive dimensions of self-report in the MMPI item pool. *Psychological Monographs, 80*(22, Whole No. 630).

Donald I. Templer, Ph.D.

Professor of Psychology, California School of Professional Psychology, Fresno, California.

MULTIPLE AFFECT ADJECTIVE CHECK LIST— REVISED

Marvin Zuckerman and Bernard Lubin. San Diego, California: Educational and Industrial Testing Service.

THE ORIGINAL OF THIS REVIEW WAS PUBLISHED IN TEST CRITIQUES: VOLUME IV (1986).

Introduction

The Multiple Affect Adjective Check List—Revised (MAACL-R) is a self-report instrument with scales of Anxiety, Depression, Hostility, Positive Affect, and Sensation Seeking. Each item, which subjects indicate as applying to them, is based on adjectives concerning one of these five dimensions.

The MAACL-R reflects three stages of development over a quarter of a century, with the first, the Affect Adjective Check List (Zuckerman, 1960), constructed as a measure of anxiety. Using it as a model, Depression and Anxiety Scales were added to expand the instrument to the Multiple Affect Adjective Check List (MAACL; Zuckerman & Lubin, 1965). Although the MAACL has been widely used and cited in the last two decades, it had psychometric weaknesses that were fortunately recognized by its authors. The three scales correlated too highly with each other to infer that the separate scales have good discriminant validity. In general, the correlations among the three scales were of about the same magnitude as the reliability of the individual scales. Additionally, the correlations with measures of response set, especially social desirability, were higher than desirable. The MAACL-R development was intended to overcome these limitations, and its mission was a success.

In the development of the MAACL-R, the 132-item MAACL was administered to several groups of normal subjects and factor analyzed. The items retained for the five scales were those that consistently loaded .30 or higher on associated factors: Anxiety, Depression, Hostility, Positive Affect, and Sensation Seeking (Zuckerman, Lubin, & Rinck, 1983; Zuckerman & Lubin, 1985).

In addition to the scores on the five scales, one may add the Anxiety, Depression, and Hostility scores to obtain a Dysphoria score. One may also add the Positive Affect and Sensation Seeking scores.

The authors, Martin Zuckerman and Bernard Lubin, are both Professors of Psychology, Diplomates in clinical psychology, and Fellows of the American Psychological Association. Both have substantial scholarly contributions in personality assessment and theory, as well as in other areas of psychology.

The MAACL-R consists of 132 adjectives that are alphabetically arranged in three columns on one side of a single sheet of paper. There are two forms: the State Form, requiring subjects to answer each item according to how they feel today, and the Trait Form, requiring them to answer according to how they generally feel. The items consist of commonly used adjectives.

The MAACL-R is untimed and requires approximately five minutes to complete. The examiner's task is to read the test instructions to the subjects, but clarification is permitted if the subject has questions. All words are at or below the eighth-grade level, but the manual cautions that an eighth-grade education does not necessarily imply an eighth-grade reading level.

The scale scores can be obtained by adding the number of adjectives on the five scales. The exception to this rule is that for Sensation Seeking four items are scored positively if they are not endorsed. The MAACL-R answers can be machine scored by Educational and Industrial Testing Service or may be locally scored by prior agreement with this testing service. Because the MAACL-R contains all 132 items of its predecessor rather than the 70 items needed to score the revised instrument, one has the option of using the three MAACL scales in addition to or instead of the MAACL-R.

Practical Applications/Uses

The MAACL-R appears to be quite useful for research involving affect and/or subjective state. This reviewer has been especially impressed by the sensitivity to change over time of the MAACL. In one study (Velebar & Templer, 1984) there were significant increases in anxiety, depression, and hostility one hour after double blind administration of the MAACL. There is no reason to believe that the revised form will prove less sensitive to change over time. In fact, the MAACL-R reflects anxiety increase over time prior to a classroom examination (Zuckerman, Lubin, & Rinck, 1983).

In this reviewer's opinion, the MAACL-R Depression Scale is often more suitable for a normal population than the standard scales of depression because most depression scales tap the classical elements of the *depressive syndrome* such as sleep difficulty, weight loss, and decreased libido, whereas the MAACL assesses *mood*. Nevertheless, research (e.g., Zuckerman, Lubin, & Rinck, 1983; Zuckerman & Lubin, 1985) has indicated that the MAACL-R can discriminate patients with various psychiatric diagnoses in the predictable direction.

The revised instrument now contains indices of both positive and negative affect and in that respect has a common element with the Affects Balance Scale (Derogatis, 1975). That instrument contains an "Affect Balance Index," which is essentially a ratio of positive to negative affect. A suggested additional score was called "total affective change" (Templer, 1985), which is essentially a sum of positive and negative affect scores and conceptualized as bearing some resemblance to the sum of color responses on the Rorschach and to the sum of the Depression Scale and Hypomania Scales T-scores of the MMPI. This reviewer is here suggesting a comparable index for the MAACL-R resulting from the summation of the Anxiety, Depression, Hostility, Positive Affect, and Sensation Seeking T-scores. Perhaps a label such as "composite affective experience" would be appropriate.

Technical Aspects

The internal consistency is good with both the state and trait forms of the MAACL-R. The stability over time of the trait form is, generally speaking, rather good. The test-retest coefficients for the state form are considerably lower. However, the lower state stability coefficients probably reflect sensitivity to changes over time rather than greater measurement error with the form. Response set does not appear to be a problem with the MAACL-R. Reasonable normative information is provided.

The construct validity of the MAACL-R is both broadly based and impressive. The MAACL-R scales correlate in the predicted direction with other measures, including the MMPI, the Profile of Mood States (McNair, Torr, & Droppelman, 1971), self-ratings, peer ratings, observer ratings, psychiatric diagnoses, self-reported health, and self-reported social activities and symptoms. Furthermore, the pattern of correlations support the differential validity of the individual MAACL-R scales. For example, the highest positive correlations of the MMPI Depression Scale tend to be with the MAACL-R Depression and Dysphoria (Anxiety + Depression + Hostility) Scales. The highest negative correlations are with the MAACL-R Positive Affect and the sum of Positive Affect and Sensation Seeking.

Critique

The MAACL-R is an instrument that is brief and easy to administer, contains both state and trait measures, assesses five dimensions of affect, and is sensitive to change over time. It has good reliability, relative independence of response set, and commendable construct validity. It has a wide range of research applications with normal and abnormal populations, perhaps especially in assessing change in affect over time.

The MAACL-R is definitely superior to the MAACL as a psychometric instrument. Some persons may prefer the older instrument because of the greater accumulation of usage and literature by which it is buttressed. Furthermore, the correlations between the old and new Anxiety, Depression, and Hostility Scales are acknowledged as low to moderate in the manual, which prudently cautions that "users of the new scales must be wary about assuming that past results with the old scales will apply to the new scales bearing the same names" (Zuckerman & Lubin, 1985, p. 10). On the other hand, because all 132 items are answered with both the MAACL and MAACL-R, the user can choose to use one or both.

References

Derogatis, L. R. (1975). *Affects Balance Scale*. Towson, MD: Clinical Psychometric Research.

McNair, D. M., Lorr, M., & Droppelman, L. F. (1971). *Profile of Mood States: Manual*. San Diego: Educational and Industrial Testing Service.

Templer, D. I. (1985). Review of Affects Balance Scale. In D. J. Keyser & R. C. Sweetland (Eds.), *Test critiques* (Vol. 2, pp. 32-34). Kansas City, MO: Test Corporation of America.

Veleber, D. M., & Templer, D. I. (1984). Effects of caffeine on anxiety and depression. *Journal of Abnormal Psychology, 93*(1), 120-122.

Zuckerman, M. (1960). The development of an affect adjective check list for the measurement of anxiety. *Journal of Consulting Psychology, 24,* 457-462.

Zuckerman, M., & Lubin, B. (1965). *Manual for the Multiple Affect Adjective Check List.* San Diego: Educational and Industrial Testing Service.

Zuckerman, M., Lubin, B., & Rinck, C. M. (1983). Construction of new scales for the Multiple Affect Adjective Check List. *Journal of Behavioral Assessment, 5,* 119-129.

Zuckerman, M., & Lubin, B. (1985). *Manual for the Multiple Affect Adjective Check List-Revised.* San Diego: Educational and Industrial Testing Service.

Carl G. Willis, Ed.D.
Counseling Psychologist, University of Missouri, Columbia, Missouri.

MYERS-BRIGGS TYPE INDICATOR

Isabel Briggs Myers and Katharine C. Briggs. Palo Alto, California: Consulting Psychologists Press, Inc.

THE ORIGINAL OF THIS REVIEW WAS PUBLISHED IN TEST CRITIQUES: VOLUME I (1984).

Introduction

The Myers-Briggs Type Indicator (MBTI) is a forced-choice, self-report inventory that attempts to classify individuals according to an adaptation of Carl Jung's theory of conscious psychological type. In essence, there is the assumption that human behavior, perceived as random and diverse, is actually quite orderly and consistent. Myers' (1962) view supposes that the observed variability is due to "certain basic differences in the way people prefer to use perception and judgment" (p. 1). The MBTI should be regarded "as affording hypotheses for further testing and verification rather than infallible expectations of all behaviors" (p. 77).

The MBTI classifies individuals along four theoretically independent dimensions. Each dimension has two dichotomous (polar) preferences with only one preference from each categorization ascribed to any one individual. The first dimension is a general attitude toward the world, either extraverted (E), in which the personal direction is actively outward to other persons and objects, or introverted (I), where the attention and energy are directed inward to focus on internal, often unspoken, representations of events. As such, the E-I dimension does not purport to measure shyness versus gregariousness, a common misconception. The second dimension, perception (how persons prefer to orient to data from their own environments), describes a function and is divided between sensation (S) and intuition (N). Sensing refers to attending to actual sensory realities and, cognitively, to facts and details. Intuition, on the other hand, is more global, even unconscious, focusing on insight and possibilities within the data a person receives. The third dimension, also a function, is that of judging. Once information is received, it is processed in either a thinking (T) or feeling (F) style. Thinking refers to a reliance on reasoning and logic in decision making. Conversely, feeling means that perceptions are compared on a value basis. In the feeling style, decisions and more personal and subjective, and are particularly attuned to how the person relates to others.

The final dimension, judging (J) versus perceiving (P), is one proposed by Myers (1962) and it serves a dual purpose in the interpretation of the test. Again, there is a distinction between two preferences. The judging attitude focuses on a

This reviewer wishes to acknowledge and thank Tom L. Ham, research associate at the University of Missouri-Columbia.

willingness to make prompt decisions, come to conclusions, and thereby excludes concurrent use of the perceiving function. Conversely, a preference for the perception (P) attitude means holding off on deciding while gathering more information and simultaneously excluding the judging function. People are seen as needing both of these attitudes, but being more comfortable in one than the other. The primary purpose of the J-P dimension, however, is to determine which one of the individual's two function preferences is dominant and which is auxiliary (i.e., S or N versus T or F).

The dominant process is the one on which the person relies the most. Theoretically, the dominant process has ascendency over the auxiliary process. The auxiliary process becomes utilized exclusively only under certain situations. These circumstances occur mainly when the extravert must react in an introverted fashion and when the introvert must respond in an extraverted fashion. The auxiliary process is not as developed as the dominant process, but is necessary to provide balance between one's introverted and extraverted attitudes. Without a developed auxiliary process, the extravert may seem to lack depth and appear superficial. The introvert who lacks an adequate auxiliary process may find dealing with the outer world uncomfortable and unpredictable. Myers (1962) has suggested that in our Western society the introvert is the one most penalized for a lack of development in the auxiliary process.

The identification of the dominant function is easily ascertained for extraverts. If extraversion is the preferred attitude, a judging (J) preference indicates that this person relies predominantly on either thinking or feeling. Thus, the auxiliary, either sensing or intuition, becomes the preferred perceptual function. Conversely, when the extravert has a perceiving (P) preference, either the sensing or intuiting process will be dominant and the thinking or feeling will be auxiliary. For example, an ESFJ type will have feeling (F) as the dominant process, dictated by the J. Thus, S (sensing) would be the auxiliary process. Another example would be an ENFP, where the P determines that the intuitive (N) process is dominant and feeling (F) would become the auxiliary.

The identification of dominant and auxiliary processes for the introvert is more complex, though still logical. The J versus P preference always indicates a person's reaction to the external world. This being the case, the introvert's J or P score points to the auxiliary process, used when assuming an extraverted role. True to the introvert's nature, the most relied upon process (i.e., the dominant one) is used on the internal world. Thus, for an INTP, the P points to intuiting (N) as the auxiliary, and thinking (T) becomes the dominant, internally focused process. Similarly, an ISFJ has feeling (F) as the auxiliary, via the judging (J), and sensing (S) is dominant.

Combinations of the four preferences exhibited by each person determine the 16 possible types. Each type defines a unique set of characteristics and tendencies in behavior. A type table was developed to visually explain the relationships among the four dimensions of each type and to aid in interpretation of the types. No one preference or type is thought of as being qualitatively superior to another. Each simply reflects differences in attitude, orientation, decision-making, and the relative importance of these functions in a person's life. Each preference and type has its strengths and implied weaknesses, though the positive perspective is encouraged throughout the manual (Myers, 1962).

Being theory based, the MBTI was, with several revisions, developed to provide a theoretical map that would order the diverse array of human behavior. To the degree in which the instrument can assess each individual's personal map, much understanding of behavior can be garnered inferentially. Although prediction of future behavior is one obvious purpose, modifying or strengthening one's behavior or coping abilities—remediation and/or development—would seem the most practical goals. In another sense, each examinee has a chance to gain more self-acceptance, other-acceptance, and an appreciation of differences in human behavior. Perhaps the moral of the MBTI story is that differences, once recognized, can be appreciated instead of scorned and complementarity in type can lead to strength rather than conflict.

The historical development of the MBTI is somewhat different than that encountered for most tests; there were two different lines of development. Carl Jung, the Swiss psychologist, published his book entitled *Psychological Types* in 1921. In this book Jung presented a theory suggesting that much apparently random behavior was actually consistent and orderly, being caused by different manners of expressing perception and judgment in human beings. He can be credited with helping the words "extravert" and "introvert" become household terms. In a sense, Jung puts the inner world of ideas and understanding on an equal basis with the external world of people and things. At this juncture between extraverting and introverting, the ego is poised to relate in either direction.

Although it was Carl Jung who developed the theory that categorized behavior into types, in the United States Katharine Briggs was at the same time raising the question, "Why are people different?" This question apparently sprang from her observations around the time of World War I as she became interested in differences and similarities of human personality. Hence, she made an extensive study of biographies. A second impetus for her pursuit of understanding human behavior came from her daughter Isabel's interest in a man whose family appeared very different from the Briggs family. After Jung's work was published, Briggs found that her thinking paralleled those types theorized by Jung. She accepted the more complete system espoused by Jung and taught it to her daughter.

Isabel Briggs and her mother continued their interest in personality assessment. During the World War II period, in response to a perceived paucity of instrumentation and to a greater respect for the power of Jung's theory in explaining and predicting human behavior, they began developing the original item pool for the MBTI. The items were based on type theory and observations, and were informally validated on friends and relatives whose type preferences seemed clear. Thus, Form A and Form B (slightly rearranged Form A) were created.

Sometime in the mid-1950s, Isabel Briggs Myers approached Educational Testing Service and gained their support and research assistance. During the late 1950s many new items, including word pairs, were tried out to develop Form D. More refinement followed until Forms E and F, identical except that F has unscored experimental items, became available. Form F, with 166 items, seemed to have some ambiguities for the user, thus a shortened Form G, without 38 experimental and two dropped items, and with nine slightly reworded items, was devised in 1975. Since the best items, statistically speaking, are the first 50 on Form G, a quick estimate of type can be obtained from these or from Form AV (Abbreviated Version), which comprises the same 50 items printed as a separate form.[1]

In 1975, the nonprofit Center for Applications of Psychological Type (CAPT) was created in Gainesville, Florida, to offer education, research, and other services to MBTI users. Extensive bibliographies, computer scoring, research assistance, continuing education, and publication of related materials are supported by CAPT.

Although there are no official non-English versions, reports in the literature indicate that the MBTI has been translated and used with Japanese and Spanish-language populations. Other translations may also have been done.

Practical Applications/Uses

The development of the MBTI is framed both from a theoretical (Jungian) perspective and from an empirical or statistical perspective. The vast repository of research reports and references describe an almost infinite array of applications. A review of the bibliography from CAPT and from other data bases provides the description of breadth of use. One or more references are included in at least 23 business journals, 30 education journals, 24 medically related journals, 37 psychology journals, 8 science journals, 3 religious journals, and 14 other specific professional journals.

Of the 14 other journals, the fields of journalism, sociology, law, pharmacy, and communications are represented. A large number of in-house reports are included in over 800 formal references by approximately 800 authors or co-authors. Over 40 theses and 260 dissertations have been completed using the MBTI. At least ten foreign journals are referenced. There are over 15 books, or chapters in books, which detail the theory or the instrument. In addition, the *Journal of Psychological Type,* now in its eighth year of publication, presents articles on the MBTI or on Jung's theory.

When considering all that has been written, the findings can be described in two categories: as an IN type (internally focused on possibilities and reliance on insight to understand) of basic or pure research, or as an ES type (externally focused on realities with a very practical observable perspective) of applied research. The manual suggests uses in educational, vocational, marriage, and personal counseling; in selection and placement in business and industry; in contrivance of optimal educational, training, and work settings; and in research and the advancement of knowledge. Although the MBTI manual is directed toward uses discerning differences among normal people, it ". . . may be found useful in psychotherapy . . ." (Myers, 1962, p. 78) as a means of normalizing individual differences. Conflict resolution can be accommodated in couples or other human relations settings when based on type differences. Teaching, learning, and communication styles can be analyzed. Specific types have been found to cluster within creative, scientific, business, and social occupations. The MBTI can be used in counseling with individuals, couples, families, and groups. Team building, leadership training, personnel selection, and teaching are examples of the many uses of the MBTI in work settings. Myers (McCaulley, 1981) believed that the application of type theory in a child's early years would give that child a chance to develop "individual gifts" and to learn to perform adequately in the less preferred functions; thus, developmental concepts and applications are suggested.[2]

From the viewpoint of a potential user, there are some strengths in the instrument. An examinee may share in the evaluation of the suggestions from the instrument since the descriptions are of normal behaviors with minimal value connotations. With the assistance of a trained interpreter, all type descriptions can be viewed as positive. There is some concern, however, that an examinee can accept in a too literal sense the description of type as a command for action rather than as another mode for self-understanding. In these cases, behavior may be viewed as static rather than developmental.

The three forms (F, G, AV) of the MBTI may be used with most populations. Although considerable discussion in the manual details use with bright junior high school students and older individuals, there may be some considerations based on the reading and concept level of actual items (more so in Form F). Average or below average students may be unable to respond appropriately. However, students are directed to omit items that they do not understand.

No general adult sample distribution is available; high school or college distributions are the comparison groups. Generally, the information available is based on school studies or on bright, career-oriented professionals. The Jungian theory would hypothesize that all human beings can be described in types, but little empirical information is available on minorities or on blue-collar workers.

All three forms (F, G, AV) of the MBTI are virtually self-administering without specific time limits. The printed directions seem satisfactory and attempt to negate the social desirability of the questions. All forms can be administered individually or in groups. Most of the items represent a forced choice between two responses. There are differential weightings for certain responses in order to offset social desirability bias.

The MBTI scoring is designed to yield two types of scores, preference and continuous. For Forms F and G, a numerical score is obtained on each preference within the four dichotomous dimensions (i.e., E versus I, N versus S, T versus F, and J versus P). The numerical scores on each of the pairs of preferences are compared and the smaller total is subtracted from the larger. The difference is converted into a preference score based on the larger numerical value. For example, a female with a T score of 21 and an F score of 4 would be typed as a T with a value (see MBTI manual for conversion) of 33. Preference scores range from 1 to an upper range of 49-67, depending on the particular preference.

The scores on the preferences within each dimension can be converted into four continuous scores where the value of 100 is the dividing point between the preferences. For E, S, T, and J, the preference scores are subtracted from 100, while I, N, F, and P preferences are added to 100. For example, a score of 40 on the E-I dimension indicates a strong preference for extraversion, converting to a continuous score of 60 (i.e., 100 − 40), while a continuous score of 140 indicates a strong preference for introversion (i.e., 100 + 40). There is a very strong statement in the manual that the only appropriate use of continuous scores is in research.

Forms F and G may be processed by computer, either through purchase of a scoring program or a commercial scoring service such as that provided by CAPT. Templates for hand scoring are also available. For self-scoring Form AV, the process is slightly different. All items are marked on a top sheet of NCR forms with the responses and scoring directions on the second sheet.

Technical Aspects

Assessment of reliability data on the MBTI draws on a body of findings largely based on post-elementary and college populations. Comparison of these findings, however, is complicated by the two data types (preference and continuous) obtained from the MBTI, the variety of statistical procedures employed, and the existence of three forms in current use (i.e., F, G, and AV).

Most reliability reports utilized Form F. Internal consistency on preference scores has been estimated by phi coefficients and tetrachoric coefficients, both with application of the Spearman-Brown prophecy formula. The phi coefficient estimation is considered to be conservative, while the tetrachoric coefficient is likely an overestimate due to its assumption of normality when applied to relatively non-normal MBTI data. Phi coefficient estimates range from .55 to .65 (E-I), .64 to .73 (S-N), .43 to .75 (T-F), and .58 to .84 (J-P). The tetrachoric coefficients range from .70 to .81 (E-I), .82 to .92 (S-N), .66 to .90 (T-F), and .76 to .84 (J-P). These reliabilities are similar to other self-report inventories, though the discrepancy between the two types of estimates are obvious (Carlyn, 1977).

Conversion of data into continuous scores yields more consistent estimates. Data obtained via two different procedures (Myers, 1962; Stricker & Ross, 1963; Webb, 1964) produced estimates of .76 to .82 (E-I), .75 to .87 (S-N), .69 to .86 (T-F), and .80 to .84 (J-P). The estimates of continuous scores retain data precision lost in the use of type-category scores, which accounts for the difference in reliabilities obtained from the two data types (Carlyn, 1977).

Several trends in these correlations are noteworthy. The T-F scale exhibits the least reliability and the S-N, generally, the most. Recent findings also show increasing reliabilities with populations of increased age and intelligence (McCaulley, 1981).

Stability in type-category scores has been reported using test-retest intervals of up to six years (McCaulley, 1981). The proportion of reclassification into the same preference ranged from 62% to 83% for E-I, 57% to 89% for S-N, 61% to 90% for T-F, and 66% to 90% for J-P. Complete agreement on all four scales on retest range from 20% to 61%; 70-88% had at least three preferences in common on retest; 10-22% had two preferences in common; 2% to 7% maintained one preference; and less than .1% changed on all four scales. Further investigation has demonstrated that test-retest reliability is related to the degree of strength on the initial preference score (Howes & Carskadon, 1979).

Stability in continuous scores is represented by fewer studies than type-category scores, probably reflecting the emphasis on type as opposed to individual scales. Nonetheless, with test intervals from five weeks to 21 months, reliability coefficients range from .73 to .83 for E-I, .69 to .87 for S-N, .56 to .82 for T-F, and .60 to .87 for J-P.

A relative paucity of information currently exists on reliability for continuous data for Forms G and AV. Two studies dealing only with a small N of college students over five- and seven-week intervals are reported for form G (Carskadon, 1982). Results are comparable to those for Form F. Nonetheless, a significant difference between males and females on T-F is reported in both studies, albeit in contradictory directions. Although minimal data are available on Form AV, the test publisher has provided some comparative information with accuracy of the type scores obtained. When the AV portion of Form G was scored compared to the total Form G on a sample of 11,615 cases, approximately 27% of the type scores on the AV portion were different than those derived from the total Form G on one or more scales. Thus, one must be very cautious in using Form AV since it is less reliable.

Critical to the examination of validity on the MBTI is whether the scales accurately measure Jung's constructs and Myers' extensions thereof. One approach to this construct validation is examination of MBI continuous scores in relation to similar constructs on other tests. The MBTI manual (1962) provides correlational data with the Allport-Vernon-Lindzey Study of Values (AVL), the Gray-Wheelwright Psychological Type Questionnaire (also known as the Jungian Type Survey), The Edwards Personal Preference Schedule (EPPS), the Personality Research Inventory (PRI), the Scholastic Aptitude Test (SAT), and the Strong Vocational Interest Blank (SVIB), among others. The Sixteen Personality Factor Questionnaire (16PF) and the Rokeach Dogmatism Scale have also been correlated with the MBTI. These numerous correlational studies indicate that ". . . a wealth of circumstantial evidence has been gathered and results appear to be quite consistent with Jungian theory" (Carlyn, 1977, p. 469). Examination of data on individual MBTI scales demonstrates the behaviors and attitudes which the MBTI appears to tap, suggesting a strong argument for construct validity.

E-I validity: Extraversion relates to preferences for action, impulsiveness, talkativeness, and gregariousness. Extraverts tend to have underlying needs of dominance, exhibition, and affiliation. Occupational interests reflect these orientations in the extravert's preferences for social service, business, sales, and ministry.

A positive relationship with measures of adjustment also exists for extraversion, whereas introversion relates similarly to maladjustment. Introversion is also associated with several measures of anxiety, though extraversion has been significantly related to at least one measure of denial of anxiety (McCaulley, 1978). Introverts tend to work alone, dislike distraction, and think before acting. Introversion is strongly related to self-sufficiency scales and tend toward occupations that allow some degree of privacy and self-reliance. Occupational examples associated with introversion include scientists, artists, farmers, and architects. The major criticism of the E-I scale is that it may be confused with the popularized conceptions of shyness and withdrawal (I) versus gregariousness (E) (Stricker & Ross, 1964).

S-N validity: Sensing has been related to a preference for facts and tangible stimuli. In particular, it correlated highly with measures of practicality. Sensors are associated with areas such as business, finance, sales, and other careers that focus on attention to concrete details.

Intuition correlates with autonomy, creativity, intelligence, and with aesthetic and theoretical values. Studies relating to creativity show especially strong relationships to intuition. Similar findings exist when comparing scores with membership in occupations such as research scientist, psychologist, writer, and musician. Intuitives seem to prefer the abstract and tolerate ambiguity.

T-F validity: Thinking tends to relate to constructs such as autonomy, order, masculine orientation, and dominance. Thinkers value the theoretical, logical, and objective aspects of situations. They tend to perform well on exams and possess positive values toward work. Research further supports these notions in the thinker attraction to scientific, technical, and business professions. Thinking is also associated with the legal field, politics, and mechanical activities.

Feelings tends to correlate with measures of nurturance and affiliation. Associated values include social and religious orientations. This appears to support the construct of feeling being associated with concern for valuing of others. Another, perhaps related, dimension is the correlation with free-floating anxiety. The suggestion is that anxiety reflects interpersonal sensitivity. Not surprisingly, feeling is associated with occupational preferences in the helping professions, particularly where children are served.

J-P validity: Judging relates to responsibility, dependability, control, and needs for order and endurance. Judging is correlated with higher school grades relative to perceiving, despite the opposite relationship vis-à-vis aptitude. Most preferred occupations for judging types cluster around those involving business and administration.

Perceptive types are associated with impulsiveness, flexibility, and tolerance for complexity. Associated needs are autonomy, exhibition, and change. Relative to judging types, perceiving is more related to procrastination and less emphasis on the value of work. Perceptive types seem most attracted to fields involving verbal skills as well as artistic endeavors.

With regard to predictive validity, much research has focused on career-related topics such as longevity and turnover. The MBTI has been shown to be moderately predictive in these instances, which is impressive considering the variability often encountered in work environments. Generally, supportive evidence also exists for prediction of retention of college students, self-description, learning, and memory tasks. Throughout most of these findings, it has been noted that data utilizing complete types or at least combinations of scales are preferable to single scale considerations.

Critique

The MBTI is a good instrument based on its substantive theoretical and empirical bases. An extensive, but dated, manual (1962) for Form F impresses the user with a wealth of information; however, a supplement (1977) for Form G seems too brief. In defense of Myers, a strong argument has been made that Forms F and G could be used interchangeably since almost no differences exist on the questions that are scored. Therefore, the data base and information on Form F could conceptually be considered as support of the use of Form G. The manual could be improved with a clearer discussion of the dominant and auxiliary processes and their delineation.

Another area of concern in using the MBTI relates to the test-taking attitude or mood of the examinee. When used in screening for a high demand situation, possible distortions of type could occur without any indices to suggest the error(s).

All of these factors—the explosion of research reports, the normality of test items and type descriptions, the positive nature of the instrument, the ease of administration and scoring, the usefulness of the theory, the development of the support organization CAPT, the publishing of a dedicated journal—have shared a role in the wide acceptance of the MBTI.

One final concern in the employment of the MBTI in all of its diverse applications relates to the skill and qualifications of the user—*you*, the reader of this evaluation. Study of the references by Myers, McCaulley, and Lawrence is important to gain a grasp of the theory, application, and interpretation of the MBTI.

Notes to the *Compendium:*

[1]It appears that Form AV has been found lacking in the quality of psychometric properties. Therefore, a new 94-item version of Form G will be released as a self-scorable edition to replace Form AV.

Recent efforts have led to the creation of another instrument, called MBTI Form J. Form J purportedly will serve to (a) create more specific preference assessments both among and *within* types and (b) assist in the resolution of type indeterminacy (i.e., near 0 preference scores). Form J expands the number of items to 290 and is considered a research-only version at present. A preliminary manual using Form J items should soon be available. This Type Differentiation Indicator (TDI) yields 27 factor-analysis-based scores. (Scores for the four traditional dimensions, however, remain readily obtainable due to inclusion of exact item content from Form F.) Scoring of the Form J scales, however, is available only through a mail-in service. The 27 scales are considered bipolar, though not necessarily opposites. Single word descriptors anchor each end of the scales. Profiles are represented within a traditional four-dimension format. Also reported is an estimate of within-subject consistency and Form G preference scores. Limited normative data in the TDI consist of both sexes from single adult and student samples. Use of Form J in the TDI methodology will require much greater psychometric sophistication on behalf of the test interpreter.

[2]In 1985, a new and thorough MBTI manual was released. This manual continues to provide a plethora of information and interpretative guides. An obvious strength of the MBTI now focuses on vocational uses. Almost one fourth of a 300-plus page manual is allotted to occupational and career counseling information. In fact, the MBTI, with over 180 different occupations represented in extensive occupation-by-type tables, compares very favorably with the SCII with 106 occupations (Hansen & Campbell, 1985) and the Kuder DD with 76 occupations (Zytowski, 1985). An *experimental* Creativity Index can also be scored. In addition, the 1985 manual provides a much clearer delineation of the dominant and auxiliary processes.

References

Carlyn, M. (1977). An assessment of the Myers-Briggs Type Indicator. *Journal of Personality Assessment, 41*,461-473.

Carskadon, T. G. (1979). Test-retest reliabilities of continuous scores on Form G of the Myers-Briggs Type Indicator. *Research in Psychological Type, 2,* 83-84.

Carskadon, T. G. (1982). Sex differences in test-retest reliabilities of continuous scores on Form G of the Myers-Briggs Type Indicator. *Research in Psychological Type, 5,* 78-79.

Hansen, J.D., & Campbell, D.P. (1985). *Manual for the SVIB-SCII* (4th ed.). Stanford, CA: Stanford University Press.

Howes, R. J., & Carskadon, T. G. (1979). Test-retest reliabilities of the Myers-Briggs Type Indicator as a function of mood changes. *Research in Psychological Type, 2,* 67-72.

Lawrence, G. (1982). *People types and tiger stripes: A practical guide to learning styles* (2nd ed.). Gainesville, FL: Center for Applications of Psychological Type, Inc.

McCaulley, M. H. (1981). *Jung's theory of psychological types and the Myers-Briggs Type Indicator.* Gainesville, FL: Center for Applications of Psychological Type, Inc.

Mendelsohn, G. A. (1965). Review of the Myers-Briggs Type Indicator. In O. K. Buros (Ed.), *The sixth mental measurements yearbook.* Highland Park, NJ: The Gryphon Press.

Myers, I. B. (1962). *Manual: The Myers-Briggs Type Indicator.* Princeton, NJ: Educational Testing Service.

Myers, I. B. (1977). *Supplementary manual: The Myers-Briggs Type Indicator.* Palo Alto, CA: Consulting Psychologists Press, Inc.

Myers, I.B., & McCaulley, M.H. (1985). *Manual: A guide to the development and use of the Myers-Briggs Type Indicator.* Palo Alto, CA: Consulting Psychologists Press.

Myers, I. B., & Myers, P. B. (1980). *Gifts differing.* Palo Alto, CA: Consulting Psychologists Press, Inc.

Stricker, L. J., & Ross, J. (1963). Intercorrelations and reliability of the Myers-Briggs Type Indicator scales. *Psychological Reports, 12,* 287-293.

Webb, S. C. (1964). An analysis of the scoring system of the Myers-Briggs Type Indicator. *Educational and Psychological Measurement, 24,* 765-781.

Zytowski, D.G. (1985). *Kuder DD manual supplement.* Chicago: Science Research Associates.

Robert E. Williams, Ed.D.

Associate Professor of Educational Psychology, University of Houston—
University Park, Houston, Texas.

Ken R. Vincent, Ed.D

Adjunct Professor of Educational Psychology, University of Houston—
University Park, Houston, Texas.

PEABODY INDIVIDUAL ACHIEVEMENT TEST

Lloyd M. Dunn and Frederick C. Markwardt, Jr. Circle Pines,
Minnesota: American Guidance Service.

THE ORIGINAL OF THIS REVIEW WAS PUBLISHED IN TEST CRITIQUES: VOLUME III
(1985).

Introduction

The Peabody Individual Achievement Test (PIAT; Dunn & Markwardt, 1970) is an individually administered measure of academic achievement that is norm-referenced. The test was designed to provide a wide-range screening measure in five content areas that can be used with students in kindergarten through the twelfth grade. The content areas covered by the PIAT are 1) mathematics, 2) reading recognition, 3) reading comprehension, 4) spelling, and 5) general information. The test materials are presented in two easel-kit volumes that present the stimulus items to the students on one side while the examiner can see both the stimulus items of the students and the examiner's instructions on the reverse side. This easel-kit format facilitates the assessment process and is easy to work with for the examiner and the student.

The authors emphasize that the PIAT was designed as a screening measure and not as a diagnostic test, a point that must be remembered by the user. Therefore, the test is useful in obtaining an overview of scholastic attainment of an individual. The results can then be used as a starting point for further, in-depth diagnostic evaluation of the individual. The test seems to have adequate reliability and validity for these purposes.

The test was developed by Lloyd M. Dunn, Ph.D., Professor of Educational Psychology, University of Hawaii, Honolulu, Hawaii, and Frederick C. Markwardt, Jr., Ph.D., consulting psychologist, Lindbom & Associates, St. Paul, Minnesota. The test was developed between 1962 and 1968 with support from many university colleagues and public school systems. The national standardization was done in 29 different, independent school systems during the 1968-1969 school year.

Included in the PIAT test materials are two volumes of test plates that contain the demonstration and training exercises, test items, and most of the instructions for the administration of the five subtests. Volume 1 includes subtest 1 (Mathematics) and subtest 2 (Reading Recognition); Volume 2 includes subtest 3 (Reading Comprehension), subtest 4 (Spelling), and subtest 5 (General Information). Packets of individual record booklets provide space for recording and scoring responses to each of the five subtests, and each has a profile. The stimulus materials to be read to the subjects for the Spelling subtest are also in the booklet. In addition to the manual, an optional training tape, which is a guide to the accepted pronunciations of the words used in the Reading Recognition and Spelling subtests, is available from the publishers. All of the test materials are contained in the easel-kits that also contain pockets where the manual and the individual record booklets are stored.

The first steps in the item development of the PIAT were 1) an extensive review of the curriculum materials pertinent to the areas being assessed and 2) consultation with curriculum experts in each particular area of study. Three times as many items as were to be used were generated in this initial step. Extensive field testing for major item selection then began and spanned seven years in six different communities in Minnesota, Tennessee, Hawaii, and Saskatchewan, with several minor selection field tests in other locations. Item difficulty and item discrimination were the two criteria used in the selection. Items were placed at grade level where roughly 50% of the students answered correctly and looked at discrimination data to select the most effective items at each grade level.

The Mathematics subtest consists of 84 multiple-choice items with four options for each item and tests skills ranging from matching and recognizing numerals to advanced concepts in geometry and trigonometry. The 84 final items were selected from an initial pool of 300 items and are of the type that can be solved without written notations and placed in a multiple-choice format, thus excluding complex and time-consuming problems.

The Reading Recognition subtest contains 84 items that range from preschool to high-school levels. In the development of the items vocabulary lists of basic reading series were surveyed. An attempt was made to balance across the analytic (look and say) and the synthetic (phonics) approach in the final selection of items. In order to control for regional differences in pronunciation, *Webster's Third New International Dictionary* (1966) was selected as the standard. The optional tape also provides acceptable pronunciations for each of the 66 words.

The Reading Comprehension subtest contains 66 items instead of 84 items contained in all other subtests. The 66 items are numbered 19-84 with item 19 corresponding in difficulty with item 19 in Reading Recognition. Each of the 66 items consists of two pages, with the first containing a passage the subject reads and the second consisting of an illustration that the subject selects as best representing what was read. The complexity of the reading passages were increased by the vocabulary level of the words used, the length of the sentences, and the complexity of the sentence structure. The final selection of the 66 items was based on the field testing of an initial pool of 250 items.

The Spelling subtest contains 84 multiple-choice items ranging from kindergarten to high school. The first 14 items require that subjects discriminate letters that are different from other symbols or the selection of a letter or word associated with a speech sound or word name. In items 15 to 84 words are pronounced and used in a sentence by the examiner, with the subject required to point out the correct spelling from four choices. *Webster's Third New International Dictionary* (1966) was again used as the standard. Words were selected from a cross section of spellers and language textbooks used in the public schools of the United States. The final 84 items were selected by item analysis from an initial pool of 1,000 items.

The General Information subtest, containing 84 questions (items) that are read to subjects and answered orally by them, measure general knowledge ranging across science, social studies, fine arts, and sports. Selected items were appropriate for students in the United States, Canada, and other English-speaking countries, not items that were specific only to the United States. Items were selected and written after a review of curriculum materials in the respective areas, and field tested only in the United States and Canada, but because the norms are based only on United States children, caution must be used when testing subjects other than those from the United States.

Practical Applications/Uses

The PIAT, a screening measure of scholastic attainment or academic achievement that can be used with children from kindergarten through high school, has several advantages as an individually administered test (as opposed to a group test of achievement): it is possible to 1) establish a personal, positive rapport with the child during testing (not possible with group tests) and 2) directly observe the child's test behavior and thus record important clinical and qualitative information about the child during testing. Davenport (1976) reports that it is useful in ascertaining the problem-solving strategies used by a student, as well as the student's attitude towards school and attitudes about particular curricular areas. Because of the test's design it is also useful for children who are physically handicapped and disadvantaged during group testing.

The test can be used with adults by using the norms for subjects aged 18 years. The results should be viewed with caution, however, as this is an extrapolation. Nevertheless, because the PIAT is not a speed test older adults are not apt to be penalized.

The test is an untimed power test with the items arranged in order of difficulty, which permits the examiner to administer only those items that can be completed by the subject successfully. The time for administration and scoring ranges from 30 to 40 minutes, thus it can be given as part of a more extensive battery of tests or given during the usual school period. According to Davenport (1976), the test takes 45-60 minutes for learning disabled students to complete. Because more than half of the items are multiple-choice in format, the subjects need only to make a pointing response. Even this response can be altered by the examiner (e.g., by having physically handicapped students who are paralyzed respond by eye blinks, thus allowing them to be assessed; Dunn & Markwardt, 1970). The test was also designed so that no academic skills other than those being tested are required for any subtest. It also requires no writing, usage of pencils, or handling of the test materials by the subject, which is an advantage in assessing those students who have problems in these areas (Dunn & Markwardt, 1970).

The widest potential use of the PIAT is in the educational setting. It can be administered by teachers in the classroom to ascertain a student's present level of academic functioning, to gain some insight into the student's attitudes toward school and problem-solving behaviors, and to know each student on an individual basis. Used in this way the PIAT can also aid in individualization of instruction and in a guidance and counseling mode. It is also an effective way of assessing the academic level of new admissions to a school and can be used as a gross measure of educational intervention techniques for individual students.

Additional potential uses identified by the authors (Dunn & Markwardt, 1970) include 1) out-patient and psychiatric inpatient diagnosis and treatment in community facilities because these facilities serve many children who are experiencing academic difficulties; 2) a variety of governmental agencies (e.g., vocational rehabilitation programs, adoption agencies and courts) that must understand and serve many clients; 3) industrial settings where personnel workers need to measure academic achievement for job placement and training; and 4) as a research tool in longitudinal studies and in demographic surveys, as well as in cross-cultural studies of academic achievement.

No formal training is necessary to administer, score, and interpret the PIAT accurately. Nonprofessional personnel can train themselves by reading the manual carefully or using the optional training tape available from the publishers. It is important that the person administering the test knows and follows the manual in a precise manner. Practice in administration and the ability to establish an open rapport with the subject is necessary in order to achieve optimal test results. If, however, clinical insights are desired, the test should be administered by the respective professional and not by a nonprofessional. The usual testing environment should be a quiet area with adequate lighting, temperature control, and tables and chairs. The five subtests should be given in the order in which they are numbered.

Administration of the test is facilitated because all of the specific instructions for the subtests—except the Spelling subtest instructions, which are in the individual record booklet—are located in the volumes on the test plates. The instructions are clear, specific, and easy to follow, and the easel-kit format helps examiners during administration because they are able to read the instructions or questions, record the answers, and see the plates on which the subject is responding.

Items for each subtest are in ascending order of difficulty. It is not expected that the subjects will be able to answer all of the questions in any given subtest or on all of the subtests. The subject is started on the Mathematics subtest at a point that the examiner considers to be an appropriate starting point for that subject. This point should not be too easy or too difficult for the subject. On subsequent subtests the starting point, according to the authors, is the item that corresponds to the subject's raw score on the preceding subtest. Wikoff (1979) recommends, however, that the starting point should be eight points lower than that suggested in the manual for each subtest. This lower starting point will facilitate testing attitude instead of backtracking from a starting point that is too high (Wikoff, 1979). It is necessary to establish basal and ceiling scores. The basal rule is five consecutive correct responses, and the ceiling rule is five errors in any seven consecutive responses. If more than one basal or ceiling score is obtained, all answers below the accepted basal are counted correct and all answers above the accepted ceiling are counted as incorrect. A subtest is discontinued after a basal and a ceiling are established.

Recording the subject's answers is easily done in the individual record booklet because of the clarity of the instructions and the format of the record booklet. To obtain the raw scores for each subtest, the examiner subtracts the errors from the number of ceiling items, except on the Reading Recognition subtest when a subject earns a score of 18 or less. In this case the Reading Comprehension subtest is not given, and the raw score obtained for Reading Recognition is repeated as the Reading Comprehension raw score. A total test raw score is then found by summing the raw scores of the five subtests. Four kinds of derived scores have been developed for the PIAT: grade equivalents, age equivalents, percentile ranks, and standard scores. Space is provided on the cover of the individual record booklet to record raw scores, derived scores, and a profile on which the scores can be plotted. Overall, neither scoring nor interpretation is difficult. Due to the clarity and ease of the format for recording the scores and developing the profile nonprofessionals should be able to interpret the scores with accuracy.

Technical Aspects

Students from the mainstream of education in public schools in the continental United States compised the standardization sample for the PIAT. tudents from Alaska and Hawaii were excluded because of the travel and expense involved in testing these students. Except for those who happened to be in the mainstream classrooms that were sampled, special education students were also excluded. These were primarily learning disabled students, but they were not identified. The subjects were from the nine geographic divisions of the U.S. that were used by the U.S. Census Bureau in 1967. The percentage of subjects used in the PIAT sample from any one geographical region was based on the percentage of children in the respective region according to the census data. The distribution of subjects did differ to a small extent from the census data to the extent of a −2.73% in the middle Atlantic states to a +1.16% difference in the Pacific region.

The total sample was composed of 2,559 students—200 from each grade (1-12), and 159 from kindergarten. These students were randomly selected from public school classrooms and balanced according to sex (50% male, 50% female), age, race (84.4% white, 11.3% black, 4.3% other), and socioeconomic level. Dean (1977) did question the underrepresentation of Mexican-American students but did find in his study that the PIAT is as reliable for Mexican-American students as it is for Anglo students. In this sample there were also small differences in the representation of students based on parents' occupational classification, but the extent of the differences is not felt to be significant.

All of the initial testing and retesting teams were trained and supervised by American Guidance Service psychologists. Overall, the standardization sample and the presentation of the information is adequate and professionally presented.

The means, medians, and a standard deviations of the raw scores of the standardization sample are essentially the same, which indicates that the distribution of items over grade levels and chronological age is roughly parallel for all students. Raw scores of the test have been converted into grade equivalent scores, age equivalent scores, percentile ranks, and standard scores; however, care must be taken when one interprets grade and age equivalent and percentile rank scores because they are ordinal data at best. For the standard scores the test has a mean of 100 and standard deviation of 15 points.

Dunn and Markwardt (1970, p. 43) report that "split-half reliability techniques were rejected as likely to result in spuriously high estimates of reliability for a test on which items have been carefully ordered in difficulty and on which the basal and ceiling procedure is utilized." Test-retest reliability coefficients (Pearson product-moment correlations) were calculated based on sample retesting of 50-75 students in Grades K, 1, 3, 5, 8, and 12. There was a month interval between testings, and no attempt was made to insure that the same examiner retested the same students. The resultant reliability coefficients ranged from .42 in kindergarten for Spelling to .94 in the third grade for Reading Recognition. The overall median reliability coefficient was .78. In terms of median coefficient values the greatest confidence in stability is in the total test (.89) and Reading Recognition (.89) and least in Reading Comprehension (.64) and Spelling (.65). The grade-level stability is greatest in Grades 1, 5, and 8, with coefficients of .80, and lowest in Kindergarten, with a coefficient of .74. Test-retest reliability coefficients are reported in the manual for selected grade levels (K, 1, 3, 5, 8, and 12). Wettler and French (1973) compared the PIAT to the Wide Range Achievement Test for 30 outpatient learning disabled children whose achievement was two grade levels below expected level. Overall, the correlations were adequate and were lowest in Arithmetic and Spelling and highest in Reading. It was concluded that both tests may be used for screening purposes.

Standard errors of measurement were calculated for the subtests of the PIAT and are presented in the manual. Caution must be taken when using the standard errors of measurement in looking for differences between subtests because of wide variability and variable interpretation.

Content validity of the PIAT was established through the selection of items described in an earlier section of this paper. Dunn and Markwardt (1970) state that content validity (whether the items reflect the curriculum) is the most important concern in the development of an achievement test. It is believed that the rigorous item selection used in the development of the PIAT supports adequate content validity.

Concurrent validity was calculated by comparing the scores of the PIAT to a measure of scholastic aptitude, the Peabody Picture Vocabulary Test (PPVT), Form A (Dunn, 1965). The resultant product-moment correlation coefficients ranged from a median of .42 in kindergarten to a median of .69 in third grade. The range for the subtest coefficients ranged from .40 in Spelling to a median of .68 in General Information. The overall coefficient for the subtests of the PIAT with the PPVT, Form A, was .57. Correlation coefficients between the PIAT and the PPVT, Form A, are presented in the manual for Grades K, 1, 3, 5, 8, and 12 by subtests.

Critique

The PIAT is designed as a screening test for assessing student achievement in five academic areas. It is very useful for this purpose. The subtest reliabilities are too low for diagnostic assessment and for making important placement decisions, but most are adequate for screening purposes. The overall standardization of the PIAT is superior to most individual achievement tests, but educational personnel who use the PIAT should make certain that the PIAT reflects the present curriculum being used in schools because the items were developed during the 1960s. The items seem pertinent, but care should be taken to make certain that the content validity is still as high as it was during development.

The PIAT is also useful in testing adults in community college and adult basic-education settings. Because of its format, it is especially useful in rehabilitation and medical settings for testing handicapped individuals.

There are several positive aspects to using the PIAT in a variety of settings. The manual is well-written, comprehensive, clear, and readable in presentation, a plus for those administering the exam. This is an important feature because the PIAT can be administered by nonprofessionals with accuracy if the manual is studied and followed. In addition, there is an optional training tape available from the American Guidance Service. PIAT scoring procedures, reference tables, and score recording procedures are also clearly presented in the manual and easily followed. Thus, few errors should be made if the procedures are followed by the examiners because of the overall quality of the manual.

The easy-to-use easel-kit format, which contains the test items, the directions for administration, and correct answers, makes the PIAT easy to administer to students.

References

This list includes text citations as well as suggested additional reading.

Baum, D. (1975). A comparison of the WRAT and the PIAT with learning disabled children. *Educational and Psychological Measurement, 35,* 487-493.

Davenport, B. M. (1976). Comparison of the Peabody Individual Achievement Test, the Metropolitan Achievement Test, and the Otis-Lennon Mental Abilities Test. *Psychology in the Schools, 13,* 291-297.

Dean, R. S. (1976). *Comparison of PIAT validity with Anglo and Mexican-American children.* (ERIC Document Reproduction Service No. ED 133 638)

Dean, R. S. (1977). Internal consistency of the PIAT with Mexican-American children. *Psychology in the Schools, 14,* 167-168.

Dunn, L. M. (1965). *Expanded manual. Peabody Picture Vocabulary Test.* Circle Pines, MN: American Guidance Service.

Dunn, L. M., & Markwardt, F. C., Jr. (1970). *Manual for the Peabody Individual Achievement Test.* Circle Pines, MN: American Guidance Service.

Harmer, W. R., & Williams, F. (1978). The Wide Range Achievement Test and the Peabody Individual Achievement Test: A comparative study. *Journal of Learning Disabilities, 11,* 58-65.

Kaufman, A. S. (1979). *Intelligent testing with the WISC-R.* New York: John Wiley & Sons.

Reynolds, C. R. (1979). Factor structure of the Peabody Individual Achievement Test at five grade levels between grades one and 12. *Journal of School Psychology, 17,* 270-274.

Reynolds, C. R., & Gutkin, T. B. (1980). Statistics related to profile interpretation of the Peabody Individual Achievement Test. *Psychology in the Schools, 17,* 316-319.

Scull, J. W., & Branch, L. H. (1980). The WRAT and the PIAT with learning disabled children. *Journal of Learning Disabilities, 13,* 64-66.

Silverstein, A. B. (1981). Pattern analysis on the PIAT. *Psychology in the Schools, 19,* 13-14.

Walden, J., Jr. (1979). A comparison of the PIAT and the WRAT: A closer look. *Psychology in the Schools, 16,* 342-346.

Wettler, J., & French, R. W. (1973). Comparison of the Peabody Individual Achievement Test and the Wide Range Achievement Test in a learning disabilities clinic. *Psychology in the Schools, 10,* 285-286.

Wikoff, R. L. (1979). Determining basals for the Peabody Individual Achievement Test. *Psychology in the Schools, 16,* 172-174.

Ysseldyke, J. E., Sabantino, D. A., & La Manna, J. (1973). Convergent and discriminant validity of the Peabody Individual Achievement Test with educable mentally retarded children. *Psychology in the Schools, 10,* 200-204.

Forrest G. Umberger, Ph.D.

Associate Professor and Director of Communication Disorders,
Department of Special Education, Georgia State University, Atlanta,
Georgia.

PEABODY PICTURE VOCABULARY TEST-REVISED

Lloyd M. Dunn and Leota M. Dunn. Circle Pines, Minnesota:
American Guidance Service.

THE ORIGINAL OF THIS REVIEW WAS PUBLISHED IN TEST CRITIQUES: VOLUME III
(1985).

Introduction

The Peabody Picture Vocabulary Test-Revised (PPVT-R) is an individually administered, norm-referenced, wide-range, power test of hearing vocabulary, available in two parallel forms—now designated L and M. Each form in this revision contains five training items, followed by 175 test items arranged in order of increasing difficulty. Each item has four simple, black-and-white illustrations arranged in a multiple-choice format. The subject's task is to select the picture that best illustrates the meaning of a stimulus word presented orally by the examiner.

The test is designed for persons 2½-40 years of age who can see and hear reasonably well and understand standard English to some degree. Testing requires only 10-20 minutes because the subject must answer only about 35 to 45 items of suitable difficulty. Items that are too easy or too hard for the subject are not administered. Scoring, which is rapid and objective, is accomplished largely while the test is being administered. Raw scores are usually converted to age-referenced norms, and grade-referenced derived scores are available from the publisher on request.

The first edition of the Peabody Picture Vocabulary Test was authored by Lloyd M. Dunn. The PPVT-R is authored by Lloyd M. Dunn and Leota M. Dunn. Lloyd M. Dunn received his training in special education and psychology from the University of Illinois. He has been Affiliate Professor of Special Education at the Manoa Campus of the University of Hawaii since 1970. Prior to his present position he served on the faculty of Peabody College for 20 years. While at Peabody he served as chairman of the special education department, and later as the first research director of the Institute on Mental Retardation and Intellectual Development. Dunn is the senior author of the Peabody Individual Achievement Test, the Peabody Language Development Kit (PLDK) series, and the Peabody Early Experiences Kit. He is past president of the Council for Exceptional Children and a fellow of the American Psychological Association.

Leota M. Dunn holds a degree in elementary education from the University of Saskatchewan at Regina. She collaborated with her husband from 1956 to 1959 in

345

developing the original Peabody Picture Vocabulary Test (PPVT). She has served as a psychometric examiner and research assistant on a number of research projects at Peabody College and Vanderbilt University. She worked on the PLDK series, serving as coordinator for Level 2, and during the 1970s she collaborated in the revisions of it and the PPVT.

The first edition of the PPVT is designed to provide an estimate of subjects' "verbal intelligence" through measuring their hearing vocabulary. In addition to its effectiveness with average subjects, it has utility with other groups. For instance, because the test items do not require the subject to read, it is nonbiased for nonreaders and remedial reading cases. Additionally, because neither a pointing nor oral response is essential, speech impaired and physically handicapped (cerebral palsy) individuals are less handicapped in the testing situation.

According to Dunn and Dunn (1981, p. 29), the procedure for the development of the original PPVT began by constructing 200 test plates from an initial pool of 3,885 words, which were selected from the 1953 edition of *Webster's New Collegiate Dictionary* (G. & C. Merriam), and the meanings of which could be clearly illustrated by black-and-white line drawings. Each plate was made up of four pictures with three stimulus words tested for each plate, for a total of 600 words. After repeated field testing and refinement of these plates, the best 300 stimulus words and their decoys were selected to construct one series of 150 plates. This series was used for both Form A and Form B of the original edition, with the stimulus words arranged in ascending order of difficulty.

Only white children and youth residing in and around Nashville, Tennessee, were included in the final standardization group. Norms were established for age equivalents (mental ages), standard score equivalents (intelligence quotients), and percentile equivalents.

During the years 1976 through 1980 the PPVT-R was developed and standardized. Selection of items began by determining how many of the original 300 stimulus words could be retained in the new item pool. Percentage passing curves and decoy action statistics, together with information from a number of research studies that reported on items that were biased culturally, regionally, sexually, and racially, were used in the selection process. Only 144 of the original 300 stimulus words were retained for further consideration. Approximately two-thirds of the stimulus words are new, and all drawings are either new or reworked for better racial, ethnic, and sex balance.

Other modifications or revisions involved in the PPVT-R include 1) expanding the number of stimulus words per form from 150 to 175, thereby increasing the sensitivity of the instrument yet keeping it short; 2) having a separate series of plates for each form, thus allowing a distractor to work for only one stimulus word, rather than for two as in the original edition; 3) using a new system for determining item concentrations, difficulty levels, and goodness of fit to the hearing vocabulary growth curve; and 4) developing an item pool for tryout that would have about twice as many items as the 350 planned for the final two forms of the test.

The authors of the PPVT-R credit advances in test construction since 1959 to the added sophistication of the present instrument. Additionally, they have responded to many suggestions for improving the test that have been accumulating during 20 years of use and research. New features that were incorporated into this revision include 1) standardization conducted on a national basis; 2) data in the norms tables that were smoothed and presented in finer increments; 3) the addition of adult norms; 4) the replacement of the terms *mental age* and *intelligence quotient* by *age equivalent* and *standard score equivalent,* respectively; 5) the spacing of items to conform to the negatively accelerating growth curve of hearing vocabulary in order to keep it equally sensitive across the range of the test; and 6) a horizontal layout and easel format to permit improved viewing of test plates and reduced interference between consecutive plates (Dunn & Dunn, 1981, pp. 1-2).

Standardization was carried out on a representative national sample of children and youth and a selected sample of adults. To insure representativeness, the sample was selected according to chronological age and sex balance and geographical, occupational, ethnic, and community-size representation. (A detailed account of the standardization of the PPVT-R is contained in the test manual and technical supplement.) The standardization procedures easily met the most rigorous expectations for test construction.

The PPVT-R test kit contains a manual for Forms L and M, a test book (Form L or M) that folds into an easel for displaying the test plates, and a supply of individual test records.

The information in the test manual is divided into two parts. Part one contains the test introduction, administration procedures, and scoring instructions. Part two describes the test construction, standardization, and norms development. The appendices contain derived scores, norms tables, a guide to pronouncing the stimulus words, and a list of facilities and coordinators participating in the PPVT-R research programs. The optional technical supplement, which elaborates on the contents of the manual, is available at extra cost from the publisher.

The test book (Form L or M) contains 175 test plates and five demonstration plates. Each plate contains four black-and-white line drawings, each of which is identified by 1, 2, 3, or 4. The back of the booklet folds out into an easel allowing the examinee to see each plate individually.

The individual test record lists the stimulus words to be used with the training plates and provides the key to the correct choices. Space is provided for recording the subject's responses, raw score, and certain derived scores, plus additional information about the subject.

The examiner's participation in the testing process is to explain the task to the subject according to predetermined suggestions. The examiner displays the training plates and explains that the subject is to point to the picture of the word that the examiner says. Once the task is understood by the subject, the examiner continues the test by turning the plates and giving the stimulus word from the individual test record. Examiners should be seated where they can see the subjects' pointing responses.

Because the test is designed for persons aged 2½-40 years, the difficulty level of the test items ranges from easy for two-year-olds to hard for superior adults.

The answer form is filled in by the examiner who records subject responses to each stimulus presentation. The answer form for the test contains the plate numbers from one to 175. Next to the plate number is the stimulus word, the number of the plate containing the correct response, a space for recording the subject's response, and seven recurring geometric figures to aid in establishing the basal and ceiling. When errors occur the examiner crosses through the corresponding geometric figure. These figures can be scanned quickly to determine when six out of eight responses are in error, thereby establishing the ceiling.

The profile is located on the back of the answer form and provides space for recording a raw score, standard score equivalent, percentile rank, stanine, and age equivalent. The percentile rank and stanine scales are laid out under the standard score equivalent. Once the standard score equivalent is marked on the scale, a perpendicular line can be drawn down through the percentile rank and stanine scale to give the subject a rating on those scales.

Practical Applications/Uses

The PPVT-R can be used across a number of disciplines and provides information along a number of dimensions. Although its primary design is to measure a subject's receptive vocabulary for Standard American English, it can also be viewed as 1) a scholastic aptitude test in that it gives a quick estimate of one aspect of verbal ability; 2) a measurement of one facet of general intelligence: vocabulary; 3) one of the best single indices of school success, due to the vocabulary; and 4) an achievement test in that it shows the extent of English vocabulary acquisition.

The test has a number of applications within the school setting, including providing the teacher with information on receptive vocabulary and, therefore, a starting level for intervention. It is helpful for screening foreign-speaking students who plan to attend English-speaking universities. The test is especially useful at the preschool level as a measure of child development because it is easy to administer to young, immature children. Finally, it is a useful indicator of scholastic aptitude as an initial screening device.

The PPVT-R's primary value as a clinical tool is that it does not require the client to read or write. Responses can be gestural, making the test appropriate for the seriously impaired, such as expressive aphasics, stutterers, and those with cerebral palsy. The test is relatively nonthreatening because little verbal interaction between the examiner and client is required. For this reason it can be used with certain autistic, withdrawn, and psychotic individuals. The illustrations are clean, bold, line drawings, enabling most moderately visually impaired persons to successfully participate. Similarly, persons with visual-perceptual impairments are able to perform on the test because the illustrations are free of fine detail and figure-ground confusions.

Because the PPVT-R has been standardized with adults it has vocational applications, including selecting persons for jobs that require comprehension of English vocabulary. It can also be used to identify individuals in need of preemployment instruction to improve English language comprehension.

For research purposes the test is useful for investigating a number of phenomena, and because it has alternate forms it is useful in studies involving pretesting and posttesting.

The PPVT-R is administered on an individual basis. The room should be in a quiet area, with the lighting arranged to avoid glare on the test plates. It is suggested that the examiner and subject sit at the corner of a table, facing each other.

The easel is placed so that the client can see only the plate being considered. The numbers at the bottom of the test plates correspond to the test-item numbers. The number appearing on the examiner's side of the easel indicates the number of the plate being administered. The examiner records the responses on the individual test record, which is placed behind the easel out of the subject's view. For subjects who use a pointing response, the examiner may choose to sit on the same side of the table as the subject in order to view the client's responses.

Although specialized course work or training is not required in order to administer the test properly, examiners should thoroughly familiarize themselves with the test materials. It is suggested that practice in administration of the test be done under the supervision of an experienced examiner. The most important requisite is that the examiner be able to pronounce each stimulus word according to *Webster's New Collegiate Dictionary* (G. & C. Merriam Co.).

Basically, the test administration instructions are straightforward and easy to understand. Test administration takes 10-20 minutes. The starting point on the test is determined according to the subject's chronological age. The basal is established when the subject makes eight or more consecutive correct responses. If the subject makes an error before making eight consecutive correct responses, the examiner simply tests backwards until eight consecutive correct responses are given. The ceiling is the lowest eight consecutive responses containing six errors. One problem that could be encountered in the administration of the test is that the inexperienced examiner may have a tendency to review the administration procedures superficially and not look beyond the basal and ceiling explanation to the examples provided. The test has definite procedures for scoring when two basals or two ceilings occur. If these procedures are not followed, the results will be invalid. Administration instructions provide information on establishing rapport, starting points, basal and ceiling rules, and include two procedures for introducing the test to clients: one for those under age eight and one for those over age eight.

The procedures for scoring the test are clear and examples are provided as a model. Scoring the test requires only a few minutes and is computed by the examiner. The procedure does not require machine or computer scoring. To score the test the examiner calculates the raw score by subtracting the errors from the ceiling item. The items below the basal are added as correct, and the items above the ceiling are considered as errors and are not considered in the calculation. The individual test record contains a space for the raw score. Using the chronological age of the client and the raw score the examiner can obtain three deviation-type age norms which are contained in the test record: standard-score equivalents, percentile ranks, and stanines. Development-type age norms and age equivalents are also reported.

The PPVT-R is based on objective scoring and norm-referenced interpretation. To aid the examiner in the accurate interpretation of derived scores the authors summarize the definition and characteristics of each of the four types of age-referenced norms in the appendices. A nice feature of the interpretation aspects of the test deals with errors of measurement. Realizing that many users often report scores as precise points when they are based on imprecise measurement devices, the authors have added a true confidence band—that range of scores in which the subject's true scores can be expected to fall 68 times in 100. The test form contains the standard-score equivalent, percentile rank, and stanine laid out horizontally such that a vertical line will cut through all three scales. The examiner marks the obtained standard-score equivalent on the top scale and then draws a vertical line through it and across the three scales. The test form also contains a table that indicates the area to be shaded on each side of the obtained standard score. This is the score-confidence band and it covers all three age-referenced norms. Realizing the tremendous responsibility that goes with interpreting derived scores, particularly the potential impact when educational placement decisions are at stake, some examiners may choose to report standard error of measurement band-width scores for age equivalents.

Technical Aspects

The reliability coefficients for the PPVT-R were provided in two forms: internal consistency (split-half) and alternate forms. Internal consistency reliability coefficients are calculated by administering one form of the test to a group of subjects and then comparing their performances on different parts of the test, usually odd versus even items (split-half procedure). The results of this procedure reflect measurement error resulting from characteristics of the particular items chosen in the construction of the test. Alternate-forms reliability coefficients represent the administration of two different test forms to a group of persons in a test-retest situation.

Reliability coefficients established for the PPVT-R consist of coefficients of internal consistency (split-half correlations), immediate retest alternate-forms reliability coefficients, and delayed retest alternate-forms reliability coefficients.

The standard procedure for split-half correlations was not appropriate for the PPVT-R because it would provide spuriously high coefficients due to the calculation including items below the basal (which are counted as correct) and items above the ceiling (which are counted as incorrect). For this reason the Rasch-Wright latent trait methodology was incorporated, thus making the split-half analysis possible. The split half correlations were based on all subjects in the standardization sample (ages 2½-40 years).

The split-half reliabilities for children and youth aged 2½-18 years are quite similar. The coefficients ranged from .67 to .88 on Form L (median .80), and from .61 to .86 on Form M (median .81). Only Form L was administered to the adults. They received a median split-half reliability of .82.

Alternate-forms reliability coefficients based on an immediate retest were calculated for 642 subjects, selected to ensure adequate representation by age, sex, occupation, and race. The subjects were selected from each age group (2½-18 years) of the standardization sample. Forms L and M were administered in counterbalanced order. Reliability coefficients for raw scores ranged from .73 to .91, with a median of .82, and from .71 to .89, with a median of .79, for standard scores.

Delayed retest alternate-forms reliability coefficients were calculated with the caveat that test reliability from such a measure is questionable because the test variables are confounded with performance changes over time. A sample of 962 subjects with adequate representational characteristics was selected from each age group of the standardization sample. Forms L and M were administered in counterbalanced order. A minimum of nine days and a maximum of 31 days elapsed between the first and second testing. The reliability coefficient for raw scores for the various age groups ranged from .52 to .90, with a median of .78. Coefficients for standard scores ranged from .54 to .90, with a median of .77. Based on median values, the reliability coefficients dropped very little from immediate retesting to delayed retesting.

In terms of sample selection and statistical application, the revised PPVT satisfies the criteria for reliability. It also exceeds the original PPVT in reliability.

Information on validity, or the degree to which the PPVT-R measures what it claims to measure, is reported for content, construct, and prediction validity. Content validity for the PPVT-R consisted of a complete search of *Webster's New Collegiate Dictionary* (G. & C. Merriam, 1953) for all words whose meanings could be represented by a picture. It was assumed that this dictionary represented the content universe for hearing vocabulary. By retaining as many stimulus words as possible from the 1959 PPVT, the PPVT-R met adequate standards for a picture vocabulary test measuring hearing vocabulary in Standard English.

Construct validity of the PPVT-R as a measure of hearing vocabulary depends on its content validity. On this dimension the test meets the criteria. The test authors' inferences of the PPVT-R's ability to measure scholastic aptitude are supported in the literature (e.g., Hinton & Knights, 1971). In summary, the literature on tests of intelligence reports that the single most important test of most batteries is the vocabulary test because it correlates more highly with full scale IQ scores than any other subtest (Wechsler, 1974, p. 47).

Evidence of construct validity is also reflected in internal consistency or how the test items will measure the trait sampled by the test. The PPVT-R test-time validity was established as the test items were selected. The selection of each item was based on a gradual increase of the subjects' correct responses to the item for each successive age group. This was accomplished by using the Rasch-Wright latent trait item analysis as refined by Woodcock and Dahl (1971) at American Guidance Service. The model enables the investigator to construct a growth curve for hearing vocabulary (the latent trait being measured by the PPVT-R) and to select items to fit the curve. The curve was drawn repeatedly on the computer, using the various tryout and calibration data. In addition, the percentage of passing data for each age was available for each item. If an item exhibited too steep or flat an item characteristic curve, it tended to be one of those not fitting the Rasch-Wright latent trait curve for hearing vocabulary.

At the time of this review predictive validity data were not available for the PPVT-R. The only concurrent validity data vailable to date is the study equating the original and revised editions. These findings reveal a good deal of overlap between the two editions. For this reason, until data are available for the new editions, justification exists for applying research findings for the PPVT to the PPVT-R. For the interested reader a number of correlational studies on the original PPVT are included in the manual of the current edition.

Critique

The PPVT-R, as a test of hearing vocabulary, is one of the most widely used instruments of its kind. It has a strong statistical foundation and the authors have continued to gather data to strengthen that foundation and to keep the test current.

One of the most significant features of the PPVT-R is its flexibility and applicability to a number of exceptional populations. The variety of purposes have already been enumerated. As an educator, this reviewer can attest to the test's value in the field of education, particularly special education. More specifically, it allows the special educator to evaluate the hearing vocabulary and, subsequently, the scholastic aptitude of multiply handicapped children or adults whose only mode of communicating may be by pointing.

It has an equally wide application as a research tool or as one test in a battery of tests on language competence. In this reviewer's opinion, anyone engaged in evaluating and prescribing educational placement should have the PPVT-R as part of their armamentarium of diagnostic tests.

References

This list includes text citations as well as suggested additional reading.

Anderson, D. E., & Flax, M. L. (1968). A comparison of the Peabody Picture Vocabulary Test with the Wechsler Intelligence Scale for Children. *Journal of Educational Research, 62*(3), 114-116.

Ando, K. A. (1968). A comparative study of the Peabody Picture Vocabulary Test and Wechsler Intelligence Scale for Children with a group of cerebral palsied children. *Cerebral Palsy Journal, 29*(3), 7-9.

Brown, L. F., & Rice, J. A. (1967). The Peabody Picture Vocabulary Test: Validity for EMRs. *American Journal of Mental Deficiency, 71*(6), 901-903.

Covin, T. M. (1976). Alternate form reliability of the Peabody Picture Vocabulary Test. *Psychological Reports, 39*, 1286.

Dunn, L. M., & Dunn, L. M. (1981). *Peabody Picture Vocabulary Test-Revised*. Circle Pines, MN: American Guidance Service.

Hinton, G. G., & Knights, R. M. (1971). Children with learning problems: Academic history, academic prediction, and adjustment three years after assessment. *Exceptional Children, 37*, 513-519.

Ivanoff, J. M., & Tempero, H. E. (1965). Effectiveness of the Peabody Picture Vocabulary Test with seventh-grade pupils. *Journal of Educational Research, 58*, 412-415.

Raskin, L. M., & Fong, L. J. (1970). Temporal stability of the PPVT in normal and educable-retarded children. *Psychological Reports, 26*, 547-549.

Wechsler, D. (1974). *Preschool and Primary Scale of Intelligence manual*. New York: The Psychological Corporation.

Woodcock, R. W., & Dahl, M. N. (1971). *A common scale for the measurement of person ability and test item difficulty* (AFS Paper No. 10). Circle Pines, MN: American Guidance Service.

Merith Cosden, Ph.D.

Assistant Professor of Special Education, University of California-Santa Barbara, Santa Barbara, California.

PIERS-HARRIS CHILDREN'S SELF-CONCEPT SCALE

Ellen V. Piers and Dale B. Harris. Los Angeles, California: Western Psychological Services.

THE ORIGINAL OF THIS REVIEW WAS PUBLISHED IN TEST CRITIQUES: VOLUME I (1984).

Introduction

The Piers-Harris Children's Self-Concept Scale is a self-report measure of children's feelings about themselves. The scale measures "self-concept" as it is addressed through children's evaluations of their behavior, intellectual and school status, physical appearance and attributes, anxiety, popularity, and satisfaction. The test consists of a series of declarative statements which describe successful or unsuccessful functioning in each of these areas; the child responds positively or negatively to each, evaluating whether or not the statement describes them *most* of the time. The test provides scores which indicate whether a child has high or low reported self-concept in relation to their peers. It is used for screening children with low self-concept, as part of clinical assessments and in research.

The current 80 test items were taken from a pool of 164 statements developed by Jersild in 1952 from his interviews with children on their likes and dislikes (Piers, 1984). A preliminary pool of items was administered to a sample of 90 students in grades 3-5. All of the items which were answered in one direction (agree/disagree) by more than 90% of the students were dropped, with the exception of certain items with high "content validity" for a self-concept scale (e.g., "my parents love me"). Additional test items were dropped after a second analysis using responses from a sample of 127 sixth-grade students. Only those items that significantly discriminated between students with overall high and low scores, and those items which were answered in the expected direction by at least half of the high scoring students, were retained.

The remaining 80-item test was conceived as a unitary measure of children's self-concept. Subsequently, a meta-analysis of interitem correlations yielded six item "clusters" (i.e., multiple group-factor scores). These clusters are based on analyses of interitem correlations for six separate samples of students, including the original sample. The item composition of each of these clusters suggests that they represent self-evaluations of: 1) behavior; 2) intellectual and school status; 3) physical appearance and attributes; 4) anxiety; 5) popularity; and 6) happiness and satisfaction. While the nature of these scales bears a close resemblance

353

to the content areas originally designated for study by Jersild, the reliability of these clusters has been questioned (e.g., Platten & Williams, 1979).

The 80-item test was normed in 1966 on a sample of 1,183 Pennsylvania school children in grades 3-12, while normative cluster scores were obtained from an independent sample of 485 students. Significant mean and standard deviation differences were not detected between students as a function of grade. Thus, scores were collapsed across grade, resulting in a mean score of 51.84 and standard deviation of 13.87 for the entire sample. The 1984 manual provides the means and standard deviations of these samples for normative use.

In addition to these scores, the means and standard deviations from 12 other studies of normal children are presented in the manual. These means and standard deviations vary from those obtained for the initial standardization samples (i.e., in studies with samples of 30 or more children reported in the manual); average Piers-Harris self-concept scores range from 51.8 to 61.4, with standard deviations of 10.2 to 12.8. These variations in mean and standard deviation scores between old and new samples are noted in the manual.

Although the cumulative N from these reported studies totals 3,692 children, these students were not selected through any broad-based or demographically stratified design. Generalization of any of these scores for normative purposes (i.e., use of these samples as a standard against which to measure other children's test scores) is not justified by their sampling designs. Rather than attempt to generate new norms based on a current, broad-based stratified sample of students, the author suggests the use of locally derived norms to assess individual deviations in self-concept.

Normative scores for special groups also differ from those of the initial sample. These differences have been noted in samples of blacks and Koreans (Chang, 1975), gifted students (Ketcham & Snyder, 1977), and behavior-disordered students (Bloom, Shea, & Ein, 1979). These score variations dictate caution in using the provided norms for a definitive assessment of low self-concept, and reinforce use of the scale as a screening rather than diagnostic device.

The test has not been revised since its publication in 1969. While experimental variations of the Piers-Harris have been suggested, such as rewriting items tapping the affective domain (Michael, Smith, & Michael, 1975) and replacing the dichotomous response format with a Likert scale (Lynch & Chaves, 1975), these suggestions have not been empirically explored.

A mail-in service for computerized scoring and interpretation of protocols has recently been made available. Western Psychological Services, the publisher of the Piers-Harris, plans to develop a microcomputer diskette of the test which would allow users to administer, score, and interpret protocols on their own microcomputers. Through this system, users could contract for a specified number of protocols, which would be programmed and sold on a diskette. While this service is further delineated in the 1984 manual, it was still under production in July, 1984.

The Piers-Harris is a paper-and-pencil test designed to be administered individually or in small groups with verbal instructions from the examiner. Children taking the test mark either "yes" or "no" in response to each of the 80 statements, reflecting whether or not the statement is true *most* of the time. The statements are constructed at a third-grade reading level, and are written with both positive and negative valences.

The direct nature of the inquiry made by this scale necessitates that the examiner maintain a positive, cooperative attitude during the session. The examiner reads the instructions aloud to all children, and continues to read each item aloud for children in the lower grades (4, 5, and 6). The examiner is permitted to define any words that students do not understand. Finally, the examiner needs to encourage children who have difficulty with the all-or-none quality of the responses, requesting that examinees respond in the way that they *usually* feel to all of the items.

While the test is easily administered to a wide range of students, the norms available in the manual allow standardized interpretation of test results only for students in Grades 4-12. Since the test is written at a third-grade reading level, one needs to assess younger students' ability to fully understand the test items. In order to insure accurate comprehension by all students, the examiner must read all questions aloud both to younger students and any who are at a lower reading level.

There are two types of answer forms available for the test. The original form is a four-page booklet in which the child circles "yes" or "no" in response to each statement. These forms are easily scored with a scoring key. More recently, an answer sheet for computerized scoring, processed through Western Psychological Services, has been developed.

A child's score is determined by the number of items checked in the direction of positive self-concept. In addition to the total score, a profile is created by analyzing the child's responses to items as they cluster around six areas of functioning: behavior; intellectual and school status; physical appearance and attributes; anxiety; popularity; and happiness and satisfaction.

Practical Applications/Uses

The Piers-Harris provides a global measure of personal satisfaction, as well as specific self-evaluations in the six areas previously described. The information obtained from this scale and the ease with which it is administered and scored make it appropriate for use in both educational and clinical settings. It is best used for screening possible at-risk children and for determining a child's self-concept in relation to peers. The test's utility in making specific assessments of children's needs is limited by subscale reliability and standardization problems. Nevertheless, the Piers-Harris provides a direct measure of children's perceptions of their skills and relationships which can be integrated with other test and interview information to provide a comprehensive picture of their social and emotional functioning.

There are three major uses for the Piers-Harris. First, it was designed as a screening device for children potentially "at risk" in terms of their social or emotional functioning. The scale is particularly suited for use in this manner because of the ease with which it can be group administered and scored. Used in this manner, the scale signals the need for further psychological evaluation and treatment. Despite its promise for use in this manner, however, few studies have studied its effectiveness in this arena.

Secondly, the Piers-Harris can be used as part of an individual assessment battery. The test directly presents the child's perceptions of his or her own functioning; this information, in conjunction with other test, observation, and interview data, can be used to generate hypotheses about the impact of the child's environment on confidence and capacity to function. Very low scores alert the examiner to potential depressive problems; very high scores suggest the need for analysis of discrepancies between the child's report and other life events, and an assessment of the child's pattern for reacting to stress.

Finally, the test has been used extensively in research, as a pre- and post-test measure in studies of factors affecting self-esteem (Wanat, 1983) and in studies evaluating the relationship of other factors to self-concept (e.g., Grayhill, 1978). The scale is frequently utilized in this manner because of the speed and ease with which it can be administered and because of the objective scoring system which quickly provides a direct measure of self-esteem. The ease with which the Piers-Harris can be used, however, may cause experimenters to rely on it for more information than is appropriate. The test does not provide reliable, detailed self-esteem measures that would be expected to change over limited periods of time. Thus, the test may not be sensitive enough to reflect the impact of short-term interventions.

The test is designed for use with children ages 8-18. It is recommended as a screening device, but is also clinically indicated for use with children who "appear" depressed or for whom there are other signs of potentially stressful life events having an impact on their self-esteem.

While the test is written at a third-grad comprehension level, and the norms extend from Grades 3-12 on a narrow demographic sample of students, the test is cited in the literature with younger children (e.g., Bloom et al., 1979) and with racial groups for whom it has not been normed (e.g., Jegede & Bamgboye, 1981). Although the simple administrative format makes the Piers-Harris adaptable for a wide range of students, caution must be exercised in interpreting scores from these younger and racially unnormed subjects. With regard to younger children, questions about comprehension of the test items, stability of scores over time, and the meaning of absolute scores inhibit analysis. Studies on various racial groups raise concerns about item comprehension, the meaning of test items within the cultural context, and the lack of appropriate normative data for test interpretation. While using the Piers-Harris with these populations may seem appropriate on the surface, interpretation of the results requires special considerations.

The test has not been adapted for use with special populations. The dual modality administration of the test (i.e., oral and visual), however, suggests that the test instructions could easily be modified for use with physically handicapped, blind, deaf, and learning disabled students.

The Piers-Harris can be administered individually or in small groups, and there are no special setting requirements. However, as in all testing situations, attention to the task at hand and optimal performance require a setting which is quiet, free of distractions, and relatively unstressful.

The test can easily be administered and scored by both professionals and paraprofessionals (e.g., psychologists, psychiatrists, educators, and teacher and child-care aides). *Interpretation* of test results, however, requires training and expertise in psychological testing and child development, including specific knowledge about children's self-concept scales and the psychometric properties of the Piers-Harris. A "User Qualification Form" must be completed before one can purchase the scale, and potential test users are evaluated by professional staff members at Western Psychological Services. In most cases, individuals with relevant master's level training or better are allowed access to the test. Students need a qualified user to sign for them in order to obtain access to the scale.

Instructions for individual and group administrations are provided in the manual. The instructions are simple and clear, and allow the examiner considerable flexibility in responding to children's requests for help. The examiner may clarify as fully as necessary the test requirements and specific test items. Motivating children to respond honestly and to provide a dichotomous response (yes/no) to statements which may seem conditional to them are the major difficulties faced by the test administrator.

The test is self-paced. When reading the items aloud, the examiner must wait for all examinees to respond before proceeding to the next item. When subjects read the items to themselves they are allowed as much time as needed to make their responses. In most instances, test administration takes 15-20 minutes.

The Piers-Harris can be hand-scored with the use of templates in approximately 30 minutes. The only difficulty in using the templates is noting which subset of questions is used to calculate each cluster score. The cluster scores do not sum to equal the total raw score, as some items "load" on more than one cluster. Each question the respondent answers in the direction of a positive self-concept is given a point; the points are then summed to determine the cluster scores and the total raw score. These scores can be compared to average and deviant scores provided in the manual. The mean and standard deviation provided in the manual were derived from children's scores across all grades (i.e., the same norms and those used for children in Grades 3-12).

After calculating these raw scores, the user can convert them to T-scores, stanines, and percentiles by mapping the raw scores on an available profile form or by using the tables provided in the manual. While the author acknowledges that the use of T-scores for scaling the cluster scores is not appropriate given the small number of questions in each cluster, these conversions are still provided. Scores between the 31st and 70th percentiles are considered average; scores considered deviant fall ± 1 standard deviation from the mean (i.e., below the 16th or above the 84th percentiles).

There is a separate Piers-Harris form which can be used for computerized scoring. The objective nature of the Piers-Harris scoring system makes either method highly reliable. The computerized scoring provided by WPS includes an interpretive report. (A discussion of the validity of this report will follow later in this review.)

The test is designed so that the interpretation of the child's self-concept is a function of the objectively obtained cumulative and cluster scores. While these scores provide the basis for interpretative use of the test, effective use (i.e., assessment of the validity and specific meaning of the test scores) relies on both psychometric data and other clinical information on the child. There are two issues that need to be addressed before one can directly interpret the Piers-Harris scale scores: 1) whether or not children's responses reflect their true feelings about themselves, and 2) whether or not existing norms are appropriate for use in interpreting test scores.

There are at least four factors, three of which are listed in the manual, that can affect the validity of the raw scores. First, a child can "fake" responses, deliberately responding to questions in a manner calculated to "look good" or "look bad." The manual states and this reviewer concurs that children have few reasons to "fake bad" and that all scores reflecting low self-concept be treated as potential indicators of distress. Attempts to "fake good," however, may reflect several underlying issues, including attempts to hide problems or a desire to look good in front of others. This tendency can be statistically detected by noting records in which children score over 1.5 standard deviation units over the mean in a positive direction. The manual notes that high scores may reflect a positive self-evaluation or a healthy desire to look good in front of others, and may not be a cover for underlying problems. Interpretation of this response set for individual children requires both psychometric knowledge of test patterns and other clinical information about the child in question.

Several other response sets may also bias the validity of test scores. Random responses to test items, or acquiescence (saying "yes" or "no" to all items), can be detected through raw score profiles described in the manual. Again, interpretation of this pattern and its meaning for the child in question relies on both clinical and psychometric expertise.

Yet another response set to note is the child's use of denial or other psychological defenses to guard against the perception of negative events in their environment. Some children may report positive self-evaluations that conflict with what one would expect, given their stressful life events. Detection of this response style requires both clinical skill and the integration of Piers-Harris test findings with other psychodiagnostic information.

The second major threat to the validity of test scores is the use of inappropriate normative data in interpreting results. While the manual provides the caveat that "some children who differ in ethnicity or socioeconomic background from the normative sample . . . may respond differently to the Piers-Harris" (Piers, 1984, p.37), the demographic characteristics of the normative sample are not clearly delineated. In lieu of the preferred but costly alternative of developing a broad-based, stratified sample of Piers-Harris scores, users need to be cautious in interpreting scores relative to the norms, using the test to *screen* deviant scores for further assessment.

The manual presents clear instructions on the procedures to follow for interpreting test results. Starting with a validity check of the test results (described earlier), the test user is instructed to calculate and determine the level of the total raw score and cluster scores. Individual item responses are then assessed before integrating this information with other test data and developing recommendations. In order to interpret test results accurately the user must understand the rationale and psychometric properties of the test. Effective use of the Piers-Harris requires a general background in assessment and the use and integration of test information, familiarity with the rules for ethical test use established in the *Standards for Educational and Psychological Tests* (1974), and an understanding of child development, psychopathology, and the relationship of children's reported self-concept to these processes.

Technical Aspects

Studies on the reliability of the Piers-Harris scale have been conducted on a variety of child populations. The internal consistency of the test as a whole is relatively high. Alpha coefficients of .90-.91 have been reported for male and female populations, and reliabilities of .88-.93 have been cited for males and females using the Kuder-Richardson formula 20 (see Piers, 1984). Similarly high internal-consistency measures have been found with special populations, including the learning disabled (Smith & Rogers, 1978) and native Americans (Lefly, 1974).

Test-retest reliabilities range from .62 to .96 in the literature with retest intervals of a few weeks to six months. These reliabilities have been established in normal populations (e.g., Shavelson & Bolus, 1982), learning-disabled students (Smith et al., 1978), and in children from different ethnic backgrounds, including black and Mexican-American children (Platten & Williams, 1979, 1981), Mexican-American migrant workers (Henggeler & Tavormina, 1979), and American Indian students (Lefly, 1974).

The discriminant validity of the scale is difficult to ascertain from the current literature. A number of studies have been conducted attempting to discriminate between two or more subject groups on the basis of Piers-Harris self-concept scores. While some have found significant between-group differences in self-concept for students who differ in achievement levels (e.g., Kanoy, Johnson, & Kanoy, 1980; Karnes & Wherry, 1981) or for handicapped and nonhandicapped children (e.g., Goldman & Hardin, 1982; Sheare, 1978), others have not found these differences (e.g., Beck, Roblee, & Hansen, 1982; Silverman & Zigmond, 1983). Few studies, however, provide explicit hypotheses which would predict or specify the nature of expected between-group differences on the self-concept dimension. More theoretically driven research in this area is needed, both to test the validity of the scale and to meaningfully interpret between-group differences when they occur.

Several validity studies both predict and find differences in children's self-evaluations as a function of other factors in the child's life. These studies are primarily correlational, and do not imply causation between self-concept and these other factors. For example, children's self-concept has been related to parental evaluation and acceptance. Smith, Zingale, and Coleman (1978) found that discrepancies between adult expectations and children's academic performance were associated with lower self-concepts in children. In a similar vein, Crase, Foss, and Colbert (1981), Grayhill (1978), and Piers (1982) all found positive relationships between children's self-concept and parental satisfaction, or perceptions of parental satisfaction.

Other studies have assessed the validity of scale scores in terms of their relationship to a child's standing with peers. Self-concept, as measured by the Piers-Harris, has been positively related to one's classroom achievement relative to one's peers (Rogers, Smith, & Coleman, 1978) and to involvement in "chum" relationships (Mannarino, 1978). Bradley (1975) reported high correlations between peer ratings and self-concept.

Few studies have attempted to validate the scale on the basis of *behavioral* characteristics associated with high self-concept. Studies along this line include Kim's (1983), in which she reported that Asian children with high self-concept scores learned English faster and made greater progress in school than did other students.

A larger body of literature has focused on the relationship of Piers-Harris self-concept scores to other personality characteristics as measured by similar paper-and-pencil scales. In particular, studies have found a relationship between self-concept and locus of control; children with lower self-evaluations report a belief in external controls (e.g., Goldman et al., 1982; Moyal, 1977), while high-esteem students report a belief in internal controls for their successes (Piers, 1977). The Piers-Harris has not been highly correlated with various other tests, such as scales of body esteem (Wheeler & Ladd, 1982) and ego strength (Gordon & Testmeyer, 1982). These studies, however, reflect the reliability problems of each scale, and do not speak to the validity of either measure.

Correlations of Piers-Harris scores with other self-concept scales are moderate. As described in the manual, correlations range from .32 on the Personal Attribute Inventory for Children to .85 on the Coopersmith Self-Esteem Inventory, which is like the Piers-Harris in its format and age range (Piers, 1984).

Internal consistency measures for the six cluster scales are provided in the manual. These internal reliability measures are based on the initial standardization sample for these scales (N = 485) and a sample of 97 children from Pennsylvania outpatient clinics. The range in their reliabilities is from .73 to .90.

Test-retest reliabilities of the cluster scales indicate that the delineated factorial structures are relatively unstable, even when given to the same group twice (Platten et al., 1979). The stability of the factors across groups has been only moderate. While most studies show partial replication of these clusters with normal (Michael et al., 1975) and special groups (Rich, Barcikowski, & Witmer, 1979; Stewart, Crump, & McLean, 1979; Wolf, Sklov, Hunter, Webber, & Berenson, 1982), somewhat different clusters have been noted in each assessment. The effective use of the reported cluster scores is limited by this instability and by the concurrent dearth of validity studies.

Western Psychological Services provides computerized scoring and detailed reports for the Piers-Harris Self Concept Scale. The report has three segments: 1) a group report; 2) an individual report; and 3) a school-oriented report.

Comparative statistics will be generated for all children for whom the group report is requested. The test user is warned not to glean too much from the individual scores at this point, as individual validity problems are not addressed in detail. Rather, the range of high and low responses to the test is presented in tabular form, along with the global mean and standard deviation. If any objective indicators of validity problems appear (i.e., either the child does not appear to have answered the questions truthfully as reflected in unusual response patterns, or the child's personal characteristics put the validity of the provided norms in question for him or her), the record is noted as having a validity consideration. Specific validity problems are not detailed in this report; the test user is referred to the individual reports for this analysis. The group report ends by noting all children whose score falls one or more standard deviation units below the mean. In sum, the group report presents a general screening of all scores with minimal interpretation.

The individual report contains identification information and a short narrative on the background and use of the scale. This information is followed by an assessment of the validity of that test based on the criteria described earlier. Whether or not the validity of the test is in question, the scores are plotted and discussed. Interpretations of the total raw score and the six cluster scores are provided. Interpretations of the global test score vary with each .5 standard deviation from the mean. Use of these particular discriminations is arbitrary; more importantly, the use of the original sample as a standard, given the fluctuations in mean and standard deviation scores across subsequent samples, limits the validity of the interpretive statements. This situation is compounded in the reported interpretations of cluster scores. Given the instability of these clusters across studies, the information derived from those item-groupings can only be considered suggestive. All scores are presented in an "ipsative table," which presents the child's scores in relation to each other with an analysis of the statistical significance of these differences. This intraindividual analysis provides an important description of the child's strengths and weaknesses, but it does not address the absolute levels of these attributes. Individual item responses are also analyzed with respect to the percentage of children in the original sample who responded in a similar manner. Use of these figures, as well as the summary report, is limited by the instability of the original sample scores on which they are based.

The school report contains the same information as the individual report, but is prepared in a briefer format suited for school needs. While this structure can be timesaving for professionals needing to make quick reports, it suffers from the same validity problems as does the individual report.

Critique

The Piers-Harris Children's Self-Concept Scale provides a direct measure of how children report they feel about themselves. Its strengths include ease of administration and scoring and a direct method of ascertaining self-concept. Two major weaknesses are the limitations in its current standardization sample and the unreliability of its factorial clusters.

The manual accurately and explicitly delineates these factors and provides a clear picture of the test's limitations, as well as its intended uses for screening within an individual assessment and in research. While the literature suggests that the test is often used for purposes which go beyond the dictates of reliability and validity considerations, the scale can provide important and interesting information on children's social and affective states when limited to its intended uses and when the obtained scores are integrated with data from additional sources.

References

American Psychological Association (1974). *Standards for educational and psychological tests* (rev. ed.). Washington, DC: Author.

Beck, M.A., Roblee, K., & Hanson, J. (1982). Special education/regular education: A comparison of self-concept. *Education, 102*(3), 277-279.

Bloom, R.B., Shea, R.J., & Eun, B. (1979). The Piers-Harris Self-Concept Scale: Norms for behaviorally disordered children. *Psychology in the Schools, 16*, 483-487.

Chang, T.S. (1975). The self-concept of children in ethnic groups: Black American and Korean American. *Elementary School Journal, 76*(1), 52-58.

Crase, S., Foss, C., & Colbert, K. (1981). Children's self concept and perception of parents' behavior. *Journal of Psychology, 108*, 297-303.

Goldman, R.L., & Hardin, V.B. (1982). The social perception of learning disabled and non-learning disabled children. *Exceptional Child, 29*, 57-63.

Gordon, M., & Tegtmeyer, P.F. (1982). The egocentricity index and self esteem in children. *Perceptual and Motor Skills, 55*, 335-337.

Grayhill, D. (1978). Relationship of material child-rearing behaviors to children's self esteem. *Journal of Psychology, 100*, 45-47.

Henggeler, S.W., & Tavormina, J.B. (1979). Stability of psychological assessment measures for children of Mexican American migrant workers. *Hispanic Journal of Behavioral Sciences, 1*(3), 263-270.

Jegede, R.O., & Bamgboye, E.A. (1981). Self-concepts of young Nigerian adolescents. *Psychological Reports, 49*(2), 451-454.

Kanoy, R.C., Johnson, B.W., & Kanoy, K.W. (1980). Locus of control and self-concept in achieving and underachieving bright elementary students. *Psychology in the Schools, 17*, 395-399.

Karnes, F.A., & Wherry, J.N. (1981). Self-concepts of gifted students as measured by the Piers-Harris Children's Self-Concept Scale. *Psychological Reports, 49*, 903-906.

Ketcham, B., & Snyder, R.T. (1977). Self-attitudes of the intellectually and socially advantaged student: Normative study of the Piers-Harris Children's Self-Concept Scale. *Psychological Reports, 40*, 111-116.

Kim, S.P. (1983). Self concept, English language acquisition, and school adaptation in recently immigrated Asian children. *Journal of Children in Contemporary Society, 15*, 71-79.

Lefley, H.P. (1974). Social and familial correlates of self-esteem among American Indian children. *Child Development, 45*, 829-833.

Mannarino, A.P. (1978). Friendship patterns and self-concept development in preadolescent males. *Journal of Genetic Psychology, 113*, 105-110.

Michael, W.B., Smith, R.A., & Michael, J.J. (1975). The factorial validity of the Piers-Harris Children's Self Concept Scale for each of three samples of elementary, junior high, and senior high students in a large metropolitan high school district. *Educational and Psychological Measurement, 35*, 405-414.

Moyal, B.R. (1977). Locus of control, self-esteem, stimulus appraisal, and depressive symptoms in children. *Journal of Consulting and Clinical Psychology, 45*, 951-952.

Piers, E.V. (1972). Parent prediction of children's self-concepts. *Journal of Consulting and Clinical Psychology, 38*, 428-433.

Piers, E.V. (1977). Children's self-esteem, level of esteem certainty, and responsibility for success and failure. *Journal of Genetic Psychology, 130*, 295-304.

Piers, E.V. (1984). *Revised manual for the Piers-Harris Children's Self Concept Scale.* Los Angeles, CA: Western Psychological Services.

Platten, M.R., & Williams, L.R. (1979). A comparative analysis of the factorial structures of two administrations of the Piers-Harris Children's Self-Concept Scale to one group of elementary school children. *Educational and Psychological Measurement, 39,* 471-478.

Platten, M.R., & Williams, L.R. (1981). Replication of a test-retest factorial validity study with the Piers-Harris Children's Self-Concept Scale. *Educational and Psychological Measurement, 41,* 453-461.

Rich, C.E., Barcikowski, R.S., & Witmer, J.M. (1979). The factorial validity of the Piers-Harris Children's Self-Concept Scale for a sample of intermediate-level EMR students enrolled in elementary school. *Educational and Psychological Measurement, 39*(2), 485-490.

Shavelson, R.J., & Bolus, R. (1982). Self-concept: The interplay of theory and methods. *Journal of Educational Psychology, 74*(1), 3-17.

Sheare, J.B. (1978). The impact of resource programs upon the self concept and peer acceptance of learning disabled children. *Psychology in the Schools, 15,* 406-412.

Silverman, R., & Zigmond, N. (1983). Self-concept in learning disabled adolescents. *Journal of Learning Disabilities, 16,* 478-490.

Smith, M.D., & Rogers, C.M. (1978). Reliability of standardized assessment instruments when used with learning disabled children. *Learning Disabilities Quarterly, 1,* 23-30.

Smith, M.D., Zingale, S.A., & Coleman, J.M. (1978). The influence of adult expectancy/child performance discrepancies upon children's self-concepts. *American Educational Research Journal, 15,* 259-265.

Stewart, D.J., Crump, W.D., & McLean, J.E. (1979). Response instability on the Piers-Harris Children's Self-Concept Scale. *Journal of Learning Disabilities, 12*(5), 71-75.

Wanat, P. (1983). Social skills: An awareness program with learning disabled adolescents. *Journal of Learning Disabilities, 16,* 35-38.

Wheeler, V., & Ladd, G. (1982). Assessment of children's self efficacy for social interactions with peers. *Developmental Psychology, 18,* 795-805.

Wolf, T.M., Sklov, M.C., Hunter, S.M., Webber, L.S., & Berenson, G.S. (1982). Factor analytic study of the Piers-Harris Children's Self-Concept Scale. *Journal of Personality Assessment, 46,* 511-513.

Rolf A. Peterson, Ph.D.

Professor and Chairman, Department of Psychology, University of Health Sciences, The Chicago Medical School, North Chicago, Illinois.

Sandra W. Headen, Ph.D.

Assistant Professor of Psychology, University of Health Sciences, The Chicago Medical School, North Chicago, Illinois.

PROFILE OF MOOD STATES

Douglas M. McNair, Maurice Lorr, and Leo Droppleman. San Diego, California: Educational and Industrial Testing Service.

THE ORIGINAL OF THIS REVIEW WAS PUBLISHED IN TEST CRITIQUES: VOLUME I (1984).

Introduction

The Profile of Mood States is an adjective rating form which assesses present mood state. Present mood state is viewed as the emotional state of an individual which is transient and responsive to changes in the environment compared with a more stable, long-term mood state which may reflect enduring personality traits. Thus, the POMS measures the momentary mood state at the time of test administration. The POMS assesses six affective or emotional dimensions: tension-anxiety, depression-dejection, anger-hostility, vigor-activity, fatigue-inertia, and confusion-bewilderment. Factor scores for each dimension are obtained from responses to 65 adjectives rated on a five-point scale ranging from zero, "not at all," to four, "extremely."

Development of the POMS began in the early 1960s when the authors, Douglas M. McNair, Ph.D., Maurice Lorr, Ph.D., and Leo F. Droppleman, Ph.D., undertook a series of factor analytic studies to determine the components of mood state and develop an assessment instrument which would measure each of the dimensions of mood. Their first step was to assemble adjectives from a wide variety of scales already in use, along with the dictionary and thesaurus providing the original pool of items. They administered adjective lists to different subject populations in a series of studies. Based on factor analytic outcomes items were discarded until the present 65-item list was obtained. Further validation studies revealed high internal consistency within mood dimensions and test-retest reliability ranging from $r = .65$ to $r = .74$. Although lower than would be expected for a stable personality trait, the test-retest reliability coefficients were deemed appropriate for assessing emotional states having transient qualities and expected to be extremely sensitive to environmental changes. Issues of predictive and construct validity were addressed in studies which assessed scale changes associated with brief psychotherapy and controlled drug trials, studies which induced emotional states, and studies correlating the POMS with other mood measures. The manual for the POMS provides norms which were primarily obtained from two sources. The first population includes 1,000 consecutive admissions to an outpatient clinic exclusive of illiterates, alcoholics, extreme psychotics, emergency admissions, and non-English-speaking individuals. Separate norms are provided for male and female patients. Means for factor scores broken down by diagnosis and treatment disposition are also provided in the manual. Norms are also provided for 340 male and 516 female college students.

The POMS in its present form was published by the Educational and Industrial Testing Service (EdITS) in 1971. It has not been revised since that time nor does the manual provide information regarding other versions of the test, e.g., alternate forms to reduce practice effects, non-English language versions, or forms designed for other special populations. However, development of a new form of the POMS to assess bipolar mood states is in progress (Lorr, McNair, & Fischer, 1982).

The test itself is contained on one side of a single 8" x 10" sheet which includes instructions for the appropriate cognitive set for completing the form, and an explanation of the conceptual meaning of the rating scale. Item difficulty is geared to a seventh-grade education level and items are presented in an easy-to-under-stand format. Since the test is a self-report measure, it can be administered individually or in groups. The examiner merely ensures that respondents under-stand the instructions and directs them to fill out the form. The form as published is set up to rate respondents' feelings, "during the past week including today," but the scale is also used with instructions for different time periods. For example, "rate your feelings during the last half hour." If time periods other than those included in the instructions are used, the examiner must be certain that respondents understand the specific time period which is being used. The one-page test format includes both items and response format. A set of six cardboard overlays is used to score each mood factor and scores are recorded on the answer form. Some caution is advised in using the overlays. Two of the factors include a correction score where the score for one item is subtracted from the total score. Because this procedure deviates from the additive method of obtaining other scale scores, care is needed to ensure that these scores are accurate. However, the forms are clear and appropriately alert the user to this feature so that normal attention to the task should result in correct scoring.

Normative profile sheets for a female and male outpatient population and a normal college population allow the transformation of raw scores into standard-ized T-scores. These scores can then be plotted to provide a line profile reflecting all six mood dimensions simultaneously. The mood state measured by each scale is relatively self-explanatory from the scale title and seems to correspond to a layman's understanding of the different moods with one exception. The tension-anxiety scale was developed to measure physical or muscular-skeletal tension rather than cognitive tension, anxiety, or the generalized feeling of discomfort.

In addition to the six mood states, a total mood disturbance score can be obtained. However, the manual does not provide norms or validity data for the total score, nor does the profile sheet provide a space for recording it. Thus, a total mood disturbance score should be used cautiously since it can only be interpreted within the context of other published studies which have reported total scores.

Practical Applications/Uses

As a research tool, the POMS has been primarily used to measure mood changes associated with brief psychotherapy, drug treatments, and psychological or physical interventions such as relaxation, stress management, or exercise. In evaluating the consistency of the results of these studies, the investigator should be alerted to the diversity of time period ratings and form used in reporting scores for the POMS. Some researchers report results in terms of average score per item rather than scale scores as indicated in the manual.

These reviewers and their students have used the POMS in a more general clinical sense to determine the present mood state of individual clients or group members and to monitor the success or lack of success of attempts to modify mood state. The POMS appears to be a convenient and quick way to assess the emotional states of individuals participating in stress management groups because of its sensitivity to changes in moods which are often the targets of training in stress management. Again, the POMS is most useful in assessing immediate, transient mood level and is probably not of value when attempting to assess severe psychiatric disorders or behavior, such as psychotic depression, manic behavior, or suicide. Also, the POMS is descriptive of present mood on six specific dimensions and was not developed for the purpose of understanding personality or environmental factors which induce mood state. It seems reasonable, however, that information on this topic can be obtained by using the POMS in conjunction with other methods. For example, when used in combination with interviews or other tests which measure recent life events, hypotheses about how these events may influence emotional state can be developed. Clinicians seeking a quick and easy psychometric assessment of mood, either for the purpose of general assessment in screening, or for the purpose of evaluating the effectiveness of the intervention provided, should give serious consideration to including the POMS as a standard pre-test and post-test as well as using it as an ongoing assessment procedure. Although actual administration and scoring can be done by an assistant, the interpretation of the results, both for research and clinical use, should be done by a professional trained in assessment and experienced with the population being assessed.

The POMS has been a sensitive and popular assessment tool for the purposes for which it was originally designed. It is frequently used in experimental studies assessing mood change and effects of psychological therapy and/or drug use on mood. Additionally, some researchers have used the POMS to obtain normative and descriptive mood profiles and mood levels for particular clinical populations. A selected set of published articles describing different uses of the POMS is provided in the reference section. Although these articles may provide some clinicians and researchers with data on comparison groups similar to the populations they are interested in, the issue of limited normative data for the POMS is problematic. The POMS was originally developed to assess mood in psychiatric outpatients and normals aged 15 and over with at least some high school education. Norms are also available in the manual for a college student population. Thus, additional normative data would be needed for non-psychiatric, non-college populations. It should also be noted that while the POMS would appear to be easily adapted for use with any desired population or format, i.e., oral presentation or a Braille form, materials are not available for such adaptations and a user would need to develop appropriate norms for this type of administration.

As stated previously, administration of the POMS as described in the manual is simple and straightforward. It can be administered to individuals or groups by a secretary or proctor under the general supervision of a psychologist or other professional trained in test development and interpretation. Hand scoring can be done quickly (two to three minutes per test) using card overlays which provide relatively clear instructions. The actual scoring, counting, and recording of the numbers, and their entry on the profile sheet, can be handled by a secretary or an assistant. At the moment, a computer scoring service is not available and in view of the ease of scoring it is probably not necessary unless an extremely large number of subjects is tested. Subject scores are easily interpreted by transforming individual scores on each mood scale to the standardized score for the appropriate normative sample. An intra-subject profile of scores on the six dimensions of the POMS relative to the normative sample is obtained by plotting the scores on the forms provided. These intra-subject profiles should be interpreted with caution, however. Validated scores indicating critical cutoff values which could be used for the purposes of individual diagnosis or intervention are not provided and use of the test in this manner is not one that is recommended in the manual. Thus, in a clinical setting, the use of the POMS to interpret emotional state and severity of problems should be restricted to psychologists trained in therapeutic technique and test interpretation and familiar with the purposes and recommended uses of the POMS.

Technical Aspects

In developing an instrument to assess their proposed multidimensional mood construct, the authors were required to demonstrate that the POMS produces scales which are conceptually valid, internally consistent, statistically independent, and replicable. A series of factor-analytic studies described in the manual show high correlations within factors, and both independence and replication of factors in the populations used.

Data on test-retest reliability obtained from a large sample of outpatient psychiatric clients at intake and again prior to entry into treatment are also provided. The median time between testing was 20 days with resulting reliability coefficients of .70 for tension-anxiety, .74 for depression-dejection, .71 for anger-hostility, .65 for vigor-activity, .66 for fatigue-inertia, and .68 for confusion-bewilderment. Since the POMS is designed to measure momentary emotional states and is sensitive to short term changes in mood, it is not clear how the test-retest reliability data reported in the manual should be interpreted. On the one hand, test-retest reliability coefficients in the .60s and .70s for a median 20-day period would appear to indicate a fairly stable mood state. Whether this means that the POMS is sufficiently sensitive to mood changes depends on the nature of the experiences of the subjects tested over the 20-day period. If little change occurred in a subject's life, the present mood state may remain highly stable until some intervention produces a mood change. However, the POMS is designed for and frequently used in research to measure changes in mood during very brief periods of time, including multiple assessments in brief periods (i.e., every five minutes for 30 minutes). Thus, more empirical information on test-retest reliability for a variety of time periods and with samples which vary in the stability of life circumstances (i.e., final exam week) is needed. On the other hand, the data from studies on drug medication and psychotherapy show significant changes in mood state in the predicted directions, indicating that the scales are highly sensitive to momentary changes in mood over periods ranging from minutes to hours, days, or weeks.

Information on the validity of the POMS comes primarily from three sources: studies involving factorial validity, score changes as a result of clinical intervention or experimentally induced conditions, and correlations with other scales that also assess aspects of mood. With six different factor analytic replications, the factor loadings of the individual items for each factor are relatively consistent and are related to the factor with sufficient strength to suggest that they are an important contributor to that factor. The results suggest good validity for the factor structure of the POMS although it cannot be assumed that these same factors nor independence of factors will be found in populations not assessed in the original developmental studies. Evidence for the construct validity of the POMS is provided by the high correlations obtained between the POMS and a variety of other scales which purport to measure emotional states of one type or another. The extent to which these criterion scales accurately assess the purported mood factor could be questioned, however. Overall, the data suggest that the POMS is a solid measure of mood state, particularly the relatively consistent findings obtained in studies assessing mood changes associated with clinical intervention or experimental manipulation.

Critique

These reviewers generally agree with the following statement by Weckowicz (1978): "In summary, the POMS is an excellent research and clinical evaluation psychometric instrument for the assessment of mood and feelings states of psychiatric outpatients. It is particularly useful for an evaluation of the effects of psychotherapy and medication. It is easy to administer and easy to score. While the validity of the test is apparently well established, some further studies of reliability are needed" (p. 1019).

Although the POMS appears to be quite adequate in assessing mood states, several cautionary statements are in order. First, norms published in the manual are rather limited and provide few comparison populations. It is also not known how well other populations, e.g., normal adults, adults exposed to normal environmental conditions, or adults under high stress, compare with the normative samples provided. It is not clear what environmental circumstances existed when the normative data were obtained. This is especially important for the sample of college students who may have been taking exams or undergoing other stressful conditions. The time and circumstances of administration might affect the degree to which the normative sample would reflect the average college student at any given time. Since the scale by its very nature is designed to assess momentary or transient mood level, the appropriate normative comparison is difficult to determine as the norms average all the different individual emotional states into a given profile. These individual states are the result of particular events which vary in uniformity and content. Therefore, for both clinical and research purposes, it is recommended that respondents' scores on the POMS be compared with those of similar people under similar conditions in order to provide a more accurate interpretation of the meaning of an individual score or change in scores.

Some problems also exist in the ease with which the POMS can be faked or answered to avoid exposure. The content of the items is generally obvious and the test does not include a lie, defensiveness, or social desirability scale. This issue is addressed briefly in the manual which reports low to moderate correlations with a social desirability scale for psychiatric outpatients. Evaluating biases due to social desirability on this population is problematic, however. Outpatients seeking treatment may not be as motivated to hide particular feelings as other groups. For example, when using the POMS for clinical or research assessment of non-volunteers or populations with special characteristics, respondents may fear the consequences of exposing their true responses to the scrutiny of others. An example of such a population would be police officers encouraged by their superiors to take a stress management course. These individuals may feel compelled to hide their true feelings from their superiors. Therefore, when the POMS is used with an individual or group which may be motivated to conceal their moods, the scores should be interpreted with extreme caution, taking into account the type of sample and the particular situation.

Another issue users of the POMS should be alerted to is the lack of specificity in meaning of different factor patterns obtained with this test. For example, the manual provides no guidelines regarding how a subject profile which shows high scores on depression, anxiety, and anger and low scores on vigor and confusion should be interpreted. It should be noted that while a particular factor pattern may be meaningful for one population, the same pattern may be less interpretable with another population. For example, in one study (Beebe et al., 1984) evaluating the impact of a behavior modification treatment program on patients with non-insulin-dependent diabetes mellitus, factor patterns showing negative associations between vigor and fatigue and elevated scores on depression, anxiety, and anger reflected the poor physiologic state of patients entering the program. Modest improvements in health status after ten weeks were associated with changes in this factor pattern in the expected direction; that is, scores on depression, anxiety, anger, and fatigue decreased and scores on vigor increased. However, in another study (Peterson et al., 1983), the factor pattern obtained was of little value in interpreting the effects of stress management training on normal supervisory staff at a facility for the developmentally disabled. In this study, the results showed that the depression scale was highly correlated with other scales: tension, $r = .90$, anxiety, $r = .75$, fatigue, $r = .74$, and confusion, $r = .89$. High intercorrelations among other scales were also observed. The vigor scale was negatively related at a relatively low level with other scales, the highest correlation being $r = -.42$ between vigor and depression. In this study at least, it appeared most appropriate to view the POMS as measuring two independent mood components, depression and vigor. Although this pattern of factor scores is similar in some ways to that obtained in the previous study, it was not useful in evaluating the expected differential impact of stress management training on ostensibly six independent components of mood.

Taking into account the comments and precautions discussed previously, the overall evidence suggests that the POMS is a relatively valid instrument for measuring momentary mood state. It is appropriate for the assessment of psychiatric outpatients and particularly useful and meaningful in studies assessing mood change associated with psychological or medication intervention. The usefulness of the scale as a general clinical tool is less clear. However, for the well-trained individual who is aware of the problems of appropriate norms and factor structure, the scale does appear to provide an important source of information in assessing present mood state and monitoring mood changes. In addition, the POMS provides an easy and convenient method for monitoring the effectiveness of treatment, at least in regard to mood changes.

References

This list includes text citations as well as suggested additional reading.

Beebe, C., Fischer, B., Headen, S., & McCracken, S. (1984). *Lifestyle change: A new approach to treatment of Type II diabetes.* Paper presented at Forty-Fourth Annual Meeting of the American Diabetes Association, Las Vegas, Nevada.

Eichman, W. J. (1978). Profile of Mood States. In O. K. Buros (Ed.), *The eighth mental measurements yearbook* (Vol. 1) (pp. 1015-1018). Highland Park, NJ: The Gryphon Press.

Gosselin, R. Y., Lubin, B., & Sokoloff, R. M. (1983). Some preliminary findings on new MAACL scales. *Psychological Reports, 53,* 1205-1206.

Jacobson, A. F., Weiss, B. L., Steinbook, R. M., Brauzer, B., & Goldstein, B. J. (1978). The measurement of psychological states by the use of factors derived from a combination of items from mood and symptom checklists. *Journal of Clinical Psychology, 34*(3), 677-685.

McNair, D. M., Lorr, M., & Droppleman, L. F. (1971). *Profile of Mood States.* San Diego: EdITS/ Educational and Industrial Testing Service, Inc.

Lorr, M., McNair, D. M., & Fisher, S. (1982). Evidence for bipolar mood states. *Journal of Personality Assessment, 46*(4), 432-436.

Peterson, R. A., Calamari, J., Greenberg, G., Giffort, D., & Schiers, B. (1983, August). *Evaluation of two strategies to reduce stress on supervisory staff in a residential facility for the DD.* Paper presented at the meetings of the Illinois American Association in Mental Deficiency, Homewood, IL.

Shacham, S. (1983). A shortened version of the Profile of Mood States. *Journal of Personality Assessment, 47*(3), 305-306.

Weckowicz, T. E. (1978). Profile of Mood States. In O. K. Buros (Ed.), *The eighth mental measurements yearbook* (Vol. 1, pp. 1018-1019). Highland Park, NJ: The Gryphon Press.

Drug Studies

Fischman, M. W., Schuster, C. R., & Rajfer, S. (1983). A comparison of the subjective and cardiovascular effects of cocaine and procaine in humans. *Pharmacology, Biochemistry and Behavior, 18,* 711-716.

Glass, R. M., Uhlenhuth, E. H., Hartel, F. W., Schuster, C. R., & Fischman, M. W. (1980), A single dose study of nabilone, a synthetic cannabinoid. *Psychopharmacology, 71,* 137-142.

Haskell, D. S., Gambill, J.D., Gardos, G., McNair, D.M., & Fisher, S. (1978). Doxepin or diazepam for anxions and anxions-depressed outpatients. *Journal of Clinical Psychiatry, 39*(2), 135-139.

Johansen, C. E., & Uhlenhuth, E. H. (1980). Drug preference and mood in humans: d-amphetamine. *Psychopharmacology, 71,* 275-279.

Health Issues and Pain

Abplanalp, J. M., Donnelly, A. F., & Rose, R. M. (1979). Psychoendocrinology of the menstrual cycle: 1. Enjoyment of daily activities and moods. *Psychosomatic Medicine, 41*(8), 587-603.

Gath, D., Cooper, P., & Day, A. (1982). Hysterectomy and psychiatric disorder: 1. Levels of psychiatric morbidity before and after hysterectomy. *British Journal of Psychiatry, 140,* 335-350.

Parker, J. C., Granberg, B. W., Nichols, W. K., Jones, J. G., & Hewett, J. E. (1983). *Journal of Clinical Neuropsychology, 5(4), 345-353.*

Shacham, S., Reinhardt, L. C., Raubertas, R. F., & Cleeland, C. S. (1983). Emotional states and pain: Intraindividual and interindividual measures of association. *Journal of Behavioral Medicine, 6(4), 405-419.*

Taylor, S. E., Lichtman, R. R., & Wood, J. V. (1984). Attributions, beliefs about control and adjustment to breast cancer. *Journal of Personality and Social Psychology, 46*(3), 489-502.

Vollhardt, B. R., Ackerman, S. H., Grayzel, A. I., & Barland, P. (1982). Psychologically distinguishable groups of rheumatoid arthritis patients: A controlled, single blind study. *Psychosomatic Medicine, 44*(4), 353-362.

Exercise

Blumenthal, J. A., Williams, S., Needels, T. L., & Wallace, A. G. (1982). Psychological changes accompany aerobic exercise in healthy middle-aged adults. *Psychosomatic Medicine, 44*(6), 529-536.

Gondola, J. C., & Tuckman, B. W. (1982). Psychological mood state in "average" marathon runners. *Perceptual and Motor Skills, 55,* 1295-1300.

Alcoholism

Freed, E. X., Ripley, E. P., & Ornstein, P. (1977). Assessment of alcoholics' moods at the beginning and end of a hospital treatment program. *Journal of Clinical Psychology, 33*(3), 887-894.

Criminal Offenders

Fagan, T. J., & Lira, F. T. (1978). Profile of mood states: Racial differences in a delinquent population. *Psychological Reports, 43*(2), 348-350.

Lira, F. T., & Fagan, T. J. (1978). The profile of mood states: Normative data on a delinquent population. *Psychological Reports, 42*(2), 640-642.

Relaxation

Gilbert, G. S., Parker, J. C., & Clairborn, C. D. (1978). Differential mood changes in alcoholics as a function of anxiety management strategies. *Journal of Clinical Psychology, 34,* 229-232.

Townshend, R. E., House, J. F., & Addaria, D. (1975). A comparison of biofeedback mediated relaxation and group therapy in the treatment of chronic anxiety. *American Journal of Psychiatry, 132,* 598-601.

Turin, A., Nirenberg, J., & Mattingly, M. (1979). Effects of comprehensive relaxation training (CRT) on mood: A preliminary report on relaxation plus caffeine cessation. *Behavior Therapist, 2*(4), 20-21.

Stress

Costantini, A. F., Braun, J. R., Davis, J., & Iervalind, A. (1973). Personality and mood correlates of schedule of recent experience scores. *Psychological Reports, 32*(3), 1143-1150.

McKinney, B., & Peterson, R. A. (1983, August). *Evaluation of stress in parents of developmentally disabled children.* Paper presented at the meetings of the Illinois American Association in Mental Deficiency, Homewood, IL.

Howard Lerner, Ph.D.

Professor of Psychology, University of Michigan, School of Medicine, Ann Arbor, Michigan.

Paul M. Lerner, Ed.D.

Faculty of Medicine, University of Toronto, Toronto, Ontario.

RORSCHACH INKBLOT TEST

Hermann Rorschach. Bern, Switzerland: Hans Huber. U. S. Distributor—Grune & Stratton, Inc.

THE ORIGINAL OF THIS REVIEW WAS PUBLISHED IN TEST CRITIQUES: VOLUME IV (1986).

Introduction

The past decade has witnessed a significant resurgence of the use of the Rorschach Inkblot Test for the study and understanding of people, a virtual revival of what Millon (1984) has referred to as its "rich heritage of the 1940's and 1950's." In the previous decade, the 1960s, both personality assessment and personality theory had fallen into disrepute, and hit hardest were the projective techniques, including the Rorschach, and psychoanalytic theory. The veritable explosion of new treatment modalities onto the psychotherapeutic marketplace, many of which stressed brevity and cost efficiency, together with the expanding roles being filled by psychologists seemed to render traditional personality assessment obsolete, if not anachronistic. Anti-test attitudes were rampant and have been well recorded in the testing literature (Holt, 1967, 1968; Shevrin & Schectman, 1973; Millon, 1984). Humanistic psychologists reviewed assessment as exploitive, as robbing clients of their individuality and dignity; those with a social perspective equated assessing with labeling and argued that the diagnostic process itself was dehumanizing and antitherapeutic; and those psychologists in academia insisted that statistically derived predictions were superior to clinically based ones.

With respect to the Rorschach the drought is over. Catapulting renewed interest in the Rorschach has been the empirical work of John Exner and his colleagues (Exner, 1974, 1978; Viglione & Exner, 1983) together with major shifts in psychoanalytic theory. With regard to the latter, the relatively recent integration of modern object relations theory, a broadened psychodynamic developmental theory, and a systematic psychology of the self into the mainstream of traditional psychoanalytic theory is now providing the conceptual basis for a more comprehensive, fruitful, and systematic Rorschach theory.

The purpose of this critique is to examine and summarize the current status of the Rorschach, beginning with a review of the contributions of Hermann Rorschach, the originator of the test. This will be followed by a discussion of the most significant issue permeating contemporary Rorschach theory and practice—conceptual versus empirical approach. One major conceptual approach, that based on psychoanalytic theory, will then be outlined. Included in the outline will be a review of the contributions of David Rapaport, together with a discussion of the impact of Rorschach research and clinical use on conceptual shifts in psycho-analytic theory, and concluding with future directions these reviewers believe that work with and on the Rorschach will take.

Contributions of Hermann Rorschach. According to Rickers-Ovsiankina (1977), "grasping the inner workings of man was a life-long concern of Rorschach's" (p. x). It was toward this broad goal that he directed his experiments with inkblots. Employing inkblots (i.e., using forms obtained through chance by folding over a piece of paper into the center of which ink had been dropped) to explore an aspect of personality did not begin with Rorschach. In 1857, Kerner (note Pichot, 1984) reported on his experiments with inkblots, and at the turn of the century Binet used chance inkblots to assess imaginative capacities in children.

Rorschach went beyond these early researchers. He not only standardized an inkblot procedure, but was also able to synthesize the procedure with Jung's work on the Word Association Test and Bleuler's notions regarding personality assessment. Jung published his research with the Word Association Test between 1906-1909 under the general title "Diagnostic Association Studies." Basic to his research were the important insights that any one particular act (i.e., handwriting, choice of clothing, etc.) could represent the whole person and that one could explore personality by systematically studying an individual's reactions to a stimulus. Bleuler had made similar observations, noting that such reactions could be taken as "an index of all mental activity so that it is only necessary to decipher them in order to know the entire man" (Pichot, 1984, p. 595).

When responses to inkblots had been used to assess imagination, emphasis was placed on the content of the responses. By contrast, Rorschach stressed not the content, but rather the formal properties of the response, and, as such, this enabled him to conceptualize the test as one of perception and not of imagination. He was then able to demonstrate that peculiarities in perception were dependent upon the nature of the underlying personality structure, including pathological deviations. In other words, by noting the intimate relationship between perceptual reactions and other psychological functions, Rorschach was able to conceptually place his technique in the middle of the assessment of total personality functioning.

Rorschach codified his theoretical understandings of the inkblot procedure by developing categories for analyzing responses—the determinants. Later Rorschach workers (Rickers-Ovsiankina, 1977; Pichot, 1984) consider this to be his most important contribution.

Contemporary Issues. As Weiner (1977) points out, historically clinicians and researchers have struggled with issues of Rorschach validity. Out of this struggle, according to Weiner (1977), three types of opinions have emerged concerning the psychometric status of the Rorschach. One extreme opinion claims that the Rorschach, despite thousands of research studies and a voluminous clinical and theoretical literature, has failed to demonstrate its validity according to strict psychometric standards. According to this point of view, the Rorschach should be discarded as an assessment procedure. A second, more clinical opinion holds that the Rorschach is not a test at all and therefore should not be judged by psychometric criteria. This point of view holds that the Rorschach is a special type of clinical interaction based upon the skills and sensitivities of the examiner rather than the psychometric characteristics of the instrument. In keeping with this point of view, psychometric studies of validity are inappropriate to the nature of the instrument and statistical studies have no bearing on its clinical utility. Weiner (1977) represents a third, more integrated point of view: ". . . some psychologists have been determined to maintain their scientist's respect for objective data without sacrificing the richness and utility of the Rorschach in which they had come to believe as clinicians. This determination has engendered thoughtful criticism of previous validity studies and proposals for new approaches to Rorschach validation that combine sophisticated research design with adequate attention to the nuances of clinical practice" (p. 576).

Evolving out of the debate concerning the validity of the Rorschach, two major approaches to using and evaluating the instrument have recently emerged: the empirical sign or psychometric approach and the conceptual approach. While these somewhat divergent approaches to the Rorschach are not mutually exclusive, issues pertaining to either a more psychometric or conceptual approach to the Rorschach underlie much of the research and clinical studies found in the contemporary literature. The evaluation of each approach will be outlined with a view toward uncovering those major issues at the forefront of Rorschach inquiry today.

Practical Applications/Uses

The empirical sign or psychometric approach to the Rorschach began with Hermann Rorschach (1921/1942) as he compared previously selected diagnostic groups and found that certain "signs" or scores occurred more frequently in some groups than in others. While this approach has spanned the Rorschach's history and has cut across theoretical orientations to the instrument, through a number of major books and publications, Exner and his colleagues (Exner, 1974, 1978; Exner, Weiner, & Schuyler, 1979) have become the most thorough, instrumental, and contemporary representatives of the Rorschach's psychometric or sign approach. What these authors term the Comprehensive System, an integration of five major Rorschach systems, synthesizes, according to psychometric criteria, the most reliable and useful indices from other systems. Supplementing the Comprehensive System are selected variables, scores, and ratios that demonstrate high interscorer reliability and correlate with other psychological variables. Much of the research conducted by Exner and his colleagues involves further delineation of the constructs associated with these variables and collection of vital normative data on the Rorschach.

According to Viglione and Exner (1983), studies based on the Comprehensive System are united by understanding the Rorschach as primarily a "problem-solving" task that activates relatively consistent styles of coping behavior. According to these authors: "The data gleaned from the subject's responses, or solutions, provide a glimpse of how the individual works with the world, how he responds to ambiguity and challenge. These data are quantified in the various scores and more indirectly expressed in the words and other behaviors that are observed. The impressive temporal consistency of the various Rorschach scores, as derived from test-retest studies (Exner, 1978; Exner, Armbruster, & Viglione, 1978), suggests that these solutions to the Rorschach problem are representative of one's relatively stable, idiographic style of coping with the world" (p. 14). The problem-solving orientation to the Rorschach, and with it an emphasis on habits and decision operations, stands in marked contrast to the instrument seen in terms of a "projective process" and wedded to a theory of personality independent of the test itself.

The foundation for Rorschach interpretation based upon the Comprehensive System is what Exner (1978) terms the "4-square," which incorporates the basic scores and ratios thought to be characteristic of one's problem-solving style. The four indices of the 4-square are 1) Erlebnistypus (EB, the ratio of human movement to weighted color responses); 2) Experience Actual (EA, the sum of human movement and weighted color responses); 3) Experience Base (eb, the ratio of non-human movement to shading and gray-black responses); and 4) Experience Potential (ep, the sum of non-human movement, shading, and gray-black responses). According to Exner and his colleagues, these four variables taken as a whole incorporate the fundamental information about the psychological habits and capacities of an individual and represent the crucial interpretative departure from previous Rorschach systems. The four variables comprising the 4-square are thought to be symmetrical, that is, two ratios (EB, eb) with a corresponding sum (EA, ep). In keeping with Rorschach's (1921) original formulations, EB generates information on individuals' tendencies toward ideational or affective coping responses to stress. By combining both sides of the EB to form the sum EA, an index is formed to represent the "amount" of organized resources available for coping (Beck, 1960). Based on the formulations of Klopfer, Ainsworth, Klopfer, and Holt (1965) that non-human movement and some shading and gray-black responses correspond to psychological activity outside of deliberate control, Exner extended Klopfer's ratio to incorporate all shading and gray-black responses and formed the "eb." Following Beck (1960), Exner added the two sides of the "eb" to produce "ep," which is thought to reflect the degree of psychological activity not readily accessible to deliberate control and frequently experienced as stimulation acting or impinging upon the individual.

Much of Exner's research based on the Comprehensive System has involved studies delving into the stimulus-input of the ten cards and the emergence of relatively consistent response styles (Exner, 1980). These studies conducted on adult populations demonstrate that many variables included in the Comprehensive System, especially the 4-square, are reliable and temporally consistent.

Most validation studies for the Comprehensive System have revolved around the 4-square interrelationships (Viglione & Exner, 1983). Research conducted by Exner and his colleagues involving temporal consistency data indicate that the EA:ep ratio stabilizes through development and achieves permanence by adulthood (Exner, Armbruster, & Viglione, 1978). Normative data indicate that EA increases relative to ep as children mature and that more normal subjects than patients have EA greater than ep. Collectively these findings demonstrate that the EA:ep relationship represents a stable, personality characteristic indicating the relative "amount" of psychological activity organized and available for "coping purposes" as opposed to more immature experiences that impinge on the person. Treatment studies reveal increases in EA, both alone and in relationship to ep, among patients who improved in psychotherapy (Exner, 1978). For subjects beginning treatment with ep greater than EA, a reversal has been demonstrated in most cases; that is, EA becomes greater than ep in retest records when intervention continues for more than a brief period of time (Exner, 1978). Exner (1978) suggests that treatment either facilitates the organization of psychological resources or relieves stress. According to Exner the occurrence of ep is not always disruptive or pathological, but rather, ep activity in moderate amounts stimulates underlying motivational processes.

While much of the Comprehensive System research has involved reliability, temporal consistency, and correlational studies of normal adult populations, Exner and his colleagues have recently initiated studies involving the usefulness of the Rorschach in differential diagnosis, particularly unraveling the relationship between borderline, schizotypal, and schizophrenic patients. In general, the findings of these studies (Exner, 1978; Viglione & Exner, 1983), particularly involving the borderline patient's emotional immaturity, excessive self-centeredness, need for closeness, intense anger, and chaotic interpersonal world, are consistent with other Rorschach studies (Spear, 1980; Lerner, Sugarman, & Gaughran, 1981; Lerner & Lerner, 1982; Lerner & St. Peter, 1984) based upon a more conceptual approach to the instrument and utilizing scales representing composite variables derived from contemporary psychoanalytic theory.

The contributions of Exner and his co-workers represent the most recent advances in a psychometric or empirical "sign" approach to the Rorschach. Based. upon major advances in computer science, data processing, and data analysis, the contribution of these researchers revolve around the psychometric features of the Rorschach and the nature of the instrument itself. As such, these contributions address directly previous psychometric critiques of the Rorschach as nonreliable and not suitable for research. In addition, Exner's careful accumulation of normative data fills an important gap in the vast Rorschach literature which contributes to our understanding of development, the Rorschach response process, and an important baseline for further research in differential diagnostic issues.

In marked contrast to the psychometric or sign approach to the Rorschach, the "conceptual" approach to using and evaluating the Rorschach evolved from the recognition that the fundamental purpose of the instrument is to assess personality processes (Weiner, 1977). According to Weiner (1977): "In the conceptual approach personality processes provide a bridge between Rorschach data and whatever condition is to be evaluated or behavior to be predicted. Viewed in this frame of reference, the Rorschach successfully identifies the presence of some condition or predicts some aspect of behavior only when the instrument accurately measures personality variables that in turn account in substantial part for the condition or the behavior" (p. 595). In contrast to the sign approach, the conceptual approach addresses issues of "construct" as opposed to "criterion-related" validity. Criterion-related validity, according to Weiner (1977), consists of the extent to which Rorschach scores correlate with some concurrent condition of the subject or a predicted aspect of behavior. On the other hand, construct validity addresses the extent to which a theoretical formulation can account for relationships between selected aspects of a Rorschach protocol and some condition or behavior. While these two forms of validity are not mutually exclusive, positive outcomes in construct validity studies of personality measures validate both the theoretical formulations concerning the personality variables studied and the tests used for assessing these variables. Personality theory, from a conceptual approach to the Rorschach, is seen as an essential guide for appropriate clinical and research applications of the instrument.

While the relationship between a conceptual approach to the Rorschach and psychoanalytic theory has a long history of mutual benefit, the most contemporary proponents of this point of view are reflected in the work of Blatt, Mayman, the Lerners, and Sugarman. Rorschach assessment from a conceptual vantage point in general and a psychoanalytic perspective in particular is markedly different from more traditional psychometric testing. Effective utilization of the Rorschach within a conceptual framework does not divorce the instrument from its clinical foundation and as Blatt (1975) notes, research requires that judgments reflect the distinctions made in clinical practice. According to Blatt: ". . . many studies have simply failed to find support for some of the most fundamental Rorschach assumptions because researchers used test scores in an undifferentiated way without understanding the basic assumptions and interpretive rationale for the procedure. Many studies have simply failed to integrate into their research methodology the finer distinctions made by experienced clinicians in interpreting a Rorschach" (p. 329). From a conceptual perspective it is important that the Rorschach as a research instrument not be used in a mechanical way; that is, the same distinctions, qualifications, and integrations made in the clinical application should carry over to research.

A conceptual utilization of the Rorschach, in contrast to the psychometric testing, maintains both a foundation in clinical application and a coherent theory of personality. As Mayman (1964, p. 53) noted, "In testing a patient for clinical purposes, we are not simply measuring: we observe a person in action, try to reconstruct how he went about dealing with the tasks we set for him, and then try to make clinical sense of this behavior" (p. 2). In pioneering this approach, Rapaport foresaw how test administration, scoring, and interpretation were inextricably interwoven. He demanded that the test administrator be thoroughly familiar with the mechanics of each test and that the examiner feel free to engage him- or herself with the subject rather than the test. Thus an integral part of the verbatim Rorschach protocol includes the examiner's own comments and the patient's spontaneous remarks.

Psychometric and clinical testing differ significantly in regard to the role and value accorded the examiner. In psychometric testing the test administrator is seen as a source of bias and error variance, hence this role is standardized and minimized as much as possible. In contrast, in clinical testing the examiner's role is maximized as his or her skill, judgment, and intuitive sensitivity are not only valued, but are regarded as the most sensitive and perceptive clinical tools available. The scope of interpersonal interactions which arise in the patient-examiner relationship are neither avoided or acted upon; rather, they are carefully observed and used as an essential aid in bringing informed meaning and understanding to the subject's behavior and attitudes.

The psychologist using the Rorschach in a clinical manner, as Schlesinger (1973) and Sugarman (1985) note, have several sources of information available. First, he or she has the subject's behavior in the standardized testing situation, the ways in which the subject interacts with the examiner, and the nature of changes in the testing relationship over time. Second, the examiner has the content of the subject's test responses, including the subject's idiosyncratic awareness of and reaction to his or her own performance. As previously noted, the examiner's own subjective reactions, stemming from what Mayman (1976) terms an "emphatic-intuitive" immersion into the testing relationship and test responses, are regarded as an important source of data that can be utilized in a disciplined and psychometrically reliable and valid way. Recently, Sugarman (1981) has applied contemporary psychoanalytic notions of countertransference derived from psychotherapy to the psychological testing situation. A final source of information the examiner can draw upon is the form aspects of the subject's test responses, including test scores and their interrelationships. As Schlesinger (1973) observes, the examiner's task is to become attuned to these various sources of data and to integrate them in a coherent and meaningful way. Each source of information must be given its due and be seen as having its own consistency and relationship with other levels of observation. The art and science of psychological testing, according to Schlesinger (1973), consists of a sensitive and informed shifting of attention from one source of data to another and careful drawing and checking of inferences throughout the course of testing and interpreting.

Along with a solid clinical foundation, the other cornerstone of Rorschach utilization from a conceptual point of view rests upon the grounding of test administration, scoring, and interpretation within a comprehensive theory of personality. Sugarman (1985) has outlined four major functions served by a comprehensive theory of personality. First, a comprehensive theory of personality serves an organizing function. An implicit theory of personality both in clinical practice and in research aids the Rorschach examiner in comprehending and organizing data that are complex, often exceedingly rich, as well as inconsistent. Second, a comprehensive personality theory goes beyond organization to integration of seemingly unrelated pieces of data. Third, a comprehensive understanding of personality theory can guide the well-trained and disciplined diagnostic clinician or researcher in filling in data gaps in an informed manner. Constriction as well as marked fluctuations in functioning on the Rorschach challenge the examiner to understand theoretical relationships on the one hand as one way of explaining the data, but caution the examiner, on the other hand, not to allow theoretical bias to distort clinical test data or results. Fourth, a theory of personality facilitates prediction. As Blatt (1975) notes, it is crucial that the examiner take into account the complex social matrix when attempting to predict on the basis of psychological variables alone; that is, in predicting, it is important that the behavior mediated is generated by the personality variables tapped by the instrument.

The contemporary trend within the conceptual approach to the Rorschach is to reduce the welter of isolated but overlapping variables and ratios based on traditional scoring categories into what Blatt and Berman (1984) term "molar variables" that integrate various Rorschach scores in a way that is psychometrically reliable and that measure central dimensions of personality. According to these authors, this conceptual approach assesses dimensions of personality independent of the test itself; that is, "These more composite variables (e.g., thought organization, quality of object representation, thought disorder, defense) are not a replacement for the traditional scoring of a Rorschach protocol, but rather offer a higher order of organization of the data. This higher order of organization is based on conceptual models and personality theories that provide a framework, external to the Rorschach itself, for integrating various aspects of responses to the Rorschach stimuli" (p. 236).

Contributions of Rapaport. Whereas Rorschach did not wed his procedure to a specific theory of personality, Rapaport did. The marriage forced by Rapaport between the Rorschach and psychoanalysis was a perfect one of technique and theory. Rapaport fashioned the relationship out of a test rationale and a specific perspective as to how the test was to be used.

Rapaport based his theoretical rationale for psychological tests, in general, and the Rorschach, in particular, on the construct "thought processes." Accordingly, the exploration of projective tests and the exploration of thought processes were considered synonymous. He conceived of thinking and its organization as the mediating process connecting behavior with its psychodynamic underpinnings on the one hand to test performance and test responses on the other. It was from this organization of thought including subprocesses such as concept formation, anticipation, memory, judgment, attention, and concentration that Rapaport derived inferences to other facets of personality functioning. This conceptualization of the predictive power of thought processes is especially well stated by Schafer (1954): "A person's distinctive style of thinking is indicative of ingrained features of his character makeup. Character is here understood as the person's enduring modes of bringing into harmony internal demands and the press of external events, in other words, it refers to relatively constant adjustment efforts in the face of problem situations. The modes of achieving this harmony are understood to consist essentially of reliance on particular mechanisms of defense and related responsiveness to stimulation associated with these defenses" (p. 17).

Rapaport envisioned the relationship between tests and theory as a two-way street. In one direction he saw how theory could provide the clinical examiner with a bedrock of conceptualizations that would allow for test inferences with remarkable depth and range. In discussing this aspect of Rapaport's work, Mayman (1976) states: "Rorschach inferences were transposed to a wholly new level of comprehension as Rapaport made a place for them in his psychoanalytic ego psychology and elevated psychological test findings from mundane, descriptive, pragmatically useful statements to a level of interpretation that achieved an incredible heuristic sweep" (p. 200). In the other direction, Rapaport also saw how the tests themselves provided a means for operationalizing concepts that were hazy, elusive, and highly abstract; how this would then permit the testing of key psychoanalytic formulations; and then, in time, how this could add to the overall evolving scope of psychoanalytic theory.

The pioneering work begun by Rapaport has been continued by others, such as Schafer, Schlesinger, Holt, Blatt, Mayman, Lerner, Sugarman, and Spear. Parallel with the evolving scope of psychoanalytic theory, both clinicians and researchers have sought to translate more recent psychoanalytic observations and formulations into empirical, test-related concepts and then, like Rapaport, employ these as tools for evaluating hypotheses generated from the theory.

Psychoanalysis has never been a static body of knowledge. Consequently, it is not a closed, tightly knit, well-integrated, totally coherent personality theory.

Rather, it is a loose-fitting composite of several complementary, internally consistent submodels, each of which furnishes concepts and formulations for observing and understanding a crucial dimension of personality development and functioning. The following section will briefly review each of these models followed by a discussion of Rorschach endeavors that are reflective of and rooted in that specific conceptual base.

Shifts in Psychoanalytic Theory—Drive Theory. In his earliest stages of theory construction, Freud was primarily interested in identifying the basic instincts. Despite changes in his theory of instincts over time, his latest writings finally settled upon two instincts: libido and aggression. Drive theory, then, refers to the instincts and their vicissitudes and changes they undergo throughout the course of development. While predating Rapaport's work, this aspect or model of psychoanalytic theory that is rooted in drive theory represents a significant contribution to the Rorschach literature.

Robert Holt (1968) developed a Rorschach scoring system for the "primary process" and "secondary process" concepts. Freud, in 1900, had first distinguished primary process thinking from secondary process thinking in his work *The Interpretation of Dreams*. Primary process, a term coined to indicate a developmentally earlier form of thinking, is organized around drive discharge and involves the formal properties of a disregard for logic and reality. As described by Lerner and Lewandowski (1975), "In primary process, ideas are fluid, they lose their identity through fusion and fragmentation, reflectiveness is abandoned and thoughts are combined in seemingly arbitrary ways. Secondary process, in contrast, is under ego control and operates in accordance with the reality principle; it is goal directed, logical, and uses delay of impulses, detours and experimental action until appropriate avenues of gratification have been realized" (p. 182).

Holt's system calls for the scoring of four sets of variables: content indices of primary process, formal indices of primary process, control and defense, and overall ratings. The part of the system most firmly grounded in drive theory, the content indices, involve ideational drive representations. A distinction is drawn between responses reflecting drives with libidinal aims and those with aggressive aims. The libidinal category is further subdivided into sections corresponding to the stages of psychosexual development. The aggression category is subdivided as well; however, these subcategories are based on whether it is the subject (aggressor), the object (victim), or the result (aftermath) that is emphasized in the destructive action. The formal indices reflect formal aspects of the response processes. Herein, categories were developed to assess the perceptual organization of the response, the thinking underlying the response, and the language used to communicate the response. The specific categories roughly parallel characteristics in thinking outlined by Freud (condensation, displacement, symbolism, etc.) and further refined by Rapaport (peculiar verbalizations, confabulations, etc.). The control and defense scores are attempts to measure the way in which the primary process material is regulated and the relative successfulness of those attempts. Two scoring aspects are included, one involving the identification of the specific defensive operation and the other involving a judgment as to whether the operation improves or further disrupts the response. The control and defense categories include remoteness, context, reflection, postponing strategies, sequence, and overtness.

The final section, overall ratings, involves summary ratings of form level, creativity, demand for defense, and effectiveness of defense. Several summary scores are also involved, including percentage of primary process, mean defense demand, mean defense effectiveness, and adaptive regression. Research involving the scoring system has primarily made use of these summary scores.

Holt's scoring system has generated an impressive amount and array of personality research. Studies involving the scoring system have included the following: 1) attempts to relate the drive and control measures to behaviors and characteristics conceptually related to primary process thinking, 2) the use of specific scores as criterion measures in studying the effects on thinking of experimentally induced or clinical conditions, and 3) attempts to find differences in the expression of and control of primary process thinking among groups differentiated on the basis of another variable such as diagnosis or level of conscience development.

Various investigators attempted to relate categories in Holt's manual to specific cognitive and perceptual variables that, on a theoretical basis, are linked to primary process forms of thinking. The variables studied include the thinking of individuals who have undergone unusual religious experiences (Allison, 1967; Maupin, 1965), the capacity to deal with cognitive complexity (Blatt et al., 1969; Von Holt et al., 1960), creativity (Cohen, 1960; Gray, 1979; Pine, 1962; Pine & Holt, 1960; Rogolsky, 1968), the capacity to tolerate unrealistic experiences (Feinstein, 1967), conjunctive empathy (Bachrach, 1968), and tolerance for perceptual deprivation (Goldberger, 1961; Wright & Abbey, 1968; Wright & Zubek, 1969). Overall, the results of these various studies indicated that several summary scores from the manual are related to a host of conceptually based cognitive and perceptual variables. Specifically, the adaptive regression summary score was found to relate to the capacity to tolerate and adaptively deal with situations in which reality contact is temporarily suspended (Zen meditation, conversion experiences, perceptual isolation). These studies further demonstrated that subjects who have the capacity to modulate drive expressions and integrate logical and illogical thoughts into acceptable Rorschach responses are better able to tolerate unrealistic experiences, are more empathic in treatment relationships, and are more effective in handling a variety of cognitive tasks (Lerner & Lewandowski, 1975). No clear-cut relationship has been found between manual scores and creativity. The inconsistency in findings that has been reported seems to be due to the sample used, the measure of creativity employed, and the specific manual score studied.

A second group of studies employed specific Holt scores to evaluate the effects on thinking of particular experimental and clinical conditions. The conditions investigated included subliminally presented aggressive stimuli (Silverman, 1965, 1966; Silverman & Candell, 1970; Silverman & Goldberger, 1966; Silverman & Spiro, 1967), drugs (Saretsky, 1966), and myxedema psychosis (Greenberg et al., 1969). From these studies the following results were reported: 1) disturbed thinking, defined by increases in the formal indices section of the manual, increased following presentation at a subliminal level of aggressive stimuli; 2) following the use of chlorpomazine there was an increase in the mean defense effectiveness score and this increase was related to an independent measure of clinical improvement; and 3) in a patient treated for myxedema psychosis, predicted changes were noted in the mean defense demand score, the mean defense effectiveness score, and three formal scores.

A third group of studies were designed to determine differences in the expression and control of primary process manifestations among groups differentiated on the basis of other variables. In these studies it was found that groups distinguished on the basis of varied factors such as diagnosis (Silverman, 1965), subdiagnosis (Zimet & Fine, 1965), and conscience development (Benfari & Calogeras, 1968) differed in the extent and quality of primary process manifestations and these differences were in the predicted directions.

Taken collectively, the above studies lend consistent and convincing support to the notion of Holt's scoring system as a valid measure of primary process thinking. As such, Holt, in keeping with a tradition begun by Rapaport, has made operational a concept (primary and secondary process thinking) basic to psychoanalytic theory and thus has provided a most valuable tool for investigating the hypothesis generated by this aspect of Freud's theorizing.

Shifts in Psychoanalytic Theory—The Structural Model. From his early concern with drive identification, Freud's interest shifted to an emphasis on studying and understanding those processes which controlled and regulated the drives and their vicissitudes. With this shift he began to outline the characteristics, synthesis, and functions of the ego with particular emphasis on the defensive function. This change in theoretical emphasis ushered in the structural model and eventuated in Freud's formulations regarding the tripartite (ego, superego, id) structure of the personality. Although Holt's system for assessing primary process manifestations (especially the section on control and defense) is theoretically rooted in this model as well as drive theory, Schafer's (1954) work on the defensive aspect of Rorschach responses best characterizes the contribution of the structural model to Rorschach theory and usage.

Based upon the pattern of formal scores, the content of the response, the nature of the patient-examiner relationship, and the attitude the patient takes toward his or her responses, Schafer outlined seven general and 36 specific types of expressions of defense and/or defended against as they might appear on the Rorschach. Also included are provisions for determining the overall success or failure of the defensive operation. The categories and expressions of defense, together with the indications of success or failure, are applied to the specific mechanisms of repression, projection, denial, regression, reaction formation, isolation, and undoing. For each defense Schafer elaborated the theoretical underpinnings, outlined expected Rorschach manifestations, and provided clinical examples.

To illustrate the above points, Schafer's treatment of the defense of repression will be presented. Based upon Fenichel (1945), Schafer defines repression as "unconsciously purposeful forgetting or not becoming aware of internal impulses or external events which, as a rule, represent possible temptations or punishments for, or mere allusions to, objectionable, instinctual demands" (p. 193). Implicit in this definition, as Schafer points out, is the aim of blocking the unacceptable impulses and their derivatives, as well as the formulation that the repressed continues to exist outside of awareness. Because repression prevents part of the personality from growing and developing, a prolonged emphasis on repression often results in marked ego restriction and various expressions of immaturity. Therefore, in a repressed individual one expects to find impulsiveness, unreflectiveness, naïvete, diffuse affects, emotional lability, superficiality, and a tendency to relate to others in a childlike way.

From these behavioral characteristics commonly associated with a strong reliance on repression, Schafer then devised Rorschach indices reflective of these traits. For example, to indicate the relative paucity of ideas and narrowness of ideation, he suggested that the test record should reveal a comparatively low number of total responses and human movement responses as well as a limited range of content categories. Long reaction times and card rejections, he further noted, were to be expected. For Schafer, formal aspects of the test record related to affect and anxiety were likely to be prominent. He anticipated a protocol with a conspicuous number of responses involving color and shading and suggested that the extent to which form entered into color and shading responses reflected the way in which the individual experienced and expressed affects. Turning from formal scores to the patient's attitude toward the examination process, Schafer suggested that one would anticipate an attitude characterized by self-centeredness, defensive vagueness, childish naïvete, and impenetrable unreflectiveness. Finally, Schafer depicted an interpersonal relationship in which the examiner was looked to for reassurance as well as permission.

In the above example, we have attempted to illustrate Schafer's methodology; how, with each defense, he began with the psychoanalytic definition of that defense, inferred from the defensive operation likely behavioral and attitudinal correlates, and then from the scores, content, and testing relationship identified Rorschach indices reflective of those correlates.

Schafer's application of the concept of defense to the Rorschach has become an indispensable part of the clinical armamentarium of the psychoanalytically oriented examiner. Concurrently, Schaefer's work has generated considerable research which, in turn, has broadened and refined the theory surrounding the concept of defense.

Contemporary Psychoanalytic View of the Rorschach. The conceptual approach to the Rorschach and psychoanalytic theory are at present in a state of evolution. New orientations in psychoanalytic theory involving the emergence and elaboration of object relations theory, self-psychology, and developmental psychoanalysis have converged into a phenomenological point of view and a developmental structural perspective that intersects with both more traditional formulations and other disciplines including cognitive, developmental, and social psychologlical theories. While traditional psychoanalytic propositions are being questioned and core concepts such as thought processes, defenses, and the impact of formative interpersonal relationships on psychological structure formation are being reconceptualized, these new developments in turn are being transformed and operationalized into a more phenomenological Rorschach test theory.

From a historical perspective, Mayman (1963) was one of the earliest Rorschach investigators to delve into more experiential and clinically relevant dimensions of personality. As Blatt and Lerner (1983a) note, he consistently stressed the need to develop a theory of psychoanalysis based on what he termed a "middle level language" as part of the "complex multileveled theory" of psychoanalysis and to distinguish among what he saw as three coordinated sets of concepts or languages. First, according to Mayman (1963, 1976), there is the language used by the therapist in transaction with the patient during the treatment hour, a language more akin to poetry than science. Outside the consultation room the clinician utilizes a "middle language" of "empirical constructs" that helps formulate clinical generalizations about an individual. A third, more abstract language consists of "systematic" or "hypothetical constructs," a system of impersonal concepts using more objective, distant, third-person terms that constitutes psychoanalytic metapsychology. Mayman argues that these three levels of abstraction should not be confused with each other but rather need to be coordinated with constant reference back to the original primary data base, the clinical material. In the remainder of this section we will discuss several of these theoretical advancements and then present research endeavors and clinical test material that reflect innovative ways of utilizing the Rorschach from a conceptual point of view.

Psychology of the Self. In a series of major publications, Kohut (1971, 1977) has laid the conceptual groundwork for a systematic psychoanalytic psychology of the self. Tolpin and Kohut (1978), in a paper with major diagnostic implications, drew important distinctions between more classical neurotic pathology and pathology of the self. Unlike the former, which is presumed to originate in later childhood and at a time when there is self-other differentiation, full structural development (id, ego, superego), and oedipal passions, self-pathology begins in earlier childhood and at a point when the psychic structures are still in formation. Because of the absence of a cohesive sense of self, in self-pathology symptoms occur when the insecurely established self is threatened by dangers of psychological disintegration, fragmentation, and devitalization. Further, in treatment, unlike the neurotic patient who develops a transference neurosis in which the therapist is experienced as a new edition of parents, patients with self-pathology develop therapeutic relationships in which the therapist is used to correct or carry out a function that is ordinarily, with optimal development, carried out intrapsychiatrically.

Based on these formulations, Paul Lerner (1979, 1981, in press), in a series of papers, has attempted to describe the test behavior and pattern of Rorschach responses of a selected subgroup of outpatients with self-pathology who, on the basis of their test performance, exhibit an identifiable self-system and mode of interpersonal relationships, definable style of thinking, and characteristic manner of experiencing affects. According to Lerner, these patients enter testing under a cloak of extreme vigilance with a readiness to be distrustful. They are described as hyperalert, highly sensitive, and extremely vulnerable. Accompanying the sensitivity is a style of compliance and accommodation. Similar to chameleons, they sensitively attune to the nuances, expectations, and anticipations of others and mold themselves and their behavior accordingly. They present what Winnicott (1960) has referred to as a "false self." They swiftly scrutinize all aspects of the examiner during the test situation, including his or her tone of voice, attire, and office furnishings. Consequently, the examiner experiences him- or herself as being viewed under a microscope and as a result may feel inhibited or as if "walking on egg shells," carefully selecting words and unconsciously fearing damage to the subject's self-esteem.

Lerner (1979) has identified a Rorschach determinant which is especially sensitive to hypervigilance and heightened sensitivity; that is, the (c) response. This score is applied to responses which are delineated and determined by variations in shading. Because the perception of variations in shading on the Rorschach are subtle, to generate a (c) response one must scrutinize the blot extremely carefully, sensitively attune to finely differentiated nuances, and feel one's way into something that is not blatantly apparent. To accomplish this, according to Schactel (1966), requires perceptual sensitivity in addition to a searching, articulating, and penetrating activity.

The tendency toward compliance becomes manifested in Rorschach imagery or content. Lifeless figures whose actions and intentions are controlled externally, such as puppets, mannequins, and robots often populate the examinees' protocols. Further, quasi-human figures distanced in time or space such as ghosts, clowns, snowmen, skeletons, and masks convey their sense of lacking substance and illusory feelings.

Lerner further describes this group of patients in terms of a specific cognitive style characterized by concreteness, passivity, and egocentricity. These individuals often exhibit a nearsighted clarity accompanied by a blatant loss of backdrop or perspective; that is, they lack reasonable objectivity and detachment. Accordingly, life events are not critically examined or placed in logical context, but rather are experienced in terms of their most obvious and immediately personal qualities. As Lerner puts it, "The present dominates and the significance of the past and of the future vanishes" (p. 25). While the (c) response vividly reflects impairments in maintaining perspective, the arbitrary and incompatible blending of color with content, as seen in the FC arbitrary score (e.g., blue monkeys, pink wolverines), and the offering of responses in which spatial or temporal relationships in the blot are concretely taken as real relationships (fabulized combinations) capture the loss of distance from the cards together with the relinquishment of a more objective, critical, and self-imposed evaluative attitude.

According to Lerner, these patients described by Kohut and his colleagues also experience distinctive and identifiable affects. Specifically, they are subject to lowered self-esteem, disintegration anxiety, and depletion depression. Disintegration anxiety, a term used by Kohut (1977), refers to a type of primal anxiety or agitated depression prompted by threats to self-cohesion and fears revolving around fragmentation and a loss of aliveness. The depressive affect involves unbearable feelings of deadness and nonexistence and a self-perception of emptiness, weakness, and helplessness. Recently, Wilson (in press) has linked these Rorschach determinants to Blatt's (1974) formulations concerning anaclitic depression, a type of pre-oedipal depression characterized by helplessness, hopelessness, and a future quest for need gratification.

Developmental Theory. A second recent advance within psychoanalysis has been the elaboration of an empirically based, dynamic, developmental theory. Mahler, Pine, and Bergman (1975) have observed the steps in the separation-individuation process, beginning with the earliest signs of the infant's differentiation or hatching from a symbiotic fusion with the mother, proceeding through the period of the infant's absorption in his or her own autonomous functioning to the near exclusion of the mother, continuing through the all-important period of rapprochement in which the infant, precisely because of his or her more clearly perceived state of separateness from the mother, is prompted to redirect attention back to the mother, and concluding with a feeling of a primitive sense of self, of individual identity, and of constancy of the object.

Based upon these formulations, Kwawer (1980) developed a Rorschach scale designed to assess early object relations, particularly self-other differentiation, in patients with severe character pathology. He first identified, theoretically and clinically, very early stages of levels of relatedness in the developmental emergence of selfhood through differentiation from a primary mothering figure and then constructed Rorschach content scores to assess these stages. An initial stage, termed "narcissistic mirroring," includes responses in which mirrors or reflections play a prominent role. Percepts such as "two men mirroring each other, two little mimes" or "somebody grabbing hold of himself in a mirror" would be included under this heading. These responses are understood as reflecting a state of self-absorption in which the other is experienced solely as an extension of the self and used for the exclusive purpose of mirroring or enhancing the self. A second stage, referred to as "symbiotic merging," consists of responses which indicate a powerful push toward merger, fusion, and reuniting. "Siamese twins joined at the stomach and they have but one hand" and "a butterfly that flew into a wall and seems to be at one with wall" are iilustrative responses included in this category. A third stage of interpersonal differentiation is found in responses conveying "separation and division" content. The Rorschach imagery is reminiscent of the biological metaphor of cell division: "These two things appear to have been once connected but broke apart . . . it's as if on the inside there was some continuity between the two." An analogous process is suggested by the following response: "It's an animal dividing going from one to two . . . it almost looks like it's breaking away into two separate objects." The fourth and final stage, according to Kwawer (1980), termed "metamorphosis and transformation," is reflective of the experience of a very early and rudimentary sense of self. Here incipient selfhood is manifested in themes of one-celled organisms, fetuses, and embryos. An example of this type of response is "Two ants dressed up in men's clothes . . . it seems like their bodies are transformed into human bodies, but their heads remained ants. They were acting like human beings."

Technical Aspects

Modern Object Relations Theory. Psychoanalytic theorists and researchers are increasingly aware of the complex interactions among early formative interpersonal relationships; the level and quality of internal psychological structures including thought processes and defensive organization; the internal representational world; and the nature of ongoing interpersonal relations and the ways they are internalized and become part of the personality. This comparatively recent perspective has provided the conceptual foundation for several innovative Rorschach studies.

One major thrust is seen in the varied attempts to assess the construct "object representation." Defined broadly, object representation refers to the conscious and unconscious mental schemata, including cognitive, affective, and experiential components of objects encountered in reality (Blatt & Lerner, 1983a, 1983b). Beginning as vague, diffuse, variable sensorimotor experiences of pleasure and unpleasure, they gradually expand and develop into differentiated, consistent, relatively realistic representations of the self and the object world. Earlier forms of representation are based more on action sequences associated with need gratification, intermediate forms are based on specific perceptual features, and higher forms are more symbolic and conceptual. Whereas these schemata evolve from and are intertwined with the developmental internalization of object relations and ego functions (Mahler et al., 1975), the developing representations provide a new organization for experiencing object relations.

Two primary research groups have contributed to the systematic study of the object representation construct by means of the Rorschach and other projective techniques through experimental procedures. These two research groups represent different but not mutually exclusive approaches to the study of object representations. Mayman and his colleagues at the University of Michigan have focused on the thematic dimension of object representations and using a variety of projective procedures (e.g., early memories, manifest dreams, written autobiographies), including the Rorschach, have studied the relationship between this construct and the severity of psychopathology, character structure, quality of object relations, and capacity to benefit from psychotherapy. Blatt and his colleagues at Yale University, by contrast, have emphasized the structural dimension of object representations. While this group has also developed independent projective measures and scales (e.g., parental descriptions, Thematic Apperception Test scales), they have studied the developmental level of object representations across a wide spectrum of normal and clinical populations. Although the contributions of both groups are conceptually rooted in an integration of ego psychology and object relations theory, Blatt and his colleagues have integrated the developmental cognitive theories of Piaget and Werner into their Rorschach scales and formulations.

Based on the theoretical contributions of Jacobson and Erickson, Mayman (1967, 1968) conceptualized object representation as templates or internalized images of the self and of others around which the phenomenological world is structured and into which the ongoing experience of others is assimilated. Methodologically, he contended that the manifest content of dreams, early memories, and Rorschach content was more than simply a screen that both expressed and concealed deeper and more significant levels of unconscious meanings. He argued that manifest content in its own right could reflect levels of ego functioning, the capacity for object relations, and the nature of interpersonal strivings. According to Mayman (1967): "When a person is asked to spend an hour immersing himself in a field of impressions where amorphousness prevails and where strange or even alien forms may appear, he will set in motion a reparative process, the aim of which is to replace formlessness with reminders of the palpably real world. He primes himself to recall, recapture, reconstitute his world as he knows it, with people, animals, and things which fit most naturally into the ingrained expectations around which he has learned to structure his phenomenal world. A person's most readily accessible object representations called up under such unstructured conditions tell much about his inner world of objects and about the quality of relationships with these inner objects toward which he is predisposed" (p. 17).

Equipped with these notions, together with a commitment to a clinical empathic-intuitive approach to the analysis of projective test data, in an early study Mayman (1967) operationalized selected Rorschach responses, particularly the human response, as expressions or facets of object representations. Building on the clinical observation that the human content of good psychiatric residents featured "warmth, openness, and a sense of contact" and those of poor residents expressed "cynicism . . . bitterness . . . fearfulness or alienation from the people they described" (p. 21), Mayman developed a technique of distilling Rorschach protocols into clusters of object representational responses independent of reference to traditional scoring categories in order to test the validity of inferences made from object representational aggregates of Rorschach responses. Mayman compared independent ratings of these responses using the Luborsky Health-Sickness Scale with similar ratings by clinicians based on clinical interviews, and found an impression correlation ($r = .86$) between the two measures for a group of patients.

Mayman's seminal contributions to the Rorschach have spawned a number of object representation scales and a host of construct validational studies conducted by his students that have further refined the concept (Urist, 1973) and have extended the thematic analysis of object representations to manifest dreams (Krohn, 1972), autobiographical data (Urist, 1973), and the capacity to benefit from insight-oriented psychotherapy (Hatcher & Krohn, 1980). The various scales designed to unearth object representational levels as specific points on a developmental continuum have been correlated with each other (Urist, 1973) and have been applied and correlated across data bases including manifest dreams, the Rorschach, early memories, and health-sickness ratings (Krohn & Mayman, 1974). These studies reflect Mayman's focal interest in thematic content, his distinct and gifted clinical approach to projective data that capitalizes on the empathic-intuitive skills of trained clinicians, and his abiding interest in variables relevant to psychoanalytic theory and treatment.

Using a scale designed by Mayman and Ryan (1972) to evaluate object representational content dimensions in early memories, Ryan (1973) found a positive relationship in neurotic patients between levels of object representation and the capacity to enter into an elementary psychotherapeutic relationship. Drawing upon Mayman and Ryan's scale, Krohn (1972), through a pilot study, developed the Object Representation Scale for Dreams. Designed to tap the degrees of wholeness, differentiation, consistency, and overall intactness of object representations with an impressionistic survey of dreams, Krohn's scale was example-anchored and intended for use by "intuitive, trained clinicians" in an empathic manner. With a view toward establishing the reliability and construct validity of the object representation concept as an empirically, researchable dimension of personality, Krohn (1972) and Krohn and Mayman (1974) applied the dream scale across a range of projective media including written manifest dream reports, early memories, and Rorschach protocols that they then demonstrated to be related to therapist-supervisor ratings of patients' overt object relations and Luborsky's health-sickness dimension. The authors report a number of important findings. Strong, statistically significant correlations were found between object representations assessed on the Rorschach, early memories, and dreams; and high correlations emerged between projective object representation scores and criterion ratings of psychopathology and level of object representation. Analyzing the data through partial correlations, the dream and early memory scores emerged as the best predictors to therapist-supervisor ratings of patients' object relations while the Rorschach ratings correlated most significantly with more global health-sickness scores. Krohn (1972) found that patients' Rorschach ratings were consistently lower; that is, less intact than scores based on either manifest dreams or early memories. "The fundamental conclusion to be drawn from this study," according to Krohn and Mayman (1974), "is that level of object representation appears to be a salient, consistent, researchable personality dimension that expresses itself through a relatively diverse set of psychological avenues ranging from a realm as private as dream life to one as interpersonal as psychotherapy. Moreover, it is not a redundant construct synonymous with level of psychopathology or severity of symptomatology" (p. 464).

Consonant with the methodological thrusts of Mayman and integrating the theoretical contributions of Kernberg and Kohut, Urist (1973), in a sophisticated construct validational study, examined the multidimensional qualitative aspects of the object representational concept by correlating several Rorschach scale ratings of 40 adult inpatients covering a broad range of psychopathology with independent ratings of written autobiographies. The specific scales designed by Urist were gauged to reflect the developmental ordering of stages in the unfolding of object relations along a number of overlapping dimensions including mutuality of autonomy, body integrity, aliveness, fusion, thought disorder, richness and complexity, and differentiation and individuation. The focus of the measures, particularly the Mutuality of Autonomy Scale, is on the developmental progression of separation-individuation from symbiosis to object constancy. Urist reports significantly high correlations among the various measures of object relations which reflect an impressive consistency among internal self and object representations across a wide range of sampled behavior. In terms of the ability of the Rorschach to tap qualitative aspects of the representational world, Urist observed, ". . . the individual's internal world of mental representations is indeed mapped into his percepts on the Rorschach" (p. 113).

In keeping with his initial hypothesis and bolstered by factor analytic techniques, Urist demonstrated that object relations are not unidimensional areas of ego functioning. A factor analysis discerned an important distinction between two related but separate structural underpinnings of object representation; that is, an integrity factor related to issues of self-other differentiation, stability, and consistency (an index of secondary narcissism), and a boundary factor related to developmental gradations in fusion-merger tendencies and thought disorder associated with the ability to maintain a cognitive-perceptual sense of boundary between self and other and between one object and another (an index of primary narcissism).

Utilizing the same data, but highlighting the Mutuality of Autonomy Scale, Urist (1977), in a further construct validational study, correlated Rorschach ratings with independent measures of the same dimension applied to the written autobiographies and behavioral ratings of ward staff. Obtaining several Rorschach scores in order to sample the range as well as the average and "best subjective overall rating," Urist reported a consistency across all variables and range of measures that point to an enduring consistency in patients' representations of relationships, and he demonstrated that the Rorschach can be utilized effectively to systematically assess aspects of mutuality of autonomy within patients' experience of self and others.

The contributions of Mayman and his colleagues at the University of Michigan, summarized by Blatt and Lerner (1983a), ". . . reflect a focal interest in thematic content—allowing the data to speak for themselves, a steadfast clinical focus which is experience-near, and which is comprehended through empathic-intuitive skills while at the same time achieving psychometrically respectable levels of validity, and finally, an abiding commitment to investigating reactive variables directly relevant to psychoanalytic theory and the treatment process" (p. 210).

A second major approach to the study and systematic assessment of object representations is represented in the work of Blatt and his colleagues. In contrast to the University of Michigan group, which stressed thematic analysis and a more clinical-intuitive methodology, the group at Yale, employing more experimental methodologies, has focused upon the structural dimension of the object representation construct.

Conceptualizing the establishment of ego boundaries between self and nonself and between fantasy and reality (inside and outside) as the initial and most fundamental stages in the development of object representations, several studies (Blatt & Ritzler, 1974; Brenneis, 1971) have involved an assessment of boundary disturbances in psychotic and borderline patients. Using three of Rapaport's classic indices of thought disorder (contamination, confabulation, fabulized combination) as measures of varying level of severity of boundary disturbances, Blatt and Ritzler (1974) found such indices to be related to a variety of ego functions (capacity for reality testing, quality of interpersonal relations, nature of object representations) in a mixed schizophrenic and borderline sample. The authors also found that poorly articulated boundaries occurred most frequently in more disturbed, chronic patients who had impoverished object relations, impaired ego functions, and a lifelong pattern of isolation and estrangement. Using the manifest content of dreams, Brenneis (1971) reported significantly more boundary disturbances in a group of schizophrenic patients as compared with other psychiatric patients having a diagnosis indicating less severe psychopathology. Johnson (1980) provided empirical support for Blatt and Ritzler's contention that patients with more articulated boundaries are more actively involved in interpersonal relations. And, more recently, Lerner, Sugarman, and Barbour (1985) have expanded the scope of research in this area by finding that independently diagnosed borderline patients can be distinguished from both schizophrenics and neurotics on the basis of boundary disturbance. These authors found support for Blatt's findings that schizophrenia involves a deficit in maintaining the distinction between self and others. In drawing developmental distinctions between groups, Lerner et al. (1985) found that borderline patients experience difficulty maintaining the inner-outer boundary as assessed through the confabulation response; that is, these patients experience difficulty discriminating between an external object and their own internal affective reaction to that object.

Building on their initial investigation of boundary disturbances, Blatt, Brenneis, Schimek, and Glick (1976) developed a highly comprehensive and sophisticated Rorschach manual for assessing object representations. Based upon the developmental theory of Werner (1948) and ego psychoanalytic theory, the system calls for the scoring of human responses in terms of the developmental principles of differentiation, articulation, and integration. Within each of these areas, categories were established along a continuum based on developmental levels. Differentiation refers to the type of figure perceived, whether the figure is quasi-human detail, human detail, quasi-human, or a full human figure. For articulation, responses are scored on the basis of the number and types of attributes ascribed to the figure. Integration of the response is scored in three ways: the degree of internality of the action, the degree of integration of the object and its action, and the integration of the interaction with another object. Responses are also second along a content dimension of benevolence-malevolence.

In an early study (Blatt et al., 1976) this particular scoring system was applied to the Rorschach protocols of normal subjects on four separate occasions over a 20-year period. In this longitudinal study of normal development the authors found a marked increase, over time, in the number of accurately perceived, well-articulated, full human figures involved in appropriate, integrated, positive, and meaningful interaction. In a second study, the records of the normal subjects obtained at age 17 were compared with the Rorschachs of a hospitalized sample of disturbed adolescents and young adults. In comparison with the normals, patients gave human figures that were significantly more inaccurately perceived, distorted, and seen as inert or engaged in unmotivated, incongruent, nonspecific, and malevolent activity. The combined studies' results lent strong support to the construct validity of the concept object representation and the manual devised to assess it.

This construct validity study had particular relevance for understanding psychopathology (Blatt et al., 1976). These researchers found that while the Rorschach responses of hospitalized schizophrenic and borderline patients were at lower developmental levels, these responses were primarily accurately perceived in terms of form level. Paradoxically, patients were found to have a significantly greater number of responses at higher developmental levels than normals on inaccurately perceived Rorschach responses; that is, hospitalized patients had a significantly greater number of developmentally more advanced responses—responses that were undistorted, intact, well articulated, integrated, and benevolent. These findings, according to Blatt and his colleagues, indicate that patients, as compared with normals, function at lower developmental levels when in contact with conventional reality but that hospitalized patients function at higher developmental levels than normals when they give idiosyncratic interpretations of reality. These findings have been replicated by Ritzler, Zambianco, Harder, and Kaskey (1980) who found this pattern to be more apparent in schizophrenic than non-schizophrenic psychotic patients.

Recently Lerner and St. Peter (1984) applied the Concept of the Object Scale (Blatt et al., 1976) to the Rorschach protocols of independently diagnosed outpatient neurotic and borderline subjects as well as hospitalized borderline and schizophrenic patients. Analyzing the results separately for perceptually accurate and inaccurate responses, Lerner and St. Peter found, consistent with previous studies, that schizophrenic patients produced significantly fewer accurate responses and portrayed realistic human figures at lower developmental levels than the other three groups. This impairment in the representation of objects serves as a distinguishing factor between schizophrenic and borderline patients and is consistent with previous clinical and research reports. Unexpectedly, however, the inpatient borderline sample functioned at the highest developmental level of differentiation, articulation, and integration for inaccurate responses and, in general, produced more inaccurate responses in these categories than the other three groups. The results, particularly in regard to the hospitalized borderline group, highlight the content dimension of object representation as being a particularly discriminating variable. The inpatient borderline subjects produced the most human responses with malevolent content and were the only group to offer inaccurately perceived malevolent responses. Whereas only 25% of all other subjects offered malevolent human responses, the inpatient borderline sample attributed malevolency to 42% of their human responses and to 94% of their quasi-human characters. These findings clearly illustrate the enormous difficulty these patients experience in managing aggression within the interpersonal relationships.

In order to assess the clinical utility of the Concept of the Object Scale, Blatt and Lerner (1983b) applied the instrument to the Rorschach records of several patients, each of whom was independently selected as a prototypic example of a specific clinical disorder. These authors not only found a unique quality of object representation for each of the clinical entities, but their findings, based on Rorschach data, were remarkably congruent with clinical expectations. In a nonparanoid schizophrenic patient the object representations were found to be inaccurately perceived and at lower developmental levels of differentiation. The representations were inappropriately articulated and seen as either inert or involved in unmotivated action. Little interaction was seen between figures and the content was basically barren. With a narcissistic-borderline patient, the object representations were found to progressively deteriorate either over time or with stress. Initially, the representations were accurate, well differentiated, and appropriately articulated; however, this gave way to representations that were inaccurately perceived, inappropriately articulated, and seen as part rather than whole figures. An infantile character with an anaclitic depression (Blatt, 1974) represented objects accurately but with little articulation. Interaction was perceived between figures, but this usually involved an active-passive transaction in which one figure was seen as vulnerable and in a depriving, rejecting, undependable relationship. In an acutely suicidal patient with an introjective depression (Blatt, 1974), the author found an alternation between representations at higher developmental levels and others at lower developmental levels. The latter representations featured activities between objects that were destructive and with malevolent intent. The representations of a delinquent adolescent were well differentiated, limitedly articulated, and characterized by activity that lacked purpose and direction. Finally, a patient diagnosed as hysteric provided representations that were clearly differentiated and highly articulated, although the articulation was primarily in terms of external physical details. The interactions ascribed between figures was mutual and reciprocal but, nonetheless, childlike.

Whereas several of the above studies involved individual cases, Spear (1980) investigated differences among diagnostic groups on the Object Representation Scale. Using a summary score, he found significantly greater impairment in representing objects in a schizophrenic group as compared with two groups of borderline patients.

Finally, other investigators have related parts of Blatt et al.'s scoring manual to psychological variables conceptually linked to object representation. Johnson (1980) reported significant correlations between the degree of articulation of the representation and more advanced developmental levels of interaction and an independent measure of field independence. He also found a significant correlation between scale measures of the integration of the object with its action and the portrayal of congruent interactions in a role-playing task. Fibel (1979), in a sample of seriously disturbed adolescent and young adult hospitalized patients, found a significantly positive relationship between scale scores and independent clinical assessments of quality of interpersonal relations.

In summary, in an earlier article Blatt (1974) made a substantial contribution to the theoretical literature on the construct object representation. In particular, he explained the relationship between level of object representation and the nature and severity of depression. Out of this theoretical groundwork, bolstered by the integration of theories of cognitive developmental psychology with psychoanalytic theory, has come a psychometrically refined scoring manual for the systematic assessment of object representation. Studies involving the development of object representations in normal subjects, differences in object representations among different individual patients representing varied clinical disorders, and investigations regarding the relationship between object representations and other theoretically linked variables have all contributed to the construct validity of the scoring manual.

A second innovation, emerging from the integration of object relations theory with more classical psychoanalytic theory, has been an attempt to investigate the more primitive defenses of the borderline patient. Kernberg (1975) has identified two overall levels of defensive organization associated with pre-oedipal and oedipal pathology. At the lower level, splitting is the core defense with a concomitant impairment of the ego's synthetic function. In addition, splitting is abetted through the related defenses of primitive idealization, primitive devaluation, denial, and projective identification. At a more advanced level, repression replaces splitting as the major defense and is accompanied by the related defensive operations of intellectualization, rationalization, undoing, and higher forms of projection.

Based upon these theoretical conceptualizations of defense advanced by Kernberg (1975) and the clinical test work of Mayman (1967), Pruitt and Spilka (1964), Holt (1968), and Peebles (1975), Lerner and Lerner (1980) devised a Rorschach scoring manual designed to evaluate the specific defensive operations presumed to characterize this developmentally lower level of defensive functioning.

This scoring manual is divided into sections on the basis of the specific defenses of splitting, devaluation, idealization, projective identification, and denial. Within each section the defense is defined, Rorschach indices of the defense are presented and clinical illustrations are offered. The sections on devaluation, idealization, and denial call for an identification of these defenses as well as a ranking of the defense on a continuum of high versus low order. In keeping with Kernberg's notion that these defenses organize (as well as reflect) the internalized object world, and with the empirical relationship found between human responses on the Rorschach and quality of object relating (Blatt & Lerner, 1983b), the system involves a systematic appraisal of the human figure response. In assessing the human percept, attention is paid to the precise figure seen, the way in which the figure is described, and the action ascribed to the figure.

To evaluate the construct validity of the manual the authors (Lerner & Lerner, 1980) conducted two separate studies. In one study Rorschach records of borderline patients were compared with those of neurotic patients, while in the second study the protocols of borderline and schizophrenic patients were compared with respect to manifestations of primitive defenses.

In the first study (Lerner & Lerner, 1980), 30 Rorschachs, 15 from borderline patients and 15 from neurotic patients, were selected from private files and scored using the proposed system. Independently obtained mental status examinations and social-developmental histories were also available on each patient. Because the testing was initially done for research purposes, it was not used in the formulating of a diagnosis. As each of the patients subsequently entered either psychotherapy or psychoanalysis, the initial diagnosis was confirmed in discussions with the patient's therapist or analyst. The patients were matched as groups on the variables of age, sex, and socioeconomic status. The Rorschach records obtained from the two groups did not differ significantly with regard to the number of total responses.

For purposes of reliability, all Rorschachs were coded, scored independently by two raters, and then the ratings were correlated. The resultant correlations were as follows: .76 for splitting, .96 for projective identification, and .58 to 1.00 for the continuum variables.

A review of the findings indicated that borderline patients used scale indices of lower level devaluation, splitting, and projective identification significantly more often than did the neurotic patients. Strikingly, measures of splitting and projective identification appeared exclusively in the borderline groups. By contrast, indices of high level devaluation and high level denial were found significantly more frequently among the neurotic patients. The authors further found that the use of excessive depreciation and idealization by the neurotics was typically mitigated through forms of higher level denial. Such was not the case with these borderlines. Their expressions of blatant devaluation and idealization were not mitigated, controlled, or regulated.

In the second study (Lerner, Sugarman, & Gaughran, 1981), which involved a comparison of borderline and schizophrenic patients, Rorschach protocols were selected from the records of a population of psychiatric inpatients hospitalized at a university teaching hospital. Patients selected for the study were between 16 and 26 years and showed no evidence of organic impairment. The schizophrenic sample, consisting of 19 patients, was selected using the Research Diagnostic Criteria (Spitzer et al., 1975). The borderline sample, which consisted of 21 patients, was selected in accordance with criteria set out in DSM-III.

Obtained reliability results, also involving interrater agreement, were comparable to those reported in the 1980 study. In this study the borderline patients were found to use test indices of primitive devaluation, primitive idealization, lower level denial, splitting, and projective identification significantly more often than the schizophrenic patients. As in the first study, scores for projective identification occurred exclusively in the borderline group.

Both studies' combined findings lend convincing support to Kernberg's (1975) contention regarding the unique and discernible defensive constellation of borderline patients as well as to the reliability and construct validity of the scoring manual developed to assess these defenses. Empirically, the scoring manual was found to be effective in distinguishing among diagnostic groups on the basis of an evaluation of defensive organization. Conceptually, implicit in the scale's development was the notion that representational capacities and each of the defensive functions are inextricably related. Because of this, the authors (Lerner & Lerner, 1982) elaborated upon their findings and advanced a developmental-structural model of defense, conceptualized within an object representational framework. More specifically, they suggested that "as development proceeds and a representational capacity is achieved, defenses take on an increasingly organizing function and protect the cohesion and integrity of poorly differentiated self and object representations. At higher developmental levels featured by enhanced affective cognitive differentiation and corresponding representational capacity, more encapsulated defenses, limited in scope and related to specific drive affective derivatives, function to protect a well structuralized ego from anxiety stemming from conflict between intrapsychic structures composed of more fully established and specific self and object representations" (pp. 35-36).

Cooper (1981, 1982) and his colleagues (Arnow & Cooper, 1984) have developed a comprehensive and sophisticated Rorschach defense scale based on a careful examination of a broad range of content, including the patient-examiner relationship, and geared toward drawing distinctions between levels of defense reflective of developmental arrest (Stolorow & Lachmann, 1980) and structural conflict. Cooper offers definitions of borderline defenses and specific scoring criteria and examples for the Rorschach assessment of splitting, devaluation, primitive idealization, omnipotence, and projective identification.

Future Directions. Recently, there has been a growing interest in the study of people by means of the Rorschach. This critique has reviewed the two major approaches to the Rorschach, the psychometric and the conceptual, and has outlined, from an historical perspective, advances in the psychoanalytic basis of Rorschach investigations. From a relatively narrow but solid clinical foundation established by Rapaport and articulated by Schafer and Holt, recent advances in the psychoanalytic understanding of primitive mental states including borderline, narcissistic, and psychotic disturbances have dramatically expanded this base by providing new formulations for significant and innovative clinical and research efforts. Examples of these developments include the investigation of core contemporary psychoanalytic variables including boundary disturbance, object representation, and defensive organization. Collectively, these endeavors have operaionalized newer, more phenomenological concepts based on a conceptual model of personality and, in turn, have proved methodologies and Rorschach scales for systematically assessing and evaluating the reliability, validity, and clinical utility of constructs generated by this expanded body of knowledge. One may expect this trend toward approaching research with higher-order variables, based on a model of personality independent of the test itself, to continue.

Blatt and his colleagues (Blatt et al., 1985) have recently presented a major study of opiate addictions based, in part, on the Rorschach. These investigators used a combination of structural and content variables that has been the focus of extensive validation and research, as well as being clinically relevant to opiate addiction. These variables were based on conventional scores, ratios, and content categories. These researchers further used well-established, reliable, and extensively validated Rorschach scales to capture significant dimensions of personality organization such as thought disorder, developmental level of object representation, and primary-secondary process thinking. By controlling for response productivity and using factor analytic statistical techniques, they developed seven composite variables that significantly distinguished a sample of opiate addicts from a control group. With regard to a conceptual approach integrated with traditional Rorschach scoring, Blatt and Berman (1984) state: ". . . these new approaches to the Rorschach based on personality theories seem to provide ways of more effectively managing what is often experienced as a confusing and overwhelming array of isolated scores and of making the Rorschach more useful for the systematic investigation of a wide range of clinical phenomena" (p. 238).

Critique

Psychoanalytic theory and the conceptual approach to the Rorschach is and has been in a constant state of evaluation. As noted previously, from an early concern with an identification of the instincts and their vicissitudes to an emphasis on studying the ego, interest has now shifted to a systematic exploration of the early mother-child relationship and its impact on the development of the self and the quality of later interpersonal relationships. Each shift in theory has provided new and stimulating conceptualizations that have served to broaden the theoretical basis for the Rorschach's clinical and research application.

There has been a resurgence of interest in projective tests, in general, and the Rorschach, in particular. C. Piotrowski (1984), in an article entitled "The Status of Projective Techniques: Or, 'Wishing Won't Make it Go Away'," analyzed the predicted decline in usefulness of projective techniques from a number of perspectives: the academic, APA's Division of Clinical Psychology members, internship centers, the applied setting, and private practitioners. On the basis of an extensive review of empirical surveys and position studies completed in the past two decades, Piotrowski found tremendous support for the utility of projective techniques in all settings with the exception of the academic. Noting that the role and function of the clinical psychologist continues to change and that the role of psychometric assessment increasingly is being passed onto technicians, the popularity and clinical acceptance of the Rorschach should, in the opinion of these reviewers, remain high and continue to grow.

References

The following selected references may be regarded as the most significant of the vast Rorschach literature reviewed in this chapter. The Society of Personality Assessment's annual convention and the *Journal of Personality Assessment* serve as the major contemporary forums for Rorschach research as well as for the discussion of issues raised in this review. Text citations follow.

Allison, J., Blatt, S., & Zimet, C. (1968). *The interpretation of psychological tests.* New York: Harper & Row.

Blatt, S., Brenneis, C., Schimek, J., & Glick, M. (1976). Normal development and psychopathological impairment of the concept of the object on the Rorschach. *Journal of Abnormal Psychology, 85,* 364-373.

Exner, J. (1974). *The Rorschach: A comprehensive system* (Vol. 1). New York: John Wiley & Sons.

Exner, J. (1976). *The Rorschach: A comprehensive system, current research and advanced interpretation* (Vol. 2). New York: John Wiley & Sons.

Holt, R. (Ed.). (1968). *Diagnostic psychological testing* (rev. ed.). New York: International Universities Press.

Kwawer, J., Lerner, H., Lerner, P., & Sugarman, A. (Eds.). (1980). *Borderline phenomena and the Rorschach test.* New York: International Universities Press.

Lerner, H., & Lerner, P. (Eds.). (in press). *Primitive mental states and the Rorschach.* New York: International Universities Press.

Rapaport, D., Gill, M., & Schafer, R. (1945-1946). *Diagnostic psychological testing* (Vols. 1 & 2). Chicago: Year Book Publishers.

Schachtel, E. (1966). *Experiential foundations of Rorschach's test.* New York: Basic Books.

Schafer, R. (1954). *Psychoanalytic interpretation in Rorschach testing.* New York: Grune & Stratton.

Weiner, I. (1977). Approaches to Rorschach validation. In M.A. Rickers-Ovsiankina (Ed.), *Rorschach psychology.* Huntington, NY: R. E. Krieger.

Allison, J. (1967). Adaptive regression and intense religious experiences. *Journal of Nervous and Mental Disorders, 145,* 452-463.

Arnow, D., & Cooper, S. (1984). The borderline patient's regression on the Rorschach test. *Bulletin of the Menninger Clinic, 48*(1), 25-37.

Bachrach, H. (1968). Adaptive regression, empathy and psychotherapy: Theory and research study. *Psychotherapy, 5,* 203-209.

Beck, S. (1960). *The Rorschach experiment: Ventures in blind diagnosis.* New York: Grune & Stratton.

Benfari, R., & Calogeras, R. (1968). Levels of cognition and conscience typologics. *Journal of Projective Techniques and Personality Assessment, 32,* 466-474.

Blatt, S. J. (1974). Levels of object representation in anaclitic and introjective depression. *Psychoanalytic Study of the Child, 29,* 107-157.

Blatt, S. J. (1975). The validity of projective techniques and their research and clinical contribution. *Journal of Personality Assessment, 39,* 327-343.

Blatt, S., Allison, J., & Feinstein, A. (1969). The capacity to cope with cognitive complexity. *Journal of Personality, 37,* 269-288.

Blatt, S., Brenneis, B., Schimek, J. G., & Glick, M. (1976). Normal development and psychopathological impairment of the concept of the object on the Rorschach. *Journal of Abnormal Psychology, 85,* 364-373.

Blatt, S., Rounsaville, B., Eyre, S., & Wilber, C. (1984). The psychodynamics of opiate addiction. *The Journal of Personality Assessment, 42,* 474-482.

Blatt, S., & Berman, W. (1984). A methodology for the use of the Rorschach in clinical research. *Journal of Personality Assessment, 48*(3), 226-239.

Blatt, S., & Lerner, H. (1983a). Investigations in the psychoanalytic theory of object relations and object representations. *Empirical Studies of Psychoanalytic Theory, 1,* 189-249.

Blatt, S. J., & Lerner, H. (1983b). The psychological assessment of object representation. *Jour-*

nal of Personality Assessment, 47, 77-92.

Blatt, S. J., & Ritzler, B. A. (1974). Thought disorder and boundary disturbances in psychosis. *Journal of Consulting Clinical Psychology, 42,* 370-381.

Brenneis, C. B. (1971). Features of the manifest dream in schizophrenia. *Journal of Nervous and Mental Disorder, 153,* 81-91.

Cohen, I. (1960). *An investigation of the relationship between adaptive regression, dogmatism and creativity using the Rorschach and dogmatism scale.* Unpublished doctoral dissertation, Michigan State University, East Lansing.

Cooper, S. (1981). *An object relations view of the borderline defenses: A Rorschach analysis.* Unpublished manuscript.

Cooper, S. (1982, March). *Restage versus defense process and the borderline personality: A Rorschach analysis.* Paper presented to the mid-winter meeting of the Division of Psychoanalysis of the American Psychological Association, Puerto Rico.

Exner, J. (1974). *The Rorschach: A comprehensive system* (Vol. 1). New York: John Wiley & Sons.

Exner, J. (1978). *The Rorschach: A comprehensive system, current research and advanced interpretation* (Vol. 2). New York: John Wiley & Sons.

Exner, J., Armbruster, G., & Vigilione, D. (1978). The temporal stability of some Rorschach features. *Journal of Personality Assessment, 42,* 474-482.

Exner, J., Weiner, I., & Schuyler, W. (1979). *A Rorschach workbook for the comprehensive system.* Bayville, NY: Rorschach Workshops.

Feinstein, A. (1967). Personality correlates for unrealistic experiences. *Journal of Consultative Psychology, 31,* 387-395.

Fenichel, D. (1945). *Psychoanalytic theory of neurosis.* New York: Norton.

Fibel, B. (1979). *Toward a developmental model of depression: Object representation and object loss in adolescent and adult psychiatric patients.* Unpublished doctoral dissertation, University of Massachusetts, Amherst.

Goldberger, L. (1961). Reactions to perceptual isolation and Rorschach manifestations of the primary process. *Journal of Projective Techniques, 25,* 287-302.

Gray, J. (1969). The effect of productivity on primary process and creativity. *Journal of Projective Techniques and Personality Assessment, 33,* 213-218.

Greenberg, N., Ramsay, M., Rakoff, V., & Weiss, A. (1969). Primary process thinking in myxoedema psychosis: A case study. *Canada Journal of Behavioral Science, 1,* 60-67.

Hatcher, R., & Krohn, A. (1980). Level of object representation and capacity for intense psychotherapy in neurotics and borderlines. In J. Kwawer, H. Lerner, & A. Sugarman (Eds.), *Borderline phenomena and the Rorschach test.* New York: International Universities Press.

Holt, R. (1967). Diagnostic testing: Present situation and future prospects. *Journal of Nervous and Mental Disorder, 144,* 444-465.

Holt, R. (1968). *Manual for scoring primary process manifestations in Rorschach responses.* Unpublished manuscript, New York University, Research Center for Mental Health.

Johnson, D. (1980). *Cognitive organization in paranoid and nonparanoid schizophrenia.* Unpublished doctoral dissertation, Yale University, New Haven.

Kernberg, O. (1975). *Borderline conditions and pathological narcissism.* New York: Jason Aronson.

Klopfer, B., Ainsworth, M.D., Klopfer, G., & Holt, R. (1954). *Developments in the Rorschach technique: Vol. I. Techniques and theory.* New York: World Book.

Kohut, H. (1971). *The analysis of the self.* New York: International Universities Press.

Kohut, H. (1977). *The restoration of the self.* New York: International Universities Press.

Krohn, A., & Mayman, M. (1974). Object representations in dreams and projective tests: A construct validational study. *Bulletin of the Menninger Clinic, 38,* 445-466.

Krohn, A. (1972). *Levels of object representations in the manifest dreams and projective tests.* Unpublished doctoral dissertation, University of Michigan, Ann Arbor.

Kwawer, J. (1980). Primitive interpersonal modes, borderline phenomena, and the Rorschach test. In J. Kwawer, H. Lerner, P. Lerner, & A. Sugarman (Eds.), *Borderline phenomena and the Rorschach test* (pp. 89-106). New York: International Universities Press.

Lerner, H., & Lerner, P. (1982). A comparative study of defensive structure in neurotic, borderline, and schizophrenic patients. *Psychoanalysis and Contemporary Thought, 1,* 77-115.

Lerner, H., Sugarman, A., & Barbour, C. (1985). Patterns of ego boundary disturbance in neurotic, borderline, and schizophrenic patients. *Psychoanalytic Psychology, 2,* 47-66.

Lerner, H., Sugarman, A., & Gaughran, J. (1981). *Borderline and schizophrenic patients: A comparative study of defensive structure.* Unpublished manuscript.

Lerner, P. (1979). Treatment implications of the (c) response in the Rorschach records of patients with severe character pathology. *Ontario Psychologist, 11,* 20-22.

Lerner, P. (1981). *Cognitive aspects of the (c) response in the Rorschach records of patients with severe character pathology.* Paper presented to the Internal Rorschach Congress, Washington, DC.

Lerner, P. (in press). Rorschach indices of the false self concept. In H. Lerner & P. Lerner (Eds.), *Primitive mental states and the Rorschach.* New York: International Universities Press.

Lerner, P., & Lerner, H. (1980). Rorschach assessment of primitive defenses in borderline personality structure. In J. Kwawer, H. Lerner, P. Lerner, & A. Sugarman (Eds.), *Borderline phenomena and the Rorschach Test.* New York: International Universities Press.

Lerner, P., & Lewandowski, A. (1975). The measurement of primary process manifestations: A review. In P. Lerner (Ed.), *Handbook of Rorschach scales* (pp. 181-214). New York: International Universities Press.

Lerner, H., & St. Peter, S. (1984). Patterns of object relations in neurotic, borderline, and schizophrenic patients. *Psychiatry: The Study of Interpersonal Process, 47,* 77-92.

Levy, M., & Fox, H. (1975). Psychological testing is alive and well. *Professional Psychology, 6,* 420-424.

Mahler, M., Pine, F., & Bergman, A. (1975). *The psychological birth of the human infant: Symbiosis and individuation.* New York: Basic Books.

Maupin, E. (1965). Individual differences in response to a Zen meditation exercise. *Journal of Consultation and Psychology, 29,* 139-145.

Mayman, M., & Ryan, E. (1972). *Level and quality of object relationships: A scale applicable to overt behavior and to projective test data.* Unpublished manuscript, University of Michigan.

Mayman, M. (1963). Psychoanalytic study of the self-organization with psychological tests. In B. T. Wigdor (Ed.), *Recent advances in the study of behavior change: Proceedings of the academic assembly on clinical psychology.* Montreal: McGill University Press.

Mayman, M. (1964). *Some general propositions implicit in the clinical application of psychological tests.* Unpublished manuscript, Menninger Foundation, Topeka, KS.

Mayman, M. (1967). Object representations and object relationships in Rorschach responses. *Journal of Projective Techniques and Personality Assessment, 31,* 17-24.

Mayman, M. (1968). Early memories and character structure. *Journal of Projective Techniques and Personality Assessment, 32,* 303-316.

Mayman, M. (1976). Psychoanalytic theory in retrospect and prospect. *Bulletin of the Menninger Clinic, 40,* 199-210.

Millon, T. (1984). On the renaissance of personality assessment and personality theory. *Journal of Personality Assessment, 48,* 450-466.

Peebles, R. (1975). Rorschach as self-system in the telophasic theory of personality development. In P. Lerner (Ed.), *Handbook of Rorschach scales* (pp. 71-136). New York: International Universities Press.

Pichot, P. (1984). Centenary of the birth of Hermann Rorschach. *Journal of Personality Assessment, 48,* 591-596.

Pine, F., & Holt, R. (1960). Creativity and primary process: A study of adaptive regression. *Journal of Abnormal and Social Psychology, 61,* 370-379.

Pine, F. (1962). Creativity and primary process: Sample variations. *Journal of Nervous and Mental Disorder, 134,* 506-511.

Piotrowski, C. (1984). The status of projective techniques: Or, "Wishing won't make it go away." *Journal of Clinical Psychology, 40,* 1495-1502.

Pruitt, W., & Spilka, B. (1964). Rorschach empathy—object relationship scale. *Journal of Projective Techniques and Personality Assessment, 8,* 331-336.

Rickers-Ovsiankina, M. (1977). *Rorschach psychology* (2nd ed.). Huntington, NY: Robert E. Krieger.

Ritzler, B., Zambianco, D., Harder, D., & Kaskey, M. (1980). Psychotic patterns of the concept of the object on the Rorschach test. *Journal of Abnormal Psychology, 89,* 46-55.

Rogolsky, M. (1968). Artistic creativity and adaptive regression in third grade children. *Journal of Projective Techniques and Personality Assessment, 32,* 53-62.

Rorschach, H. (1942). *Psychodiagnostics.* Bern: (Hans Huber).

Ryan, E. R. (1973). *The capacity of the patient to enter an elementary therapeutic relationship in the initial psychotherapy interview.* Unpublished doctoral dissertation, University of Michigan, Ann Arbor.

Saretsky, T. (1966). Effects of chlorpromazine on primary process thought manifestations. *Journal of Abnormal Psychology, 71,* 247-252.

Schactel, E. (1966). *Experiential foundations of Rorschach test.* New York: Basic Books.

Schafer, R. (1954). *Psychoanalytic interpretation in Rorschach testing.* Boston: Grune & Stratton.

Schlesinger, H. (1973). Interaction of dynamic and reality factors in the diagnostic testing interview. *Bulletin of the Menninger Clinic, 37,* 495-518.

Shevrin, H., & Shechtman, F. (1973). The diagnostic process in psychiatric evaluation. *Bulletin of the Menninger Clinic, 37,* 451-494.

Silverman, L., & Candell, P. (1970). On the relationship between aggressive activation, symbiotic merging, intactness of body boundaries and manifest pathology in schizophrenia. *Journal of Nervous and Mental Disorders, 150,* 387-399.

Silverman, L., & Goldberger, A. (1966). A further study of the effects of subliminal aggressive stimulation on thinking. *Journal of Nervous and Mental Disorders, 143,* 463-472.

Silverman, L., & Spiro, R. (1967). Further investigation of the effects of subliminal aggressive stimulation on the ego functioning of schizophrenics. *Journal of Consulting Psychology, 31,* 226-232.

Silverman, L. (1965). Regression in the service of the ego. *Journal of Projective Techniques and Personality Assessment, 29,* 232-244.

Silverman, L. (1966). A technique for the study of psychodynamic relationships: The effects of subliminally presented aggressive stimuli on the production of pathological thinking in a schizophrenic population. *Journal of Consulting Psychology, 30,* 103-111.

Spear, W. (1980). The psychological assessment of structural and thematic object representations in borderline and schizophrenic patients. In J. Kwawer, H. Lerner, P. Lerner, & A. Sugarman (Eds.), *Borderline phenomena and the Rorschach* (pp. 321-342). New York: International Universities Press.

Spitzer, R., Endicott, J., & Robbins, E. (1975). Research diagnostic criteria. *Psychopharmacological Bulletin, 11,* 22-24.

Stolorow, R., & Lachmann, F. (1980). *Psychoanalysis of developmental arrest.* New York: International Universities Press.

Sugarman, A. (1981). The diagnostic use of countertransference reactions in psychological testing. *Bulletin of the Menninger Clinic, 45,* 473-490.

Sugarman, A. (1985). The nature of clinical assessment. Unpublished manuscript.

Tolpin, M., & Kohut, H. (1978). The disorders of the self: The psychopathology of the first years of life. In G. Pollock & S. Greenspan (Eds.), *Psychoanalysis of the life cycle* (NIMH Publication). Washington, DC: U. S. Government Printing Office.

Urist, J. (1973). *The Rorschach test as a multidimensional measure of object relations.* Unpublished doctoral dissertation, University of Michigan, Ann Arbor.

Urist, J. (1977). The Rorschach test and the assessment of object relations. *Journal of Personality Assessment, 41,* 3-9.

Von Holt, H., Sengstake, C., Sonoda, B., & Draper, W. (1960). Orality, image fusion and concept formation. *Journal of Projective Techniques, 24,* 194-198.

Viglione, D., & Exner, J. (1983). Current research in the comprehensive Rorschach systems. In J. Butcher & C. Spielberger (Eds.), *Advances in personality assessment* (Vol. 1). Hillsdale, NJ: Lawrence Erlbaum.

Weiner, I. (1977). Approaches to Rorschach validation. In M. A. Rickers-Ovsiankina (Ed.), *Rorschach psychology.* Huntington, NY: Robert E. Krieger.

Werner, H. (1948). *Comparable psychology of mental development.* New York: International Universities Press.

Wilson, A. (in press). Depression in primitive mental states. In H. Lerner & P. Lerner (Eds.), *Primitive mental states and the Rorschach.* New York: International Universities Press.

Winnicott, D. (1960). Ego distortion in terms of true and false self. In D. Winnicott (Ed.), *The maturational processes and the facilitating environment.* London: Hogarth Press.

Wright, N., & Abbey, D. (1965). Perceptual deprivation tolerance and adequacy of defense. *Perceptual and Motor Skills, 20,* 35-38.

Wright, N., & Zubek. J. (1969). Relationship between perceptual deprivation tolerance and adequacy of defenses as measured by the Rorschach. *Journal of Abnormal Social Psychology, 74,* 615-617.

Zimet, C., & Fine, H. (1965). Primary and secondary process thinking in two types of schizophrenia. *Journal of Projective Techniques and Personality Assessment, 29,* 93-99.

Merith Cosden, Ph.D.

Special Education Program, Graduate School of Education, University of California-Santa Barbara, Santa Barbara, California.

ROTTER INCOMPLETE SENTENCES BLANK

Julian B. Rotter. San Antonio, Texas: The Psychological Corporation.

THE ORIGINAL OF THIS REVIEW WAS PUBLISHED IN TEST CRITIQUES: VOLUME II (1985).

Introduction

The Rotter Incomplete Sentences Blank (RISB) is a projective test in which examinees are asked to complete forty sentence "stems" to create statements that reflect their feelings about themselves and others. The test is designed for use with adolescents and adults. Unlike many other sentence completion tasks the RISB has both qualitative and quantitative scoring procedures, with the goal of assessing the types of stressors and level of adjustment experienced and reported by the examinee.

The RISB was revised from a form that was used by Rotter and Willerman in the army and that had been a revision of a variety of then extant sentence completion blanks (Rotter & Rafferty, 1950). The RISB was designed to provide specific "projective" information while also being economical to administer and score. Rotter and Rafferty intended the instrument to serve a purpose similar to that of a lengthy structured interview. Although the major function of the test was not seen as "exposing" deep layers of personality, the test is frequently used to obtain information which examiners feel could not be as well obtained through direct interview.

Three similar RISB forms—High School, College, and Adult—have been developed. These forms can be objectively scored by assigning an empirically derived numerical value to each completed sentence. Responses are scaled on the basis of the level of conflict or adjustment reflected in each statement. The test can also be subjectively interpreted through qualitative analysis of the needs and dynamics projected into the subjects responses.

Practical Applications/Uses

The RISB, like other sentence completion forms, is a projective test that is structured to identify areas of personal adjustment and maladjustment. By placing some "distance" between the examinee and the examiner, the test format allows the examinee to respond more freely than might be expected through direct interview. At the same time, the structure of the test, i.e., its direct inquiry into different areas of functioning, makes it possible for the examiner to use test responses at face value and rapidly develop at least a surface understanding of some of the psychological issues facing the individual.

Unlike many other sentence completion tests, the RISB has a standardized scoring system that allows the examiner to synthesize responses and assess the general level of adjustment presented by the individual. Thus, the RISB provides an efficient method for obtaining information about the individual's adjustment. This method of assessment allows the individual to respond rather directly to specific issues while eliciting more projective material than would be expected through direct interview.

This test, frequently used with adult clients in the early stages of an assessment, was found in a 1961 review of clinical assessments (Sundberg, 1961) to be the 13th most frequently used clinical test instrument and the 2nd most used test for group personality instruments. It is popular partly because it can be easily administered and quickly interpreted at "face value" as well as through more in-depth analysis of dynamic patterns.

Although many see the purpose of the test as serving as a projective device for gathering information that could not be as well obtained through direct interview, there have been questions raised about the level of projective information obtained through sentence completion methods. The RISB provides a relatively direct type of inquiry in comparison to other projective devices. Further, there is empirical evidence to suggest that individuals respond in a qualitatively different manner to this type of test than to a less structured projective test, such as the Thematic Apperception Test (TAT). Newmark and Hetzel (1974), for example, found that subjects experienced significantly greater levels of state anxiety after unstructured tests such as the Rorschach and TAT than after more structured tests like the RISB or Minnesota Multiphasic Personality Inventory (MMPI). Additionally, Murstein and Wolf (1970), in a study of the effects of different "levels" of stimulus structure, found the predicted inverse relationship between pathology and stimulus structure for normal subjects but not for psychiatric subjects using the RISB as an example of a relatively structured projective measure. This study suggests that, at least for normal subjects, less pathology would be expected to emerge on the RISB than on more unstructured projective tests, while psychiatric subjects could be expected to project as much pathological material on this type of test as on other projective devices.

Related to the issue of the "depth" of the projective information elicited by this test is the question of whether subjects' responses reflect conscious or unconscious processes and whether or not subjects can control their responses. Of specific concern is the extent to which responses reflect the underlying needs and dynamics of individuals, or the conscious attempts of the individuals to present themselves in a given manner. Clinical criticisms of this type of test describe the more conscious response process tapped by this procedure, arguing that this same information could be obtained through direct interview. Goldberg (1965), however, in his review of sentence completion tests, notes that the reliability and validity of test information inversely varies with the depth of level of the assessment. Thus, although the RISB might provide less projective information than other types of projective devices, the reliability and validity of that information can be viewed as more stable.

The RISB can be administered to any size group of examinees. The instructions are simple, i.e., the examiner asks the subjects to respond to all items, to express their real feelings, and to make complete sentences. The examiner is advised not to present the sentence stems in any but the designated format as this would disrupt use of the standardized scoring system. Altering the response context by increasing the space provided to complete sentences, for example, would prohibit the effective use of that part of the scoring system which differentiates unusually long responses from other responses. Studies have noted that changing the complexity or structure of sentence stems can have a significant effect on subject responses (Cromwell & Lundy, 1965; Turnbow & Dana, 1981). Other factors being equal, Meltzoff (1951) reported that the tone of instructions as well as the stimuli can have a strong impact on responses. Although the emotional tone of the instructions may have a significant impact, emphasizing speed (Cromwell et al., 1965) or administering items verbally (Flynn, 1974) have not been found to alter written output.

Test administration itself requires little experience or training. The test can be administered in approximately thirty minutes but is self-paced and there is no time limit.

The scoring systems for all three forms of the RISB are based on the scaled responses of college freshmen used to develop the Incomplete Sentences Blank-College Form. Although this form was later adapted to create the High School and Adult Forms through modification of the wording of certain RISB stems to make them more appropriate for their target populations, no additional normative data were collected. The manual places the responsibility on "competent clinical workers" to make the necessary transitions to appropriately use the college school scoring system with high school and adult populations. Given the broad nature of the questions, however, there is at least strong "face validity" to utilizing the scoring system with all three populations.

The information obtained through sentence completions can be summarized or "reduced" for purposes of interpretation through both qualitative and quantitative procedures. Both approaches are detailed, with case examples, in the manual (Rotter & Rafferty, 1950, pp. 14-50).

In the objective scoring system each completed sentence receives an "adjustment" score. To obtain this score, the examiner assigns each sentence one of four possible categorical codes: 1) omission (no response after the sentence stem or a response too short to provide a meaningful sentence), 2) conflict response (a response that reflects hostility or unhappiness), 3) positive response (a statement which reflects positive or hopeful attitudes), or 4) neutral response (a simple declarative statement which does not imply either a positive or negative affect). Both conflict and positive statements are given a weight (1-3) that reflects the degree of sentiment expressed by the response. The manual provides examples for each scoring category, with responses for men and women listed separately.

Each code is scaled from 0-6, with higher values given to more negative responses. A summative adjustment score with a possible range of 0-240 is obtained in this manner. The average adjustment score reported in the manual is 127 with a standard deviation of 14. A cutoff score of 135 is frequently used to differentiate subjects who are "maladjusted" (i.e., in need of psychotherapy) from those who are well adjusted. The effectiveness of this cutoff score is discussed in the Technical Aspects section of this review.

The coding system is relatively simple, with a limited number of exclusive categories to be considered. Nevertheless, wide response variations are anticipated on a projective test of this sort, mandating reliance on the manual and its coding exemplars to ensure high reliability. The manual does not state the time needed for scoring each protocol but this reviewer has found the scoring time to be approximately 45 minutes depending on the length and complexity of the protocol.

A comprehensive but dated review of sentence completion forms by Goldberg (1965) notes that the two distinguishing characteristics of the RISB are its quantified scoring system and its single variable method of analysis. Test interpretation based on quantitative analysis of the level of adjustment reflected in one's responses is fairly straightforward. Each statement is scored and summed to provide a measure of that individual's overall adjustment. This system is touted by investigators both as the test's strength, i.e., it provides an efficient and reliable manner for interpreting test scores, and as its weakness, i.e., it limits the scope and depth of test interpretation to one variable. Goldberg summarizes these findings by concluding that the RISB sacrifices scope for efficiency in its scoring system.

Other methods for coding the RISB have been explored with some success. Rogers (1978) conducted a computerized content analysis of the words used in the RISB. Using five content categories, he was able to predict more accurately certain types of deviant behaviors than was possible with regular RISB scores. Albert (1970), on the other hand, tested a method for scoring response differences in the RISB over time. He was able to show that difference scores in this system predicted therapeutic gains. Neither of these models, however, have been advanced in the literature.

Qualitative analysis of subject responses, although allowing for a greater range and scope of interpretations, relies solely on the clinical expertise and knowledge of the test user for its reliability and validity. The manual suggests content analysis of the RISB in a manner similar to that utilized in analyzing the TAT. This procedure carries with it the problems found in all subjective analyses of projective materials, i.e., questionable reliability and validity of test interpretations. While little research has been conducted to validate the use of this type of interpretive procedure, Starr and Katkin (1969) found that subjective interpretation of the RISB was subject to the same "illusory correlations" between responses and patient symptoms as were other projective devices. That is, test users are likely to posit stronger relationships between certain types of statements on the RISB and behavior patterns than is warranted by observations of these relationships.

Technical Aspects

Reported reliability coefficients for the RISB are high. In the normative sample, Rotter, Rafferty, and Schachtitz (1965) reported interscorer reliabilities of .96 for the women tested in their sample and .91 for the men. Split-half reliability coefficients were found to be .83 for women and .84 for men. Although these reliability coefficients were obtained from researchers trained by the test developers, other investigators have reported similar high reliabilities. Churchill and Crandall (1965), for example, found interscorer reliabilities of .94 to .95 using assistants with minimal psychological expertise who were trained using the manual. While Lah and Rotter (1981) continue to report high interrater reliabilities, they also note that current student scores differ significantly from those obtained in the initial sample.

A number of studies have assessed the validity of the RISB as a screening device using a "maladjustment" cutoff score. A cutoff score of 135 has been found to correctly identify delinquent youths 60% of the time while screening nondelinquent youths correctly 73% of the time (Fuller, Parmelee, & Carroll, 1982); to identify students scoring low in manifest anxiety defenses 86% of the time and pinpoint those high in defenses 75% of the time (Milliment, 1972); to identify students in counseling 54-66% of the time and differentiate those not in counseling 64% of the time (Churchill et al., 1965); and to identify severe drug users 80-100% of the time (Gardner, 1967). These validation studies suggest that the maladjustment cutoff score does successfully screen those in crisis from those not in crisis on more than a chance basis, but the screen may not be sufficiently high to utilize alone without fear of missing too many other individuals in need of psychological help.

The subjects in the above studies had been identified in need of services prior to administration of the RISB, with the RISB administered to test its effectiveness in detecting their problems. The RISB has also been used to evaluate stresses when this information was not immediately available through other sources. Scores on the RISB have been shown to correlate with the level of difficulty experienced by individuals going through new vocational experiences during mid-life career changes (Vaitenas & Wiener, 1977). The RISB has also been successful in identifying depression in adult clients even when depression was not initially evident in overt behaviors (Robbins, 1974). There have been moderate correlations found between RISB scores and both anxiety (Richardson & Soucar, 1971), and disparities in one's self-image (De Mann, 1983). The RISB has also been used to identify students with high levels of physical complaints as measured by utilization of medical facilities (Getter & Weiss, 1968).

The RISB has been used to assess a wide range of anxiety and depressive disorders. It does not, however, appear effective in assessing psychological adjustment over short periods of time as measured by pre- and post-test differences (e.g., Krouse & Krouse, 1981; Rosenheim & Dunn, 1977). Albert (1970) found significant adjustment differences as a function of therapeutic progress on pre- and post-test RISB administrations using his own scoring system. The RISB has also had equivocal results in screening for drug use, with only heavy drug users scoring significantly in the maladjusted range (Cross & Davis, 1972; Gardner, 1967; Pascale, Hurd, & Primavera, 1980).

The validity of the RISB as an adjustment index has been questioned by some on the basis of its high loading on social desirability. Some investigators (e.g., Janda & Galbraith, 1973) feel that adjustment scores may reflect awareness of socially desirable responses rather than a lack of disorder. Janda and Galbraith argue that the high correlations between the social desirability of responses and their adjustment scores confound the direct interpretation of the RISB. Other investigators

(Banikiotes, Russell, & Linden, 1971; McCarthy & Rafferty, 1971), however, report low to moderate linear relationships between social desirability and adjustment as measured by the RISB. These investigators do not believe that adjustment scores on the RISB can be as easily interpreted as social desirability scores; rather, they posit a more limited and complex relationship between social desirability and adjustment, acknowledging that some stems on the RISB tap self-concept that, in turn, is related to social desirability.

Critique

The RISB is a relatively structured projective instrument designed for use with high school and college students and adults. It differs from other incomplete sentence blanks in that it allows both qualitative and quantitative assessment of responses. This assessment provides information on the types of stressors and level of adjustment experienced and reported by the individual. The extent to which this test taps conscious processes (i.e., attempts of the individual to provide the investigator with certain types of information) or unconscious processes (i.e., underlying needs and dynamics that the individual is not aware of expressing) is yet unclear.

As discussed by Goldberg (1965), it is the structure of the test that is both the test's strength and weakness. On the one hand, the test provides a relatively fast, efficient method for obtaining direct information on how individuals are functioning in many different aspects of their lives. Use of the empirical scoring system further allows the examiner to quickly synthesize information and obtain a picture of how well the individual is functioning overall. By the same token, this type of format may not provide the same sort of rich information obtained through other projective techniques. For some subjects it may not provide much more information than would be obtained through direct interview. Further, to the extent that this information reflects the conscious processes of the individual it is subject to the same types of bias as other self-report measures, and interpretation has to take into consideration what the individual wants the examiner to know.

The RISB is probably most effective in the early stages of an assessment with adults who are either having a difficult time defining their problems or who need some minimal distance from the examiner in order to express themselves more openly. In this way, the test can be used to help clarify for the examiner the specific areas in which individuals are feeling stressed as well as indicate the general extent of their problems and their level of adjustment.

References

Albert, G. (1970). Sentence completions as a measure of progress in therapy. *Journal of Contemporary Psychology, 3*, 31-34.

Banikiotes, P. G., Russell, J. M., & Linden, J. D. (1971). Social desirability, adjustment and effectiveness. *Psychological Reports, 29*, 581-582.

Churchill, R., & Crandall, V. J. (1965). The reliability and validity of the Rotter Incomplete Sentences Test. In B.I. Murstein (Ed.), *Handbook of projective techniques* (pp. 873-882). New York: Basic Books.

Cromwell, R. L., & Lundy, R. M. (1965). Productivity of clinical hypothesis on a sentence completion test. In B.I. Murstein (Ed.), *Handbook of projective techniques* (pp. 883-890). New York: Basic Books.

Cross, H., & Davis, G. (1972). College students' adjustment and frequency of marijuana use. *Journal of Counseling Psychology, 19*, 65-67.

De Man, A. F. (1983). Self image disparity and adjustment in young adult females. *Psychological Reports, 52*, 78.

Flynn, W. (1974). Oral vs. written administration of the incomplete sentences blank. *Newsletter for Research in Mental Health and Behavioral Sciences, 16*, 19-20.

Fuller, G. B., Parmelee, W. M., & Carroll, J. L. (1982). Performance of delinquent and nondelinquent high school boys on the Rotter Incomplete Sentence Blank. *Journal of Personality Assessment, 46*, 506-510.

Gardner, J. (1967). The adjustment of drug addicts as measured by the sentence completion test. *Journal of Projective Techniques and Personality Assessment, 31*, 28-29.

Getter, H., & Weiss, S. (1968). The Rotter Incomplete Sentences Blank adjustment score as an indicator of somatic complaint frequency. *Journal of Projective Techniques and Personality Assessment, 32*, 266.

Goldberg, P. (1965). A review of sentence completion methods in personality assessment. In B.I. Murstein (Ed.), *Handbook of projective techniques* (pp. 777-822). New York: Basic Books.

Janda, L., & Galbraith, G. (1973). Social desirability and adjustment in the Rotter Incomplete Sentences Blank. *Journal of Consulting and Clinical Psychology, 40*, 337.

Krouse, H. J., & Krouse, J. H. (1981). Psychological factors in postmastectomy adjustment. *Psychological Reports, 48*, 275-278.

Lah, M. I., & Rotter, J. B. (1981). Changing college student norms on the Rotter Incomplete Sentences Blank. *Journal of Consulting and Clinical Psychology, 49*, 985.

McCarthy, B. W., & Rafferty, J. E. (1971). Effect of social desirability and self-concept scores on the measurement of adjustment. *Journal of Personality Assessment, 35*, 576-583.

Meltzoff, J. (1951). The effect of mental set and item structure upon response to a projective test. *Journal of Abnormal and Social Psychology, 46*, 177-189.

Milliment, R. (1972). Support for a maladjustment interpretation of the anxiety-defensiveness dimension. *Journal of Personality Assessment, 36*, 39-44.

Murstein, B., & Wolf, S. (1970). Empirical test of the "levels" hypothesis with five projective techniques. *Journal of Abnormal Psychology, 75*, 38-44.

Newmark, C., & Hetzel, W. (1974). The effects of personality tests on state and trait anxiety. *Journal of Personality Assessment, 38*, 17-20.

Pascale, R., Hurd, M., & Primavera, L. (1980). The effects of chronic marijuana use. *Journal of Social Psychology, 110*, 273-283.

Richardson, L., & Soucar, E. (1971). Comparison of cognitive complexity with achievement and adjustment: A convergent-discriminant study. *Psychological Reports, 29*, 1087-1090.

Robbins, P. R. (1974). Depression and drug addiction. *Psychiatric Quarterly, 48*, 374-386.

Rogers, G. (1978). Content analysis of the Rotter Incomplete Sentences Blank and the Prediction of Behavior Ratings. *Educational and Psychological Measurement, 38*, 1135-1141.

Rosenheim, H. D., & Dunn, R. W. (1977). The effects of rational behavior therapy in a military population. *Military Medicine, 142*, 550-552.

Rotter, J. B., & Rafferty, J. E. (1950). *Manual for the Rotter Incomplete Sentences Blank: College Form*. New York: The Psychological Corporation.

Rotter, J. B., Rafferty, J. E., & Schachtitz, E. (1965). Validation of the Rotter Incomplete Sentences Test. In B.I. Murstein (Ed.), *Handbook of projective techniques* (pp. 859-872). New York: Basic Books.

Starr, B. J., & Katkin, E. S. (1969). The clinician as an aberrant actuary: Illusory correlation and the incomplete sentences blank. *Journal of Abnormal Psychology, 74*, 670-675.

Sundberg, N. D. (1961). The practice of psychological testing in clinical services in the United States. *American Psychologist, 16,* 79-83.

Turnbow, K., & Dana, R. (1981). The effects of stem length and directions on sentence completion. *Journal of Personality Assessment, 45,* 27-32.

Vaitenas, R., & Wiener, Y. (1977). Developmental, emotional, and interest factors in voluntary mid-career change. *Journal of Vocational Behavior, 11,* 291-304.

Howard Tennen, Ph.D

Associate Professor of Psychiatry, University of Connecticut School of Medicine, Farmington, Connecticut.

Glenn Affleck, Ph. D.

Associate Professor of Psychiatry, University of Connecticut School of Medicine, Farmington, Connecticut.

Sharon Herzberger, Ph.D.

Associate Professor of Psychology, Trinity College, Hartford, Connecticut.

SCL-90-R

Leonard R. Derogatis. Riderwood, Maryland: Clinical Psychometric Research.

THE ORIGINAL OF THIS REVIEW WAS PUBLISHED IN TEST CRITIQUES: VOLUME III (1985).

Introduction

The SCL-90-R (the Symptom Check List-90-Revised) is a 90-item, self-report inventory designed to assess the psychological symptoms of psychiatric and medical patients. The inventory measures somatization, obsessive-compulsive symptoms, interpersonal sensitivity, depression, anxiety, phobic anxiety, psychoticism, paranoid ideation, hostility, and global indices of psychopathology.

The Clinical Psychometrics Research Unit of Johns Hopkins University devised the SCL-90-R, which evolved from the Hopkins Symptom Checklist (HSCL). The HSCL was designed to assess psychopathological symptomatology, but has several drawbacks. First, it had not been designed for use with individual patients, but was a research instrument. Second, the items failed to cover all of the important dimensions of symptomatology identified by Derogatis and his colleagues. Finally, the scale had no companion instrument to provide reliable coassessment of the patient from a clinician's viewpoint. With these deficiencies in mind, work on the SCL-90-R began.

Five primary symptom dimensions from the HSCL were retained, some items were dropped, and four new dimensions (composed of 45 new items) were added. All items were set to a 5-point response scale. An early version of this scale (SCL-90) was tested, modified, and then validated in the present form as the SCL-90-R. Validation procedures revealed that the subscales have high levels of internal consistency and high convergent validity. The subscales also show ade-

The reviewers wish to thank John Chapman for his assistance in verifying references for this review.

quate factorial invariance across subsamples and the results of confirmatory factor analyses demonstrated considerable correspondence between the theoretically and empirically derived factor structures (see technical aspects section for details).

Our review of the literature revealed that investigators continue to use the SCL-90 instead of the SCL-90-R. This may be due, in part, to the similarities between these two measures. Eighty-seven of the 90 items are identical and only one item was changed completely. The SCL-90 asks the respondent "how much were you bothered by . . ." while the SCL-90-R asks "how much were you distressed by . . . ." Both measures use the same five-point response scale. Nonetheless because the SCL-90-R is a revised measure and because only the SCL-90-R has norms, these reviewers strongly recommend that investigators and clinicians use the revised measure. In the studies reviewed below, only nine used the SCL-90-R, and when referring to these studies the SCL-90-R applies. The SCL-90 applies when reviewing a study that used the earlier version.

A shorter version of the test, the Brief Symptom Inventory (BSI), has also been devised (Derogatis & Melisaratos, 1983). Its 53 items are also on 5-point scales. The BSI includes the same 9-symptom dimensions and global indices. Because, according to the manual (Derogatis, 1977), the BSI and SCL-90-R correlate highly, the BSI may be used when the longer scale is impractical.

The SCL-90 version of the scale has been translated into various languages, including Hebrew (Roskin & Dasberg, 1983), Farsi (Siassi & Fozouni, 1982), and Dutch (Arrindell & Ettema, 1981).

The SCL-90-R consists of 90 self-description items that list symptoms related to various aspects of psychopathology. Respondents note the degree to which they are distressed by the symptom on a scale from 0 (not at all) to 4 (extremely). The test administrator sets a time period for which the symptoms are assessed, typically the last week.

The instructions are simple and straightforward, and the items are easy to comprehend. During test development, the Lorge-Thorndike *Word Book* (Thorndike & Lorge, 1944) was used to ensure that the items were simple to read and comparable in difficulty level across symptom dimensions.

Subtests include somatization, obsessive-compulsive, interpersonal sensitivity, depression, anxiety, hostility, phobic anxiety, paranoid ideation, and psychoticism. Furthermore, three global indices provide an assessment of the intensity of perceived distress, the number of symptoms experienced, and a summary measure combining intensity and a number of symptoms. Finally, seven additional items that load onto several symptom dimensions, but are not scored as a subscale, are included for their clinical relevance.

Practical Applications/Uses

The SCL-90-R is designed for psychiatric and other medical patients. It can be used with adolescents (age 13 years or above), but should not be used with delusional clients or those for whom a self-report measure would be inappropriate. The SCL-90-R has been used with a wide range of patients, including alcoholics, drug users, students, cancer and heart patients, and those with sexual disorders.

The test manual recommends that the measure be administered by someone who provides a positive impression of the benefits of psychological assessment. Administration is straightforward, however, and can be done by a technician. The technician should be familiar with the test content and vocabulary and be prepared to answer definitional questions when necessary. If a patient has difficulty responding to the test due to physical limitations, the administrator should provide a written version of the response scale and read the items aloud. The respondent can then answer verbally to each item. The test takes 10-30 minutes (10 minutes longer for verbal administration).

A score profile booklet is included with the test to aid scoring. The booklet lists each item on the relevant subscale, leaving spaces for noting the patient's response. Scoring requires simple addition and division, and conversion of raw scores to T-scores standardized according to the appropriate patient norm. Separate norms are given for males and females. Finally, a symptom profile may be plotted to facilitate interpretation. Detailed information about scoring and handling missing data is given in the test manual.

Interpretation of the subject's responses must be done by a clinician who is familiar with the symptom dimensions. Interpretation should include an examination of the global indices of distress, the standardized scores on each symptom dimension, and response to individual items.

Normative scores for a heterogeneous psychiatric sample, normals, and adolescent psychiatric patients are provided in the test manual, as are endorsement frequencies for each item within each sample. Furthermore, profiles from a wide range of other samples (e.g., weight reduction clients, alcohol rehabilitation clients, cancer patients) are provided, but should be used with caution. These samples are typically small and the norms have not been formally investigated.

Two other instruments are provided by Derogatis and his research group to complement the SCL-90-R. The use of these scales permits an examination of correspondence between clinician and patient assessments and between clinician assessment and that provided by other health professionals. The Hopkins Psychiatric Rating Scale (HPRS) is designed for psychiatrists and other clinicians. It contains 18 symptom dimensions, nine matching those on the SCL-90-R and nine that are not amenable to self-observation. The SCL-90 Analogue Scale may be completed by health professionals who lack sophisticated knowledge of psychopathology. It should be noted, however, that the correlation between scores on the SCL-90-R and the Analogue measure has been low in some samples.

Technical Aspects

Studies investigating the reliability of the SCL-90-R have used one of three strategies: 1) measuring the homogeneity of items of the SCL-90, 2) using a test-retest design to determine scale score changes over time, and 3) assessing interrater reliability by having the respondent complete the SCL-90-R and a clinician complete an analogue scale.

Derogatis, Rickels, and Rock (1976) assessed the homogeneity of each of the nine symptom dimensions in a group of 219 symptomatic volunteers who completed the SCL-90-R. All of the coefficient alphas were appropriate, ranging from .77 for the psychoticism dimension to .90 for depression. Clark and Friedman (1983a, 1983b) also reported extremely high internal consistency for the SCL-90, but found that the symptom subscales were highly intercorrelated.

In a study of a Dutch version of the SCL-90, Arrindell and Ettema (1981) assessed the responses of normal nonpatients and phobic outpatients. These authors report relatively high internal consistency as measured by item-total score correlations.

Derogatis (1977) presents test-retest coefficients from a sample of 94 psychiatric outpatients who completed the SCL-90-R during an initial evaluation and once again before their first psychotherapeutic hour. One week elapsed between assessments. Test-retest coefficients ranged from .78 for the hostility dimension to .90 for phobic anxiety.

In a study comparing the stability of five adjustment scales in a nonpatient sample, Edwards (1978) administered the SCL-90 and four other measures of psychological distress at three points in time, two weeks apart. There was no evidence of systematic changes in scores due to multiple administrations of the measures. The SCL-90 showed not only adequate test-retest stability, but "only the SCL-90 showed promise for allowing reliable assessment of individual change over time" (p. 275). Thus, the SCL-90 not only demonstrates adequate reliability, but appears to be more reliable and sensitive than other measures of psychological distress.

Finally, a longitudinal study of psychiatric symptomatology among drug abusers (Rhoads, 1983) has relevance to the issue of the test-retest reliability of the SCL-90-R. Individuals who had recently been discharged from heroin detoxification programs completed the SCL-90-R at monthly intervals for three months. Participants who reentered treatment (methadone maintenance) had decreased depression and anxiety scores on the SCL-90-R, suggesting that the measure was sensitive to treatment effects. More important with regard to stability was the finding that subjects who were not in treatment showed no significant changes over time in their SCL-90-R scores.

In summary, studies of the test-retest reliability of the SCL-90-R and the SCL-90 suggest that scores are fairly stable over time, yet they are sensitive to treatment effects. Respectable test-retest reliability coefficients have been reported in normal populations as well as in psychiatric samples.

Yet another method of determining the reliability of a measure is to administer alternate forms of the test and complete the correlation between scores on the two forms. As Derogatis and Melisaratos (1983) point out, the SCL-90-R and the BSI are not alternate forms of the same test in the strict sense of the term. Both measures do, however, tap the same symptom dimensions and the 53 items on the BSI are also on the longer SCL-90-R. Thus correlations between these measures are of interest with regard to reliability, but should be interpreted conservatively. In a sample of 565 psychiatric outpatients, correlations betweem SCL-90-R scores and BSI scores ranged from .99 for the hostility to .92 for psychoticism (Derogatis & Melisaratos, 1983).

A test's reliability can also be ascertained by determining how well two raters can agree when using the test. Because the SCL-90-R is a self-report measure interrater reliability assessments cannot be conducted. However, the SCL-90 Analogue (Derogatis, 1977) provides an opportunity to compare the degree of agreement between self-ratings and clinician ratings. Such comparisons can be viewed as measures of construct validity or as measures of interrater reliability using alternate forms. When two clinicians use the Analogue to rate the same target person, reliability coefficients range from .78 for the psychoticism dimension to .94 for somatization (Derogatis, 1977). Reports comparing self-ratings and clinician ratings, however, have produced mixed results. In one study, Turner, McGovern, and Sandrock (1983) found significant agreement among the ratings of schizophrenic clients, hospital clinicians, and independent evaluators.

In two other studies, clinicians' ratings did not conform to patients' self-reports, as measured by the SCL-90 Analogue and the SCL-90-R, respectively. Nahmias, Beutler, Crago, Osborn, and Hughes (1983) asked individuals entering the emergency room of a university hospital to complete the SCL-90-R. Following treatment in the E.R., the house physician and the nurse completed the SCL-90 Analogue. Three sets of comparisons were presented: physician-nurse ratings, which provide an index of interrater reliability for the Analogue, and physician-patient ratings and nurse-patient ratings, which test agreement between self-report and clinical assessment.

Although physicians and nurses showed some correspondence on most dimensions, the correlations were quite modest, with significant correlations ranging from .20 for interpersonal sensitivity to .50 for anxiety. This level of agreement is considerably lower than that reported in the manual. The reasons for this discrepancy are unclear, as the manual reports that the Analogue was designed for use by health professionals "without detailed training or knowledge in psychopathology" (p. 33). One reasonable possibility, worthy of further investigation, is that the valid use of the analogue requires the clinician to probe into those areas measured by the SCL-90-R. Otherwise, ratings can reflect mere speculation. Nonetheless, the modest physician-nurse agreement noted by Nahmias et al. (1983), using measures of agreement that fail to control for chance (Fleiss, 1981), should be cause for some concern. Whether these modest agreement rates reflect some problem in the SCL-90-R, the Analogue, or the need for more psychological-psychiatric training of emergency room staff is still to be determined.

In another study of psychiatric emergency room patients who were also in psychotherapy, Kass, Skodol, Buckley, and Charles (1980) computed a psychopathology recognition index in which clinicians' ratings were compared to SCL-90 ratings from the patient. They found that therapists were able to recognize depression (94% of the cases) and anxiety (89% of the cases), but were unable to adequately recognize psychotic symptoms (35% recognition) or obsessive-compulsive symptoms (16% recognition). The total recognition score for all SCL-90 scales was 61%, which is modest or moderate at best whether one considers this agreement rate as an index of interrater reliability or a measure of construct validity.

Once again, one might question the adequacy of therapists' ratings as a benchmark for comparing SCL-90 or SCL-90-R scores. Indeed, Derogatis (1983) raises just this issue in response to Kass et al.'s (1983) recent report of an unacceptably low correlation ($r = .17$) between the SCL-90 and the SCL Analogue. Kass et al. (1983) attribute this discordance to the underreporting of symptoms by patients due to fear and distrust or a "denying" characterological style, and to overreporting by patients with a "dramatizing" style. Derogatis (1983) suggests that these findings may be due to the investigators' use of the SCL-90, the earlier prototype of the SCL-90-R, their reliance on a study sample that was heavily skewed toward low SES patients, their failure to establish interrater reliability for the Analogue, and the use of inexperienced clinicians. To date, there is no solid evidence regarding the concordance between the SCL-90-R and the SCL Analogue when rated by experienced clinicians.

In summary, the SCL-90-R has consistently demonstrated high levels of internal consistency and impressive stability coefficients. The relationship between SCL-90-R scores and clinician ratings using the SCL Analogue requires further investigation. We should note, however, that many self-report rating scales with sound psychometric qualities fail to show a high level of correspondence to clinician ratings. Until studies comparing self-report measures and clinician reports incorporate stringent requirements for interclinician reliability, we believe that the current evidence favors the use of the SCL-90-R for screening as well as measuring symptom change.

Derogatis and Melisaratos (1983) provide reliability data on the BSI, which, like the longer SCL-90-R, reflects the nine primary symptom dimensions as well as the three global indices of distress. The internal consistency of the BSI was established in a sample of 1,002 outpatients. Coefficient alphas ranged from .71 on the psychoticism dimension to .85 for depression. This range of internal consistency coefficients parallels those found in the longer SCL-90-R.

Test-retest reliability coefficients for the BSI symptom dimensions were derived from a sample of sixty nonpatients with two weeks between test administrations (Derogatis, 1977). Test-retest values ranged from .68 for the somatization dimension to .91 for phobic anxiety. The test-retest coefficients for the GSI (Global Severity Index), PSDI (Positive Symptom Distress Index), and PST (Positive Symptom Total) were .90, .87, and .80 respectively. These are respectable values for measures of symptom constructs. As Derogatis (1977) and Derogatis and Melisaratos (1983) note, stability indices for psychopathological syndromes usually fall between those for stable personality characteristics such as intelligence and state-dependent characteristics such as mood.

The SCL-90, but not the SCL-90-R, has been used in numerous studies of psychological symptomatology and distress in clinical and nonclinical populations. This literature gives substantial evidence of the SCL-90's concurrent, predictive, and construct validity. Investigators have 1) studied the covergence of SCL-90 scores with those of measures of similar constructs; 2) incorporated SCL-90 scores as predictors of response to treatment; 3) used the SCL-90 as an outcome variable in studies of the efficacy of drug treatment, psychotherapy, and behavioral management programs; 4) examined the SCL-90 as a discriminator of psychological symptoms in various clinical samples; 5) used the SCL-90 as a measure of distress associated with aversive experiences; and 6) attempted to replicate SCL-90 factor subscales.

Several investigators have reported significant correlations between SCL-90 symptom subscale scores and measures of analogous constructs. These include significant relationships between SCL-90 subscales or global scores and similar symptom dimensions measured by the MMPI (Derogatis, Rickels, & Rock, 1976; Dinning & Evans, 1977), the Beck Depression Inventory (Dinning & Evans, 1977), the Denver Community Mental Health Questionnaire, and Personal Adjustment and Role Skills Inventory (Turner, McGovern, & Sandrock, 1983). In one study, the SCL-90 correlated with measures of dissimilar constructs in psychiatric inpatients, suggesting low discriminant validity for this population (Dinning & Evans, 1977).

A few studies have been conducted on the usefulness of the SCL-90 as a predictor of response to therapy. Patients whose hypertension was relatively resistant to a self-management program reported higher levels of psychological distress on the SCL-90 prior to treatment (Egan et al., 1983). SCL-90 scores obtained at the beginning of treatment were also found to predict duration of involvement in a rehabilitation program as well as functional outcome following hospital admission of patients with spinal cord injuries (Malec & Neimeyer, 1983) and to correlate with length of stay in an outpatient program for drug abusers (Egan et al., 1983).

Numerous researchers have used the SCL-90 as an outcome variable in drug efficacy studies. These studies show that the SCL-90 is sensitive to changes associated with drug therapy with depressed patients (Davidson et al., 1983; Davidson et al., 1981b), patients with anxiety neuroses (Binstok, Foster, & Mullane, 1984), or chronic anxiety disorders (Kathol et al., 1980), and in samples with mixed diagnoses (Davidson et al., 1981a). There is also some indication that symptom improvement as assessed by the SCL-90 is associated with biological markers such as MAO inhibition (Davidson et al., 1981a).

The SCL-90 has also been used as a criterion measure in evaluating psychotherapy or behavioral treatments. Positive changes in SCL-90-R dimensions have been demonstrated in a comparison of different modes of group therapy for psychiatric inpatients (Beutler et al., 1984), in a support program for families of suicide victims (Rogers et al., 1982), in a small group intervention program for people who had recently experienced multiple life stressors (Roskin, 1982), in a comparative study of the effects of drug counselling and psychotherapy with addicts (Woody et al., 1984), and in the evaluation of a psychiatric outpatient service (Speer & Swindle, 1982). Training normal subjects in relaxation techniques (Shapiro & Lehrer, 1980) and chronic pain patients in biofeedback (Hendler et al., 1977) has also reduced symptom distress as reported on the SCL-90. Depressive symptoms recorded on the SCL-90 were also lowered by involvement in a running program for patients with moderate or mild depression (Greist et al., 1979).

Another type of validity study concerns the test's ability to discriminate between clinical groups or between clinical and nonclinical populations. When compared with nonpatient norms or control subjects, greater levels of psychological symptomatology have been reported on the SCL-90 by burn patients (Blank & Perry, 1984), head trauma patients (Sbordone et al., 1984), patients with acoustic and vestibular disorders (Bastecky, Boleloucky, & Skovronsky, 1981), patients seeking treatment for sexual dysfunction (using the SCL-90-R) (Derogatis, Meyer, & King, 1981), people suffering from recurrent nightmares (Kales et al., 1980a), clients on a methadone maintenance program (Jacobs, Doft, & Koser, 1981), and patients admitted to an oncology research unit (Craig & Abeloff, 1974). The SCL-90 also successfully discriminated between patients with low back versus other types of chronic pain (Pelz & Merskey, 1982), between heroin addicts seeking different types of treatment (using the SCL-90-R) (Steer, 1982), between current and previous somnambulists (Kales et al., 1980b), and between women who were seeking first-time versus repeat abortions (Freeman et al., 1980). Abeloff and Derogatis

(1977) found that breast cancer patients had a distinct SCL-90 symptom profile when compared with that of women with other forms of cancer.

Many investigators have also used the SCL-90 to measure responses to stressful life events or chronic strains in nonpatient samples. For example, elevated symptom subscales have been shown for women seeking abortions (Freeman et al., 1980), spouses of chronic pain patients (Shanfield et al., 1979), parents of children who had died in traffic accidents (Shanfield & Swain, 1984), people bereaving the death of a parent (Horowitz et al., 1984), survivors of a catastrophic fire (using the SCL-90-R) (Green et al., 1983), relatives of suicide victims (Rogers et al., 1982), and rape victims (Kilpatrick, Resick, & Veronen, 1981). A recent study of people living near the Three Mile Island nuclear power plant showed lower symptom distress to be a function of the use of certain coping strategies and social supports in the aftermath of the accident (Baum, Fleming, & Singer, 1983). Another study linked parents' perceptions of their young infants' difficult temperament to higher scores on the depression and anxiety subscales (Ventura, 1982). The ability of the SCL-90 to discriminate between people who sought treatment for pathological effects of bereavement and those who had experienced the loss of a loved one but did not seek treatment is further demonstration of the test's ability to identify individual differences in adaptation to aversive events (Horowitz et al., 1984).

Finally, there are data from several studies that bear on the validity of the SCL-90's hypothesized factorial structure of nine symptom dimensions. Derogatis and Cleary (1977a) described results of a confirmatory factor analysis on data from over 1,000 psychiatric outpatients. They interpreted their findings to show a reasonably good match between empirical findings and the hypothesized symptom structure in eight of the nine areas measured by the test. Generally positive confirmatory factor analyses have also been reported in studies of neurotic patients (Jerabek, Klimpl, & Boleloucky, 1982), psychiatric inpatients (Holcomb, Adams, & Ponder, 1983), and outpatients (Evenson et al., 1980). Other studies of psychiatric inpatients (Clark & Friedman, 1983a) and outpatients (Hoffmann & Overall, 1978) have identified discrepancies between the hypothesized dimensional structure of the SCL-90 and the results of factor analysis.

Most of the validity studies cited used the SCL-90 rather than the SCL-90-R. There are some additional studies in the literature that seem to support the validity of the SCL-90-R, but as a group, they do not provide much additional evidence. Thus, clinicians and researchers using the scale should be aware that much of the available validity data concerns the earlier prototype of the SCL-90-R.

Critique

The SCL-90-R is a widely used self-report measure of psychopathology with sound psychometric properties, including high levels of internal consistency and temporal stability. The SCL-90 version of the scale appears sensitive to clinical change and has been used successfully in psychotherapy outcome studies, epidemological studies, investigations of drug treatment, studies of victimized individuals, and studies of psychopathology in medical patients.

The weaknesses of the SCL-90-R are similar to those of many self-report instruments: 1) there is an assumption that the research participant or patient will accurately describe symptoms and behavior (Derogatis & Melisaratos, 1983; Wilde, 1972); 2) the measure is based on the premise that response bias and social desirability (Crowne & Marlowe, 1960; Edwards, 1957) do not contribute to scores in a systematic fashion (Derogatis & Melisaratos, 1983); and 3) there is equivocal concordance between patient reports and clinician ratings. Future investigations of the psychometric properties of the SCL-90-R should include the contribution of social desirability and the benefits of nine separate scales compared to the exclusive use of the global measures of distress. Based on current knowledge, the SCL-90-R is a sound self-report assessment instrument of psychopathology.

References

Abeloff, M. D., & Derogatis, L. R. (1977). Psychological aspects of the management of primary and metastatic cancer. *Proceedings of the International Conference on Breast Cancer.* New York: A. R. List Inc.

Arrindell, W. A., & Ettema, H. (1981). Dimensional structure, reliability and validity of the Dutch version of the Symptom Checklist (SCL-90): Data based on a phobic and a "normal" population. *Nederlands Tijdschrift Voor de Psychologie en Haar Grensgebieden, 36,* 77-108.

Bastecky, J., Boleloucky, Z., & Skovronsky, O. (1981). Psychotropic drugs in acoustic and vestibular disorders. *Activitas Nervosa Superior, 23,* 187-188.

Baum, A., Fleming, R., & Singer, J. (1983). Coping with victimization by technological disaster. *Journal of Social Issues, 39,* 117-138.

Beutler, L. E., Frank, M., Schieber, S. C., Calvert, S., & Gaines, J. (1984). Comparative effects of group psychotherapies in a short-term inpatient setting: An experience with deterioration effects. *Psychiatry, 47,* 66-76.

Binstok, G., Foster, L. G., & Mullane, J. F. (1984). Propranolol and the depression component of anxiety neurosis. *Current Therapeutic Research, 35,* 423-432.

Blank, K., & Perry, S. (1984). Relationship of psychological processes during delirium to outcome. *American Journal of Psychiatry, 141,* 843-847.

Clark, A., & Friedman, M. J. (1983a). Factor structure and discriminant validity of the SCL-90 in a veteran psychiatric population. *Journal of Personality Assessment, 47,* 396-404.

Clark, A., & Friedman, M. J. (1983b). Nine standardized scales for evaluating treatment outcome in a mental health clinic. *Journal of Clinical Psychology, 39,* 939-950.

Craig, T., & Abeloff, M. (1974). Psychiatric symptomatology among hospitalized cancer patients. *American Journal of Psychiatry, 131,* 1323-1327.

Crowne, D. P., & Marlowe, D. (1960). A new scale of social desirability independent of psychopathology. *Journal of Consulting Psychology, 29,* 349-354.

Davidson, J., Linnoila, M., Raft, D., & Turnbull, C. D. (1981a). MAO inhibition and control of anxiety following amitriptyline therapy. A pilot study. *Acta Psychiatry Scand, 63,* 147-152.

Davidson, J. R., McLeod, M. N., Turnbull, C. D., & Miller, R. D. (1981b). A comparison of phenelzine and imipramine in depressed inpatients. *Journal of Clinical Psychiatry, 42,* 395-397.

Davidson, J., Miller, R., Van Wyck, F. J., Strickland, R., Manberg, P., Allen, S., & Parrott, R. (1983). A double-blind comparison of bupropion and amitriptyline in depressed inpatients. *Journal of Clinical Psychiatry, 44,* 115-117.

Derogatis, L. R. (1977). *SCL-90: Administration, scoring & procedures manual for the revised version.* Baltimore: Clinical Psychometric Research.

Derogatis, L. R. (1983). Misuse of the Symptom Checklist 90. *Archives of General Psychiatry, 40,* 1152.

Derogatis, L. R., & Cleary, P. A. (1977a). Confirmation of the dimensional structure of the SCL-90: A study in construct validation. *Journal of Clinical Psychology, 33,* 981-989.

Derogatis, L. R., & Cleary, P. A. (1977b). Factorial invariance across gender for the primary symptom dimensions of the SCL-90. *British Journal of Social and Clinical Psychology, 16,* 347-356.

Derogatis, L. R., & Melisaratos, N. (1983). The brief symptom inventory: An introductory report. *Psychological Medicine, 13,* 595-605.

Derogatis, L. R., Meyer, J. K., & King, K. M. (1981). Psychopathology in individuals with sexual dysfunction. *American Journal of Psychiatry, 138,* 757-763.

Derogatis, L. R., Rickels, K., & Rock, A. F. (1976). The SCL-90 and the MMPI: A step in the validation of a new self-report scale. *British Journal of Psychiatry, 128,* 280-289.

Dinning, W. D., & Evans, R. G. (1977). Discriminant and convergent validity of the SCL-90 in psychiatric inpatients. *Journal of Personality Assessment, 41,* 304-310.

Edwards, A. L. (1957). *The social desirability variable in personality assessment and research.* New York: Holt.

Edwards, D. W. (1978). Test-taking and the stability of the adjustment scales: Can we assess patient deterioration? *Evaluation Quarterly, 2,* 275-291.

Egan, K. J., Kogan, H. N., Garber, A., & Jarrett, M. (1983). The impact of psychological distress on the control of hypertension. *Journal of Human Stress, 9,* 4-10.

Evenson, R. C., Holland, R. A., Mehta, S., & Yasin, R. (1980). Factor analysis of the Symptom Checklist-90. *Psychological Reports, 46,* 695-699.

Fleiss, J. L. (1981). *Statistical methods for rates and proportions* (2nd ed.). New York: John Wiley & Sons.

Freeman, E. W., Rickels, Huggins, G. R., Garcia, C. R., & Polin, J. (1980). Emotional distress patterns among women having first or repeat abortions. *Obstetrics and Gynecology, 55,* 630-636.

Green, B. L., Grace, M. C., Lindy, J. D., Tichener, J. L., & Lindy, J. G. (1983). Levels of functional impairment following a civilian disaster: The Beverly Hills Supper Club fire. *Journal of Consulting and Clinical Psychology, 51,* 573-580.

Greist, J. H., Klein, M. H., Eischens, R. R., Faris, J., Gurman, A. S., & Morgan, W. P. (1979). Running as treatment for depression. *Comprehensive Psychiatry, 20,* 41-54.

Hendler, N., Derogatis, L., & Long, D. (1977). EMG biofeedback in patients with chronic pain. *Diseases of the Nervous System, 38,* 505-509.

Hoffman, N. G., & Overall, P. B. (1978). Factor structure of the SCL-90 in a psychiatric population. *Journal of Consulting and Clinical Psychology, 46,* 1187-1191.

Holcomb, W. R., Adams, N. A., & Ponder, H. M. (1983). Factor structure of the Symptom Checklist-90 with acute psychiatric inpatients. *Journal of Consulting and Clinical Psychology, 51,* 535-538.

Horowitz, M. J., Weiss, D. S., Kaltreider, N., Krupnick, J., Marmar, C., Wilner, N., & DeWitt, K. (1984). Reaction to the death of a parent: Results from patients and field subjects. *Journal of Nervous and Mental Disease, 172,* 383-392.

Jacobs, P. E., Doft, E. B., & Koger, J. (1981). A study of SCL-90 scores of 264 methadone patients in treatment. *International Journal of the Addictions, 16,* 541-548.

Jerabek, P., Klimpl, P., & Boleloucky, Z. (1982). Factor analysis of the SCL-90 inventory. *Activitas Nervosa Superior, 24,* 183-184.

Kales, A., Soldatos, C. R., Caldwell, A. B., Charney, D. S., Kales, J. D., Markel, D., & Cadieux, R. (1980a). Nightmares: Clinical characteristics and personality patterns. *American Journal of Psychiatry, 137,* 1197-1201.

Kales, A., Soldatos, C. R., Caldwell, A. B., Kales, J. D., Humphrey, F. J., Charney, D. S., & Schweitzer, P. K. (1980b). Somnambulism: Clinical characteristics and personality patterns. *Archives of General Psychiatry, 37,* 1406-1410.

Kass, F., Charles, E., Klein, D. F., & Cohen, P. (1983). Discordance between the SCL-90 and therapists' psychopathology ratings. *Archives of General Psychiatry, 40*, 389-393.

Kass, F., Skodol, A., Buckley, P., & Charles, E. (1980). Therapists' recognition of psychopathology: A model for quality review of psychoptherapy. *American Journal of Psychiatry, 137*, 87-90.

Kathol, R. G., Noyes, R., Slyman, D. J., Crowe, R. R., Clancy, J., & Kerber, R. E. (1980). Propranolol in chronic anxiety disorders. *Archives of General Psychiatry, 37*, 1361-1365.

Kilpatrick, D. G., Resick, P. A., & Veronen, L. J. (1981). Effects of a rape experience: A longitudinal study. *Journal of Social Issues, 37*, 105-122.

Malec, J., & Neimeyer, R. (1983). Psychologic prediction of duration of inpatient spinal cord injury rehabilitation and performance of self-care. *Archives of Physical Medicine and Rehabilitation, 64*, 359-363.

Nahmias, L. M., Beutler, L. E., Crago, M., Osborn, K., & Hughes, J. H. (1983). Use of the SCL-90R to assess health care staff's perceptions of patient's psychological states. *Perceptual and Motor Skills, 57*, 803-806.

Pelz, M., & Merskey, H. (1982). A description of the psychological effects of chronic painful lesions. *Pain, 14*, 293-301.

Rhoads, D. L. (1983). A longitudinal study of life stress and social support among drug abusers. *International Journal of the Addictions, 18*, 195-222.

Rogers, J., Sheldon, A., Barwick, C., Letofsky, K., & Lancee, W. (1982). Help for families of suicide: Survivors support program. *Canadian Journal of Psychiatry, 27*, 444-449.

Roskin, M. (1982). Coping with life changes: A preventive social work approach. *American Journal of Community Psychology, 10*, 331-340.

Roskin, M., & Dasberg, H. (1983). On the validity of the Symptom Check List-90 (SCL90): A comparison of the diagnostic self-ratings in general practice patients and 'normals', based on the Hebrew version. *International Journal of Social Psychiatry, 29*, 225-230.

Sbordone, R. J., Kral, M., Gerard, M., & Katz, J. (1984). Evidence of a "command performance syndrome" in the significant others of the victims of severe traumatic head injury. *International Journal of Clinical Neuropsychology, 6*, 183-185.

Shanfield, S. B., Heiman, E. M., Cope, D. N., & Jones, J. R. (1979). Pain and the marital relationship: Psychiatric distress. *Pain, 7*, 343-351.

Shanfield, S. B., & Swain, B. J. (1984). Death of adult children in traffic accidents. *Journal of Nervous and Mental Disease, 172*, 533-538.

Shapiro, S., & Lehrer, P. M. (1980). Psychophysiological effects of autogenic training and progressive relaxation. *Biofeedback and Self-Regulation, 5*, 249-255.

Siassi, I., & Fozouni, B. (1982). Psychiatry and the elderly in the Middle East: A report from Iran. *International Journal of Aging and Human Development, 15*, 107-120.

Speer, D. C., & Swindle, R. (1982). The "monitoring model" and the morality treatment interaction threat to validity in mental health outcome evaluation. *American Journal of Community Psychology, 10*, 541-552.

Steer, R. A. (1982). Symptoms discriminating between heroin addicts seeking ambulatory detoxification or methadone maintenance. *Drug and Alcohol Dependence, 9*, 335-338.

Thorndike, E. L. & Lorge, I. (1944). *The teacher's word book of 30,000 words.* New York: Bureau of Publications.

Turner, R. M., McGovern, M., & Sandrock, D. (1983). A multiple perspective analysis of schizophrenics' symptoms and community functioning. *American Journal of Community Psychology, 11*, 593-607.

Ventura, J. N. (1982). Parent coping behaviors, parent functioning, and infant temperament characteristics. *Nursing Research, 31*, 269-273.

Wilde, G. J. S. (1972). Trait description and measurement by personality questionnaire. In R. B. Cattell (Ed.), *Handbook of modern personality theory* (pp. 69-103). Chicago: Aldine Publishing Co.

Woody, G. E., McLellan, A. T., Luborsky, L., O'Brien, C. P., Blaine, J., Fox, S., Herman, I., & Beck, A. T. (1984). Severity of psychiatric symptoms as a predictor of benefits from psychotherapy: The veterans administration—Penn study. *American Journal of Psychiatry, 141,* 1172-1177.

Jack L. Bodden, Ph.D.

Psychologist, Olin E. Teague Veterans Center, Temple, Texas.

THE SELF-DIRECTED SEARCH

John L. Holland. Odessa, Florida: Psychological Assessment Resources, Inc.

THE ORIGINAL OF THIS REVIEW WAS PUBLISHED IN TEST CRITIQUES: VOLUME V (1987).

Introduction

The Self-Directed Search (SDS) is a self-administered, self-scored, and self-interpreted vocational counseling measure. It consists of four main parts involving the use of a number of scales and ratings. As a whole, the SDS attempts to simulate the vocational counseling experience.

The SDS was created by John L. Holland, Ph.D., representing an outgrowth of his theory of vocational choice (Holland, 1959, 1966). Holland began with the premise that people view the world in terms of occupational stereotypes, and these stereotypes reveal something about the individual and his or her career choices. Over a period of more than 35 years Holland has revised not only his theory of vocational choice but also the SDS. His theory and the SDS have been utilized and refined by an extensive body of empirical research.

Holland's purpose in developing the SDS was to create a "scientifically sound and practical simulation of the vocational counseling experience" (Holland, 1985b, p. v). He hoped to accomplish this goal by developing a self-administered, self-scored, and self-interpreted vocational assessment booklet and a compatible file of occupational possibilities. Part of his motivation to develop the SDS was his desire to develop an inventory that would avoid the problems involved in separate answer sheets, mailing, scoring, and so on.

The origin of the SDS can be traced to the Vocational Preference Inventory (VPI), which Holland developed in 1953, prior to the first presentation of his theory of vocational choice (1959). The VPI was used to define six theoretical personality types (Realistic, Investigative, Social, Conventional, Enterprising, and Artistic). The next step in the process was defining occupational environments according to VPI profiles (1966). In 1969, Holland published his "Hexagonal Ordering" concept, which showed that the correlation between the six theoretical typologies could best be represented by a hexagon so that distances between the six points on the hexagon were inversely proportional to the correlations among the six types). These preliminary steps led to the publication of the first version of the SDS in 1971. Some format changes, a simplified scoring procedure, the inclusion of the seventh revision of the VPI, and the addition of 50 job titles to the Occupations Finder comprised the revision published in 1977. The 1985 edition, which is the subject of this review, is the latest refinement of the SDS. This edition includes updated items, rewritten directions for scoring the test (designed to improve scoring accuracy), and a doubling of the number of occupations in the Occupations Finder (from 500 to more than 1,100). There is also a new form (Form E), which was developed for adolescents and adults with limited reading skills. Form E has simplified scoring procedures and is shorter than the standard SDS. In addition, a special version of the SDS has been developed for use with the blind.

Research investigations with the SDS have been done in a number of foreign countries. These studies have reportedly demonstrated the validity of the theory as well as the utility of the instrument (Holland, 1985b). However, there are no foreign language versions available to the general test user.

The SDS consists of two main components: 1) the Assessment Booklet and 2) the Occupations Finder. There is also a supplementary booklet entitled *You and Your Career* (Holland, 1985c), which provides additional background on the instrument, the theory, and suggestions for enriching the experience provided by the SDS.

The heart of the SDS is the Assessment Booklet (which incorporates the VPI). The booklet includes four sections: 1) Activities (six scales of 11 items each), 2) Competencies (six scales of 11 items each), 3) Occupations (six scales of 14 items each), and 4) Self-Estimates (two sets of six ratings, each rating corresponding to a type). There is a fifth part to the Assessment Booklet, entitled "Occupational Daydreams," which does not actually contribute to the formal scoring of the SDS; however, the occupations listed in the "Daydreams" section are coded and can be compared with the SDS personality code.

The Occupations Finder includes 1,156 occupational titles (comprising 99% of the workers in the United States). The occupations are arranged according to personality types and subtypes. Each occupational subtype is arranged according to the level of general educational development (GED) that an occupation requires according to the *Dictionary of Occupational Titles* (Department of Labor, 1977).

Each of the four main parts of the Assessment Booklet contain six clusters of items, which correspond to and determine the score for each of Holland's personality types. The personality types (based on Holland's early work on vocational titles, occupational stereotypes, and personality) follow:

1) *Realistic:* a practical, mechanically inclined individual who may be somewhat lacking in social skills;

2) *Investigative:* a scientifically oriented, intellectual person who may lack leadership ability;

3) *Artistic:* an imaginative, creative person who may lack clerical, practical skills;

4) *Social:* a people-oriented individual whose strength is his or her social skills;

5) *Enterprising:* an individual with sales and leadership skills; and

6) *Conventional:* a person who likes orderly pursuits but may lack artistic ability.

The Activities section includes items to which the user responds with "L" (like) or "D" (dislike). Items include interests such as "Fix electrical things" and "Sell something." The Competencies section asks the user to indicate whether or not he or she can perform a given activity well. Items include examples like "I can type 40 words per minute." The Occupations section consists of occupational titles (and is actually the VPI). Users indicate whether they like or are interested in a particular occupation. Self-Estimates include two scales, on which the user rates his or her perceived ability on dimensions such as Manual Skill or Scientific Ability.

The participation of the examiner or counselor in the testing process with the SDS is intended to be nominal. Depending on the age, education, and sophistication of the group or individual using the SDS, the person who administers the instrument may do no more than mail out the instrument or simply serve as a monitor for groups, answering questions and assisting with scoring. Holland encourages monitoring counselors to be trained vocational counselors or psychologists, knowledgeable about psychological assessment and vocational counseling. For the most part the examiner assists in administration or scoring problems and provides vocational guidance to the few who may require additional assistance. It should be noted that Holland never proposed that the SDS would replace vocational counselors or other vocational assessment instruments; rather, he hoped to develop an instrument that, under ideal conditions, would require little or no attention and time from the highly trained professional. The SDS provides an experience that may in itself be sufficient vocational counseling for some individuals; however, it may also stimulate others to ask for more information or even professional counseling.

Although no clear-cut upper and lower age limits have been established, Holland (1985b) states that the SDS can be used for persons 15 years of age and older. The test may be taken individually (and in private) or in groups having one monitor for every 25-30 persons. The reading difficulty level is estimated to be at the seventh- to eighth-grade level using a standard readibility formula.

When the user completes the SDS, he or she is guided through the scoring process, which is actually a rather simple process of adding scale subtotals or rating values. The answer sheet is the booklet itself. The scoring process results in a summary or profile code, which is comprised of the three highest values for the six types (e.g., Investigative, Social, and Artistic, or I-S-A). After determining the profile code, the user looks up the occupations listed under the corresponding code in the Occupations Finder (which for an I-S-A would include occupations such as physician, nurse, and educational psychologist). For most people, the process is relatively simple and takes less than 50 minutes.

Practical Applications/Uses

The SDS is designed to multiply the number of people a vocational counselor can serve and to provide a vocational counseling experience for people who may not need or have access to a professional counselor. The SDS eliminates the need—or at least the time—required to administer, mail, score, interpret, and give unnecessary feedback/counseling.

The SDS is intended to give a user an indication of his or her personality type (or style) and a list of occupations that are congruent or compatible with that personality style. The test does not, as Holland points out, tell an individual what jobs he can or should do. Some awareness of Holland's theory can be helpful to the potential user and this information is given in simple terms in *You and Your Career* (Holland, 1985c). The counselor could also provide explanations to those who want an understanding of the rationale of the test.

Holland (1985b) proposes that the SDS can be used with a very wide range of persons. He states that it could serve perhaps as many as 50% of the students and adults who have only a minimal need for vocational assistance. It can and has been used in schools (high school and college), adult centers, correctional institutions, employment offices, women's centers, business, and industry. All of these uses have been evaluated by research and found appropriate.

The SDS is suited to the needs of vocational counselors, counseling and clinical psychologists, and other mental health professionals who have some training in psychometrics, personality theory, and vocational theory. As the examiner serves primarily in a back-up role, the amount of professional training he or she needs depends largely on how the test is to be used.

The SDS has been used in many settings, countries, and by a wide variety of users. The only limitations to its use would be with subjects less than 15 years of age or persons who have no interest in vocational self-exploration. This reviewer suspects that the SDS will always find its greatest use in college counseling centers and high school guidance offices. A considerable amount of research indicates that the SDS is a highly adaptable and pragmatic instrument.

The SDS is suitable for administration individually, in groups, or through the mail. The setting can be anyplace that is relatively quiet. Holland (1985b) states that private settings appear to be conducive to greater individual involvement with the SDS.

Unlike the majority of psychological tests, which are designed to provide data about the user and must be handled from administration to interpretation by a highly skilled professional examiner, the SDS is entirely self-administered, self-scored, and self-interpreted. The instructions to the examinee are quite clear and easy to follow. Form E is especially simple. It is this reviewer's opinion that very few individuals would have difficulty understanding how they are supposed to respond to the SDS.

It is in the area of scoring where most questions and concern have arisen. The booklet gives succinct scoring instructions, the required operations are quite simple (nothing more complicated than adding small numbers), and the scoring procedures, when followed correctly, lead the examinee to a determination of his or her 3-point summary of personality code. However, Holland (1985b) and other reviewers (e.g., Gelso, Collins, Williams, & Sedlacek, 1973) have expressed concern about scoring errors. The manual cites one study of college students in which 60% of the men and 63% of the women made no scoring errors; however, about 83% of both men and women correctly identified their 3-point summary code (even if they made an error along the way). Whether scoring errors are significant enough to raise serious questions as to how truly self-scoring and self-administering the SDS actually is cannot be answered unequivocally. This reviewer believes the scoring instructions are very clear, but monitoring and double-checking by a counselor or monitor is still advisable.

All scoring is accomplished by counting scale totals; no templates are needed. A machine version has been developed for large-scale research studies, but this version is less desirable because self-scoring is an integral part of the SDS's effort to simulate the vocational counseling experience.

In order to interpret the SDS, the examinee is instructed to look up the occupations listed in the Occupations Finder under his or her summary code. Finally, the examinee is given some suggestions for further reading and exploration. The entire process should require no more than 45-50 minutes. The only limitation is that users who are curious about the "hows and whys" of the summary code and the occupational grouping that corresponds to their code may be left with some unanswered questions. Some of these questions could be answered if the student has access to *You and Your Career.* Other questions, such as what the examinee should do if he or she obtains a rare or an undifferentiated summary code (i.e., little or no difference between the top three scores that comprise the code), may require additional input from the psychologist or counselor.

Technical Aspects

According to Holland (1985b), the current version of the SDS is more reliable than the 1977 revision. Measures of internal consistency (alpha) generally range from about .70 to .93. Test-retest reliability (1-4 weeks) for the 1977 edition (using a sample of high school students) ranged from .56 to .95. The manual does not report test-retest data on the 1985 edition but states that it should be good because of the high intercorrelations between the 1977 and 1985 editions.

In general, the 1985 version of the SDS has shown adequate evidence of its reliability with a diverse group of subjects, and it will probably prove to be as stable as the 1977 edition once test-retest correlations are reported.

Holland (1985b) refers to more than 400 studies of construct validity on the 1971 edition alone. A comparison of concurrent validities of the 1977 and 1985 editions, which involved comparing the examinee's summary code with his or her stated vocational choice, yielded agreement scores ranging from 44% to 70% for the 1977 edition and from 48% to 62% for the 1985 edition. Holland suggests that the concurrent validity of the two editions is probably comparable.

Predictive validity studies based upon the 1971 edition showed moderate support. These validity studies compared SDS summary codes with expressed vocational choice one and three years after taking the SDS. Kappa values ranged from .24 ($p < .001$) to .33 ($p < .001$). Percent correct prediction values ranged from 26.6% to 72.4% in the studies cited by Holland (1985b).

An in-press study cited by Holland (1985b), in reviewing concurrent and predictive validity studies, indicates that most interest inventories have "hit rates" (i.e., predicting stated vocational choice from test profile scores) in the range of 40% to 55% in a six-category scheme. If so, the SDS is comparable to other interest measures in this regard.

In addition to reliability and validity data, Holland (1985b) cites what he calls "effects and outcome data." Studies reported in this section of the manual show that the SDS has the expected positive effect on its user's self and vocational understanding, and also creates more career choice options. Moreover, several studies have found the SDS to "equal the influence of professional counselors" (p. 51).

The SDS appears to possess reliability and validity that are on a level with other vocational interest measures. In addition, a number of studies have also demonstrated its apparent utility.

Critique

This reviewer has been familiar with the theory and research conducted by John Holland for a number of years and has been impressed with the utility of the SDS and the theory on which it rests. As when it was first introduced, the revised SDS fills a genuine need in the vocational counseling field.

John Holland is to be commended for the thorough and thoughtful way in which the SDS manual is constructed. Not only does it cover the theory and rationale upon which the instrument was built, it goes on to discuss administration, scoring, and practical applications. The manual anticipates commonly asked questions regarding the SDS and includes a particularly good section on interpretation and uses of test results. In short, this manual could well serve as a model for other developers of vocational interest inventories.

The SDS is strongly recommended for its practical as well as research value. Holland's theory of career development is probably one of the more viable theories and the SDS is a logical outgrowth of his theory.

References

This list includes text citations and suggested additional reading.

Department of Labor. (1977). *Dictionary of occupational titles* (4th ed.). Washington, DC: U.S. Government Printing Office.

Gelso, C. J., Collins, A. M., Williams, R. O., & Sedlacek, W. E. (1973). The accuracy of self-administration and scoring of Holland's Self-Directed Search. *Journal of Vocational Behavior, 3,* 375-383.

Gottfredson, G. D., Holland, J., & Ogawa, D. K. (1982). *Dictionary of Holland occupational codes.* Palo Alto, CA: Consulting Psychologists Press.

Holland, J. L. (1959). A theory of vocational choice. *Journal of Counseling Psychology, 6,* 35-45.

Holland, J. L. (1966). *The psychology of vocational choice.* Waltham, MA: Blaisdell.

Holland J. L. (1973). *Making vocational choices: A theory of careers.* Englewood Cliffs, NJ: Prentice-Hall.

Holland, J. L. (1985a). *Making vocational choices: A theory of vocational personalities and work environments.* Englewood Cliffs, NJ: Prentice-Hall.

Holland, J. L. (1985b). *The Self-Directed Search, professional manual.* Odessa, FL: Psychological Assessment Resources, Inc.

Holland, J. L. (1985c). *You and your career.* Odessa, FL: Psychological Assessment Resources, Inc.

Holland, J. L., & Gottfredson, G. D. (1976). Using a typology of persons and environments to explain careers: Some extensions and clarifications. *The Counseling Psychologist, 6,* 20-29.

Osipow, S. H. (1973). *Theories of career development.* New York: Appleton-Century-Crofts.

Raymond G. Johnson, Ph.D.

Professor and Co-Chair, Department of Psychology, Macalester College, St. Paul, Minnesota.

SHIPLEY INSTITUTE OF LIVING SCALE

Walter C. Shipley. Los Angeles, California: Western Psychological Services.

THE ORIGINAL OF THIS REVIEW WAS PUBLISHED IN TEST CRITIQUES: VOLUME V (1987).

Introduction

The Shipley Institute of Living Scale (SILS), more commonly known as the "Shipley-Hartford," has been extensively used by clinical psychologists for nearly half a century. The test consists of two parts, a group of 40 multiple-choice vocabulary items and a set of 20 open-ended series that require the person to abstract a rule with which to determine the next element in the list. A Vocabulary and an Abstraction score may be obtained; intellectual impairment is purportedly indicated when the Abstraction score is considerably lower than expected from the Vocabulary score. The paper-and-pencil test takes less than 20 minutes and may be self-administered and scored by a clerk. A new version is available for microcomputers; the computer administers the test, scores it, and produces a multipage report.

The test was devised by Walter C. Shipley and was originally known as the Shipley-Hartford Retreat Scale. After the institution modified its name, Shipley changed the test title to the Shipley Institute of Living Scale. Articles in the literature are evenly divided between the two titles. The test was published by Shipley and his family for several decades. In recent years the publisher has been Western Psychological Services.

Shipley (1953) states that his test was "inspired" by the work of Harriet Babcock, who had devised an individually administered scale to assess mental deterioration. Babcock (1930) assumed that vocabulary is relatively more resistant to deterioration than speed and performance tests, and used the vocabulary list from the Stanford-Binet as a measure of intellectual level. She compared a person's tested vocabulary level with 24 sets of items. These ranged from memory for designs to learning paired associates, from memory of a paragraph to picture recognition, and from Knox Cubes to timed responses to general information questions. On the basis of the vocabulary score, an expected score is projected for each of the other subtests. Deterioration or impairment may be inferred when actual scores fail to reach the expected level. Babcock's tests are clinical instruments administered

---

The reviewer wishes to thank Robert A. Zachary, former director of clinical assessment at Western Psychological Services, for graciously allowing me to examine a prepublication draft of the revised manual for the Shipley Scale. It is due to be published in the fall of 1986. The final version of that manual may be somewhat different from what is cited in this review.

individually, allowing the psychologist to observe and make qualitative judgments about the person's performance in addition to the quantitative comparison.

Shipley's goal was to follow Babcock's model "with a view to measuring the same sort of thing on a group-test basis" (1953, p. 751). He wanted a self-administered test that could be given quickly and objectively (1940). He realized that it would be difficult to devise group tests of recent memory; therefore, he chose to use abstract thinking as the ability manifesting rapid deterioration, taking as his guide the sorting tests devised by Kasanin and Hanfmann. Shipley (1953, p. 752) describes his abstraction items as requiring the subject "to induce some principle common to a given series of components and then to demonstrate his understanding of this principle by continuing the series."

Each Abstraction item is a series of letters or numbers followed by blanks indicating the number of characters in the answer. Three items similar to those in the test follow:

1. A   C   E   G   I __
2. bib bib   part trap   269 _ _ _
3. 234   162   45   81   315   43_

Instructions are to complete each series and to place the appropriate answer in the blanks. (Answers to the above items are K, 962, and 2.) The individual must infer a rule that governs placement of the elements in the list and then use that rule to determine the next item. The Abstraction score is the number of items correct multiplied by two; thus, scores on this subtest can range from 0 to 40.

The Vocabulary section consists of 40 items. The individual's task is to select the synonym of a word from a set of four alternatives. Vocabulary items have this form:

1. SHIP   animal   tree   knife   boat
2. INANE   slender   timely   silly   damp

The Vocabulary score is the number correct plus one for every four omitted to provide a guessing correction, and scores can range from about 10 to 40.

A total score is the sum of the Vocabulary and the doubled Abstraction score. The original 1940 manual provides tables to derive Vocabulary Age, Abstraction Age, Mental Age, and Conceptual Quotient. A new manual being developed by the publisher (Zachary, 1986) reports these scores but discourages their use. The new manual provides age-corrected T-scores for Vocabulary and Abstraction based on a new normative sample of psychiatric patients. It also provides a new impairment index, the Abstraction Quotient (AQ). The new manual gives an extensive review of relevant literature and many suggestions for appropriate use of the instrument.

Practical Applications/Uses

Each part of the scale has a 10-minute time limit, but because an individual rarely needs that much time the test is often considered a power test. A number of psychiatric institutions have used the Shipley as part of an admissions test battery that often includes the MMPI and a sentence completion device. The psychologist may have the results from these instruments before ever seeing the patient.

The Shipley-Hartford Vocabulary score is usually interpreted as a rough estimate of the individual's intellectual level. The general population mean appears to be in the upper 20s, while bright normals score in the low 30s. The Vocabulary test has a low ceiling and does not distribute persons well at the upper end, and the range is restricted at the lower end as well. A person not answering any item or guessing blindly at all items will obtain a score of 10. Obviously the examinee needs to be able to read some English to earn points. Low scores may be due to language problems, inability to concentrate or follow directions, low intelligence, or other factors. In the original manual, Shipley cautions against the use of the deterioration index (the Conceptual Quotient) with persons of subnormal intelligence; he feels the index should not be used when the Vocabulary score is less than 23. Garfield (1947) maintains that the Shipley-Hartford, and especially the Abstraction scale, is too difficult for persons of below average intelligence. He tested 350 men in a disciplinary barracks who had Wechsler-Bellevue IQs ranging from 50 to 106 and found that mental age scores based on the Shipley-Hartford were lower than those based on the Wechsler.

The Abstraction scale has a low ceiling. In this reviewer's experience using the test as a demonstration instrument in undergraduate classes, Abstraction always yields a negatively skewed distribution. The vagaries of the Shipley distribution probably matter little when used in conjunction with other information to develop clinical hypotheses, but when used in place of the WAIS to estimate IQ or to assess "brain damage" of an individual, the Shipley by itself is inadequate.

Motivation may be a problem in test performance. When high school students were grouped by responses to the question, "How important is it that you do well?", there were no group differences in performance. Poorly motivated chronic schizophrenics, on the other hand, had lower scores than well-motivated patients (Schalock & Wahler, 1968).

An Abstraction score considerably lower than a Vocabulary score may reflect reduced intellectual efficiency, though such an interpretation must take into account other information developed about the individual, such as age, social history, and motivation, as well as anxiety level, psychotic mentation, or possible brain damage.

The Shipley is frequently used in research studies, sometimes as a selection or matching variable, less often as a dependent variable. In recent years it has been particularly popular in the alcoholism literature.

Technical Aspects

In constructing his test, Shipley (1953) used students rather than psychiatric or neurological patients. His initial sample comprised 462 youth from three educational levels: high school freshmen, high school juniors and seniors, and college upperclassmen. This sample was given preliminary forms containing 24 abstraction items and one of three vocabulary lists containing about 40 items each. An item analysis of the abstraction items was performed to select the 20 items that best differentiated students from the three educational levels. Each item in the abstraction test was then matched with two vocabulary items that matched the abstraction item in percent passing at each educational level. The result was a 40-item vocabulary test and a 20-item abstraction test of approximately equal difficulty for this sample that differentiated on the basis of educational level and age.

The Conceptual Quotient is not quite the same as the ratio of Abstraction Age to Vocabulary Age times 100. Shipley continued to follow Babcock's (1930) procedure and constructed a CQ table by first finding the median abstraction age that accompanied each vocabulary age in the normative group. This predicted abstraction age is divided by the obtained abstraction age, then multiplied by 100 to obtain the CQ.

In the new Shipley manual, Zachary (1986) summarizes criticism of age-equivalent and ratio scores. While the concept of mental age allowed Alfred Binet to develop his revolutionary instrument, the notion has outlived its usefulness. Age-equivalent scores are not appropriate for adults, the regression procedures used are arbitrary, variability around the age curve is ignored, no normative information is provided, and the concept of mental age contains misleading connotations. The new manual, however, continues to report Shipley's original Vocabulary Age and Abstraction Age.

Zachary also notes the deficiency in ratio IQs; namely that they may have different variances at various age levels, and that deviation IQs are used in the Wechsler tests and more recently in the Stanford-Binet. The new manual continues to report the CQ, while at the same time strongly criticizing it and recommending that it not be used—an undesirable compromise between selling and science that the publisher feels compelled to make. Zachary has devised a new impairment index, the Abstraction Quotient, which will be described later in this review.

Scherer (1949) found no differences between individual and group administration of the Shipley. Zachary (1986) reports that no differences were obtained between a computer-administered Shipley and the paper-and-pencil version.

An alternate form of the Abstraction scale has been devised by Horlick and Monroe (1954). While the items contain letters and numbers that are different from the original, the rules to be induced are identical to Shipley's version. It is of little use; practice effects remain.

In the literature, Shipley results have been reported in a variety of ways: raw scores for Vocabulary, Abstraction, and total; mental age equivalents of each; IQ transformations; and Conceptual Quotients.

To develop mental-age norms Shipley (1940, 1946) used a sample of 1,046 students who had previously been given standardized group intelligence tests. This group included 572 grammar school pupils in Grades 4 through 8, 257 high school students in Grades 9 through 12, and 217 college students. The college students were given the Otis test, while school records provided mental-age scores from various tests for the other students. On the basis of these results, tables of Vocabulary Age, Abstraction Age, and Conceptual Quotient were developed.

Table 1 lists several studies that have reported mean raw scores. Examination of the table suggests that persons who score above 30 on both subtests may be considered to be of above-average intelligence. Persons in the lower 20s and below on each subtest have difficulty with these kinds of tests and related tasks. The data in Table 1 are not a representative sample of the normal population but include more educated groups. Across these samples median Vocabulary raw score is 31.5 and median Abstraction raw score is 28.3. To the extent that we can generalize from these samples, we may expect normally functioning adults to score on average

Table 1

Shipley-Hartford Mean Raw Scores for Several Nonpsychiatric Samples

Sample	Voc (SD)	Abs (SD)
812 British "normals"[1]	22.8 (8.5)	20.7 (9.0)
485 jr.-sr. high school students	26.4 (5.4)	28.3 (7.3)
65 psychiatric aides[3]	28.1 (5.3)	22.5 (8.9)
43 student nurses[4]	30.5 (2.8)	33.5 (4.3)
56 student nurses[5]	32.5 (2.9)	35.1 (3.8)
200 hospital staff[6]	33.8 (4.2)	28.3 (6.5)
40 medical students[7]	34.7 (3.5)	36.4 (3.3)
60 teachers[8]	34.9 (3.4)	28.3 (4.1)
MEDIAN	31.5	28.3

Sources: [1]Slater (1943); "Approximately random sample of the normal adult population" of Britain, "old men and imbeciles not included"; [2]Palmer (1964); [3]Sines (1958); [4]Shaw (1966); [5]Ruiz and Krauss (1967); [6]Kraus, Chalker, and Macindor (1967), includes nurses, clerks, technicians, and domestic staff in Australia; [7]Jones (1974), medical school students volunteering to participate in alcohol ingestion study; and [8]Garfield and Blek (1952), median of three groups of unmarried female teachers ages 20-30, 40-50, 60-70.

about 3 points higher on the Vocabulary than on the Abstraction portion. Such a figure is consistent with the difference psychiatric patients show on discharge, as will be noted.

Spearman rank-order correlations between means and standard deviations for the samples in Table 1 are .62 for Vocabulary and .99 for Abstraction. Thus as the means get higher, the standard deviations get lower because of the low test ceiling and the resulting piling up at the top of the curve. High as well as low scores on the Shipley are more subject to distortion than scores in the midrange.

The new norms presented in the revised manual (Zachary, 1986) are based on a sample studied by Paulson and Lin (1970) comprising 290 mixed psychiatric patients drawn from three sources: 160 clinic patients from the Neuro-Psychiatric Institute of the UCLA Medical Center (a low SES group with some neurological cases); 60 patients from private psychiatric practices in Los Angeles (an upper-middle-class group); and 70 patients from private files of psychologists at NPI-UCLA (also upper-middle class). Each group contained both out- and inpatients, but the proportions were not reported. The only criterion for selection was that each person had taken both the Shipley and the WAIS. The sample contains approximately equal numbers of men and women. The youngest examinees in the sample are 16 years old and a small proportion are over 54 years. The mean age is 34.9, but, reflecting the positive skew, the median is about 30 and Q is roughly equal to 10. This sample obtained a Vocabulary mean of 29.2 (SD = 6.0) and an Abstraction mean of 22.0 (SD = 10.1).

The new manual has tables of age-corrected T-scores for Vocabulary and

Abstraction based on this sample. It has age-based tables for estimating WAIS and WAIS-R IQs. The Abstraction Quotient (AQ), a new impairment index, based on the difference between Vocabulary and Abstraction scores, is also reported. To obtain the AQ, a predicted Abstraction score is compared to the actual Abstraction score. The predicted score is derived from a regression equation that uses Vocabulary score, age, and educational level and is reported as a standard score with a mean of 100 and an SD of 15. The AQ is an improvement over the CQ, because the latter was strongly influenced by age (e.g., Garfield & Fey, 1948).

Because the new norms are based on a sample of psychiatric patients, the Abstraction T-scores and the AQ will tend to overestimate an individual's functioning. Zachary and Huba (1985) have developed a table, reprinted in the manual, listing the statistical uniqueness of the various T-score combinations, correcting for the correlation between the two subtests. A more useful table for clinicians would be a normative table for predicted Abstraction scores. (I could not find the SE(est) for Abstraction score predictions in the manual [Zachary, 1986]).

Shipley assessed the reliability of his scales with a sample of 322 Army recruits. The odd-even correlations were .87 (Vocabulary), .89 (Abstraction), and .92 (total score). Manson and Grayson (1947) obtained similar figures; they calculated Spearman-Brown reliabilities for 1,262 prisoners as .82 for Vocabulary and .92 for Abstraction. Thus, internal consistency appears satisfactory.

Retest reliabilities are more difficult to interpret; a practice effect should be evident over short intervals. Most studies of retest reliability have used small samples and intellectually homogeneous groups (nurses or college students) whose mean scores are over 30 on each Shipley subtest. The resulting restriction of range should attenuate reliability coefficients. Retests of three samples of nurses (Ns of 43, 56, and 19) over a 3-month period yielded coefficients for Vocabulary of .54, .77, and .60, and for Abstraction of .47, .63, and .61 (Shaw, 1966; Ruiz & Krauss, 1967; Goodman, Streiner, & Woodward, 1974). Reliabilities of the total score were .62, .74, and .63. Another study using 40 undergraduates found total score reliability of .80 over a 45-day period (Martin, Blair, Stokes, & Lester, 1977).

The practice effect found in these studies (Shaw, 1966; Ruiz & Krauss, 1967; Goodman et al., 1974) tended to be less than one point for Vocabulary but about two to four points for Abstraction. The latter differences could be greater in more heterogeneous samples; with Abstraction means over 37 for student nurses on 3-month retest, the test ceiling left little room for the practice effect.

A study of 181 state hospital patients who retook the Shipley on readmission following an average interval of 14 months found a retest reliability coefficient of .73 (Stone, 1965b). The estimated IQs rose from 103.8 to 105.7. Moreover, Schalock and Wahler (1968) administered the Shipley weekly to chronic schizophrenics and to high school students. Over repeated testings the students' scores improved while the patients' scores did not.

Thus, while internal consistency reliability measures are satisfactory, coefficients of stability reach only moderate levels. Interpretation of these data is complicated because of the homogeneity of some of the samples used, the low test ceiling, practice effects, and possible motivation problems among patient groups.

Several studies have reported the intercorrelations of the Vocabulary and the Abstraction tests in samples ranging from high school students to "normal" British adults, to psychiatric and nonpsychiatric patients. The results ranged from .31 to .71 with a median correlation of .53 (Eisenthal & Harford, 1971; Kish, 1970; Mason & Ganzler, 1964; Palmer, 1964; Paulson & Lin, 1970; Pishkin, Lovallo, Lenk, & Bourne, 1977; Salzman, Goldstein, Atkins, & Babigian, 1966; Slater, 1943).

A number of studies have shown that results on the Shipley-Hartford are strongly related to age. Among adolescents there is a positive correlation between age and Vocabulary and Abstraction scores, reflecting Shipley's strategy of selecting items that differentiated students of various grades or ages. In his study of junior and senior high school students, Palmer (1964) found correlations with age of .58 for Vocabulary and .47 for Abstraction.

The results reported by Tarter and Jones (1971) indicate the different trends among adults. VA patients under 45 years have Vocabulary scores of 27.4, and patients between 46 and 60 have a score of 29.3. On Abstraction, the younger patients score 19.4 and the older patients score 17.6. Hoffmann and Nelson (1971) grouped alcoholics into three age groups, 18-44, 45-54, and 55-67, and found decreasing Conceptual Quotients of 84, 74, and 67. Another study at the same state hospital (Jansen & Hoffmann, 1973) showed that CQ also increases with IQ and with educational level. Their ANOVA results were highly significant and no interaction was found. Corotto (1966) tested several hundred psych-tech trainees. His ANOVA showed no interactions, but main effects of age of Vocabulary, Abstraction, and CQ.

Salzman et al. (1966) report a correlation of -.45 between age and Abstraction. They found that when they controlled for age the Shipley-Hartford differences between various neurotic and psychotic admissions to a university psychiatric unit washed out. Mason and Ganzler (1964) obtained a correlation of -.29 between Abstraction and age, with a positive correlation of .18 between Vocabulary and age. Eisenthal and Harford (1971) obtained age correlations with Vocabulary and Abstraction of .07 and -.33 in 100 hospitalized VA patients.

To examine age differences on the Shipley in a nonpsychiatric group, Garfield and Blek (1952) studied a group of unmarried women who were or had been teachers. They had three groups, each with 20 persons, aged 20-30, 40-50, and 60-70. Vocabulary means were higher in the older groups (31.3, 34.9, and 37.6), while Abstraction remained constant or showed some decrease (31.6, 28.3, 29.2).

Any interpretation of the Shipley-Hartford must consider the factor of age. Among adults we may expect that older samples will have slightly higher Vocabulary scores and considerably lower Abstraction scores and CQs. The T-score values presented in the new manual (Zachary, 1986) are age corrected, as is the Abstraction Quotient.

Stone and Chambers (1965) found no gender effect among over 2,500 state hospital patients; Paulson and Lin (1970) found none in the 290 mixed psychiatric patients used for the new norms. But at least two studies have reported gender differences. Pauker (1975) began with 54 males and 62 females. Both inpatients and outpatients were represented. He matched them by age and WAIS Full Scale IQ and found that females had higher Shipley total scores than males: 47.4 to 42.2. Palmer (1964) tested almost 500 junior and senior high school students and found the girls to be slightly but consistently higher than the boys, especially on Abstraction. Corotto (1966) found no gender differences in his sample of psych-tech trainees, and by and large there is an assumption implicit in the literature that gender differences are not significant.

Contrary to the expectation of some, educational level seems more related to Abstraction than to Vocabulary. Kish and Ball (1969) randomly selected subjects from VAH files, omitting chronic brain syndrome patients. They divided the sample into persons with no more than a grade school education and those who have gone to high school. The groups did not differ in age. Vocabulary scores for the two groups were identical (24.5 and 25.2), but the groups differed on Abstraction (11.4 and 16.6) and CQ (72.7 and 81.0).

A recent study at the Institute for Living in Hartford obtained means considerably higher than those found at state and VA hospitals. The authors (Phillips, Phillips, & Shearn, 1980) had two groups of patients, 23 schizophrenics and 36 nonschizophrenics. The Vocabulary means for these groups were 28.5 and 32.1, and the Abstraction means were 27.0 and 27.2. The higher mean scores as compared with state hospital and VA groups probably reflect the higher educational level, lower age, and better functioning of this more select group.

Eisenthal and Harford (1971) obtained Vocabulary and Abstraction correlations with social class of .44 and .41 on a sample of VAH patients. Among high school students Palmer (1964) found correlations of .29 and .20.

Clearly, scores on the Shipley reflect demographic variables, particularly age and SES. Such a condition is not unique to the Shipley-Hartford. Demographic variables (age, gender, race, education, occupation) of the WAIS normative group explain over half the variance of WAIS Full Scale IQs (Wilson et al., 1978). Shipley-Hartford scores must always be interpreted with respect to age and social history.

A number of studies have calculated the correlation between Wechsler-Bellevue or WAIS Full Scale IQs and Shipley-Hartford total scores (Garfield & Fey, 1948; Lewinski, 1946; Monroe, 1966; Sines, 1958; Sines & Simmons, 1959; Stone & Ramer, 1965; Suinn, 1960; Wahler & Watson, 1962; Weins & Banaka, 1960; Wright, 1946). These studies used psychiatric inpatients and outpatients as well as aides. These correlations range from .65 to .90 with a median of .76. Thus, the Shipley total score typically explains over one-half the Wechsler Full Scale IQ variance, even though the range has been restricted both by omitting those low scoring individuals who cannot read the Shipley, and by the low ceiling of the test.

Prado and Taub (1966) tested 59 VA patients and almost as many employees or applicants. They recommend the Shipley as a screening instrument, noting that of the persons who had at least an average Shipley, all had at least an average WAIS.

A number of investigators have developed regression equations and tables, some with age corrections, to estimate WAIS Full Scale IQs from Shipley total scores. As Zachary (1986) points out, these equations are most accurate for midrange scores. Most will tend to underestimate IQs above 115.

Dennis (1973) did a cross-validation of a number of these prediction equations on a sample of 37 psychiatric patients. He found correlations between Shipley-predicted and actual WAIS FSIQs to range from 0.716 to 0.792, and the SE(est) ranged from 7.68 to 8.79. The best of these estimates is age corrected, but even it provides 95% confidence intervals (1.96 × 7.68) that range about two standard deviations of IQ [(X-15.05) < X < (X + 15.05)]. With so much error, pinpoint estimates are inappropriate.

The most recent attempts to develop WAIS FSIQ and now WAIS-R FSIQ estimates have been by the publisher, apparently to be used with a computer scoring and interpretive program and as part of a new manual (Zachary, Crumpton, & Spiegel, 1985; Zachary, Paulson, & Gorsuch, 1985; Zachary, 1986). They have brought the SE(est) down to 5.75 for the WAIS-R on a cross-validation sample providing 95% confidence intervals of ± 11.3 or about 1.5 standard deviations of IQ. The authors recognize the problem when they state that while the Shipley is "useful for general intellectual screening, it should not be used to make more fine-grained discriminations in the higher and lower IQ scores" (p. 537). They still provide a table with increments as low as one IQ point. Instead of reporting that the estimated WAIS-R IQ of a 40-year-old man with a Shipley Total Score of 60 is 102, why not have the table show the 95% confidence limits of 91 and 113? The table would then be less subject to misuse. If Zachary et al. and the publisher are concerned about the deficiencies at the top and bottom of the Shipley, their energies would be better spent revising the instrument to obtain better spread at the extremes.

The conversion tables provided in the Zachary articles go down to a Shipley Total Score of zero. With the multiple-choice format and guessing correction on the Vocabulary test, a "zero" score is 10. What do scores less than 10 mean?

Responsible clinicians will use the WAIS when a more refined individual assessment is needed. Converting the Shipley score to an estimated WAIS IQ should not be encouraged in clinical practice, although such conversion may have usefulness in research. The new manual and scoring programs will probably stimulate more WAIS estimates.

Palmer (1964) found the Shipley Vocabulary and Abstraction subtests to correlate .77 and .64 with the STEP-SCAT academic achievement tests in 151 high school students. Eisenthal and Harford (1971) correlated the Shipley Vocabulary and Abstraction scores with the Raven Progressive Matrices in 100 VA hospital patients and obtained coefficients of .57 and .67. The Shipley has been found to correlate approximately .4 with the Quantitative score of the Porteus Mazes (Bennett, 1956; Sutker, Moan, & Swanson, 1972); Bennett found nonsignificant correlations with Porteus Qualitative scores. Martin, Blair, Sadowski, and Wheeler (1981) obtained a .39 correlation between total score on the Shipley and the Intellectual Efficiency scale of the California Psychological Inventory.

Table 2 lists mean Vocabulary and Abstraction raw scores for several psychiatric samples. It is a heterogeneous bunch, ranging from state hospital patients of the 1940s and soldiers referred for evaluation to contemporary private hospital patients. Both inpatients and outpatients of varying ages are represented. The studies are ranked by Vocabulary means. From the table one may infer poor functioning among city hospital alcoholics, deterioration and motivational problems in long-time state hospital patients, lack of deterioration in servicemen referred largely for behavior problems, and higher functioning among the younger, better-educated patients at a private psychiatric facility. The median across the samples in Table 2 is 28.0 for Vocabulary and 18.7 for Abstraction, a difference of about 9 points.

Three studies compared Shipley-Hartford scores taken at admission and later at discharge of psychiatric patients (Kobler, 1947; Lewinsohn & Nichols, 1964; Gold-

Table 2

Shipley Scale Mean Raw Scores for Several Psychiatric Samples

Sample	Voc (SD)	Abs (SD)
376 alcoholic inpatients[1]	24.2 (nr)	15.5 (nr)
50 state hospital patients[2]	24.8 (6.2)	12.5 (9.3)
977 naval psychiatric patients[3]	24.8 (6.9)	18.2 (nr)
251 VAH psychiatric patients[4]	25.8 (7.1)	16.6 (10.0)
100 VAH psychiatric patients[5]	27.7 (nr)	15.6 (nr)
99 VA MH clinic patients[6]	27.7 (6.0)	20.3 (10.8)
886 state hospital admissions[7]	28.0 a	18.0 a
100 psychological referrals[8]	28.0 a	29.5 a
86 psychiatric outpatients[9]	28.4 (7.2)	21.4 (10.2)
23 schizophrenic patients[10]	28.5 (8.5)	27.0 (9.6)
150 VAH psychiatric patients[11]	28.8 (6.6)	19.4 (9.7)
290 mixed psychiatric patients[12]	*29.2 (6.0)*	*22.0 (10.1)*
44 psychiatric outpatients[13]	29.4 (5.4)	18.1 (10.6)
36 non-schizophrenic patients[14]	32.1 (5.2)	27.2 (8.8)
MEDIAN	28.0	18.7

a: Scores reported as Vocabulary Age and Abstraction Age.
Sources: [1]Dalton and Dubnicki (1981), city hospital patients; [2]Magaret and Simpson (1948), patients aged 40-50 years; [3]Wright (1946), neuropsychiatric unit of naval hospital; [4]Sines (1958); [5]Zachary, Crumpton, and Spiegel (1985); [6]Sines (1958); [7]Stone (1965a), all new admissions to state hospital; [8]Lewinski (1946), men referred for psychological evaluation; [9]Zachary, Paulson, and Gorsuch (1985); [10]Phillips, Phillips, and Shearn (1980), private hospital, young, educated; [11]Eisenthal and Harford (1971); [13]*Paulson and Lin (1970), the new normative sample;* [13]Zachary, Paulson, and Gorsuch (1985); and [14]Phillips, Phillips, and Shearn (1980), private hospital, young, educated.

stein & Salzman, 1967). The results are shown in Table 3. From admission to discharge median Vocabulary scores went from 27.5 to 28.8, while median Abstraction scores went from 20.6 to 26.3. Psychiatric patients in the acute phase of their disorder on admission show a diminution of intellectual efficiency as represented in lowered Abstraction scores. At discharge these Abstraction scores rise to normal levels. The Vocabulary scores at discharge have increased only a point or two, while the Abstraction scores have increased 3 to 6 points.

In a study of factors related to improvement of mental hospital patients, Lewinsohn (1967) used a variety of measures and outcome criteria. He found that Shipley Vocabulary scores were not related to the outcome measures, but that improvement in Abstraction was related to financial security and father's occupational level. Jansen and Nickles (1973) compared Shipley scores of patients with two or more admissions with patients who had been admitted but once; they found no differences in IQ or CQ.

Table 3

Shipley-Hartford Mean Raw Scores on Admission and Discharge for Several Psychiatric Samples

	Admission		Discharge	
Sample	*Voc(SD)*	*Abs(SD)*	*Voc(SD)*	*Abs(SD)*
100 Army psychiatric patients[1]	26.5 (6.4)	20.1 (9.8)	27.7 (6.3)	25.4 (9.9)
45 psychiatric patients[2]	28.1 (nr)	21.0 (nr)	28.3 (nr)	27.1 (nr)
45 nonschizophrenic patients[3]	29.1 (6.1)	24.2 (9.4)	30.2 (6.2)	27.6 (8.8)
44 schizophrenic patients[4]	26.9 (8.3)	18.1 (9.3)	29.4 (8.0)	21.6 (10.5)
MEDIAN	27.5 (6.4)	20.6 (9.4)	28.8 (6.3)	26.3 (9.9)

Sources: [1]Kobler (1947); [2]Lewinsohn and Nichols (1964); [3]Goldstein and Salzman (1967); [4]Goldstein and Salzman (1967).

Braff and Beck (1974) compared depressed patients with schizophrenic patients in a run of consecutive admissions. On Vocabulary and Abstraction the depressed patients scored 28.0 and 18.7, while the schizophrenics had 23.0 and 10.8. Normal controls from a local church with similar SES had means of 31.5 and 32.3. The depressed patients were considerably older than the schizophrenics and the controls, making more salient the impairment of the schizophrenics as represented by their low Abstraction score. Abrams and Nathanson (1966) studied intellectual deterioration by retesting schizophrenics who had been hospitalized for over 6 years since the initial testing. Both Vocabulary and Abstraction showed statistically significant decreases, Vocabulary by 2-3 points and Abstraction by 4-5 points. It seems too short a time for aging to have a major effect, but motivational change may well have occurred together with or as part of the schizophrenic process.

Murray, Page, Stotland, and Dietze (1970) found a correlation of -.52 between Shipley vocabulary scores and the Phillips scale of process-reactive schizophrenia. Lewinsohn (1963) used a regression procedure to control for age and vocabulary level and found that schizophrenics, especially chronics, tended to score lower than psychiatric and normal controls. Pishkin, Lovallo, Lenk, and Bourne (1977) studied cognitive dysfunction in schizophrenics. The Shipley scores had a correlation of about -.5 with errors on the Whitaker Index of Schizophrenic Thinking, and correlations over -.7 with errors on a rule learning task. Knight, Epstein, and Zielony (1980) also compared the Shipley with the Whitaker Index. They found correlations of -.32 with Vocabulary and -.43 with Abstractions.

Psychiatric patients have lowered Abstraction scores on admission that are on average improved at discharge. The results lend support to the construct validity of the test as a measure of intellectual impairment. Schizophrenics show lower scores than other patients on both Vocabulary and Abstraction.

Shipley and Burlingame (1941) ranked Conceptual Quotients for various diagnostic groups and found that psychoneurotics had a mean CQ of 93, which was similar to normals (CQ = 100). The CQs of functional psychotics ranged from 75 to 85, while CNS syphilitics had the lowest score (CQ = 58).

Other empirical studies of brain-damaged patients do not lend much support for the Shipley as a subtle measure of impairment. Robinson (1946) found the CQs of lobotomized schizophrenics to be no different from control schizophrenics. Winfield (1953) estimated IQs from General Classification Tests taken by men entering service with IQs estimated from the Shipley taken after they suffered brain trauma and found no differences. Nine Klinefelter's syndrome (a congenital endocrine condition) cases were found in the military and had a mean Conceptual Quotient of 103.5 (Barker & Black, 1976). No test showed deficit or dysfunction in these individuals who as servicemen were a highly selected sample from the Klinefelter's population. Canter (1951) challenged the validity of the CQ when he found no measured deterioration in multiple sclerosis patients. Ross and McNaughton (1944) examined 90 head-injured patients and reported that the Shipley "in our series showed no relation to severity of the injury or to the electroencephalographic or pneumoencephalographic evidence of cerebral damage except in cases of extreme injury" (p. 259).

The expectation that psychological test scores should have a linear relationship with EEG or morphological measures is not justified, especially when extreme cases have been omitted. It is impossible to say whether the aforementioned negative results represent test misses or the recuperative function of the brain. Negative results may indicate that the Shipley is insensitive to impairment resulting from brain malfunction or injury. Negative results may also reflect that no significant intellectual impairment has occurred. To validate a test of intellectual impairment, independent evidence of deterioration (psychometric, behavioral, social history) is needed.

Several studies have looked at the ability to abstract in selected groups of subjects with hypothetically impaired abstracting ability. Malerstein and Beldon (1968) matched 10 Korsakoff's syndrome (polyneuritic psychosis) patients with 10 alcoholic controls on age, gender, education, and WAIS Vocabulary. The controls had Shipley Vocabulary and Abstraction means of 31.3 and 23.0, while the Korsakoff's patients had means of 31.8 and 10.7. Black (1973) found that patients with diffuse brain damage had lower Conceptual Quotients than patients with penetrating missile wounds, 79.2 to 92.7. WAIS Full Scale IQs showed a similar difference. Black (1976) found that subjects with frontal lesions had better scores on a number of WAIS measures than subjects with posterior lesions. Shipley-Hartford CQs were in the same direction (94.6 and 88.09), but were not significantly different. Lyle and Gottesman (1977) followed up the offspring of Huntington's disease patients and attempted to ascertain premorbid psychometric indicators of the Huntington's gene. They found that Shipley-Hartford total score had a point-biserial correlation of .41 with the criterion (still normal, late onset, early onset). Over these groups Vocabulary declined slightly, but there was a marked difference in Abstraction between the still-normal group and those who showed the Huntington's symptoms.

Aita, Armitage, Reitan, and Rabinovitz (1947) compared 70 veterans with brain damage with 61 controls who were on the same wards. While 46% of the brain-injured patients had pathological CQs, so did 26% of the controls. The groups had similar Vocabulary scores, but differed on Abstraction (22.2 and 17.7). The authors found little relationship between the Shipley and the Hunt-Minnesota Test and concluded that the Shipley was of "doubtful validity" for diagnosing brain damage. Armitage (1946) pointed out that the Conceptual Quotient is not valid for diagnosing brain damage; both false negatives and false positive are produced.

Among the general run of psychiatric patients there appears to be no relationship between Shipley scores and number of Bender figures recalled (Aaronson, Nelson, & Holt, 1953). However, while comparing "brain-damaged" patients with schizophrenics, Watson (1968) found correlations with the Shipley of .58 with the Bender, .60 with the Graham-Kendall, and .46 with BVRT correct. Parker (1957) tested 30 brain damaged soldiers and compared their results with control patients. The mean age of each group was in the mid-20s. He used a variety of tests: Bender-Gestalt, Goldstein-Scheerer sorting, Wechsler Memory Scale, and Block Design. Only Block Design differentiated the groups—on the Shipley the controls had a lower CQ than the brain-damaged subjects! Garfield and Fey (1948) found no relationship between Shipley Conceptual Quotient and a Wechsler-Bellevue Hold-No Hold Index; the age corrected correlation was .13.

Inferring brain damage on the basis of psychological test results involves using some scores as indicators of the client's premorbid level of functioning and then comparing other scores to that level. There are a large number of possible comparison standards and many potential measures of functioning. The Shipley provides only one of each. Yates (1954, 1956) questions the assumption that vocabulary level may be used as the measure of previous intellectual functioning. He argues that spurious organic scores may be obtained when using such a procedure. He also questions the "rather curious (assumption) that organic deterioration is similar to the deterioration accompanying age" (1954, p. 368).

From the studies done with neurological patients and those performed with psychiatric patients, one must conclude that low Abstraction scores and low Conceptual Quotients are not pathognomic of brain damage. In clinical situations the Shipley is useful as a screening instrument, but one cannot diagnose brain damage on the basis of this test alone. This reviewer is not the first to point out that vague terms such as "brain damage" or "organicity" need much more specific definition in order to clarify the relationships between brain lesions and psychological functioning.

Critique

The Shipley-Hartford is a useful device to make a rough assessment of intellectual functioning. It taps some of the same abilities as the Wechsler tests. Hunt (1949, p. 456) found the Shipley to be

> . . . as satisfactory as, if not more so than, the other available measures. Brevity and ease of administration are definitely in its favor. The answers on the Abstraction Test also offer relatively rich material for clinical interpretation despite the pencil-and-paper nature of the test. Some weakness develops at the lower ranges of intelligence, suggesting the need for further extension at this level. (p. 456)

Ives (1949) concludes her review,

> . . . for screening purposes or to supplement information from other tests, the Shipley scale provides valuable information regarding impairment in abstract thinking when restricted to the select group for which it is suited— above average in intelligence, reasonably well educated with no language handicaps, test-sophisticated, not too disturbed to be cooperative, and prefer- ably young. (p. 45)

Armitage (1946) concludes that "as a screening device is the only possible way in which [the Shipley-Hartford] can profitably be employed. . . . To determine the presence or absence of intellectual impairment, regardless of its cause, the test is of value" (p. 16). Filskov and Leli (1981) feel that short tests such as the Shipley can be used only as "adjunctive" measures "since validity data regarding their sensitivity to brain deficits have not been evaluated" (p. 558).

Yates (1972) concludes his review of the test by saying

> The test, used with due caution because of possible confounding factors listed above [failure to control in the standardization data, for confounding variables such as age, sex, educational level, and the effects of slowness or responding], remains a useful screening device or indicator of change where more inten- sive or direct experimental investigation is not possible (p. 322).

The new Shipley manual is a major improvement; all users of this test should obtain a copy. It provides a comprehensive review of the Shipley literature and is replete with cautions about the appropriate use of the instrument.

Users of the Shipley need to be aware of the characteristics of the new normative sample when making inferences on the basis of T-scores. These norms are based on a group of psychiatric patients who may be slightly above average in SES, but who manifest some performance deficit in Abstraction. The decision to use psychiatric patients as the normative group may not have been wise. Clinicians are accustomed to interpreting roughly equal Vocabulary and Abstraction scores as showing no evidence of deterioration. The new norms tend to minimize impair- ment. Consider a 35-year-old man with a high-school education: suppose his scores were the same as the medians in Table 3 as he entered and left the hospital. His admission raw scores for Vocabulary and Abstraction of 28 and 21 correspond to T-scores of 45 and 48. His discharge scores of 29 and 26 have T-scores of 47 and 53. The Abstraction Quotient on admission would be 99 and on discharge it would be 106. A more accurate picture of his functioning would have been an admission AQ in the low 90s and approximately 100 on discharge. Clinicians will need to look to their own experience to see whether their interpretations of Shipley results are greatly modified on the basis of these new norms.

The computerized version of the Shipley will now print out a 6-page report that, despite the caveats of the manual, are certain, I fear, to be misused.

In the new manual, Zachary lists several directions for future research on the Shipley. They include study of the AQ, establishing its norms, conducting validity studies with criterion groups and correlations with other measures, and relating the Shipley to brain scans. Goals for the Shipley should be more modest. It cannot replace the Halstead-Reitan, the Michigan, or the idiosyncratic test batteries used by clinicians, and validity data for these batteries are difficult to come by.

Clinical and research users should be cautioned against overinterpreting results from the Shipley and consider the following:

1. The results are affected by age. Older samples have slightly higher Vocabulary scores but considerably lower Abstraction scores; test development was done with samples of students. Yates (1954, p. 367) calls the standardization of the Shipley "completely inadequate" because only young, intelligent normals were used.

2. The new norms are age and education corrected, but are based on psychiatric patients rather than normals; thus, an Abstraction T-score of 50 may imply some impairment. The T-scores and the AQ should be used with caution.

3. The test is inappropriate for low-IQ persons or individuals with language handicaps.

4. The Shipley also has a low ceiling and will not spread high IQ subjects.

5. Shipley results may be lowered by poor motivation of the examinee. The test is self-administered, and motivational level cannot be evaluated by the clinician.

6. Retest reliabilities are only moderate. This may be because of the shortness of the test, low ceiling and high base, uncontrolled motivational variables, or the homogeneity of the samples used.

7. Although correlations with WAIS FSIQ are about .7 to .8, estimates based on the Shipley should be used with caution. Under the best of circumstances, the 95% confidence intervals of the IQ estimates are 1.5 standard deviations of IQ.

8. Relatively low scores on the Abstraction section may reflect diminished intellectual efficiency. Schizophrenics, for example, tend to score lower on Abstraction than other psychiatric patient groups, and chronic schizophrenics lower yet.

9. A relatively low Abstraction score or low Conceptual Quotient or low Abstraction Quotient is *not* pathognomic of intellectual impairment resulting from brain damage.

The Shipley-Hartford is useful as a screening instrument, but it cannot replace more intensive individual devices to assess individual intellectual functioning. We can do no better than conclude with Shipley's own caution:

> [The application of the scale] is pretty well restricted to use with individuals of at least average intelligence who are reasonably well educated and are without language handicaps. (1953, p. 755)

The scale should not be used as a final, definitive measure of intellectual impairment. Instead, it should be used as a preliminary screening device to call to the attention of clinicians those individuals who may be functioning at a lowered level in an important area of their thinking and therefore need to be studied more carefully. Where impairment is indicated in the test score, only further clinical evaluation will determine whether the impairment is genuine and, if so, to what it is attributable.

References

Aaronson, B. S., Nelson, S. S., & Holt, S. (1953). On a relation between Bender-Gestalt recall and Shipley-Hartford scores. *Journal of Clinical Psychology, 9,* 88.

Abrams, S., & Nathanson, I. A. (1966). Intellectual deficit in schizophrenia: Stable or progressive. *Diseases of the Nervous System, 27,* 115-117.

Aita, J., Armitage, S. G., Reitan, R. M., & Rabinovitz, A. (1947). The use of certain psychological tests in the evaluation of brain injury. *Journal of General Psychology, 37,* 25-44.

Armitage, S. G. (1946). An analysis of certain psychological tests used for the evaluation of brain injury. *Psychological Monographs, 60* (Whole No. 277).

Babcock, H. (1930). An experiment in the measurement of mental deterioration. *Archives of Psychology, 18* (Whole No. 117).

Barker, T. E., & Black, F. W. (1976). Klinefelter syndrome in a military population. *Archives of General Psychiatry, 33,* 607-610.

Bennett, H. J. (1956). The Shipley-Hartford Scale and the Porteus Maze test as measures of functioning intelligence. *Journal of Clinical Psychology, 12,* 190-191.

Black, F. W. (1973). Cognitive and memory performance in subjects with brain damage secondary to penetrating missile wounds and closed head injury. *Journal of Clinical Psychology, 29,* 441-442.

Black, F. W. (1976). Cognitive deficits in patients with unilateral war-related frontal lobe lesions. *Journal of Clinical Psychology, 32,* 366-372.

Braff, D. L., & Beck, A. T. (1974). Thinking disorder in depression. *Archives of General Psychiatry, 31,* 456-459.

Canter, A. H. (1951). Direct and indirect measures of psychological deficit in multiple sclerosis. *Journal of General Psychology, 44,* 3-50.

Corotto, L. V. (1966). Effects of age and sex on the Shipley-Institute of Living Scale. *Journal of Consulting Psychology, 30,* 179.

Dalton, J. L., & Dubnicki, C. (1981). Sex, race, age, and educational variables in Shipley-Hartford scores of alcoholic inpatients. *Journal of Clinical Psychology, 37,* 885-888.

Dennis, D. M. (1973). Predicting Full Scale WAIS IQs with the Shipley-Hartford. *Journal of Clinical Psychology, 29,* 366-368.

Eisenthal, S., & Harford, T. (1971). Correlation between the Raven Progressive Matrices Scale and the Shipley Institute of Living Scale. *Journal of Clinical Psychology, 27,* 213-215.

Filskov, S. B., & Leli, D. A. (1981). Assessment of the individual in neuropsychological practice. In S. B. Filskov & T. J. Boll (Eds.), *Handbook of clinical neuropsychology* (pp. 545-576). New York: Wiley.

Garfield, S. L. (1947). The Shipley-Hartford Retreat Scale as a quick measure of mental status. *Journal of Consulting Psychology, 11,* 148-150.

Garfield, S. L., & Blek, L. (1952). Age, vocabulary level, and mental impairment. *Journal of Consulting Psychology, 16,* 395-398.

Garfield, S. L., & Fey, W. F. (1948). A comparison of the Wechsler-Bellevue and Shipley-Hartford scales as measures of mental impairment. *Journal of Consulting Psychology, 12,* 259-264.

Goldstein, R. H., & Salzman, L. F. (1967). Cognitive functioning in acute and remitted psychiatric patients. *Psychological Reports, 21,* 24-26.

Goodman, J. T., Streiner, D. L., & Woodward, C. A. (1974). Test-retest reliability of the Shipley-Institute of Living Scale: Practice effects or random variation. *Psychological Reports, 35,* 351-354.

Hoffmann, H., & Nelson, P. C. (1971). Personality characteristics of alcoholics in relation to age and intelligence. *Psychological Reports, 29,* 143-146.

Horlick, R. S., & Monroe, H. J. (1954). A study of the reliability of an alternate form of the Shipley-Hartford Abstraction scale. *Journal of Clinical Psychology, 10,* 381-383.

Hunt, W. A. (1949). Shipley-Institute of Living Scale for Measuring Intellectual Impairment. In O.K. Buros (Ed.), *The third mental measurements yearbook* (p. 456). Highland Park, NJ: The Gryphon Press.

Ives, M. (1949). Shipley-Institute of Living Scale for Measuring Intellectual Impairment. In O.K. Buros (Ed.), *The third mental measurements yearbook* (pp. 456-457). Highland Park, NJ: The Gryphon Press.

Jansen, D. G., & Hoffmann, H. (1973). The influence of age, intelligence, and educational level on Shipley-Hartford Conceptual Quotients of state hospital alcoholics. *Journal of Clinical Psychology, 29*, 468-470.

Jansen, D. G., & Nickles, L. A. (1973). Variables that differentiate between single- and multiple-admission psychiatric patients at a state hospital over a 5-year period. *Journal of Clinical Psychology, 29*, 83-85.

Jones, B. M. (1974). Cognitive performance of introverts and extraverts following acute alcohol ingestion. *British Journal of Psychology, 65*, 35-42.

Kish, G. B. (1970). Alcoholics' GATB and Shipley profiles and their interrelationships. *Journal of Clinical Psychology, 26*, 482-484.

Kish, G. B., & Ball, M. E. (1969). Low education level as one factor producing a verbal-abstract disparity on the Shipley-Institute of Living Scale. *Journal of Clinical Psychology, 25*, 183-184.

Knight, R. A., Epstein, B., & Zielony, R. D. (1980). The validity of the Whitaker Index of Schizophrenic Thinking. *Journal of Clinical Psychology, 36*, 632-639.

Kobler, F. J. (1947). The measurement of improvement among neuropsychiatric patients in an Army convalescent facility. *Journal of Clinical Psychology, 3*, 121-128.

Kraus, J., Chalker, S., & Macindoe, I. (1967). Vocabulary and chronological age as predictors of "abstraction" on the Shipley-Hartford Retreat Scale. *Australian Journal of Psychology, 19*, 133-135.

Lewinski, R. J. (1946). The Shipley-Hartford Scale as an independent measure of mental ability. *Educational and Psychological Measurement, 6*, 253-259.

Lewinsohn, P. M. (1963). Use of the Shipley-Hartford Conceptual Quotient as a measure of intellectual impairment. *Journal of Consulting Psychology, 27*, 444-447.

Lewinsohn, P. M. (1967). Factors related to improvement in mental hospital patients. *Journal of Consulting Psychology, 31*, 588-594.

Lewinsohn, P. M., & Nichols, R. C. (1964). The evaluation of changes in psychiatric patients during and after hospitalization. *Journal of Clinical Psychology, 20*, 272-279.

Lyle, O. E., & Gottesman, I. I. (1977). Premorbid psychometric indicators of the gene for Huntington's disease. *Journal of Consulting and Clinical Psychology, 45*, 1011-1022.

Magaret, A., & Simpson, M. M. (1948). A comparison of two measures of deterioration in psychotic patients. *Journal of Consulting Psychology, 12*, 265-269.

Malerstein, A. J., & Beldon, E. (1968). WAIS, SILS, and PPVT in Korsakoff's syndrome. *Archives of General Psychiatry, 19*, 743-750.

Manson, M. P., & Grayson, H. M. (1947). The Shipley-Hartford Retreat Scale as a measure of intellectual impairment for military prisoners. *Journal of Applied Psychology, 31*, 67-81.

Martin, J. D., Blair, G. E., Sadowski, C., & Wheeler, K. J. (1981). Intercorrelations among the Slosson Intelligence Test, the Shipley-Institute of Living Scale, and the Intellectual Efficiency Scale of the California Psychological Inventory. *Educational and Psychological Measurement, 41*, 595-598.

Martin, J. D., Blair, G. E., Stokes, E. H., & Lester, E. H. (1977). A validity and reliability study of the Slosson Intelligence Test and the Shipley Institute of Living Scale. *Educational and Psychological Measurement, 37*, 1107-1110.

Mason, C. F., & Ganzler, H. (1964). Adult norms for the Shipley Institute of Living Scale and Hooper Visual Organization Test based on age and education. *Journal of Gerontology, 19*, 419-424.

Monroe, K. L. (1966). Note on the estimation of the WAIS Full Scale IQ. *Journal of Clinical Psychology, 22*, 79-81.

Murray, M. D., Page, J., Stotland, E., & Dietze, D. (1970). Success on varied tasks as an influence on sense of competence. *Journal of Clinical Psychology, 26*, 296-298.

Palmer, J. O. (1964). A restandardization of adolescent norms for the Shipley-Hartford. *Journal of Clinical Psychology, 20*, 492-495.

Pauker, J. D. (1975). A gender difference and a caution in predicting WAIS IQ from Shipley-Hartford scores. *Journal of Clinical Psychology, 31,* 94-96.

Paulson, M., & Lin, T-T. (1970). Predicting WAIS IQ from Shipley-Hartford scores. *Journal of Clinical Psychology, 26,* 453-461.

Phillips, W. M., Phillips, A. M., & Shearn, C. R. (1980). Objective assessment of schizophrenic thinking. *Journal of Clinical Psychology, 36,* 79-89.

Pishkin, V., Lovallo, W. R., Lenk, R. G., & Bourne, L. E., Jr. (1977). Schizophrenic cognitive dysfunction: A deficit in rule transfer. *Journal of Clinical Psychology, 33,* 335-342.

Prado, W. M., & Taub, D. V. (1966). Accurate prediction of individual intellectual functioning by the Shipley-Hartford. *Journal of Clinical Psychology, 22,* 294-296.

Robinson, M. F. (1946). What price lobotomy? *Journal of Abnormal and Social Psychology, 41,* 421-436.

Ross, W. D., & McNaughton, F. L. (1944). Head injury: A study of patients with chronic post-traumatic complaints. *Archives of Neurology and Psychiatry, 52,* 255-269.

Ruiz, R. A., & Krauss, H. H. (1967). Test-retest reliability and practice effect with the Shipley-Institute of Living Scale. *Psychological Reports, 20,* 1085-1086.

Salzman, L. F., Goldstein, R. H., Atkins, R., & Babigian, H. (1966). Conceptual thinking in psychiatric patients. *Archives of General Psychiatry, 14,* 55-59.

Schalock, R. L., & Wahler, H. J. (1968). Changes in Shipley-Hartford scores with five repeated test administrations: Statistical conventions vs. behavioral evidence. *Psychological Reports, 22,* 243-246.

Scherer, I. W. (1949). The psychological scores of mental patients in an individual and group testing situation. *Journal of Clinical Psychology, 5,* 405-408.

Shaw, D. J. (1966). The reliability of the Shipley-Institute of Living Scale. *Journal of Clinical Psychology, 22,* 441.

Shipley, W. C. (1940). A self-administering scale for measuring intellectual impairment and deterioration. *Journal of Psychology, 9,* 371-377.

Shipley, W. C. (1946). *Shipley-Institute of Living Scale: Manual of directions and scoring key.* Hartford: Institute of Living.

Shipley, W. C. (1953). Shipley-Institute of Living Scale for measuring intellectual impairment. In A. Weider (Ed.), *Contributions toward medical psychology: Theory and psychodiagnostic methods* (Vol. 2, pp. 751-756). New York: Ronald.

Shipley, W. C., & Burlingame, C. C. (1941). A convenient self-administering scale for measuring intellectual impairment in psychotics. *American Journal of Psychiatry, 97,* 1313-1325.

Sines, L. K. (1958). Intelligence test correlates of Shipley-Hartford performance. *Journal of Clinical Psychology, 14,* 399-404.

Sines, L. K., & Simmons, H. (1959). The Shipley-Hartford Scale and the Doppelt Short Form as estimators of WAIS IQ in a state hospital population. *Journal of Clinical Psychology, 15,* 452-453.

Slater, P. (1943). Interpreting discrepencies. *British Journal of Medical Psychology, 19,* 415-419.

Stone, L. A. (1965a). Recent (1962-1964) psychiatric patient validation norms for the Shipley-Institute of Living Scale. *Psychological Reports, 16,* 417-418.

Stone, L. A. (1965b). Test-retest stability of the Shipley-Institute of Living Scale. *Journal of Clinical Psychology, 21,* 432.

Stone, L. A., & Chambers, A., Jr. (1965). The distribution of measured adult intelligence in a state psychiatric hospital population. *Psychology, 2,* 27-29.

Stone, L. A., & Ramer, J. C. (1965). Estimating WAIS IQ from Shipley Scale scores: Another cross-validation. *Journal of Clinical Psychology, 21,* 297.

Suinn, R. M. (1960). The Shipley-Hartford Retreat Scale as a screening test of intelligence. *Journal of Clinical Psychology, 16,* 419.

Sutker, P. B., Moan, C. E., & Swanson, W. C. (1972). Porteus Maze test qualitative performance in pure sociopaths, prison normals, and antisocial psychotics. *Journal of Clinical Psychology, 28,* 349-352.

Tarter, R. E., & Jones, B. M. (1971). Absence of intellectual deterioration in chronic alcoholics. *Journal of Clinical Psychology, 27,* 453-454.

Wahler, H. J., & Watson, L. S. (1962). A comparison of the Shipley-Hartford as a power test with the WAIS verbal scale. *Journal of Consulting Psychology, 26,* 105.

Watson, C. G. (1968). The separation of NP hospital organics from schizophrenics with three visual motor screening tests. *Journal of Clinical Psychology, 24,* 412-414.

Wiens, A. N., & Banaka, W. H. (1960). Estimating WAIS IQ from Shipley-Hartford scores: A cross-validation. *Journal of Clinical Psychology, 16,* 452.

Wilson, R. S., Rosenbaum, G., Brown, G., Rourke, D., Whitman, R., & Grisell, J. (1978). An index of premorbid intelligence. *Journal of Consulting and Clinical Psychology, 46,* 1554-1555.

Winfield, D. L. (1953). The Shipley-Hartford vocabulary test and pre-trauma intelligence. *Journal of Clinical Psychology, 9,* 77-78.

Wright, M. E. (1946). Use of the Shipley-Hartford test in evaluating intellectual functioning of neuropsychiatric patients. *Journal of Applied Psychology, 30,* 45-50.

Yates, A. (1954). The validity of some psychological tests of brain damage. *Psychological Bulletin, 51,* 359-379.

Yates, A. (1956). The use of vocabulary in the measurement of intellectual deterioration—A review. *Journal of Mental Science, 102,* 409-440.

Yates, A. (1972). Shipley-Institute of Living Scale for Measuring Intellectual Impairment. In O.K. Buros (Ed.), *The seventh mental measurements yearbook* (p. 322). Highland Park, NJ: The Gryphon Press.

Zachary, R. A. (1986). *Shipley-Institute of Living Scale manual* (revised). Pre-publication draft.

Zachary, R. A., Crumpton, E., & Spiegel, D. E. (1985). Estimating WAIS-R IQ from Shipley Institute of Living Scale. *Journal of Clinical Psychology, 41,* 532-540.

Zachary, R. A., & Huba, G. J. (1985). Simplified formula and table for the unusualness of combinations of subtest scores on the Shipley Institute of Living Scale. *Journal of Clinical Psychology, 41,* 832-833.

Zachary, R. A., Paulson, M. J., & Gorsuch, R. L. (1985). Estimating WAIS IQ from the Shipley Institute of Living Scale using continuously adjusted age norms. *Journal of Clinical Psychology, 41,* 820-831.

Brent Edward Wholeben, Ph.D.

Senior Graduate Faculty and Associate Professor, Department of Educational Leadership and Counseling, University of Texas at El Paso, El Paso, Texas.

SIXTEEN PERSONALITY FACTOR QUESTIONNAIRE

Raymond B. Cattell and IPAT Staff. Champaign, Illinois: Institute for Personality and Ability Testing, Inc.

THE ORIGINAL OF THIS REVIEW WAS PUBLISHED IN TEST CRITIQUES: VOLUME IV (1986).

Introduction

The Sixteen Personality Factor Questionnaire (16PF) is an objective test of 16 multidimensional personality attributes arranged in omnibus form. In general, it provides normed references to each of these attributes (the primary scales). Four additional factor scores, second-order scales (Cattell, Eber, & Tatsuoka, 1970), are also computable—based on linear combinations of the 16 primary scales. Conceptualized and initially developed by Raymond B. Cattell in 1949 as a broad, multipurpose measure of the "source traits" of individual personality, the 16PF is appropriate for a wide range of multifaceted populations. It provides a global representation of an individual's coping style, the person's reactive stance to an ever-fluid and transactional environment, and that individual's ability to perceive accurately certain specific environmental requisites for personal behavior. Unlike such instrumentation as the Minnesota Multiphasic Personality Inventory (MMPI), the 16PF attempts to measure personality attributes and behavioral styles of a more "normal" rather than a "pathological" population, although a more clinical use of the 16PF may be appropriate (Karson & O'Dell, 1976).

The 16PF assesses a total of 16 indices, or attributes, of the human personality, attempting to convey a map of the individual's "personality sphere" as originally intended by Cattell. Based on the subject's reaction to certain situations, namely, individual interpretations based on certain questions, a profile of that subject's personality is constructed based on each of the following sixteen factors:

Warmth (Factor A): detached, critical, cool, impersonal vs. outgoing, participating, interested in people, easygoing;

Intelligence (Factor B): concrete-thinking vs. abstract thinking, bright;

Emotional Stability (Factor C): emotionally less stable, easily upset, changeable vs. mature, faces reality, calm, patient;

Dominance (Factor E): mild, accommodating, easily led, conforming vs. aggressive, authoritative, competitive, stubborn;

Impulsivity (Factor F): prudent, serious, taciturn vs. impulsively lively, enthusiastic, heedless;

Conformity (Factor G): disregards rules, feels few obligations vs. persevering, proper, moralistic, rule-bound;

Boldness (Factor H): restrained, threat-sensitive, timid vs. socially bold, uninhibited, spontaneous;

Sensitivity (Factor I): self-reliant, realistic, no-nonsense vs. intuitive, unrealistic, sensitive;

Suspiciousness (Factor L): adaptable, free of jealousy, easy to get along with vs. opinionated, hard to fool, skeptical, questioning;

Imagination (Factor M): careful, conventional, regulated by external realities vs. careless of practical matters, unconventional, absent-minded;

Shrewdness (Factor N): natural, genuine, unpretentious vs. calculating, socially alert, insightful;

Insecurity (Factor O): self-assured, confident, secure, self-satisfied vs. self-reproaching, worrying, troubled;

Radicalism (Factor Q_1): respecting established ideas, tolerant of traditional difficulties vs. liberal, analytical, likes innovation;

Self-sufficiency (Factor Q_2): a joiner and sound follower vs. prefers own decisions, resourceful;

Self-discipline (Factor Q_3): careless of protocol, follows own urges vs. socially precise, following self-image, compulsive; and

Tension (Factor Q_4): tranquil, torpid, unfrustrated vs. frustrated, driven, restless, overwrought.

(Aiken, 1976; Cattell, Eber, & Tatsuoka, 1970; Karson & O'Dell, 1976; Krug, 1981; Lanyon & Goodstein, 1982)

The four second-stratum measures of the subject's personality are constructed based on factor loadings from each of the 16 profile aspects of human behavior. These second-order aggregates are identified and described as follows:

Extraversion (Factor Q_1): introversion vs. extraversion; principally accounted for by the four primary factors of warmth (high A), impulsivity (high F), boldness (high H), and group dependence (low Q_2);

Anxiety (Factor Q_{II}): low anxiety vs. high anxiety; principally accounted for by the six primary factors of emotional instability (low C), threat sensitivity (low H), suspiciousness (high L), guilt (high O), low integration (low Q_3), and tension (high Q_4);

Tough Poise (Factor Q_{III}): sensitivity, emotionalism vs. tough poise; principally accounted for by the three primary factors of detachment (low A), tough-mindedness (low I), and practicality (low M); and

Independence (Factor Q_{IV}): dependence vs. independence; principally accounted for by the three primary factors of dominance (high E), rebelliousness (high Q_1), and self-sufficiency (high Q_2).

(Aiken, 1976; Cattell, Eber, & Tatsuoka, 1970; Institute for Personality and Ability Testing, Inc., 1970, 1972, 1979, 1985b; Karson & O'Dell, 1976; Krug, 1981)

The 16PF is constructed in five forms requiring an administration time of 30-60 minutes, depending on the form. Separate forms vary in readability requirements from a seventh-grade level for Forms A and B to a third-grade level for Form E. The target population for administration is high-school senior through adult. The test is objective, forced-option, and composed of 6-13 items per each of the surveyed 16 personality attributes. Most items are declarative elicitors requiring the subject to give only a short, reactive response to a generally described situation or milieu. The test is usually untimed and, due to its ease of hand scoring, can often be interpreted for the subject immediately after the administration.

In the early 1940s, Cattell's consuming interest in the concept of the "personality sphere" (the global specification of all personality factors or traits that can be measured directly and then subsequently analyzed to understand better the phenomenon of human behavior and interpersonal coping) led him to build on the earlier work of Allport and Odbert (1936). Allport and Odbert identified 17,954 "trait names" relative to human behavior, reduced them to a list of 4,504, and called them "real traits." Cattell further reduced this list to 171 terms by simply eliminating synonyms; through a cluster analysis of peer ratings he identified several clusters that he named "surface traits." Using the then-revolutionary technique of factor analysis, Cattell eventually identified a total of 20 distinct factors that he called "source traits." The result was the composition of the 16 primary personality factors measured in the 16PF instrument originally published in 1949 (Lanyon & Goodstein, 1982).

Subsequent editions of the 16PF, including the construction of seemingly parallel forms (possibly better referred to as extended, repeated measurement forms), were published in 1956-57, 1961-62, and 1967-69. A total of five forms now exist: Forms A, B, C, D, and E. Recently (1985), Form E has been renormed for highly diverse populations, including prison inmates, culturally disadvantaged, physical rehabilitation clients, and limited schizophrenic patients. The norms and interpretive data for the use of the remaining forms (A-D) are based on updated validity and reliability studies conducted in 1970 (Forms A-B) and 1972 (Forms C-D). Normative data comparisons are based on the demographic characteristics of gender and age for senior high school, college, and more general populations, but age corrections can be applied when widely age-ranging populations are being compared.

The 16PF has been translated, or structurally adapted, into over 50 languages in Europe, South America, Africa, and Asia. Most of these translations were initiated by the actual user of the test. Over the past several years, a series of structured analyses have been conducted concerning the effect of the "administration language" on the resulting measurements of individual personality by the instrument (see Cattell, Schroder, & Wagner, 1969; Kapoor, 1965; Nowakowska, 1970; Rodriguez, 1981). Normative comparison tables for culturally disadvantaged populations based on the administration of Form E in the United States (Institute for Personality and Ability Testing, Inc., 1985b) may eventually assist a resolution of this concern for the probability of uncontrolled bias in results based on the language of administration. However, Form E is administered in English, albeit at a reduced literacy competency of a third-grade reading level.

Each of the five forms of the 16PF contain a collection of declarative elicitors that require the subject to respond to a specific situation by choosing from among three forced-choice options in Forms A, B, C, and D or from among two forced-choice options in Form E. Because the number of items differ by form (187 items in each of Forms A and B; 105 items in each of Forms C and D; 128 items in Form E), the administration times are relationally affected; approximately one hour is required for Forms A, B, and E, whereas 30-45 minutes may be sufficient for Forms C and D.

According to the Institute for Personality and Ability Testing, Inc. (1970, 1972, 1985b), reading grade levels are 7.5 for Forms A and B, 6.5 for Forms C and D, and 3.3 for Form E. However, Forms C and D may replace A and B whenever administration time must be limited due to other constraints (Cattell, Eber, & Tatsuoka, 1970).

Items used in the 16PF may be of ordinal or nominal form. For example, in Forms A, B, C, and D (as given in the "examples" section of Form A, 1967-68 Edition R) an ordinal-form item would state: "People say I'm impatient," with options being a) true, b) uncertain, and c) false, whereas a nominal-form item would state: "Adult is to child as cat is to —," with the options being a) kitten, b) dog, and c) baby. In Form E (as given in the "examples" section of Form E, 1967 Edition) subjects would respond to an ordinal-form item such as "Do you like to play" with either "jokes on people" or "do you not like to do that," whereas on a nominal-form item they would indicate their preference by marking either "Would you rather play baseball" or "go fishing."

Subjects respond by checking the item number of their appropriate response on an accompanying answer sheet. Scoring templates, wherein specific choices from among the available options are tallied for producing summative, raw weights, promote a relatively time-efficient and accurate scoring framework for each individual instrument. The raw weights are then converted to sten scores, half-interval standard score equivalents, based on appropriate normative tables supplied with the instrument. These sten scores are subsequently depicted graphically on a subject profile sheet for each of the 16 primary personality factors suggested by the subject's responses.

Second-order factor scores (Extraversion, Anxiety, Tough Poise, and Independence) are easily extracted from the individual primary scale results by the use of a hand-calculating worksheet. Sten scores for the second-order factors are entered on the individual profile sheet but are not graphically depicted. In the event that repeated measurements are indicated (e.g., administering both Forms A and B on separate occasions), aggregated measures and combinatorial sten scores can be computed and entered on the profile sheet for subsequent comparison.

Practical Applications/Uses

The attributes of human personality are three-fold: 1) what of a given situation one is able to perceive, 2) how one perceives the situation and subsequently interprets it, and 3) what one does in response to that situation based on this interpretation. The approach of the 16PF to personality assessment—the predictable reactive interpretation of a subject based on the variable indices of individual personality—is an evaluation of the subject's projective view of "self" rather than a reactive view toward the intentions of "others."

The 16PF is described as appropriate for ages 16 and above. The items in the various forms are representative of situations that middle-school children and adults encounter daily in our modern, complex, and sophisticated world. In addition, although considerable research still needs to be accomplished, the 16PF may provide a vehicle for better understanding the dynamics of personality development, change, and influence throughout the various maturational benchmarks of the adolescent child.

Aside from a purely clerical function, the administrator or examiner is simply a convenor and distributor of materials. For some populations, the administrator may wish to read aloud the simple, one-page instructions prior to the beginning of testing. However, most subjects will read the instructional page privately and commence testing at their volition. Although the administrator's manual for the 16PF (Institute for Personality and Ability Testing, Inc., 1979, p. 14) suggests that approximately every ten minutes the administrator remark to respondents: "Most people are now doing question —," this reviewer does not advise doing so. Subjects may begin to view their performances as substandard (e.g., in the case of a particular subject being slower in terms of response time than others) and thus elicit a personological response that would uncontrollably influence subsequent responses to the items in the instrument.

According to Lanyon and Goodstein (1982), the quantity of references for the 16PF is second only to the MMPI. Of the over 2,000 research citations that exist today concerning the formal application of the 16PF as a personality assessment instrument (e.g., Buros, 1978; Institute for Personality and Ability Testing, Inc., 1977, 1979, 1985a; Mitchell, 1985), the major thrust of application has recently been in terms of career guidance, vocational exploration, and occupational testing. Of the seven machine (computer) generated reports that are based entirely on the 16PF, three are linked directly to career, vocational, and/or occupational assessment: 16PF Narrative Scoring Report, Personal Career Development Profile, and Law Enforcement Assessment and Development Report. As a more direct example, for nearly ten years (1971-80) participants in the National Institute of Education's research and development effort to rehabilitate the hard-core unemployed in Montana, Idaho, Wyoming, Nebraska, and North and South Dakota (Mountain-Plains Education and Economic Development Program, Inc.; later, Family Training Center, Inc. in Glasgow, Montana) were evaluated by the 16PF to better understand the personological dynamics of this disadvantaged population. Unfortunately, few, if any, of these data records were formally analyzed in aggregate or subsequently disseminated in published form.

The 16PF has provided information in a multitude of settings, from the study of personality and motivational factors related to potential aviation accidents among U.S. Naval Academy graduates to the study of personality differences between U.S. Olympic national and nonnational swimmers (Institute for Personality and Ability Testing, Inc., 1985a). Research professionals interested in personality relationships to such areas as cancer treatment, cross-cultural acclimation, teachers of the emotionally disturbed, juvenile delinquency, religion, child abuse, imminent death anxiety, air traffic control, nutrition behavior, controlled substance abuse, long-term effects of concentration camp internment, attitudes towards variable physiological functioning, the screening of seminarians for the priesthood, and child adoption have selected the 16PF as their instrument for the study of personality (see Buros, 1978; Institute for Personality and Ability Testing, Inc., 1977, 1979, 1985a; Mitchell, 1985).

As mentioned in a previous section of this review, the changing mores and standards of an increasingly complex society have provided an opportunity to utilize the 16PF on many diverse populations. Moreover, the demands for leadership and excellence in all occupations, education, and formal training identify our needs to understand better human personality and the reactive behavior associated with it.

The 16PF identifies ages 16 years through adulthood as the primary targets for personality assessment by this instrument. For widely ranging population comparisons in terms of age, score corrections for age are available.

The current norms for Forms A, B, C, and D were constructed for high school juniors and seniors, college students, and a general nation-wide population of age and income levels commensurate with then current U.S. Bureau of Census figures. These original norms have been updated to reflect the dramatic societal changes since the late 1960s. For example, Form E, which has recently (1985) entered its second edition, provides newly composed norms for culturally disadvantaged, rehabilitation, and prison inmate populations that have attempted to assist better normative interpretations based on present societal attitudes and perceptions toward human behavior.

As with most tests, skillful modification of recommended administration procedures make the 16PF useful for special populations. For example, the publisher has made available video cassettes for substituting a visual testing in American sign language for the Form A version. With caution, the test could also be administered verbally to special populations for whom reading and/or writing are precluded (e.g., the learning disabled and/or multiple handicapped). Although not available at this time, a braille reader or Kurzweil text-character interpreter could also facilitate individual administration with visually handicapped populations without sacrificing validity or reliability of the subject's responses.

The 16PF can also be utilized to complement other measures typically associated with more generally psychopathological populations, although this might be contested by Butcher (1985). Experience on the part of this reviewer demonstrates that this test can add to the information gained through such administrations as the MMPI to cross-reference other results, as well as validate the responses of the more "nonnormal" clientele populations.

The administration of the 16PF is straightforward and simple, requiring little or no training on the part of the examiner. Subjects respond individually and directly to each item in the test booklet by marking the appropriate blank on the answer form. With the possible exception of the examiner reading aloud the brief directions that appear on the front page of the test booklet, there is no interaction between examiners and examinees. For this reason, the examiner could be a secretary or other untrained staff person; in the case of some examinees, it is plausible that individuals could enter the examination room, begin the test at their volition, and return the completed test to a predesignated area. In such open exit situations, an examiner may not be necessary at all.

The 16PF may be administered individually or in a large group setting. With some preparation and training, an examiner could administer the test orally to a large group, with participants recording their individual responses on separate answer forms. This could be particularly helpful for English-language-deficit populations for which even the Form E (third-grade reading level) might be precluded.

The administrator's manual (Institute for Personality and Ability Testing, Inc.,

1979) is equally straightforward and simple to follow. Step-by-step instructions are presented (pp. 14-15) for direct use by the examiner. Combined with the directions printed on the front cover of the test booklet for the examinee to read during test preparation, little confusion or ambiguity typically results. As mentioned in a previous section of this review, the recommendation by the test developers that the examiner report general progress to the examinee group as a whole is not recommended by this reviewer. The manual also provides information regarding the various profile scales of the 16PF, including a listing of relevant validity and reliability coefficients and directions for hand scoring and interpreting final results.

Generally, the time required for adequate administration of the 16PF, assuming the particular form and associated reading level have been properly selected, is approximately 60 minutes for Forms A, B, and E. For Forms C and D, 30-45 minutes have usually been found to be sufficient for adequate completion of all items. Because the test measures underlying facets of human personality and the intrinsic motivations associated with personal behavior, it is essential that the administration and testing environment not introduce extraneous stimulants that will bias the response behavior of the subject.

Scoring of the 16PF may be accomplished by hand or optical scanning device. Scoring templates (overlays) are available for hand scoring. In addition, computer scoring, including computer-generated interpretations for the gestalt of the subject's responses, is also available from the publisher. With the exception of Form E, a subject's raw score for each of the 16 primary factors is obtained through a weighted procedure where particular responses count as "1" summatively toward the final raw score, whereas others count as "2." Form E counts all appropriate responses as a "1" towards the unweighted sum. These weighted or unweighted sums are then compared to the desired normative score tables in the particular tabular supplement (Institute for Personality and Ability Testing, Inc., 1970, 1972, 1985b) where a particular sten score is identified based on the magnitudinal range of the response and the individual normative demographics of the respondent. This sten score is entered on the profile form and subsequently depicted graphically for ease of interpretation.

A separate worksheet is available for calculating by hand each of the four second-order (higher-order) scales. This involves transforming each of the various 16 sten scores as weighted factor loadings of standardized variables in a regression equation for each of the second-stratum scales. The arithmetic result is the interpretable sten score for the particular second-order scale. Simple integer multiplication, addition, and subtraction are required. However, the worksheet is such that the procedure is straightforward and uncomplicated.

Scoring requirements in terms of time depend specifically on the scorer. With a little practice, total scoring for one test can be accomplished easily within ten minutes. This time estimate includes conversion of raw measures to sten scores and entering them graphically on the interpretive profile. Additional scoring requirements are associated with determining individual validity measures for "probable faking" by the subject, which is discussed in the technical aspects section of this review.

Interpretation of the 16PF may be surmised through study of the individual profile sheet or through the use of one of the many computer-generated interpretive reports available from the publisher. Of course, interpretation in each of these situations should be construed differently.

As with all standardized tests concerned with the assessment of human personality, the terminology of interpretation provides an inherent bias that is often uncontrollable. For example, when considering the term *impulsivity*, which describes a scale for the extent to which a subject is prudent or serious as opposed to enthusiastic, many individuals think of the adjective "impulsive" as representing a negative condition of the human spirit. Yet, if one were to use the juxtaposed descriptors "prudent vs. enthusiastic," it might be difficult to determine which represented the more "preferred" state or condition. The use of terminology to describe various personological substates that might be interpreted incorrectly because of their colloquial usage has always been a major difficulty associated with personality assessment. This is not, by any means, a select criticism of the 16PF, for all tests are susceptible to this charge; however, the 16PF seems to be easy enough to understand that an untrained observer might draw erroneous conclusions based on a misinterpretation of the terminology used.

The computer-generated interpretive reports, the 16PF Narrative Scoring Report, 16PF Single-Page Report, Personal Career Development Profile, Marriage Counseling Report, Karson Clinical Report, Law Enforcement Assessment and Development Report, and Human Resource Development Report provide easy-to-read summative narrative based on individual subject responses to the 16PF. For the professional untrained in test interpretation, such reports can be very helpful as they consolidate the technical record of each subject into readable summaries based on sten-score ranges. On the other hand, such machine-scoring opportunities can also provide the environment for generalized misuse and abuse and once again restate the need for only "qualified professionals" to have access to such data.

One of the strengths of the 16PF is its ease of usage for structuring the interpretive interview with the client or subject. The profile sheet is easy to read and quickly understood (cautiously, of course, as mentioned above) in the presence of the trained professional. Discussing the profile with the client will often illuminate future points or issues for further discussion, especially if the test were administered in preparation for upcoming counseling or therapy sessions with the client.

Technical Aspects

The various technical reports available from the publisher (Cattell, Eber, & Tatsuoka, 1970; Institute for Personality and Ability Testing, Inc., 1970, 1972, 1979, 1985b) divide their validity assessment efforts into direct and indirect construct (or concept) comparisons. Similarly, reliability assessments are test-retest analyses for what the authors call "dependability and stability." In addition, equivalence coefficients between separate forms have also been evaluated.

Direct construct validity is reported in the form of multiple correlation coefficients, representing the degree of relationship between each of those items that "load" the particular personality factor (the particular primary factor of the 16PF) and the magnitude of the factor itself. Forms A and B are reported to have the greatest total direct validity where each form has seven scales with validity coefficients of at least .70 magnitude. At the same time, Form A has the lowest direct validity coefficients across all four forms, with .35, .41, and .44 correlational magnitudes for Intelligence, Shrewdness, and Imagination, respectively. For Form E, direct construct validity coefficients fall below a .70 magnitude in six areas: Warmth (.66), Conformity (.65), Suspiciousness (.66), Imagination (.41), Shrewdness (.21), Radicalism (.59), and Self-discipline (.67). These coefficients for Form E were derived from a sample of 914 male convicts.

Indirect construct validities for Forms A, B, C, and D are also reported in the form of multiple correlation coefficients, representing the degree of relationship between each primary scale magnitude and the total remaining primary scale magnitudes in the 16PF. The authors refer to this form of investigation as relating each specific factor with all other factors, namely, comparing all of what is "A" with all of what is determined as "not-A." As might be anticipated, correlational coefficients fall below a .80 magnitude in only two instances: .63 for Shrewdness and .74 for Imagination. No indirect validities are reported for Form E.

Dependability coefficients, reliability coefficients calculated by test-retest with short intervals (single or multiple day) between administrations, demonstrate relatively acceptable coefficients, with only sporadic instances of a scale falling below a .70 magnitude. For stability coefficients, test-retest administrations conducted over long intervals (several weeks), magnitudes are expectedly reduced. These results are available for Forms A and B only. For Form E, only stability coefficients are reported, and only the Sensitivity Scale lies above a .80 magnitude. The sample in this exercise was limited to 32 male and female rehabilitation clients, with a test-retest, median interval of six years, four months.

Equivalence coefficients, intercorrelations between primary factor scales generated from different test forms, are generally low. Few equivalence coefficients are greater than .50 magnitude when Forms A and B are compared. Fewer coefficients of .50 or more magnitude exist for Forms C and D. However, when combined administrations of A and C are compared with similar administrations of B and D, correlational magnitudes increase dramatically where 12 of the 16 scales display equivalence coefficients greater than .50 magnitude. When Form E results are compared with combined Forms C and D administrations, equivalence magnitudes generally fall within the .50 to .69 magnitudinal range.

Unfortunately, most of the validity and reliability analyses for all forms except Form E are based on data collected almost 15-20 years ago and are representative of populations who might have viewed their behavioral responses and, therefore, their personality in a different light than might be representative of today's societal mores. In addition, much of the data was collected by second parties not under the direct, standardized control of the publisher.

Internal respondent validity is also discernible for only Form A administrations using a publisher-supplied validity (appropriate response verification) key for determining "faking good" and "faking bad" potential when the subject may wish to appear more normal or less normal, respectively. Based on this supplementary analysis, a correction for distortion is calculable (see Krug, 1978). In addition, a similar analysis is available through the computing of the "motivational distortion scale" for Forms C-D (Institute for Personality and Ability Testing, Inc., 1972).

Critique

The 16PF provides an inventory of personal "source traits" that relate to individual perception, human behavior, and reactive (coping) potential. The test has a wide variety of uses ranging from career guidance to precounseling (intake) assessment. Originally developed for individuals aged 16 years or older, the primary factor scales, which are calculated from subject responses, provide a useful tool for better understanding personological differences between subjects and, furthermore, might provide a highly valid technique for explaining the underlying dynamics of cultural norms and ethnicity-based mores.

The 16PF is best used in individual personality assessment as a means of comparing individual and group mean differences. Its use as an instrument for determining actual trait-specific personality factors (that a subject definitively portrays one behavioral aspect as opposed to another) has been in dispute since the test's original publication (Anastasi, 1976; Bloxom, 1978; Bolton, 1978; Butcher, 1985; Lanyon & Goodstein, 1982; Mossholder, 1985; Walsh, 1978; Zuckerman, 1985). These concerns range from mild remonstrations regarding the use of the test for certain purposes to harsh criticism concerning the use of the instrument for any purpose whatsoever.

The low validity and reliability coefficients for the various testing forms do little to instill consistent faith in the instrument. It is somewhat disconcerting that more research and validation studies have not been initiated by the developer to enhance the utility of the test. However, criticisms regarding potential misuse of the instrument often overshadow the real value of the test in other areas.

Much of the criticism of the 16PF could be dispelled through an updated standardization of the test and its various forms. This would include, of course, a regeneration of norms to identify the dynamics of current societal populations better. An equally strict analysis of validity and reliability coefficients, based on clearly identifiable subpopulations, would serve to enhance the correlational magnitudes that presently exist as marginal. Much of the validity and reliability coefficients formally reported for the test are based on second-party studies not under the direct control of the developer. Although combined administrations of different forms of the test may improve overall validity and reliability measures considerably, the significant increase in terms of administration time and the transient-situational aspects of personality itself may preclude such an option.

These criticisms notwithstanding, the 16PF is a valuable means for identifying differential behavioral attributes in individuals and groups. The type of items and their wording, the situational relevancy described by these items, and the simplicity of response offer a potentially high value for this test in behavioral research and attitudinal exploration with diverse populations. The renewed emphasis on cross-cultural research in multiethnic populations also holds significant promise for the future application of this instrument.

References

Aiken, L. R. (1976). *Psychological testing and assessment* (2nd ed.). Boston: Allyn and Bacon.

Allport, G. W., & Odbert, H. S. (1936). Trait names: A psycho-lexical study. *Psychological Monographs, 47* (Whole No. 211).

Anastasi, A. (1976). *Psychological testing* (4th ed.). New York: Macmillan.

Bloxom, B. M. (1978). Review of the Sixteen Personality Factor Questionnaire. In O. K. Buros (Ed.), *The eighth mental measurements yearbook* (pp. 1077-1078). Highland Park, NJ: The Gryphon Press.

Bolton, B. F. (1978). Review of the Sixteen Personality Factor Questionnaire. In O. K. Buros (Ed.), *The eighth mental measurements yearbook* (pp. 1078-1080). Highland Park, NJ: The Gryphon Press.

Butcher, J. N. (1985). Review of the Sixteen Personality Factor Questionnaire. In J. V. Mitchell, Jr. (Ed.), *The ninth mental measurements yearbook* (Vol. I). Lincoln, NE: Buros Institute of Mental Measurements. (BRS Document Reproduction Service No. AN 0910-679)

Cattell, R. B., Eber, H. W., & Tatsuoka, M. M. (1970). *Handbook for the Sixteen Personality Factor Questionnaire (16PF).* Champaign, IL: Institute for Personality and Ability Testing, Inc.

Cattell, R. B., Schroder, G., & Wagner, A. (1969). Verification of the structure of the 16PF Questionnaire in German. *Psychol. Forsch., 32,* 369-386.

Institute for Personality and Ability Testing, Inc. (1970). *Tabular supplement no. 1 to the 16PF handbook: Norms for the 16PF, Forms A and B (1967-68 edition).* Champaign, IL: Author.

Institute for Personality and Ability Testing, Inc. (1972). *Tabular supplement no. 2 to the 16PF handbook: Norms for the 1969 edition of Forms C and D.* Champaign, IL: Author.

Institute for Personality and Ability Testing, Inc. (1977). *16PF research bibliography 1971-1976.* Champaign, IL: Author.

Institute for Personality and Ability Testing, Inc. (1979). *Administrator's manual for the 16PF* (2nd ed.). Champaign, IL: Author.

Institute for Personality and Ability Testing, Inc. (1985a). [Compilation of research citations of 16PF use]. Unpublished raw data.

Institute for Personality and Ability Testing, Inc. (1985b). *Manual for Form E of the 16PF* (2nd ed.). Champaign, IL: Author.

Kapoor, S. D. (1965). Cross-validation of the Hindi version of the 16PF test (VKKJ). *Indian Journal of Psychology, 40*(3), 115-120.

Karson, S., & O'Dell, J. W. (1976). *Guide to the clinical use of the 16PF.* Champaign, IL: Institute for Personality and Ability Testing, Inc.

Krug, S. E. (1978). Further evidence on 16PF distortion scales. *Journal of Personality Assessment, 42*(5), 513-518.

Krug, S. E. (1981). *Interpreting 16PF profile patterns.* Champaign, IL: Institute for Personality and Ability Testing, Inc.

Lanyon, R. I., & Goodstein, L. D. (1982). *Personality assessment* (2nd ed.). New York: John Wiley & Sons.

Mitchell, J. V., Jr. (Ed.). (1985). *The ninth mental measurements yearbook.* Lincoln, NE: Buros Institute of Mental Measurements.

Mossholder, K. M. (1985). Review of the Personal Career Development Profile. In J. V. Mitchell, Jr. (Ed.), *The ninth mental measurements yearbook* (Vol. I). Lincoln, NE: Buros Institute of Mental Measurements. (BRS Document Reproduction Service No. AN 915-2082)

Nowakowska, (1970). Polish adaptation of the Sixteen Personality Factor Questionnaire (16PF) of R. B. Cattell. *Educational Psychology, 13*(4), 478-500.

Rodriguez, F. N. (1981). Normas Argentinas del test 16 PF para sujetos de 17 y 20 anos. *Acta Psiquiatrica y Psicologica de America Latina, 27*(3), 219-226.

Walsh, J. A. (1978). Review of the Sixteen Personality Factor Questionnaire. In O. K. Buros (Ed.), *The eighth mental measurements yearbook* (pp. 1081-1083). Highland Park, NJ: The Gryphon Press.

Zuckerman, M. (1985). Review of the Sixteen Personality Factor Questionnaire. In J. V. Mitchell, Jr. (Ed.), *The ninth mental measurements yearbook* (Vol. I). Lincoln, NE: Buros Institute of Mental Measurements. (BRS Document Reproduction Service No. AN 0910-679)

Maria M. Llabre, Ph.D.

Associate Professor of Educational Psychology, University of Miami, Coral Gables, Florida.

STANDARD PROGRESSIVE MATRICES

J. C. Raven. London, England: H. K. Lewis & Co. Ltd. U.S. Distributor—The Psychological Corporation.

THE ORIGINAL OF THIS REVIEW WAS PUBLISHED IN TEST CRITIQUES: VOLUME I (1984).

Introduction

The Standard Progressive Matrices (SPM) was designed to measure a person's ability to form perceptual relations and to reason by analogy independent of language and formal schooling, and may be used with persons ranging in age from 6 years to adult (Raven, Court, & Raven, 1978). It is the first and most widely used of three instruments known as the Raven's Progressive Matrices, the other two being the Coloured Progressive Matrices (CPM) and the Advanced Progressive Matrices (APM). All three tests are measures of Spearman's g. This review focuses specifically on the SPM. The CPM and APM are reviewed elsewhere in this volume.

The SPM resulted from the work of a British psychologist, John C. Raven, and British geneticist Lionel Penrose (Penrose & Raven, 1936). Raven was a student of Charles E. Spearman. Their work was based on Spearman's two-factor theory of intelligence, which postulated that every cognitive test measures a general factor (g) that is common to all cognitive tests and a specific factor (s) that is unique for each test. Conceptually, g was defined by Spearman as the "eduction of relations and correlates."

The idea of using perceptual analogy items based on simple geometric patterns is credited to Cyril Burt who, along with his students, factor analyzed these items together with more conventional items of reasoning ability and noted their high loadings on g. Penrose and Raven translated the idea into a two-dimensional array or matrix, where the analogy could be viewed from simultaneous transformations of the rows and columns of the matrix. Raven published the first test in 1938 and worked on extensions of the test throughout his productive career. After his retirement in 1964 he continued to work with colleagues, in particular Dr. J. H. Court of Flinders University in South Australia and his own son, John Raven of the Scottish Council for Research in Education, who are jointly responsible for consolidating the three Matrices and the associated Vocabulary Scales. J. C. Raven died in 1970, but impressive work on the tests has continued. Raven's SPM is

The reviewer wishes to acknowledge and express appreciation for the assistance of William Summers.

probably the most widely used and researched culture-reduced test, its uses extending to diverse cultures throughout the world.

The current SPM is essentially the same as the 1938 edition. In 1947 a correction was made to one of the original items. During that same year, the CPM was published in order to extend the applicability of the test to young children as well as very old people and mentally retarded adults. The APM also came into restricted circulation in 1947, and further extended the applicability of the test to superior adults.

The 1956 revision of the SPM involved a reordering of items and of the incorrect alternatives. However, no major revisions of the original scale have been done. This version is currently available with a 1958 copyright.

Guides for using the test were published in 1956 and the bibliography was revised in 1960. The successor to the guides is the manual, first published in 1976 and revised in 1978 and 1983. Two research supplements to the manual are currently available: the first contains the 1972 Irish standardization and the 1979 British standardization of the SPM and the Mill Hill Vocabulary Scales (MHV) (Raven, 1981); the second contains summaries of additional reliability, validity, and normative studies (Court & Raven, 1982).[1]

A researcher's bibliography with short summaries of research studies is independently published and periodically updated by J. H. Court. The most recent update was in 1982 and brings the number of references to over 1,500.

The SPM consists of 60 items arranged in five sets (A, B, C, D, & E) of 12 items each. Each item contains a figure with a missing piece. Below the figure are either six (sets A & B) or eight (sets C through E) alternative pieces to complete the figure, only one of which is correct. Each set involves a different principle or "theme" for obtaining the missing piece, and within a set the items are roughly arranged in increasing order of difficulty. The first item in each set is relatively obvious, providing guidance to the examinee by elucidating the principle used to work the subsequent problems in that set. The diversity of figures and the arrangement of items in terms of difficulty level both serve to maintain examinees' interest and motivation.

The principles involved in the five sets may be described as: 1) completion of a pattern in a continuous figure, 2) figural analogy in a two-by-two matrix, 3) systematic alteration of a pattern in a three-by-three matrix, 4) systematic permutations and alterations of figures in a three-by-three matrix, and 5) systematic decomposition and synthesis of figural parts in a three-by-three matrix.

All items are contained in a test booklet, with one item per page. The booklet is attractive and durable. (Sets A and B are also available in board form). The figures are presented boldly in black-and-white within a rectangle located in the upper half of the page. Below each figure, two rows of numbered alternatives are presented. The choices are equally well drawn and of adequate size to ease possible eye strain. The correct responses are accurately drawn. Since responses are not registered in the test booklet, it is constructed for multiple use. Only one form of the SPM is available.

Answer sheets are available for both hand and machine scoring. The scoring key is provided in the manual; alternatively, a stencil scoring key for easier scoring is available. Machine scoring is done by Document Reading Services Ltd., who also provide a computing service.

The 1978 version of the manual (Raven, Court, & Raven) is comprised of eight sections which may be purchased separately or together. Sections 1, 3, and 7 are relevant to the SPM. Other sections pertain to the CPM, APM, MHV, and the Crichton Vocabulary Scale (CVS). Section 3 was revised in 1983 to incorporate information found in the research supplements (Raven, Court, & Raven, 1983). Section 1 contains background information on the development of the scales and serves as a general overview (an updated version is currently in press). Section 3 is essential for using the SPM, since it has the instructions for administration and scoring, normative data, and information on reliability and validity. Additional research on reliability, validity, and norms is documented in Section 7. This last section also contains the references.

Practical Applications/Uses

The SPM was designed to cover a broad range of mental ability and to be equally applicable across age groups. The nonverbal nature of the SPM makes it useful for testing persons from different linguistic backgrounds as well as those with communication disorders or limited language proficiency. It can serve as a screening device of intellectual ability for selecting students for gifted programs. The amount of research data available from countries around the world facilitates the determination of applicability to specific cultural groups. However, the SPM is not a useful diagnostic tool for purposes of remediation, since the perceptual processes involved are not measured by the current scoring system. Some practitioners have tried to overcome this limitation by questioning examinees on the reasons for their response selection after the completion of a conventional format. More work along these lines is desirable, since the authors suggest that even the examinees' incorrect responses are guided by hypotheses (Raven, Court, & Raven, 1983).

Test instructions are simple, allow for either individual or group administration, and could even be pantomimed. The examiner's involvement is minimal and no special training is required for test administration. The setting may be a classroom or any other quiet room where examinees can be free of distraction. Everyone is given the same items in a fixed order with no time limit. Most examinees are able to complete the test in less than 45 minutes. Examiners are encouraged to repeat the instructions, which include demonstrating the first item, when examinees do not appear to have understood the nature of the task. Examinees tested individually may respond simply by pointing to the answer while the examiner records the response. In group administration the examinees record their responses on the separate answer sheets. The flexibility in administration procedures and the lack of time constraints facilitate the use of the SPM with examinees who have severe communication disorders or physical disabilities.

A person's score is the number of correct responses. The raw score is typically converted to a percentile rank by using the appropriate norms. The norm groups included in the manual are 1) British children between the ages of 6 and 16 tested in groups in 1979; 2) Irish children between the ages of 6 and 12 tested in 1972; 3) Colchester children between the ages of 6 and 14 tested individually in 1943; 4) military and civilian subjects between the ages of 20 and 65 tested in the early 1940s; and 5) deaf British children, 15 to 16 years of age, tested in 1979. Norms obtained in the 1940s may be inadequate for present comparisons, since the authors have noted an upward trend in scores. A large scale normative study is currently being conducted in the United States.[2]

Normative studies from other parts of the world abound in the literature. For example, data from studies done in Canada, the United States, Newfoundland, and East and West Germany are compared to the original and recent norms in the research supplement to the manual. The measurement literature also contains results from normative studies done with children, adults, or special subgroups in diverse places including Australia, Belgium, Canada, Colombia, the Congo, Cuba, France, Hong Kong, India, Iraq, Iran, Italy, Rhodesia, Rumania, Spain, South Africa, Tanzania, Uruguay, and the United States. Users should note the use of time limits in many of these studies, a practice not recommended by the author. In addition, the majority of these studies did not use representative samples.

Technical Aspects

Approximately 50 reliability studies attest to the stability and internal consistency of the SPM. These studies focus on diverse groups with respect to age, nationality, and physical and psychological conditions. The results generally support the reliability of the SPM. A thorough review of reliability studies is provided in the manual.

Internal consistency studies using either the split-half method corrected for length or KR-20 estimates result in values ranging from .60 to .98, with a median of .90. The extreme values were reported by Georgas (1970); the .60 was with children 6 years of age for whom the CPM would have been more appropriate. In his review of the SPM, Burke (1958) questioned the test's adequacy with respect to reliability. His subsequent calculation of a .96 split-half estimate with American veterans (Burke, 1972) led him to conclude that "reliability seems quite adequate" (p. 254).

Test-retest correlations range from a low of .46 for an eleven-year interval to a high of .97 for a two-day interval. The median test-retest value is approximately .82. Coefficients close to this median value have been obtained with time intervals of a week to several weeks, with longer intervals associated with smaller values. Sinha (1968) reported a value as high as .96 for a six-month interval. Keir (1949) reported two-year interval values of .54 and .74 for small samples of 11-year-old boys and girls respectively. Raven provided test-retest coefficients for several age groups: .88 (13 yrs. plus), .93 (under 30 yrs.), .88 (30-39 yrs.), .87 (40-49 yrs.), .83 (50 yrs. and over) (reported in Raven, Court, & Raven, 1983).

Raven stated that the ability measured by the SPM "increases rapidly during childhood, appears to have reached its maximum somewhere about the age of 14, stays relatively constant for about 10 years, and then begins to decline, slowly but with remarkable uniformity right through old age" (Raven, Court, & Raven, p. G4). Thus, one can expect stability in SPM scores for older children and young adults. This is in fact supported by the evidence for intervals of up to one year.

Given the dependence of classical reliability coefficients on group heterogeneity, it is surprising to find that the more stable standard error of measurement was not reported in the manual for any of the studies. Also absent from the manual were reports of the standard deviations of the norm groups.

Numerous validation studies of the SPM can be found in the literature. Most common are correlational studies assessing concurrent and predictive validity with diverse groups and criteria. Construct validity has typically been investigated through traditional factor analytic methods in order to determine the factorial composition of the test. As was the case with reliability, the manual includes a synthesis of validation studies and a discussion of the relevant issues.

Spearman (1938) considered the test to be the best measure of g, and Vernon (1947) described it as one of the purest. After reviewing validity studies, Burke (1958) stated that he was not convinced the SPM is a pure measure of g and expressed doubts as to the feasibility of obtaining such a measure, a concern shared by other researchers. When evaluated by factor analytic methods which were used to define g initially, the SPM certainly comes as close to measuring it as one might expect. The majority of studies which have factor analyzed the SPM along with other cognitive measures in Western cultures report loadings higher than .75 on a general factor. On the other hand, some studies have shown SPM scores to be influenced by a spatial factor. The spatial factor, however, explains a lesser proportion of test variance and is not as stable across studies.

The factorial composition of a test is linked to its cultural loading. This issue is important when it comes to the SPM because of the presumed lack of cultural influences on it. This reviewer considers cultural loading to be a relative rather than an absolute characteristic. When evaluated against other measures of reasoning ability the SPM fares well in minimizing cultural influences. Studies have also demonstrated the suitability of norms in Western societies. In non-Western cultures, however, educational influences have been shown to play an important role in test performance (Sinha, 1968), and factor loadings vary considerably, indicating that cross-cultural applications of the SPM do have limitations.

Techniques other than factor analysis have been used to determine whether or not the SPM measures g. For example, Hunt (1974) sought to identify the specific cognitive processes involved in solving APM items. Using computer technology he identified two qualitatively different processes, gestalt and analytic, both of which could lead to similar scores. On this basis he questioned the suitability of scores as measures of g. Some support for the existence of two distinct strategies for solving SPM items lies in the bimodal distributions of SPM scores presented in the manual for seven- to nine-year-olds. Similar approaches to understanding the meaning of test scores have been taken by others (e.g., Bereiter & Scardamalia, 1979) and should prove very useful in increasing the diagnostic potential of the SPM. Kirby and Lawson (1983) have provided experimental data which also support Hunt's analyses and, like Guinaugh (1971), demonstrate that the strategies can be taught. Although Hunt was generally critical of the test, he demonstrated its usefulness as a research tool in cognitive psychology.

Concurrent validity coefficients between the SPM and the Stanford-Binet and Weschler scales for English-speakers range between .54 and .88, with the majority in the .70s and .80s. Correlations are slightly lower with other nonverbal and performance tests of intelligence, and lower yet with tests of verbal intelligence or vocabulary, where the typical value is below .7 (Burke, 1958; Raven, Court, & Raven, 1983). The specific values vary as a function of age, group heterogeneity, and criterion used, but generally demonstrate reasonable concurrent validity for English-speakers. Coefficients tend to be lower for non-English-speakers and fluctuate more widely across studies.

Not surprisingly, the SPM's ability to predict scholastic or occupational performance is not outstanding, although some researchers (e.g., Irvine, 1966) have reported validity coefficients as high as .90. The large variability in these coefficients emphasizes their dependence on group and criterion characteristics. Since most criterion measures used in predictive studies are heavily dependent on acquired information, the more "pure" SPM, by itself, predicts less well than other measures of scholastic aptitude.

Before ruling out the SPM as a useful predictor of academic performance, one must note that in the manual the authors repeatedly stress its use in conjunction with a measure of acquired information, such as the MHV scales. In fact, both measures are presented together as a package in the manual, although it is possible to use other vocabulary measures. Unfortunately, most predictive studies report a single correlation as evidence of validity rather than investigating the test's usefulness in the context of a prediction equation. Vernon (1949) suggested that a test battery which included a measure such as the SPM and a verbal test would provide useful predictions for most occupations. The relative "purity" of the SPM should prove advantageous in the context of multiple-regression equations.

Analyses of the 1979 standardization data showed that SPM scores correlated with age (.68) and that this relationship is similar for boys and girls. Court (1983) reviewed studies in which the scores of females and males were compared and concluded that most demonstrate no sex differences; where differences occurred, they were related to other variables.

Item analysis information based on the 1972 and 1979 standardization is presented in a research supplement to the manual (Raven, 1981). Item characteristic curves are shown which indicate that the test discriminates between children at all ability levels except the most able (p. 34). Item difficulties were computed separately for different socioeconomic status groups and correlated between pairs of groups. These resulted in coefficients higher than .95, indicating that the order of item difficulties was similar across all socioeconomic groups. Keir (1949) and Banks and Sinha (1951) had found a preponderance of items of medium difficulty and several items having low discrimination indices. Jensen (1980) commented on the gaps in item difficulties which prevented total scores from yielding smooth distributions. Certainly more attention needs to be given to item scaling. Recent developments in item calibration using latent-trait models should be useful for scaling existing as well as future items, and could eventually lead to an item pool large enough to result in a second form of the test.

Critique

The SPM is a very useful nonverbal measure of a person's ability to form perceptual relations and reason by analogy. The nature of the items and the flexibility in administration make it useful for testing individuals with communication disorders or physical disabilities. The stability and internal consistency reliabilities of the SPM have been adequately established.

Although not a strictly "pure" measure of Spearman's *g*, it comes as close as many consider possible. The construct validity of the test has been established in Western societies, where any psychologist or educator wishing to get a measure of intellectual ability should give the SPM serious consideration. In particualr, those working with non-native speakers of English, the deaf, or the physically handicapped should consider the SPM.

In non-Western cultures the factor loadings and validity coefficients are less stable; however, sufficient research data are available to enable users to decide its appropriateness cross-culturally. Although in its present form the SPM is not useful diagnostically, the items lend themselves to meaningful task and error analyses. The fact that the SPM has not proven to be a good predictor of practical criteria may not be so much a problem with the test as a problem with the manner in which it has been used. The author's recommendation that for prediction purposes the test be used in conjunction with a measure of acquired knowledge has largely been ignored, particularly in the United States.

Notes to the *Compendium:*

[1]Now three research supplements to the manual are available. The third is a compendium of normative data for the RPM and the MHS in the United States (Raven, 1987).

[2]U. S. norms are now available in the previously mentioned 1987 research supplement to the manual.

References

Banks, C., & Sinha, U. (1951). An item analysis of the Progressive Matrices test. *British Journal of Psychology (Stat. sect.), 4,* 91-94.

Bereiter, C., & Scardamalia, M. (1979). Pascual-Leone's M construct as a link between cognitive-developmental and psychometric concepts of intelligence. *Intelligence, 3,* 41-63.

Burke, H.R. (1958). Raven's Progressive Matrices: A review and critical evaluation. *Journal of Genetic Psychology, 93,* 199-228.

Burke, H.R. (1972). Raven's Progressive Matrices: Validity, reliability and norms. *Journal of Psychology, 82,* 253-257.

Court, J.H. (1983). Sex differences in performance on Raven's Progressive Matrices: A review. *Alberta Journal of Educational Research, 29,* 54-74.

Court, J.H., & Raven, J. (1982). *Manual for Raven's Progressive Matrices and Vocabulary Scales* (Research Supplement No. 2). London: H.K. Lewis.

Guinaugh, B.J. (1971). An experimental study of basic learning ability and intelligence in low socioeconomic-status children. *Child Development, 42,* 27-36.

Hunt, E. (1974). Quote the Raven? Nevermore. In L.W. Gregg (Ed.), *Knowledge and cognition.* Potomac, MD: Lawrence Erlbaum.

Irvine, S.H. (1966). Towards a rationale for testing attainments and abilities in Africa. *British Journal of Educational Psychology, 36,* 24-32.

Jensen, A.R. (1980). *Bias in mental testing,* New York: The Free Press.

Keir, G. (1949). The Progressive Matrices as applied to school children. *British Journal of Psychology (Stat. sect.), 2,* 140-150.

Kirby, J.R., & Lawson, M.J. (1983). Effects of strategy training on Progressive Matrices performance. *Contemporary Educational Psychology, 8,* 127-140.

Penrose, L.S., & Raven, J.C. (1936). A new series of perceptual tests: Preliminary communication. *British Journal of Medical Psychology, 16,* 97-104.

Raven, J. (1981). *Manual for Raven's Progressive Matrices and Mill Hill Vocabulary Scales* (Research Supplement No. 1). London: H.K. Lewis.

Raven, J. (1987). *A compendium of American normative data for Raven's Progressive Matrices and Mill Hill Vocabulary Scales* (Research Supplement No. 3). San Antonio: The Psychological Corporation.

Raven, J.C., Court, J.H., & Raven, J. (1978). *Manual for Raven's Progressive Matrices and Vocabulary Scales.* London: H.K. Lewis.

Raven, J.C., Court, J.H., & Raven, J. (1983). *Manual for Raven's Progressive Matrices and Vocabulary Scales* (Section 3). London: H.K. Lewis.

Sinha, U. (1968). The use of Raven's Progressive Matrices test in India. *Indian Education Review, 3,* 75-88.

Spearman, C. (1938). Measurement of intelligence. *Scientia, 64,* 75-82.

Vernon, P.E. (1947). The variations of intelligence with occupation, age, and locality. *British Journal of Psychology (Stat. sect.), 1,* 52-63.

Vernon, P.E. (1949). Recent developments in the measurement of intelligence and special abilities. *British Medical Bulletin, 6,* 21-23.

Raymond H. Holden, Ed.D.

Professor of Psychology, Rhode Island College, Providence, Rhode Island.

STANFORD-BINET INTELLIGENCE SCALE: FORM L-M

Lewis M. Terman and Maud A. Merrill. Chicago, Illinois: The Riverside Publishing Company.

THE ORIGINAL OF THIS REVIEW WAS PUBLISHED IN TEST CRITIQUES: VOLUME I (1984).

Introduction

The third revision of the Stanford-Binet Intelligence Scale by Lewis M. Terman and Maud A. Merrill was published in 1960 by the Houghton Mifflin Company. It is a wide-range individual test, assessing intelligence from age two through the superior adult level. It is an age scale, requiring subjects to solve problems, give definitions, memorize new material, and use some visual-motor skills at various age levels.

The history of the Binet Intelligence Scales is almost as long as the history of mental measurement itself. Alfred Binet (1857-1911), working with Theophilus Simon (one of the few historical examples of a psychiatrist working as secondary collaborator to a psychologist), published the first practical measure of intelligence in 1905 in Paris, France. Binet's test was designed so that children of approximately average ability could solve about 50% of the problems at their own age level and most of the tasks at lower age levels. A "mental age" was then assigned: a ten-year-old who responded like an average eight-year-old achieved a mental age of eight.

In 1916 Lewis Terman (1877-1956) produced the first widely accepted American version of the Binet Scale at Stanford University. Terman adopted the "intelligence quotient" or IQ (mental age/chronological age x 100) from Wilhelm Stern, a German behavioral scientist. The test items used were of the same type used by Binet, but increased from 54 to 90 items. The items were varied and heterogeneous to test "general intelligence," conceived of as the sum of the thought processes involved in mental adaptation.

The second Stanford revision, the most extensive and comprehensive thus far, was published in 1937 (Terman and Merrill). Based on years of experience with the first revision, research, and additional standardization, it provided a wider sampling of the same kinds of mental abilities and a measure of general intelligence.

There evolved two comparable forms, Form L and Form M, with 129 subtests each. These test items included analogies, opposites, comprehension, vocabulary, similarities and differences, verbal and pictorial completions, absurdities, reproducing designs, and memory for meaningful memory and unmeaningful

466

digits. Verbal tasks were far more heavily weighted than performance tasks, the latter often being eliminated because they contributed little or nothing to the total score. After all, the Binet is a verbal test.

The third (1960) revision combined the most discriminating test items from Forms L and M into a single instrument, Form L-M. The authors justified the elimination of a second form because of less need (following the publication of the Wechsler Intelligence Scale for Children in 1949) and the possibility of adding an alternate subtest at each age level. Changes in the revision included 1) the relocation of test items that were found to have changed significantly in ease or difficulty since the 1937 editions (e.g., the vocabulary words "Mars" and "juggler" were found to be considerably easier than in the 1937 test); 2) clarification of directions for administration and scoring; 3) substitution of some items not suitable because of cultural changes; and 4) revision and extension of IQ tables so that standard score IQs are comparable at all age levels.

The 1960 Stanford-Binet Form L-M test material comes in a durable carrying case measuring 11" x 15" x 4". Included are the manual, large printed card material, small printed card material, three-hole formboard, blocks, beads, small boxes, and small toy test items such as a doll, car, railroad engine, scissors, spoon, fork, dog, ball, chain, cup, and flag. There are two record forms available: the record booklet, which has much more space for examiner's notes and comments, and an abbreviated record form for scoring. As stated previously, the test items range from age II to Superior Adult III level. There are six items at each half-year age level from II to V, six items at each one-year age level from VI to XIV, eight items at the Average Adult level, and six items each at Superior Adult I, II, and III.

Some sample items include identifying six parts of the human body on a picture doll (age II-6); identifying 14 items by name (Picture Vocabulary (age IV); copying a diamond (age VII): making rhymes (age IX); repeating complex sentences (age XII); defining abstract words (Average Adult); providing the meaning of proverbs (Superior Adult II).

Practical Applications/Uses

At present, the most suitable use for this test is with young children between the ages of two and five to determine their intellectual or developmental level, or with those children suspected of developmental deviation below or above average. It is not very useful for school-age children, normal adolescents, or adults, since the Wechsler scales take priority. The use of this test in institutions to determine the precise level of retardation in adults is more or less an academic exercise. Subjects of any age with language impairment or cultural diversity may be unduly and unfairly handicapped. However, with clinical judgment an examiner can administer selected items in pantomime and the child can point to indicate answers. Two adaptations for the visually impaired, the Perkins-Binet and the Hayes-Binet Scale, have also been published.

The Stanford-Binet is designed to be administered individually in a quiet, secluded room by a school or clinical psychologist who has had specific training and experience in giving this test. Rapport, as usual, is expected to be developed appropriately before testing begins and maintained throughout the test session. The examiner must explicitly follow the printed directions for scores to be valid, but must also be empathetically sensitive to the needs of the subject. The authors discourage serial testing (the presentation of all the subtests of one type of item consecutively, such as memory for sentences or digits, or abstract words), as well as "adaptive testing" (the alternation of hard and easy items successively). Nevertheless, they do not present adequate research data to discourage these procedures (Terman & Merrill, 1973, p. 48). The time for administration can vary from 30 to 90 minutes, depending on the successes and failures of the subject.

The manual is very clear in its instructions for scoring. All answers are scored " + " or "−." However, on many items marginal answers may be obtained. The manual provides many actual expamples of plus, minus, and marginal responses to facilitate scoring. The examiner may query marginal or "Q" responses to determine if they may be changed to a " + " score; otherwise, they become a "−."

A "basal age" is established when the subject has passed all the items at a given age level (e.g., 3 years, 6 months) and a "ceiling" is established when the subject fails all test items at a higher mental age level (e.g., six years). The examiner must establish the basal age of the child and continue testing until a ceiling is reached. The test is then completed and the examiner scores it by hand. No machine or computer scoring is available.

Interpretation is based primarily on objective scoring, and not necessarily the clinician's judgment. The IQ distributions for the 1937 standardization group, carried over to the 1960 revision, are given in Table 1.

Table 1

IQ Classifications for the 1937 Stanford-Binet Intelligence Scale
and Percent of Subjects Represented

IQ	Percent	Classification
140-169	1.3	Very superior
120-139	11.3	Superior
110-119	18.1	High average
90-109	46.5	Average (normal)
80-89	14.5	Low average
70-79	5.6	Borderline
30-69	2.6	Mentally defective

Technical Aspects

Reliability of the scale varies at different ages and different IQ ranges. From age 2½ to 5½, the reliability coefficients range from .83 for IQs 140-149 to .91 for IQs

60-69. For ages 6 to 13, the coefficients are .91 to .97 respectively, and for ages 14-18 the coefficients range from .95 to .98 respectively.

Validity of the scale depends on three sources: 1) the choice of items according to mental age on the 1937 scale assures that the new scale is measuring the same thing that was measured in the original 1916 scale; 2) regular increases in mental age from one age to another agreed with increases in percent passing from one chronological age to the next in both forms of the 1937 scale; and 3) biserial correlations were computed for each item of Forms L and M of the 1937 scale. The retention of an item for the 1960 scale partly depended on its correlation with the total score. The mean biserial correlation for the 1960 scale is .66 (range from .61 at young age levels to .73 for the adult levels).

Critique

In perusing the reviews of the Stanford-Binet Scale in subsequent editions of the *Mental Measurements Yearbooks* (Buros, 1953, 1959, 1965, 1972), it is interesting to observe the reviews becoming increasingly more negative over the years. In the fifth edition (Buros, 1959) Mary Haworth states: "In spite of its old fashioned tinge . . . the Stanford-Binet continues to serve as the standard and generally accepted criterion against which other tests are validated" (p. 595). Norman Sundberg in the same edition reports: "The grand old test has had to bow to an upstart (Wechsler Intelligence Scale for Children)" (p. 596). In the sixth edition (Buros, 1965), Elizabeth Fraser notes that despite some minimal changes in the 1960 version, and a criticism that changes were made with no restandardization, "the Stanford-Binet remains very much its old self." In the seventh edition (Buros, 1972, p. 772-773), David Friedes has this to say:

My comments . . . are not very different from those made by F.L. Wells 32 years ago in the *1938 Mental Measurements Yearbook*. The Binet scales have been around for a long time and their faults are well known . . .

The *Stanford-Binet Intelligence Scale* is an old, old vehicle. It has led a distinguished life as a pioneer in the bootstrap operation that is the assessment enterprise. Its time is just about over. *Requiescat in pace.*

In the eighth edition (Buros, 1978), no new review of the Stanford-Binet is presented.

The greatest criticism of the 1960 revision lies in the lack of restandardization of the new scale. Over 52% of the subjects used in comparing scores in the early 1950s with the 1937 Binet scores of subjects in the 1930s came from California, mainly near the area of Stanford University. Only 18% came from either New Jersey or Minnesota; approximately 2% each from Iowa, New York, and Massachusetts. There was no representative sampling by socioeconomic status throughout the U.S. (Terman & Merrill, 1973, p. 21), and no children with any psychological referral problems were included. Only white, native-born American children were sampled; no mention is made of any other ethnic backgrounds or any minority groups.

The Riverside Publishing Company has just announced a new fourth edition of the Stanford-Binet to be published in late 1985, authored by Elizabeth Hagan, Jerome Sattler, and Robert L. Thorndike (Riverside Publishing Co., 1984). For the first time, separate scores will be provided for verbal reasoning, visual-spatial reasoning, quantitative reasoning, and short-term memory, with a resulting composite general ability factor considered representative of general intelligence. Time will tell if a truly representative standardization will be accomplished.

Let us hope that with the publication of the fourth edition of the Stanford-Binet Intelligence Scale, we may scotch the adage that "You can't teach an old dog new tricks."

References

Buros, O.K. (1953). *The fourth mental measurements yearbook.* Highland Park, NJ: The Gryphon Press.

Buros, O.K. (1959). *The fifth mental measurements yearbook.* Highland Park, NJ: The Gryphon Press.

Buros, O.K. (1965). *The sixth mental measurements yearbook.* Highland Park, NJ: The Gryphon Press.

Buros, O.K. (1972). *The seventh mental measurements yearbook.* Highland Park, NJ: The Gryphon Press.

Buros, O.K. (1975). *Intelligence tests and reviews.* Highland Park, NJ: The Gryphon Press.

Terman, L.M. (1916). *The measurement of intelligence.* Boston: Houghton Mifflin Co.

Terman, L.M., & Merrill, M. (1937). *Measuring intelligence.* Boston: Houghton Mifflin Co.

Terman, L.M., & Merrill, M. (1937). *Stanford-Binet Intelligence Scale: Manual for the third revision.* Boston: Houghton Mifflin Co.

Jean Spruill, Ph.D.

Associate Professor and Director, Psychological Clinic, The University of Alabama, Tuscaloosa, Alabama.

STANFORD-BINET INTELLIGENCE SCALE, FOURTH EDITION

Robert L. Thorndike, Elizabeth P. Hagen, and Jerome M. Sattler. Chicago, Illinois: The Riverside Publishing Company.

THE ORIGINAL OF THIS REVIEW WAS PUBLISHED IN TEST CRITIQUES: VOLUME VI (1987).

Introduction

The Stanford-Binet Intelligence Scale: Fourth Edition (Stanford-Binet/Fourth Edition) is an individually administered intelligence test based on a three-level hierarchical model of cognitive abilities. Taking into consideration the ways that clinicians and educators have used previous editions of the Stanford-Binet, the developers of the hierarchical model were influenced by current theories and research in cognitive psychology. Figure 1 illustrates the model and the tests that measure each of its factors. At the top of the model is *g*, a *general reasoning* factor used by individuals to solve problems they have not been taught to solve. The next level divides *g* into three broad factors: Crystallized Abilities, Fluid-Analytic Abilities, and Short-Term Memory. The Crystallized Abilities factor represents the cognitive factors necessary to acquire and use information to deal with verbal and quantitative concepts in order to solve problems. Highly influenced by education, crystallized abilities also represent more general verbal and quantitative problem-solving skills acquired through a variety of learning experiences, both formal (school) and informal. At the third level of the model, Crystallized Abilities are further divided into verbal and quantitative reasoning. The fluid-analytic abilities factor represents the cognitive skills necessary to solve new problems involving nonverbal or figural stimuli. General life experiences are considered more important than formal education in the development of these abilities. Fluid-analytic abilities are measured by abstract/visual reasoning.

The third factor at the second level of the cognitive abilities model, Short-Term Memory, is included because of its positive relationship to more complex tasks of cognitive performance. This is a measure of the individual's ability to retain information until it can be stored in long-term memory and to hold information drawn from long-term memory so that the individual may use it for solving problems. Short-Term memory is not further differentiated at the third level of the model.

The term *intelligence* has been replaced by *cognitive development*. The terms *intelligence, IQ* and *mental age* are not used in reference to the Fourth Edition anywhere in the administration or technical manual. Instead of IQ, the term *Standard Age Score* (SAS) is used. The Fourth Edition is intended to provide a clinically useful profile of an individual's cognitive abilities in addition to the overall level of cognitive development. Another goal was to de-emphasize verbal skills from previous editions. To accomplish these goals, a wide variety of item types (tests) were used

Figure 1

Model of the theoretical structure of the Stanford-Binet Intelligence Scale, Fourth Edition

$$g$$
General Reasoning

CRYSTALLIZED ABILITIES		FLUID-ANALYTIC ABILITIES	SHORT-TERM MEMORY
Verbal Reasoning	*Quantitative Reasoning*	*Abstract/Visual Reasoning*	———
Vocabulary	Quantitative	Pattern Analysis	Bead Memory
Comprehension	Number Series	Copying	Memory for Sentences
Absurdities	Equation Building	Matrices	Memory for Digits
Verbal Relations		Paper Folding & Cutting	Memory for Objects

to assess each area of cognitive abilities in the theoretical model. Although the Fourth Edition is a revision of earlier editions of the test, "it is a thoroughly modern test in terms of content, scales, and testing procedures" (Thorndike, Hagen, & Sattler, 1986b, p.3).

Extensive reviews of the Binet scales can be found in the various editions of the *Mental Measurements Yearbook* (e.g., Buros, 1972), and only a brief sketch of their voluminous history is given here.

The history of modern intelligence testing began with the publication of the first Binet test in 1905 in Paris, France. Alfred Binet, collaborating with Theophilus Simon, devised 30 objective tests with the goal of differentiating mentally retarded from normal children. Binet revised his test twice before his death in 1911. Refinements of the test included grouping the tests into age levels, introducing the concept of mental age, eliminating tests that were heavily dependent upon reading and writing, and including adults in the standardization sample.

The Binet-Simon scales received wide acceptance in the United States. In 1916, Lewis Terman of Stanford University published the American version of the test (Terman, 1916). Major changes included the use of the intelligence quotient (IQ), adopted from Wilhelm Stern's work, and perhaps the first attempt at a truly representative sample for standardization. The 1916 edition was called the Stanford Revision and Extension of the Binet-Simon Intelligence Scale. Although inadequate by modern standards, the 1916 version was standardized on 1,000 children and 400 adults, representing a very large sample for that time.

The 1937 revision by Terman and his colleague at Stanford, Maude Merrill, resulted in two forms of the test, Form L and Form M, and improved standardization. There were still problems with the test, namely, in the scoring and difficulty level of items, lack of adequate discrimination at the upper end of the intelligence distribution, a heavy emphasis on verbal and rote memory items, and, although improved over prior versions, still inadequate standardization.

A third revision of the Stanford-Binet was published in 1960. Although Terman's name appears on the revision, it was carried out primarily by Maude Merrill, because Dr. Terman died in 1956. The best subtests from the L and M Forms of the 1937 scale were incorporated into a single scale; some tests were relocated, dropped, or rescored; items were updated; a new group of children was used to check changes in test difficulty; scoring principles were clarified; and age 18 rather than 16 was used as a ceiling level. Also, the Deviation IQ was substituted for the ratio IQ (Terman & Merrill, 1960). Although the revision was an improvement over earlier versions, many of the criticisms of the 1937 edition remained true of the 1960 revision.

In 1971, Robert Thorndike undertook a revision of the norms of the 1960 Form L-M. Unlike the earlier versions, which sampled only white, English-speaking individuals, the 1972 norms included black, Mexican-American, and Puerto Rican-American English-speaking subjects (Terman & Merrill, 1973). Insufficient information is provided in the publication of the 1972 norms to adequately assess the representativeness of the normative sample.

The 1972 revision was only an updating of the test norms; prior to this Fourth Edition, the test itself had not been revised since 1960. The present revision of the test was undertaken because of the vast social and cultural changes in the United States that have taken place since 1960. "These changes, together with new research in cognitive psychology," led Thorndike, Hagen, and Sattler (1986c, p. 8) to revise Form L-M. These authors of the Fourth Edition are all well-known psychologists with a great deal of expertise in the areas of intelligence and test development.

Although the Fourth Edition is markedly different from Form L-M, it has retained some of its predecessor's features. Cognitive abilities related to progress in school continue to be a major factor. The Fourth Edition covers the same age range, requires the examiner to establish a basal and ceiling for each subject, and includes a few of the items from Form L-M. However, these features seem to be the only similarities between the two tests. In keeping with the development of other individual tests of intelligence (e.g., the Wechsler scales and the Kaufman Assessment Battery for Children), the Fourth Edition has a point scale format rather than an age scale format and is devised so that an individual's pattern of abilities, as well as overall cognitive development, can be assessed. A wide variety of item types (tests) are used to measure each area of cognitive ability.

To be included in the Fourth Edition, each item type had to be 1) a measure of verbal, abstract/visual, or quantitative reasoning, or short-term memory, 2) reliably scored, 3) relatively free of ethnic or gender bias, and 4) able to function over a relatively wide age range. Many new item types were constructed, and in 1979 the authors began preliminary tryouts of the items. After the tryouts, the item types were reduced to 23 for possible inclusion in the final test. Items were arranged according to difficulty level within each type, and field-test booklets were prepared. Because 23 tests were too many to give to any one person, three forms were prepared. Each form (A, B, and C) had at least two tests that were designed to measure each of the four cognitive areas being tested.

Initial testing was done on a small group of subjects during 1981-82. Based on the results of the field testing, additional revisions were made, and a second field test was carried out on a larger sample during late 1983 and early 1984. After the second field trial, the additional revisions in administration and scoring took place, and the final form of the test was developed. At each stage of development, examiners were asked to critique the items, and standard item analyses and Rasch analyses were run on each item. A panel of judges made decisions about any potential ethnic and gender biases of items. In the final test, 15 item types were retained. The field tryouts and development of the test appear to have been done very well.

Standardization was carried out from January through July, 1985. The standardization sample of 5,013 people was stratified according to six variables: geographic region, community size, race/ethnic group, gender, parental occupation, and parental education. School districts that met the stratification variables were selected throughout the country. Children in representative classes were sent home with parental permission slips for testing. Each examiner was responsible for locating children (from those whose parents gave permission) who met certain specifications of the sampling design. The technical manual (Thorndike et al., 1986c) describes how examiners were selected and trained and gives examples of how specific subjects were selected for testing. Under-school-aged subjects were children enrolled in some type of preschool or day care or were siblings of school-aged children tested. Adult subjects aged 18 to 23 were siblings of school-aged children. The standardization sample originally included a group of subjects 24 to 32 years of age but did not include them in the normative data because they were not a representative sample.

The standardization sample was roughly representative of the 1980 U.S. census population for all variables except parental occupation and parental education. The sample was overrepresented for the college-graduate-or-beyond education group and for the managerial/professional occupation group. This was taken into consideration and the sampling bias presumably reduced by weighting the data in the development of the scale scores. The weighting of scores assumed the differences among the subjects were only quantitative, not qualitative, and this is a questionable assumption. The weighting procedure was described only in general terms. The authors state: "Each child from an advantaged background was counted as only a fraction of a case (as little as 0.28), while each child from a less advantaged background was counted as more than one case" (Thorndike et al., 1986c, p. 25). What is meant by "advantaged" and "disadvantaged" backgrounds is not explained, nor is the weighting procedure further described.

The Fourth Edition comes in a sturdy carrying case that contains the test materials, the administration and scoring manual, and the technical manual. The record book must be ordered separately. Most of the test items are contained in four item books arranged in an easel-kit format that allows the examiner to see the test directions and scoring key at the same time the child sees the items. In addition to the item books, there are four boxes containing test materials. Two boxes contain blocks; in one are the nine green blocks used in the copying test and in the other are black and white cubes with varying designs to be used in the Pattern Analysis test. The third box contains beads used in the Bead Memory test, and the last box contains dice used in the Quantitative test. Additional materials are a modern picture of a child and a new version of the Form Board.

For each of the 15 tests of the Fourth Edition, the items are arranged in levels of increasing difficulty designated by the letters A through Y. Each level has two items of approximately equal difficulty. For each test, the examiner must establish a basal and a ceiling age, and the administration manual gives detailed instructions on doing so (Thorndike, Hagen, & Sattler, 1986b, p. 12).

Testing with the Fourth Edition uses a multistage format. The Vocabulary test, administered first, is a routing test to determine the entry level for all other tests. Entry level for the Vocabulary test is determined by the examinee's chronological age. Using the results of the Vocabulary test, the examiner determines the entry level for the remaining tests by using the "Entry-Level Chart" on the back cover of the record booklet. The rows of the chart are the chronological ages, and the columns are the highest pair of vocabulary items administered. The intersection of the appropriate row and column determines the entry level for the remaining tests.

Test 1: Vocabulary. The Vocabulary test contains 46 items; the first 14 are pictures of common objects, and the remaining 32 are words the examinee is asked to define. This test is administered to all examinees.

Test 2: Bead Memory. The examinee is shown a pattern of beads for 5 seconds and is then asked to reproduce the pattern from memory by using beads of three colors and four shapes. Bead Memory is administered for all entry levels; the sample items and tasks required differ depending upon entry level. For example, entry levels A-G (items 1-10) merely require the examinee to point to the beads they have been shown. Entry levels H-Q (items 11-26) require the examinee to copy a pattern of beads.

Test 3: Quantitative. The Quantitative test requires examinees to place counting blocks (dice) in the counting-blocks tray to match, count, add, subtract, or form logical series of numbers for items 1-12. For items 13-40, the examinee is asked to solve quantitative problems presented either visually or orally. The Quantitative test is administered at all entry levels, and examinees may use paper and pencil in answering. There are no sample items.

Test 4: Memory for Sentences. In the Memory for Sentences test, the examinee repeats sentences after the examiner. Sample items are given, and the test is administered to all entry levels. The examiner is instructed to read at a steady pace and to drop his or her voice at the end of each item.

Test 5: Pattern Analysis. This is the only test that is timed; it is administered to all entry levels and has sample items except for Levels A-C. The first six items use the Form Board; items 7-42 use the black and white cubes. For some items, the examiner uses the cubes to construct a model that the examinee then copies. For other items, the examinee is shown a picture that he or she is expected to copy.

Test 6: Comprehension. The first six items require the examinee to point to various body parts on a card showing a picture of a child. Items 7-42 require a verbal response to questions about common events (e.g., "Where do people buy books?"). Comprehension is administered at all entry levels; there are no sample items.

Test 7: Absurdities. The Absurdities test is administered only for examinees with entry levels A-L. The examinee is asked to tell what is wrong with absurd pictures; there are examples for entry levels A-B.

Test 8: Memory for Digits. The examinee is read a series of digits and asked to repeat them, first forwards, then backwards. Digits are read at the rate of one per second, and the examiner's voice is dropped at the end of each item. This test is administered only to examinees with entry levels I or above.

Test 9: Copying. This test is given only for entry levels A-J; no sample items are administered. For items 1-12, the examinee is asked to duplicate the examiner's design made from blocks. Items 13-28 require the examinee to draw the designs shown. Erasing is allowed.

Test 10: Memory for Objects. This test is administered to examinees with an entry level of I or above. Examinees are shown pictures of one or more objects and then asked to identify the objects in the correct order of their appearance. Both order and choice of object must be correct. Objects are shown at the rate of one per second, and sample items are given.

Test 11: Matrices. In this test, the examinee is asked to fill in the missing object in a series, first using a multiple choice format (items 1-22) and then using a written format (items 23-26). The examinee must figure out the rule or pattern to determine the correct response.

Test 12: Number Series. The examinee is shown a series of numbers that are arranged according to a certain rule. The examinee is asked to supply the next two numbers according to the rule. This test is given only to examinees with entry levels of I or above. The examinee may use paper and pencil, and both numbers must be correct to receive credit. Although the test is not timed, a limit of 2 minutes per item is suggested.

Test 13: Paper Folding and Cutting. This test is administered only to examinees with entry levels M or above. For the sample items, the examiner folds and cuts pieces of paper, and the examinee, using a multiple-choice format, picks the picture that demonstrates how the paper would look if unfolded. For the test items, the examinee looks at a series of pictures that illustrate a piece of paper being folded and cut, then picks the picture that shows how it would look when unfolded.

Test 14: Verbal Relations. This test is administered only to examinees with entry levels M or above. The examiner names four things, and the examinee tells how the first three things are alike and how they are different from the fourth.

Test 15: Equation Building. In this test, given to entry levels M or above, the examinee is given several numbers and operational symbols (+, -, =, etc.) to use in building a meaningful mathematical relationship. This test is not timed, but a limit of 2 minutes per item is suggested. The examinee may use paper and pencil in figuring out the answers.

No more than 13 of the above tests are given to any one examinee. The complete battery ranges from 8 to 13 tests, depending upon the age of the examinee and the entry level established by the Vocabulary routing test. There are several abbreviated versions recommended by the authors, and these are detailed in the administration manual (Thordike et al., 1986b, pp 34-36).

The record booklet is 40 pages long and at first glance appears quite intimidating. The first page contains information about the examinee and his or her scores on the test. The administration manual gives clear guidelines for completing the front page and obtaining the Standard Age Scores. Five Standard Age Scores are obtained, one for each of the four areas measured (Verbal Reasoning, Abstract/ Visual Reasoning, Quantitative Reasoning, and Short-Term Memory) and a Test Composite. The Composite SAS is a deviation score; it appears to be simply another name for the Deviation IQ score in the previous editions of the Stanford-Binet.

The second page contains a profile analysis, which is used to determine the strengths and weaknesses of the examinee's cognitive abilities. The last page in the booklet is the Entry-Level Chart. Pages 3-39 are for recording verbatim the examinee's answers in the spaces provided for each test. Each test has the entry level items marked, and the sample items, if any, that accompany each entry level are designated. The raw score for a test is the item number of the highest item administered minus the number of items that were failed.

Practical Applications/Uses

The Stanford-Binet Intelligence Scale, Fourth Edition was constructed as a measure of the cognitive abilities of individuals from 2 to 23 years of age. According to the authors, additional purposes of the test are 1) to differentiate between mentally retarded and learning disabled students, 2) to aid in understanding why a particular student is having difficulty in learning in school, and 3) to identify gifted students.

Although the Fourth Edition covers the age range from 2 to adult, the use of the term *adult* is misleading. Normative data are available only for individuals 18 to 23 years of age, limiting the usefulness of the Fourth Edition for adults. It is most appropriate for the purposes given previously, which relate primarily to school-aged children.

At the present time it is not clear if the Fourth Edition will be useful in institutions for the mentally retarded. The lowest SAS that can be obtained for any area or composite is 36, making the test inappropriate for individuals classified in the severe or profound range of retardation.

The Fourth Edition is designed to be individually administered by an experienced examiner in a setting that is free from distractions. This test may be used by several professional groups, among them school psychologists, educational specialists, and clinical psychologists. To administer and interpret the test, graduate-level training in test administration in general and in the Fourth Edition in particular is necessary. The administration manual gives many helpful suggestions, such as how to 1) tailor tests to specific individuals, 2) develop and maintain rapport, and 3) test preschool children.

One of the major criticisms of Form L-M was the difficulty in administration. Unfortunately, difficulty in administration also is likely to be a criticism of the Fourth Edition. Much practice with the Fourth Edition is needed to achieve a smooth transition from test to test.

Instructions for administration appear in the item books and, in general, are clear and easy to follow. Finding the correct starting place is not always easy; the tabs in the item books are not labeled. The administration guide recommends administering the tests in a certain order, changing the order only if circumstances dictate. Although the order of administration of the tests may be changed if necessary, the order of items within each test should never be changed.

Whereas most administration instructions are clear and easy to follow, there is a problem in the administration of the Bead Memory test. In Bead Memory, items 1-10 involve one type of response and items 11-42 another. Individuals starting with item 11 (or above) are given sample items. The problem arises when an examinee starts the test below item 11 and continues beyond it. The directions do not specify the administration of a sample item or an explanation that the task has now changed. In fact, the directions explicitly state that sample item 1 is given only to those examinees starting the test at entry levels H-K (Item Book 2, p. 15). Without an explanation of the task and a sample item, examinees starting below entry level H will have difficulty understanding the changed nature of the task. Apparently, at workshops on administering the Fourth Edition, examiners are being told to give Sample 1 to all examinees who are administered item 11, regardless of their entry Level. Surely, later printings of Item Book 2 (Bead Memory) will correct this problem.

The administration of the Fourth Edition is lengthy. To keep the testing time to 60 to 90 minutes, the authors recommend that the examiner use one of the batteries of tests suggested for different age groups; however, the most reliable and valid measure of intellectual functioning is obtained with the complete battery. Thus, for some individuals, testing may have to be split over two sessions. Other reasons for not administering an abbreviated battery are given below.

Instructions for scoring the responses are contained in the item books, generally on the same page as the instructions for administering the items. Only 5 of the 15 tests require using an expanded scoring guide from the administration manual. Few problems with scoring items are noted; one exception is the scoring for the Pattern Analysis test. In this test, the examiner shows the subject a picture of a design to copy. The examiner also sees the design that the subject is supposed to make. However, for items numbered 26, 28, 29, 30, Sample 5, 31, 32, 33, 34, 35, 36, 37, and 42, the pattern seen by the examiner as the "supposedly" correct answer is wrong—typically, the patterns are rotated 180x either vertically or horizontally. Presumably, this error will be corrected in future printings of Item Book 1. The record booklet also shows the pattern for each item, and these illustrations are accurate, so the examiner can use the record book to score the item.

Potential problems also exist in scoring the Equation Building test. Although the scoring key in the item book lists many correct solutions, there are other correct solutions not listed. An examiner who forgets that there are other correct answers, or whose mathematical skills are limited, may incorrectly score some answers.

Using the tables in the administration manual, raw scores are transferred to the front page and converted to Standard Age Scores. Although scoring of individual items is easy, obtaining test raw scores, test SASs, Area SASs, and the Composite SAS is more complex; therefore, to ensure accuracy, scoring should be gone over at least twice.

All test raw scores are converted to SASs with a mean of 50 and a standard deviation of 8. The Area and Composite SASs have a mean of 100 and a standard deviation of 16. The standard deviation was kept at 16 to make comparisons with Form L-M scores easier and to maintain historical continuity with Form L-M. The technical manual also has a table converting the scores to a distribution with a mean of 100 and a standard deviation of 15 in order to facilitate comparisons with scores from tests with a mean of 100 and a standard deviation of 15.

For older and/or very bright examinees, some tests have a ceiling that is much too low. For example, a perfect score on Copying could result in an SAS of 55 for a 12- to 13-year-old child. An opposite problem occurs with the mentally retarded subjects. In testing subjects ranging in ability from mild to moderate retardation, this reviewer has had difficulty in establishing a basal level for several tests. In addition, when starting the Bead Memory test at the level dictated by the Vocabulary routing test, this reviewer has had to go backwards an average of 5 levels to establish a basal.

Interpretation is based on objective scoring and clinical judgment. At the present time, the information provided for the interpretation of the Fourth Edition is inadequate. The *Expanded Guide for Interpreting and Reporting Fourth Edition Results* (Thorndike, Hagen, & Sattlerr, in press) has not been published, although it has been advertised since the publication of the test. There is very little information, other than technical data, in either the administration or technical manual to assist in interpreting the results. The technical manual does provide descriptive classifications for the Composite SAS. The terminology is similar to that associated with the IQ scores of Form L-M, except that the terms *Mentally Retarded* and *Slow Learner* have replaced *Mentally Defective* and *Borderline Defective,* respectively. In order to conform more closely with current classification practices, the score ranges for the categories were changed. In fact, when converted using a scale with a standard deviation of 15 (Table D.1 in the technical manual), the score ranges are the same as those used in the Wechsler scales and the K-ABC.

It has become increasingly common to report confidence intervals around standard (IQ) scores. Confidence intervals are not presented in the technical manual; however, if one knows how, they can be calculated easily from the standard errors of measurement. In addition to using confidence intervals around SAS scores, interpretation of test performance is aided by determining the examinee's strengths and weaknesses. Table F.2 in the technical manual (Thorndike et al., 1986c, p. 134) presents the differences between SASs for the areas and composite required for statistical significance at the 15% and 5% levels of confidence. However, these values were computed without taking into consideration the fact that multiple comparisons are being made. Although the problems involved in making multiple pairwise tests of significance have been recognized for years, it is only recently that test developers and clinicians have considered the problems caused by multiple comparisons when doing a profile analysis of test scores.

Sattler (1982) and Silverstein (1982), in their calculations of the differences required for significance among Wechsler subtests, used the Bonferroni *t* inequality to control for the increased error rate caused by multiple comparisons on the same data. Similarly, Naglieri (1982) corrected the differences required for significance for the McCarthy Scales using the Bonferroni *t*. Kaufman and Kaufman (1983), in their development of the K-ABC, made the appropriate corrections for their comparisons among test scores. The failure of the developers of the Fourth Edition to apply currently accepted statistical procedures in their analysis of differences represents a major error.

Technical Aspects

Reliability of the scores varies as a function of the different tests and the different age groups. For the most part, the reliability coefficients reported are based on Kuder-Richardson Formula 20 (KR-20) and are well within the typical range for individual tests of intelligence. However, as pointed out in the technical manual, to use the KR 20 formula it is necessary to assume that all items below the basal level were passed and all items above the ceiling level were failed. Because the assumption of failure above the ceiling level is not likely to be strictly met, the resulting correlation coefficients may be somewhat inflated and therefore should be considered as upper bound estimates of reliability.

As would be expected, the most reliable score, at all ages, is the Composite SAS; reliability coefficients ranged from .95 to .99. Reliability coefficients for the area scores depended upon the number of tests making up the score and ranged from .80 to .97. Not surprisingly, the more tests given in a particular area, the more reliable the Area SAS. The individual test reliabilities, with the exception of Memory for Objects, are in the .80s and .90s. Memory for Objects is also a short test (14 items), which contributes to its lower reliability. Reliability coefficients tend to be higher for older age groups.

Test-retest reliability was computed for two samples, ages 5 and 8. For most of the tests, the coefficients reported, although lower than the internal consistency measures based on KR-20, are in the acceptable range. Exceptions are the very low test-retest correlations of .28 for the Quantitative test, .46 for the Copying test, and .51 for the Quantitative Reasoning area. The authors speculate that a restricted range of scores in their small sample of 8-year-olds ($n=57$) resulted in the low retest correlations. However, the authors did not correct for restricted range, nor did they explain why the range of scores may have been restricted. It would have been helpful to see some test-retest data for adolescents and adults.

With the exceptions mentioned, the reliabilities reported in the technical manual (Thorndike et al., 1986c, pp. 38-49) compare favorably with those reported for the WISC-R, WAIS-R, K-ABC, and other individually administered tests.

Standard errors of measurement are reported for the various age groups for each test, each area score, and the composite score. Standard errors appear to be of reasonable magnitude given the reliability of the tests and considering that they are based on scales with a standard deviation of 8.

To understand what the Fourth Edition measures, the authors focused their research on three areas: 1) the internal structure of the test (i.e., factor analysis and intercorrelations among tests); 2) correlations between the Fourth Edition and other measures of intellectual ability; and 3) the performance of individuals identified by other measures as being high or low in intellectual ability.

A variant of confirmatory factor analysis was used to investigate the internal factor structure of the Fourth Edition. The results of the factor analysis were reported "to provide positive support for the rationale underlying the battery" (Thorndike et al., 1986c, p. 55). Unfortunately, the authors do not explain how this "variant of confirmatory factor analysis" was carried out, making it difficult to evaluate the accuracy of their statement. Except for the loadings on the g, or general factor, the factor loadings reported in the technical manual are smaller than desirable. Clearly, each test's highest loading is on the *g*, or general ability factor. Not surprisingly, Vocabulary had the highest loading on *g* (.76), whereas Memory for Objects had the lowest (.51). Some tests (e.g., Memory for Sentences) show loadings on factors other than those proposed by the model, and other tests fail to load on the group factors dictated by the model (e.g., Matrices and Copying). Test specificity for most tests is high, indicating the tests are strong measures in their own right.

When additional factor analyses were computed on various age groups, the model did not fare well. According to the theoretical model, four factors in addition to the *g* factor should occur. Separate analyses were conducted for three age groups. For the age group 2 through 6, only two factors in addition to *g* emerged: a Verbal factor and an Abstract/Visual factor. The Short-Term Memory and Quantitative factor did not appear. Because only one quantitative test is given at these ages, a common Quantitative factor was not expected. There are two memory tests given at ages 2 through 6; Bead Memory had a high loading on the Abstract/Visual factor, and Memory for Sentences had a high loading on the Verbal factor. The Quantitative test also loaded on the Abstract/Visual factor; in fact, its factor loading of .25 was higher than the factor loadings of two of the tests (Pattern Analysis and Copying) that supposedly are measuring abstract/visual reasoning.

Factor analysis for the age group 7 through 11 resulted in the identification of three factors: Verbal, Memory, and Abstract/Visual. A Quantitative factor did not occur, even though two quantitative tests were given in this age range. Only three of the four memory tests loaded on the Memory factor. Bead Memory had a factor loading of .05 on Memory, and Memory for Sentences had a loading of .19 on the Verbal factor.

The four factors predicted by the theoretical model did emerge for "age group 12 through 18-23." However, the Quantitative factor was not very strong; factor loadings were .10 for the Quantitative test, .20 for Number Series, and .40 for Equation Building. Matrices also had a weak loading on the Abstract/Visual factor, perhaps because it had such a high loading on *g*.

Some support for the theory underlying the construction of the Fourth Edition is found in a study by Keith and his colleagues at the Lindquist Center at the University of Iowa (Keith, Cool-Hauser, Novak, White, & Pottebaum, 1987). Keith et al. used a LISREL VI computer program to perform a first order, confirmatory factor analysis (CFA) using the entire standardization sample and the same three age groups used by the authors of the Fourth Edition in their factor-analytic studies. The CFA approach is somewhat different from programs typically used and is thought to provide a stronger test of the underlying structure of a test than does exploratory factor analysis. Keith et al. conducted a systematic series of analyses, checking the fit of various models to the proposed structure of the Fourth Edition. Based on their analyses, Keith et al. found some support for the four factors underlying the structure of the Fourth Edition. However, neither theirs nor Thorndike et al.'s factor-analytic procedure found a Memory factor for the ages 2 through 6 group.

The conclusion by Thorndike et al. (1986c) that the factor structure of the Fourth Edition both conformed to the theoretical framework used to construct the test and provided good support for the theoretical rationale underlying the test is questionable. Clearly, the results support the existence of a strong g component; the results are less supportive of the second level of the model in general, especially for the younger age groups.

The technical manual reports a number of studies conducted to obtain data on the validity of the Fourth Edition. Most of the subjects selected for the studies were not part of the standardization sample. Two kinds of samples were used for these studies: 1) groups of examinees designated by their schools or institutions as members of an exceptional group (gifted, mentally retarded, or learning disabled), and 2) nonexceptional samples consisting of groups of subjects that were in a regular educational setting.

When comparing the Fourth Edition Composite SAS with the Full Scale IQ scores of the Wechsler tests (WISC-R, WAIS-R, and WPPSI) and the Mental Processing Composite score of the K-ABC, the correlations ranged from .81 to .91, and the differences between the mean SASs and Full Scale IQ or Mental Processing Composite scores ranged from a low of .4 (K-ABC) to a high of 5.0 (WPPSI). These results were for the nonexceptional samples and clearly indicate that, for these groups, the Fourth Edition Composite SAS is equivalent to other test scores.

The authors had predicted certain patterns of correlations between the Fourth Edition SASs and the various standard scores of the other tests. The results were mixed. Overall, the studies were well done and the results generally supportive of this aspect of the validity of the Fourth Edition.

Using exceptional samples, that is, samples identified by their school system or institution as "gifted, learning disabled, or mentally retarded," the Fourth Edition was compared to the Form L-M, WISC-R, WAIS-R, and K-ABC. Two samples of gifted students were studied, one using the WISC-R and one using Form L-M. Using gifted students restricted the range of scores on the two tests, and hence reduced the possible correlation between the two. An additional factor that might reduce the correlation is the different emphasis on verbal skills for the two tests, particularly as gifted individuals are generally considered to be higher in verbal skills than in other areas. Nevertheless, the correlation of .27 between the Composite SAS of the Fourth Edition and the IQ score of Form L-M is quite low, particularly considering that the sample size was 82. The mean Composite SAS was 121.8 (SD = 9.0); the mean Form L-M IQ was 135.3 (SD = 9.7). The difference in overall scores is probably a function of the differing verbal emphasis between the two tests.

When comparing the Fourth Edition with the WISC-R in a sample of 19 gifted students, the mean Composite SAS was 116.3 (SD = 16.4), and the Mean Full Scale IQ was 117.7 (SD = 12.1). The correlation between the Composite SAS and the Full Scale IQ score was .69. It is interesting that the mean WISC-R IQ score for this "gifted" sample was 117.7, only slightly more than one standard deviation above the population mean IQ—hardly what is typically thought of as gifted.

Using learning disabled subjects to compare the Fourth Edition with the WISC-R, Form L-M, and the K-ABC, the authors found that the Fourth Edition Composite SAS scores are comparable to scores obtained on the other tests. The mean differences between the Fourth Edition SAS and the total standard scores for the other tests ranged from 1.7 to 3 points; the correlations between the Composite

SAS and the total standard scores for the other tests ranged from .66 to .87. Specific predictions concerning patterns of relationships between the various test scores were only partially supported; however, the overall results indicate that the Fourth Edition is comparable to other tests for this group of subjects.

Studies using subjects identified as mentally retarded showed similar results to the studies using learning disabled subjects. In general, when testing groups of retarded subjects, the Fourth Edition Composite SAS is very similar to the Form L-M IQ score (50.9 vs. 49.5 respectively) and the WISC-R Full Scale IQ score (66.2 vs. 67.0 respectively). When compared to the WAIS-R Full Scale IQ Score of 73.1, the mean Composite SAS was 63.8, which is almost 10 points lower. However, the difference between the WAIS-R and Fourth Edition scores is not surprising; the WAIS-R overestimates the IQ scores of the retarded (Spruill & Beck, in press), and the Fourth Edition has a slightly lower floor than the WAIS-R.

At present, there is little published information about the Fourth Edition in the professional literature, primarily because of the short time since its publication; however, there are many studies being done. Carvajal and Weyand (1986) compared the General Purpose Abbreviated Battery of the Fourth Edition with the WISC-R and found a correlation of .783 between the Composite SAS and the Full Scale IQ score. It is likely that the correlation would be even higher had they given the complete Fourth Edition. Carvajal and his colleagues (Carvajal, McVey, Sellers, Weyand, & McKnab, 1986) also compared the General Purpose Abbreviated Battery of the Fourth Edition to the Peabody Picture Vocabulary Test-Revised (PPVT-R) and found a correlation of .601; this is similar to correlations found between the PPVT and other individual measures of intelligence.

Several studies comparing the Fourth Edition and various other measures of intellectual ability were reported at the 1987 meeting of the National Association of School Psychologists. Livesay and Mealor (1987) presented the results of their study comparing the scores obtained on the Fourth Edition with Form L-M using a sample of subjects referred for evaluation for the gifted programs in their schools. Their results were similar to those found by the authors when studying gifted children; the IQ scores on the Form L-M were significantly higher than the Composite SAS (130 vs. 122), and the correlations between the two tests were all equal to or higher than those reported for similar studies in the technical manual. One reason for Livesay and Mealor's higher correlations could have been a broader range of scores in their sample. Because the subjects were referred for a determination of eligibility for a gifted program, they probably had a broader range of scores than would be found in a group of subjects identified as gifted.

Krohn and Lamp (1987), using a sample of low income preschool children, compared the K-ABC, Form L-M, and the Fourth Edition. There were no significant differences among the means of composite scores for the total group on all three measures. In general, the correlations found by Krohn and Lamp were slightly lower than those reported in the technical manual, probably because of the low socioeconomic status of the subjects.

Sims (1987), in her investigation of the relation between the Fourth Edition and the WISC-R, concluded that the Fourth Edition was a strong measure of general reasoning ability, providing a Composite SAS very similar to the WISC-R Full Scale IQ. The correlations between the two tests, except for Composite SAS and

Full Scale IQ, were smaller than those reported in the technical manual for similar studies.

In conclusion, there is a strong theory underlying the development of the Fourth Edition. The extent to which the test matches the theory is yet to be determined. The evidence presented thus far is inconclusive. In particular, the factor-analytic results are not adequately explained, and the factor structure of the theoretical model is not supported for all age groups. Correlational data between the Fourth Edition Composite SAS and the other major individual tests of cognitive ability is acceptable; however, specific predicted patterns of results between the various scores of the Fourth Edition and óther tests have not always been borne out. Clearly, validity of the Fourth Edition as a measure of *g*, or a general reasoning, cognitive ability factor, has been adequately demonstrated. At the present time, the validity of the other levels of the theoretical model is undetermined. Much more needs to be done in this area to ascertain what other factors are measured by the test.

Critique

During the last two decades, the Stanford-Binet Intelligence Scale, Form L-M has declined in popularity (Lubin, Larsen, & Matarazzo, 1984), and the reviews in the various editions of the *Mental Measurements Yearbook* (e.g., Buros, 1972) have become increasingly negative. The development of the Fourth Edition was clearly aimed at regaining the Stanford-Binet's prominence in the area of assessing cognitive abilities. The Fourth Edition is, in most respects, a completely new version of a very old test. Unfortunately, the test appears to have had significant problems from the start.

The publication of the Fourth Edition seemed somewhat rushed. In retrospect, it would have been better to have delayed the publication date. The test was published without accompanying technical data to allow the user to judge the appropriateness and technical adequacy of the instrument. This is a violation of Standard 5.1 in the *Standards for Educational and Psychological Testing* (American Educational Research Association, American Psychological Association, & National Council on Measurement in Education, 1985). Even more serious than the failure to provide technical information is the fact that the first printing of the *Guide for Administering and Scoring the Fourth Edition: Stanford-Binet Intelligence Scale* (Thorndike et al., 1986b) contained errors in the norms tables. The errors were corrected, and a new administration manual was sent to all known purchasers of the test. Hopefully, no examiners are still using the original administration manual. The *Expanded Guide for Interpreting and Reporting Fourth Edition Results* is not yet available. The test has been out for well over 1 year, and materials crucial to its use still are not available.

However, the problems with the Fourth Edition started earlier than the development of the manuals. One consistent criticism of the Binet test and its subsequent revisions has been inadequate standardization. This continues to be a problem. In the Fourth Edition, the standardization sample contained a larger percentage of high-socioeconomic-status subjects than in the population at large. The adequacy of the weighting procedure used to correct for sample bias in the data is not clear.

The Fourth Edition also does not include adults over age 23 in its normative data.

There are problems associated with the administration of the Bead Memory test and in the scoring of the Pattern Analysis test. Furthermore, the statistical data presented in the technical manual do not fully support the factor structure of the model underlying the construction of the test.

The National Association of School Psychologists did not endorse the use of the Fourth Edition when it was first published because of the lack of accompanying technical information. As of the time of this writing (June, 1987), some states have yet to approve the test for use in making decisions about placement in educational programs.

In spite of the preceding negative comments, the Fourth Edition is an exciting addition to the array of tests used to measure intellectual abilities. It has a strong theory based on recent research in cognitive psychology and provides a broader coverage of the cognitive skills of examinees than previous editions. The administration is more flexible than is true of many tests, and children, especially younger ones, find the items challenging and fun.

Although less emphasis on verbal skills is present in the Fourth Edition, the Stanford-Binet continues to be a very good assessment of cognitive skills related to academic progress. In spite of numerous problems, the Fourth Edition will be around for many years—after all, it is still the Binet.

References

This list includes text citations and suggested additional reading.

American Educational Research Association, American Psychological Association, & National Council on Measurement in Education. (1985). *Standards for educational and psychological testing.* Washington, DC: American Psychological Association.

Buros, O. K. (Ed.). (1972). *The seventh mental measurements yearbook.* Highland Park, NJ: The Gryphon Press.

Carvajal, H., & Weyand, K. (1986). Relationships between scores on Stanford-Binet IV and Wechsler Intelligence Scale for Children-Revised. *Psychological Reports, 59,* 963-966.

Carvajal, H., McVey, S., Sellers, T., Weyand, K., & McKnab, P. (1986). Relationships between scores on the General Purpose Abbreviated Battery of Stanford-Binet IV, Peabody Picture Vocabulary Test-Revised, Columbia Mental Maturity Scale, and Goodenough-Harris Drawing Test. *The Psychological Record, 1,* 127-130.

Holden, R. H. (1984). Stanford-Binet Intelligence Scale: Form L-M. In D.J. Keyser & R.C. Sweetland (Eds.), *Test Critiques: Volume I* (pp. 603-607). Kansas City, MO: Test Corporation of America.

Kaufman, A. S., & Kaufman, N. L. (1983). *Kaufman Assessment Battery for Children.* Circle Pines, MN: American Guidance Service.

Keith, T. Z., Cool-Hauser, V. A., Novak, C. G., White, L. J., & Pottebaum, S. M. (1987). *Confirmatory factor analysis of the Stanford-Binet Fourth Edition: Testing the theory—test match.* Unpublished manuscript.

Kennedy, W. A. (1973). *Intelligence and economics: A confounded relationship.* Morristown, NJ: General Learning Press.

Krohn, E. J. & Lamp, R. E. (1987, March). *Validity of K-ABC and Binet-Fourth Edition for low income preschool children.* Paper presented at the annual convention of the National Association of School Psychologists, New Orleans.

Livesay, K., & Mealor, D. J. (1987, March). *A comparison of the Stanford-Binet: Fourth Edition and Form L-M for gifted referrals.* Paper presented at the annual convention of the National Association of School Psychologists, New Orleans.

Lubin, B., Larsen, M., & Matarazzo, J. (1984). Patterns of psychological test usage in the United States: 1935-1982. *American Psychologist, 39,* 451-454.

Naglieri, J. A. (1982). Interpreting the profile of McCarthy Scale indexes: A revision. *Psychology in the Schools, 19,* 49-51.

Sattler, J. M. (1982). *Assessment of children's intelligence and special abilities* (2nd Ed.). Boston: Allyn & Bacon.

Silverstein, A. B. (1982). Pattern analysis as simultaneous statistical inference. *Journal of Consulting and Clinical Psychology, 50,* 234-240.

Sims, L. (1987, March). *Concurrent validity of the new Stanford-Binet and WISC-R.* Paper presented at the annual convention of the National Association of School Psychologists, New Orleans.

Spruill, J., & Beck, B. (in press). Comparison of the WAIS and WAIS-R: Different results for different IQ groups. *Professional Psychology: Research and Practice.*

Terman, L. M. (1916). *The measurement of intelligence.* Boston: Houghton Mifflin Company.

Terman, L. M., & Merrill, M. A. (1937). *Measuring intelligence.* Boston: Houghton Mifflin Company.

Terman, L. M., & Merrill, M. A. (1960). *Stanford-Binet Intelligence Scale.* Boston: Houghton Mifflin Company.

Terman, L. M., & Merrill, M. A. (1973). *Stanford-Binet Intelligence Scale: 1972 norms edition.* Boston: Houghton Mifflin Company.

Thorndike, R. L., Hagen, E., & Sattler, J. (1986a). *Stanford Binet Intelligence Scale: Fourth Edition.* Chicago: The Riverside Publishing Company.

Thorndike, R. L., Hagen, E., & Sattler, J. (1986b). *Guide for administering and scoring the fourth edition: Stanford Binet Intelligence Scale.* Chicago: The Riverside Publishing Company.

Thorndike, R. L., Hagen, E., & Sattler, J. (1986c). *Technical manual: Stanford-Binet Intelligence Scale, Fourth Edition.* Chicago: The Riverside Publishing Company.

Thorndike, R. L., Hagen, E., & Sattler, J. (in press). *Expanded guide for interpreting and reporting fourth edition results.* Chicago: The Riverside Publishing Company.

Oliver C. S. Tzeng, Ph.D.

Professor of Psychology and Director of the Osgood Laboratory for Cross-Cultural Research, Purdue University, School of Science at Indianapolis, Indianapolis, Indiana.

STRONG-CAMPBELL INTEREST INVENTORY

E. K. Strong and D. P. Campbell. Stanford, California: Distributed exclusively by Consulting Psychologists Press, Inc.

THE ORIGINAL OF THIS REVIEW WAS PUBLISHED IN TEST CRITIQUES: VOLUME II (1985).

Introduction

In measuring an individual's interest in different types of occupations, educational disciplines, personality associations, and recreational activities, one of the most prominent instruments in the history of psychometrics is the Strong-Campbell Interest Inventory (SCII), the current edition of the Strong Vocational Interest Blank (SVIB). Historically, the SCII represents the culmination of over 50 years of scientific endeavors that involved hundreds of thousands of people in diverse occupations. Since the publication of the initial work by Strong in 1927, the SCII and its earlier editions have enjoyed vast popularities not only in career counseling and personnel selection, but also in scientific research communities. Literally over one thousand articles have been published on the inventory, and countless master's and doctoral dissertations have used it as the major source. Therefore, it seems almost impossible to conduct an exhaustive review of all relevant materials in this critique. For this reason the focus will be on the evaluation of its theories and psychometric properties.

As its name suggests, the SCII is intended to measure an individual's interest, not aptitude or intelligence, in various occupations. The theoretical foundation for developing the SCII is from typology and trait psychology. That is, individuals in the same occupation will display similar interests and personality characteristics, whereas people in different occupations will display different types of interests and personality characteristics. Based on the nature of homogeneity within the same occupation and heterogeneity across different occupations, the following inferences of trait attribution become the basic assumptions in the development and application of the SCII: 1) Each occupation has a desirable pattern of interests and personality characteristics among its workers. The ideal pattern is represented by successful people in that occupation. 2) Each individual has relatively stable interests and personality traits. When such interests and traits match the desirable interest patterns of the occupation the individual has a high probability to enter that occupation and be more likely to succeed in it. 3) It is highly possible to differentiate individuals in a given occupation from others-in-general in terms of the desirable patterns of interests and traits for that occupation.

The SVIB was first published in 1927 with 420 items in differentiating male certified accountants from other occupational groups (Strong, 1927). Since the first distribution of the manual in 1928, the SVIB has undergone many major revisions and expansions: 1) the publications of the women's SVIB in 1933 and its manual in 1935; 2) major revision of the men's form in 1938; 3) revision of the women's form in parallel with the men's form in 1946; 4) first publication of modern SVIB manual by Consulting Psychologists Press in 1959; 5) major revisions of the men's form in 1966 with 399 items; 6) major revision of the women's form in 1969 with 398 items; 7) the publication of the handbook for the SVIB by Campbell in 1971; 8) publication of the 1974 edition of the SVIB, designated as the SCII (Form T325) that includes the merging of the men's and women's forms into a single booklet and the introduction of a theoretical framework to guide the organization and interpretation of scores (Campbell & Hansen, 1981)[1]; 9) publication of the 1984 user's guide for the SVIB-SCII (Hansen, 1984).

For the present review, the major focus will be on three contemporary sources that are vital for application of the SCII: the 1971 handbook, the 1981 SVIB-SCII manual (Form T325), and the 1984 user's guide.

The basic materials for the SVIB-SCII include three parts: the inventory booklet (Form T325) and answer sheet, a profile form, and an interpretive report.

The inventory booklet and answer sheet contain 325 items grouped under the following seven domains of measurement:

Occupations: (131 items). Respondents indicate how they would feel about doing a particular kind of work by marking each occupation (e.g., Actor/Artist, Advertising executive, Architect) as either "Like" (L), "Indifferent" (I), or "Dislike" (D).

School Subjects: (36 items). Respondents indicate their interest in each subject (e.g., Agriculture, Algebra, Arithmetic) even though they may not have studied it by marking each item L, I, or D.

Activities: (51 items). Respondents indicate their interest in a number of diverse activities that include individual behaviors (e.g., cooking, living in a city), social interactions (e.g., discussing politics, contributing to charities), and work-related functions (e.g., operating machinery, interviewing job applicants) by marking each item L, I, or D.

Amusement: (39 items). Respondents indicate their interest in nonwork-related, leisure activities (e.g., golf, fishing, concerts) by marking each item L, I, or D.

Types of People: (24 items). Respondents indicate how they would feel having day-to-day contact with each type of people (e.g., babies, aggressive people, musical geniuses) by marking each item L, I, or D.

Preference Between Two Activities: (30 items). Respondents express their preference between two "opposing" activities or occupations (e.g., taxicab driver vs. police officer, reading a book vs. watching TV or going to a movie) by marking their preference "R" if it is on the right side, "L" if it is on the left side, or " = " if they like or dislike both equally or if they are undecided.

Your Characteristics: (14 items). Respondents show what kind of person they are, according to whether or not various statements (e.g., win friends easily, put drive into an organization) describe them, by marking each item "Y" if it does, "N" if it does not, or "?" if they are undecided.

Answers to all 325 items will yield raw scores for each respondent from which various scale scores were derived by combining different items into various organizational groups (scales). The results of such derived scores are entered by computer on the profile form that contains the following five types of scale scores:

General Occupational Themes: (6 themes). Holland's (1973) six occupational types—Realistic, Investigative, Artistic, Social, Enterprising, and Conventional—were adopted to derive six theme scales. Each theme was measured by 20 marker items that were selected based on the descriptions given by Holland (1966). The raw scale score on each theme is a simple summation of the individual item scores across the 20 markers (e.g., L = 1, I = 0, D = 1). The respondent's raw scale score on each theme is further transformed to a standard T-score with respect to the general reference sample of 600 adults of both sexes. Such a standard score becomes the theme scale score printed on the profile for interpretation. The scores on the six themes provide a parsimonious view of the respondent's interests with respect to both the total lifestyle and the global occupational environment.

Basic Interest Scales: (23 scales). Based on the analysis of inter-item correlations across all items, 23 clusters were obtained and identified as the Basic Interest Scales. Each scale has 5 to 24 marker items. Psychometrically, the items within each cluster are more homogeneous than the items across different clusters in representing one specific area of activities. Therefore, a respondent's score on each scale is further transformed into standard T-scores for comparisons with the general reference sample. For example, a high score on a given scale would indicate the consistency of answering "Like" to the activities in that area. These 23 scales are further clustered on the profile into the six General Occupational Theme categories. That is, the Realistic Theme contains Agriculture, Nature, Adventure, Military Activities, and Mechanical Activity; the Investigative Theme contains Science, Mathematics, Medical Science, and Medical Service; the Artistic Theme contains Music/Dramatics, Art, and Writing; the Social Theme contains Teaching, Social Service, Athletics, Domestic Arts, and Religious Activities; the Enterprising Theme contains Public Speaking, Law/Politics, Merchandising, Sales, and Business Management; and the Conventional Theme contains Office Practices.

The Occupational Scales: (162 items). The Profile carries 77 common occupations for both sexes and eight specific occupations, four for males and four for females. For example, the common occupations include Air Force Officer, Farmer, and Chemist; the four male-specific occupations are Agribusiness Manager, Skilled Crafts, Investment Fund Manager, and Vocational Agriculture Teacher; and the four female-specific occupations are Dental Assistant, Dental Hygienist, Home Economics Teacher, and Secretary. Each scale contains between 50 to 70 marker items, and each item has maximal discriminability (large response-percentage differences) between the reference group (people-in-general) and the criterion groups ("successful" employees of a specific occupation). For each scale, the respondent's answers to the markers are checked against the sex-normal keys (either "Like" or "Dislike"). A congruent response is scored as +1, whereas an incongruent response (e.g., a subject's "Like" vs. the norm group's "Dislike" key) is scored as −1. Summing over such comparison scores across all markers yields an Occupational Scale Score for each of the 85 occupations which is further transformed to the standard T-score with reference to the mean and standard deviation of the respective criterion groups (males and females). As a result, each respondent's standard scores, in the range of −44 to +96, on all 85 Occupational Scales for each sex are entered in the Profile.

For facilitating interpretation, such standard scores are further broken down into seven categories, from very dissimilar (standard score of 12 or below), dissimilar (13-21), moderately dissimilar (22-27), mid-range (28-39), moderately similar (40-45), similar (46-54), to very similar (55 or higher). A high score will indicate the respondent's interests are similar to the interests of people in that occupation (i.e., the criterion group).

In addition, the 162 occupational scales of both sexes are coded by the six General Occupational Themes based on the response scores of the criterion samples on the six themes. These codings provide easy integration of the General Occupational Themes, Basic Interest Scales, and Occupational Scales (Hansen notes in the user's guide and in a personal communication [January 18, 1985] that in the 1985 revision, the Occupational scales will contain about 207 items that will include scales for female investment managers and vocational agriculture teachers, and both female and male carpenters, etc. The new 1985 manual by Hansen and Campbell will be available July 1, 1985. The major changes and features of the 1985 revision can be found in Hansen, in press.)[2]

Special Scales (Academic Comfort and Introversion-Extroversion): The Academic Comfort Scale (AC) is designed to serve as an indicator of 1) degree of comfort in an academic setting, 2) degree of interest in intellectual endeavors, and 3) degree of persistence toward either investigative and/or artistic vocations (high scores) or toward organizational and/or action-oriented problems (low scores).

The same procedure used for constructing the Occupational Scales was used in developing the Academic Comfort Scale: 1) A large sample of college students at the University of Minnesota's College of Liberal Arts was categorized into two groups (good students vs. poor students) based on the combination of their GPA's and achievement test scores. 2) By treating these good vs. poor student groups as two criterion samples, the SCII items differentiating the two samples were then used as the marker items to construct the AC scale. 3) The respondent's endorsements of the marker items emphasizing the interests in Investigative and Artistic Themes are scored positively (+ 1s), whereas the endorsements of the marker items in the interests of Enterprising and Realistic themes are scored negatively (−1s); each respondent's AC score (in the range of −14 to 92) is the sum of weighted scores on all marker items. 4) The AC scale, normed across 198 Occupational Samples, will reflect a hierarchical relationship between the educational backgrounds of the individuals in various occupations and their AC scores (Ph.D.s score about 56 to 65; professional degrees (e.g., lawyers, dentists, optometrists) score about 50 to 55; master's and bachelor's degrees about 45 to 55; associate or vocational/technical 2-year degree about 35 to 44; and high-school diplomas about 34 or lower. 5) The AC scale is also normed so that for both bachelor's and master's degrees, occupations with Investigative or Artistic interests score higher (50-55) than those with interests in the Social or Enterprising themes (45-50). Both in the 1981 manual and 1984 user's guide, mean scores on the AC Scale for a variety of occupational samples are listed for counseling purposes; for example, Biologist (female) with a mean of 65, Chemists (male) with a mean of 62, etc.

The Introversion-Extroversion (I-E) Scale is designed to discriminate between people-oriented and nonpeople-oriented occupational interests. Low scores on the I-E Scale are earned by extroverted individuals who enjoy working with others, especially in social-service settings, whereas high scores are earned by introverted individuals who would prefer to work with ideas and things. This scale is developed under the same procedure used in developing the AC Scale: 1) A large sample of University of Minnesota students who scored in either the extroverted or introverted direction on the Minnesota Multiphasic Personality Inventory (MMPI) was categorized into two criterion (extroverted and introverted) groups. 2) The SVIB-SCII items that differentiated these two criterion groups were used as the markers for the I-E Scale. 3) Items emphasizing working with ideas or things are scored in the introverted direction (+ 1, high scores); items emphasizing working with people are scored in the extroverted direction (–1, low scores). 4) Each respondent's I-E Scale score, in the range of 18 to 94, is the sum of individual introverted or extroverted scores on all marker items. 5) The I-E Scale is normed so that the entire scale score range is trichotomized into three successive categories: scores of 45 or below reflect the extroverted direction, scores between 46 and 54 reflect a balance between extroverted and introverted interests, and scores of 55 or higher reflect the introverted direction. In the manual and user's guide, I-E mean scores for 200 occupational samples are listed for counseling purposes.

The Administrative Indexes (3 statistics): The Administrative Indexes (AI) are routine computer analyses of each respondent's answer sheet on all 325 items. Three types of indexes are reported on the profile: The Total Response, Infrequent Response, and Relative Frequency Distribution on Response Alternatives.

The Total Response Index indicates the total number of item responses read by the computer from the answer sheet. If every item was answered by the respondent and correctly read in by the computer, the Total Response Index printed on the profile should be 325. Less than 325 indicates that either omissions of item responses by the subject or mechanical problems in computer data processing have occurred. When the index drops below about 310, the answer sheet should be checked before accepting the resulting profile.

The Infrequent Response Index is a general check of subject response validity. Based on prior analyses of item-response percentages of two General Reference Samples, along with several other samples, unpopular (infrequent) items, each with 6% or less of endorsement (e.g., "Like") from all respondents, were identified. As a result, the maximal numbers of infrequent responses are 6 for women and 11 for men. These numbers are used as the baseline for computing each respondent's score on the Infrequent Response Index. That is, for each endorsement of the unpopular items, 1 point is subtracted from this baseline. Therefore, the maximal scores of 6 (for female respondents) and 11 (for males) on this index would indicate no endorsement of any unpopular items. (Note, in the 1984 user's guide, the marker items for this Index include several popular items such as taking responsibility. Under such circumstances, the computation method will include the subtraction of the number of rejecting popular items from the baseline.) Generally speaking, a very low (especially negative) score on this index would indicate a potential problem of the respondent in answering the SCII items (e.g., misunderstanding the instructions, or lack of cooperation).

The Relative Frequency Distribution on Response Alternatives contains 24 numbers. As indicated earlier, the 325 items in SCII are categorized in the test booklet under seven sections of measurement. For all items, a common three-step response format is used: the Like-Indifferent-Dislike format for the first five sections, the Left-Neutral-Right format for the measurement of Preference Between Two Activities, and the Yes-Undecided-No format for checking personality characteristics. Within each section, frequency distribution of items on these three response alternatives is computed for each respondent and converted into percentages for reporting on the profile. In addition, across all seven sections, an overall frequency distribution of the 325 items is also obtained and reported in percentage form. As a result, a total of 24 numbers is reported on the profile. These figures are intended to reveal the respondent's response style in filling in the inventory. For example, an excessively high response percentage (60% or higher) on the first alternative (''Like,'' ''Left,'' and ''Yes'') will result in high score on all six General Occupational Themes and most of the Basic Interest Scales: people with high percentages on the second alternative tend to be indecisive, apathetic, or confused about their future occupations.

Furthermore, the indexes of Total Responses and Infrequent Responses are used to check the accuracy of the scale scores on the profile prepared by various commercial scoring series. It was reported that these checks have reduced mechanical inaccuracies in computing profile scale scores and increased the consistency and reliability of profile reporting (Hansen, 1982).

Finally, on the issue of faking (invalidity) in responses, it was concluded that most people answer the SCII items honestly even in highly competitive selection situations. However, in order to check a possible faking, the recommendation was made in the manual to check the inconsistencies between scores on the Basic Interest Scales and on the Occupational Scales, and to follow the check-and-balance system provided by using homogeneous and heterogeneous scales.

In addition to the basic, graphic profile form on which the respondent's scores can be printed directly, a descriptive form, the Interpretative Report (or Interpretative profile), is generated and printed directly by the computer. Each report includes scale scores and interpretations of each score for the individual respondent. Such a computer-interpretive profile is useful for one-to-one interactions between the examiner and examinee in various education/counseling situations.

Practical Applications/Uses

The SCII 132 items are easy to administer and can be given individually or in groups, in person or by mail. The respondent needs only a place to write, a computer-scored test booklet, and a dark lead pencil. It will take an average of 25 to 35 minutes to complete the entire test, including five demographic items.

The prospective subjects who will benefit from taking the SCII include high-school and college students, and adults seeking guidance on job entry or continued education. Therefore, the SCII is appropriate for individuals in a life span of 50 years or more. Because most students' interests have not developed enough to be differentiated before the eighth grade (ages 13-14), the SCII is not recommended for application below high school level.

The SCII items are written at about the sixth-grade level to ensure the understanding of all prospective subjects, including the slower readers with eighth- and ninth-grade education backgrounds. When a few items (e.g., calculus, botany) are unfamiliar to some respondents, the administrator is encouraged to offer definitions or explanations such that respondents can estimate accurately their preference to the items.

Before administration of the SCII, pretest orientation is needed for all respondents, including such concepts as 1) career-planning is a lifelong activity, 2) the SCII is a measure of interest, and 3) the inventory is to help the organization of existing interests, rather than to provide solutions to occupational problems.

Completed answer sheets are normally sent to one of the commercial scoring agencies for scoring. Currently, there are three commercial agencies that will handle scoring and distribution of the SCII materials. They are CPP (P.O. Box 60070, Palo Alto, California, 94306), CPP (5100 No. Roxboro Rd., Durham, North Carolina, 27703), and CPP (P.O. Box 944, Minneapolis, Minnesota, 55440).[3]

The SCII is reported to have diverse applicabilities. The common applications can be summarized into three broad areas: 1) counseling the individual student or employee for such purposes as high-school and college curriculum-planning, mid-career evaluation and change, occupational rehabilitation, and leisure counseling; 2) conducting field research on groups for such purposes as studying characteristic interests of particular occupations, studying indigenous cultural characteristics and cross-cultural influences; 3) conducting basic research programs that will identify interpersonal homogeneity in various social interactions, investigate the nature and process of career development, study the impacts of similar interests on human relationships (e.g., marital, sibling, and parent-child relations), and apply scientific knowledge for various occupational improvements.

For these applications, a variety of professionals who will conduct the above studies or who are in the position of assisting clients can use the SCII.

Both the manual and user's guide devote a chapter, as well as other space, to the description of procedures for interpreting the SCII results to the respondent. These procedures include 17 steps in sequence and each step is explained in detail.

Many case studies are also included to provide the foundation necessary for an understanding of the psychometric principles of the SCII and its applicational mechanisms.

Technical Aspects

The utility of a psychological measurement inventory depends on the psychometric evidence of reliability and validity at the levels of items, scales, and the entire test as a whole. This section of the review focuses on the evaluation of such psychometric properties of the SCII and its earlier edition, the SVIB, especially at the item and scale levels.[4]

Three types of statistical analyses at the item level are reported in the manual. The first type is item-response distribution for all of the SCII items. That is, for each item, the percentages of ''Like'' responses from 438 occupational samples were plotted in an item-response distribution. The distributions of all 325 items were then used to assess various psychometric properties, such as the differentiality across various occupations, the relationship between the content of each item and the occupations endorsing the item, the stability of distributions over time (some over four decades), and the identification of sex differences. (These distributions are available in the computer-accessible storages at the University of Minnesota.)

The second type is between-group comparisons on item-response percentages. That is, for each item, the "Like" percentages from two groups (the occupational criterion and general reference samples) were compared. The differences on all 325 items were rank-ordered, and the items that exhibited large response differences were then selected as the markers (in an average of 60 items) for that occupation in the Occupational Scales.

The third type is inter-item correlations. That is, for the selection of marker items for the 23 Basic Interest Scales, intercorrelations between the 400 items of the SVIB were calculated for the men's and women's booklets separately. Based on frequency distributions of the correlations, a cutoff point of .25 was used to select the marker items for each of the Basic Interest Scales. In the SCII scales, the male and female SVIB sets were merged by combining the parallel scales of both sexes.

Using scale scores as basic units in analyses at the scale level three types of statistics are reported; the first type is inter-group mean differences. That is, the mean scores on a given Occupational Scale were compared between the reference sample and a new respondent sample, and also between different occupational samples. The minimal differences were established as criteria to evaluate mean differences for psychometrical significance (concurrent validity). In addition, the SVIB has a long history of validity study in predicting occupational choices. Detailed information can be found in the handbook and the manual.

Concurrent validity of the Basic Interest Scales was supported by numerous comparisons among people currently in different occupations. However, the predictive validity was found not as good as concurrent validity, due to both interscale and interpersonal differences (some scales and persons are less predictable than others).

The second type of analysis is on the relationship between the Basic Interest Scales and the Occupational Scales. The manual reports that about 20% of the SVIB profiles had one or more apparent inconsistencies between these two types of scales (e.g., mathematics vs. mathematician). Such discrepancies are due to the different procedures involved in selecting their respective marker items.

The third type involves test-retest correlations for the Basic Interest Scales over various time intervals and for various age samples. The test-retest correlations and the stability of the means were also computed for the Occupational Scales. In general, the correlations are reasonably high, ranging from .60s to .90s, demonstrating that the SCII scales are quite stable over time (especially over short periods).

Critique

It should be noted that in assessing the above psychometric properties of the SCII items, not a single test of statistical significance has been used and reported in the manual. Instead, Campbell (1974) emphasized magnitude and replicabilities of differences and consistency of trends. This reviewer readily agrees with Campbell that, in many cases, statistical significance does not automatically imply psychological significance. Numerous studies that involve between-group comparisons in means or between-item correlations within such content areas as human intelligence, implicit personality theories, and semantic differential ratings have been tricked by the notion of ''statistical significance'' (Tzeng, 1982; Tzeng & Tzeng, 1982; Tzeng, Powers, & Schliessmann, 1984).

The approaches used in the SVIB and SCII have many desirable characteristics, but also bear many weaknesses.

There are five major strengths about the development and presentations of the SVIB-SCII:

1) The development of the seven item-response contents includes diverse occupations, school subjects, activities, amusements, important-others, preferred activities, and personality characteristics. This approach guarantees the comprehensiveness of measurement issues and concerns that are directly and indirectly related to the development of individual occupational interests and preferences. In this regard, the vision of Strong in the 1920s encompasses all four personality research paradigms that were prevalent during the last 50 years: typology, trait psychology, situationism, and interactionism.

2) The handbook provides complete documents on the theories, methods, and procedures for the development and revision of the SVIB-SCII, and the manual and user's guide synthesize very well the materials that are necessary for the users and interpreters of the test.

3) The uses of means, standard deviations, percentages, correlations, and standard scores as the major statistics are easy to understand by average researchers and prospective users.

4) Extensive studies are presented on three types of validity—concurrent validity, predictive validity, and postdictive validity—within and across various forms. Also, extensive studies on reliability (mostly test-retest) are reported.

5) The two interpretation materials—the graphic profile and the descriptive interpretation report—are easy to understand and are complementary to each other. The combination of these two will provide a comprehensive picture of a respondent's occupational interests and preferences.

Regarding the measurement properties of the SCII and its earlier edition, there are several issues that have never been assessed. As a result, the lack of such information generates many unexplainable properties of scale profiles that are interpreted as ''special properties'' of the SCII in the manual and user's guide. These issues are centered around the measurement concept of construct validity for the development of three types of derived measures: six General Occupation Themes, 23 Basic Interest Scales, and 162 different Occupational Scales.

Throughout the 325 SCII items, a uniform response format with trichotomous options (Like/Indifferent/Dislike, Left/ = /Right, Yes/?/No) is used. Psychometrically, the response on each item should represent an isomorphic mapping (quantification) between the empirical response continuum of the three steps and the underlying trait continuum to be measured (e.g., extroversion/introversion) (cf. Tzeng, 1983). However, the trichotomous format is qualitative (rather than quantitative) in nature, and it can detect neither the relative differentiations between two endorsed items on the same construct nor the true relative saturations of the two constructs (both with 100% Like responses) for a single respondent. Under such circumstances, the equal number of endorsed items for an individual on two different scales does not necessarily indicate an equal interest level on these two psychological constructs. Similarly, the equal numbers of endorsed items for two persons on the same scale do not automatically imply an equal occupational interest to both individuals. Therefore, improvement of such a trichotomous response format would seem necessary in order to establish the quantitative evidence of construct validity in the future. (For detailed discussions and illustrations of comparative results across different response formats see Tzeng, 1983.)

The organizations of the six General Occupational Themes and the 23 Basic Interest Scales from the same 325 items are hierarchically inclusive in the theoretical postulations for the 1974 version of the SVIB. However, in the actual process of constructing the marker items, these two types of scales were initially obtained by two separate procedures, and only later linked by computing the correlations between their scale scores. Given the availability of these correlations (see Manual, Table 5–3), the assignment of the 23 Basic Interest Scales into the six General Themes was still rather arbitrary. The obvious problems include 1) the treatment of multiattributions of many individual Basic Interest Scales as a single, isolated attribution (e.g., Mathematics has apparent coattributions under the Realistic, Investigative, and Conventional Themes, but is clustered under the Investigative Theme only); 2) The use of inconsistent criteria to cluster individual Basic Interest Scales (e.g., the high correlation coefficients of .57 [male] and .64 [female] did not warrant the inclusion of Business Management under the Conventional Theme, whereas the low correlations of .28 [male] and .40 [female] were used to include Medical Service under the Investigative Theme); 3) the repeated reference in the manual to the psychometric concepts of "factors" and "clustering" in identifying markers for different scales. However, the various multivariate techniques that will perform the designated purposes of identifying factors and clusters were never used in the empirical treatment of the data. Therefore, partial and/or biased assignments of the marker items become inevitable.

To illustrate the third point, the two matrices of intercorrelations between the 23 Basic Interest Scales obtained from the two General Reference Samples of both sexes (see Manual, Table 5-2) were factor analyzed by this reviewer. The results of three to eight dimensions were rotated through Varimax criterion. The six-dimensional solutions should in theory recapture completely the categorizations of the 23 Basic Interest Scales into the six General Occupational Themes.

Table 1 presents the factor structures of both sexes with only the salient loadings being presented in the table. Generally speaking, the present results provide only partial support for the construct validity of heirarchical organizations between the Interest Scale level and the General Occupation Theme level.

Specifically, in the table the present factor solutions suggest the following: 1) For both male and female groups, six empirical factor solutions seem reasonable and each accounts for over 89% of the respective total input variance. 2) In comparisons between male and female structures, there appears roughly a one-to-one correspondence across all six factors. 3) Based on salient loadings, especially those coded with asterisks, the SCII clusterings of Basic Interest Scales into the six General Themes are not totally supported. In fact, only the five Basic Interest Scales under the Enterprising Theme appear to be relatively homogeneous and "simple" (uniquely loaded on two dimensions) for the male reference group. 4) There exist important sex differences in interrelationships between various Basic Interest

Table 1

Rotated Factor Structures for Male and Female General Reference Groups[a]

General Themes	Basic Interest Scales	Male Factor Structure						Female Factor Structure					
		I	II	III	IV	V	VI	I	II	III	IV	V	VI
Realistic	Agriculture	—[b]	-.70*	.58	—	—	—	-.30	-.76*	—	—	—	—
	Nature	-.44	-.50*	—	.43	.48	—	-.40	-.68*	—	—	.48	—
	Adventure	.52*	—	—	—	—	-.92*	—	—	—	—	—	.86*
	Military Activities	—	—	.49	—	—	—	—	—	.46	—	—	.73*
	Medical Activities	—	-.76*	—	-.43	—	—	-.30	—	—	-.81*	—	—
Investigative	Science	-.60	—	—	-.62*	.33	—	-.50	—	—	.74*	.40	—
	Mathematics	—	-.37	—	-.89*	—	—	-.31	—	.37	-.83*	—	—
	Medical Science	-.47	—	—	—	.77*	—	-.33	—	—	-.30	-.84*	—
	Medical Service	—	—	—	—	.93*	—	—	—	.45	—	.70*	—
Artistic	Music/Dramatics	-.42	.36	-.70*	.34	—	—	—	.31	-.87*	—	—	—
	Art	-.48	.55	-.69*	.37	—	—	—	—	-.94*	—	—	—
	Writing	-.45	—	-.60*	—	—	—	—	.37	-.88*	—	—	—
Social	Teaching	.44	.61*	—	—	.37	—	—	.70*	-.32	—	—	—
	Social Service	—	.73*	—	.48	.30	—	—	.49	—	.75*	—	—
	Athletics	—	—	.82*	—	—	—	—	—	.60*	—	—	.48
	Domestic Arts	—	—	-.41	.58*	.50	—	—	-.31	.43	.71*	—	—
	Religious Activities	—	—	—	.34	—	.77*	-.31	—	.52	.59*	—	—
Enterprising	Public Speaking	—	.91*	—	—	—	—	.54	.65*	-.42	—	—	—
	Law/Politics	—	.92*	—	—	—	—	.48	.67*	-.42	—	—	—
	Merchandising	.91*	—	—	—	—	—	.88*	—	—	—	—	—
	Sales	.91*	—	—	—	—	—	.87*	—	—	.35	—	—
	Business Management	.90*	—	—	—	—	—	.88*	.32	—	.31	—	—
Conventional	Office Practices	.79*	—	—	—	—	.50	.35	—	.81*	—	—	—

[a]Correlation matrices, reported in the 1981 SVIB-SCII Manual (p. 37), were analyzed separately for males and females through principal components analysis with varimax rotation.

[b]Factor loadings with an absolute value below .30 were not reported in this table. For each Basic Interest Scale, its highest loading was coded with an asterisk (*) for interpretation purposes.

Scales. For example, to females, the interests of Military Activities and Athletics were found to be highly congruent with the interest of Adventure. For males, on the other hand, Military Activities were not related to Adventure, but rather related to popular occupations for general male adults (i.e., Sales, Merchandising, Business Management, Office Practice, and Teaching as opposed to Science, Medical Science, and Artistic interests). 5) The salient interests on the six empirical factors suggest that Holland's six general themes may not be apropriate for assigning the six distinctive "clusters" from the 23 Basic Interests. For example, the third factor of the male structures indicates the dimensional nature with two opposite characteristics in behavioral contexts—overall body functions (Athletics, Agriculture, and Military Activities) vs. fine mental activities (Music, Art, Dramatic Arts, and Writing).

Psychometrically, the derivation of the 23 Basic Interest Scales and identification of their respective markers can be and should be obtained from simultaneous assessment (e.g., factor analysis) of the intercorrelations among all 325 items. Unfortunately, such intercorrelations are not available in the handbook for psychometric verification of the clusterings. Unless such analysis is completed, many unexplainable problems will remain a mystery in the literature of the SCII (e.g., only one Basic Interest Scale, Office Practice, fits into Holland's Conventional category; cf. Manual, p. 37).

The method for selecting marker items for each Vocational Scale was based on the relative between-group discriminabilities among items. This strategy bears several potential difficulties in the context of construct validity. First, from the sheer mathematical rule of combinations, 60 items were selected from the same 325 item pool for some eighty different occupations. It would be inevitable to generate a lot of overlapping marker items for different occupation scales, thus yielding many equally appealing occupations for each respondent. Secondly, in the contemporary society individuals tend to exhibit gender interests and involvements in various activities, amusements, interdisciplinary school subjects and occupational skills, etc. As a result, it is highly probable for a respondent to be characterized as having high interests in many and diverse occupations. Under both conditions, construct validity of the SCII and its power for predicting an individual's future occupational choice will be arbitrarily inflated.

In addition to the above weaknesses many other issues have been raised by various reviewers of the SCII and reported in the *Mental Measurements Yearbooks* (Buros, 1972, 1978). Among them, the major criticisms include 1) the application difficulty of the SCII to the respondent who is either a "yes sayer" or "nay sayer"; 2) the inconsistency between the Basic Interest and Occupational Scales; 3) the sex-restrictive nature of the SCII, due to separate sex scoring keys for the common Occupation Scales, and separate sex norms for the homogeneous scales (It should be noted that although the SCII is sex-restrictive, it meets all of the NIE guidelines on sex-fairness for interest inventories. In addition, data in the manual [chap. 7] indicate that separate-sex scales are less restrictive than combined-sex scales in the new development of interests of women and men. Also, the combined-sex norms are available for homogeneous scales in the manual.); 4) the omission of the report of marker items for individual SCII scales in the manual; 5) the omission of the report of means and standard deviations for each occupational criterion group on all Occupational Scales and the omission of the report of intercorrelations among the Occupational Scales for both sexes.

It should be noted that many of these and related issues have been discussed in depth by various scholars. Unfortunately, no consensus has been reached. The basic reason underlying such controversies seems to center around the lack of rigorous scientific studies on construct validity of the SCII at the item, scale, and profile levels. Therefore, unless all psychometric properties are thoroughly evaluated and documented in the manual or handbook, the utilities and interpretabilities of the SCII will probably continue to be questioned by some critics.

Overall, the new SCII is well constructed from the revision of its preceding edition, the SVIB. It appears to be one of the best vocational interest inventories available and is widely used for educational guidance and vocational counseling. The 1971 handbook, 1981 manual, and 1984 user's guide provide complete information for all individuals who might be interested in using the SCII either for practical purposes or for research work by or with this inventory. Generally speaking, the users should be cautious about many unique aspects of the SCII construction in interpretation of scale scores and profiles. The researchers, especially those who plan to employ the SCII scores as indexes for prediction and/or discrimination purposes, should take into consideration many unexplainable and uncertain properties of the SCII scales (i.e., interdependence and confounding influences of individual scales that are within and/or across the three types of scale scores reported on each profile). Finally, as far as the perfection of the inventory is concerned, there is still some room for improvement, especially in the area of scientific evidence of its construct validity.

Notes to the *Compendium:*

[1]See also Hansen and Campbell (1985).

[2]The new 1985 manual by Hansen includes 264 scales: 6 General Occupational Themes that measure Holland's vocational types, 23 homogeneous Basic Interest Scales, 207 Occupational Scales representing 106 occupations, two special scales to measure Academic Comfort and Introversion-Extroversion, and 26 Administrative Indexes.

[3]From the scoring agency, each customer will receive two copies of the computerized interpretive report and the standard profile for the SVIB-SCII (see Hansen, 1987, for an illustrative layout).

[4]It should be noted that this review makes reference to the 1981 manual rather than the 1985 manual because the psychometric properties of the last two SVIB-SCII revisions remain unchanged.

References

Buros, O. K. (1972). *The seventh mental measurements yearbook* (Vols. 1 & 2). Highland Park, NJ: The Gryphon Press.

Buros, O. K. (1978). *The eighth mental measurements yearbook* (Vols. 1 & 2). Highland Park, NJ: The Gryphon Press.

Campbell, D. P. (1971). *Handbook for the SVIB.* Stanford, CA: Stanford University Press.

Campbell, D. P., & Hansen, J. C. (1981). *Manual for the Strong-Campbell Interest Inventory* (3rd ed.). Stanford, CA: Stanford University Press.

Hansen, J. C. (1982). *Scale score accuracy of the 1981 SCII.* Unpublished manuscript, University of Minnesota, Center for Interest Measurement Research, Minneapolis.

Hansen, J. C. (1984). *User's guide for the SVIB-SCII.* Stanford, CA: Stanford University Press.

Hansen, J. C. (1987). Computer-assisted interpretation of the Strong-Campbell Interest Inventory. In J. N. Butcher (Ed.), *Computerized psychological assessment.* New York: Basic Books.

Hansen, J. C. (in press). Strong Vocational Interest Blank/Strong-Campbell Interest Inventory. In W. B. Walsh and S. H. Osipow (Eds.), *Advances in vocational psychology: The assessment of interest.*

Hansen, J.C., & Campbell, D. P. (1985). Manual for the SVIB-SCII (4th ed.). Stanford, CA: Stanford University Press.

Holland, J. L. (1966). *The psychology of vocational choice.* Waltham, MA: Blaisdell.

Holland, J. L. (1973). *Making vocational choices: A theory of careers.* Englewood Cliffs, NJ: Prentice-Hall.

Strong, E. K., Jr. (1927). *Vocational Interest Blank.* Stanford, CA: Stanford University Press.

Tzeng, O. C. S. (1982). The artificial dispute over implicit personality theory. *Journal of Personality, 50,* 251-260.

Tzeng, O. C. S. (1983). A comparative evaluation of four response formats in personality ratings. *Educational and Psychological Measurement, 43,* 935-950.

Tzeng, O. C. S., Powers, K., & Schliessmann, R. (1984, November). *Cross-measurement and cross-statistical comparisons on studies of person perceptions: Measuremental fantasy vs. statistical ghost.* Paper presented in Seventh Midwest Psychometric Conference, Bloomington, IN.

Tzeng, O. C. S., & Tzeng, C. (1982). Implicit personality theory: Myth or fact? An illustration of how empirical research can miss. *Journal of Personality, 50,* 223-239.

James A. Walsh, Ph.D.
Professor of Psychology, University of Montana, Missoula, Montana.

TENNESSEE SELF CONCEPT SCALE

William H. Fitts. Los Angeles, California: Western Psychological Services.

THE ORIGINAL OF THIS REVIEW WAS PUBLISHED IN TEST CRITIQUES: VOLUME I (1984).

Introduction

The Tennessee Self Concept Scale (TSCS) consists of 100 self-descriptive items by means of which an individual portrays what he or she is, does, likes, and feels. The scale is intended to summarize an individual's feeling of self-worth, the degree to which the self-image is realistic, and whether or not that self-image is a deviant one. As well as providing an overall assessment of self-esteem, the TSCS measures five external aspects of self-concept (moral-ethical, social, personal, physical, and family) and three internal aspects (identity, behavior, and self-satisfaction). In addition, crossing the internal and external dimensions results in the mapping of 15 "facets" of self-concept.

Because each of 90 TSCS items taps both an internal and an external dimension as well as one facet, the item sets for the several dimensions overlap, but in a completely balanced way. The other 10 items are taken from the MMPI Lie Scale and provide a measure of a subject's defensiveness in responding to the TSCS. Numerous other indices of response sets and styles are generated from a subject's responses and are used in interpreting the scores on the self-concept scales.

William H. Fitts, a psychologist, did his doctoral work at Vanderbilt University and completed his dissertation on the role of self-concept in social perception in 1954. He has pursued parallel careers as a psychotherapist, mental health consultant, and researcher. According to his own description, Fitts was ". . . continuously involved in self concept research of some kind" from 1953 to 1965 (Fitts, 1972).

Early in his career Fitts became convinced that ". . . the self concept is a central and critical variable in human behavior and that an adequate measure of the self concept should contribute to the criterion problem [*sic*] in behavioral science" (Fitts, 1972, p. 1), He also believed ". . . that the measurement problem in self concept research was critical" (p. 1). From 1955 to 1965 his primary focus was on the development of better instruments for the measurement of self concept. The original mimeographed version of the Tennessee Self Concept Scale (TSCS) was published in 1956 and the present version appeared in 1965. The new version differs from the old primarily in terms of 1) the improved formats for the test booklet and answer sheet, and 2) the means for the majority of the scales. It is possible without great difficulty to translate most data obtained with the old form to that used on the new and vice versa.

In 1965 Fitts outlined a program of research devoted to understanding the way in which self-concept is influenced by psychotherapy, developmental factors, social influences, and the other independent variables that behavioral and social scientists have employed in studying self-concept. The strategic retreat which inevitably followed such an ambitious thrust led Fitts to focus on two target populations (public offenders and the mentally ill) in order to establish a manageable research agenda. Even so, the large volume of research in which other investigators employed the TSCS eventually drew Fitts away from his heavy load of direct involvement and gradually led him into the roles of methodological consultant and facilitator of scientific communication regarding self-concept. In 1968 this "clearinghouse" function resulted in the first of a series of monographs (there are now seven) and bibliographies (including a major one in 1980). The monographs review and attempt a broad synthesis of research involving self-concept and, as major foci, delinquency, interpersonal competence, self-actualization, psychopathology, and performance.

The enormous volume of work to which Fitts has contributed in his several roles has produced mountains of data about self-concept, but as will be explained later much too little of it touches directly on the structural validity of the TSCS. While various studies have produced special versions of the TSCS to this reviewer's knowledge they are not generally available to test users.

Two forms of the TSCS are in use and a third is under development. The "regular form" consists of a reuseable 6-page booklet of test items and a consumable "combination packet." Each packet includes an answer sheet, a piece of carbon paper, a scoring matrix, and a profile page. The item responses are transferred to the scoring matrix by means of the carbon and scores, once totaled from the matrix, are transferred by the scorer to the profile page. Depending on the intended use of the TSCS, one of two different profile pages can be ordered: the counseling form or the clinical and research form. Because several additional scales can be scored for clinical and research purposes, a set of scoring templates is required in addition to the scoring matrix.

The computer form consists of a consumable two-page booklet, the first page of which is an answer and identification sheet. The second page contains the self-descriptive items, and after testing is returned to the publisher for scoring. The answer sheet from the regular form can also be returned to the publisher for computerized scoring. Computerized handling of either form results in a printout that contains scores on 29 variables together with a set of standard punched cards that can be used as input in further analyses.

A one-page interpretive test report form of the TSCS merges questions and answer sheet and after computer handling will provide a textual interpretation of the test record in addition to test scores. The test report form was not operational at the time of this writing.

The TSCS is self-administering and may be given to individuals or groups. The instructions are simple and, in this reviewer's experience, most college student subjects understand them without any amplification from the examiner. The scale was, however, designed for use with subjects as young as 12 years of age, and while a fifth- or sixth-grade level of reading ability seems sufficient to insure understanding of the items, younger or less literate subjects do occasionally have trouble with the numbering system used in the regular form test booklet and answer sheet. This happens because in order to make scoring a test record reasonably easy, consecutive items are not numbered in sequence in the booklet. While the same numbering system is maintained on the answer sheet, a subject moves down columns in the usual way in recording his or her answers, but starts with the far right column and moves back towards the left in proceeding through the test. Younger subjects also occasionally have some difficulty in comprehending the 5-point response scale, which runs from "completely false" to "completely true."

Both profile sheets for the TSCS are well designed and easy to read. However, so much more information regarding response styles, as well as scores on a set of empirical scales, is presented on the clinical and research form that details can become (literally) blurred unless care is taken in drawing the profile. As is usual on profile sheets, raw score values are numbered along vertical lines below each scale name and horizontal lines intersect the vertical ones to connect to T-score values at the margins of the sheet and percentile values along the midline. It should be noted that while scores on the internal and external dimensions of self-concept are profiled, scores on the 15 facets are not. Given the brevity of the facet scales and their subsequently lower reliabilities, this is appropriate.

Practical Applications/Uses

The optimal clinically oriented use of the TSCS is probably in counseling young adults who do not have severe clinical problems but who may be lacking in self-esteem or have distorted self-images. A profile from the counseling form of the scale can provide a takeoff point for several sessions of extended discussion concerning the reality and/or appropriateness of a client's self-image. The TSCS will also undoubtedly continue to be widely employed in research and especially, as its author intended, ". . . for tying together many research and clinical findings" (Fitts, 1965, p. 1). The instrument seems to be especially useful in generating hypotheses about relatively specific effects of psychotherapeutic regimens and psychosocial interventions.

Fitts designed the TSCS to be ". . . simple for the subject, widely applicable, well standardized . . ." and to provide a ". . . multidimensional . . . description of the self concept" (1965, p. 1). It was intended ". . . for a variety of purposes—counseling, clinical assessment and diagnosis, research in behavioral science, personnel selection, etc." (Fitts, 1965, p. 1). At a somewhat higher level of abstraction, it was intended as a means of synthesizing clinical and research findings through their common relationship(s) with self-concept.

The TSCS measures the overall level of self-esteem and the previously mentioned internal and external dimensions of self-concept as well as a set of variables that are of primary importance for interpreting the self-concept scores. The following scores are derived on the counseling form: 1) the Self Criticism scale, which consists of 10 items from the MMPI Lie Scale and measures test-taking defensiveness in the sense of giving socially desirable but improbable responses; 2) the three Variability scales (one for internal, one for external, and one total for both), which provide indices of the amount of variability subjects display with respect to the range of their scores across dimensions, and which appear to be related to the degree of integration of self-concept; 3) the Distribution scale, which summarizes the degree to which subjects use all five categories on the response scale rather than favoring only one or two, and for which extreme scores are usually associated with severe psychological disturbances; and 4) the Time score, which is simply a measure to the nearest minute of how long a subject takes to complete the TSCS. Most individuals with the required reading skills complete the scale in less than 20 minutes, but psychiatric patients in general take more time than non-patients.

The clinical and research form is scored for the following scales in addition to those just described: 1) the True-False Ratio score, which serves primarily as a measure of acquiescence response set, but which may also, within self-theory, be taken as an indicator of the degree to which subjects define themselves by what they are rather than by what they are not; 2) the three Conflict scores (acquiescence, denial, and total), which reflect the degree ". . . to which an individual's responses to positive items differ from, or *conflict* with, his responses to negative items in the same area of self perception" (Fitts, 1965, p. 3); 3) the six "empirical" scales, Defensive-Positive, General Maladjustment, Psychosis, Personality Disorder, Neurosis, and Personality Integration; and 4) the Number of Deviant Signs score, which is the sum of deviant features from all of the other scales, and which Fitts indicates is an excellent indicator of "psychological disturbance."

Because the empirical scales resemble a mini-MMPI, their development may deserve some elaboration. Items were selected for given scales on the basis of distinguishing members of one criterion group from those of all the others. Because an item can make several such distinctions, and because Fitts did not limit any items to a single scale, the empirical scales overlap somewhat. Six criterion groups were used: psychotics (n = 100); neurotics (n = 100); subjects with diagnosed personality disorders (n = 100); subjects labelled "defensive positive" (n = 100); subjects called the "personality integration" group (n = 75); and subjects comprising the general norm group (n = 626).

The Psychosis scale consists of 27 items that, on the basis of chi-squared statistics, differentiate psychotic subjects from all others. Similarly, the Neurosis scale (23 items) and the Personality Disorders scale (27 items) distinguish members of their respective criterion groups from members of all other groups. The Defensive Positive scale is composed of the 29 items that distinguished between 100 psychiatric patients with total self-esteem scores above the TSCS general norm group mean and the members of all the other groups. The rationale for the Defensive Positive scale lies in a hypothesis of self-theory regarding discrepancies between general self-concept and specific self-description. The Personality Integration scale is composed of the 25 items that distinguished 75 subjects ". . . who, by a variety of criteria, were judged as average or better in terms of level of adjustment of degree of personality integration" (Fitts, 1965) from all of the other subjects in the six criterion groups. The General Maladjustment scale contains 24 items that separated psychiatric patients from non-patients, but which did not distinguish between members of the various patient groups. Thus, it is presumably a more general measure of psychopathology or lack of integration than are the other scales.

From comprehensive samplings (Reed, Fitts, & Boehm, 1980) of the literature generated by the TSCS, it appears that although clinical and counseling applications are not negligible components of the test's pattern of utilization, their volume is minor in comparison with that of use in research. Many high school students, clients of college counseling centers, psychiatric inpatients and outpatients, incarcerated individuals, and disabled persons are administered the TSCS, but the test appears to be employed primarily for the information it gives the researcher rather than for the insight it might provide the subjects of the investigations in which it is used.

Given the wide variety of guidance, counseling, and mental health professionals who employ the TSCS, such as school counselors, psychologists, psychiatrists, rehabilitation counselors, corrections specialists, and substance abuse counselors, it would seem that its potential for use by still other professionals would be quite limited. However, should the factor structure of the instrument continue to receive the confirmation that more focused analyses (McGuire & Tinsley, 1981; Walsh, Wilson, & McLellarn, 1984) have begun to provide, there is a distinct possibility that it could become an instrument routinely used in personnel selection and in diagnostic categorization. The more technical aspects of the factor-analytic studies will be discussed later.

Fitts has placed very few limitations on the types of subjects for whom the TSCS is appropriate beyond minimum age (12 years) and reading ability (sixth-grade). Although there may be at least one indication that young people under 16 years of age are not suitable subjects for the TSCS (McGuire & Tinsley, 1981), the behavioral science literature (Reed, Fitts, & Boehm, 1980) attests to its use with individuals of nearly every age and demographic characteristic. This reviewer has recorded the TSCS items and instructions for nonliterate and blind subjects, and adaptations for other groups of handicapped individuals would appear to offer no special challenge.

Administration of the TSCS requires only a quiet room equipped with a clock, so that subjects can record their starting and finishing times, and a secretary or clerk to hand out testing materials and collect them when the examinees are finished. Subjects can be tested individually or in groups. In this reviewer's experience, the instructions on the test booklet are generally sufficient and few questions are asked except by subjects under 16 years of age. Those questions that are asked will usually have to do with the numbers of the items in the booklet and/or in the answer sheet. Most subjects will require about 15 minutes to complete the TSCS, but individuals with poor reading ability may take 45 minutes or slightly longer.

The instructions for and illustrations of the scoring procedure in the TSCS manual are quite clear, even though the process is a lengthy one. The scoring matrix in the regular form version reduces much of the work to routine, but considerable concentration is required the first half-dozen or so times one scores a TSCS simply to insure that no steps have been overlooked.

The first five or six times the average professional hand scores a TSCS, 15 minutes or more will probably be required, but that time will soon be reduced to an average of 7 or 8 minutes, even for the longer clinical and research form. Scoring is primarily a matter of counting weighted keyed responses, often according to simple geometric or spatially segregated patterns that are drawn onto the scoring matrix. No clinical judgment or interpretations of any kind are needed.

In this reviewer's experience, machine scoring other than by computer is not feasible. The computer-scoring service provided by the publisher is good, but there is the inevitable time lag due to mail pick up and delivery. Unless the volume of tests processed is large, and especially for the counseling form, most users of the TSCS will probably wish to hand score their tests.

As exemplified in the manual and in Fitts' other writings on the TSCS, interpretation of test profiles is a matter first of reference to scale norms, second of attention to extreme scores and possible response sets and biases, and third of a substantial infusion of clinical knowledge, experience, and judgment. These elements are blended in proportions that appear to be unique to each test profile. A thorough grounding in the theory of personality, and especially in self-theory, is almost a necessity if complete use is to be made of the information provided by the TSCS profile. Indeed, it would be a very useful addition to the test manual if Fitts were to add an appendix that included what he believes to be the most important elements of self-theory in relation to the TSCS.

Although many, perhaps the majority, of test profiles from nonclinical subjects might be straightforward and amenable to analysis by an M.A.-level counselor, many test records require the expertise of a doctoral-level clinical or counseling psychologist if they are to be adequately interpreted. As a rule of thumb, interpretation should be done by doctoral-level practitioners.

Technical Aspects

The generation of items for the TSCS is perhaps the most appropriate place to begin a consideration of its technical aspects. As is usual Fitts began with a pool of items larger than he expected to use. He took them from previously developed self-concept measures and from written "self-descriptions" of samples of psychiatric patients and nonpatients. Based on this pool of items, ". . . a phenomenological system was developed for classifying items on the basis of what they themselves were saying" (Fitts, 1965). This evolved into the internal/external dimensions schema with equal numbers of positively and negatively worded items selected for each of the 15 facets. Then seven clinical psychologists, blind to the initial item classification, were asked to place each item into the three-by-five dimensional schema and to judge whether the item was positive or negative in content. Six items on which there was perfect agreement by these judges were retained for each facet.

While such a procedure is defensible it has the disadvantage that a particular sample of items was allowed to determine the overall structure of the scale. If the item sample was exhaustive in its representation of the dimensions or facets of self-concept, there would be no problem. But apparently no attempt was made, for example, to work backward and determine how adequate a mapping of self-concept had been achieved. Were there, for example, sets of written self-descriptions that were excluded from the scale because they did not correspond to the internal-external dimensions schema that a majority of the items fit into? The manual does not address this issue, but it is an important one.

The primary norm group for the TSCS was a sample of 626 people who varied in age from 12 to 68 years. The group was composed of approximately equal members of men and women and ranged over a variety of educational, social, and economic levels. There is a reasonable degree of racial and geographic diversity, but younger white subjects, especially students, are overrepresented.

Although the data have long been available to do so, Fitts has not constructed norms for different groups. There are two reasons for this: "First, it has been apparent that samples from other populations do not differ appreciably from the norms, provided they are large enough samples (75 or more). Second, the effects of such demographic variables as sex, age, race, education, and intelligence on the scores of this scale are quite negligible. With large samples (n = 100 or more) a few scattered scores will correlate significantly with these variables, but these correlations are usually in the .20s and thus account for very little of the variance" (Fitts, 1965, p. 13). Data from various studies (Sundby, 1962; Hall, 1964) tend to support these contentions.

The raw score distributions of almost all of the variables of the TSCS are approximately normal in form. The one exception, Number of Deviant Signs, has been subjected to a normalizing tranformation. All of the T-scores are based on a transformation that will result in distributions that are very close to normal if the raw score distributions are similar to those with which Fitts worked.

The reliability estimates for all TSCS scales are retest coefficients based on a sample of 60 college students over a two-week period and in general range from .60 to .90. These stability coefficients are, unfortunately, inappropriate. The assumption underlying all but the empirical scales is that they are homogeneous, and thus a measure of internal consistency such as coefficient alpha is what is called for. It is often argued that the retest coefficient provides a lower bound estimate of reliability, but as both Nunnally (1972) and Edwards (1970) have demonstrated, retest coefficients have *no* necessary relationship to reliabilities, being based on fundamentally different models. While retest coefficients must be a part of every comprehensive experimental design for the estimation of measurement error (Nunnally, 1972), they are, given the lack of alternate forms for TSCS scales, sensitive only to short-term change with respect to the variables in question. Adequacy of item samples, values of the standard error of measurement, and other important questions about the TSCS cannot be answered based on the reliability data given. The size and narrowness of the sample on which the retest coefficients are based are also liabilities. College students are not a good representation of the population as a whole, and a sample of size 60 (probably from one introductory psychology class) will not even provide an adequate picture of the population of college students. The question of the reliability of TSCS scales is an open one, but at the very least it can be said that the available data do not justify use of this instrument as an important component of any major decision about choices in a particular individual's life.

Numerous efforts at both convergent and discriminant validation of the TSCS have been undertaken, although the stated purpose may have been more specific. For the most part, these studies have suffered from a lack of the sort of focus that would be possible if there were a commonly accepted definition of self-concept and its components. As it is, the usual strategy has been to select particular variables of interest to an individual experimenter and to relate them correlationally to TSCS variables. As a perusal of any reasonably large sample of studies from Fitts' bibliography (Reed, Fitts, & Boehm, 1980) makes clear, there are a great many variables that correlate significantly with TSCS variables and a great many others that do not. It has been rare, although not unheard of, that a priori predictions have been made on the basis of either self-theory or some more general personality theory. Generally, correlational results have been justified on the basis of post hoc explanations and no consistent pattern of results that supports the *structure* of self-concept asserted by Fitts has emerged.

This is not to say good studies supplying evidence for discriminant validity have not been done, and Fitts cites several in the manual. They speak tellingly to the issue of discriminating between patient and non-patient groups, between various patient groups, and of the generalizability of these results. However, confirmation of self-theory predictions that the self-concepts of psychiatric patients or those of juvenile offenders will be less favorable and more disorganized than those of non-patients is not totally convincing, especially when detailed *and comprehensive* predictions with respect to the structure of self-concept as mapped by the TSCS seldom appear.

Concerning convergent and discriminant validity from a more purely psychometric point of view, Fitts (1965) presents correlations between the TSCS scales and those of the MMPI, the Edwards Personal Preference Schedule, and several other well-known measures. A great many of these correlations are significant; so many in fact where the MMPI is concerned it appears the two inventories must be nearly completely overlapping. Taken together with the fact that many of the TSCS scales are highly correlated with each other, one suspects that response-set variance must be a major determiner of responses to the TSCS. This seems especially likely since Fitts did not use the procedure pioneered by Edwards (1966) and Jackson (1974) of eliminating items with extreme social-desirability scale values from his item pool. In fact, Bentler (1972) has surveyed this evidence and concluded that the TSCS taps no more than two or three dimensions.

Regarding the TSCS structure as highly limited and redundant is not the only tenable viewpoint, however. McGuire and Tinsley (1981) reviewed a number of factor-analytic studies and concluded that none supported Fitts' specific hypothesis of 15 dimensions of self-concept. They also pointed out faults in many of these studies, especially the fact that most of them were oriented toward condensation of variance rather than testing structural hypotheses. Their own study (where it concerned college students) as well as one by this reviewer and colleagues (Walsh, Wilson, & McLellarn, 1984) provide some support for both the internal and external dimensions when the intercorrelations among the scales are examined by means of a multiple-group factor analysis. The most likely conclusion to draw from the factor studies is that the items of the TSCS do tap the internal and external dimensions envisioned by Fitts, but that the use of items that have a high degree of social-desirability saturation masks these dimensions through their confounding with response-set tendencies. In such a case one would tend to see the high correlations with scales such as those of the MMPI (Edwards, 1970) which are actually observed.

Critique

The TSCS is the most ambitiously and comprehensively conceived scale of self-concept that exists today. It is employed with some frequency in clinical and counseling settings and has figured in a great variety of research. While the process of its construction was well ordered and reasonable, constraints of unknown magnitude were imposed by the author's item selection procedure. The normative data provided for the TSCS are satisfactory, but the reliability measures are inappropriate and inadequate. From a research point of view, the TSCS has proven its utility as an instrument that frequently distinguishes between different groups, particularly clinical and nonclinical ones, but studies pertaining to the validity of its structure as presently mapped have provided mixed results. Thus, counseling college students or other subjects with respect to specific internal and external dimensions seems to be unjustified at this time, and the TSCS should probably be used primarily as a focal point for initiating discussion about a client's self-concept.

References

Below are listed not only references cited in this review, but also some older studies and some more recent that address the general issues of applied use of the TSCS, syntheses of evidence regarding its use, and its dimensional structure. Reading these studies should provide a reasonably broad frame of reference within which to consider strengths and weaknesses of the TSCS.

Bentler, P. M. (1972). Review of Tennessee Self Concept Scale. In O. K. Buros (Ed.), *The seventh mental measurements yearbook* (pp. 366-367). Highland Park, NJ: The Gryphon Press.

Boyle, E. S., & Larson, P. C. (1981). Factor structure of the Tennessee Self Concept Scale for an institutionalized, disabled population. *Perceptual and Motor Skills, 52,* 575-582.

Edwards, A. L. (1966). *Manual for the Edwards Personality Inventory.* Chicago, IL: Science Research Associates.

Edwards, A. L. (1970). *The measurement of personality traits by scales and inventories.* New York: Holt, Reinhart, & Winston.

Ezeilo, B. N. (1982). Cross-cultural utility of the Tennessee Self Concept Scale. *Psychological Reports, 51,* 897-898.

Fitts, W. H. (1965). *Manual: Tennessee Self Concept Scale.* Los Angeles, CA: Western Psychological Services.

Fitts, P. H. (1972). *The self concept and behavior: Overview and supplement* (Monograph VII). Nashville, TN: Counselor Recordings and Tests.

Fitzgibbons, D. J., & Cutler, R. (1972). The factor structure of the Tennessee Self Concept Scale among lower-class urban psychiatric patients. *Journal of Clinical Psychology, 28,* 184-186.

Gellen, M. I., & Hoffman, R. A. (1984). Analysis of the subscales of the Tennessee Self Concept Scale. *Measurement and Evaluation in Counseling and Development, 17*, 51-55.

Jackson, D. N. (1974). *Personality Research Form manual.* Goshen, NY: Research Psychologists Press.

Lund, N. L., Carman, S. M., & Kranz, P. L. (1981). Reliability in the use of the Tennessee Self Concept Scale for educable mentally retarded adolescents. *The Journal of Psychology, 109*, 205-211.

McGuire, B., & Tinsley, H. E. A. (1981). A contribution to the construct validity of the Tennessee Self Concept Scale: A confirmatory factor analysis. *Applied Psychological Measurement, 5*, 449-457.

Nunnally, J. C. (1978). *Psychometric theory* (2nd ed.). New York: McGraw-Hill.

Reed, P. F., Fitts, W. H., & Boehm, L. (1980). *Tennessee Self Concept Scale: Bibliography of research studies* (rev. ed.). Nashville, TN: Counselor Recordings and Tests.

Resnick, H., Fauble, M. L., & Osipow, S. H. (1970). Vocational crystallization and self-esteem in college students. *Journal of Counseling Psychology, 17*, 465-467.

Sherman, D. S. (1983). *Psychosocial correlates of adolescent substance abuse.* Unpublished doctoral dissertation, University of California, Los Angeles.

Stanwyck, D. J., & Garrison, W. M. (1982). Detection of faking on the Tennessee Self Concept Scale. *Journal of Personality Assessment, 46*, 426-431.

Stewart, D. W. (1976). Measuring self-concept: A multitrait-multimethod look. *Catalog of Selected Documents in Psychology, 6*, 41.

Vacchiano, R., & Strauss, P. (1968). The construct validity of the Tennessee Self Concept Scale. *Journal of Clinical Psychology, 24*, 323-326.

Walsh, J. A., Wilson, G. L., & McLellarn, R. W. (1984, April). *A confirmatory factor analysis of the Tennessee Self Concept Scale.* Paper presented at the meeting of the Rocky Mountain Psychological Association, Las Vegas.

Weinbaum, J., Fayans, A., & Gilead, S. (1982). Consistency across modalities of self/nonself-segregation. *Psychological Reports, 50*, 835-838.

Robert H. Bauernfeind, Ph.D.

Professor of Education, Northern Illinois University, DeKalb, Illinois.

TEST OF STANDARD WRITTEN ENGLISH

The College Board. New York, New York: The College Board Publications.

THE ORIGINAL OF THIS REVIEW WAS PUBLISHED IN TEST CRITIQUES: VOLUME VI (1987).

Introduction

The Test of Standard Written English (TSWE) is an objective (multiple-choice) test of editing skills designed for use with high school seniors and college freshmen. Students' raw scores are converted to scale scores ranging from a low of 20 to a high of 60+.

Initial forms of the TSWE were developed at Educational Testing Service in the early 1970s. The newer forms, carrying 1980s copyrights, consist of 50 five-choice items. These newer forms are distributed by the College Board in two ways. First, TSWE can be given along with the College Board SATs through the national Admissions Testing Program of the College Board. Second, TSWE can be purchased by high schools and colleges for local institutional use.

The test materials are visual and oriented to appropriate use of the English language. Thus, we should expect no adaptations for individuals who are visually impaired, nor translations into other languages.

The TSWE test booklet consists of 12 pages—a cover page, one page of directions, a page showing a sample answer sheet, six pages of test questions, and three pages that are blank. The sample test sent for review is printed black-on-white—not especially attractive, but certainly readable.

The TSWE consists of two item types. The first, called "Usage," has the student attempt to identify an error in a short sentence. Example:

Prefabricated housing <u>is economical</u> <u>because</u>
A   B

<u>they reduce</u> labor costs <u>considerably.</u> <u>No error</u>
C*   D   E

This kind of skill is tested in items 1-25 and 41-50.

The second item type, called "Sentence Correction," asks the student to select which of five parallel phrases best completes a short sentence. Example:

<u>Eddie was as angry as Linda was</u> when he discovered that thieves had stripped her car.

- *(A) Eddie was as angry as Linda was
- (B) Eddie had anger like Linda's
- (C) Eddie's anger was like Linda was
- (D) Eddie's anger was as great as Linda
- (E) Eddie had an anger as great as Linda

This kind of skill is tested in items 26-40.

511

TSWE will be quite easy for students who are good writers, but difficult for those who have not tuned in to the "logical flow" of good English and/or who are poor readers.

The examiners' responsibilities simply involve making certain that each student understands the test problems and the use of the separate answer sheet. The examiners also are responsible for calling time after exactly 30 minutes.

Practical Applications/Uses

The authors intend that TSWE will be used with high school seniors and with college freshmen. Noting that many colleges provide two semesters of English composition courses in the freshman year, Breland (1977a) suggested that TSWE scale scores be used to divide incoming freshmen into four groups:

Category 1. Students scoring 60+ might be exempted from freshman composition, provided they also pass an actual essay-writing test provided by the local college's English faculty. (Breland's studies found that only 4% of college-bound high school seniors scored 60+ on the TSWE.)

Category 2. Students scoring 45-59 might take only the second of the two writing courses. (Breland's studies found that 46% of high school seniors scored in the 45-59 range.)

Category 3. Students scoring 35-44 would be required to take both writing courses. (Breland's studies found that 27% of high school seniors scored in the 35-44 range.)

Category 4. In Breland's (1977a) opinion, students scoring below 35 would need remediation *before* enrolling in a college writing course. (Breland's studies found that 23% of high school seniors scored in the 20-34 range.)

Individual colleges will, of course, wish to make categories based on their own experience, but Breland's suggestions give each a model from which to make local adaptations.

TSWE is a group test, very easily administered. In a single English class it could be administered just by the teacher. In an all-school testing program, there should probably be one proctor for every 30 students. The 30-minute time limit means that TSWE can be easily administered in the usual 40-50 minute high school or college class period.

The College Board provides a complete scoring service. In addition, the authors provide an answer key (one correct answer per item) that is readily adaptable for hand-scoring or machine-scoring. One disturbing aspect of the TSWE program is that the authors use the antiquated correction-for-guessing formula in arriving at each student's final raw score. While TSWE was developed and normed at Educational Testing Service, so also were the Pre-Professional Skills Tests:

> Your score on any of the PPST is based on the number of questions you answer correctly. . . . On these tests, there is no subtraction for incorrect answers. Therefore, if you are uncertain about the answer to any of the multiple-choice questions . . . you should guess at the answer rather than not respond at all. You risk nothing by guessing; but if you do not respond, you lose the chance to

raise your score by making a lucky guess. (Educational Testing Service, 1985, p. 8)

And the Law School Admission Test:

Answer every question—guess if necessary. When we score your answer sheet, we will give you credit for every correct answer but deduct nothing for wrong ones. (Educational Testing Service, 1974, p. 12)

The right hand knoweth not. . . .

The correction-for-guessing formula imposed on TSWE scores is disturbing for at least two reasons. First, it is unnecessary and probably counterproductive (Bauernfeind, 1978). Secondly, it requires *three* additional steps in the scoring process, making it extremely error-prone in high schools and colleges that do their own scoring. Schools using the College Board's scoring services will have to live with the correction-for-guessing procedures. But for schools doing their own scoring, this reviewer strongly urges use of the kinds of directions cited above, to be followed by local norms and local cutoffs on the items-right raw scores.

TSWE is a straightforward test of ability to recognize correct standard English. It is further intended to distinguish among students who need help in this skill. It is *not* intended to distinguish among students whose command of standard English is "better than average" (College Board, 1984b).

Technical Aspects

The TSWE package sent to test reviewers contains a strange assortment of materials. We are given (in my opinion) an excellent test of editing skills and a summary of Breland's (1977a) interesting article. The Breland article makes repeated references to 50% common variance with ratings of actual essays, thus suggesting correlations around .70 between TSWE scores and essay ratings. Such correlations, not corrected for attenuation, would be remarkably high. But, we are given no scoring key, no information about the 20-60+ scale, no information about the percentile rank norming group, no information about interform equating studies, and not a single report of a reliability coefficient.

The review package *does* tell us the following:

1. The percentile rank scores "are based on the scores of candidates who took the TSWE in 1977-78" (College Board, n.d.). The word "candidates" suggests that the percentile rank scores are based on scores of college-bound high school seniors.
2. TSWE is fairly easy. The norms suggest a median number-correct raw score of about 33 for Form WWE6, and about 34 for Forms WWE1 and WWE2.
3. The wide spread of raw scores indicates that these scores are quite reliable. This reviewer estimates the KR-21 reliability for the percentile-rank norms group to be .88. However, there is a skew to the left in the raw score distribution. This means that the percentile-rank scores are more reliable with low-scoring students, less reliable with high-scoring students.

Reviewers who write for a conventional technical report are sent only a copy of the original project report by Breland (1977b). Although the Breland report is *not* a conventional technical report, it does show some interesting correlation data involving TSWE scores:

	r	N
SAT Verbal	.73	7013
SAT Mathematical	.53	7013
Essay, early in Freshman year	.63	770
Essay, later in Freshman year	.58	766
Essay, total of three	.72	261
TSWE Posttest (Fall)	.83	789
TSWE Posttest (Spring)	.84	493

Source: Breland (1977b, pp. 38, 47)

In "Essay, total of three," the TSWE scores were correlated with the sums of ratings of three student essays gathered at different times during the college freshman year. Correlations with TSWE posttests constitute reliability studies of a sort, but there were intervening treatments between the TSWE testings, making it very difficult to interpret these data.

The Breland (1977b) study was focused on *uses* of the TSWE, not on the test itself. Nowhere in the Breland report does one find an explanation of the 20-60 + scale, a description of the percentile rank norming group, or a single report of an internal consistency reliability coefficient.

One final effort by this reviewer to obtain such basic technical information brought forth a 225-page technical report that had been published in 1984 (College Board, 1984a). This totally unpublicized technical handbook is one of the best I have ever seen. Among its revelations:

1. The TSWE scale *is* equated to the SAT 200-800 scale, except that TSWE deletes the final zero on the SAT scale. Thus, TSWE 40 is equivalent to SAT 400, TSWE 50 is equivalent to SAT 500, and so on (p. 20).
2. The TSWE is fairly easy for good editors. Rather than following a bell-shaped curve, raw scores are somewhat bunched at the top, skewed at the bottom. For this reason it was not possible to set the scores to the bell-shaped 20-80 scale; rather, the bunched high scores are simply given a scale value of 60 + (p. 20).
3. Each new form is anchored to previous forms using conventional linear methods. SAT Verbal scores are also referenced in scaling new forms of the TSWE (p. 20).
4. TSWE correlations with SAT Verbal and SAT Mathematical have been running slightly higher than those reported in the Breland study. Across eight studies, TSWE scores showed a median correlation of .78 with SAT Verbal, .64 with SAT Mathematical (p. 81).
5. In studies in 25 colleges, TSWE scores showed a mean correlation of .37 with grades in English courses (pp. 166-167). Note, however, that "English courses" often include skills other than writing/editing.
6. TSWE is essentially a power test. Typically around 85% of students complete the entire test (pp. 80-81).
7. TSWE reliabilities are quite satisfactory for the groups tested. Across eight

studies the median KR-20 coefficient was .88; across six studies the median test-retest coefficient was .82 (p. 82).

An incidental note: While 500 was at one time the mean scale score for 12th-graders taking the SAT, the mean had fallen to V = 427, M = 468 during the 1977-78 school year (Educational Testing Service, 1978). The mean of those two means is 447.5, which accounts for the fact that the median 20-80 scale score on TSWE is about 44, not 50 (College Board, 1984b).

Critique

This reviewer would offer three suggestions to the authors of the TSWE. First, redesign the review package to make clear the fact that the original Breland study and a full technical handbook are in print and available. Secondly, develop number-correct norms for the benefit of high schools and colleges doing their own scoring. In addition, update the percentile rank norms. While the 1977-78 norms are probably very close to current norms (because the SAT means are very similar), it would look better to students and test critics if the norms appeared to be current.

However, those considerations are peripheral. The "heart" of any test embodies the test items themselves, and the heart of TSWE is in good shape. Although each prospective test user will have to judge the "content validity" of the TSWE, this reviewer considers it very high. The items reflect real-life communication problems: the authors do *not* test for *laid/lain, can/may, better/best,* or other "errors" that in fact do *not* interfere with clear communications. While the task is clearly one of "editing," the test items come so close to real life that I will not quarrel with those who call this a test of "writing skills."

The test is relatively easy. It focuses on low-scoring students and the kinds of errors they make. The raw scores of good editors are bunched at the top of the test; these high-scoring students will be told that they are good with standard English and, as suggested by Breland, they might be excused from one or more college English composition courses.

I gave this 50-item test to five people who work as "copy editors" for newspapers, magazines, test publishers, or book publishers. The raw scores were 49, 49, 49, 50, and 50. Four of the five volunteered that this is "a very good test." The usefulness of the TSWE could be greatly enhanced if the authors would generate raw score distributions for successful copy editors, successful secretaries, and successful writers (James Reston and Mike Royko types). These latter kinds of data could be extremely helpful to high school and college students who want something more than a norm-referenced score within a local norms group. In short, TSWE may have a much greater variety of uses than its authors have proclaimed.

References

Bauernfeind, R. H. (1978). *Building a school testing program* (2nd ed.). Bensenville, IL: Scholastic Testing Service.

Breland, H. M. (1977a). *Can multiple-choice tests measure writing skills?* (The College Board Review No. 103). New York: College Board.

Breland, H. M. (1977b). *A study of college English placement and the Test of Standard Written English.* Princeton, NJ: Educational Testing Service.

Cohn, S. J. (1985). Review of the College Board Scholastic Aptitude Test and Test of Standard Written English. In J. V. Mitchell (Ed.), *The ninth mental measurements yearbook* (pp. 360-362). Lincoln, NE: The Buros Institute of Mental Measurements.

College Board. (n.d.). *Guidelines for TSWE administrations.* New York: Author.

College Board. (1980). *Test of Standard Written English—Form M-WWE6.* New York: Author.

College Board. (1984a). *The College Board technical handbook for the Scholastic Aptitude Test and Achievement Tests.* New York: Author.

College Board. (1984b). *Information for candidates about the Test of Standard Written English.* New York: Author.

Cronbach, L. J. (1985). Review of the College Board Scholastic Aptitude Test and Test of Standard Written English. In J. V. Mitchell (Ed.), *The ninth mental measurements yearbook* (pp. 362-364). Lincoln, NE: The Buros Institute of Mental Measurements.

Educational Testing Service. (1974). *Law school admission bulletin 1974-1975.* Princeton, NJ: Author.

Educational Testing Service. (1978). *1977-1978 update for Table 2 in* On Further Examination. Princeton, NJ: Author.

Educational Testing Service. (1985). *Bulletin of information for the Pre-Professional Skills Tests of Reading, Writing, and Mathematics.* Princeton, NJ: Author.

Richard M. Ryan, Ph.D.
Assistant Professor of Psychology, University of Rochester, Rochester, New York.

THEMATIC APPERCEPTION TEST

Henry A. Murray. Cambridge, Massachusetts: Harvard University Press.

THE ORIGINAL OF THIS REVIEW WAS PUBLISHED IN TEST CRITIQUES: VOLUME II (1985).

Introduction

The Thematic Apperception Test (TAT) is, along with the Rorschach, among the most widely used, researched, and taught projective tests in existence (Wade, Baker, Morton, & Baker, 1978). It is also among the most controversial devices, with proponents citing its manifold clinical uses and long history of provocative and sometimes impressive research findings, and critics describing it as a psychometrician's quicksand (Varble, 1971; Entiwisle, 1972; McClelland, 1980). What becomes evident following any comprehensive review is that both views have some validity. Used in theoretically appropriate contexts and with regard to relevant criteria, the TAT can demonstrate remarkable sensitivity and utility. Nonetheless, when subjected to the criteria developed for and appropriate to many objective test instruments, the TAT provides little firm ground upon which the traditional psychometrician could stand.

The TAT is now at its golden anniversary, being introduced in 1935 by Morgan and Murray with only minor changes since that inception. It consists of a series of pictures of relatively ambiguous scenes to which subjects are requested to make up stories or fantasies concerning what is, has, and is going to happen, along with a description of the thoughts and feelings of the various characters depicted. The test protocol thus provides the examiner with a rich source of data, based on the subject's perceptions and imagination, for use in the understanding of the subject's current needs, motives, emotions, and conflicts, both conscious and unconscious (Murray, 1943). The data from the TAT can be scored according to a variety of existing quantitative systems. However, more commonly in clinical use the stories are interpreted in accord with general principles of inference derived from psychodynamic theory. Its use in clinical assessment is generally part of a larger battery of tests and interview data, which provide the background and convergent information necessary for appropriate interpretation.

As a research device, the TAT can be best understood as a method of eliciting a sample of perceptual and verbal material in response to a set of standard stimuli, which can lend itself to a plethora of empirical uses. It can be rated on nearly any clearly defined dimension or criterion, content based or formal in character, that can be reliably assessed. Its status then, both as a clinical and research instrument, is based primarily on its capacity to elicit an abundant, complex, albeit somewhat variable, response from varied subjects to a common situation. Accordingly, the strength of the "test" is no better than the properties of the chosen scoring system

517

or the interpretive skills of the examiner in dealing with the material this situation generates. In this sense the TAT is more a method than a "test."

Aside from its own specific properties and popularity, the TAT is also important as a projective paradigm. Since its release 50 years ago, the technique of using relatively ambiguous pictures to elicit free response stories has been widely emulated. Close relatives include the Children's Apperception Test (CAT) and the Senior Apperception Test (SAT) (Bellak, 1975), but its lineage can also be traced to a still growing extended family that includes the Blacky Pictures Test (Blum, 1950), the Rosenzweig Picture-Frustration Study (Rosenzweig, 1976), the Michigan Pictures Test-Revised (Hutt, 1980), the Tasks of Emotional Development (TED) (Cohen & Weil, 1975), the Symonds (1948), the Picture Arrangement Test (PAT) (Tomkins & Miner, 1957), and undoubtedly numerous others. Although these descendants may have applications to specific populations or may "pull" for issues and themes not covered by the TAT, they all rely to some extent on the assumptive basis of their originator. Before explicating the specific applications and scoring systems of the TAT, the projective assumptions that underlie it and all of its imitators should be presented.

The major uses and interpretation of the TAT are based on several key interrelated principles herein subsumed under the general rubric of the projective hypothesis or viewpoint. The projective viewpoint is widespread among practicing clinicians and personality psychologists and is intimately interwoven with psychodynamic theory, yet it has never fully achieved acceptance with mainstream academic psychology (Karon, 1981). Indeed projective psychology has often been described as dissident, or a "psychology of protest" (Abt, 1952) with respect to more behavioral or traditional scientific approaches, embracing values and beliefs not shared by that wider community (Dana, 1982). This has undoubtedly influenced the selection of individuals who comfortably employ a technique like the TAT.

Projective psychology assumes as a starting point that individuals' perceptions or behavior cannot be understood merely as an outcome of impinging, objectively describable stimuli. Rather, that which determines psychological and behavioral processes is assumed to be at least in part an outcome of subjective interpretation of what the environment affords. Perception is conceived as an active, constructive, and selective process, the organization of which is influenced by the person's unique capacities and history, and more importantly by current drives or needs that energize such processes. In short, perception is never neutral but rather selective and motivated.

Without doubt, however, projective psychology recognizes that strong situational determinants lead predictably to similar perceptual and behavioral responses across most subjects. Skinner (1957) called such potent situational determinants "mands." However, most situations in life are more equivocal and subject to multiple meanings, a fact that leads to individual differences in responses of all kinds. Indeed the more ambiguous the situation, the more a subject's own current needs, motives, or propensities are likely to play a role in determining its meaning and the actions that occur with respect to it.

The rationale of projective tests follows from this reasoning. By providing relatively ambiguous stimuli along with minimal constraints or structuring of the response that follows, the assumption is that a person's needs, motives, sentiments, and interests can be allowed maximal play in the formation of a response. Perceptions or responses elicited in the projective situation are thus evaluated in terms of the underlying dynamic processes in the subject's "inner world" that are assumed to have shaped them. Projective tests are thus of most significant value when used in the assessment of motives, needs, and other organizing influences on perception and behavior.

Nonetheless, the emphasis is on the *relative* ambiguity of the stimuli and nonstructuring of response in projective methods because a wide variety of research has shown that even projective stimuli mand or "pull" for certain types of responses (Peterson & Schilling, 1983) and the testing situation always places some constraints upon response. Card pull and response structuring are a recognized and important element in projective test construction and scoring, and in the TAT in particular.

Test Description. The TAT materials consist of 30 pictures with a variety of subject matters and themes portrayed. The majority of the cards depict life situations involving one or more persons, based on the assumption that this facilitates the subject's projection of motives, emotions, and attitudes. Each card is coded on the reverse side of the picture with numbers and letters to designate the order of administration and to whom it should be administered, i.e., whether it is appropriate to adult males (M), adult females (F), boys (B), girls (G), or some combination. Thus, for example, the code 3BM is designed for use with boys and adult males, whereas 3GF would be used with either girls or adult females. Ten cards merely have a number without letter codes, indicating their appropriateness for subjects of all ages and both gender. Using this coding system there is a series of 20 sequentially arranged cards appropriate for use with any given group of subjects.

Murray (1943) originally advocated that subjects receive all 20 cards divided over two sessions. In actual practice, however, few examiners administer all 20 cards. Generally, both clinicians and researchers select some smaller subset of cards ranging from 6-10 in number. Some authors have identified subsets of cards that they advocate for use with all subjects, such as Arnold (1962). In her "story sequence analysis" method she recommended 13 cards (1, 2, 3BM, 4MF, 6BM, 7BM, 8BM, 10, 11, 13MF, 14, 16, and 20) regardless of the subject's gender. Dana (1956) also selected a subset of cards (2, 3GF, 4, 6GF, and 7GF for females; and 2, 3BM, 4 6BM, and 7BM for males) on which he collected extensive normative data. Finally, the popular Menninger Institute method employs the following sequence: 1, 5, 15, 3BM, 14, 10, 7GF, 13MF, 18GF, and 12M.

The most common practice, however, particularly in clinical applications, is to preselect cards that elicit themes that are thought to be pertinent to a given examinee's conflicts or concerns. Such selection, of course, will vary from person to person and depends on what is called the "pull" of the card (i.e., the properties of the card that tend to evoke particular affective and/or thematic responses across subjects). Presumably it is the manner in which subjects respond or adjust to the pull of each card that provides the information about their own idiosyncratic motives, conflicts, and capacities.

A number of investigators have researched the particular properties of the cards with regard to pull or response, eliciting properties using normative analyses of content or themes (Eron, 1950; Murstein, 1963; Campus, 1976), formal characteristics (Dana, 1955, 1956), or semantic differential techniques (Goldfried & Zax, 1965). Murstein (1963) has gone so far as to state that the "stimulus is by far the most important determinant of a TAT response" (p. 195). Although not everyone would agree with this point, it nonetheless behooves the examiner to be familiar with the properties of the cards and the typical responses they elicit. What follows is a list of the 31 cards (30 pictures and 1 blank), along with a brief description of their features and more suggestions as to typical themes and issues they elicit. The interested examiner should beware that this list of themes is in no way intended to be exhaustive. The card description is based on Murray's (1943) original test manual.

Card 1: A young boy is contemplating a violin which rests on a table in front of him. Common themes and issues: needs for achievement; autonomy, particularly with respect to parents and/or authorities; self- versus other-motivation.

Card 2: Country scene—in the foreground is a young woman with books in her hand; in the background a man is working in the fields and an older woman is looking on. Common themes and issues: family relations; separation and individuation; achievement values and aspirations; pregnancy issues.

Card 3BM: On the floor against a couch is the huddled form of a boy with his head bowed on his right arm. Beside him on the floor is a revolver. Common themes and issues: depression, helplessness, suicide; guilt; impulse control; handling of aggression.

Card 3GF: A young woman is standing with downcast head, her face covered with her right hand. Her left arm is stretched forward against a wooden door. Common themes and issues: depression; loss; suicide; guilt.

Card 4: A woman is clutching the shoulders of a man whose face and body are averted as if he were trying to pull away from her. Common themes and issues: male-female relationships; sexuality; infidelity; interpersonal control, dominance, and conflict.

Card 5: A middle-aged woman is standing on the threshold of a half-opened door, looking into a room. Common themes and issues: attitude toward mother or wife; guilt; autonomy issues; fear of intruders; paranoia.

Card 6BM: A short elderly woman stands with her back turned to a tall young man. The latter is looking downward with a perplexed expression. Common themes and issues: mother-son relations; loss and grief; separation-individuation.

Card 6GF: A young woman sitting on the edge of a sofa looks back over her shoulder at an older man with a pipe in his mouth who seems to be addressing her. Common themes and issues: daughter-father or male-female relationships; heterosexual relationships; interpersonal trust; employer-employee relationships.

Card 7BM: A gray-haired man is looking at a younger man who is sullenly staring into space. Common themes and issues: father-son relationships; employer-employee relationships; authority issues.

Card 7GF: An older woman is sitting on a sofa close beside a girl, speaking or reading to her. The girl, who holds a doll in her lap, is looking away. Common themes and issues: mother-daughter relationships; rejection issues; child-rearing attitudes and experiences.

Card 8BM: An adolescent boy looks straight out of the picture. The barrel of a rifle is visible at one side, and in the background is the dim scene of a surgical operation, like a reverie-image. Common themes and issues: aspirations and achievement; handling of aggression; guilt; fears of being harmed; oedipal issues.

Card 8GF: A young woman sits with her chin in her hand, looking off into space. Common themes and issues: because of card ambiguity it elicits very diverse themes, aspirations; sense of future possibilities often noted.

Card 9BM: Four men in overalls are lying on the grass, taking it easy. Common themes and issues: homosexuality; male-male relationships; work attitudes; social prejudice.

Card 9GF: A young woman with a magazine and a purse in her hand looks from behind a tree at another young woman in a party dress running along a beach. Common themes and issues: female-female relationships; rivalry; jealousy; sexual attack; trust versus suspicion; suicide.

Card 10: A young woman's head rests against a man's shoulder. Common themes and issues: marital or parents' relationships; intimacy; loss or grief.

Card 11: A road skirts a deep chasm between high cliffs. On the road in the distance are obscure figures. Protruding from the rocky wall on one side is the long head and neck of a dragon. Common themes and issues: unknown, threatening forces; attack and defense; aggression. (Good card for assessing imaginative ability.)

Card 12M: A young man is lying on a couch with his eyes closed. Leaning over him is the gaunt form of an elderly man, his hand stretched out above the face of the reclining figure. Common themes and issues: health; homosexuality; father-son relationships; issues of control; response to psychotherapy.

Card 12F: The portrait of a young woman. A weird old woman with a shawl over her head is grimacing in the background. Common themes and issues: mother or mother-in-law relationships; guilt and superego conflicts; good and evil.

Card 12BG: A rowboat is drawn up on the bank of a woodland stream. There are no human figures in the picture. Common themes and issues: loneliness; nature; peace; imaginal capacities; suicide.

Card 13MF: A young man is standing with downcast head buried in his arm. Behind him is the figure of a woman lying in bed. Common themes and issues: sexual conflict and attitudes; heterosexual relationships; guilt; handling of provocative stimulus; aggression.

Card 13B: A little boy is sitting on the doorstep of a log cabin. Common themes and issues: loneliness; abandonment; childhood memories. (Stories are extremely diverse due to ambiguity of picture.)

Card 13G: A little girl is climbing a winding flight of stairs. Common themes and issues: childhood memories; loneliness. (Themes extremely varied due to ambiguity of picture.)

Card 14: The silhouette of a man (or woman) against a bright window. The rest of the picture is totally black. Common themes and issues: wishes and aspirations; depression; suicide; loneliness; burglary; intrapsychic concerns.

Card 15: A gaunt man with clenched hands is standing among gravestones.

Common themes and issues: death; religion; fantasy; aggression.

Card 16: Blank card. Common themes and issues: extremely varied—handling of unstructured situation; imaginal capacities; optimism versus pessimism.

Card 17BM: A naked man is clinging to a rope. He is in the act of climbing up or down. Common themes and issues: achievement and aspirations; homosexuality; optimism and pessimism; danger, escape, competitiveness.

Card 17GF: A bridge spans over water. A female figure leans over the railing. In the background are tall buildings and small figures of men. Common themes and issues: loneliness; suicide; intrapsychic concerns.

Card 18BM: A man is clutched from behind by three hands. The figures of his antagonists are invisible. Common themes and issues: alcoholism, drunkenness; homosexuality; aggression; paranoia; helplessness.

Card 18GF: A woman has her hands squeezed around the throat of another woman, whom she appears to be pushing backward across the banister of a stairway. Common themes and issues: aggression, particularly mother-daughter; rivalry; jealousy; conflict.

Card 19: A weird picture of cloud formations overhanging a snow-covered cabin in the country. Common themes and issues: highly varied; imaginal capacities.

Card 20: The dimly illuminated figure of a man (or woman) in the dead of night leaning against a lamppost. Common themes and issues: loneliness; fears; aggression.

Administration. In addition to a selection of cards for the subject, the examiner should come prepared with a stopwatch, pencils, and paper. The stopwatch is used to time response latency, and the pencils and paper are used for response recording. Some examiners prefer an audio recording device.[1] Nonetheless, paper and pencils should be available to note any specific behavioral or emotional reactions of the subject not accessible through the voice recording.

The single most important consideration in TAT administration is the creation of a psychological atmosphere in which the examiner feels relaxed, comfortable, and freely available to respond to the situation. Examiner rigidity or lack of friendliness or evaluative attitude is likely to result in constriction or discouragement of the subject, an unwanted situational contribution to the results. A quite, comfortable physical setting is most suitable for the session. Murray advocated that when possible the examinee be seated with his or her back to the examiner, presumably to facilitate free responding. However, many test users find that this standard arrangement is somewhat unnatural and instead sit across from or adjacent to the subject. The particulars of the seating arrangement are less important than maximizing rapport and freedom of response.

Appropriate subjects are those who range in age from 4 through adult. However, since projective story tasks more pertinent to childhood issues are available, the TAT is most widely used for subjects who are late adolescents or older.

Instructional sets have been varied by many examiners; thus, in actual practice there is not one standard phrasing. However, it is very important to introduce the following elements: 1) the subject is to make up a dramatic or imaginative story; 2) the story should include a description of what is occurring in the picture, what led up to it, and what will occur; and 3) the story should include something about what the various characters are thinking and feeling. Murray's (1943) manual used the following set for adults:

> I am going to show you some pictures, one at a time; and your task will be to make up as dramatic a story as you can for each. Tell what has led up to the event shown in the picture, describe what is happening at the moment, what the characters are feeling and thinking; and then give the outcome. Speak your thoughts as they come to your mind. Do you understand? Since you have fifty minutes for ten pictures, you can devote about five minutes to each story. Here is the first picture. (p. 3)

Murray's instructions also included a statement that this was "a test of imagination, one form of intelligence" (p. 3). However, because such a statement can create an ego-involved approach to the task (Ryan, 1982) it is not advocated by the present author. Murray also had a separate set of instructions for children:

> I have some pictures here that I am going to show you, and for each picture I want you to make up a story. Tell what has happened before and what is happening now. Say what the people are feeling and thinking and how it will come out. You can make up any kind of story you please. Do you understand? Well, then, here is the first picture. You have five minutes to make up a story. See how well you can do. (p. 4)

At times subjects may need these instructions repeated, in full or in part, particularly if the examiner is interested in obtaining a complete story with all the elements for each picture. In any case, it is noteworthy when subjects, once having grasped the task, omit certain elements either consistently or in response to certain cards.

After the initial instructions are presented, the examiner hands the first card of the selected series to the subject. The examiner then begins timing, recording the number of seconds that pass from the presentation of the card to the point at which the subject actually begins the true response. Some comments, exclamations, or halting verbalizations given by the subject may not be the beginning of the story, although they should be noted. The measure of time from card presentation to response is called the *response latency,* and is useful in gauging the subject's handling of the specific card. For example, very long response latencies may reflect a struggle with conflictual material. On the other hand, extremely short responses may suggest an impulsive or perhaps counter-phobic approach to the card material.

The examiner should record all of the subject's verbalizations, including spontaneous verbalizations, extra-test comments, utterings, laughter, etc., as well as the story response itself. In addition, it is extremely helpful to note specific behavioral and affective reactions such as fidgeting, facial expressions, sighs, gestures, etc., as these may reflect aspects of the subject's psychological response.

The examiner should interfere with a subject's responses as little as possible. However, at times it may be important to remind subjects about parts of the tasks being omitted, or to repeat something they miss. If the subject asks for more detailed instructions as to what to do, or seeks feedback or input from the examiner, it is best to respond openly but noncommittally, such as stating "You may make it anything you please." Discussion is to be avoided. Occasionally certain subjects, particularly children, may require encouragement, but this should be applied judiciously.

Following the administration of all cards, many examiners like to obtain further information from the subject about specific responses, a procedure often called the *inquiry*. This may be introduced by saying "I have a few more questions I'd like to ask," or alternatively by providing a more elaborate rationale concerning one's interest in the factors that enter into imaginal construction as Murray did. In any case, the inquiry should be accomplished in a way that minimizes evaluation apprehension. Typical foci of the inquiry are the subject's possible sources for the themes that emerged, associations to the cards or stories, or emotional reactions to particular stimuli. Other examiners have asked subjects to recall their most and least favorite cards or stories. This is especially useful if a day or more has passed since the actual administration. Finally the inquiry may be useful in "testing the limits" of certain responses. Thus the examiner can inquire about aspects of the pictures that may have been conspicuously omitted or distorted in the stories, or encourage elaborations to themes or stories that were particularly barren or sparse.

Murray's Theory and Its Relation to the TAT Interpretation. The assumptions associated with the projective hypothesis are given perhaps their most elegant expression in the personality theory of Murray (1938), whose constructs and formulations formed the basis through which the TAT was originally developed and interpreted. Murray articulated a dialectical theory of behavior that considers both the psychobiological and environmental determinants of human action. He provided a system that describes both how people are influenced by external forces and how they select and organize their own actions based on current needs and values.

The two central constructs in his psychology were those of *needs* and *press*, both of which the TAT was initially developed to assess. According to Murray, needs represent a force or energy that "organizes perception, apperception, intellectualization, connotation and action" (1938, p. 123). Needs play a role in the selection of those aspects of the world that are perceived and what meaning is given to them. Furthermore, they energize behavior in the direction of their satisfaction. Needs can be classified as either primary or biological (e.g., hunger, thirst, sex), or secondary or psychogenic (e.g., achievement, affliction, autonomy, dominance). Specific needs can be more or less dominant at any given time, or alternatively weak. They can also conflict, fuse, or become subsidiary to one another. Needs can also either arise spontaneously or be evoked by certain environmental circumstances. In any case, needs represent the most important internal determinants of behavior.

In contrast, press refers to the power of events in the environment to affect or influence a person. Murray described two kinds of press, which he designated as "Alpha" and "Beta." Alpha press is the influence on the person of objective or "real" external forces. Beta press concerns the subjective components of those forces, or pushes and pulls, that the individual perceives to be affecting him or her in the world. What is forceful and salient to one individual is not necessarily so for another.

Needs and press interact to form *thema,* dynamic combinations or "molar units" that reflect significant patterns of behavior (Maddi & Costa, 1972). Thema may be discreet or complex and can vary in importance. Fantasy productions will display various thema since here, in the subjects's verbal behavior, needs and press of the stimulus will combine to produce characteristic reactions.

This skeletal presentation of Murray's theory is introduced merely to illustrate how individuals' behavior (and TAT stories) can be conceptualized within a framework that includes consideration of needs, press, and thema. These bare bones, however, fail to capture the real body of the theory, which is indeed intricate and comprehensive. Nonetheless, it provides description of some of the major dimensions upon which a subject's TAT stories can be evaluated.

In interpreting a TAT protocol the initial task is to identify in each story the protagonist, who Murray called the *hero.* It is assumed that, in developing their fantasies, subjects identify with the hero and are therefore likely to project onto this figure needs, motives, beliefs, or emotions that are actually the subjects' own. At times subjects may identify with more than one figure, in which case attributes of other figures might be considered.

The next step is to analyze and quantify the needs that the hero expresses or exhibits. For this purpose Murray developed a list of 36 needs, both primary and secondary, that are rated by the examiner on a scale from 1-5 in terms of the intensity and centrality of their expression within the story. A similar procedure is used for rating press as revealed by the stories. Press includes such forces as physical dangers, interpersonal rejections, deaths, and other significant forces in the environment that may affect the hero.

Needs and press then must be put together in a meaningful way that corresponds to the consideration of thema and story outcomes. For instance, does the hero achieve success or failure? Does he or she do right or wrong? Does the hero engage in or flee from relationships? Does he or she fall into depression or turn to optimism in reaction to adversity? What interests or endeavors does the hero typically form in the force of needs or press?

From both these quantitative and qualitative considerations interpretation follows. Here, undoubtedly, the examiner's theoretical bias plays a significant role in the organization and integration of the data into some useful formulation, as well as the use to which that formulation will be put. Useful guidelines for using Murray's system are provided by several sources, the most up-to-date and comprehensive of which is offered by Stein (1981).

Although Murray's own "personological" theory undoubtedly spawned the development of the instrument and his scoring system, soon after its release a variety of researchers and clinicians used the TAT method as a way of studying their own specific interests or followed divergent pursuits. Some of these resulted in quantitative systems, others in qualitative ones, and many, like Murray's own, combined elements of both. In addition, some systems are oriented primarily toward the scoring or interpretation of the stories' content, while others focus on formal aspects (i.e., the manner in which the story is constructed or its form). Among some of the more noteworthy of these various approaches are those presented by Eron (1950); Dana (1955, 1982); McClelland et al. (1953); Veroff (1958); Arnold (1962); Cramer (1983); Bellak (1975); Bunin (1978); Rapaport, Gill, and

Schafer (1968); and Wyatt (1947). Some of these will receive further elaboration later in this review.

Practical Applications/Uses

Clinical Use. Despite the fact that the clinical use of the TAT appears in many contexts to be essentially ad hoc and idiosyncratic, the vast literature pertaining to its interpretations contains a fairly consistent core of strategies and lore. Although not intended in any way to be exhaustive or authoritative, what follows is a distillation of some of these common themes of the interpretive literature. There is no replacement for direct contact with the empirical and clinical literature in the topic. For heuristic purposes, *formal* considerations and interpretation of *content* will be examined separately here, although the two often overlap. Formal analyses are those that stem from the study of *how* a subject constructs or presents his or her responses; content analyses emerge from the study of *what* the subject provides (Holt, 1958).

TAT stories are, before all else, samples of the subject's verbal behavior. As such they can be used to assess aspects of the subject's language fluency, degree of concreteness versus abstractness, coherence of thought, and intellectual capacities. Assessing such formal aspects is often a useful starting point for interpretation.

There are three levels of possible responses to the card stimuli that can be used as gauges for the examinee's current functional capacities. The most primitive style of response is one of *enumeration,* in which the subject simply lists or names various elements or characters in the picture without elaborations or connections between them. A "story" is not developed. For instance, a subject might say, in response to Card 1, "A boy, a violin. Nothing else in the picture." Somewhat more complex is a *descriptive* style of response, which occurs when the subject describes the card and even what may be occurring, but does not complete a story or elaborate beyond what is explicitly presented. Elements may be linked together, but there is no attempt to move far beyond the stimulus characteristics per se. Finally, and most usually, subjects will offer an *interpretive* response to the card, incorporating the stimulus elements into a fantasy or story that extends well beyond the presented elements into a rich network of projections.

Adult subjects who, if not neurologically impaired, merely enumerate or describe the explicit elements of the cards may be markedly defensive or constricted, which prevents them from more freely responding to the evocative material at hand. Such a situation, although minimal in terms of interpretive possibilities, is helpful in determining the person's functional level. Interpretive responses, on the other hand, are most common and more useful to the examiner because as they afford more complex dynamic interferences and interpretations.

Also noteworthy in formal approaches is the subject's level of fluency and productivity. Fluency can be related to aspects of intelligence, social background, and developmental level. Productivity is a useful variable in considering the energy and inner resources available to the test taker. Depressed subjects, or those who are markedly constricted or guarded for other reasons, often provide sparse or impoverished records. Finally, the degree to which stories have manifest internal coherence or consistency can be an important issue and help in the assessment of the integrity and control of thought processes.

Another level of formal investigation involves examining the degree to which the subject has given each of the requested elements of the story (i.e., has described the situation, what preceded and followed it, and the thoughts and feelings of the characters). According to Rapaport et al. (1968) the failure to provide specific elements may result from the fact that what comes to the subject's consciousness is conflictual and therefore suppressed, or perhaps repressed altogether. At any rate, the general issue of the way in which the subject complies with the task instruction is seen frequently as having dynamic significance.

Formal analyses have also been directed to more microscopic levels, as in examining linguistic characteristics such as number and type of verbs used relative to other elements (Henry, 1956; Wyatt, 1947). A related technique of considerable clinical interest was suggested by Hutt (1980) with reference to another projective test (MPT-R) in which stories are evaluated for the direction of forces depicted. A story is rated as *centrifugal* when the hero acts upon his world and as *centripetal* when the hero is acted upon. Hutt reports evidence suggesting that this direction of forces construct can be useful for assessing adjustment levels.

Despite the fact that formal considerations bear upon interpretation, unquestionably the TAT is most often interpreted primarily in terms of the content of the responses. The central consideration of content scoring harks back to Murray's theory and its "hero" assumption. The hero is the protagonist of the subject's story and it is assumed that this is the character with whom the subject most clearly identifies. Accordingly, the hero's attributes, needs, strivings, and feelings are often interpreted as having significance in the storyteller's own life. Conversely, those thoughts, feelings, or actions that the hero avoids or denies may represent areas of conflict for the subject. For example, a depressed and potentially self-destructive client gave the following response to Card 3BM:

> Looks like . . . I can't tell if it's a girl or boy. Could be either. I guess it doesn't matter. This person just had a hard physical workout. I guess it's a her. She's just tired. No trauma happened or anything. She was sitting around a table with friends and she got real tired. She's not in a health danger or anything. These are her keys. Her friends drag her back to her room and put her to bed. She's O.K. the next day. No trauma. She's tired physically, not mentally.

Note particularly the repetitive denial of danger or trauma. However, later in the TAT series, in response to the blank card (16), the same subject told a story of a young man, traumatized at school, who takes his car down to the river:

> He sees the bridge, he's really down. He remembers that he's heard stories about people jumping off and killing themselves. He could never understand why they did that. Now he understands, he jumps and dies . . . he should have waited 'cause things always get better sometime. But he didn't wait, he died.

This rather dramatic case example illustrates one way in which the hero assumption operates. Impulses of the characters are assumed to be impulses of the storyteller. Additionally, impulses, actions, or emotions that are repeatedly denied or otherwise defended against are seen as conflict areas even if not directly or literally expressed. The verbal material is interpreted in accord with general principles of inference drawn from psychodynamic theory. Such interpretation needs to be tempered by an appropriate hypothesis-testing framework, such as that outlined by Deinhardt (1983).

A question that follows from these considerations is the problem of levels of prediction (Karon, 1981). For instance if the hero "thinks" about doing something versus actually doing something versus denying that something is done, does this suggest differing predictions about the subject's own propensities? Tomkins (1947) argued, for example, that story behavior predicts subject's behavior, character's thoughts predict subject's thoughts, and manifest defenses of the characters predict manifest defenses of the subject. Others have argued that all of these levels should be used only to index the subject's motives rather than to predict behavior. Research by Kagan and Leaser (1961) suggested that responses to earlier cards in the series are more highly correlated with overt behavior, and stories later in the series less so. Needless to say, predictions of behavior from the TAT are risky and should be backed up by convergent data from other methods. The TAT is most appropriate for looking at psychological processes that may or may not have direct behavioral correlates.

Using the hero assumption, one can form hypotheses by fully assessing the appropriate story characters. It might be important to ask, for example: How adequately does the hero deal with the situation at hand? Is he she active or passive? What coping mechanisms does he or she employ? What types of out-comes are expected and achieved? What is the nature of the interpersonal relations depicted? etc. Through such a procedure one tries to gain a description of the *lebenswelt* of the hero and, by implication, of the subject. For the most part the validity of this description derived by the interpreter will require outside confirmation or support.

In addition to the hero assumption, a second central aspect of content interpretation involves consideration of the stimulus pull of the cards. As previously noted each card has a very definite content upon which subjects elaborate. Some aspects of a story will then have close correspondence to the stimulus materials and others will reflect more subjective, interpretive elements. Bellak (1975) labels these subjective components the *apperceptive distortion,* which is the dynamically significant part of the protocol. The concept of "distortion" is not meant to suggest pathology, but rather merely a departure from the objectively given. The interpretive activity of sorting out the apperceptive distortion from the more directly stimulus-bound material was once described by Murray as "separating the wheat from the chaff," the wheat having the most value in evaluating the subject's inner processes. This sorting process on the part of the examiner requires great familiarity with the literature on card pull and personal experience with the test.

A number of excellent articles exist that pertain to clinical interpretation. Among the foremost are the psychoanalytic treatments of Rapaport, Gill, and Schafer (1968), the ego-analytic approach of Bellak (1975), and the personological approach by Stein (1981). These and other approaches to clinical interpretation undoubtedly have many pitfalls due to their ideographic assumptions and personalized style of application. As Rapaport et al. (1968) state:

These techniques are not hard and fast rules like those of scoring other tests, they are rather like viewpoints for looking upon the TAT stories that must become ingrained in the examiner, so that he can use them flexibly and judiciously (p. 490).

Research Use. The TAT method has been used for a variety of research purposes. In these cases data collected through standard or only slightly modified procedures are subjected to analyses or ratings tailored to detect specific variables or dimensions that can be reliably assessed. Data pertaining to the reliability and validity of these analyses vary from case to case and system to system; thus no attempt will be made here to present such data. Rather, an overview of some fo the more prominent systems is presented.

McClelland and Atkinson have provided what is perhaps the most famous and well-researched method of scoring the TAT for empirical use. Together with Clark and Lowell they published a book in 1953 entitled *The Achievement Motive,* which outlined a theory and measurement strategy for the study of achievement motives and other dimensions of motivation. They developed a scoring system with excellent reliability, but which takes considerable training and experience to master. McClelland (1980) has recently presented an elegant argument for the validity of his approach, as well as theoretical justification for gauging validity issues on criteria more appropriate to projective (or, as he calls them "operant") techniques than those traditionally applied to structured, "respondent," measures.

May (1966) developed a formal approach to the TAT that he used to measure aspects of the subject's gender identity. This approach assesses what he termed the Deprivation-Enhancement pattern of the stories. Men tend to give stories that begin with enhancement (e.g., success, fame, happiness) and end with deprivation (e.g., disaster, failure, unhappiness), whereas the tendency of women is to develop stories that move in the opposite direction. He has a detailed manual used to score such patterns, which has good interrater reliability. A number of studies have had success in employing this approach to the study of gender identity in adult and child subjects.

Cramer (1983) has recently developed a manual for the assessment of defenses from TAT stories. Her scoring system reliably identifies three major categories of defenses—denial, projection, and identification. She has also demonstrated with her system that these three defenses show a definite developmental course, with denial being prominent in young children, projection more prominent in late childhood and early adolescence, and identification more characteristic in late adolescence. Gender differences are also apparent.

Sutton and Swensen (1983) have provided evidence that the TAT can be reliably scored for the subject's level of ego development using priniciples derived from Loevinger and Wessler's (1970) manual for scoring sentence completion tests. Concurrent validity for their technique with Loevinger's sentence completion technique is impressive.

Technical Aspects

These few examples of empirical uses for the TAT are not meant to be exhaustive, as numerous other scoring systems exist. They do, however, illustrate that the TAT method can be used for special purposes and can provide credible, sometimes impressive, results. It also should suggest that summary conclusions about *the* reliability or *the* validity of the TAT are necessarily overgeneralizations. One must speak more specifically about the properties of specific approaches. It also lends more credence to the view that the TAT is not a test but rather a method, one with considerable adaptability and generative capacities.

The data that emerge from the story production procedure are varied and intricate, and thus lend themselves to many analytic techniques. The usefulness of the data is thus equivalent to the validity of any given scoring system that the examiner chooses to apply, or alternatively to the skill of the examiner in interpreting the complex products of the TAT situation. Several of the existing scoring systems have demonstrable reliability and construct validity, while most interpretive methods and lore have not been adequately tested or researched.

Critique

The existing literature on the TAT is voluminous. However, comprehensive reviews appear periodically that can help the test user to orient to various TAT scoring, interpretive, or empirical uses. Some of the more authoritative reviews include Varble (1971), Vane (1981), Stein (1978, 1981), Dana (1982), and Deinhardt (1983). In addition, several excellent papers exist on projective theory and TAT theory in particular. These include Abt (1952), Rabin (1981), McClelland (1980), and Bellak (1975).

The proliferation of systems of approach to material elicited with the TAT method attests to its flexibility and richness. One of the TAT's major assets is that it can be subjected to so many constructs and techniques; however, this is also a significant liability in many respects. Practitioners interpreting the TAT are likely to use different systems, an idiosyncratic combination of systems, or no system at all. This is the bane of the psychometrician, and it also suggests that in common usage the interpretation of the TAT is based on strategies of unknown and untested reliability and validity, a potentially dangerous outcome. That this is a reality rather than speculation is attested to by the results of a recent survey (Wade & Baker, 1977) that revealed that 81.5% of projective test users employ "personalized" procedures for interpretation. The impact of this upon the effectiveness and value of these tests is presently unknown.

Notes to the *Compendium:*

[1]Tape recordings may increase the guardedness of subjects (Kahn & Taft, 1983). Recent issues in malpractice cases suggest that an audio device is ill advised as a standard procedure because the effects of recording on the subject are unknown and courts may deem such practice to be below an appropriate "standard of care" (Kahn & Taft, 1983). In addition, the paper-and-pencil approach facilitates recording of behavioral and emotional reactions not accessible through a voice recording.

References

Abt, L. E. (1952). A theory of projective psychology. In L. E. Abt & L. Bellak (Eds.), *Projective psychology* (pp. 33-66). New York: Alfred A. Knopf.

Arnold, M. B. (1962). *Story sequence analysis.* New York: Columbia University Press.

Bellak, L. (1975). *The T.A.T., C.A.T. and S.A.T. in clinical use* (3rd ed.). New York: Grune & Stratton.

Blum, G. S. (1950). *The Blacky Pictures.* Cleveland: The Psychological Corporation.

Bunin, A. I. (1978). *Structural analysis of the TAT.* Unpublished manuscript, Florida State University, Talahasee.

Campus, N. (1976). A measure of needs to assess the stimulus characteristics of TAT cards. *Journal of Personality, 40,* 248-258.

Cohen, H., & Weil, G. R. (1975). *Tasks of emotional development.* Brookline, MA: TED Assoc.

Cramer, P. (1983, August). *Defense mechanisms: A developmental study.* Paper presented at the annual meeting of The American Psychological Association, Anaheim, CA.

Dana, R. H. (1955). Selection of abbreviated TAT sets. *Journal of Clinical Psychology, 11,* 401-403.

Dana, R. H. (1956). Selections of abbreviated TAT sets. *Journal of Clinical Psychology, 12,* 36-40.

Dana, R. H. (1982). *A human science model for personality assessment with projective techniques.* Springfield, IL: Charles C. Thomas.

Deinhardt, C. L. (1983). *Personality assessment and psychological interpretation.* Springfield, IL: Charles C. Thomas.

Entiwisle, D. R. (1972). To dispel fantasies about fantasy-based measures of achievement motivation. *Psychological Bulletin, 77,* 377-391.

Eron, L. D. (1950). A normative study of the Thematic Apperception Test. *Psychological Monographs, 69.*

Goldfried, M. R., & Zax, M. (1965). The stimulus value of the TAT. *Journal of Projective Techniques, 29,* 46-58.

Henry, W. E. (1956). *The analysis of fantasy: The thematic apperception technique in the study of personality.* New York: John Wiley.

Holt, R. R. (1958). Formal aspects of the TAT: A neglected resource. *Journal of Projective Techniques, 22,* 163-172.

Hutt, M. L. (1980). *The Michigan Picture Test-Revised.* New York: Grune & Stratton.

Kagan, J., & Leaser, G. (1961). *Contemporary issues in thematic apperception methods.* Springfield, IL: Charles C. Thomas.

Kahn, M., & Taft, G. (1983). The application of the standard of care doctrine to psychological testing. *Behavioral Science and the Law, 1,* 71-84.

Karon, B. P. (1981). The Thematic Apperception Test (TAT). In A. I. Rabin (Ed.), *Assessment with projective techniques* (pp. 85-120). New York: Springer Publishing Co.

Loevinger, J., & Wessler, R. (1970). *Measuring ego development* (Vol. 1). San Francisco: Jossey-Bass.

McClelland, D. C. (1980). Motive dispositions: The merits of operant and respondent measures. In L. Wheeler (Ed.), *Review of personality and social psychology* (pp. 10-41). Beverly Hills, CA: Sage Publications.

McClelland, D. C., Atkinson, J. W., Clark, R. A., & Lowell, E. L. (1953). *The achievement motive.* Englewood Cliffs, NJ: Prentice-Hall.

Maddi, S. R., & Costa, P. T. (1972). *Humanism in personology.* Chicago: Aldine Publishing Co.

May, R. (1966). Sex differences in fantasy patterns. *Journal of Projective Techniques and Personality Assessment, 30,* 576-586.

Morgan, C. D., & Murray, H. A. (1935). A method for investigating phantasies: The Thematic Apperception Test. *Archives Neurology and Psychiatry, 34,* 289-306.

Murray, H. A. (1938). *Explorations in personality.* New York: Oxford University Press.

Murray, H. A. (1943). *Thematic Apperception Test-Manual.* Cambridge: Harvard University.

Murstein, B. I. (1963). *Theory and research in projective techniques (emphasizing the TAT).* New York: John Wiley.

Perterson, C. A., & Schilling, K. M. (1983). Card pull in projective testing. *Journal of Personality Assessment, 47,* 265-275.

Rabin, A. I. (Ed.). (1981). *Assessment with projective techniques: A concise introduction.* New York: Springer Publishing Company.

Rapaport, D., Gill, M. M., & Schafer, R. (1968). *Diagnostic psychological testing.* New York: International Universities Press.

Rosenzweig, S. (1976). *Manual for the Rosenzweig Picture-Frustration Study, Adolescent Form.* St. Louis, MO: Roma House.

Ryan, R. M. (1982). Control and information in the intrapersonal sphere: An extension of cognitive evaluation theory. *Journal of Personality and Social Psychology, 43,* 450-461.

Skinner, B. F. (1957). *Verbal behavior.* New York: Appleton-Century-Crofts.

Stein, M. I. (1978). The Thematic Apperception Test and related methods. In B. B. Wolman (Ed.), *Clinical diagnosis of mental disorders: A handbook* (pp. 179-236). New York: Plenum Publishing Corp.

Stein, M. I. (1981). *Thematic Apperception Test* (2nd ed.). Springfield, IL: Charles C. Thomas.

Sutton, P. M., & Swensen, C. H. (1983). The reliability and concurrent validity of alternative methods for assessing ego development. *Journal of Personality Assessment, 47,* 468-475.

Symonds, P. M. (1948). *Symonds Picture-Story Test.* New York: Columbia University, Bureau of Publications.

Tomkins, S. S. (1947). *The Thematic Apperception Test.* New York: Grune & Stratton.

Tomkins, S. S., & Miller, J. B. (1957). *The Tomkins-Horn Picture Arrangement Test.* New York: Springer Publishing Company.

Vane, J. R. (1981). The Thematic Apperception Test: A review. *Clinical Psychology Review, 1,* 319-336.

Varble, D. L. (1971). Current status of The Thematic Apperception Test. In P. McReynolds (Ed.), *Advances in psychological assessment* (pp. 216-235). Palo Alto, CA: Science & Behavior Books.

Veroff, J. (1958). A scoring manual for the power motive. In J. W. Atkinson (Ed.), *Motives of fantasy, action and society.* Princeton, NJ: VanNostrand.

Wade, T. C., & Baker, T. B. (1977). Opinions and use of psychological tests. *American Psychologist, 32,* 874-882.

Wade, T. C., Baker, T. B., Morton, T. L., & Baker, L. J. (1978). The status of psychological testing in clinical psychology: Relationships between test use and professional activities and orientation. *Journal of Personality Assessment, 42,* 3-10.

Wyatt, F. (1947). The scoring and analysis of The Thematic Apperception Test. *Journal of Psychology, 24,* 319-330.

Raymond H. Holden, Ed.D.

Professor of Psychology, Rhode Island College, Providence, Rhode Island.

VINELAND ADAPTIVE BEHAVIOR SCALES

Sara S. Sparrow, David A. Balla, and Dominic V. Cicchetti. Circle Pines, Minnesota: American Guidance Service.

THE ORIGINAL OF THIS REVIEW WAS PUBLISHED IN TEST CRITIQUES: VOLUME I (1984).

Introduction

The Vineland Adaptive Behavior Scales assess personal and social adaptability of individuals from birth to adulthood. Like the original scale, the present Vineland is not a test in the usual sense of the word. It is a scorable structured interview, conducted by a trained interviewer, with a respondent who is familiar with the subject's everyday behavior. Adaptive behavior is defined as the performance of daily activities required for personal and social sufficiency. The scales are applicable to both handicapped and non-handicapped individuals.

Edgar A. Doll, author of the Vineland Social Maturity Scale (1935, 1965), was a major pioneer in the objective assessment of adaptive behavior. His view was that social competency should be compared with evidence of intellectual fuctioning, measured by instruments like the Binet Scales. In his six criteria of mental deficiency, Doll listed social incompetence as the first and most important. He also broadened the concept of adaptive behavior to include a wide range of areas or domains. He classified eight different categories on his scale: self-help general, self-help dressing, self-help eating, self-direction, socialization, locomotion, and occupation.

From the 1930s to the 1960s, IQ scores dominated the classification of mental retardation. However, in 1973, 1977, and 1983 the American Association of Mental Deficiency published several revised editions of its manual, which formally included deficits of adaptive behavior as well as subaverage intelligence as criteria for a diagnosis of mental retardation (Grossman, 1983). The importance of the adaptive behavior construct was also enhanced by the passage of the Education for All Handicapped Children Act of 1975 (Public Law 94-142, 1975) promoting the education, employment, and training of all children regardless of handicap. Stringent guidelines for the assessment of handicapped children, including adaptive behavior, were clearly specified in the law. In the last ten years, legislation and litigation have resulted in the further development of adaptive behavior scales.

There are three versions of the newly revised Vineland: 1) Interview Edition, Survey Form; 2) Interview Edition, Expanded Form; and 3) Classroom Edition. Both the Survey and Expanded Form are administered during a semi-structured interview with parents or caregivers of individuals ranging in age from birth to 18 years, 11 months. The Survey Form is the most similar in content to the original

Vineland scale, includes 297 items, and provides a general assessment of adaptive behavior.

The Survey Form includes a record booklet available in English or Spanish and a manual. The record booklet is used by the interviewer during the assessment to record item passes and failures, and includes incidental informal information considered pertinent as well. There is a score summary page for recording and profiling derived scores. The manual presents detailed guidelines for administering, scoring, and interpreting test results. Criteria for scoring the items are specified and must be adhered to by the examiner. There is voluminous technical information about the development and standardization of the scale as well as normative tables supplying derived scores for the Survey Form.

The Expanded Form contains 577 items, including 297 from the Survey Form, in English and in Spanish. This longer version provides not only a more comprehensive and detailed assessment of adaptive behavior, but also a systematic basis for preparing educational, habilitative, and treatment programs through use of a separate booklet, the program planning report.

The Classroom Edition is a questionnaire completed by a regular school or special education teacher for children aged 3 to 12 years, 11 months.

There are supplementary materials available, including an audiocassette tape providing sample interviews with parents or other caregivers as a possible training guide for examiners. Microcomputer software programs are provided for rapid score conversion, profile displays, and record management. A Technical and Interpretive Manual containing detailed statistical data concerning standardization, reliability, and validity studies will soon be available.

Practical Applications/Uses

The most important use of the Vineland Adaptive Behavior Scale continues to be provision of up-to-date standards for social and adaptive behavior of children from ages 2 to 18. Social competency scores can be compared to intellectual development on such tests as the Bayley Mental and Motor Scales, the revised Stanford-Binet, the WPPSI, and the WISC-R. The revised Vineland is not limited to use with mentally retarded individuals but can be useful in assessing physically handicapped, emotionally disturbed, hearing-impaired, and visually handicapped children. Each of these groups is represented in special national samples providing supplementary norms for the scale.

The Vineland can be used to develop individual educational, habilitative, and treatment programs. Strengths and weaknesses in specific areas can be pinpointed in order to select the most suitable program for the individual or to zero-in on specific activities in the program which might be most beneficial.

The Survey Forms should be administered in a quiet room or office after suitable rapport has been developed with the adult respondent. A clinical or school psychologist, social worker, or other graduate professional with specific training and experience in individual assessment and test interpretation is expected to administer the scale. The individual being assessed must *not* be present.

Administration time for the Survey Form ranges from 20 to 60 minutes, the Expanded Form 60 to 90 minutes, and the Classroom Edition approximately 20 minutes for nonhandicapped individuals. The starting point for administration of items is usually based on chronological age, but *not* rounded upward to the next year level. For retarded or other handicapped individuals, the starting point is usually the best estimate of either mental age or social age.

There are four general areas or "domains" to be sampled: Communication (receptive and expressive language); Daily Living Skills (self-care activities of eating, dressing, washing, etc.); Socialization (interpersonal relations, play, and leisure); and Motor Skills (gross and fine coordination). These four domains are summed together to obtain an Adaptive Behavior Composite.

A Maladaptive Behavior domain is optionally available to the examiner, and there are two parts. Part 1 describes minor maladaptive behaviors; Part 2, more serious maladaptive behaviors that may be rated "Serious" or "Moderate" in degree of frequency of occurance.

Instructions for scoring each item are very clearly presented in the Appendix of each instructional manual. Scores and their indications are as follows: 2—the activity is performed satisfactorily and habitually; 1—there is emerging performance of an activity, or adequate but not habitual performance; 0—the individual is too young or immature to perform the activity, or can perform it but rarely does; N—there is "no opportunity" to perform the activity; and DK—the respondent does not have opportunity to observe whether the activity is performed or not. A basal age is established in each domain when seven consecutive items are scored 2, and a ceiling is reached when seven consecutive items are scored 0.

Interpretation is based on objective scores derived from a national standardization sample. Understanding the interpretation requires the examiner to be familiar with standard scores and standard error, a level of difficulty above many school teachers and residential workers and most nonprofessionals. The Report Booklet for parents requires an understanding of percentile rank and stanines. Concessions are made to parents that the ratings be grouped under the headings Above Average, Average, and Below Average, but to understand how *far* above or how *far* below average, the reader must know how to interpret the "Stanine."

Technical Aspects

21,876 parental participation forms provided a pool from which a sample of 3,000 standardization subjects were randomly selected by computer to correspond to 1980 U.S. Census figures by age, sex, community size, geographic region, parental education, and racial or ethnic representation. The standardization group comprised 100 subjects in each of 30 age groups between birth and 18 years, 11 months. It consisted of 1,500 males and 1,500 females from four geographic regions (Northeast, North Central, South, and West) and four racial groups: white, black, Hispanic, and other. Items for the standardization of the scale were selected from 529 items used in a national pre-test sample.

Three types of reliability estimates were developed: split-half (or internal consistency), test-retest, and interrater reliability. Split-half correlations range from a low of .83 for the Motor Skills domain to .94 for the Adaptive Behavior composite scores on the Survey Form. On the Expanded Form, median split-half correlations are considered very satisfactory. Test-retest reliability obtained by testing 484 individuals from ages 6 months through 18 years, 11 months at 2 to 4 week intervals (mean: 17 days) varied by age groups with correlations in the .90's for ages 6 months through 6 years, 11 months, and correlations in the .80's for ages 7 through 18 years, 11 months.

Over the whole age range, however, domain test-retest correlations ran .98 to .99, considered to be excellent. Interrater reliability for 160 individuals interviewed twice by different interviewers resulted in correlations ranging from .96 to .99 for different domains, also excellent.

Construct validity was demonstrated in three ways: 1) mean raw scores increased with age for all four major domains; 2) factor analysis of the general Adaptive Behavior Composite resulted in one significant factor, accounting for 55-70% of the variance at all yearly age levels; and 3) correlations of Vineland scores with tests of intelligence yielded low but positive correlations, as predicted. For example, in 2,018 cases correlations between the Vineland and the Peabody Picture Vocabulary Test ranged from .12 to .37.

Critique

There are no current reviews available of the new Vineland Adaptive Behavior Scales. However, previous reviews of the Vineland Social Maturity Scale have generally been favorable, while pointing out certain flaws and shortcomings. In *The Third Mental Measurements Yearbook* (Buros, 1947), C.M. Loutitt praised the methodology of the scale while urging more adequate standardization (p. 210). John Rothney (p. 211) cautioned that only the most thoroughly trained interviewers would obtain similar results, since he felt the scoring was highly subjective. In the fourth edition of Buros (1953), William Cruickshank praised the development of such a scale, but opted for a new standardization and a new manual (p. 94). Florence Teagarden was impressed by the extensive research done with the scale, but favored the instrument as a clinical interview of the parent or surrogate (p. 94).

The new scale remedies all the faults indicated by previous reviewers. The authors are to be commended on their thorough and elaborate standardization based on 1980 U.S. census data, and for their use of elegant sampling techniques and sophisticated statistical analysis. However, the new users' manuals are now overweighted with statistical tables beyond the need and usefulness of most practitioners. It seems unlikely that the usual clinician could digest the findings of 117 pages of tables devoted to domain standard scores, adaptive levels of subdomains, age equivalents, and supplementary norms. It appears that a tabular overkill is manifest.

Without computer assistance, hand-scoring of a multitude of domains, subdomains, standard error, standard error of the mean, percentile rank, stanine, supplementary norm ranking, adaptive level and age equivalents, along with drawing a score profile for the form domains, amounts to a lengthy, involved, tedious, and time-consuming process.

For easier administration, Appendix C of the manuals (Scoring Criteria) should be off-printed in a separate booklet.

Despite these objections, we can look forward with eager anticipation to the clinical usefulness of this carefully restandardized version of a well-known instrument that measures contemporary social and personal development.

References

Buros, O.K. (1947). *The third mental measurements yearbook.* Highland Park, NJ: The Gryphon Press.

Buros, O.K. (1953). *The fourth mental measurements yearbook.* Highland Park, NJ: The Gryphon Press.

Doll, E.A. (1935). A genetic scale of social maturity. *American Journal of Orthopsychiatry, 5,* 180-188.

Doll, E.A. (1953). *The measurement of social competence: A manual for the Vineland Social Maturity Scale.* New York: Educational Test Bureau.

Doll, E.A. (1965). *Vineland Social Maturity Scale.* Circle Pines, MN: American Guidance Service.

Grossman, H.J. (Ed). (1983). *Classification in mental retardation* (1983 revision). Washington, DC: American Association on Mental Deficiency.

Public Law 94-142. The Education of All Handicapped Children Act, 20 U.S.C. 1401-1461.

Sparrow, S.S., Balla, D.A., & Cicchetti, D.V. (1984). *Vineland Adaptive Behavior Scales: Survey Form manual, Interview edition.* Circle Pines, MN: American Guidance Service.

Sparrow, S.S., Balla, D.A., & Cicchetti, D.V. (1984). *Vineland Adaptive Behavior Scales: Expanded Form manual, Interview edition.* Circle Pines, MN: American Guidance Service.

Sparrow, S.S., Balla, D.A., & Cicchetti, D.V. (1984). *This is your sampler of the Vineland Adaptive Behavior Scales.* Circle Pines, MN: American Guidance Service.

Jean Spruill, Ph.D.

Associate Professor and Director, Psychological Clinic, The University of Alabama, University, Alabama.

WECHSLER ADULT INTELLIGENCE SCALE-REVISED

David Wechsler. San Antonio, Texas: The Psychological Corporation.

THE ORIGINAL OF THIS REVIEW WAS PUBLISHED IN TEST CRITIQUES: VOLUME I (1984).

Introduction

The Wechsler Adult Intelligence Scale-Revised (WAIS-R) is designed as a broad-based sampling of a wide range of abilities believed to reflect one's overall capacity for intelligent behavior. While there is not a uniformly agreed upon definition of intelligence, historically intelligence has been most frequently defined in terms of abilities, such as the ability to reason abstractly, to learn, to adapt, etc. Wechsler believed that intelligence was multidetermined and multifaceted, calling for an overall competency, or "global capacity," which ". . . enables a sentient individual to comprehend the world and deal effectively with its challenges" (WAIS-R Manual, p. 8).

The WAIS-R, published in 1981, is the latest revision in a series of tests that have become the most widely used individual measures of adult intelligence in the world. First introduced in 1939 as the Wechsler-Bellevue Intelligence Scale, the forerunner to the Wechsler Adult Intelligence Scale (WAIS) enjoyed unparalleled success. It is impossible to imagine anyone involved in the testing of adults who is not familiar with the WAIS. It is expected that the WAIS-R, like its predecessors, will also become *the* standard against which all other tests for adults are compared.

David Wechsler was born in Lespedi, Romania, in 1896 and emigrated to New York City at the age of six. Upon graduating in 1917 with an M.A. in psychology from Columbia University, he entered the army. There he had the opportunity to work with E.G. Boring in scoring and evaluating the newly developed Army Alpha Test. He also spent much time assessing new recruits on the Stanford-Binet, Yerkes Point Scale, and the Army Individual Performance Scales. It was during this time that he became aware of the shortcomings of the current tests as measures of adult intelligence.

After World War I Wechsler received his doctorate from Columbia University (1925) and was employed in a variety of clinical settings. In 1932 he became the chief psychologist at Bellevue Psychiatric Hospital in New York. A major part of his responsibility was to provide psychological assessments of adult patients, most of whom were emotionally disturbed adults, many from New York City's non-English-speaking immigrant population. The Stanford-Binet lacked proper adult norms, many of the items seemed childish to adults, and the Binet subtests at the upper levels were predominantly verbal. For these reasons the Binet was deemed not suitable for the assessment of adult intelligence, and in 1934 Wechsler began the development of a test to replace the Binet.

During the development of the Wechsler-Bellevue Intelligence Scale, Wechsler experimented with a large number of individual tests, primarily by trial and error and this culminated in 11 tests that constituted the Wechsler-Bellevue, Form 1. A twelfth test, Cube Analysis, was discarded because of large sex differences and a low ceiling. The test was widely adopted by clinics, hospitals, and other facilities that conducted intellectual testing of adults. Part of the success of the Wechsler-Bellevue was due to the zeitgeist of the times, as Wechsler's test became available at a time when clinical psychology was emerging as a profession and testing was the prime domain of the clinician. In fact, at one time, the diagnosis of general intelligence constituted the principal function of most psychologists employed by clinics, hospitals, and other agencies.

In 1944 Form II of the Wechsler-Bellevue was published, again focusing on adolescent and adult populations. In 1949 Wechsler published the Wechsler Intelligence Scale for Children (WISC), which was for the most part a downward extension of the Wechsler-Bellevue, Form II. By primarily adding easier items to each subtest, this allowed the testing of children as young as five years of age. The WISC was standardized independently from the Wechsler-Bellevue, and was revised in 1974 and called the WISC-R. In 1967 Wechsler extended the WISC downward for children ages 4½ to 6½ years, calling this test the Wechsler Preschool and Primary Scale of Intelligence (WPPSI).

In the early 1950s Wechsler had begun work on an extensive revision and restandardization of the Wechsler-Bellevue. The new test, published in 1955 and designated the Wechsler Adult Intelligence Scale (WAIS), quickly became the most widely used individual measure of adult intelligence in the U.S. It is expected that the WAIS-R will follow in the footsteps of the WAIS.

Because the WAIS is so well known and has been reviewed extensively in the past, no attempt will be made to duplicate the already published accounts of its history, development, and research. The interested reader is referred to the previous several editions of the *Mental Measurements Yearbook* and Matarazzo's book *Wechsler's Measurement and Appraisal of Adult Intelligence* (1972). The WAIS was a substantial improvement over the Wechsler-Bellevue with respect to standardization but represented little change in the actual subtests themselves. The primary changes in subtests were in terms of 1) adding items, thus increasing the range in scores of some of the subtests, which in turn increased the upper range of IQ scores, and 2) adding, modifying, or deleting items to amend problems in scoring. The reliability of both the subtests and total scales was improved, and the WAIS proved to be more discriminating at the higher level of ability than the Wechsler-Bellevue.

The WAIS-R, like its predecessor, has 11 subtests, six verbal and five performance. The subtests may be administered together or separately to yield Verbal, Performance, and Full Scale IQ scores. This permits the use of the test with visually, motorically, or verbally handicapped individuals. In developing the

WAIS-R, two new subtests were tried as possible additions to the battery. One, a measure of nonintellective ability called Level of Aspiration, was deemed promising but in need of more research before being added as an integral part of the test battery, and the other, a new measure of spatial ability, was too highly correlated with Block Design to warrant inclusion. Thus, the WAIS-R continues to reflect the basic design of the original Wechsler-Bellevue but with substantial modifications in the content of some tests and in the quantity and quality of the standardization group.

Changes in the WAIS-R have focused on modifying or eliminating dated WAIS items, increasing the cultural diversity of the items, and deleting items that were either too hard or too easy. The order of administration of the subtests was changed by alternating verbal and performance subtests to help maintain the subject's interest in the test. There were also some slight modifications in the instructions and scoring for some of the items. The standardization group consisted of 1,880 Americans, equally divided by sex and stratified according to the 1970 U.S. census data, with the controlled variables being age, race, sex, geographic location, urban-rural residence, education, and occupation. There were nine age groups used, ranging from 16 to 74 years of age.

The standardization was thorough and carefully done. There were 115 testing centers located in 39 states, with the field supervisors selecting, training, and supervising the examiners who administered the tests. Great care was taken to insure that the examiners were qualified and experienced in test administration. The sample did not include institutionalized mentally retarded persons, individuals with known brain damage, or persons with severe behavioral or emotional problems; it was limited to individuals who could physically respond to all the items and whose primary language was English or who were able to adequately speak and understand the English language. Only one member of a family was tested. This represents the best attempt at obtaining a representative sample of the U.S. population of any test currently published. The WAIS-R manual gives a very good description of the standardization sample and procedures used to derive the scores and IQ tables.

Although the WAIS-R is too new to have been translated into other languages, four subtests (Information, Comprehension, Arithmetic, and Digit Span) have been adapted and translated into Hindi. However, the scores obtained were affected to a serious degree by educational level, and therefore more work is needed before the translation can be considered a valid measure of intellectual functioning. The WAIS was translated into Spanish and standardized on a Puerto Rican population in 1965. The research on the Spanish version reflects adequate reliability and validity, and it is extensively used in Spanish-speaking countries as well as in many parts of the U.S. It is expected that the WAIS-R also will be translated into Spanish, particularly since such a large subgroup of the U.S. population is Spanish-speaking. Readers wishing to obtain a Spanish version of the WAIS are referred to The Psychological Corporation, which publishes the Escala de Inteligencia Wechsler para Adultos. It should be noted that the Spanish and English materials for the WAIS are not interchangeable.

Administration of the WAIS-R or WAIS requires approximately 60 to 90 minutes, and many attempts have been made to shorten both tests. These attempts vary from omitting certain subtests entirely to eliminating items within subtests and prorating the scores. Short forms typically involve selecting various verbal and performance subtests that correlate highly with the Full Scale IQ score and prorating the scores. Research on short forms suggests that they are adequate *if* one is looking for a general measure of intelligence. The major advantage of any short form lies in its use as a screening device. If the individual scores 90 or above on the short form, there is little likelihood that the individual is functioning below average. However, if the short form IQ score is less than 90, the examiner should consider additional testing for proper diagnostic findings.

The WAIS-R's subtests consist of six Verbal Scale subtests and five Performance Scale subtests. The content of the WAIS-R is essentially the same as the WAIS; about 80% of the items are the same or have only slight modifications. The following represents a brief description of the WAIS-R subtests. More complete information can be found in Sattler's *Student's Manual to Accompany Assessment of Children's Intelligence and Special Abilities* (1982).

The Information subtest consists of questions that sample a broad range of general knowledge. The questions generally can be answered with a simply stated fact and represent items that individuals with average opportunity should be able to acquire for themselves. For example, questions such as "How many days are there in a month?" or "At what temperature does water freeze?" are typical of the questions found on this subtest.

On the Digit Span subtest the subject listens to a series of numbers given orally by the examiner and then is asked to repeat the digits. This test is primarily a measure of short-term memory and attention.

The Vocabulary Subtest consists of words arranged in order of difficulty, and the subject is asked to explain orally the meaning of each word (e.g., "What does _____ mean?"). This subtest involves a variety of cognitive functions and is often considered the best single measure of a subject's general intellectual capacity.

The Arithmetic subtest contains 14 orally presented problems and subjects are required to answer orally. All of the problems are timed and reflect various kinds of skills, from direct counting of objects to the operations of addition, subtraction, multiplication, and division, as well as more complex problem-solving skills.

The Comprehension subtest contains 16 questions covering a wide range of problem situations involving such issues as health, social mores, interpersonal relations, and laws. A typical question on this subtest would be similar to the following: "Why do people who want a divorce have to go to court?" This test generally measures an individual's social judgment, common sense, grasp of social conventions, and ability to use knowledge appropriately.

The Similarities subtest contains 14 pairs of words, and the subject is asked to explain the similarity between each pair. This subtest requires subjects to perceive common elements within the terms. To receive maximum credit, subjects must be able to bring these similarities together to form a concept; thus Similarities is a measure of the capacity for verbal concept formation or abstract thinking.

The Picture Completion subtest consists of a series of 20 drawings of common objects or scenes (e.g., door, frog, tennis game), from which some important element is missing. This is a test of the subject's ability to differentiate essential from nonessential details and requires both recognition of the object and some reasoning and memory ability.

The Picture Arrangement subtest requires the subject to arrange a series of pictures in a logical order, much like a comic strip which has been cut into segments and rearranged. The individual pictures are placed in a prearranged order by the examiner and the subject is requested to put them in the "right" order to tell a story that makes sense. This requires the ability to grasp the essence of the story and to anticipate and understand the possible antecedents and consequences of events, particularly social events.

In the Block Design subtest the subject is required to use blocks to reproduce a design from a model. The designs are two-dimensional, red-and-white abstract designs arranged in order of increasing difficulty. All design tasks are timed and involve the ability to perceive and analyze forms by breaking down the whole (the design) into its component parts and then reassembling the parts into an identical design. This is a nonverbal concept formation task that requires perceptual-visual organization and abstract conceptualization.

The Object Assembly subtest consists of four jigsaw problems. The pieces of each problem are laid out in a prearranged manner by the examiner, and the subject is requested to "put the pieces together to make something." The jigsaw problems are timed and require both visual-motor coordination and perceptual organization ability. To put the pieces together correctly, the subject is required to grasp a whole relationship or pattern from the individual parts.

The Digit Symbol subtest consists of nine symbols paired with numbers, the symbol appearing below each number. The subject's task is to fill in 93 blank boxes, matching the appropriate symbol with the number in the box. This test is timed and requires the ability to learn an unfamiliar task and involves speed and accuracy of eye-hand coordination and short-term visual memory.

Practical Applications/Uses

The WAIS-R is the standard against which all other measures of adult ability are compared. The paragon of intelligence tests, it is used in a wide variety of settings requiring an accurate assessment of a person's abilities. However, intellectual assessment in isolation is of little value. When used in conjunction with other information relating to personality and to the educational, social, and occupational opportunities available to the individual, the IQ score provides profuse information in the assessment process.

In educational settings measures of intelligence are very highly correlated with educational level. The question arises of whether years of schooling predict one's IQ or whether one's IQ predicts the number of years of schooling one will complete. Research seems to indicate the latter; IQ contributes significantly to one's grade-point average. Numerous studies show that the correlation between IQ scores and academic success in the classroom is about .50. This correlation occurs at all levels of education, grade school through graduate and professional schools. Thus the WAIS-R is frequently used in settings which require the prediction of academic success.

In vocational settings the WAIS-R can be a valuable aid in the prediction of whether an individual has the general intellectual ability required for different types of occupations. Most of the studies of IQ and vocational attainment show a correlation of about .50, similar to that found between academic success and intelligence. However, attempts to correlate intelligence and "vocational success" have produced much lower correlations, averaging about $r = .20$. Part of the lowered correlation stems from the difficulty in defining vocational success. It is simply much easier to define academic success than to define success in work; neither production rates nor income are generally good measures of success, and thus quantification of success is difficult. Supervisors' ratings often reflect many nonintellectual factors, such as interpersonal relationships, reporting to work on time, etc. Therefore, a "successful" rating may depend on factors other than work ability.

Most assessments of organic problems begin with the use of the WAIS-R. The IQ scores and the various patterns of responses for the subtests are used to identify areas of intellectual deficit. Because our intellectual abilities are typically rather even in their development, subtest scores that are substantially lower than average are indicative of significant deficits and possible organic defects.

Beginning with the development of the Binet Scales, psychologists have attempted to use intelligence tests in the area of personality to gather more information about an individual than just an IQ score. For example, Binet observed that psychotic patients "scatter" their passes and failures on the Binet Scale more than retarded individuals. Wechsler also believed that his scales (Wechsler-Bellevue, WAIS, WISC) provided more information than a simple IQ score. He firmly believed that psychopathology has an impact on intellectual functioning and further that different types of disorders produced different patterns of responding (profiles). The first belief, that psychopathology has an impact on intellectual functioning, has received ample support from the literature. Unfortunately, hundreds of studies have not borne out the belief that different disorders have differential patterns or profiles. Nevertheless, research in this area continues, fueled by some positive studies. It does appear that when one carefully specifies the personality variable, there is a greater chance of finding a relationship. For example, if one distinguishes between state and trait anxiety rather than using a relatively crude measure of anxiety, then there is a relationship between anxiety and patterns of responding. Although no research in the area of personality variables and the WAIS-R is yet reported, it is expected that some will be forthcoming. Refinement of personality measures has made this area a fruitful endeavor for further exploration.

The WAIS-R is appropriate for individuals ages 16-0 to 74-11. The norms were based on groups considered representative of the U.S. adult population and a stratified sampling plan was used to insure that representative proportions of various classes of individuals would be included. The 1970 U.S. census report, updated as information became available, was used to determine the stratification. The variables controlled were age, sex, race (white-nonwhite), geographic region, occupation, education, and urban-rural residence. The manual states that sex was stratified, but although there are actually more women than men at all age levels except ages 16 through 17, equal numbers of men and women were used for each age group. However, the total numbers did vary among the age groups depending on the proportion in the population.

One of the attractive features of the WAIS-R is that it can be used in part with many handicapped individuals. Visually and physically impaired subjects can be given the verbal portions of the test and individuals with auditory problems can be administered the performance portions. Also, most verbal subtests can be typed on cards and presented to deaf individuals who can read and write. Sattler's text (1982) has an excellent section on administering the Wechsler tests to handicapped individuals.

Examiners may question whether to use the WAIS-R or the WISC-R for sixteen-year-olds because there is an overlap between the two tests in the age range 16-0 to 16-11-30. One study (Wechsler, 1981) compared the WAIS-R and WISC-R for normal IQ subjects in this age group. The results showed similar mean IQs on both tests and similar standard errors of measurement, indicating that either test would be appropriate. However, it should be noted that these subjects were in the normal range of intelligence. Similar findings occurred when the WAIS and WISC were compared for normal IQ subjects, but when subjects in the borderline range of intellectual functioning were compared, the WAIS produced IQ scores approximately 6 to 8 points higher than the WISC. A similar study comparing the WAIS-R and WISC-R for borderline subjects has not been reported. However, it is worthwhile at this point to note that research has consistently shown the WISC-R to yield IQ scores approximately 6 to 8 points below WISC scores; similar findings have occurred with respect to WAIS-R and WAIS scores. Thus one needs to be very cautious when comparing scores made on the various tests. When testing sixteen-year-olds it is recommended that the WISC-R be used unless there is some specific reason to use the WAIS-R.

This test is always administered individually and requires a highly skilled and professionally trained examiner to administer and interpret the results. Typically the examiner has at least a master's degree in psychology or special education. The manual gives excellent instructions for the administration of each subtest and for the recording and scoring of the items. Considerable practice is required for a smooth administration and the rules for starting, discontinuing, repeating items, and probing answers are laid out fairly clearly in the manual.

Because the administration of the full battery of tests is a lengthy 60-90 minutes, various short forms of the WAIS-R are often used as previously noted. For many purposes a brief (rough) intellectual assessment is appropriate, and one or two subtests of the WAIS-R serve this purpose quite well. However, the abbreviated versions do not have the same reliability as the standard version and may not adequately measure IQ at the upper and lower ranges of the scale. The most successful WAIS short form that involves eliminating items within subtests is that proposed by Satz and Mogel (1962) and modified by Silverstein (1982). Their model has been tested with the WAIS-R and obtained approximately the same results; administration time is cut in half and the reliabilities, while lower, are still respectable (.94, .89, and .95 for Verbal, Performance and Full Scale IQs respectively). When using a short form of the WAIS-R, if the individual's score is low (89 or less) it is recommended that complete testing be carried out to accurately assess the individual's ability.

Many of the problems reported in scoring items on the WAIS have been corrected in the WAIS-R. Answers to the subtests are recorded on an answer sheet which is designed well for recording answers but leaves little room to write behavioral observations of the subjects while testing. The criteria for discontinuing each subtest are now printed on the answer sheet, an improvement over the WAIS. Nevertheless, arriving at a score for some subtests is not simple and often requires judgment, particularly when ambiguous responses are given. A careful study of the scoring criteria, using the guidelines and examples given in the manual, is vital to correct scoring. The record booklet facilitates the scoring of responses and should be carefully checked. It is important to confirm one's arithmetic; addition errors are the most frequent types of errors found in scoring.

The front of the record booklet provides a space for summarizing the raw scores, converting them to standard scores, and recording the corresponding IQ scores. The raw number of correct answers on each subtest is converted into a standard score. The standard scores for the Verbal subtests are added together to yield a Verbal Scale IQ score; similarly the standard scores for the Performance subtests yield a Performance Scale IQ score. The Full Scale IQ score is derived from the addition of the Verbal and Performance standard scores. The range of IQ scores is from 46 to 150, which is not sufficient for the moderate and below classification of retardation nor does it adequately cover the extremely gifted range of intellectual ability.

As in administration and scoring, interpretation of the WAIS-R requires professional training. In general the Verbal, Performance, and Full Scale IQ scores are first compared to the population norms, and the scores are classified in terms of functioning (e.g., moderate range of mental retardation, high average, superior classification of intelligence, etc.). Any discrepancy between Verbal and Performance Scale IQ scores is discussed, and then the strengths and weaknesses of the individual within the Verbal and Performance areas are discussed. Interpretation of results does not occur in isolation but rather in conjunction with all other available information about the person.

Technical Aspects

Reliability of a test refers to consistency of measurement; that is, the extent to which scores obtained from two or more administrations of the test on the same subjects agree. There are several methods of calculating test reliability; where appropriate, the split-half method, corrected by the Spearman-Brown formula, was used for the WAIS-R subtest reliabilities. For tests involving speeded items, test-retest reliability was calculated. While reliability coefficients for the subtests varied from a low of .52 (Object Assembly, ages 16 to 17) to a high of .96 (Vocabulary, for six of nine age groups), the average reliability coefficients for the Verbal, Performance, and Full Scale IQ scores are consistently higher (.97, .93, and .97 respectively). Separate reliability coefficients were calculated for each age group and are presented in the WAIS-R manual. As mentioned earlier, the normative sample did not contain individuals known to have psychiatric or neurological problems. A study by Ryan, Prifitera, and Larsen (1982), using psychiatric patients, found approximately the same reliability coefficients and standard errors of measurement for Verbal, Performance, and Full Scale IQ scores and for all the subtests except the Arithmetic subtest, as was found in the standardization sample. It was hypothesized that the significant difference between the WAIS-R reliability coefficient for the Arithmetic subtest and the estimate of Ryan et al. was due to concentration difficulties which are frequently noted in psychiatric patients. Thus their results indicated the WAIS-R was a reliable instrument when used in the evaluation of hospitalized patients exhibiting psychiatric, neurological, or mixed symptomatology.

Other reliability studies of the WAIS-R have not been reported in the current literature. Research on the reliability of the WAIS has been extensive, and results have shown the WAIS to be a highly reliable instrument for assessing adult intelligence in a variety of settings and with a variety of populations. Because of the similarity between the WAIS and WAIS-R and the careful standardization of the WAIS-R, it is expected that research on the reliability of the WAIS-R will produce similar results.

The standard errors of measurement for the subtests and IQ scores for each group are reported in the WAIS-R manual. The standard error of measurement allows us to place a band of "error" around a theoretical or true IQ score. In other words, it is an estimate of the variability of IQ scores obtained if a person could be tested hundreds of times, with the "true" IQ being the average IQ score obtained from these hundreds of testings. The standard errors for the Performance Scale are somewhat larger than those for the Verbal Scale at all age levels. The standard error for the Full Scale IQ score is, with one exception (ages 16-17), smaller than either the Verbal or Performance standard errors. In looking at the standard errors presented in the manual, one may note that the standard errors for the IQ scores are larger than those for the individual subtests. This is because the standard errors for the subtests are expressed in terms of scale score units (mean of 10, standard deviation of 3) and standard errors of the IQ scores are expressed in terms of IQ units (mean of 100, standard deviation of 15). As noted, similar standard errors were also found in a study of patients with psychiatric and neurological symptomatology indicating the appropriateness of the WAIS-R for this group, which had been excluded from the standardization sample.

Standard errors of measurement were also used to calculate the differences between scaled scores necessary for significance at the 15% probability level. For most purposes a scaled score difference of 3 or more points represents a significant difference between subtest scores. Exact figures are presented in tables in the WAIS-R manual, along with samples explaining the use of the tables.

Wechsler reported the differences between Verbal and Performance IQs required for statistical significance; as little as 6.48 points (ages 45-54) is significantly different at the 15% level, with the average significant Verbal-Performance discrepancy being 7.15 points. Yet it should be noted that 50% of the subjects in the standardization group had Verbal-Performance discrepancies of 8 or more points (Grossman, 1983). Thus interpreting a significant difference without also some knowledge of the frequency of this difference in the normative population is misleading. As a general rule, a difference of 15 or more points warrants further investigation, as a difference of this magnitude only occurs in 10% or less of the normal population (Grossman, 1983).

Verbal-Performance discrepancies are often used as support for findings of neurological dysfunction in patients. Bornstein (1983) used the WAIS-R with a sample of patients with neurodiagnostically confirmed unilateral or bilateral cerebral disease. He found that patients with left-hemisphere disease obtained significantly lower Verbal IQ than Performance IQ scores and patients with right-hemisphere or bilateral disease obtained lower Performance than Verbal IQ scores. These findings are consistent with those found in numerous studies with the WAIS, and according to Bornstein confirm the utility of the WAIS-R in assisting in the diagnosis of neurological impairment. Bornstein does caution against the use of Verbal-Performance IQ discrepancies in isolation and states that they should be used only in the context of a complete neurological examination and with relevant medical and educational background information. It should be pointed out that while Bornstein's research was well designed and carefully carried out, the differences he found, though statistically significant, occur many times in the population at large. At least 50% of the standardization population have Verbal-Performance discrepancies of 8 or more points; the patients in Bornstein's study had differences ranging from 4.9 for the left-hemisphere group to 10.7 for the right-hemisphere group. Most of the research concerning Verbal-Performance discrepancies on the WAIS has ignored the frequency of such discrepancies occurring in the population at large.

The validity of any test refers to the extent to which that test measures what it purports to measure. Since the WAIS-R measures the same abilities as the WAIS and has approximately 80% of the same content as the WAIS, it is expected that the validity studies of the WAIS-R will be similar to those of the WAIS. Because the literature on the WAIS is so extensive (over 1,300 publications were cited in *The Eighth Mental Measurements Yearbook* [Buros, 1978]), no attempt will be made to summarize the research reported. Guertin and his colleagues have published several review articles of the research literature on the WAIS, the latest in 1971, and Zimmerman and Woo-Sam (1973) have an excellent book on the research focusing on the clinical use of the WAIS. The validity of the WAIS as a measure of adult intelligence has been soundly established, and by extension the WAIS-R is also considered to be a valid instrument for assessing adult intelligence.

At present, validity studies on the WAIS-R are few. There have been two recent studies (Smith, 1982; Mishra & Brown, 1983) comparing WAIS and WAIS-R IQs, both using a college population. Both studies were well designed and found results similar to those reported by Wechsler (1981) in the WAIS-R manual. Thus three studies to date have found that IQ scores on the WAIS are approximately 7 to 8 points higher than IQ scores on the WAIS-R. These results are consistent with the data reported in earlier studies comparing revisions of the WISC and the Stanford-Binet with the older scales. Whether this discrepancy also holds at the lower end of the intelligence scale is not known. A study (Ryan & Rosenburg, 1983) reporting on the correlation between the WAIS-R and the Wide Range Achievement Test for a sample of mixed patients demonstrated that the WAIS-R is a valid indicator of achievement test scores, with the Verbal and Full Scale IQ scores being better predictors of the achievement test scores than the Performance Scale IQ. The pattern of relationship shown for the WAIS-R was essentially the same as previously found for the WAIS and WISC-R. Similarly, Traub and Spruill (1982), in comparing the WAIS-R and Quick Test IQ scores, found essentially the same results for the WAIS-R and Quick Test that have been found for the WAIS and the Quick Test.

Several factor-analytic studies of the WAIS-R have been conducted, with results similar to those found with the WAIS. Most of the studies give strong support to the separation of the WAIS-R into Verbal and Performance Scales. Typically three basic factors have been identified: a "verbal comprehension" factor, a "perceptual organization" factor, and a "memory/freedom from distractability" factor. In general, the verbal subtests make up the verbal comprehension factor, with the performance subtests making up the perceptual organization factor. There is some overlap in these factors; that is, subtests from the verbal scale correlate moderately with the perceptual organization factor and vice versa. Thus these factors are not pure. Verbal comprehension seems to measure both knowledge of content of the verbal items and the mental processes the individual goes through to answer the questions (comprehension). Perceptual organization describes the ability under-lying the factor for both item content (perceptual) and mental process (organiza-tion). Freedom from distractability seems to measure processes related to concentration, memory, and attention. The two major subtests for this factor are Digit Span and Arithmetic, followed by Digit Symbol. In addition to the three basic factors identified above, the WAIS-R subtests are all relatively good mea-sures of the general factor (g) of intelligence, with the verbal subtests being better measures of g than the performance subtests.

Critique

The WAIS-R continues in the tradition of its predecessors and will undoubtedly serve in the future as a standard of comparison for tests of ability. It is very similar to the WAIS and appears to have the same statistical properties as its predecessor with respect to reliability, standard error of measurement, validity, etc. The stan-dardization is excellent, including both white and nonwhite individuals in pro-portion to the population distribution. The administration procedures have been changed slightly and the modifications represent an improvement over the WAIS. The manual is clear, easily read, and provides good scoring directions.

The major limitations of the WAIS-R, which were also true of the WAIS, are its limited floor and ceiling and the nonuniformity of scaled scores. The test allows subjects who have a raw score of 0 on a subtest to receive 1 scaled-score point on that subtest. In addition, the range of Full Scale IQ scores (45 to 150) is not sufficient to allow for the assessment of individuals who are extremely gifted or severely retarded, a distinct disadvantage when working with individuals at the lower end of the intelligence scale. The range of scaled scores is not uniform for each subtest, so that some subjects reach a ceiling on certain subtests more quickly than others. For example, the highest scaled score that can be obtained on the

Vocabulary subtest is 19 but only 17 for the Arithmetic subtest. This makes it difficult to use profile analysis, particularly for the extremely gifted subjects.

In summary, the WAIS-R has very little competition in the measurement of adult intelligence and will likely be as well received as was the WAIS. The principal contribution of the WAIS-R is its standardization and updating of some of the subtest items.

References

Buros, O. K. (Ed.). (1978). *The eighth mental measurements yearbook.* (Vol. 1). Highland Park, NJ: The Gryphon Press.

Bornstein, R. A. (1983). Verbal IQ-Performance IQ discrepancies on the Wechsler Adult Intelligence Scale-Revised in patients with unilateral and/or bilateral cerebral dysfunction. *Journal of Consulting and Clinical Psychology, 51*(5), 779-780.

Grossman, F. M. (1983). Percentage of WAIS-R standardization sample obtaining verbal-performance discrepancies. *Journal of Consulting and Clinical Psychology, 51*(4), 641-642.

Guertin, W. H., Ladd, C. E., Frank, C. H., Rabin, A. I., & Hiester, D. (1971). Research with WAIS: 1965-1970. *Psychological Record, 21,* 289-339.

Matarazzo, J. D. (1972). *Wechsler's measurement and appraisal of adult intelligence.* Baltimore, MD: Williams & Wilkins.

Mishra, S. P., & Brown, K. H. (1983). The comparability of WAIS and WAIS-R IQs and subtest scores. *Journal of Clinical Psychology, 39*(5), 754-757.

Ryan, J. J., Prifitera, A., & Larsen, J. (1982). Reliability of the WAIS-R with a mixed patient sample. *Perceptual and Motor Skills, 55*(3), 1277-1278.

Ryan, J. J., & Rosenberg, S. J. (1983). Relationship between WAIS-R and Wide Range Achievement Test in a sample of mixed patients. *Perceptual and Motor Skills, 56*(2), 623-626.

Sattler, J. (1982). *Assessment of children's intelligence and special abilities.* Boston: Allyn & Bacon, Inc.

Sattler, J. (1982). *Student's manual to accompany assessment of children's intelligence and special abilities.* Boston: Allyn & Bacon, Inc.

Satz, P., & Mogel, S. (1962). An abbreviation of the WAIS for clinical use. *Journal of Clinical Psychology, 18,* 77-79.

Silverstein, A. B. (1982). Validity of the Satz-Mogel-Yudin-type short forms. *Journal of Consulting and Clinical Psychology, 50*(1), 20-21.

Smith, R. S. (1983). A comparison study of the Wechsler Adult Intelligence Scale and the Wechsler Adult Intelligence Scale-Revised in a college population. *Journal of Consulting and Clinical Psychology, 51*(3), 414-419.

Traub, G. S., & Spruill, J. (1982). Correlations between Quick Test and WAIS-R. *Psychological Reports, 51*(1), 309-310.

Wechsler, D. (1981). *WAIS-R manual: Wechsler Adult Intelligence Scale-Revised.* New York: The Psychological Corporation.

Zimmerman, I. L., & Woo-Sam, J. M. (1973). *Clinical interpretation of Wechsler Adult Intelligence Scale.* New York: Grune & Stratton.

Philip A. Vernon, Ph.D.

Assistant Professor of Psychology, The University of Western Ontario, London, Ontario.

WECHSLER INTELLIGENCE SCALE FOR CHILDREN-REVISED

David Wechsler. San Antonio, Texas: The Psychological Corporation.

THE ORIGINAL OF THIS REVIEW WAS PUBLISHED IN TEST CRITIQUES: VOLUME I (1984).

Introduction

The Wechsler Intelligence Scale for Children-Revised (WISC-R) is one of the best known and most widely used individually administered tests of intelligence. Despite a number of problems which will be described, it is generally and deservedly regarded as a standard measure of general intelligence, against which the validity of other tests of mental ability can be evaluated. Like its predecessor the WISC, the WISC-R consists of six verbal and six performance subtests, of which ten comprise the test proper and two are supplementary or optional, and is scored to provide verbal, performance, and full scale IQs. The subtests measure a diverse set of abilities, reflecting their author's position that "Intelligence can manifest itself in many forms . . . [and] is best regarded not as a single unique trait but as a composite or global entity" (Wechsler, 1974, pp. 5-6).

David Wechsler (1896-1981), an American clinical psychologist, is highly regarded for his work in the area of intelligence testing. The WISC-R, published in 1974, is one of a series of tests of intelligence which he developed, beginning with the Wechsler-Bellevue (W-B) in 1939, the WISC in 1949, the Wechsler Adult Intelligence Scale (WAIS) in 1955, the Wechsler Preschool and Primary Scale of Intelligence (WPPSI) in 1967, and the WAIS-Revised (WAIS-R) in 1981. Together the tests allow the measurement of intelligence from four to seventy-four years of age.

The 1949 WISC was patterned closely after the W-B, which had been designed for use with adults, and was often criticized for being too adult-oriented. This notwithstanding, the WISC established itself as a valuable diagnostic instrument and was widely used for the assessment of both normal children and those with learning or intellectual disabilities. The revision of the WISC was undertaken with the goals of preserving the positive features of the original test and, as far as possible, enhancing these with updated items and improved procedures for administration and scoring. The revision was carried out over a period of approximately three years and included soliciting critical comments and suggestions from practicing psychologists, developing new or modified items, materials, and

550

directions for administration and scoring, and administering the revised test to an appropriately aged representative sample of the U.S. population.

The standardization sample for the WISC-R consisted of 2,200 subjects between the ages of 6½ to 16½. At each year of age within this range, 100 boys and 100 girls were tested within 6 weeks of their midyear. Based on the 1970 U.S. Census, the sample was stratified with respect to geographic region, urban or rural residence, occupation of head of household, and race. Eighty-five percent of the sample (n = 1870) was white; the remaining 15% included blacks (n = 305), American Indians, and Orientals. Only one child in any family was tested. Institutionalized mental defectives and children with severe emotional problems were not included in the sample. Testing was carried out by 202 examiners in 32 states (including Hawaii) and Washington, D.C. The publishers of the WISC-R, The Psychological Corporation, are to be commended for the thoroughness and extensiveness of their standardization procedure. It has resulted in a sample which has been described as being "more nearly representative of the U.S. population within the designated age limits than is any other sample employed in standardizing individual tests" (Anastasi, 1982, p. 254).

The items within each of the subtests are arranged in order of difficulty, generally starting with very simple questions or problems and extending up into items which are difficult or highly complex. The test provides a measure of IQs for 6- to 16-year-olds within the range from 40 to 160, or over 8 standard deviations, and is thus applicable for the great majority of the population in this age range. The 12 subtests of the WISC-R, in the order they are presented to subjects, are as follows:

1. *Information:* a 30-item verbal test of general knowledge and information. Items are scored 0 or 1. Testing is discontinued after five consecutive failures.

2. *Picture Completion:* a 26-item performance test in which subjects are shown pictures of common objects or familiar scenes. Each picture has one component missing and the subject's task is to identify and name or point out the missing part within 20 seconds. Items are scored 0 or 1. Testing is discontinued after four consecutive failures.

3. *Similarities:* a verbal test in which subjects are given 17 pairs of words and asked to describe how each pair is alike or similar. The first four pairs are scored 0 or 1; the remaining 13 items are scored 0, 1, or 2, depending on the quality of the subject's answer. The test is discontinued after three consecutive failures.

4. *Picture Arrangement:* a 12-item performance test (plus one example item) in which subjects are given three to five cards with pictures on them and instructed to arrange the pictures so that they tell a sensible story. The first four items each have a 45-second time limit and are scored 0, 1, or 2, the latter being awarded if the subject performs an item correctly without any assistance from the examiner. The remaining eight items have time limits of 45 seconds (items 5 to 8) or 60 seconds (items 9 to 12), and bonus scores up to a maximum of 5 points are applied if the items are completed within shorter designated time limits. Testing is discontinued after three consecutive failures.

5. *Arithmetic:* an 18-item verbal test of arithmetic, ranging in difficulty from simple counting and addition to fairly complex problems involving fractions and numerical reasoning. The items have time limits of 30 seconds (items 1 to 13), 45

seconds (items 14 and 15), or 75 seconds (items 16 to 18), but no bonus points are awarded for quick solutions. Items 5 to 15 are read aloud by the examiner and may be repeated once if the child requests it. Items 16 to 18 are presented in a booklet, and the child is asked to read each question aloud before attempting to solve it. Items are scored 0 or 1 and testing is discontinued after three consecutive failures.

6. *Block Design:* an 11-item performance test in which subjects are shown pictures of designs in a booklet and are provided with plastic blocks with which they are to copy or match the designs. All the blocks are the same, having two red sides, two white sides, and two half-red half-white sides divided along the diagonal. The items have time limits of 45 seconds (items 1 to 4), 75 seconds (items 5 to 8), or 120 seconds (items 9 to 11). Items 1 to 3 are scored 0, 1, or 2; bonus points up to 7 are awarded on items 4 to 11 if each is completed within shorter designated time limits. The first 8 items use 4 blocks, the remaining items use 9 blocks. Testing is discontinued after two consecutive failures.

7. *Vocabulary:* a 32-item verbal test in which the child is asked to provide definitions of words read aloud by the examiner. Responses are scored 0, 1, or 2 depending on their quality. Testing is discontinued after five consecutive failures.

8. *Object Assembly:* a 4-item performance test (plus one example item) in which subjects are required to put together pieces of a puzzle to form familiar objects within time limits ranging from 120 to 180 seconds. The puzzles consist of from 6 to 8 pieces and partial scores are awarded for partially correct solutions. Bonus points can also be earned for completing the puzzles correctly within shorter designated time limits. All four items are administered.

9. *Comprehension:* a 17-item verbal test in which the examiner asks what the child would do in a variety of situations, or why certain things are done the way they are. The test is designed to measure the child's understanding (or comprehension) of different social situations. Responses are scored 0, 1, or 2 depending on their quality or extensiveness. Testing is discontinued after four consecutive failures.

10. *Coding:* There are two forms to this test. Form A, used with children under age eight, consists of five geometric shapes, each of which has another symbol (or "code") associated with it. Example items show the codes within their respective shapes. Below these, 50 of the shapes appear in random order and the child's task is to draw the appropriate code in as many shapes as possible within a 120-second time limit. Form B, for children eight or older, is similar but the codes are associated with the digits 1 to 9. Below the example items, 93 digits appear above empty boxes and the child has 120 seconds to fill in as many of the boxes with their appropriate codes as possible. In both forms, each item correctly filled in is awarded 1 point.

11. *Digit Span:* a supplementary verbal test which can be administered in addition to the tests already described or as a substitute if one of the regular verbal tests is invalidated (e.g., if testing was interrupted) or cannot be given as a result of some special circumstance. It consists of two parts—digits forward and digits backward. In digits forward, 14 strings of digits, each containing between 3 to 9 digits, are read aloud to the child and the task is to repeat the digits in the order they were presented. In digits backward, 14 strings containing between 2 to 8 digits are read aloud and the child's task is to repeat them in reverse order. Items

are scored 2 (if both strings of a given length are correctly recalled), 1 (if only one string of a given length is recalled), or 0. Testing is discontinued when a child fails to recall both strings of a given length.

12. *Mazes:* a supplementary performance test which may also be administered in addition to the other tests or if necessary as a substitute for one of the regular performance tests. The manual suggests that Mazes may also be substituted for Coding at any time if the examiner chooses to do so, and recommends this substitution for children below age eight because Mazes is more reliable than Coding in this age range. The test consists of 9 mazes (plus one example item) or various sizes and complexity, through which the child is required to trace without crossing lines or entering blind alleys within time limits ranging from 30 to 150 seconds. Partial scores are awarded if the child completes a maze within its designated time limit but makes some errors, the number of permissible errors being specified.

The test materials are conveniently packaged in a case about the size of a briefcase. The complete kit contains the booklets, cards, puzzle pieces, and blocks required for the performance subtests. Red pencils are also provided for Coding and Mazes, which are printed in a separate form. The manual contains complete instructions for administering the subtests, including the precise wording of the verbal questions, and the record form includes the test items for Similarities, Vocabulary, and Digit Span. The front page of the record form contains space for recording subjects' scores, drawing of profile of subtest scores, and for making notes. A separate, full-page profile form is also provided. Within the record form, space is provided for recording subjects' scores on each item of each test; for recording their actual responses (e.g., on Similarities, Arithmetic, Vocabulary, and Comprehension); and for recording their times on the performance tests and on Arithmetic. A number of the subtests have different starting points depending on the subject's age, and these are marked in the record form. Finally, the kit contains a scoring key for Coding, cards which are used in Arithmetic, and a layout shield for Object Assembly. The shield shows the way the examiner is to lay out the Object Assembly puzzle pieces on the table and also serves as a barrier to prevent the child from seeing the pieces before the examiner is ready. The examiner must provide a stopwatch.

Practical Applications/Uses

It is recommended that the subtests be given in the order described earlier, although the order may be changed to suit the needs of a particular child. According to the manual, the regular battery of ten subtests will require approximately 50 to 75 minutes to administer, and it is recommended that the entire battery be given in a single session. The testing room should be quiet, have good ventilation and lighting, be free from interruptions and distractions, and contain two chairs and a table. Generally, it is not advisable for anyone to be in the room except the child and the examiner, although special circumstances might warrant the presence of another person.

Before an examiner can give the test, it is mandatory that one be completely familiar with the procedures for administration and scoring. The manual provides detailed instructions for each subtest but considerable practice is required in order to learn how to administer the test properly. The instructions and test items must be presented precisely as they appear in the manual (although some leeway is allowed in asking follow-up questions), timing must be adhered to closely, and the examiner must score each item during testing in order to know whether to discontinue a subtest. The sorts of probes and questions the examiner is permitted to make are clearly specified in the manual, as are examples of the sorts of responses which earn different points on some of the verbal tests (e.g., Similarities, Comprehension, and Vocabulary). The ambiguity of some of these scoring criteria has been noted by at least one critic (Freides, 1978), but in most cases one would anticipate high interrater agreement.

Once the test has been administered, subjects' total raw scores on each subtest are computed and recorded on the front page of the record form. The raw scores are then converted to scaled scores (with a mean of 10 and a standard deviation of 3), using tables in the manual which are provided for every four-month interval between the ages of six years and sixteen years-eleven months. The sum of the child's scaled scores on the five regularly administered verbal subtests is the Verbal Score. The Performance Score is the sum of scaled scores on the five regularly administered performance subtests. The child's Full Scale Score is the sum of Verbal and Performance Scores. Once these have been computed, they are converted to Verbal, Performance, and Full Scale IQs (with a mean of 100 and a standard deviation of 15), using a single table for all ages.

If Mazes and Digit Span have been administered *in addition* to the ten regularly administered subtests, their scaled scores can be recorded but are not included in computing Verbal, Performance, or Full Scale Scores or IQs. If either was used *instead* of one of the regular subtests, then its scaled score should be included in computing IQ scores. If for any reason less than five verbal or five performance subtests are administered, the sum of the scaled scores on those subtests that were given can be prorated before obtaining Full Scale and IQ Scores. However, Verbal and Performance Scores should each be based on at least four subtests and Full Scale Scores should be based on at least four verbal *and* four performance subtests.

A number of tables are provided in the manual to assist in the interpretation of a child's performance on the WISC-R. Verbal, Performance, and Full Scale IQs, as well as subtest scaled scores, can be converted to percentile ranks or evaluated quantitatively in terms of their deviation from the mean in standard deviation units. For example, a child with a Full Scale IQ score of 130 is 2 standard deviations above the mean and scores at the 98th percentile. Another table allows a qualitative assessment of a child's IQ score. For example, 130 and above is described as "very superior" and between 80 to 89 is "low average" or "dull." Raw scores on each subtest can be converted into what the manual refers to as "test ages" or "equivalent mental ages" to facilitate the interpretation of a child's performance from a developmental perspective or to allow the comparison of WISC-R scores with age norms (or mental ages) from other tests. For the latter purpose, the child's mean or median test age can be computed across all the subtests.

A particularly useful feature of the manual is the inclusion of a table reporting the minimum differences between subtest scaled scores, and between Verbal and Performance IQs, which are statistically significant at the 15% or the 5% level. For example, if a child's scaled score on Information is 3 or more points higher than the score on Similarities, this is significant at the 15% level and may represent an important difference. Similarly, a difference of about 12 points between Verbal and Performance IQs is significant at the 5% level and is probably indicative of a real difference in these abilities. Users of the test who wish to analyze subtest profiles should consult this table, since it will assist them in deciding whether score differences are merely chance fluctuations or are indicative of actual ability differences.

With respect to profile analysis, it should be pointed out that this remains more of an art than a science (Graham & Lilly, 1984). Kaufman (1979) reports that quite large differences between subtest scores are not infrequent and should not be overinterpreted. Hirshoren and Kavale (1976) refer to WISC-R profile analysis as a "continuing malpractice." Clinicians should use their own judgment, and naturally the valildity of their assessments of subtest score differences will depend on their training, experience, and expertise. Kaufman (1979) provides an excellent discussion of profile analysis and other clinical and diagnostic uses of the WISC-R. It is strongly recommended that users of the test consult this book.

Technical Aspects

Internal consistency (split-half) reliability coefficients and test-retest stability coefficients are reported in the manual for all subtests (except Coding and Digit Span, for which only retest coefficients are appropriate), and for Verbal, Performance, and Full Scale IQs. The reliability coefficients are reported for each year of age from 6½ to 16½ and are generally satisfactory and higher than the reliabilities of the WISC subtests. Subtest reliabilities range from .62 to .92, while the average reliabilities of Verbal, Performance, and Full Scale IQs across all ages are .94, .90, and .96, respectively. Retest (stability) correlations are reported for three age groups: 6½-7½ ($n = 97$), 10½-11½ ($n = 102$), and 14½-15½ ($n = 104$); the interval between tests was 3 to 5 weeks for the majority of the children. The subtest stability coefficients are also satisfactorily high, ranging from .65 to .88 averaged across ages, while the average Verbal, Performance, and Full Scale stability coefficients are .93, .90, and .95, respectively. The manual reports that some practice effect was observed on the second testing, amounting to gains of 3½ points in Verbal IQ, 9½ in Performance IQs, and 7 in Full Scale IQs, and notes that this effect should be taken into consideration if a child is retested with the WISC-R after a relatively short interval.

Extensive longitudinal test-retest studies of the WISC-R have not yet been conducted (Graham & Lilly, 1984), but those studies that are available show high stability coefficients for IQs. Vance, Blixt, Ellis, and Debell (1981), for example, retested 75 exceptional children after a two-year interval and reported stability coefficients of .80, .91, and .88 for Verbal, Performance, and Full Scale IQs, respectively. Tuma and Applebaum (1980) reported correspondingly high stability coefficients of .95, .89, and .95 for Verbal, Performance, and Full Scale IQs, respectively, testing 45 normal children over a six-month interval. They also observed some practice effect. Overall, the reliability and stability of the Full Scale scores are as high as or higher than those of other tests of intelligence. The lower reliability of the subtest scores is to be expected, but consideration of these and the subtests' standard errors of measurement (also reported in the manual) should be given if a clinician wishes to analyze differences between a child's subtest scores.

Information pertaining to the validity of the WISC-R is somewhat limited in the manual, although a considerable number of studies have now been conducted which address this topic. The manual presents the intercorrelations among the subtests and Verbal, Performance, and Full Scale scores for each year of age between 6½ and 16½; another table contains the average intercorrelations across all ages. The average subtest intercorrelations are all positive and range from .19 (between Coding and Picture Completion) to .69 (between Vocabulary and Information). On an average, Verbal and Performance scores correlate .67, indicating that these scales measure quite a lot in common, but as Anastasi (1982) points out this correlation is low enough to justify the use of separate scales.

A factor analysis of the intercorrelations would have been useful, but does not appear in the manual. However, Kaufman (1975) performed a factor analysis on the standardization sample data and found evidence for three factors underlying the WISC-R at each age level from 6½ to 16½. He interpreted the factors as representing verbal comprehension, perceptual organization, and freedom from distractability. A similar factor structure was reported among a sample of gifted children (Karnes & Brown, 1980) among samples of Mexican-Americans and whites (Dean, 1980), and among whites and blacks (Gutkin & Reynolds, 1981). Dean (1977, 1979) also reported that the reliability and predictive validity of the WISC-R subtest scores and IQs were similar for his white and Mexican-American samples.

The manual reports three studies in which WISC-R scores were correlated with scores on other tests of intelligence. In the first, 50 six-year-old children were given the WISC-R and the WPPSI in counterbalanced order, yielding a correlation of .82 between Full Scale IQs. WISC-R verbal subtests showed higher correlations with WPPSI Verbal IQ than with WPPSI Performance IQ, and WISC-R performance subtest scores were more highly correlated with WPPSI Performance IQ than with WPPSI Verbal IQ. On an average the WPPSI IQ scores were only 2 points higher than were the corresponding WISC-R IQs. In the second study, 40 children aged 16 years-11 months were administered the WISC-R and the WAIS in counterbalanced order. Correlations between Verbal, Performance, and Full Scale IQs on these tests were .96, .83, and .95, respectively. Finally, four groups of children aged 6 (*n* = 33), 9½ (*n* = 29), 12½ (*n* = 27), and 16½ (*n* = 29) were given the WISC-R and the Stanford-Binet (Form L-M, 1972 norms) in that order. Since each sample is fairly small, the average correlations across age groups are the most reliable (though the manual reports correlations for all four groups). The average correlations between Stanford-Binet IQs and WISC-R Verbal, Performance, and Full Scale IQs were .71, .60, and .73, respectively. These are similar in magnitude to those reported in several studies between the Stanford-Binet and the 1949 WISC (see Zimmerman & Woo-Sam, 1972). On the average, there was about a 2-point difference between WISC-R Full Scale IQs and Stanford-Binet IQs, indicating that both tests can be expected to yield similar IQs for normal children between the ages of 6 and 16 years.

Other studies have found correlations ranging from .82 to .95 between the WISC-R and the Stanford-Binet, and, as in the study reported in the manual, mean IQs on these tests have been similar (Brooks, 1977; Kaufman & Van Hagen, 1977; Raskin, Bloom, Klee, & Reese, 1978). Swerdlik (1977) reviewed 12 studies comparing WISC-R and WISC IQs: WISC IQs were always higher on an average, presumably because the 1949 WISC norms are now outdated (Graham & Lilly, 1984), and in most studies correlations between Verbal, Performance, and Full Scale IQs were in the .70s and .80s. Samples included both normal and mentally retarded children and special education students, and the interval between tests was as much as three years.

The WISC-R has also been correlated with other tests of intelligence and achievement, yielding correlations of about .60 to .80 for Verbal and Full Scale IQs and somewhat lower correlations for Performance IQ (Graham & Lilly, 1984). As yet, not many studies have investigated the predictive validity of the WISC-R, but those which have report relatively good prediction of grades and scholastic achievement (Graham & Lilly, 1984). Typically, Verbal IQs are better predictors of grades than are Performance or Full Scale IQs (e.g., Dean, 1979).

Overall, studies provide strong support for the WISC-R's construct validity. Its content validity is harder to assess since Wechsler's definition of intelligence is so broad (Littell, 1960; Graham & Lilly, 1984), but the diverse nature of the subtests reflects his global conception. In addition, as Graham and Lilly (1984) point out, the test was constructed carefully, attention was given to the suggestions of practicing psychologists, and the items and materials seem appropriate for the age range for which the test was designed. More studies of the WISC-R's predictive validity would be useful and will no doubt be forthcoming given the test's extensive use.

Critique

Generally, the revisions incorporated in the WISC-R have resulted in an improved test. Nearly three-quarters of the original WISC items have been retained, but those that were changed or modified have made the test more child-oriented and less ambiguous, and contain more pictures of females and blacks. Several subtests were lengthened, resulting in higher reliability. The sequence of subtest administration has been changed, with verbal and performance tests now given alternately, and this should make the test session more interesting. The role of the examiner has been changed considerably, now playing a much more active part in terms of probing, demonstrating, and asking follow-up questions. As Whitworth (1978) has pointed out, this is a distinct improvement over the 1949 WISC, but also requires more technical competence and evaluative judgment on the part of the examiner. The 1949 WISC was designed for ages 5 to 15, while the WISC-R is for 6- to 16-year-olds. According to the manual, this change was effected in order to reduce age overlap with the WPPSI (designed for 4- to 6½-

year-olds), but Whitworth (1978) suggests that it might have been more appropriate to incorporate the WPPSI items in the WISC-R and thus have a single test for 4- to 16-year-olds.

Not all of the changes have been for the better. Freides (1978) criticizes the "irresponsible or downright antisocial themes" of a large number of the Picture Arrangement stories. He also questions why a separate record form is required for Coding and notes the distracting effect of Mazes showing through from the other side of the form. Other critics are supportive of the change in Block Design to blocks colored only red and white, but note that since they are plastic they are more likely to break than the original wooden blocks. As mentioned previously, the scoring criteria for some of the verbal subtests are still somewhat ambiguous or inconsistent, and a report of interrater reliability in the manual would have been useful (Graham & Lilly, 1984).

As stated previously, the WISC-R is widely considered to be one of the best available individual tests of general intelligence for its age range. It is used extensively by clinicians, school psychologists, and in psychological research. The manual provides sufficient information to administer and score the test, although most users of the WISC-R will require extra training and practice in order to put the test to its best use. The items and test materials are appropriate for 6- to 16-year-olds, and the test spans a wide enough range of difficulty to make it usable with the great majority of children between these ages. Research with the WISC-R has supported its construct validity as a test of a wide range of mental abilities, reflecting its author's global concept of general intelligence. In school settings the WISC-R is a relatively good predictor of academic achievement, and it is an excellent instrument for the diagnosis of intellectual retardation. Examiners familiar with the 1949 WISC will find several improvements in the WISC-R, but they will also probably have to unlearn quite a lot of the earlier administration procedures. A number of articles and books are available to assist clinicians in their interpretations of children's WISC-R performance; of these, Kaufman's 1979 book is again particularly recommended.

References

Anastasi, A. (1982). *Psychological testing* (5th ed.). New York: Macmillan Publishing Co.

Brooks, C.R. (1977). WISC, WISC-R, S-B L & M, WRAT: Relationships and trends among children ages six to ten referred for psychological evaluation. *Psychology in the Schools, 14,* 30-33.

Dean, R.S. (1977). Reliability of the WISC-R with Mexican-American children. *Journal of School Psychology, 15,* 267-268.

Dean, R.S. (1979). Predictive validity of the WISC-R with Mexican-American children. *Journal of School Psychology, 17,* 55-58.

Dean, R.S. (1980). Factor structure of the WISC-R with Anglos and Mexican-Americans. *Journal of School Psychology, 18,* 234-239.

Freides, D. (1978). Review of WISC-R. In O.K. Buros (Ed.), *The eighth mental measurements yearbook.* Highland Park, NJ: The Gryphon Press.

Graham, J.R., & Lilly, R.S. (1984). *Psychological testing.* Englewood Cliffs, NJ: Prentice-Hall.

Gutkin, T.B., & Reynolds, C.R. (1981). Factorial similarity of the WISC-R for white and black children from the standardization sample. *Journal of Educational Psychology, 73,* 227-231.

Hirshoren, A., & Kavale, K. (1976). Profile analysis of the WISC-R: A continuing malpractice. *The Exceptional Child, 23,* 83-87.

Karnes, F.A., & Brown, K.E. (1980). Factor analysis of the WISC-R for the gifted. *Journal of Educational Psychology, 72,* 197-199.

Kaufman, A.S. (1975). Factor analysis of the WISC-R at eleven age levels between 6½ and 16½ years. *Journal of Consulting and Clinical Psychology, 43,* 135-147.

Kaufman, A.S. (1979). *Intelligent testing with the WISC-R.* New York: John Wiley & Sons, Inc.

Kaufman, A.S., & Van Hagen, J. (1977). Investigation of the WISC-R for use with retarded children: Correlation with the 1972 Stanford-Binet and comparison of WISC and WISC-R profiles. *Psychology in the Schools, 14,* 10-14.

Littell, W.M. (1960). The Wechsler Intelligence Scale for Children: Review of a decade of research. *Psychological Bulletin, 57,* 132-156.

Raskin, L.M., Bloom, A.S., Klee, S.H., & Reese, A. (1978). The assessment of developmentally disabled children with the WISC-R, Binet, and other tests. *Journal of Clinical Psychology, 34,* 111-114.

Swerdlik, M.E. (1977). The question of the comparability of the WISC and WISC-R: Review of the research and implications for school psychologists. *Psychology in the Schools, 14,* 260-270.

Tuma, J.M., & Applebaum, A.S. (1980). Reliability and practice effects of WISC-R IQ estimates in a normal population. *Educational and Psychological Measurement, 40,* 671-678.

Vance, H.B., Blixt, S., Ellis, R., & Debell, S. (1981). Stability of the WISC-R for a sample of exceptional children. *Journal of Clinical Psychology, 37,* 397-399.

Wechsler, D. (1974). *Manual for the Wechsler Intelligence Scale for Children-Revised.* New York: The Psychological Corporation.

Whitworth, R.H. (1978). Review of the WISC-R. In O.K. Buros (Ed.), *The eighth mental measurements yearbook.* Highland Park, NJ: The Gryphon Press.

Zimmerman, I.L., & Woo-Sam, J.M. (1972). Research with the Wechsler Intelligence Scale for Children. *Psychology in the Schools, 9,* 232-271.

Elaine M. Heiby, Ph.D.

Assistant Professor of Clinical Psychology, University of Hawaii at Manoa, Honolulu, Hawaii.

WECHSLER MEMORY SCALE

David Wechsler and C.P. Stone. San Antonio, Texas: The Psychological Corporation.

THE ORIGINAL OF THIS REVIEW WAS PUBLISHED IN TEST CRITIQUES: VOLUME I (1984).

Introduction

The Wechsler Memory Scale (WMS) is a rationally derived, individually administered test designed to measure short-term and long-term memory deficits.

The test's developer, David Wechsler, was born in Lespedi, Romania, on January 12, 1896. His family moved to New York City in 1901. He received his undergraduate degree from the College of the City of New York in 1916 and his M.A. from Columbia University (under Robert Woodworth) in 1917. During World War I he worked with E.G. Boring on validation of the Army Alpha Test. Following the end of the war, he returned to Columbia University to complete his Ph.D., again under Woodworth, in 1925. His dissertation represents the first major work on the psychogalvanic reflex response. From 1934, Wechsler devoted his professional efforts to developing and validating a variety of measures of intellectual capacity, including the Wechsler Adult Intelligence Scale (WAIS) and the Wechsler Intelligence Scale for Children (WISC).

The Wechsler Memory Scale was developed in response to the need for a test measuring the memory ability of psychiatric patients that would not be highly correlated with intelligence. The earliest memory test was developed by Wells and Martin in 1923 (Wechsler, 1945). This instrument was difficult to administer and not well standardized. In 1940 the Babcock Test of Mental Efficiency was published. However, this instrument was both highly correlated with intelligence and time-consuming to administer (Erickson & Scott, 1977). In an attempt to resolve these problems Wechsler published the WMS in 1945 after approximately ten years of development while at Bellevue Hospital in New York City.

In developing normative data for the WMS, Wechsler chose 200 non-hospitalized male and female subjects (*Ss*) between the ages of 20 and 50 years. The sex ratio and intellectual capacity of these *Ss* has never been published. Because many memory deficits develop after the age of 50, Hulicka (1966) attempted to correct this possible deficit in the instrument by developing normative scores for five different age levels (15-17, 30-39, 60-69, 70-79, 80-89). Sex differences for normative data have not been reported.

The reviewer would like to acknowledge and thank James D. Becker for his editorial assistance.

There are currently two forms of the WMS: Form I and the more recent Form II (Stone & Wechsler, 1946). All major studies involving the WMS have utilized Form I, and there appear to be no norms or standardization data available for Form II (Lezak, 1983). One study (Prigatano, 1978) does report that WMS scores tend to remain stable across forms. Lezak (1983) states that in her opinion the two forms may not be interchangeable. Until more rigorous standardization and normative data for both forms are available, this empirical question remains unanswered.

Since the development of the original forms of the WMS, research in the area of memory has demonstrated that there is more than one "global" memory process. For example, Milner (1968) demonstrated that a major aspect of verbal memory is located in the left temporal lobe while figural memory is located within the right temporal lobe. Additionally, the concepts of short-term and long-term memory have been developed since the original WMS was published. Russell (1975), wanting to develop a clinical memory test that would reflect recent concepts in memory research, developed the Revised Wechsler Memory Scale (RWMS) (Russell, 1975, 1981, 1982). The RWMS consists of six separately scored tests: verbal short-term memory, verbal long-term memory, verbal retention, short-term figural memory, long-term figural memory, and figural retention. Thus, the revised WMS more accurately reflects the current conceptualization of multiple memory processes than does the original WMS, which reflects a more unitary view of memory. However, the RWMS is not yet widely available and is still considered experimental.

As of this writing there are no special forms of the WMS for the visually, hearing, or physically handicapped or those for whom English is not the primary language.

The two forms of the WMS consist of seven subtests. In Form I the first two subtests consist of questions common to most mental status examinations. Subtest 1, Personal and Current Information, asks the subject six questions concerning age, date of birth, and identification of current and recent public officials (such as the current and past president of the United States). Subtest 2, Orientation, contains five questions about time and place. Subtest 3, Mental Control, consists of three items concerning simple conceptual tracking (counting by threes from 1 to 40) and automatisms (alphabet). Immediate recall of ideas, presented orally in two paragraphs, is evaluated in Subtest 4, Logical Memory. This subtest is intended to measure both logical memory and immediate recall. Subtest 5, Memory Span, was borrowed from the Wechsler Adult Intelligence Scale with modifications. The three- and nine-digit trials of Digits Forward and the two- and eight-digit trials of Digits Backward are omitted from the original WAIS Digit Span. Subtest 6, Visual Reproduction, is an immediate visual memory task which involves drawing from memory four simple geometrical figures exposed for periods of 10 seconds each. Two of the figures were borrowed from the Army Performance Scale and the others from the Stanford-Binet. Subtest 7, Associate Learning, is a test of verbal retention and consists of ten paired-associates which the subject is asked to learn in three trials.

The examiner sums the scores on each of the seven subtests to produce a raw score. A constant is added to this score based on the age group of the subject. A table of constants is provided in the manual and this new sum is the subject's weighted or corrected memory score. Using the weighted sum, the score is then converted to a Memory Quotient (MQ) by referring to another table provided in the manual.

Form II of the WMS also consists of seven subtests along the same format as Form I. Subtests 1 and 2 are identical to those in Form I. Subtest 3 is similar but has a different order of item presentation and the counting item is by fours rather than by threes. Subtest 4 involves two different passages that were also borrowed from the Army Performance Scale. Subtest 5 involves different digits that were taken from the sets of equivalent forms used in a preliminary form of the Army Alpha. Subtest 6 involves designs of approximately equal difficulty as those used in Form I. Form I designs were taken from the 1937 revision of the Stanford-Binet Form L while Form II designs were taken from the 1937 revision of the Stanford-Binet Form M. Subtest 7 involves new items for the associate learning task that were selected from Wechsler's original list and appear to be of approximately equal difficulty as those in Form I.

For both Forms I and II the examiner presents the items orally and must record errors and the time required for most responses. Standardized instructions are provided for each subtest.

The WMS is intended for individuals between the ages of 20 and 64 years. The test is easy for most individuals as it is designed to identify basic memory impairments and not to discriminate among individual differences in normal memory skills.

The answer sheet for Form I and II is composed of two sides of a single 8½"x11" sheet. The first side provides space for scoring Subtests 1-5 and the second side for Subtests 6 and 7. The examiner must remember to fold the answer sheet to cover up Subtest 7 while the subject is performing the reproductions for Subtest 6. There is a section of the answer sheet for indicating subtest scores, total raw scores, age correction, corrected or weighted score, and the MQ.

Practical Applications/Uses

There are several appropriate practical uses of the WMS. Its primary use assesses general brain dysfunction in a neuropsychological and neurological examination or as an adjunct to an intellectual assessment involving intelligence testing. The WMS may be useful in the specification of functional memory problems during a battery of tests designed to assess psychopathology or general personality. The WMS may be of value in an assessment of the effects of treatments such as electroconvulsive shock therapy or chemotherapy on basic memory functions. It can be used to document changes in brain function after injury, during a degenerative disease, or in the aging process. Finally, the WMS is of value in evaluating the effects of rehabilitative attempts in the brain-damaged individual.

The WMS is inappropriate for use with individuals under the age of 20 and over the age of 64. It is also inadequate as a global measure of brain damage since numerous brain functions are not evaluated.

The WMS provides a measure of overall memory function and does not measure specific types of memory function such as immediate, short-term, or remote memory. This limitation is partly due to the fact that there are no standardized scores provided for the subtests of the WMS, rendering it impossible to interpret performance on any individual items or groups of items.

The WMS is most often used by clinical psychologists and neuropsychologists in both applied and research settings. Potential applications of the WMS include the evaluation of various hypotheses regarding the nature of memory as posited by cognitive psychologists. For example, Russell (1981) suggests that memory involves a three-stage system including sensory, short-term, and long-term recent and remote memory. Items on the WMS could be rationally and factorially analyzed to evaluate whether the test measures these various types of memory. If the WMS were shown to assess these different types of memory, then the scale could be organized along these dimensions and yield scores reflecting functioning for each type of memory.

The WMS is individually administered to subjects between the ages of 20 and 64 years who have use of the dominant hand and who are not deaf or visually impaired. It is a brief instrument, requiring approximately 15 minutes to administer. The examiner must time most responses and provide individual feedback to the subject during Subtest 7. The test can be administered by a doctoral-level clinical psychologist or neuropsychologist or by a psychological testing technician under supervision.

The WMS is easy to score by hand and the instructions are clear and simple. As of this writing, no machine- or computer-scoring services are available. There is no information provided in the manual, or by other research evaluating the WMS, that indicates the effects of changing the sequence of items or subtests. The manual also fails to provide the examiner with general instructions for the introduction of the test to the subject. There is also no information on whether the examiner can repeat items or whether items can be paraphrased.

The instructions for scoring the WMS are clearly presented in the manual and can be learned after one or two administrations. It takes about 5 minutes to score the test once the administration is mastered, although several items do present some difficulty. On both forms of the WMS, there are questions in Subtest 1 and 2 that assume the subject resides in an urban environment by inquiring about the city's name and mayor. This is obviously problematic for subjects from rural locations. Subtest 4 on both forms is also difficult to score. The examiner is asked to score the number of ideas reproduced from each of the two passages presented to the subject, but what constitutes an idea is not defined.

Some investigators (Power, Logue, McCarty, Rosenstiel, & Ziesat, 1979) have suggested the following guidelines to facilitate judgment on this subtest: 1) allow one half-credit for the substitution of one or more synonyms that do not violate the basic idea presented in the passage and 2) allow one half-credit for the omission of an adjective, adverb, or article that only slightly changes the basic idea in the passage. These seem like reasonable suggestions since Wechsler's original intent was to score concepts remembered and not literal reproductions of the passage. These passages are also somewhat stilted and are conducive to paraphrasing. Power et al. evaluated their proposed scoring guidelines and found interscorer reliability of .97 for Logical Memory immediate recall and .96 for Logical Memory delayed recall.

A final concern in scoring either form of the WMS regards Subtest 6, Visual Reproduction. Although the manual presents written descriptions on how to score the answers to these items, it would be helpful to provide the examiner some examples of acceptable, partially acceptable, and unacceptable responses.

Interpretation of the WMS is as objective as the interpretation of most IQ test scores, however, the standard deviation of the Memory Quotient (MQ) derived from the WMS is not provided and this makes it difficult to interpret deviations from the mean. For example, how much below the mean must a score be in order to suspect brain damage or dysfunction? While the failure to provide the standard deviation of scores places severe limitations on the interpretation of the MQ, the fact that the MQ can be compared to an individual's IQ facilitates interpretation in terms of relating memory functioning to general intellectual functioning. This makes interpretation of the MQ considerably more complex, and interpretation should be limited to doctoral-level clinical psychologists with training in neuropsychology.

Technical Aspects

Several types of test reliability are relevant to the WMS. First, because the scale is presented as a measure of global memory functioning, one would expect internal consistency (split-half reliability) to be high for each subtest as well as for the overall test. Second, because the WMS is designed to measure brain damage, one would expect consistency of test scores over time, given no intervening treatment or recovery (test-retest reliability). Third, because scoring of some of the items requires judgment, it is necessary to evaluate the consistency of such judgments by providing measures on interscorer reliability. Finally, one would expect each form of the WMS to provide approximately the same score (alternate form reliability). Surprisingly, to this author's knowledge there has been no research evaluating internal consistency, temporal consistency, or interscorer agreement of either form of the WMS. Alternate form reliability, in contrast, has been evaluated.

Wechsler (1945) evaluated agreement between the two forms of the WMS by administering the two scales to 60 college students in counterbalanced order with a two-week interval between administrations. He does not report the correlation between the two sets of scores, but he did conduct a *t*-test (to evaluate the difference between mean scores on the two forms) and found that the two sets of scores did not differ significantly. More recently, Prigatano (1978) reported an alternate form reliability correlation of .80 for the two forms, suggesting acceptable equivalence.

Validity of the WMS has been somewhat more thoroughly investigated than has reliability. The types of validity relevant to the WMS include concurrent, construct, and factor validity. Concurrent validity is evaluated by correlating scores on the WMS with other measures of memory functioning and with clinical judgment of such functioning. Perhaps because the WMS is the only widely used memory test available, concurrent validity of the scale has not been evaluated. Nevertheless, scores on the WMS could be related to clinical evaluation, such as that based upon the result of a mental status examination. In addition, scores on the WMS could be related to the status of individuals with known memory problems.

The second type of relevant validity, that evaluating the construct of memory as measured by the WMS, has been investigated more thoroughly than has concurrent validity. One would expect that an individual with dysfunctional memory would also evidence an impairment in overall intellectual functioning. Prigatano (1978) reports that for brain-damaged, mentally retarded, and low IQ subjects, there is a .80 correlation between an MQ obtained from the WMS and a Full Scale IQ obtained from the Wechsler Adult Intelligence Scale. Another way to look at construct validity of the WMS is to evaluate the ability of WMS scores to discriminate between left- and right-hemisphere lesions. Milner (1968) found lower MQ scores in subjects with left temporal lobectomy than in subjects with right temporal lobectomy and this finding was replicated by Black (1973). This difference would be expected since the WMS is an orally and verbally administered test. Finally, if the WMS is a measure of memory functioning one would expect scores on the WMS to decrease with advancing age, as has been found (e.g., Prigatano, 1978). Further evaluation of construct validity of the WMS would involve relating the site of a brain lesion to specific memory dysfunction.

Factor validity studies evaluate the third type of validity relevant to the WMS, which is whether the items on the WMS cluster in ways that are consistent with current theorizing regarding the nature of the types of memory. Recent reviews of studies evaluating the factor structure of the WMS have found that a three-factor solution is consistently obtained across a variety of populations (Larrabee, Kane, & Schuck, 1983; Skilbeck & Woods, 1980). Factor I, Immediate Learning and Recall, has loading from Subtest 4 (Logical Memory), Subtest 6 (Visual Reproduction), and Subtest 7 (Associate Learning). Factor II, Attention and Concentration, has loading from Subtest 3 (Mental Control) and Subtest 5 (Memory or Digit Span). Factor III, Orientation and Long-term Information Recall, has loading from Subtest 1 (Personal and Current Information) and Subtest 2 (Orientation). These factors roughly correspond to the current theorizing that posits three types of memory (e.g., Russell, 1981). It has been suggested that the WMS be scored according to these three factors rather than in terms of a total score as proposed in the test manual (Erickson & Scott, 1977). Of course, such scoring changes would require the development of norms and standard scores for each of the three memory measures. Meanwhile, Kear-Colwell (1973) conducted a stepwise regression analysis and developed the following formulae for the conversion of subtest scores to the three factor scores: 1) Factor I = −0.27 + (0.25 x Logical Memory Score) + (0.19 x Visual Reproduction Score) + (0.21 x Associate Learning Score); 2) Factor II = −1.80 + (0.28 x Information Score) + (−0.34 x Orientation Score) + (0.43 x Mental Control Score) + (0.45 x Memory Span Score); and 3) Factor III = −9.68 + (0.78 x Information) + (2.34 x Orientation).

The strongest source of support for the validity of the WMS is from studies evaluating its factor structure. Much work is still needed to establish concurrent and construct validity.

Critique

The WMS has several distinct assets. This test is the best standardized memory scale available even though there are many problems with its norms and instructions. The WMS is easy to use, as it requires only about 15 minutes to administer. It makes allowance for variations with age and the MQ is somewhat comparable to the IQ, which facilitates interpretation. Finally, the stable factor structure of the WMS suggests that further refinement of the scale is possible to yield memory scores that are consistent with current theorizing on types of memory (e.g., Russell, 1981).

The problems of the WMS include both the lack of support for the assumption that memory is a unidimensional construct and the failure to revise the test to accommodate the findings that it actually measures three types of memory, as has been found in factor-analytic studies. The subtests lack norms, making it impossible to inspect intratest variability in performance. Similarly, the entire WMS is lacking norms based on a large representative sample with consideration of various demographic and diagnostic variables. It would be helpful to revise the presentation of the test to accommodate persons with expressive and receptive verbal deficits as well as those with motor dysfunction in the dominant hand. It would also be useful to have forms of the test for non-English-speaking subjects and those with visual and auditory handicaps. Scoring difficulties mentioned earlier could be remedied by providing alternative questions in Subtests 1 and 2 for administrations of the test in rural communities and by providing scoring examples for Subtests 4 and 6.

It is also necessary to establish further reliability and validity for the WMS, as well as measures of internal and temporal consistency. It would also be helpful to provide the standard error of measurement, which would facilitate the interpretation of borderline scores. Further evidence of the validity of the WMS is also needed. Concurrent validity would help demonstrate the accuracy of the WMS in identifying distinct diagnoses. Construct validity studies would clarify our understanding of the role of memory in various dysfunctions. Finally, all evidence in support of the WMS has been based on Form I. Form II still awaits the establishment of norms as well as support of its reliability and validity. Until further psychometric support for the WMS is provided, it is recommended that the application of the WMS be restricted to Form I.

References

Black, F. (1973). Memory and paired-associate learning of patients with unilateral brain lesions. *Psychological Reports, 33,* 912-922.

Erickson, R., & Scott, M. (1977). Clinical memory testing: A review. *Psychological Bulletin, 84,* 1130-1149.

Hulicka, I.M. (1966). Age differences in Wechsler Memory Scale scores. *Journal of Genetic Psychology, 109,* 135-145.

Kear-Colwell, J. (1973). The structure of the Wechsler Memory Scale and its relationship to "brain damage." *British Journal of Social and Clinical Psychology, 12,* 384-392.

Larrabee, G.L., Kane, R.L., & Schuck, J. (1983). Factor analysis of the WAIS and Wechsler Memory Scale: An analysis of the construct validity of the Wechsler Memory Scale. *Journal of Clinical Neuropsychology, 5,* 159-168.

Lezak, M.D. (1983). *Neuropsychological assessment.* New York: Oxford University Press.

Logue, P., & Wyrick, L. (1979). Initial validation of Russell's Revised Wechsler Memory Scale: A comparison of normal aging versus dementia. *Journal of Consulting and Clinical Psychology, 471,* 176-178.

Milner, B. (1968). Visual recognition and recall after right temporal lobe excision in man. *Neuropsychologia, 6,* 191-209.

Power, D., Logue, P., McCarty, S., Rosenstiel A., & Ziesat, H. (1979). Inter-rater reliability of the Russell revision of the Wechsler Memory Scale: An attempt to clarify some ambiguities in scoring. *Journal of Clinical Neuropsychology, 1,* 343-345.

Prigatano, G. (1978). Wechsler Memory Scale: A selective review of the literature. *Journal of Clinical Psychology, 34,* 816-832.

Russell, E. (1975). A multiple scoring method for the assessment of complex memory functions. *Journal of Consulting and Clinical Psychology, 43,* 800-809.

Russell, E. (1981). The pathology and clinical examination of memory. In S. Filskov & T. Boll (Eds.), *Handbook of clinical neuropsychology.* New York: John Wiley & Sons, Inc.

Russell, E. (1982). Factor analysis of the Revised Wechsler Memory Scale tests in a neuropsychological battery. *Perceptual and Motor Skills, 54,* 971-974.

Skilbeck, C., & Woods, R. (1980). The factorial structure of the Wechsler Memory Scale: Samples of neurological and psychogeriatric patients. *Journal of Clinical Neuropsychology, 2,* 293-300.

Stone, C., & Wechsler, D. (1946). *Wechsler Memory Scale Form II.* New York: The Psychological Corporation.

Wechsler, D. (1945). *Wechsler Memory Scale.* New York: The Psychological Corporation.

Jean C. Elbert, Ph.D.

Assistant Professor, Child Study Center, Department of Pediatrics, University of Oklahoma Health Sciences Center, Oklahoma City, Oklahoma.

E. Wayne Holden, Ph.D.

Post-Doctoral Fellow, Pediatric Psychology Division, Department of Psychiatry and Behavioral Sciences, University of Oklahoma Health Sciences Center, Oklahoma City, Oklahoma.

WECHSLER PRESCHOOL AND PRIMARY SCALE OF INTELLIGENCE

David Wechsler. San Antonio, Texas: The Psychological Corporation.

THE ORIGINAL OF THIS REVIEW WAS PUBLISHED IN TEST CRITIQUES: VOLUME III (1985).

Introduction

The Wechsler Preschool and Primary Scale of Intelligence (WPPSI; Wechsler, 1967) is a widely used test of global intelligence designed for preschool children ages 4 to 6½ years. Consistent with the previous Wechsler scales, the WPPSI contains 11 subtests (six verbal, five performance), which are divided into a Verbal Scale and a Performance Scale, yielding Verbal, Performance, and Full Scale IQ scores scaled to a mean of 100 and a standard deviation of 15. Intelligence quotients are calculated with the deviation IQ method rather than the ratio IQ method utilized by more traditional intellectual measures, such as the Stanford-Binet Intelligence Scale.

The WPPSI was developed by David Wechsler in the tradition of the Wechsler Adult Intelligence Scale (WAIS), the Wechsler Intelligence Scale for Children (WISC), and their subsequent revised versions (WAIS-R and WISC-R). Three years were spent modifying and extending the WISC scales to create a test based on Wechsler's model of intelligence for use with younger children. After preliminary studies on 300 children, the WPPSI was standardized on 1,200 children, 100 boys and 100 girls in each of six age groups ranging in half years from 4 to 6½. The 1960 U. S. Census data were used to generate quotas for geographic regions, urban-rural distribution, race, and socioeconomic status. As with the other Wechsler scales, the WPPSI has the advantage of having an excellent standardization sample. Neither a revised version with restandardization nor the development of alternative versions for special populations have occurred since the initial publication of the WPPSI. However, considerable investigation of the WPPSI has been undertaken since its publication, and 250 references are cited in Buros' *Seventh* and *Eighth Mental Measurements Yearbooks*, as well as *Tests in Print II* and *Tests in Print III*.

Wechsler had assumed that mental abilities are continuous and that the same or similar tasks could be used across age levels to assess these abilities. Consequently, five of the WPPSI verbal subtests and three of the five performance subtests are downward extensions of the previously established WISC subtests; at the higher levels, specific WPPSI items overlap with the WISC. Two WPPSI subtests were developed as analogues to the WISC Digit Span and Coding subtests: the supplementary Sentences and Animal House subtests, respectively. A third addition, Geometric Design, replaces Object Assembly and Picture Arrangement in the previously developed Wechsler scales. As in all of Wechsler's scales, items within each subtest are ordered in terms of difficulty with ceiling points established by a maximum numbr of consecutive items failed. Individual WPPSI subtests are scaled to a mean of 10 and a standard deviation of 3. The subtests in their order of presentation are as follows:

1. Information: a 23-item test of general verbal factual information and knowledge of the environment. Twelve of the items, with minor changes, were taken from the WISC Information subtest. Scoring is 1 or 0, and the subtest is discontinued after five consecutive failed items.

2. Animal House and Animal House Retest: a 20-item timed pegboard-like task in which the child must associate animal pictures with colored pegs by placing the appropriately colored pegs in the corresponding holes. It is a test of perceptual-motor speed and sign-symbol association in which both speed and accuracy are assessed in the scoring. The Animal House Retest is a supplementary performance subtest, which is identical to the original subtest but which has separately normed scaled scores reflecting the effect of prior learning. It is not used in calculating the Performance IQ when administered as a sixth Performance Scale subtest.

3. Vocabulary: a 22-item test of general word knowledge in which the child must provide word definitions. Fourteen of the items are taken directly from the WISC. Items are scored 2, 1, or 0, depending on the quality and degree of abstraction of the verbal definition, and the subtest is discontinued after five consecutive failures.

4. Picture Completion: a 23-item performance task involving the identification of a single missing element in pictures of common objects. It involves attention to visual detail, understanding of part-whole relationships, and the ability to discriminate essential from unessential detail. Scoring is 1 or 0, verbal or pointing responses are allowed, and testing is discontinued after five consecutive failures.

5. Arithmetic: a 20-item test of quantitative reasoning involving object counting, comprehension of quantitative concepts, verbal problem-solving, and addition and subtraction. Six of the items are taken from the WISC, and items 9-20 have a time limit of 30 seconds. Children who are six years and older begin with item 7 and must establish a basal level of two consecutive correct items, whereas children below six years, or who are suspected to be mentally retarded, begin with item 1. The test is discontinued after four consecutive failures.

6. Mazes: a ten-item maze-tracing task, the last seven items of which are taken from the WISC. A visual-motor response is required, and motor planning, perceptual-motor speed, and visual organization skills are believed to be tapped by this task. Errors consist of entries into "blind alleys," time limits range from 45 to 135 seconds, and the test is discontinued after failure on two consecutive mazes.

7. Geometric Design: a ten-item geometric design copying task in which figures range from simple lines to more complex visual patterns. The test is untimed, and scores based on a judged degree of accuracy range from 0-2 (items 1-5), 0-3 (items 6-7), and 0-4 (items 8-10). Testing is discontinued after failure on two consecutive items.

8. Similarities: a 16-item verbal task including six items taken from the WISC. The first ten items involve verbal analogies requiring a specific word association, whereas the remaining items involve word pairs for which the child must describe the conceptual similarity. The verbal analogies are scored 1 or 0, whereas scoring for the remaining word pairs is 2, 1, or 0, depending on the quality and level of abstraction of the response. Testing is discontinued after four consecutive failures.

9. Block Design: a ten-item performance test involving perceptual analysis and manipulation in which the child is asked to reproduce block patterns from a model (items 1-7) and from cards (items 8-10). Blocks are flat and two-sided (as opposed to the cubic blocks used for the other Wechsler scales), with solid red or white blocks used in items 1 and 2, and diagonally divided red/white blocks used for the remaining items. Time limits range from 30 to 75 seconds with no time-bonus credits. Items consist of two trials; the first trial is demonstrated by the examiner on items 1, 3, 4, 5, and 8, and failure on the first trial is followed by demonstration on the second trial. The test is discontinued after failure on two consecutive items.

10. Comprehension: a 15-item verbal subtest consisting of questions evaluating such areas as general understanding of the environment, social situations, common sense judgment and ability to utilize past experience in responding to hypothetical problem situations. Responses are scored 2, 1, or 0, depending on the quality of content and degree of elaboration of the response, and testing is discontinued after four consecutive failures.

11. Sentences: a 13-item supplementary verbal subtest in which verbatim sentence repetition is required after oral presentation by the examiner. The sentences are of increasing length, ranging from two to eighteen words, and scoring ranges from 0-1 (items A and B), 0-2, (items C, 1-4), 0-3 (items 5-6), and 0-4 (items 7-10). Testing is discontinued after three consecutive failures. Although Sentences may be included as an alternate if one of the Verbal subtests cannot be administered, it is not used in calculating the Verbal IQ when administered as a sixth Verbal Scale subtest.

The WPPSI test materials include two booklets (for presenting items on the Picture Completion, Arithmetic, Geometric Design, and Block Design subtests), blocks, colored cylinders and board, test record forms, and response forms for Mazes and Geometric Design. The test forms provide space for recording verbatim responses to the Verbal subtest items, and specific criteria for discontinuing subtests as well as time limits for individual Performance Scale items.

Practical Applications/Uses

Since its initial publication the WPPSI has been used extensively for clinical evaluation, classification, placement, and individualized program planning for preschool children. With the increased attention devoted to early childhood education programs in recent years, the WPPSI has received considerable attention as a measure for identifying children who may be at risk for developmental delays or who may be intellectually gifted. Consideration will be given to 1) general ease of administration and scoring, 2) use of the WPPSI with special populations of exceptional children, 3) use of short forms of the WPPSI, and 4) interpretation of WPPSI performance.

As with other formal individual tests of intelligence, the WPPSI requires extensive specialized training in proper administration, scoring, and interpretation; consequently, administration of the scale is generally restricted to psychologists or those in the field of special education and/or early childhood education who have been trained in the measurement of intelligence. In most graduate training programs, the WPPSI is taught as one of several individually administered tests of intelligence, with prerequisite training in theories of intelligence often required.

The prescribed order of WPPSI subtests was designed so as to alternate verbal and nonverbal tasks as well as difficult and easier tasks. This balance appears effective in maintaining interest in the young child (Yule, Berger, Butler, Newhem, & Tizzard, 1969); however, changes in the prescribed order are allowed as clinical needs dictate. Although the nonverbal subtests are generally intrinsically appealing to most young children, it is often difficult to maintain interest and optimal effort on verbal subtests such as Vocabulary and Comprehension.

Administration time has been found to range between 60-90 minutes (Sattler, 1982). In the standardization sample it was noted that 10% of the children assessed required 90 minutes or longer to complete the test. This is a decided disadvantage when one is assessing very young or handicapped children whose short attention spans may preclude the completion of the scale in one session. No data are available to aid in determining whether alterations in standard administration procedures, such as dividing the administration time over two or more sessions, may alter a child's score. Investigation has been made of effects of reward on WPPSI subtest performance, and preliminary evidence has been found for a relationship between reward and test performance as a function of individual task strategies (Moran, McCullers, & Fabes, 1984).

In general, instructions for administration procedures are clearly specified in the WPPSI test manual. Despite overall clarity, however, it has been noted that some ambiguity exists in both procedures and scoring criteria for certain subtests. On the Information subtest, for example, there are insufficient examples of acceptable responses on items having multiple acceptable answers, thus introducing an element of subjectivity into the scoring. Other Verbal subtests (e.g., Vocabulary, Similarities, Comprehension) similarly require a degree of subjectivity in scoring, and poor quality of verbalizations may influence an examiner's judgment of content. Of the Performance subtests, Geometric Design has the most subjectivity in scoring in that the examiner relies to a large extent on visual comparisons of a child's reproduction with sample drawings.

Despite similarity to the WISC-R for the majority of subtests, the examiner must not assume that administrative procedures are identical for the two test instruments. On Mazes, for example, when children make an error on Mazes 2 and 3 they are interrupted, and the maze replaced with the next item. Such indications of a child's failure have been criticized by several reviewers who recommend allowing the child the opportunity to complete a maze on which errors have been made

(Sattler, 1982; Yule et al., 1969). In contrast to the WISC-R, the WPPSI Block Design subtest involves demonstration by the examiner on many initial trials (items 1, 3, 4, 5, and 8) and on all second trials where failure has occurred on the first trial. A final note in comparing the WPPSI and WISC-R is the overlap in the 6 to 6½ year age range. Because a more thorough sampling of ability can be obtained from a greater number of WPPSI than WISC-R items, the WPPSI scale is preferable for use with this overlapping age group of children. In addition, comparison of children in this age group indicates that the WPPSI tends to yield higher Full Scale IQ scores than the WISC-R (Sattler, 1982).[1]

Although the standardization sample for the WPPSI is very broad and extensive, no clinical populations of children were evaluated during the development of the test. Because practical use of the WPPSI has been primarily with clinical samples of children, considerable research has emerged regarding the efficacy of its use as a measure of identifying children with special education needs. First, as with any global measure of intelligence, the WPPSI has been used in the diagnosis of mental retardation. However, because of the levels of maturation presumed, the WPPSI is noted for its limited sensitivity and discrimination, particularly at the lower age limits. Although IQ scores in the manual range from 45 to 155, the range is not applicable until the 5½-year-old level. A child of 4 years can fail all items on a subtest, yet obtain scaled scores of 4 on six of ten subtests. The absence of a sufficient number of easy items thus results in inflated IQ ratings for low functioning children. The manual cautions that four-year-old children with suspected IQ < 75 should not be evaluated with the WPPSI. The limited floor for this test would suggest that it may not be the measure of choice in evaluating moderately mentally retarded children across the entire age range.

Because of similar limitations at the test ceiling, the appropriateness of the WPPSI for identifying intellectually gifted children has also been questioned (Rellas, 1969; Sattler, 1982; Silverstein, 1968b). In a more recent study of 306 children with WPPSI IQs > 120 it was shown that no ceiling was obtained for at least half of the subjects on six subtests: Similarities, Comprehension, Vocabulary, Picture Completion, Mazes, and Information (Hawthorne, Speer, & Buccellato, 1983). As Eichorn (1972) observes, gifted children may earn a sum of scaled scores higher than those for which IQ equivalents are available in the test manual.[2]

The need for identifying children for Head Start and other supplemental preschool programs has generated some research on the use of the WPPSI with culturally disadvantaged, linguistically different, minority, and other clinical samples of children. Although approximately 14% of the WPPSI standardization sample was composed of "nonwhite" children, the relative representation of Hispanic, black, American Indian, and other minority groups is not available in the manual. Kaufman (1973) found significant differences on the Performance Scale in favor of white over black children, which were greatest at the lowest age level. In a follow-up study (Kaufman, 1978) it was shown that test instructions for several of the WPPSI Performance subtests assume the ability to comprehend a number of basic verbal concepts. Relatively poor performance may therefore be a function of communication problems, and the valid assessment of nonverbal abilities may be questionable for those children with poor language stimulation or communication disorders. Sewell (1977) found significantly higher mean WPPSI IQs in lower SES black children when compared to their Stanford-Binet IQs, and the WPPSI was suggested as a more appropriate measure of cognitive ability for this minority group. Hispanic children have been shown to have poorer Verbal IQs than either Performance or Full Scale IQs on the WPPSI (Gerken, 1978; Valencia, 1984). In their work with Mexican-American five-year-olds from economically depressed areas, Henderson and Rankin (1973) found an 18-point difference between Verbal and Performance IQ (74 vs. 92). Consequently, concern has been raised regarding the efficacy of educational classification of these children on the basis of WPPSI Full Scale IQ.

Finally, with regard to the appropriateness of the WPPSI for handicapped children, no adaptation of the test for use with hearing or visually impaired children has as yet been published. However, the distribution of subtests into verbal and nonverbal tasks does allow for the administration of a single scale to a visually or motorically handicapped child.

Following test administration, total raw scores are recorded and converted to scaled scores using tables provided in the manual. These encompass each three-month interval between the age of 3 years, 10 months, 16 days (4 years) and 6 years, 7 months, 15 days (6½ years). The sum of scaled scores from the five verbal subtests and the five performance subtests are converted to Verbal IQ and Performance IQ, respectively, and the Full Scale IQ is computed using the sum of all ten subtests. The WPPSI manual allows for prorated IQ scores in the event that a verbal or performance subtest is invalid or not administered. In this case, prorated IQ scores can be calculated on the basis of four verbal or four performance subtests, and/or Sentences can be used as an alternate fifth verbal subtest if only four have been administered. Tables are provided for converting IQ scores into percentiles (though interpolation is required because not all possible IQ scores have equivalent percentile ranks) and for converting raw scores into test age equivalents. Subsequently, a mean or median test age can be computed. The latter conversion was included to permit comparison of WPPSI scores to age norms of other scales. Because of the limited floor and ceiling of the WPPSI, however, test age equivalents cover only the age range of the standardization sample (4 to 6½ years); thus test ages of < 4 or > 6½ are unobtainable.

To reduce administration time and provide a valid and reliable screening measure, a number of WPPSI short forms have been developed.[3] Construction of short forms has differed on both empirical and conceptual grounds, and relatively little formal data have been collected on their utility. Sattler (1982) and Reynolds and Clark (1983) provide ample discussion of the merits of various short form versions. Depending on the examiner's needs, short forms range from two subtests (Silverstein, 1970) to five subtests (Sattler, 1982).[4]

Considerable attention has been given to interpretation of WPPSI scaled scores and IQs. (See Anastasi, 1982; Reynolds & Clark, 1983; Sattler, 1982 for representative reviews.) The interpretation of an individual's performance on any of Wechsler's scales has generally involved the following types of pattern analysis: 1) classification of an individual's overall intellectual status based on statistical derivations from a normal distribution; 2) comparison of Verbal and Performance Scale IQs; 3) comparison of individual subtest scores to the mean scaled scores from the Verbal, Performance, and/or Full Scales (intersubtest scatter); 4) comparison of recategorized scaled scores based on theoretical and/or factor analytic classifications; and 5) comparison of successes and failures within subtests (intrasubtest scatter).

Reynolds and Gutkin (1981a) performed normative analyses of the WPPSI standardization sample in order to determine the justification for interpretation of WPPSI subtest scatter, and their findings are relevant for those using the WPPSI for clinical purposes. Their analyses indicated that approximately one third of the children showed a significant VIQ-PIQ discrepancy at $p = .05$ (11 IQ points), and one fourth showed a significant discrepancy at $p = .01$ (14 IQ points). What were previously believed to be unusual degrees of variability in performance for individual children were found to characterize the profiles of many normal preschool children; therefore, caution against overinterpreting VIQ-PIQ discrepancies is recommended. With regard to the direction of the discrepancy, the frequency of VIQ > PIQ was the same as VIQ < PIQ in children between −1 and +1 SD from the mean (FSIQ = 85-115). With children of FSIQ > 115 the VIQ > PIQ pattern occurred significantly more frequently; conversely, with children of FSIQ < 85, the opposite pattern of PIQ > VIQ was significantly more frequent. In general, subtest scatter appears to be more common on the WPPSI than on the WISC-R; a 5- to 7-point scaled score range is reported to be about average, and Reynolds and Gutkin (1981a) advise that ranges of at least ten scaled score points be required prior to inferring statistical abnormality in a child's performance.

One of the most frequent uses of the WPPSI has been in the early identification of children with specific learning disabilities.[5] Although the use of separate Verbal and Performance Scales provides valuable information in the identification of a child's cognitive strengths and weaknesses, the utilization of pattern analysis as the sole means for diagnosing specific learning disability is not recommended.

Finally, recategorization of WPPSI subtests into Bannatyne's theoretical groupings (Spatial, Conceptual, and Sequential) have been performed on data from the standardization sample, and minimum deviation values for statistical significance have been derived (Reynolds & Gutkin, 1981b).[6] However, it must be cautioned that interpretations based on such recategorizations do not have factor analytic support, because only two factors have emerged from various factor analyses of the WPPSI.

Technical Aspects

Psychometric theory provides a scientific framework for evaluating the technical aspects of intelligence tests. Reliability and validity represent two psychometric indices that determine the adequacy of test measurement. Reliability reflects consistency of measurement, whereas validity reflects accuracy of measurement. Although both criteria are important to the evaluation of test instruments, reliability assumes a superordinate position in relation to validity because consistency of measurement places limits on the accuracy of measurement.

Reliability estimates derived from the standardization sample were provided across age levels for summary scores and individual subtests in the WPPSI manual (Wechsler, 1967). Except for the Animal House subtest, which was evaluated by computing an immediate test-retest coefficient, internal consistency indices were utilized to evaluate summary score and subtest reliabilities. Coefficients for the total sample were quite acceptable for Full Scale scores (.96), Performance Scale scores (.93), and Verbal Scale scores (.94). Subtest reliabilities ranged from .77 for the Animal House subtest to .87 for the Mazes subtest. Minimal variability was obtained when coefficients were recalculated separately for subjects within six-month age levels between 4 and 6½ years. Additionally, the standard error of measurement for each subtest and summary score at each age level were reported. Standard errors of measurement varied widely across subtests and age levels indicating that the degree of error variability should be considered when interpreting subtest score differences.

Test-retest reliabilities were computed within a sample of 50 five-year-old kindergarten children who were a part of the standardization group. As expected, coefficients were somewhat lower than those obtained with internal consistency methods, but well within the acceptable range for intellectual measures. Summary score test-retest coefficients were particularly impressive: Full Scale score (.92), Performance Scale score (.89), and Verbal Scale score (.86). Furthermore, intercorrelations of subtest scores were calculated for a representative subpopulation of the standardization sample. An acceptable degree of internal consistency for the overall WPPSI was apparent. This is not surprising in light of the fact that the WPPSI was constructed based on a model of intelligence that has received substantial support over the last thirty years.

Further support for WPPSI reliability has been obtained in subsequent investigations. Internal consistency of Full Scale scores has been found to be equivalent to estimates obtained from the standardization sample in Mexican-American children, gifted children, mentally retarded children, and English children (Sattler, 1982). Internal consistency for summary measures from a short form of the WPPSI constructed by Kaufman (1972) have also proven to be comparable to the original internal consistency estimates.[7] Additional data on test-retest reliability have been quite limited subsequent to the initial test publication. Full Scale score test-retest reliability for a small group of black, lower SES children was not substantially different from data published in the WPPSI manual (Croake, Keller, & Catlin, 1973). WPPSI reliability in general, however, has been found to be somewhat higher than WISC-R reliability and substantially higher than Stanford-Binet reliability (Reynolds & Gutkin, 1981b).

In contrast to reliability, scant attention was given to validity in the WPPSI manual. Correlations between WPPSI Full Scale IQs and the Stanford-Binet (.75), Peabody Picture Vocabulary Test (.58), and Pictorial Test of Intelligence (.64) within a sample of 98 five- to six-year-old children were reported. Relationships with these tests were somewhat lower than reported by authors who have subsequently reviewed investigations of the criterion related validity of the WPPSI (Anastasi, 1982; Reynolds & Clark, 1983; Sattler, 1982). Similar to other Wechsler intelligence scales, however, Verbal Scale scores correlated significantly higher than Performance Scale scores with Stanford-Binet IQ. Because the mean Full Scale score for the 98 children who were subjects in the initial validity study was about ten points below that expected for the general population of children the same age, generalizability of validity coefficients reported in the manual is somewhat limited.

Subsequent investigations, however, have provided substantial support for the criterion-related validity of the WPPSI. In reviewing investigations published since the WPPSI manual, Sattler (1982) reports that a median correlation of .82 was obtained between the WPPSI Full Scale IQs and the Stanford-Binet. Furthermore,

Verbal Scale scores correlated significantly higher than Performance Scale scores with the Stanford-Binet IQ across investigations. A median correlation similar in magnitude was obtained from investigations that examined the relationship between the concurrent administration of the WPPSI and the WISC in samples between the ages of 5 and 6½ years.[8] Slightly lower coefficients were reported between WPPSI summary indices and WISC-R summary indices when tests were administered four years apart in an English sample (Bishop & Butterworth, 1979). Yule, Gold, and Busch (1982), however, found a .86 correlation between Full Scale WPPSI scores at age 5½ and Full Scale WISC-R scores at age 16½. Correlations with the General Cognitive Index from the McCarthy Scales have been somewhat lower than with other intellectual measures (Arinoldo, 1982; Phillips, Pasewark, & Tindall, 1978).[9] The decrement in magnitude of these relationships, however, might be expected, given that the McCarthy GCI is based on a summary score reflecting performance on verbal, perceptual-performance, and quantitative subtests.

Evidence for the criterion-related validity of WPPSI short forms is available. IQ estimates from Kaufman's (1972) short forms correlated highly with WPPSI Full Scale scores. Furthermore, the length of the short form has been identified as an important variable in the estimation of validity. It was reported that the validity of short forms increased as the number of subtests increased (Bishop, 1980). Moreover, the validity of short forms comprised of randomly selected subtests rivaled the validity of short forms constructed on conceptual grounds (Silverstein, 1983).

Investigations of WPPSI validity have also utilized achievement test scores as criterion measures. These data are especially important in evaluating the WPPSI, as a general consensus exists that intelligence tests are most useful in predicting academic achievement.

Wechsler (1967) failed to provide data on the relationship between WPPSI scores and academic achievement. It is unclear why this oversight occurred within the otherwise exemplary development of the test. Subsequent support for predictive validity has been found in a number of studies investigating Caucasian populations and employing several different achievement measures (Sattler, 1982). Data collected on culturally disadvantaged and lower socioeconomic groups, however, has been less impressive. It should be noted that the predictive validity of WPPSI summary scores rivaled the predictive validity of WISC-R summary scores in separate school samples referred for psychological evaluation (Reynolds, Wright, & Dappen, 1981). Moreover, sex differences in predictive validity were not found in the above sample (Wright, Reynolds, & Dappen, 1980). A prospective study provides further support for the predictive validity of the WPPSI. Correlations between WPPSI summary scores and math/reading achievement test scores remained consistent across a four-year span (Feshbach, Adelman, & Fuller, 1977). Sex differences were apparent, however, with predictive validity indices higher for males than females.

Factor analysis is a multivariate statistical technique that provides estimates of the common and specific variances that test components measure. The results of factor analyses are one index of the overall validity of intelligence tests. In addition, they provide variables that can be used to evaluate the validity of test components. Estimates of the specific and error variances represented by subtest scores can be constructed from individual factor loadings.

Separate Verbal and Performance factors have emerged in factor analytic investigations of the WPPSI (Anastasi, 1982; Reynolds & Clark, 1983; Sattler, 1982). Almost universally, Verbal Scale subtests have loaded highly on the verbal factor, and Performance Scale subtests have loaded highly on the performance factor. These results have been obtained in a number of populations utilizing differing factor analytic techniques. They contrast strongly with the three factor solutions (e.g., Verbal, Performance, and Freedom from Distractibility) typically obtained in factor analyses of WISC-R data.[10] Some minor discrepancies to this pattern of results, however, have been reported. The Arithmetic subtest has loaded highly on both factors in two investigations (Haynes & Atkinson, 1984; Reynolds & Clark, 1983), a separate factor for the Picture Completion subtest has been identified (Heil, Barclay, & Endres, 1978), and it has been reported that two-factor solutions deteriorate significantly in children below age five (Hollenbeck & Kaufman, 1972). In general, however, factor analytic research has provided substantial support for Wechsler's model of intelligence and the separation of the WPPSI into Verbal and Performance Scales.

Subtest specificity has been estimated for the eleven WPPSI subtest scores in factor analytic investigations. Authors examining data from the standardization sample report that adequate specificity exists for all subtests except Information and Comprehension (Carlson & Reynolds, 1981; Kaufman, Daramola, & DiCuio, 1977). Recently, however, these results were challenged in a factor analytic investigation of the WPPSI test scores of children evaluated in a mental health clinic setting (Haynes & Atkinson, 1984). It was reported that error variance exceeded specific variance in approximately one half of the subtests. Based on these results, it was concluded that profile analysis of WPPSI subtest scores may be invalid in clinical settings.

Perusal of the literature on the technical aspects of the WPPSI indicates that substantial attention has been devoted to the evaluation of both reliability and validity. Clearly, the WPPSI rivals most child and adult intellectual measures in the amount of data amassed on its psychometric properties. Two striking weaknesses, however, plague the literature and should be addressed in future research. First, more data is needed on test-retest reliability at different age levels and with groups varying with respect to demographic characteristics (e.g., socioeconomic status, race). Second, both standard errors of measurement and specific variances of subtest scores are in need of extensive evaluation. Interpretation of profile configuration and subtest differences is currently hampered by inadequacies and inconsistencies in the data compiled on these latter topics.

Critique

A well-standardized and technically sound measure of global intelligence, the WPPSI continues to be widely considered as one of the best available choices for cognitive assessment of the preschool child. It is used extensively by clinicians, school psychologists, and early childhood specialists, as well as in research with preschool populations. At present, the McCarthy Scale of Children's Abilities is the only other comprehensive measure of cognitive abilities in preschool children that rivals the WPPSI in its conceptual and technical adequacy. The Kaufman Assessment Battery for children is a more recently developed measure that can be used with the preschool age group, but it has not as yet had extensive research validation.

Despite its obvious technical qualities, however, several issues warrant consideration by those using the WPPSI as an assessment instrument with preschool children. First, the obvious advantage for continuity of intellectual assessment and follow-up across the developmental age span is suggested by the surface similarity of tasks on all of Wechsler's scales. However, it is questionable whether the tasks are measuring identical cognitive processes at different age levels. Since the WPPSI's inception, there has been a large accumulation of information regarding the cognitive development of preschool children. Such an extended data base reinforces a need to question the conceptual validity of a downward extension of tasks originally developed for older children and/or adults. As an example, the dissimilarities in factor structure between the WISC-R and WPPSI (e.g., Freedom from Distractibility factor) suggest that attentional variables account for differing proportions of variance in performance on the two tests. This issue is most salient when considering the WPPSI for use with both the younger age groups and with clinical populations for whom attentional processes are quite variable.

Second, the major current use of the WPPSI is apt to be with clinical populations of children, despite the absence of data regarding the performance of such children in the original test standardization sample. The limited floor and ceiling of the WPPSI restrict the range of its use with children at the extremes of the distribution of IQ scores, namely, mentally retarded and gifted children. The degree of verbal comprehension required on the Performance Scales also limits one's ability to validly assess nonverbal intelligence in communication disordered, bilingual, and, in some cases, minority children. Moreover, the length of administration time places restrictions on the applicability of the WPPSI to the youngest age group and to children displaying severe attentional disorders.

Third, the need for caution in interpretation of WPPSI test performance previously stressed in this review should be reiterated. More variability in performance is apparent on the WPPSI than either the WISC-R or WAIS-R. Therefore, larger discrepancies are required between Verbal and Performance Scale scores on the WPPSI before valid interpretation can be made of the differences. Moreover, because of the larger standard errors of measurement on individual subtests relative to the WISC-R, a greater between subtest difference is needed in order to make intersubtest comparisons. Therefore, although scaled scores of individual WPPSI subtests may be an apparent advantage over other methods of subtest clustering, caution against unwarranted interpretation of scatter must be recommended.

In view of the emphasis on early childhood evaluation since the WPPSI's development in 1967, it is likely that normative standards may have changed in the population of preschool children, and a restandardization of the WPPSI may be timely.[11] In the meantime, the authors' personal experience and the literature reviewed here suggest that the WPPSI may be most appropriate for the five- to six-year-old children who are not expected to fall at either extreme of the distribution of intelligence but who may be displaying specific information-processing deficiencies. Administration of the WPPSI by a skilled clinician can provide considerable information regarding both a child's cognitive strengths and weakness as well as the opportunity for observing individual problem-solving strategies. When used in conjunction with other measures, data from the WPPSI provide a solid basis for diagnostic formulation and remediation planning.

Notes to the *Compendium:*

[1]A direct comparison of the WISC, WISC-R, and WPPSI scales has been performed in order to test the comparability of the scales. In a counterbalanced design, 72 5- to 6-year-old children were administered all three scales (Quereshi & McIntire, 1984). Results indicated that the Verbal IQs were comparable, but the Performance and Full Scale IQs were not. Among the subtests, only Comprehension, Arithmetic, Picture Completion, and Mazes could be unequivocally considered comparable across the three scales. Much of the discrepancy in Performance IQ was apparently due to great divergence in content; the WPPSI shares only three same-name subtests with the WISC and WISC-R.

[2]Data suggest that the WPPSI measures similar intellectual constructs for males and females. However, sex differences have been shown favoring Animal House and Sentences in females and Mazes in males (Kaiser & Reynolds, 1985). These investigators found that previous work showing better female performance on memory and psychomotor tasks and better male performance on spatial tasks also was supported in sex differences on the WPPSI.

[3]Both selected-item (King & Smith, 1972; Silverstein, 1968, 1982) as well as selected-subtest versions (Haynes & Atkinson, 1983; Kaufman, 1972; King & Smith, 1972; Silverstein, 1968a) have been evaluated.

[4]In a review, Watkins (1986) reported that each of the short form approaches possessed potential flaws and that none is a valid abbreviated IQ measure. Watkins thus advised that short forms be used only as screening instruments.

[5]See Badian (1984), Hagin, Silver, and Corwin (1971), and Raviv, Rahmani, and Ber (1986).

[6]Badian's (1984) sample of 5-year-old children identified 3 years later as being poor readers demonstrated a Conceptual > Spatial > Sequential profile. The abilities important in the acquisition of reading skills may differ considerably from those associated with later good reading, and such differences may explain the difference in Bannatyne's groupings found in older LD children (Spatial > Conceptual > Sequential profile).

[7]The reliabilities of Verbal-Performance IQ differences, differences among subtest scores, and differences between subtest scores and overall IQ were reported to be higher than for either the WISC-R or the WAIS (Feingold, 1985).

[8]In contrast, Qureshi and McIntire (1984) reported that Verbal IQs were comparable, but Performance and Full Scale IQs were not, in an investigation comparing performances on the WPPSI, WISC, and WISC-R in 5- to 6-year-old children.

[9]See also Valencia and Rothwell (1984).

[10]Similarly, a recent cluster analysis of the WPPSI standardization data yielded separate Verbal and Performance clusters, providing further support for the results of factor analytic investigations (Silverstein, 1986).

[11]A revised version of the WPPSI (Wechsler, 1987) currently is undergoing field testing with the collection of standardization data scheduled to begin in late 1987. The goals of the revision are to 1) extend the age range covered by the test downward to age 3 and upward to age 7, 2) increase the appeal of the test materials for children, 3) delete items that are included on the WISC-R, 4) revise test materials that are dated or biased, 5) include an object assembly subtest, and 6) update the normative data base. The 11 original subtests from the WPPSI have been retained in the WPPSI-R, but have been modified to meet the goals of the revision. Extensive psychometric analyses are planned that will address the technical limitations noted in this review of the WPPSI (C. Burns, personal communication, April, 1987). It is anticipated that the WPPSI-R will be available for clinical use beginning in 1989.

References

Anastasi, A. (1982). *Psychological testing* (5th ed.). New York: Macmillan Publishing Co.

Arinoldo, C. G. (1982). Concurrent validity of McCarthy's scales. *Perceptual and Motor Skills, 54,* 1343-1346.

Badian, N. A. (1984). Can the WPPSI be of aid in identifying children at risk for reading disability. *Journal of Learning Disabilities, 17,* 583-587.

Bishop, D. (1980). Predictive validity of short forms of the WPPSI. *British Journal of Social and Clinical Psychology, 19,* 173-175.

Bishop, D., & Butterworth, G. E. (1979). A longitudinal study using the WPPSI and WISC-R with an English sample. *British Journal of Educational Psychology, 49,* 156-168.

Carlson, L., & Reynolds, C. R. (1981). Factor structure and specific variance of the WPPSI subtests at six age levels. *Psychology in the Schools, 18,* 48-54.

Croake, J. W., Keller, J. F., & Catlin, N. (1973). WPPSI, Rutgers, Goodenough, Good-enough-Harris I.Q.'s for lower socioeconomic, black, preschool children. *Psychology, 10,* 58-65.

Eichorn, D. H. (1972). Review of the WPPSI. In O. K. Buros (Ed.)., *The seventh mental measurements yearbook* (pp. 806-807). Highland Park, NJ: Gryphon Press.

Feshbach, S., Adelman, H., & Fuller, W. (1977). Prediction of reading and related academic problems. *Journal of Educational Psychology, 69,* 299-308.

Feingold, A. (1985). Reliability of score differences on the Wechsler Preschool and Primary Scale of Intelligence. *Psychological Reports, 57,* 663-664.

Gerken, K. G. (1978). Performance of Mexican American children on intelligence tests. *Exceptional Children, 44,* 438-443.

Hagin, R. A., Silver, A. A., & Corwin, C. G. (1971). Clinical-diagnostic use of the WPPSI in predicting learning disabilities in grade 1. *Journal of Special Education, 5,* 221-232.

Hawthorne, L. W., Speer, S. K., & Buccellato, L. (1983). Appropriateness of the Wechsler Preschool and Primary Scale of Intelligence for gifted children. *Journal of Consulting and Clinical Psychology, 51,* 463-469.

Haynes, J. P., & Atkinson, D. (1983). Validity of two WPPSI short forms in outpatient clinic settings. *Journal of Clinical Psychology, 39,* 961-964.

Haynes, J. P., & Atkinson, D. (1984). Factor structure of the WPPSI in mental health clinic settings. *Journal of Clinical Psychology, 40,* 805-808.

Heil, J., Barclay, A., & Endres, J. M. (1978). A factor analytic study of WPPSI scores of educationally deprived and normal children. *Psychological Reports, 42,* 727-730.

Henderson, R. W., & Rankin, R. J. (1973). WPPSI reliability and predictive validity with disadvantaged Mexican-American children. *Journal of School Psychology, 11,* 16-20.

Hollenbeck, G. P., & Kaufman, A. S. (1972). Factor structure of the WPPSI. *Proceedings of the Annual Convention of the American Psychological Association, 7.*

Kaiser, S. M., & Reynolds, C. R. (1985). Sex differences on the Wechsler Preschool and Primary Scale of Intelligence. *Personality and Individual Differences, 6,* 405-407.

Kaufman, A. S. (1972). A short form of the Wechsler Preschool and Primary Scale of Intelligence. *Journal of Consulting and Clinical Psychology, 39,* 361-369.

Kaufman, A. S. (1973). Comparison of the performance of matched groups of black children and white children on the Wechsler Preschool and Primary Scale of Intelligence. *Journal of Consulting and Clinical Psychology, 41,* 186-191.

Kaufman, A. S. (1978). The importance of basic concepts in the individual assessment of preschool children. *Journal of School Psychology, 16,* 207-211.

Kaufman, A. S., Daramola, S., & DiCuio, R. F. (1977). Interpretation of the separate WPPSI tests for boys and girls at three age levels. *Contemporary Educational Psychology, 2,* 232-238.

King, J. D., & Smith, R. A. (1972). Abbreviated forms fo the Wechsler Preschool and Primary Scale of Intelligence for a kindergarten population. *Psychological Reports, 30,* 539-542.

Moran, J. D., McCullers, J. C., & Fabes, R. A. (1984). Developmental analysis of the effects of reward on selected Wechsler subscales. *American Journal of Psychology, 97,* 205-214.

Phillips, B. L., Pasewark, R. A., & Tindall, R. C. (1978). Relationship among McCarthy Scales of Children's Abilities, WPPSI, and Columbia Mental Maturity Scale. *Psychology in the Schools, 15,* 352-356.

Quereshi, M. Y., & McIntire, D. H. (1984). The comparability of the WISC, WISC-R, and WPPSI. *Journal of Clinical Psychology, 40,* 1036-1043.

Raviv, A., Rahmani, L., & Ber, H. (1986). Cognitive characteristics of learning-disabled and immature children as determined by the Wechsler Preschool and Primary Scale of Intelligence Test. *Journal of Clinical Child Psychology, 15,* 241-247.

Rellas, A. J. (1969). The use of Wechsler Preschool and Primary Scale of Intelligence (WPPSI) in the early identification of gifted students. *California Journal of Educational Research, 20,* 117-119.

Reynolds, C. R., & Clark, J. H. (1983). Assessment of cognitive abilities. In K. D. Paget & B. A. Bracken (Eds.), *The psychoeducational assessment of preschool children* (pp. 163-189). New York: Grune & Stratton.

Reynolds, C. R., & Gutkin, T. B. (1981a). Test scatter on the WPPSI: Normative analyses of the standardization sample. *Journal of Learning Disabilities, 14,* 460-464.

Reynolds, C. R., & Gutkin, T. B. (1981b). Statistics for the interpretation of Bannatyne recategorizations of WPPSI subtests. *Journal of Learning Disabilities, 14,* 464-467.

Reynolds, C. R., Wright, D., & Dappen, L. (1981). A comparison of the criterion-related validity (academic achievement) of the WPPSI and the WISC-R. *Psychology in the Schools, 18,* 20-23.

Sattler, J. M. (1982). *Assessment of children's intelligence and special abilities* (2nd ed.). Boston: Allyn and Bacon, Inc.

Sewell, T. E. (1977). A comparison of the WPPSI and Stanford-Binet Intelligence Scale among lower SES black children. *Psychology in the Schools, 14,* 158-161.

Silverstein, A. B. (1968a). Validity of WPPSI short forms. *Journal of Consulting and Clinical Psychology, 32,* 229-230.

Silverstein, A. B. (1968b). WISC and WPPSI IQ's for the gifted. *Psychological Reports, 22,* 1168.

Silverstein, A. B. (1968c). Cluster analysis of the Wechsler Preschool and Primary Scale of Intelligence. *Journal of Psychoeducational Assessment, 4,* 83-86.

Silverstein, A. B. (1970). Reappraisal of the validity of the WAIS, WISC, and WPPSI short forms. *Journal of Consulting and Clinical Psychology, 34,* 12-14.

Silverstein, A. B. (1983). Validity of random short forms: III. Wechsler's intelligence scales. *Perceptual and Motor Skills, 56,* 572-574.

Valencia, R. R. (1984). Concurrent validity of the Kaufman Assessment Battery for Children in a sample of Mexican-American children. *Educational and Psychological Measurements, 44,* 365-372.

Valencia, R. R., & Rothwell, J. G. (1984). Concurrent validity of the WPPSI with Mexican-American preschool children. *Educational and Psychological Measurement, 1984, 44,* 955-961.

Watkins, C. E. (1986). Validity and usefulness of WAIS-R, WISC-R, and WPPSI short forms: A critical review. *Professional Psychology: Research and Practice, 17,* 36-43.

Wechsler, D. (1967). *Wechsler Preschool and Primary Scale of Intelligence.* Cleveland, OH: The Psychological Corporation.

Wechsler, D. (1987). *Wechsler Preschool and Primary Scale of Intelligence-Revised (Standardization Edition).* San Antonio, TX: The Psychological Corporation.

Wright, D., Reynolds, C. R., & Dappen, L. (1980). Criterion-related validity of three common preschool assessment instruments for boys and girls. *Psychological Reports, 47,* 1291-1296.

Yule, W., Berger, M., Butler, S., Newham, V., & Tizzard, J. (1969). The WPPSI: An empirical evaluation with a British sample. *British Journal of Educational Psychology, 39,* 1-13.

Yule, W., Gold, D., & Busch, C. (1982). Long-term predictive validity of the WPPSI: An 11 year follow up study. *Personality and Individual Differences, 3,* 65-71.

Robert C. Reinehr, Ph.D.

Assistant Professor of Psychology, Southwestern University, Georgetown, Texas.

WIDE RANGE ACHIEVEMENT TEST

Joseph F. Jastak and Sarah Jastak. Wilmington, Delaware: Jastak Associates, Inc.

THE ORIGINAL OF THIS REVIEW WAS PUBLISHED IN TEST CRITIQUES: VOLUME I (1984).

Introduction

The Wide Range Achievement Test (WRAT) is a test of reading, spelling, and arithmetic skills. It is intended primarily for individual administration, although instructions are included for the administration of some subtests to small groups of older children or adults.

The 1984 revision of the test, the WRAT-R, is the sixth edition since the test first appeared in 1936. Originally developed by Joseph F. Jastak, Ph.D., as a supplement to tests of general ability and to group tests of achievement, the 1936 and 1946 editions had only one scale, extending from kindergarten to college level on each of the three subtests. Later editions co-authored with Sarah Jastak, Ph.D., have been divided into two levels; one for use with children 5 to 11 years of age and the other for use with individuals 12 to 75 years.

Although similar in content to previous editions, the WRAT-R incorporates several significant changes from the 1978 version: 1) new recording forms with larger print and more space; 2) new norms, based on a sample of 5,600 subjects, stratified by age, geographic region, sex, race, and urban/rural residence; 3) new methods of calculating test scores; 4) minor item changes in the arithmetic subtest, designed to extend the lower level of usage; 5) an extension of upper level norms to include individuals through age 75; and 6) provision of two separate manuals, an Administration Manual and a Diagnostic and Technical Manual.

Practical Applications/Uses

The WRAT-R is designed for use in clinical and school settings as a screening measure of academic achievement in reading, spelling, and arithmetic. The only materials required for test administration are an Administration Manual and a Record Form, although a supplementary card containing the reading and subtest items is provided for the convenience of the examiner. The subtests may be administered in any order. The Record Form and Manual present the tests in a different order, a minor annoyance during administration.

The Reading subtest consists of asking the subject to pronounce aloud each word from a list ranging from simple to difficult. For subjects unable to complete the first line without an error, a pre-reading section is provided. Pre-reading consists of naming or recognizing individual letters printed on the Record Form or supplementary card. A time limit of 10 seconds per word is suggested but need not be strictly adhered to in situations where the examiner feels it appropriate to ask the subject to repeat words or to otherwise deviate slightly from usual procedure. The raw score obtained is the total number of words correctly read aloud plus any credit received on the pre-reading section, if administered.

The Spelling subtest consists of printing or writing 45 words from dictation. A pre-spelling section consists of a coding task and printing or writing of the subject's name. All writing is recorded on the Record Form. Time limits for dictation are 15 seconds per word. The raw score is the total number of correct words plus all credit received on the pre-spelling section.

The Arithmetic subtest consists of written arithmetic computation problems plus an oral section for use with subjects who are unable to read the written problems. The time limit for the written section is 10 minutes. The raw score is the total number of correct answers from the two sections. Full credit for the oral section is given to all subjects who are able to read the written problems.

Raw scores may be converted to Grade Equivalents with the use of a table provided on the Record Form and to Standard Scores with the use of a table provided in the Administration Manual. The manual suggests that Grade Equivalents be used only as an aid in explaining test results to parents or for estimating approximate instructional needs. It is suggested that Standard Scores be used for most interpretations of test data.

Standard Scores are scaled much like Intelligence Quotients, with a mean of 100 and a standard deviation of 15. A table of classification based on these scores is presented in a format which implies that the Standard Score is analogous to an Intelligence Quotient (the Average Range is 90-109, the Superior Range 120-129, etc.). It should be emphasized very strongly that these scores are *not* Intelligence Quotients by any stretch of the imagination. They are based on extremely limited samples of intellectual performance, but the method of presentation in the manual (they are even referred to as "deviation quotients") makes it very likely that this distinction will not be apparent to non-professional users. The Grade Equivalents may be a more valuable and valid indication of test performance, emphasizing as they do the fact that the intent of the test is to measure academic skills rather than general intelligence.

Technical Aspects

Previous reviews of the WRAT have criticized its standardization and the presentation of reliability and validity information (Merwin, 1972; Thorndike, 1972). Standardization was much improved in the 1978 version and reliability and validity was deemed adequate at that time by Helton, Workman, and Matuszek (1982).

The WRAT-R appears to be further improved with respect to standardization. The authors still have an unfortunate tendency to present exotic and unusual statistical information which is confusing and sometimes misleading, particularly to the non-technically oriented readers for whom the Administration Manual is primarily intended. It does appear that both internal consistency and test-retest reliabilities are adequate, however. Concurrent validity, as reflected in the correlation of the WRAT-R with other tests of scholastic achievement, also appears to be adequate. No validity studies are reported employing the WRAT-R itself, but it is very similar in content to previous editions of the test, and results of previous studies should still be applicable. Some comparisons of the WRAT-R and earlier editions of the WRAT are provided, but these appear on close reading to be of questionable appropriateness.

The factors which contribute to the high internal consistency coefficients sharply limit the value of the test as a diagnostic tool. Item content is much too limited to allow meaningful generalizations about specific skill deficits, particularly in reading or arithmetic. Neither is there any assurance that the modes which are tested are those most critical to development of these skills. The very important dimension of reading comprehension is specifically excluded from consideration, for example. Subtests were highly correlated with each other in the 1978 edition. Although these data are not reported in the 1984 manual (and certainly should be), it seems entirely reasonable to suppose that similar high relationships will be found between the subtests of the WRAT-R. This suggests, of course, that each subtest score is to a substantial extent a measure of some global achievement dimension rather than achievement in distinct modalities.

Critique

The WRAT-R has several very desirable features. It can be administered and scored easily and quickly by teachers or psychometricians. It is an acceptable alternative to group-administered achievement tests when clinical considerations are involved or when it seems likely that the subject will perform at an achievement level below that sampled by group instruments. However, it should be used only as a screening instrument for the determination of a global achievement level. Restricted item content and high intercorrelations among the subtests render it unsuitable for use as a diagnostic tool in the identification of specific skill deficits. The Grade Equivalent score is perhaps the best vehicle for communicating test results in a useful and appropriate context. The discussion and presentation of standard score information in the manual are somewhat misleading and are likely to encourage misuse of these scores by non-professionals.

References

This list includes text citations as well as suggested additional reading.

Hallahan, D., & Kauffman, J. (1976). *Introduction to learning disabilities: A psycho-behavioral approach*. Englewood Cliffs, NJ: Prentice-Hall, Inc.

Helton, G., Workman, E., & Matuszek, P. (1982). *Psychoeducational assessment*. New York: Grune & Stratton, Inc.

Jastak, J., & Jastak, S. (1978). *Wide Range Achievement Test manual of instructions*. Los Angeles: Western Psychological Services.

Jastak, S., & Wilkinson, G. (1984). *The Wide Range Achievement Test-Revised administration manual.* Wilmington, DE: Jastak Associates Inc.

Merwin, J. (1978). Review of the Wide Range Achievement Test. In O. K. Buros (Ed.), *The seventh mental measurements yearbook* (p. 35). Highland Park, NJ: The Gryphon Press.

Salvia, J., & Ysseldyke, J. *Assessment in special and remedial education.* Boston: Houghton Mifflin Company.

Thorndike, R. (1978). Review of the Wide Range Achievement Test. In O. K. Buros (Ed.), *The seventh mental measurements yearbook* (p. 36). Highland Park, NJ: The Gryphon Press.

Paul C. Hager, Ph.D.

Associate Dean for Academic Affairs, Berea College, Berea, Kentucky.

WOODCOCK-JOHNSON PSYCHO-EDUCATIONAL BATTERY

Richard W. Woodcock and Mary Bonner Johnson. Allen, Texas: DLM Teaching Resources.

THE ORIGINAL OF THIS REVIEW WAS PUBLISHED IN TEST CRITIQUES: VOLUME IV (1986).

Introduction

The Woodcock-Johnson Psycho-Educational Battery (WJPEB) is a comprehensive, individually administered set of standardized tests designed to measure the concepts of cognitive ability, scholastic aptitude, academic achievement, scholastic/nonscholastic interests, and independent functioning. It is intended for both handicapped and nonhandicapped populations from infancy through age 60 years and over. For the most part, the WJPEB is norm referenced rather than criterion referenced. Woodcock states that the major uses of the battery include ''. . . individual assessment, selection and placement, individual program planning, guidance, appraising gains or growth, program evaluation, and research'' (1978, p. 4). The battery consists of four parts: Tests of Cognitive Ability (I), Tests of Achievement (II), Tests of Interest Level (III), and Scales of Independent Behavior (IV).

The WJPEB scales were developed by a team of specialists, with Richard W. Woodcock as senior author. The first three parts were developed by Woodcock and Mary Bonner Johnson. Woodcock received his Ed.D. in psycho-education and statistics from the University of Oregon. He served as Professor of Special Education, senior researcher and acting director of the Institute on Mental Retardation, and director of the Research Group on Sensori-Motor Disorders and Adaptive Behavior at George Peabody College of Vanderbilt University. Since 1972, Woodcock has been director of Measurement/Learning/Consultants. Johnson is assistant director of Measurement/Learning/Consultants.

Part IV (Scales of Independent Behavior) was developed by Woodcock and three other authors: R. H. Bruininks, R. F. Weatherman, and B. K. Hill. Bruininks, who received his Ph.D. in special education from Peabody/Vanderbilt, is presently chair of the educational psychology department and a professor at the University of Minnesota. He has served as director of the Developmental Disability Planning Office, Minnesota State Planning Agency. R. F. Weatherman, who holds an Ed.D. in special education from Michigan State University, is currently Professor of Educational Psychology at the University of Minnesota and principal investigator for the Minnesota Severely Handicapped Delivery System Project. B. K. Hill, who received his M.A. and is a doctoral student in educational psychology at the University of Minnesota, has worked in a number of professional organizations concerned with the handicapped.

The WJPEB was developed to meet the need for a comprehensive, wide-age range set of measures that would be useful in a variety of settings. According to Woodcock (1978), the broad criteria serving as guides to the instrument's development included a broad set of measures that could provide a comprehensive psycho-educational evaluation from preschool through adult levels; high technical quality with APA standards followed at all times; packaged materials for ease of administration by persons with minimal psychometric experience; subtests and clusters designed to be administered separately if a comprehensive assessment is not required; and reporting procedures designed so that individual scores would be presented with due regard to commonly accepted score precision (i.e., in terms of SE_m and percentile bands).

Data for Parts I-III were gathered from 4,732 subjects from 49 communities selected from 18 states in 1976 and 1977. Subjects aged 6-17 years (N = 3,577) were chosen from 42 communities. Preschoolers (N = 555) and adults aged 18-65 + years (N = 600) were selected from seven additional communities. The norms for all three parts are based on the same sample of subjects, allowing for direct comparison of groups or individuals across all parts of the battery. The standardization sample for Part IV is based on a different sample of 1,700 subjects from over 40 communities tested in the early 1980s. The communities and settings for all four parts were selected so as to approximate the rural/urban, race and sex composition, regional representation, and socioeconomic characteristics of the 1970 census data as closely as possible. Although the samples for Parts I-III did not include severely handicapped subjects unless they were "mainstreamed," Part IV norms incorporated extensive samples of handicapped subjects, including the moderately-severely handicapped (Woodcock, 1978; Bruininks, Woodcock, Weatherman, & Hill, 1984).

After battery objectives were determined, specifications were developed to meet these objectives. Pools of items for each subtest were developed and preliminary forms were prepared and administered. The first preliminary form was administered to a few subjects at all levels. From the data gathered, a calibration-norm battery was prepared and administered to 1,000 subjects chosen from the norm samples. As a result of an extensive data analysis, some items and two subtests were dropped because they contributed little or nothing to the battery's discriminating power. Additional items were added to eight subtests, primarily at the lower and upper ranges. Validity studies with normal and clinical samples were also carried out at this stage. Additional redundant items were dropped after another 2,000 subjects were tested and data reanalyzed using the Rasch model. Because the Rasch procedure 1) allows the researcher to identify items that do not fit the test model and need to be discarded or revised and 2) specifies the probability that subjects will correctly answer an item based on their abilities, item difficulty and subject ability are measured on the same scale, permitting the development of equal-interval scales of test items. In general, 5-10% of the items were dropped in the final version (Hessler, 1984). The final correlation, multiple-regression, factor, and other analyses were carried out on the total sample of over 4,000 subjects, with a similar procedure followed for Part IV. At the time of this review, there have been no revisions of the battery, except for the addition of Part IV in 1984 and the development of a Spanish form (Woodcock-Johnson Spanish Educational Bateria) for Parts I and II covering ages 6-18 years.

The complete battery, consisting of three flip-page easel books designed to stand on the table during administration, is individually administered in its entirety or as single tests or clusters to meet specific appraisal needs. Each book contains the items for each subtest and instructions for administration. Separate manuals provide instructions for scoring, administrator training materials, and the tables necessary for score interpretation. The parts may be purchased separately except for II-III, which are printed together. A prerecorded cassette tape for Part I (Cognitive Ability), recommended for standardized administration of the three subtests for which exact pronunciation and presentation is required, may be used as a pronunciation guide and training aid. Only three subtests (Visual Matching, Calculation, and Dictation) require written responses from the subject, although the subject may use a type of scratch paper for Applied Problems if desired. Some subtests contain sample items to be used as teaching devices to introduce controlled-learning tasks. Both timed and untimed formats are used, with the untimed predominating. The use of basal and ceiling levels makes it possible to match the difficulty level of subtests to the individual being tested.

A response booklet is provided for each of the four test parts in which the administrator records responses, summarizes results, and interprets test performance. Technical manuals are purchased separately. Examiners must furnish pencils, watch or clock (stopwatch preferred), and a cassette player with good reproduction. Microcomputer scoring is available.

Parts I-II are suitable for subjects aged 3-65 years and beyond; Part III is most suitable for Grade 5-65 + years; and Part IV may be administered from infancy-29 years (and older, in certain cases).

The four parts of the battery, which consist of a total of 41 subtests and a variety of clusters or scales, are described as follows:

Part I. Tests of Cognitive Ability (see Table 1): Consists of 12 subtests and 11 scales covering three areas that provide a general measure of cognitive ability (Area I), measure cognitive structures necessary to academic functioning (Area II), and provide an estimate of examinees' "expected" achievement in the four most common areas of academic achievement (Area III). The differences observed between "expected" achievement as measured by the scales in Part I and "actual" achievement as measured by the achievement tests in Part II may be significant in diagnosing a student's learning difficulties (Hessler, 1984).

In Area I, the *Full Scale* (primarily verbal) is intended to cover the full range of functions from the lower to the higher mental processes (Woodcock, 1978); the *Preschool Scale* (primarily verbal) is most appropriate for preschool aged children but provides norms through the adult level (Woodcock, 1978) and can be used as an alternative to the Full Scale when a brief verbal ability assessment is needed (Hessler, 1984); the *Brief Scale,* requiring approximately 15 minutes to administer and score, correlates highest with the Full Scale cluster score (Woodcock, 1978) and provides a fast screening assessment of vocabulary and quantitative skills.

In Area II, the four cluster scales were derived statistically through factor, cluster, and multiple-regression analyses, with *Verbal Ability* measuring concrete vocabulary comprehension and expression and minimizing nonverbal abstract reasoning by using the Analysis-Synthesis score as a suppressor variable (deriving the cluster score by adding the two vocabulary scores and subtracting the

Table 1

Part I. Tests of Cognitive Ability

Areas	Cluster Scales	Picture Vocabulary	Spatial Relations	Memory for Sentences	Visual-Auditory Learning	Blending	Quantitative Concepts	Visual Matching	Antonyms-Synonyms	Analysis-Synthesis	Numbers Reversed	Concept Formation	Analogies
I. Broad Cognitive Abilities	Full	X	X	X	X	X	X	X	X	X	X	X	X
	Preschool	X	X	X	X	X	X						
	Brief						X		X				
II. Cognitive Factors	Verbal Ability	X							X	X			
	Reasoning Ability								X	X		X	X
	Perceptual Speed		X					X					
	Memory			X							X		
III. Scholastic Aptitude	Reading Aptitude				X	X			X				X
	Mathematics Aptitude							X	X	X		X	
	Written Language						X	X	X		X		
	Knowledge Aptitude			X			X		X				X

Analysis-Synthesis score); *Reasoning Ability* measuring the ability to carry out problem-solving strategies, nonverbal abstract reasoning, and utilization of instruction and decreasing the effects of verbal ability by employing Antonyms-Synonyms as a suppressor; *Perceptual Speed* measuring visual-perceptual fluency and accuracy and problem-solving strategies; and *Memory* measuring short-term auditory memory and the ability to manipulate or reorganize information and understand and follow directions (Woodcock, 1978).

In Area III, *Reading Aptitude* assesses verbal processing and how well examinees are expected to read rather than how well they actually read; *Mathematics Aptitude* is a composite of visual-perceptual and verbal/nonverbal abstract reasoning processes with visual and nonverbal predominating (Hessler, 1984); *Written Language* measures visual perceptual, verbal, and mathematical processing abilities and provides an expectation of the examinee's ability to understand (process) written language; and *Knowledge Aptitude* emphasizes verbal processing and assesses mathematical and verbal processing abilities necessary to learn in the sciences, social studies, and humanities.

All of the 12 subtests, most of which are untimed, begin with easier or less complex items and progress through the more difficult or complex ones. Unless noted otherwise in the following descriptions, responses to items are oral, and the starting point for each subtest is determined by the basal-level five consecutive correct responses and the ceiling-level five consecutive incorrect responses. The subtests are described as follows (Woodcock, 1978; Hessler, 1984):

Picture Vocabulary (37 items). Measures primarily expressive vocabulary skills. Requires examinee to identify pictured objects or actions.

Spatial Relations (31 items). Measures efficient problem-solving strategies and visual-perceptual and nonverbal conceptual ability. Examinee is given three minutes to make a complete shape by selecting from a series of shapes (2 or 3 components per item). After two sample items, all examinees start with the first item. Speed is a factor in the final score, but increasing difficulty of items and scoring make this equally a power test.

Memory for Sentences (22 items). Measures short-term memory and expressive syntax abilities by testing the ability to remember and repeat a sentence. The first three items are two- or three-word phrases presented orally by the examiner; the remaining 19 are on tape and are increasingly complex. The entire subtest can be read by the examiner to young examinees or those who need interaction.

Visual-Auditory Learning (seven stories with a 26-word vocabulary and two word endings). Measures ability to associate unfamiliar visual symbols (rebuses) with familiar words, translate these sequences into meaningful phrases, and possibly differentiate poor readers from good readers. Each story is preceded by teaching the names of four new rebuses to the examinee who then begins with the first story and proceeds until all stories are completed or a preestablished cumulative number of errors occur.

Blending (25 items). Measures auditory synthesis ability by testing ability to integrate, identify, and verbalize whole words (from two syllables to those composed of several single phonemes) after hearing syllables and/or phonemes of the word sequentially. Test material can be presented by examiner or prerecorded tape, depending on the examinee's need. All examinees begin with the first item and continue until five consecutive items are missed.

Quantitative Concepts (46 items). Assesses understanding of mathematical concepts, symbols, and vocabulary, rather than the ability to use them in practice, and predicts difficulty with mathematics achievement. Examinee responds to questions about quantitative and mathematical concepts, symbols, and vocabulary. No calculations or applications are required.

Visual Matching (30 items). Measures visual-perceptual fluency and accuracy and, for normal subjects, perceptual processing. A paper-pencil test requiring the examinee to find two identical words in a row of six numbers. Although two minutes are given to complete as many items as possible, this is also a power test (Woodcock, 1978).

Antonyms-Synonyms (49 items). Measures examinee's knowledge of word meanings and assesses receptive and expressive vocabulary, word definition, and (to some degree), cognitive flexibility. Requires examinee to respond with antonyms (26 items) or synonyms (23 items), with only one-word responses acceptable. Words are presented both orally and visually.

Analysis-Synthesis (30 items). Measures ability to learn symbols and their logical manipulation, abstract reasoning, and problem-solving by assessing the ability to learn symbolic formulations, use higher-level mental processing and reasoning, and apply them to problem solving. Skills measured are those required to learn mathematics and other areas requiring symbolic learning and manipulation (e.g., physics, chemistry, logic). Inability to identify four colors (yellow, black, blue, and red) correctly results in a score of zero and examinee is not tested. Examinees who correctly identify the first item proceed until all items are answered or predetermined cutoff points are reached.

Numbers Reversed (21 items). Measures short-term memory and ability to perceptually reorganize data by assessing ability to hold a sequence of numbers in memory and reorganize that sequence. Items (organized in groups of three) are composed of three to eight random numbers, with the first six items read by the examiner and the remaining presented by the examiner or tape, depending on examinees' needs. Basal level is established when all three items in a group are answered correctly; ceiling is established when all three items in a group are missed.

Concept Formation (32 items). Measures primarily nonverbal abstract reasoning and rule-learning ability, and possibly cognitive flexibility, analytic problem-solving strategies, and the ability to profit from instruction. A nonverbal subtest requiring examinee to identify rules underlying concepts when given examples of the concept (drawings varying along four dimensions: red or yellow, large or small, round or square, single or in pairs) and to identify instances when the concept does not apply. One or more boxes containing one or more combinations of the drawings are presented to the examinee, who determines which combination is appropriate in order for the drawings to be included in the box(es). All examinees start at the beginning and continue until all items are attempted or a predetermined cutoff point is reached.

Analogies (35 items). Assesses conceptualization, comprehension, and active word expression. Requires the examinee to complete an incomplete verbal phrase with a single word. Stimulus material is visible to the examinee and read by the examiner. Nine items are quantitative in nature.

Part II. Tests of Achievement (see Table 2). Consists of ten subtests and five cluster scales, with the first four cluster scales measuring achievement in the four most common areas of instruction in school: *Reading* (measuring basic reading skills, primarily sight-word vocabulary and phonic/structural analysis, but including some literal reading comprehension [Hessler, 1984]); *Mathematics* (measuring the most common achievement areas in mathematics); *Written Language* (measuring achievement skills in conventional and linguistic components of spelling, punctuation, capitalization, and word usage); and *Knowledge* (providing an estimate of the examinee's general knowledge across the major areas of the curriculum). The fifth scale, *Preschool Achievement,* is a combination of the three easiest skills subtests from the Reading, Mathematics, and Written Language Scales. It is designed

Table 2

Part II. Tests of Achievement and Part III. Tests of Interest Level

Cluster Scales	Subtests
TESTS OF ACHIEVEMENT	
Reading	Letter-Word Identification Word Attack Passage Comprehension
Mathematics	Calculation Applied Problems
Written Language	Dictation Proofing
Knowledge	Science Social Studies Humanities
Preschool Achievement	Letter-Word Identification Applied Problems Dictations
TESTS OF INTEREST LEVEL	
Scholastic Interest	Reading Interest Mathematics Interest Written Language
Nonscholastic Interest	Physical Interest Social Interest

to be used at the preschool level, but adult norms make it appropriate for administration to other special populations, such as mentally retarded adults.

All of the subtests in Part II are untimed and (with the exception of the Word Attack subtest which begins with the first item and ends when five consecutive items are missed) have basal and ceiling levels established by five consecutive correct and five consecutive failed items, respectively.

The Reading Scale consists of the following three subtests:

Letter-Word Identification (50 items). Measures ability to recognize and pronounce letters and words (printed in large type in the test book) and assesses sight-letter or word vocabulary, not reading comprehension. There is no expectation that word meaning is known. The first seven items consist of single letters.

Word Attack (26 items). Measures phonic and structural-analysis concepts. Requires the examinee to read nonsense words or infrequently used English words, beginning with simple consonant-vowel-consonant trigrams and progressing to multisyllable words. All phonemes in the English language are used (Woodcock, 1978).

Passage Comprehension (26 items). Measures literal reading comprehension. Requires the examinee to identify and supply an appropriate word to complete a short passage after silently reading the passage (a form of the "cloze" procedure). More than one response is acceptable for several of the items; seven of the 26 items include illustrations.

Calculation (42 items). Measures the given calculation skills of addition, subtraction, division, and multiplication. Includes whole numbers, fractions, and decimals, with the more complex items involving the use of trigonomic, logarithmic, geometric, and calculus operations. Procedures are specified, and no application skills are required. Items are completed on a special page in the test response booklet.

Applied Problems (45 items). Measures mathematical reasoning skills. Requires the examinee to solve practical problems in mathematics by recognizing the correct procedure, identifying the relevant data (many problems contain superfluous information), and performing the calculations. Problems are presented visually or read to the examinee to minimize the effect of reading skills. Scratch paper is permitted if requested.

The Written Language Scale consists of the following two subtests:

Dictation (40 items). Measures conventional and language components of written expression, including spelling. The examinee uses a special page from the response booklet to respond in writing to a variety of spoken instructions involving letters of the alphabet, forms, spelling, punctuation, capitalization, and language usage (e.g., contractions, abbreviations, plurals).

Proofing (29 items). Assesses conventional language usage. The examinee indicates correct usage of punctuation, capitalization, spelling, or word usage after reading passages in which errors occur.

The Knowledge Scale consists of the following three subtests:

Science (39 items). Measures knowledge of the physical and biological sciences, rather than an understanding and application of scientific methods. Items (printed only on the examiner's side of the test booklet) are read aloud by the examiner. Some illustrations and formulas are used.

Social Studies (37 items). Assesses knowledge of the social sciences (e.g., government, economics, geography). Items (many of which are printed on the examinee's side of the test book) are read aloud by the examiner. Some illustrations are used.

Humanities (36 items). Measures knowledge of the visual arts, literature, and music (including genres and styles), as well as facts and the ability to read a short musical passage. Items are read aloud and many include visual stimuli (e.g., poems, illustrations, musical passages).

Part III. Tests of Interest Level (see Table 2). Consists of five subtests and two cluster scales: Scholastic Interest (assessing the examinee's levels of interest in reading, mathematics, and written language) and Nonscholastic Interest (assessing the examinee's levels of interest in physical and social activities). Although Part III is most appropriate for Grades 5 and above, norms are provided down to Grade 3.

Each subtest item consists of two activities, one of which is relevant to the area of interest being tested. Examinees read the items from the test book (or the examiner may read items aloud if examinees have difficulty with reading) and choose which of the two activities they prefer.

The Scholastic Interest Scale consists of three subtests, each of which contains 25 items: *Reading Interest* (assessing preference for participating in activities involving reading, such as preference between going to a party or to a bookstore); *Mathematics Interest* (assessing preference for activities involving the learning or application of mathematics, such as preference between attending a party or making up mathematics games); and *Written Language* (assessing preference for various forms of activities requiring written language, such as preference between reading a newspaper or writing a newspaper story).

The Nonscholastic Interest Scale consists of two subtests, each of which contains 35 items: *Physical Interest* (assessing levels of interest in individual and group physical activities, such as choosing between mountain climbing and reading a story) and *Social Interest* (assessing levels of interest in activities involving other people, such as planning a party).

Part IV. Tests of Independent Behavior (see Table 3). Consists of four cluster scales (Motor Skills, Social Interaction and Communication Skills, Personal Independence, and Community Independence Skills) containing 14 subtests formed from 226 items, plus the Broad Independence, Short Form, Early Development, and Problem Behaviors Scales.

All items (behaviors or tasks) are printed in the interviewer's and respondent's testbook. Respondents may be the individual being examined, someone who knows the person well enough to respond, or the examiner. Typically, the respondent will be another person for very young children or persons with fairly severe emotional handicaps or retardation.

The Motor Skills Cluster Scale assesses fine- and gross-motor skills and consists of two subscales, each of which contains 17 items: *Gross-Motor Skills* (sampling skills involving the large muscles from infancy, such as sitting without support, to adult behaviors, such as regular strenuous exercise) and *Fine-Motor Skills* (assessing performance on typical hand-eye coordination skills, such as picking up small objects with the hand in infancy or small-part assembly tasks for adults).

The Social Interaction and Communications Skills Cluster Scale assesses the behaviors necessary to interact and communicate effectively with others; it consists of three subscales: *Social Interaction* (assessing performance on 16 tasks involving social interaction with others from infancy, such as distinguishing friends from strangers, to adulthood, such as participation in social activities); *Language Comprehension* (assessing the level of deriving information from spoken and written language on 16 tasks ranging from recognizing one's name to the use of reference materials); and *Language Expression* (assessing the ability to express

Table 3

Part IV. Scales of Independent Behavior

	Cluster Scales	Subscales
Broad Independence Scale	Motor Skills	Gross-Motor Skills Fine-Motor Skills
	Social Interaction and Communication Skills	Social Interaction Language Comprehension Language Expression
	Personal Independence	Eating and Meal Preparation Toileting Dressing Personal Self-Care Domestic Skills
	Community Independence Skills	Time and Punctuality Money and Value Work Skills Home/Community Orientation
Short Form Scale Early Development Scale		32 items from above subscales
Problem Behaviors Scale (Eight categories)		Presence of Problems Frequency of Occurrence Severity of Problems

one's self orally, in writing, or through other devices such as language boards on 17 tasks ranging from simple "yes/no" responses to the delivery of complex reports).

The Personal Independence Cluster Scale assesses performance in areas needed to function independently primarily within the home and consists of five subscales: *Eating and Meal Preparation* (evaluating eating skills in infancy and food preparation behaviors in adults on 16 tasks); *Toileting* (evaluating bathroom and toilet and related behaviors from infancy through childhood on 14 tasks); *Dressing* (evaluating performance in dressing on 18 tasks ranging from very simple to complex, such as the selection and maintenance of clothing); *Personal Self-Care* (evaluating performance in personal self-care on 15 tasks ranging from use of toothbrush to seeking professional help in illness); and *Domestic Skills* (evaluating skills necessary for the maintenance of a functional home environment on 16 tasks ranging from returning dishes to the kitchen to selecting appropriate housing).

The Community Independence Skills Cluster Scale assesses the behaviors necessary to function in the community, primarily outside the home, and consists of four subscales: *Time and Punctuality* (evaluating time concepts on 15 tasks ranging from telling time to making and keeping appointments); *Money and Value* (evaluates skills in determining the value of items and in use of money on 17 items); *Work Skills* (evaluating work tasks through certain prevocational behaviors on 16 tasks more developmentally advanced than other subscales); and *Home/Community Orientation* (assessing the behaviors necessary to effectively get around the home, the neighborhood, and the home community on 16 tasks).

The Broad Independence Scale (consisting of all 226 items and all 14 subscales) provides an assessment of the full range of behaviors necessary to function in everyday life.

The Early Development Scale (32 items) is designed for subjects whose developmental level is below approximately two and one-half year of age and is particularly appropriate for very young children and severely or profoundly handicapped children and adults.

The Short Form Scale (32 items) is appropriate when a brief screening or evaluation of independent functioning is needed.

All tasks on the Early Development, Short Form, and Broad Independence Scales are arranged in developmental order, with the easiest tasks presented first. Basal and ceiling levels are established, and because the tasks cover a range from infancy to adulthood, interviewers select the beginning task based on their assessment of the subjects' "operating range" (Bruininks, Woodcock, Weatherman, & Hill, 1984). Each task contains the stem asking whether or not the subject does or could do tasks completely without help or supervision. Tasks are rated by respondent, according to subject's ability, on a four-point scale (0 = never or rarely; 1 = not well or $\frac{1}{4}$ of the time; 2 = fairly well or about $\frac{3}{4}$ of the time; and 3 = very well, always, or almost always).

The Problem Behaviors Scale measures three dimensions of problem behaviors: the presence, frequency of occurrence, and severity. Problem behaviors are defined as those which, if frequently exhibited, may limit personal and social adjustment (Bruininks et al., 1984). The eight major categories of problem behaviors are hurtful to self, hurtful to others, destructive to property, disruptive behavior, unusual or repetitive habits, socially offensive behavior, withdrawal behaviors, and uncooperativeness. Each category contains several examples of such behaviors. This scale has two rating scales: frequency (scored on a continuum ranging from "less than once a month" to "one or more times an hour") and severity (scored on a five-level continuum ranging from "not serious" to "extremely serious"). The respondent indicates which of those behaviors occur, how often they occur, and which are the most serious within that category.

Practical Applications/Uses

Because of the wide range of cognition, achievement, interest, and behaviors assessed and its wide age range, the WJPEB has a large number of uses—and its potential is even greater. Woodcock (1978) and Hessler (1984) recommend four general areas of usage: 1) individual assessment, including screening, identification, and diagnosis of individual strengths and weaknesses in cognitive, academic achievement, and interest areas; individual program planning for general educational instructional objectives; and the planning of short- and long-term educational and vocational goals; 2) evaluation of individual growth and program effectiveness; 3) as a valuable instrument for research into a wide range of educational, developmental, and behavioral factors, due to the range of subtests, cluster scores, and norming; and 4) as a useful instrument for psychometric training purposes, due to the individualized nature of the administration, the use of a variety of scores, and technically superior materials.

The battery lends itself to extensive applications in both educational and non-educational settings. For example, identification of individuals with special learning problems or disabilities; diagnosis of cognitive abilities, achievements, or interests; selection or placement of subjects into appropriate groupings; and individual program planning or guidance. The age range over which it is normed makes it possible to chart individual or group growth from infancy to senior adulthood. The subtests are well-designed for research purposes in a variety of settings (Woodcock, 1978). The recent addition of the Scales of Independent Behavior (Bruininks, Woodcock, Weatherman, & Hill, 1984) makes it possible to assess the ability of individuals to function successfully in a variety of social, educational, and personal settings.

Parts I-III (Cognitive Ability, Achievement, and Interest) introduce an innovation that has been sorely needed. It is the first major individualized assessment instrument to report standardized cognitive and achievement measurements on the same samples from childhood through adulthood. This avoids the uncontrolled variance inherent in separate norms groups. The Scholastic/Nonscholastic cluster for children normed on the same sample provides a much needed assessment of children's interests (Estabrook, 1983).[1]

Woodcock (1978) recommends Parts I-III for identifying and evaluating the behaviorally and learning disabled, but the norms groups were small and the evidence not conclusive. Part IV, which was developed with nonhandicapped subjects and with larger and more representative retarded, behavioral, and learning-disabled groups, appears to be a valid diagnostic instrument for evaluating independent functioning. The preliminary evidence (Bruininks et al., in press) is impressive.

The ease of administration, wide age range, and broad conceptual coverage make the WJPEB useful in a great variety of settings, such as preschool-high school, special and adult education programs, private practices, higher education, research studies, and training of psychometrists.

The battery can be administered by anyone who has had a basic introduction to measurement; however, the materials providing specific training exercises for administration are recommended for all users, regardless of the level of their training. Administration time for the entire battery is approximately three hours, with Part I requiring about one hour, Part II about 50 minutes, Part III about 15 minutes, and Part IV 45-60 minutes. Certain subjects will require more time than others. Although the battery may be given in four different sessions, it is recommended that no more than two days be used to complete it. As with any test, especially if individually administered, the WJPEB must be given in a room with a minimum of distractions and ample working space.

The scoring is complex, perhaps to the point that most errors of administration may occur at this point. Raw scores for each subtest are converted to part scores, which are summed to produce cluster scores and further summed to obtain total cluster scores. After a grade score, age score, and instructional grade range are obtained from tables for each cluster score, an expected average cluster score for the age or grade group of the subject is entered. A difference score is calculated between the observed score and the expected cluster score. The percentile ranks for cluster scores are based on this difference score (with a zero difference representing the 50th percentile), entered on the summary sheet, and transferred to the profiles in terms of confidence bands based on SE measurement. Subtest profiles are prepared using raw score confidence bands. From other tables, the Relative Performance Index (RPI) is determined and entered, and the cluster difference scores are interpreted by functioning levels ranging from ''very superior'' to "severely deficient." Other types of scores (e.g., deviation IQ, T-scores, stanines) are also available from tables, but are not ordinarily entered on the profile. Although the technical manuals (Woodcock, 1978; Hessler, 1984) state that the user determines which type of score is most useful, the scorer must complete the summary sheet through difference scores for the clusters to be able to obtain percentile ranks and standard scores. Percentile ranks are given in a separate book of norms (Marston & Ysseldyke, 1984b).

The response booklet for Part I provides a profile on which the Full Scale and subtest raw scores are plotted. An academic achievement vs. aptitude profile may be used to show the measured differences between expected learning and that actually observed by the test. The percentile rank profile provides a graphic presentation of the relative levels among the broad cognitive ability, the four aptitude scales, and the four cognitive scales.

Parts II-III response booklets provide profiles based on percentile ranks for each of the five achievement and two interest scales. A profile may also be drawn depicting relative levels for each of the ten achievement subtests, the five related subtests from Part I, and the five interest subtests. An ''instructional implications'' profile provides information about the subject's actual performance relative to others of the same grade or age with a suggested instructional level in the broad areas of reading, mathematics, written language, and general knowledge.

The Part IV response booklet contains five profiles. The first is based on percentile ranks for the Broad Independence Scale and the four major independence clusters. If available, percentile ranks from related Cognitive Ability Scales may also be plotted for comparison purposes. Another percentile rank profile may be completed based on both age and cognitive abilities. The third is a profile of scores from the 14 subscales and the Full Scale. A ''training implications'' report may be prepared depicting the relative performance of the subject compared to others the same age. A fifth profile depicts by stanine the frequency and severity of problem behaviors.

A microcomputer scoring package, COMPUSCORE (Hauger, 1984), is available for both the Apple II series and the IBM-PC. It provides convenient scoring from raw scores for each of the 27 subtests from Parts I-III. The system allows missing scores and has considerable error checking built in. Scoring the entire battery and completing the summary sheet and profile will require between one and two hours; the COMPUSCORE computer scoring should reduce the time to approximately 30 minutes. In addition, the chance for scorer errors will be greatly reduced.

Interpretation of the battery is based on a series of objective scores rather than on clinical judgment. Various criterion- and norms-related data are provided for each subtest and cluster score. A comprehensive understanding of both the content and processes measured by each task is important, especially if the examinee's scores are significantly high or low. Further, failure to fully understand the relationships between and the meanings of individual subtest scores may result in misinterpretation of cluster scores because clusters are based on combinations of the subtests. Before attempting interpretation of the battery, users should generally have completed courses in both statistics and introductory psychometrics, as well as have had considerable experience in the interpretation of standardized tests.

Technical Aspects

The technical manual (Woodcock, 1978) provides extensive information about the methodological procedures used in norming the Woodcock-Johnson, including data on reliability and validity for Parts I-III. Preliminary data available for Part IV are adequate for this review (Bruininks et al., in press).

Reliability. Reliability is an estimate of the stability of test scores. It refers to the consistency of the scores obtained by the same person over time when reexamined with the same test, or with different sets of items from the same or equivalent test, or under variable testing conditions. Test reliability indicates the extent to which individual differences in test scores are attributable to "true" differences in the characteristic being measured and the extent to which differences are due to chance, unrelated errors. The higher the correlation coefficient on a scale of –1.00 to 1.00, the closer the relationship. Generally, coefficients of .80 or higher are indicative of test stability. Split-half reliability estimates are a measure of the internal consistency of a test (i.e., the ability of the items throughout the test to consistently measure the concept). Test-retest reliability is a measure of consistent measurement over time. Woodcock (1978) and Bruininks et al. (1984) report both.

Reliability estimates for Parts I-III were obtained by the split-half method (corrected for length by the Spearman-Brown procedure) for untimed subtests and for cluster scores. Test-retest reliabilities were obtained only for the two timed subtests. Split-half (corrected by Spearman-Brown) reliability estimates are reported for both nonhandicapped and handicapped groups for Part IV (Scales of Independent Behavior). Test-retest reliabilities were also reported for two groups of elementary-school children.

Reliabilities were reported for ages 20-39, 40-64, and 65 + years and Grades 1, 3, 5, 8, and 12 for 11 of the 25 subtests in Parts I-II. In addition to the above, estimates were also obtained for ages three and four and Kindergarten for 11 more subtests. Only elementary and secondary grades were included for the Visual Matching subtest. Median subtest reliabilities for the two parts ranged from .65 to .95, with only two falling below .80. Median cluster scores ranged from .70 (Perceptual Speed) to .97 (Full Scale). Seven of the 16 cluster-score reliabilities were .90 or above. Reliabilities were reported across all 11 norms groups for the Preschool, Knowledge, and Skills clusters and for the three adult age groups and the elementary and secondary grade levels for the remaining 13.

Reliability estimates were reported for Grades 5, 8, and 12, and ages 20-39, 40-64, and 65+ years for Part III (Interest). Median subtest reliability estimates ranged from .79 (Social) to .88 (Reading and Written Language). The two Interest cluster-score reliabilities were .93 (Scholastic) and .88 (Nonscholastic).

Part IV (SIB) subscale median internal consistency estimates ranged from .69 to .86 for three levels of normal populations. The range of cluster-score estimates was .83 to .96, with a .96 for the Broad Independence cluster. Reliability estimates for four groups of handicapped persons at two age levels (moderately/severely retarded, mildly retarded, behavior-disordered, and learning-disabled in children and adolescents/adults) and for moderately retarded adults were also reported. The median subscale reliabilities ranged from .88 to .95. Median cluster reliabilities for these groups ranged from .94 (Motor Skills) to .99 (Broad Independence). However, the sizes of the handicapped samples were generally small (N = 15 to 86), which may have inflated the estimates.

Part IV test-retest reliabilities were obtained for two small nonhandicapped samples of children (ages 6-8 and 10-11 years). The scales were administered by the same examiner over a one- to four-week period. Cluster scale test-retest reliabilities ranged from .71 to .96, with 11 of the 12 estimates in the .80 and .90 range. Maladaptive index estimates ranged from .71 to .96. Subscale reliabilities were generally high with 16 of the 20 in the .80s and .90s.

In general, reliability estimates were acceptable, with especially high reliabilities being reported for cluster and Full Scale scores. One shortcoming is the lack of test-retest data for the untimed subtests and clusters for Parts I and II, especially at the younger levels where retesting is often necessary.

Validity. To be valid, a test must measure what the author claims it measures, especially when the test results will be used to make decisions that will have long-term effects on the examinee. No test is valid in and of itself. It may be appropriate for some purposes and not for others. In a battery, some tests may be more valid than others. It is the task of the test author to present evidence of validity, that is, of the usefulness of the test for the purposes to which it will be put (Anastasi, 1982). Woodcock has presented impressive amounts of validity data. The technical manual (Woodcock, 1978) presents evidence of four types of validity: construct, criterion-related (concurrent and predictive), and content.

Construct validity is not directly measurable, but is inferred from observed relationships to other measures that appear to have the same underlying theoretical construct. For example, a measure of "quantitative ability" would be expected to show a high correlation with other measures of mathematics. In addition, it would be expected to show a low relationship with measures intuitively unrelated to mathematics, such as word attack skills. Woodcock's examples present good evidences of construct validity. For example, the Quantitative Concepts subtest correlates fairly high with Calculation (.63) and Applied Problems (.68) and lower with Word Attack (.44), a reading measure. Other data for both subtests and cluster scores show moderate to high evidences of construct validity. Unfortunately, the evidence presented is for normal populations, and no separate evidence for special populations is given.

Concurrent validity is a direct measure between a test and some other known measure of the same criterion, usually another test. Concurrent studies were reported for both normal and special populations between various parts of the WJPEB and a number of well-known instruments, e.g., Stanford-Binet, WISC-R, Iowa Tests of Basic Skills (ITBS), Peabody Individual Achievement Test (PIAT), Peabody Picture Vocabulary Test (PPVT), Wide Range Achievement Test (WRAT), KeyMath Diagnostic Test, Minnesota Behavioral Scales (MDPS), and Woodcock Reading Mastery Test (WRMT). The small special populations represented the severe learning disabled (SLD), the severe learning and behavior problems (SLBP), and the trainable mentally retarded (TMR), with 20, 30, and 33 subjects, respectively.

For normal groups, concurrent validity coefficients were in the moderate range. For example, the correlation between the Cognitive Ability Full Scale and the PPVT ranged from .58 (Grade Three) to .64 (Grade Five). The highest reported relationships were between the WJPEB Reading Cluster score and other reading measures (.75 to .90) than for either mathematics or written language which generally ran in the .70s and .80s.

Concurrent relationships groups fell into the moderate range for the TMR between the Preschool Cognitive Ability Scale and MDPS scales and Language Word Flow Chart (median, .69). Coefficients for the TMR sample between the Achievement clusters and the MDPS scales range between .56 and .86, with a median correlation of .71. The correlations between the achievement clusters and the PIAT, KeyMath, and WRAT Spelling are even lower for the SLD sample (.28 to .84, median of .55).

Bruininks, Woodcock, Weatherman, and Hill (in press) report concurrent validity coefficients between the Broad Cognitive Ability score of Part I and the cluster scores of Part IV for two age levels of handicapped and nonhandicapped groups. These range from high for both groups of children (high .70s and .80s) to low for nonhandicapped adults (.20s to low .40s). This pattern is replicated in the correlations reported between chronological age and the Scales of Independent Behavior, where only at the lower age levels is a significant relationship found. These findings tend to support the hypothesis of the scale's authors (Bruininks et al., 1984) that functional independence does vary with age and stage of development. On the other hand, low relationships reported between the maladaptive behavior scales and chronological age support the assumption that such behaviors are nondevelopmental in nature. Good discriminant validity for the Problem Behaviors Scale, in general, is shown by the relatively high mean scores of the behavior-disordered children and the zero scores for normal, high-ability, and learning-disabled subjects.

Content validity is the extent to which a test samples the ability under investigation. It is established by systematically examining the test content to determine whether it is a representative sample of the behavior domain to be measured (Anastasi, 1982). As with other types of validity, Woodcock (1978) provides adequate information on which to judge. The achievement, interest, and adaptive behavior domains appear to be adequately covered; intellectual or ability are somewhat less so, especially as they relate to special groups.

Predictive validity is the ability of a test to predict performance on some other measure of the same concept, usually other test scores, grades in specific classes, or overall academic performance. All of these predictions are important, especially when testing subjects in an academic setting. Unfortunately, at the time of this review only predictive studies involving performance on other tests have been reported for the WJPEB.

Woodcock (1978) reports the results of studies of the relationships between the four Part II achievement clusters (Reading, Mathematics, Written Language, and Knowledge) and three measures of academic aptitude (PPVT, WISC-R, and the WJPEB, Part I cluster scales). For normal groups, the relationships among the various measures are moderately high (primarily, .60s to .80s), with the highest relationships reported between the Part I and Part II clusters. This last finding is not unexpected because both were normed on the same groups. It should be noted that the Broad Cognitive Scale (Part I) is as good a predictor of performance on any of the achievement clusters as are corresponding aptitude cluster scores.[2]

Prediction of achievement cluster scores from aptitude cluster scores proved more troublesome for the three special groups. Although moderately high predictive validity coefficients were reported (.63 to .86, median of .76), there was a significant discrepancy between the expected scores on the achievement clusters and those predicted, with predicted scores averaging almost 20 points below those expected. Later research has confirmed that problems may exist with using Part I with exceptional groups. McGue, Shinn, and Ysseldyke (1982) found that neither cognitive nor aptitude factor clusters appeared to be valid for assessing learning-disabled fourth-graders. Reeve, Hall, and Zakreski (1979) and Ysseldyke, Shinn, and Epps (1981) found significant discrepancies between Part I of the WJPEB and the WISC-R Full Scale for learning-disabled students.[3]

For a behavior-disordered population (some of whom were also learning-disabled) Phelps, Rosso, and Falasco (1984) found a high concurrent validity between the Broad Cognitive score (Part I) and the WISC-R Full Scale IQ score, with only a 2.33 point average difference in means. However, an analysis of the differences between WISC-R subtests and Part I cluster scores indicated discrepancies in what was being measured—discrepancies that may call into question the applicability of Part I to this population. Other researchers have also raised questions about the use of Part I with handicapped populations (e.g., Algozzine, Ysseldyke, & Shinn, 1982; Breen, 1983; Marston, 1980; Marston & Ysseldyke, 1984a). Woodcock (1984a, 1984b) counters that the discrepancy is methodological rather than actual. However, when Thompson and Brassard (1984a) controlled for the methodological errors claimed by Woodcock, discrepancies between Part I and WISC-R still existed. Thompson and Brassard concluded that lower Part I scores for LD populations were a function of the instrument's heavy emphasis on achievement rather than systematic error in the Part I norms.

On the other hand, both Harrington (1984) and Weaver (1984) found Part I useful and "adequate" in the assessment of the achievement of mentally retarded children, and Mira (1984) found both Parts I and II useful in the assessment of hearing-impaired children. Because of the oral nature of the WJPEB's presentation, Fewell (1983) suggests that parts of both Parts I and II are useful in evaluating visually impaired preschool children. In a study of learning-disabled college students, Gregg and Hoy (1985) found no differences in performance on the WAIS-R and the Part I cluster scores.

Woodcock (1978) analyzed the relationship among the aptitude and achievement cluster scores, sex, and race. (Sex and race membership were carefully matched to 1970 census data.) The prediction of achievement cluster scores from aptitude clusters did not appear to be generally affected by either race or sex. The only exception was the finding that older boys were more likely to be inappropriately identified as having a deficit in written language skills than were older girls, for whom the opposite was true.[4] Arffa, Rider, and Cummings (1984) found no significant differences between mean scores on the Stanford-Binet and the Preschool Cognitive Scale (Part I) for a sample of 60 black preschoolers. However, the correlation between the two tests was a modest .45, indicating that the two instruments are probably measuring different concepts. The Knowledge Scale (Part II) mean for the group was significantly lower than either the Stanford-Binet or the Preschool Cognitive scores. Both findings may suggest that the cultural loadings of the WJPEB may be biased somewhat towards white, middle-class, preschool children. Further research with culturally distinct groups is needed in this area, Arffa, Rider, and Cummings (1984) suggest.

The "mean-score" discrepancy also raises important psychometric issues with regard to the usefulness of the WJPEB with special populations (Estabrook, 1983). The purpose of this test is to evaluate and predict more accurately in order that more valid placement or treatment decisions can be made. If Part I does not predict performance on other measures accurately, especially for exceptional groups, its use may be more harmful than helpful. The sources of the mean-score discrepancies, therefore, need to be identified. Several possible sources have been suggested. Reeve et al. (1979) suggest that the problem may lie with differences in difficulty between the aptitude and achievement tests. Shinn et al. (1982) and Thompson and Brassard (1984a) believe the source to be the high achievement-oriented content of Part I when compared to other measures (WISC-R or Stanford-Binet). Higher verbal saturation of the Part I tasks may also be the culprit (Cummings & Moscato, 1984a).[5] The resolution of the psychometric problems (both theoretical and practical) of Part I is essential if this instrument is to reach its full potential.

Critique

The Woodcock-Johnson Psycho-Educational Battery is a significant addition to the American psychometric "pantheon," even taking into consideration its shortcomings with regard to exceptional groups, cost, and difficult scoring procedures. The technical information presented in the various support publications is comprehensive and psychometrically sound. The concurrent norming of Parts I-III (and, it is hoped by this reviewer, Part IV in the near future) is an innovation long needed. The norms are generally representative of the nonhandicapped population of the United States. In short, the battery compares very favorably with other tests in use in this country and in Canada. It deserves careful consideration when setting up or adding to a comprehensive testing program.

Notes to the *Compendium:*

[1]McGrew (1986) introduced a new and useful tool for evaluation through his "Grouping Strategy Strength/Weakness Worksheet." Using the worksheet, the examiner can identify client strengths and weaknesses on the WJTCA across four specific hypothetical models of ability: Luria-Das (simultaneous/successive), Right/Left Brain, Verbal/Nonverbal, and Cattell Fluid/Crystallized. This technique allows an analysis of strength/weakness groupings simultaneously rather than in isolation. McGrew's book is highly recommended to anyone who plans to use the WJPEB; in fact, is a "must"!

[2]Reilly, Drudge, Rosen, Loew, and Fischer (1985) have since reported that WJPEB Cognitive Scale scores of first-graders are predictive of teacher ratings (and Full-Range Achievement Test scores) in the third grade. Kroft, Ratzlaff, and Perks (1986) found the WJPEB Achievement Battery to be predictive of underachievement in first-graders. Researchers continue to report significant correlations between WJPEB and WISC-R (Reilly et al., 1985; Mather & Udall, 1985; Ingram & Hakari, 1985; Hutton & Davenport, 1985; Phelps, Rosso, & Falasco, 1985; Lyon & Smith, 1985), although Coleman and Harmer (1985) and Ingram and Hakari (1985) found that the WJPEB-Part I is more heavily loaded with verbal factors than is the WISC-R.

[3]Hoy and Gregg (1986) found that subtest scores were better predictors of learning skills than were Part I (Cognitive) cluster scores for learning disabled college students. Breen (1986) found that several discrete or unique cognitive patterns for three subtypes of learning disabled students were present, although some overlap existed. Salvia and Salvia (1985) administered Part II to 100 handicapped college students. Their results appeared to show that pronounced ceiling effects and other psychometric problems raised questions about the reliability estimates of both the subtest and cluster scores for this population.

[4]Significant sex differences on the reasoning section on the Test of Cognitive Reasoning were found for learning disabled adults by Buchanan and Wolf (1986), differences that apparently were present in childhood.

[5]Other researchers (e.g., Reeve et al., 1979; Ysseldyke, Shinn, & Epps, 1981) have offered the opinion (and perhaps statistical evidence) that the observed (.5 to 1 SD) mean differences between WJTCA Cognitive Ability scores and WAIS-R are due to the higher verbal or crystallized *content* (emphasis mine) of the WJTCA. McGrew (1986) concludes that the differences *are* due to content, but in other domains. He found that the two instruments are equal in verbal and crystallized factors (about 26% shared variance) and speed (about 4% shared variance). The remaining 70% of unshared variance account for the observed differences.

The differences between the two instruments are an important strength, rather than weakness, of the WJTCA. As McGrew (1986) states, "In the final analysis the choice of intellectual instruments will largely depend on the philosophical beliefs of clinicians" (p. 274). If the clinician is interested primarily in predicting academic success or evaluating scholastic abilities, the WJPEB-Part I is the proper instrument. If an estimate of a client's intelligence within the context of a theoretical model is desired, the Wechsler scales (or K-ABC) are more appropriate.

References

This list includes text citations as well as suggested additional reading.

Algozzine, B., & Ysseldyke, J. E. (1981). An analysis of difference score reliabilities on three measures with a sample of low-achieving youngsters. *Psychology in the Schools, 18,* 133-38.

Algozzine, B., Ysseldyke, J. E., & Shinn, M. (1982). Identifying children with learning disabilities: When is a discrepancy severe? *Journal of School Psychology, 20,* 299-305.

Anastasi, A. (1982). *Psychological Testing* (5th ed.). New York: Macmillan

Arffa, S., Rider, L. H., & Cummings, J. A. (1984). A validity study of the Woodcock-Johnson Psycho-Educational Battery and the Stanford-Binet with black pre-school children. *Journal of Psychoeducational Assessment, 2,* 73-77.

Bracken, B. A., Prasse, D., & Breen, M. (1984). Concurrent validity of the Woodcock-Johnson Psycho-Educational Battery with regular and learning-disabled students. *Journal of School Psychology, 22,* 185-192.

Breen, M. J. (1983). A correlational analysis between the PPVT-R and Woodcock-Johnson Achievement cluster scores for non-referred regular education and learning disabled students. *Psychology in the Schools, 20,* 295-97.

Breen, M. J. (1984). The temporal stability of the Woodcock-Johnson Test of Cognitive Ability for elementary-aged learning disabled children. *Journal of Psychoeducational Assessment, 2,* 257-261.

Breen, M. J. (1985). The Woodcock-Johnson Tests of Cognitive Ability: A comparison of two methods of cluster scale analysis for three learning disability subtypes. *Journal of Psychoeducational Assessment, 3,* 167-174.

Breen, M. J. (1986). Cognitive patterns of learning disability sub-types as measured by the Woodcock-Johnson Psycho-Educational Battery. *Journal of Learning Disabilities, 19,* 86-90.

Bruininks, R. H., Woodcock, R. W., Weatherman, R. F., & Hill, B. K. (1984). *Scales of Independent Behavior.* Allen, TX: DLM Teaching Resources.

Bruininks, R. H., Woodcock, R. W., Weatherman, R. F., & Hill, B. K. (in press). *Development and standardization of the Scales of Independent Behavior.* Allen, TX: DLM Teaching Resources.

Buchanan, M., & Wolf, J. S. (1986). A comprehensive study of learning disability adults. *Journal of Learning Abilities, 19,* 34-38.

Coleman, M. C., & Harmer, W. R. (1985). The WISC-R and the Woodcock-Johnson Tests of Cognitive Ability: A comparative study. *Psychology on the Schools, 22,* 127-132.

Cummings, J. A. (in press). Review of the Woodcock-Johnson Psycho-Educational Battery. In J. V. Mitchell, Jr. (Ed.), *The ninth mental measurements yearbook* (pp. 1759-1762). Lincoln, NE: Buros Institute of Mental Measurements.

Cummings, J. A., & Moscato, E. M. (1984a). Research on the Woodcock-Johnson Psycho-Educational Battery: Implications for practice and future investigations. *School Psychology Review, 13,* 33-40.

Cummings, J. A., & Moscato, E. M. (1984b). Reply to Thompson and Brassard. *School Psychology Review, 13,* 45-48.

Cummings, J. A., & Sanville, D. (1983). Concurrent validity of the Woodcock-Johnson tests of cognitive ability: EMR children. *Psychology in the Schools, 20,* 298-303.

Epps, S., Ysseldyke, J. E., & Algozzine, B. (1983). Impact of different definitions of learning disabilities on the number of students identified. *Journal of Psychoeducational Assessment, 1,* 341-352.

Estabrook, G. E. (1983). Test review: The Woodcock-Johnson Psycho-Educational Battery. *Journal of Psychoeducational Assessment, 1,* 315-319.

Estabrook, G. E. (1984). A canonical correlation analysis of the Wechsler Intelligence Scale for Children-Revised and the Woodcock-Johnson Tests of Cognitive Ability in a sample referred for suspected learning disabilities. *Journal of Educational Psychology, 76,* 1170-1177.

Fewell, R. R. (1983). Assessment of visual functioning. In K. D. Paget & B. A. Bracken (Eds.), *The psychoeducational assessment of preschool children* (pp. 85-103). New York: Grune & Stratton.

Gregg, N., & Hoy, C. (1985). A comparison of the WAIS-R and the Woodcock-Johnson Tests of Cognitive Ability with learning-disabled college students. *Journal of Psychoeducational Assessment, 3,* 267-274.

Hall, R. J., Reeve, R. E., & Zakreski, R. S. (1984). Validity of the Woodcock-Johnson Test of Achievement for learning disabled students. *Journal of School Psychology, 22,* 193-200.

Harrington, R. G. (1984). Assessment of learning disabled children. In S. J. Weaver (Ed.), *Testing children: A reference guide for effective clinical and psychoeducational assessments* (pp. 85-103). Kansas City, MO: Test Corporation of America.

Hauger, J. (1984). *COMPUSCORE for the Woodcock-Johnson Psycho-Educational Battery.* Allen, TX: DLM Teaching Resources.

Hessler, G. L. (1984). *Use and interpretation of the Woodcock-Johnson Psycho-Educational Battery.* Allen, TX: DLM Teaching Resources.

Hoy, C., & Gregg, N. (1986). The usefulness of the Woodcock-Johnson Psycho-Educational Battery cognitive cluster scores for learning disabled college students. *Journal of Learning Disabilities, 19,* 489-491.

Hutton, J. B., & Davenport, M. A. (1985). The WISC-R as a predictor of Woodcock-Johnson Achievement cluster scores for learning-disabled students. *Journal of Clinical Psychology, 41,* 410-414.

Ingram, G. F., & Hakari, L. J. (1985). Validity of the Woodcock-Johnson Tests of Cognitive Ability for gifted children: A comparison study with the WISC-R. *Journal for the Education of the Gifted, 9,* 11-23.

Kampwirth, T. J. (1983). Problems in use of the Woodcock-Johnson suppressors. *Journal of Psychoeducational Assessment, 1,* 337-340.

Kaufman, A. S. (in press). Review of the Woodcock-Johnson Psycho-Educational Battery. In J. V. Mitchell (Ed.), *The ninth mental measurements yearbook* (pp. 1762-1765). Lincoln, NE: Buros Institute of Mental Measurements.

Kroft, S. B., Ratzlaff, H. C., & Perks, B. A. (1986). Intelligence and early academic under-achievement. *British Journal of Clinical Psychology, 25,* 147-148.

Lyon, M. A., & Smith, D. K. (1985, April). *Referred students' performance on the K-ABC, WISC-R, and Woodcock-Johnson.* Paper presented at the annual convention of the National Association of School Psychologists, Las Vegas.

Marston, D. (1980). *An analysis of subtest scatter on the tests of cognitive ability from the Woodcock-Johnson Psycho-Educational Battery* (Research Rep. No. 46). Minneapolis: University of Minnesota, Institute for Research on Learning Disabilities. (ERIC Document Reproduction Service No. ED 203 591)

Marston, D., & Ysseldyke, J. E. (1984a). Concerns in interpreting subtest scatter on the tests of cognitive ability from the Woodcock-Johnson Psycho-Educational Battery. *Journal of Learning Disabilities, 17,* 510-591.

Marston, D., & Ysseldyke, J. E. (1984b). *Derived subtest scores for the Woodcock-Johnson Psycho-Educational Battery.* Allen, TX: DLM Teaching Resources.

Mather, N., & Udall, A. J. (1985). The identification of gifted underachievers using the Woodcock-Johnson Psycho-Educational Battery. *Roeper Review, 8,* 54-58.

McGrew, K. S. (1983). Comparison of the WISC-R and the Woodcock-Johnson Tests of Cognitive Ability. *Journal of School Psychology, 21,* 271-276.

McGrew, K. S. (1984a). Normative based guides for subtest profile interpretation of the Woodcock-Johnson Tests of Cognitive Ability. *Journal of Psychoeducational Assessment, 2,* 141-148.

McGrew, K. S. (1984b). An analysis of the influence of the Quantitative Concepts subtest and the Woodcock-Johnson scholastic aptitude clusters. *Journal of Psychoeducational Assessment, 2,* 325-332.

McGrew, K. S. (1985). Investigation of the verbal/nonverbal structure of the Woodcock-Johnson: Implications for subtest interpretation and comparisons with the Wechsler scales. *Journal of Psychoeducational Assessment, 3,* 65-71.

McGrew, K. S. (1986). *Clinical interpretation of the Woodcock-Johnson Tests of Cognitive Ability.* New York: Grune & Stratton.

McGue, M., Shinn, M., & Ysseldyke, J. E. (1979). *Validity of the Woodcock-Johnson Psycho-Educational Battery with learning disabled students* (Research Rep. No. 15). Minneapolis: University of Minnesota, Institute for Research on Learning Disabilities. (ERIC Document Reproduction Service No. ED 185 759)

McGue, M., Shinn, M., & Ysseldyke, J. E. (1982). Use of cluster scores on the Woodcock-Johnson Psycho-Educational Battery with learning disabled students. *Learning Disability Quarterly, 5,* 274-287.

Mira, M. (1984). Psychological evaluation of hearing-impaired children. In S. J. Weaver (Ed.), *Testing children: A reference guide for effective clinical and psychoeducational assessments* (pp. 121-136). Kansas City, MO: Test Corporation of America.

Nisbet, R. (1981). *A comparison of the WISC-R and Woodcock-Johnson with a referral population.* Unpublished master's thesis, Moorhead State University, Moorhead, MN.

Phelps, L., Rosso, M., & Falasco, S. L. (1985). Multiple regression data using the WISC-R and the Woodcock-Johnson Tests of Cognitive Ability. *Psychology in the Schools, 22,* 46-49.

Phohl, W. F., & Enright, B. E. (1981). A review of the Woodcock-Johnson Psycho-Educational Battery. *Diagnostique, 6*(2), 8-15.

Pieper, E. L., & Deshler, D. D. (1980). *Analysis of cognitive abilities of adolescents learning disabled specifically in arithmetic computation.* (Research Rep. No. 26). Lawrence, KS: University of Kansas, Institute for Research in Learning Disabilities. (ERIC Document Reproduction Service ED 217 638)

Reeve, R. E., Hall, R. L., & Zakreski, R. S. (1979). The Woodcock-Johnson Tests of Cognitive Ability: Concurrent validity with the WISC-R. *Learning Disability Quarterly, 2,* 63-69.

Reilly, T., Drudge, O. W., Rosen, J. C., Loew, D. E., & Fischer, M. (1985). Concurrent and predictive validity of the WISC-R, McCarthy Scales, Woodcock-Johnson, and academic achievement. *Psychology in the Schools, 22,* 380-382.

Salvia, J., & Salvia, S. A. (1985). Use of the Woodcock-Johnson Psycho-Educational Battery, Part II: Tests of Achievement, with a college population. *Diagnostique, 11,* 3-8.

Shinn, M. R. (1980). Review of the Woodcock-Johnson Psycho-Educational Battery. *School Psychology International, 1,* 20-22.

Shinn, M., Algozzine, B., Marston, D., & Ysseldyke, J. E. (1980). *A theoretical analysis of the performance of learning disabled students on the Woodcock-Johnson Psycho-Educational Battery* (Research Rep. No. 38). Minneapolis, MN: University of Minnesota, Institute for Research on Learning Disabilities. (ERIC Document Reproduction Service No. ED 203 612)

Shinn, M. R., Algozzine, B., Marston, D., & Ysseldyke, J. E. (1982). A theoretical analysis of learning disabled students on the Woodcock-Johnson Psycho-Educational Tests Battery. *Journal of Learning Disabilities, 15,* 221-226.

Skrtic, T. M. (1980). *Formal reasoning abilities for learning disabled adolescents: Implications for mathematics instruction* (Research Rep. No. 7). Lawrence, KS: Kansas University, Institute for Research in Learning Disabilities. (ERIC Document Reproduction Service No. ED 217 624)

Stein, W., & Brantley, J. (1981). Woodcock-Johnson Psycho-Educational Battery—Test review. *Journal of School Psychology, 19,* 184-187.

Thompson, P. L., & Brassard, M. R. (1984a). Validity of the Woodcock-Johnson Tests of Cognitive Ability: A comparison with the WISC-R in learning disabled and normal elementary students. *Journal of School Psychology, 22,* 201-208.

Thompson, P. L., & Brassard, M. R. (1984b). Cummings and Moscato soft on Woodcock-Johnson. *School Psychology Review, 13,* 41-44.

Weaver, S. J. (1984). Assessment of mentally retarded children. In S. J. Weaver (Ed.), *Testing children: A reference guide for effective clinical and psychoeducational assessments* (pp. 50-70). Kansas City, MO: Test Corporation of America.

Woodcock, R. W. (1978). *Development and standardization of the Woodcock-Johnson Psycho-Educational Battery*. Allen, TX: DLM Teaching Resources.

Woodcock, R. W. (1982, March). *Interpretation of the Rasch ability and difficulty scales for educational purposes*. Paper presented at the annual meeting of the National Council on Measurement in Education, New York, NY. (ERIC Document Reproduction Service ED 223 673)

Woodcock, R. W. (1984a). A response to some questions raised about the Woodcock-Johnson: I. The mean score discrepancy issue. *School Psychology Review, 13,* 342-354.

Woodcock, R. W. (1984b). A response to some questions raised about the Woodcock-Johnson: II. Efficacy of the aptitude clusters. *School Psychology Review, 13,* 355-362.

Woodcock, R. W., & Johnson, M. (1978). *Woodcock-Johnson Psycho-Educational Battery*. Allen, TX: DLM Teaching Resources.

Ysseldyke, J. E., Algozzine, B., & Shinn, M. R. (1981). Validity of the Woodcock-Johnson Psycho-Educational Battery for learning disabled youngsters. *Learning Disability Quarterly, 4,* 244-249.

Ysseldyke, J. E., Shinn, M., & Epps, S. (1980). *A comparison of the WISC-R and the Woodcock-Johnson Tests of Cognitive Ability* (Research Rep. No. 36). Minneapolis: University of Minnesota, Institute for Research on Learning Disabilities. (ERIC Document Reproduction Service No. ED 203 610)

Ysseldyke, J. E., Shinn, M., & Epps, S. (1981). Comparison of WISC-R and the Woodcock-Johnson Tests of Cognitive Ability. *Psychology in the Schools, 18,* 15-19.

AUTHOR/REVIEWER INDEX

611

PUBLISHERS/DISTRIBUTORS INDEX

SUBJECT INDEX

ABOUT THE EDITORS

Daniel J. Keyser, Ph.D. Since completing postgraduate work at the University of Kansas in 1974, Dr. Keyser has worked in drug and alcohol rehabilitation and psychiatric settings. In addition, he has taught undergraduate psychology at Rockhurst College for 15 years. Dr. Keyser specializes in behavioral medicine—biofeedback, pain control, stress management, terminal care support, habit management, and wellness maintenance—and maintains a private clinical practice in the Kansas City area. Dr. Keyser co-edited *Tests: First Edition, Tests: Supplement,* and *Tests: Second Edition* and has made significant contributions to computerized psychological testing.

Richard C. Sweetland, Ph.D. After completing his doctorate at Utah State University in 1968, Dr. Sweetland completed postdoctoral training in psychoanalytically oriented clinical psychology at the Topeka State Hospital in conjunction with the training program of the Menninger Foundation. Following appointments in child psychology at the University of Kansas Medical Center and in neuropsychology at the Kansas City Veterans Administration Hospital, he entered the practice of psychotherapy in Kansas City. In addition to his clinical work in neuropsychology and psychoanalytic psychotherapy, Dr. Sweetland has been involved extensively in the development of computerized psychological testing. Dr. Sweetland co-edited *Tests: First Edition, Tests: Supplement,* and *Tests: Second Edition.*